Bottled in Illinois

Embossed Bottles and Bottled Products of Early Illinois Merchants from Chicago to Cairo 1840–1880

ILLINOIS

University of Illinois at Urbana-Champaign
Studies in Archaeology
Number 6

Thomas E. Emerson, Series Editor
Kenneth B. Farnsworth, Volume Editor

FRONT COVER IMAGES.
Examples of 1840–1880 Illinois commercial product bottles.
Top row: 1840–1860s bottles.
Center and bottom rows: 1860–1870s bottles.
During each decade, embossed bottles like these were used by a greater number and wider variety of Illinois bottlers statewide. Over time, the bottles themselves also became increasingly colorful and decorative as more and more bottlers competed to draw consumer attention to their products.

BACK COVER IMAGES.
(1) Hamlin's early oval *Wizard Oil* bottle from Chicago (*Figure 306*)
(2) *Dr. Winchell's Teething Syrup* advertisement by Snyder & Eilert of Chicago (*Figure 256*)
(3) Hand-colored print of the interior of A. Parent's beer and ale dept in Peoria (*Figure 919*)
(4) A colored pressed-paper George A. Miller advertising sign for placement in store windows in Quincy (*Figure 974*)
(5) Trade card for Eilert's *Extract of Tar and Wild Cherry* distribted by Emmert Proprietary Co., Chicago (*Figure 257*)
(6) Woodard's medicine-bottle paper labels reproduced on the back of scenic trade cards distributed by the Woodard Company in Bloomington (*see pg. 155*)
(7) Hamlin's *Wizard Oil* poster promoting his hugely popular traveling shows (*Figure 317*)

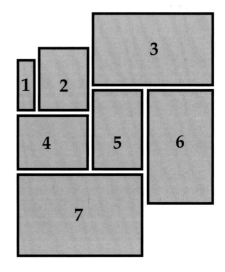

Cover Design by Linda Alexander, Sarah Boyer, Corinne Carlson, Mike Lewis, and Valerie Alexander Vallese
Book Design by Sarah Boyer
ISAS Production

Bottled in Illinois

Embossed Bottles and Bottled Products of Early Illinois Merchants from Chicago to Cairo 1840–1880

Kenneth B. Farnsworth
John A. Walthall

With Historical Research Contributions by

David Beeler	*David A. Jackson*	*Thomas Miller*	*James Searle*
Frank Donath	*Ray Komoroski*	*Eva Mounce*	*Tim Wallace*
Scott Garrow	*Robert Kott*	*Ben Oertle*	*Paul Welko*
Anthony Green	*Curtis Mann*	*John Panek*	*John Wilson*

Volume Production and Design
Sarah Boyer

ILLINOIS STATE
ARCHAEOLOGICAL SURVEY
PRAIRIE RESEARCH INSTITUTE

Studies in Archaeology No. 6
Illinois State Archaeological Survey
University of Illinois
Urbana, Illinois

Studies in Archaeology

Series Editor
Thomas E. Emerson

The Illinois State Archaeological Survey's (ISAS) mission is to investigate, preserve, and interpret the archaeological heritage of Illinois within the context of long-term public needs and sustainable economic development through its scientific research, public service, education, and outreach activities. To accomplish this mission, the Survey partners with public and private organizations as well as federal, state, and local government entities.

ISAS and the University of Illinois have enjoyed a more than half-century cooperative relationship with the Illinois Department of Transportation (IDOT) focused on supporting the preservation, protection, and scientific investigation of the state's important archaeological resources. This relationship began under Charles J. Bareis with the newly created Illinois Archaeological Survey (1959–1980), continued when the statewide survey program was transferred to the Resource Investigation Program (1980–1993), and through the creation of the Illinois Transportation Archaeological Program (1994–2009) under Thomas E. Emerson. The establishment of ITARP was the result of IDOT's interest in developing a centralized program to facilitate its cultural resources protection efforts. In 2010, the IDOT archaeology programs were incorporated into the newly created Illinois State Archaeological Survey (ISAS). During this half-century, IDOT-sponsored research has transformed our understanding of Illinois's native and European inhabitants and put Illinois researchers in the forefront of North American archaeological practitioners.

One of ISAS's primary goals, as a state scientific survey and research unit of the Prairie Research Institute (University of Illinois) is the dissemination of information of archaeological and historical topics to both professional and public audiences. This externally peer reviewed series, Studies in Archaeology, is designed to bring scholarly works of national significance on topical and regional research to the broadest possible audience.

The production of these volumes is accomplished through the efforts of the ISAS Publication Production Office under the direction of Mike Lewis (Production Manager); Corinne Carlson, Sarah Boyer, and Angie Patton (Production Coordinators); and Linda Alexander (Illustrator/Photographer/Graphic Designer). Additional copies of this report may be obtained online at: www.isas.illinois.edu or by faxing ISAS at 217-244-7458.

ISBN: 978-1-930487-23-9
© 2011 University of Illinois Board of Trustees. All rights reserved.
For permissions information, contact the Illinois State Archaeological Survey.
Printed by Authority of the State of Illinois.
Printed in the United State of America.

Publication of this volume was funded by the Illinois Department of Transportation (IDOT) and the Illinois State Archaeological Survey (ISAS) at the University of Illinois. The contents and opinions do not necessarily reflect the views or policies of IDOT, ISAS, or the University of Illinois.

Contents

List of Tables ix

List of Figures xi

Acknowledgments xv

Author Biographies xix

Dedication To Paul Welko and Eva Mounce xxi

The Embossed and Stamped Bottles Used by Illinois Merchants from 1840 to 1880
    Study Focus 1
    Sources of Information 2
    Sources of Misinformation 3
    Study Organization, Format, and Terminology 4
    Descriptive Format 5
    Abbreviated Bottle-Technology Terms 7
    Abbreviated Terms for Commercial-Data Sources 8
    Terms Used to Describe Bottle Color, Shape, and Manufacturing Technology 9
        Colors 9
        Embossing and Stamping 9
        Measurements and Volumes 10
        Manufacturing-Technology Terms 11
        Body Mold Lines 12
        Body Shape and Style Terms 13
        Base Shapes, Kick-Ups, "Decorations," and Mold Lines 22
        Shoulder Shape and Style Terms 22
        Neck Shape and Style Terms 23
        Lip Styles 24
        Closures 26
    Study Limits: The 40-Year Period Bracketing the Civil War 29
        Why Begin the Study at ca. 1840? 29
        Why End the Study at ca. 1880? 34
    Products Packaged in Embossed Glass and Stamped Stoneware Bottles 41
    Study Completeness: Late Additions to the Text Bottle Listings 43
    Recently Discovered Bottles 44
        Aurora 44
        Belleville 45
        Bloomington 46
        Cairo 47
        Chicago 47
        DeKalb 52
        Dixon 52
        LaSalle 53
        Mendota 54
    Glass-Factory Maker's Marks on Illinois Embossed Bottles 54
        Pre-1860 Illinois Bottle Manufacturers 56
        Post-1860 Illinois Bottle Manufacturers 56

Changes in Embossed-Bottle Manufacturing Technology from 1840–1880 71
Pattern of Embossed-Bottle Use in Illinois from 1840–1880 74
    Assemblage Patterns 74
    Antebellum Product-Use Pattern 75
References 80

A 85
    Albion 85
    Alton 86
    Arlington Heights 103
    Aurora 104

B 107
    Beardstown 107
    Belleville 110
    Belvidere 120
    Bloomington 120
    Blue Island 158
    Braidwood 162
    Breese 163

C 165
    Cairo 165
    Capron 172
    Carlinville 172
    Carlyle 174
    Carrollton 176
    Centralia 179
    Centreville [Centerville] 181
    Champaign 182
    Charleston 183
    Chester 184
    Chicago 186
    Colehour 462
    Collinsville 464
    Columbia 465

D 467
    Danville 467
    Darmstadt 469
    Decatur 471
    Dixon 473
    Dunton 475
    DuQuoin 476

E 479
    East St. Louis 479
    Edwardsville 487
    Elgin 487
    Evanston 492

F 495
    Farmington 495
    Forreston 496

- Franklin Grove ... 497
- Freeburg ... 499
- Freeport ... 499
- Frogtown ... 501

G ... 503
- Galena ... 503
- Galesburg ... 507
- Galva ... 510
- Greenville ... 511

H ... 515
- Highland ... 515

J ... 521
- Jacksonville ... 521
- Jerseyville ... 534
- Joliet ... 539
- Jonesboro ... 543

K ... 545
- Kankakee ... 545
- Kewanee ... 546

L ... 547
- Lacon ... 547
- LaSalle ... 548
- Lebanon ... 553
- Lemont ... 554
- Lewistown ... 557
- Lincoln ... 558
- Litchfield ... 559
- Louden City ... 563

M ... 565
- Maryland ... 565
- Mascoutah ... 566
- Mattoon ... 568
- Mendota ... 569
- Metamora ... 570
- Minot ... 571
- Monee ... 572
- Morris ... 574
- Mt. Vernon ... 575
- Murphysboro ... 576

N ... 579
- Naperville ... 579
- Naples ... 585
- Nelson ... 586
- New Athens ... 587
- Nokomis ... 588

- O 589
  - O'Fallon 589
  - Olney 591
  - Oquawka 592
  - Ottawa 593

- P 599
  - Pana 599
  - Pekin 600
  - Peoria 603
  - Perry 646
  - Peru 648
  - Pittsfield 649
  - Pontiac 651
  - Prentice 651
  - Princeton 653

- Q 657
  - Quincy 657

- R 687
  - Red Bud 687
  - Rockford 688
  - Rock Island 690

- S 697
  - Sandwich 697
  - Sheldon 698
  - Springfield 699
  - Staunton 730
  - Streator 731

- V 733
  - Virden 733

- W 735
  - Waterloo 735
  - Waukegan 738
  - West Urbana 740
  - Western Saratoga 741
  - Wilmington 742
  - Woodstock 743

- BIBLIOGRAPHY 745

- ABRIDGED INDEX 757

List of Tables

I.1. County distribution of 102 bottling towns and 494 bottlers in 60 of 102 Illinois counties....................4–5
I.2. Glass factory maker's marks on pre-1880 Illinois embossed bottles 55

LIST OF FIGURES

A.1.	ISAS staffers Kjersti Emerson, Corinne Carlson, and Sarah Boyer re-create the arduous task of preparing thousands of bottles for publication in this volume	xvi
D.1.	Photo of Paul Welko and his wife, Lenore, on their 25th wedding anniversary	xxi
D.2.	Paul Welko at an early 1970s Chicago bottle show	xxi
D.3.	Eva Mounce studying a collection of Alton, IL, stoneware and earthenware in 2002 with Ken Farnsworth	xxii
I.1.	Illinois county distribution map of 1840–1880 bottlers using embossed or stamped bottles	6
I.2.	Variable neck lengths of soda bottles used in Cairo, IL, during the 1860s and 1870s by the Andrew Lohr and Henry Breihan bottling works	10
I.3.	Examples of the open glass-rod pontil scars seen on smaller embossed bottles and the iron-pontil scars seen on larger embossed bottles used by Illinois bottlers during the 1840s and 1850s	11
I.4.	Varieties of paneled and fluted bottle decoration	13
I.5.	Shape varieties of larger-size whiskey bottles and decanters used by Illinois bottlers during the study period	14
I.6.	Shape varieties of smaller whiskey flasks used by Illinois bottlers during the study period	14
I.7.	Shape and size varieties of quart- and pint-size ale bottles used by Illinois bottlers during the study period	15
I.8.	Shape and size varieties of half-pint porter bottles used by Illinois bottlers during the study period	16
I.9.	Examples of mold-engraved and slug-plate embossed pint and quart lager beer bottles used by Illinois bottlers during the 1870s	16
I.10.	Examples of 1860s and 1870s beveled-edge square quart bitters bottles used by Illinois bottlers	16
I.11.	Examples of high-shoulder and low-shoulder pint sodas, slope-shoulder quart sodas, long-neck vs. squat Chicago-style quart sodas used by Illinois bottlers	17
I.12.	Examples of "dumpy" and large-sized round-base and semi-round ginger-ale–style bottles (including "medicated aerated waters") used by Illinois bottlers during the 1870s	17
I.13.	Examples of Saratoga-spring style and Selters-spring style mineral-water bottles used by Illinois bottlers in the post Civil War era	18
I.14.	Examples of early and late-style ring-necked cider quarts used by Illinois bottlers during the 1870s	18
I.15.	Embossed Illinois ink-bottle shapes of the 1860s and 1870s	19
I.16.	Round, cylindrical and oval, lens-shaped medicine bottle shapes used by Illinois bottlers during the study period	19
I.17.	Beveled-edge flat-paneled and sunken-paneled square and rectangle medicine bottle shapes used by Illinois bottlers during the study period	20
I.18.	Wide-mouthed salve and unguent medicine bottle shapes used by Illinois bottlers during the study period	20
I.19.	Citrate-style cylinder medicine bottle shape used by Illinois bottlers during the study period	20
I.20.	Illinois preserve-jar styles of the 1860s and 1870s	21
I.21.	Stamped stoneware bottle shapes used by Illinois bottlers during the study period	21
I.22.	Round, semi-round, and tenpin-shaped bottle bases	22
I.23.	Examples of raised-glass embossed "dimples" on the bases of Illinois quart ales of the 1860s and 1870s	22
I.24.	Examples of bottle-base mold seams	23
I.25.	Examples of bottle shoulder decorations	23
I.26.	Examples of decorative neck styles and finishes	24
I.27.	Examples of 1840s lip-finish styles used by Illinois medicine bottlers	24

I.28.	Earlier and later tapered-collar styles, earlier and later shouldered-blob styles, and oval blob and lozenge-shaped lip styles of soda and mineral-water bottle lip finishes used by Illinois bottlers during the study period	25
I.29.	Preferred style of iron swing-stopper used in Illinois to hold corks in place on blob-top sodas of the antebellum and Civil War era, and later 1870s collared-blob-style soda lip finish soda and mineral-water bottle lip finishes used by Illinois bottlers during the study period	25
I.30.	Ring-neck tapered-collar lip finishes and tapered-collar lip finishes found on embossed ales, whiskey, and bitters bottles used by Illinois bottlers during the study period	25
I.31.	Examples of medicine-bottle lip finishes found on embossed Illinois bottles dating to the later 1850s, 1860s, and 1870s	25
I.32.	Patent illustrations for swing stoppers and cork bales used to secure corks in blob-top bottles during the study period	26
I.33.	Patent illustrations for swing stoppers and cork bales used to secure corks in ring-neck bottles during the study period	27
I.34.	An example of a protective wire mask worn by bottlers during the study period to protect them against injury from exploding glass sherds while filling carbonated-beverage bottles	28
I.35.	1860s patent illustrations for the Matthews gravitating-stopper bottle closures used by Illinois bottlers	28
I.36.	A patent illustration for the Christin gravitating-stopper bottle closures used by Illinois bottlers	28
I.37.	1870s patent illustrations for Kelley gravitating-stopper bottle closures used by Illinois bottlers	28
I.38.	Unembossed, paper-labeled, thin-walled "puff"-style pharmacy bottle used during the 1840s by Geneseo, IL, druggist N. L. Lutz	30
I.39.	1870 distribution map showing the density of foreign immigrant settlement in the upper Great Lakes-Riverine area	31
I.40.	1870 distribution map showing the density of German immigrant population in the eastern U.S., concentrated in the upper Great Lakes-Riverine area	32
I.41.	1870 distribution map showing the wealth in the eastern U.S., concentrated in the upper Great Lakes-Riverine area	33
I.42.	Illustration commissioned for the 1868 Chicago City Directory showing Liberty amid the fruits of creativity and capitalism in the foreground of the consumer Mecca of the Midwest	36
I.43.	Post-1878 Hutchinson-style soda bottles and their patented internal rubber stoppers	36
I.44.	Four styles of mass-produced embossed druggist bottles patented/used prior to 1880	38
I.45.	Four styles of mass-produced embossed druggist bottles patented/used prior to 1880	38
I.46.	Four styles of mass-produced embossed druggist bottles patented/used prior to 1880	38
I.47.	Examples of post-1880 embossed, stamped, and etched bottle styles used in Illinois too late for inclusion in the present 1840–1880 study: Weiss beers	39
I.48.	Example of post-1880 embossed, stamped, and etched bottle style used in Illinois too late for inclusion in the present 1840–1880 study: half-pint stoneware small beers	39
I.49.	Examples of post-1880 embossed, stamped, and etched bottle styles used in Illinois too late for inclusion in the present 1840–1880 study: flint-glass milks	39
I.50.	Examples of post-1880 embossed, stamped, and etched bottle styles used in Illinois too late for inclusion in the present 1840–1880 study: cigar jars	39
I.51.	Examples of post-1880 embossed, stamped, and etched bottle styles used in Illinois too late for inclusion in the present 1840–1880 study: poisons	40
I.52.	Examples of post-1880 embossed, stamped, and etched bottle styles used in Illinois too late for inclusion in the present 1840–1880 study: flint-glass whiskey flasks	40
I.53.	Examples of post-1880 embossed, stamped, and etched bottle styles used in Illinois too late for inclusion in the present 1840–1880 study: drugstore citrates	40
I.54.	Examples of post-1880 embossed, stamped, and etched bottle styles used in Illinois too late for inclusion in the present 1840–1880 study: colored and flint-glass seltzers	40
I.55.	Examples of blob-top and Hutchinson-style sodas still filled with product contents but lacking paper labels	41

List of Figures

I.72.	Illinois county map showing the distribution of embossed bottles documented by our study that have **C&I** maker's marks	61
I.73.	Illinois county map showing the distribution of embossed bottles documented by our study that have **IGCo** maker's marks	65
I.74.	One of Charles Yockel's early 1870s ads touting his experience in the production of metal "proprietary" embossed-bottle molds for private bottlers	72
I.75.	Grouping of flint-glass pharmacy bottles produced ca. 1875–1915	73
I.76.	Grouping of pharmacy bottles produced for Illinois druggists ca. 1875–1915	73
I.77.	Illinois county map showing the distribution of embossed medicine bottles documented by our study that were made and used during the late 1840s	76
I.78.	Illinois county map showing the distribution of embossed soda and mineral water bottles documented by our study that were made and used during the 1850s	77
I.79.	Illinois county map showing the distribution of embossed medicine bottles documented by our study that were made and used during the 1850s	78
8.	The 1873 exterior and interior views of the Illinois Glass Company's fledgling five-pot glass-factory building at the original Belle Street location	91
11.	Stenciled shipping crate for national-brand patent medicine (*Dr. Pierce's Golden Medical Discovery*) used in Alton by Quigley, Hopkins, & Lea during the late 1860s or early 1870s	96
22.	Late nineteenth-century photo of "Pop" Muller and his soda-water delivery wagon in Arlington Heights	104
65.	Photographic portrait of Dr. Howe from the *Biographical Record of McLean County, Illinois*	128
105.	Early image of Wakefield's medicine factory published in his second annual *Western Farmer's Almanac* in 1862	146
106.	Later image of Wakefield's expanded medicine factory, used on his company's letterhead during the 1870s	146
107.	Portraits of Cyrenius Wakefield commissioned between the late 1860s and the early 1880s	147
144.	Hand-drawn illustration of Smith's "Celebrated" Old Style Bitters" embossed variety	178
146.	Rubbing of side-embossed lettering taken from restored excavated sherds of a fragmentary Besant blob-top bottle	180
147.	Rubbing made by Ron Ridder of side-embossed lettering on a blob-top bottle previously in his possession	180
184.	Both sides of a printed trade card distributed in the 1870s by the Chicago branch of Barrett & Barrett cider company	199
216.	Printed design of a pasteboard-box container for *Christie's Ague Balsam* bottles	218
218.	Arthur Christin embossed *Belfast Ginger Ale* varieties drawn by Robert Kott from fragments recovered by Paul Welko	220
226.	Cory's coat-of-arms logo and "**BIBE ET VIVE**" banner slogan as it appeared in the company's 1859 *Stomachic Bitters* ad	225
334.	A 1 3/4" glass portion of internal gravitating stopper used in the 1871 Kelly-patent soda bottle produced for Wm. A. Hausburg	283
347.	The small-lettered version of the Hedlund & Co. bottle includes both a cobalt variety (with an iron-pontil scar on the base) and a teal variety (with an open-glass-rod scar on the base)	288
370.	Late-variety W. H. Hutchinson and W.H.H. 12-sided Merrill-patent stoneware bottles	301
466.	Comparative view of standard-size and large-size cobalt Lomax soda bottles	344
713.	Examples of Evans Brothers embossed black-glass ales used in St. Louis during the late 1840s and early 1850s	504
872.	Ca. 1880 historical photo of James E. Eaton and soda-factory workers in front of R. E. Hickey's Soda Water Factory	612
873.	Ca. 1880 historical photo of James E. Eaton and two of his soda factory employees (including Alex Hickey) on one of his soda wagons, at Dye's Practical Horse Shoer shop in Peoria	613
880.	Detail from 1844 Peoria city plat map showing the location of W. B. & H. G. Farrell's Linseed near the Illinois River	619

919.	Hand-colored print of the interior of A. Parent's beer and ale depot at 321 Main Street in Peoria	640
971.	Detail from ca. 1856 half-plate daguerreotype photo by Thomas M. Easterly of the recently established St. Louis Glass Works at the N.W. corner of Broadway and Monroe streets	681
986.	Broken examples of the distinctive sharp-shouldered blob-top Codd Marble gravitating-stopper sodas produced by an unknown U.S. glass maker for three Davenport, IA, companies during the late 1870s (and perhaps early 1800s)	693
1024.	Daguerreotype photo of Alfred North's elaborate 1850s patent-medicine sales and delivery wagon	723
1029.	1876 Springfield city directory image of the *Illinois Hominy and Spice Mills* at 301-303-305-307-309-311 South 4th Street, operated by J. C. and J. J. Conkling and S. A. Slemmons	726

Acknowledgments

An encyclopedic book project like *Bottled in Illinois* can be an important building block for the growing discipline of Illinois historical archaeology. But the ability to see past its potential value to a successful end product requires focused leadership and management. Dr. Thomas Emerson, Illinois State Archaeological Survey (ISAS) director, is a practical leader of extensive experience who knows how to organize staff and resources to get this kind of job done. *Bottled in Illinois* was produced by the ISAS research and publication system that Tom has designed, built, and nurtured during his 17 years as director of the Illinois Transportation Archaeological Research Program (now ISAS). ISAS publications manager, Mike Lewis, and publication coordinator, Sarah Boyer, were the two people responsible for transforming our *Bottled in Illinois* manuscript into a high-quality publication. More than once, Mike has saved authors and production staff alike from impending computer disasters, and he has always provided calm, reassuring cyber-psychology advice to those of us for whom computers are a poorly mastered second language. The beautiful book design and page layout for *Bottled in Illinois* are the work of Sarah Boyer. Her design work and long labors to prepare countless figures and tables has so dramatically increased book's impact, that we have come to view her as a coauthor of the volume. She was ably assisted in the daunting task of freeing Farnsworth's photographic bottle images from their original backgrounds by a "surgical" team (*Figure A.1*) including ISAS staffers Corinne Carlson, Kjersti Emerson, and Linda Alexander. Sarah, Linda, and Valerie Alexander Vallese also worked together to design the impressive front and back covers for the book. Publication coordinator Corinne Carlson also applied her extensive design skills to create a beautiful full-color poster to promote initial bottle book sales. Corinne's poster design dramatically illustrates the range of colors, shapes, and sizes of some of the most impressive embossed bottles used by Illinois bottlers during the decades of the 1840s, 1850s, 1860s, and 1870s.

Over the course of data gathering for the volume, we recorded many embossed bottles we could not effectively photograph for the book—primarily because they were too dark ("black glass") or because well-embossed original examples of the bottles could not be located for photography. Three skilled artists documented many of these bottles as line drawings. We are very grateful to Liz Hansen, Laurel Norton, and Valerie Alexander Vallese for their beautiful line art work. In a book that incorporates many historical oddities of spelling, grammar, and terminology, final copyediting is a difficult proposition. But Carol McGillivray did a great job navigating her way through the text and quotations, making sure that we meant what we said and said what we meant. We would also like to thank Sue Klefstad for her careful indexing of the final volume.

Some research projects would be impossible to complete without the help of a great many people. *Bottled in Illinois* is a prime example of this kind of collaborative project. Our book could never have been completed without the enthusiastic help of literally dozens of Illinois-area history enthusiasts, bottle collectors, archaeologists, and professional archivists, librarians, and historians who live and work and collect historical artifacts in cities, towns, villages, and rural areas extending from the southern Great Lakes to the confluence of the Mississippi and Ohio Rivers. Our study was further improved by historical information and examples of rare bottles made available to us by historical researchers and regional collectors residing

Figure A.1. ISAS staffers Kjersti Emerson (left), Corinne Carlson (middle), and Sarah Boyer (right) re-create the arduous task of preparing thousands of bottles for publication in this volume. (Photo by Linda Alexander.)

as far from the Illinois heartland as Florida, Ohio, Wisconsin, and Virginia. As our study progressed, many of the regional collectors who helped us assemble the first draft of the volume also banded together to form a "distant early warning system" alerting us to newly discovered Illinois embossed bottle styles that surfaced, which insured their inclusion in the book. The tireless help of many of these friends and advisors is detailed in the paragraphs below, but phone conversations and bottle show discussions with many others will no doubt "slip through the net" of our acknowledgments. We sincerely hope that everyone who helped realizes how grateful we are for their shared knowledge, encouragement, and advice during the (often painfully slow) process of assembling the building blocks of our *Bottled in Illinois* volume.

In addition to this informal "network" of collectors, archivists, and students of Illinois history, our study greatly benefitted from several key earlier publications, and from some important modern glass studies being assembled on the Internet. We were heavily influenced and aided by the pioneering research of George and Helen McKearin and Kenneth M. Wilson (*American Glass* [1941], *Two Hundred Years of American Blown Glass* [1950], *American Bottles and Flasks and their Ancestry* [1978]), by John Riley's *History of the American Soft Drink Industry* [1958], by James Harvey Young's history of early patent medicines (*The Toadstool Millionaires* [1961]), by the works of Julian Harrison Toulouse (*A Primer on Mold Seams* [1969]; *Bottle Makers and their Marks* [1971], by John Paul and Paul Parmalee's pioneering Illinois State Museum study on *Soft*

Drink Bottling [1973], by Lowell Innes' 1976 volume on *Pittsburgh Glass: 1797–1891*, by several important glass-technology studies by Olive Jones published between 1971 and 1983 in *Historical Archaeology* and the *Journal of Glass Studies*, and by Richard Fike's 1987 patent-medicine guide entitled *The Bottle Book*. We also benefitted greatly from recent detailed studies of individual bottle styles (in particular, several editions of *For Bitters Only* and *Bitters Bottles* by Carlyn Ring and William Ham), from studies of embossed-glass products in nearby states (especially Mike Burggraaf and Tom Southard's wonderfully detailed *Antique Bottles of Iowa: 1846–1915*), from a monumental ongoing study of the technological history and chronology of glass-bottle manufacture (Bill Lindsay's massive and ever-expanding *Historic Glass Bottle Identification and Information Website*, hosted by the Society for Historical Archaeology), and from a growing body of knowledge about regional embossed-glass-bottle manufacturing houses and their maker's marks (including a wonderfully detailed series of articles on the histories of individual nineteenth-century glass-manufacturing houses, published by Bill Lockhart and his colleagues under the general title "The Dating Game" in *Bottles and Extras* magazine).

We are also greatly indebted to the authors/editors of three series of important, privately produced/distributed reference volumes on embossed Illinois bottles that greatly benefitted and speeded up our work. Their regional studies—carefully researched and lovingly assembled over many decades—include Thomas Miller's "*St. Clair County, Illinois, Soda and Related Beverage Bottles*," Ben Oertle's "*Central Illinois Bottles and Glasses*" (for a several county area of west-central Illinois centered on Pekin and Peoria), and Robert Kott's "*Chicago Sodas and Mineral Waters*" and his "*Ex-Urbus Soda and Mineral Waters*" (for the Chicago area). The help and advice of the authors themselves, as well as their cheerfully granted access to several editions of their privately printed research studies and their research data files, is gratefully acknowledged.

During the course of the *Bottled in Illinois* study, we met several regional Illinois-glass collectors who in their spare time over the years had taken time to study newspaper microfilm files and to conduct focused searches of collections and records at local historical societies, seeking information about the business histories of nineteenth-century bottlers who used embossed bottles they collected, and about the bottled products that had been put-up and sold in the bottles themselves. When these historical sleuths learned of our project, they readily made their study files available to us for incorporation into our book. Their historical discoveries about several early Illinois bottlers gave us an important "head start" on our own (somewhat daunting) statewide archival research task. All of these self-motivated regional researchers are listed on the title page of the present volume as *Historical Research Contributors* to the volume.

As noted above, Tom Miller's St. Clair County studies, Ben Oertle's west-central Illinois embossed-bottle lists, and Bob Kott's Chicago area bottle documentation and bottler-history information had already been assembled in tabular or narrative form before our study began. But the research raw-data files of several other collectors were also rich information sources. Over a span of many years, Dave Beeler has conducted wide-ranging searches of Illinois newspaper microfilm and county historical society documents to learn all he could about pioneer Illinois medical beliefs and bottled medicinal and pharmaceutical products. Dave cheerfully loaned us his ring-bound notebooks and notecard files to use in our *Bottled in Illinois* study. Along with his focused medicinal product studies, Dave had also gathered information on other bottled products as the opportunity allowed.

Several other researchers we met had sought out historical information on the embossed bottles that had been used in Illinois towns near where they lived. This information was enthusiastically shared with us as well. In the Pekin/Peoria/Galesburg area, Frank Donath, Ben Oertle, and Jim Searle shared unpublished information they had gathered on local historical bottlers. In the Rockford–to–Chicago area, Scott Garrow and John Wilson gave us historical bottler information. For the bottler-rich Chicago metropolitan area itself, we benefitted from earlier bottling-house research by Paul Welko and John Panek, in addition to Bob Kott's extensive bottle-documentation files. Ray Komorowski's detailed study records on John A. Lomax, the largest and most prolific Chicago bottler of the later nineteenth century, were also of great help in sorting out the history of the Lomax bottling works. Peter Maas, who has done extensive urban-archaeology and local history research in Wisconsin, also corresponded with us to help sort out the northern Illinois vs. southern Wisconsin connections of three prominent early bottling families: the Hickey brothers, the Taylor brothers, and the Eaton family. For the Bloomington area, we benefitted from historical society research information and from historical advertising materials gathered by Anthony Green. For Springfield and surrounding areas of west-central Illinois we benefitted from Tim Wallace's historical photographs and records of early brewery and liquor-product dealers.

Two very skilled and intuitive professional historical archivists spearheaded our own historical research efforts for the *Bottled in Illinois* volume. *Illinois State*

Archaeological Survey archivist Eva Mounce conducted work during the first two years of the bottle book study. But Eva was in poor health, and passed away before she could see her archival research efforts come to fruition (*see the dedication section*). During the final months of assembling the volume, we were very lucky indeed to secure the services of Curtis Mann, manager of the Sangamon Valley Collection at the Lincoln Library in Springfield. Curt's exceptional archival research skills helped us shepherd the book to final completion.

Many interested bottle collectors across the state also graciously invited us into their homes to photograph and record their collections of Illinois historical bottles. This book would not have been possible without their hospitality and their patience. For allowing us to study their collections, and for making it possible for us to so effectively illustrate our *Bottled in Illinois* volume, we would like to express our sincere thanks Theo Adams, John Bauman, Bill Beckman, Dave and Paul Beeler, Mike Burggraaf, Frank Bradbury, Fred Brown, Jeff Cress, Frank Donath, Jeremy Erp, Tom Feltman, Scott Garrow, Anthony Green, Jim Hall, Dave Hast, Joe Healy, Mike Henrich, Randy Heutsch, Gale Joyce, Steve Kehrer, Ray Komorowski, Bob Kott, Rich Kramerich, Keith Leeders, Tom Majewski, Tom Miller, Ron Neumann, Ben Oertle, John Panek, Dan Puzzo, Bob Rhineberger, Jim Searle, Brandon Smith, Tom Southard, Tim Wallace, Greg Watt, Paul Welko, John Wilson, and John Wolf. The Chicago section of the volume, in particular, would have been far less comprehensive had we not had access to Paul Welko's astonishing bottle collection and his store of historical knowledge about Chicago bottlers and their bottled products. At times the task of recording and researching all the 1840–80 Chicago bottlers and bottling houses seemed almost insurmountable, but Paul opened up his home to us and kept our noses to the grindstone. During the course of the study, Paul was diagnosed with cancer and, sadly, he did not survive to see the finished product he helped us create, but his efforts live on in the pages of this study (*see the dedication section*).

We also benefitted from the use of historical and archaeological resources and collections curated at several Illinois institutions and libraries. Illinois State Archaeological Survey curator, Laura Kozuch, made excavated historical collections available for study, as did laboratory and excavation staff at the ISAS American Bottom Field Station labs and Western Illinois Field Station labs. We are particularly grateful to field director, Pat Durst, for granting us study access to artifacts and photographs from his ongoing East St. Louis historical excavations, and to Western Illinois Survey coordinator, Dave Nolan, for study access to historic artifacts and excavation records from his recent Chenowith site excavations near Macomb. Staff members at the McLean County Historical Society and Museum of History (MCMH) in Bloomington were similarly helpful and supportive of our research project. We are particularly grateful to librarian/archivist, Bill Kemp, and registrar, Terri Clemens, for research insights and for their help locating and copying library materials and archival records relating to our study. Curator Susan Hartzold allowed study access to MCMH historical bottle collections and provided us with a photographic image of an original painting of Cyrenius Wakefield in the MCMH collections for use in use in our book. Staff members at the Peoria Historical Society Collection facility at Bradley University Library were also very helpful guiding us through research collections, city directories, and photographic archives curated there. At the Illinois State Museum, staff photographer, Doug Carr, went out of his way to search for original photographic material from the 1973 Paul and Parmalee *Soft Drink Bottling* publication, and with the help of author John Paul, successfully located and made available the original illustration we were seeking. The website of the Arlington Heights Historical Museum (AHHM) was an important resource for developing short histories of early bottlers in Dunton and Arlington Heights. The AHHM itself is actually located in the original Arlington Heights "Pop" Muller bottling works building, and its bottling history is highlighted on the museum website. Finally, we were able to only lightly sample a potentially highly valuable archival resource curated by the Harvard Business School's Baker Library. This resource is the massive, largely hand-written, mid- to late-nineteenth-century R. G. Dun credit rating report files housed there (*see entries for Buell & Schermerhorn in the Jacksonville section of this volume*). We are very grateful to staff of the Baker Library for helping us investigate the value of the records by locating and sending us the entries for some of out post–Civil War Jacksonville bottlers.

Authors' Biographies

Kenneth B. Farnsworth

Ken Farnsworth's previous archaeological research projects include numerous studies and publications on Illinois' later Archaic and Woodland-period native populations (5,000 B.C.–A.D. 1,100), a co-authored study of early European trade goods found in Native American contexts (1995a), and many fieldwork studies at Illinois sites occupied by nineteenth-century historic immigrants. His previous Illinois historical field research includes numerous remote-sensing surveys (in advance of historic/pioneer site excavations statewide) at rural farmsteads, as well as military, industrial, and town sites dating from the early 1800s to the Civil War–era. Farnsworth has also conducted a salvage excavation project at the 1855–80 Wilhelms pottery-kiln site in Alton, Illinois (for which analyses are under way). His historic research studies in Illinois also include excavations at a rural mid-nineteenth-century utopian community residence in the lower Illinois Valley (1999) and at the Civil War–era Perry Springs mineral water resort in upland Pike County (1995b). An archival research project with Eva Mounce resulted in a published volume on late-nineteenth-century liquor and mineral water bottling operations at the Gravel Springs site near Jacksonville, Illinois (1996). *Bottled in Illinois* is Ken's third study of Illinois historical bottling works.

Selected Bibliography

1995a Trade and Transition: European Trade Goods and Late Historic Mortuary Sites in Illinois (with Jodie O'Gorman). *Illinois Archaeology* 7:109–147.

1995b *Sulphur Spring Excavations, Perry Springs, Pike County, Illinois* (with K. Atwell and R. Perkins). Report of Investigations submitted by the Center for American Archeology to the J. L. Wade Museum, Griggsville, IL.

1996 *A Short History of the Gravel Springs Distillery and Bottling Works* (with K. Atwell and R. Perkins). Kampsville Studies in Archeology and History No. 2. Center for American Archeology, Kampsville, IL.

1999 *Evaluation of an 1825–1850 Pioneer Homestead Occupation at the Levis Site: 1998 Excavation Update* (with K. Atwell and R. Perkins). Report of Investigations submitted by the Center for American Archeology to the Illinois Valley Cultural Heritage Society and the Illinois Historic Preservation Agency.

John A. Walthall

Dr. Walthall's distinguished archaeological career includes extensive research and publications on Illinois' Paleo-Indian, Archaic, Woodland, Mississippian, and Proto-Historic native populations (12,000 B.C.–A.D. 1700), as well as studies of the state's more recent historic immigrants. His published studies of eighteenth-century French occupations in Illinois include articles on French-colonial ceramics and material culture (1991a, 2001, 2007), a monograph on excavations at Cahokia's *River L'Abbe Mission* (1987), and two edited overview volumes on Illinois French settlers and Indian interaction: *French Colonial Archaeology* (1991b) and *Calumet and Fleur-de-Lys* (1992). His nineteenth-century Illinois historical research has resulted in reports on stoneware production at Vermillionville (1989), vessel styles made at two stoneware potteries in Upper Alton (1991c), and stoneware bottles used by Chicago bottlers (1993). *Bottled in Illinois* is John's first venture into Illinois historical glass studies.

Selected Bibliography

1987   *The River L'Abbe Mission: A French Colonial Church for the Cahokia Illini on Monks Mound* (with Elizabeth D. Benchley). Studies in Illinois Archaeology No. 2. Illinois Historic Preservation Agency, Springfield.

1989   Lighting the Pioneer Homestead: Stoneware Lamps from the Kirkpatrick Kiln Site, LaSalle County, Illinois (with Floyd R. Mansberger and Eva Dodge Mounce). *Illinois Archaeology* 1:69–81.

1991a   Faience in French Colonial Illinois. *Historical Archaeology* 25:80–105.

1991b   *French Colonial Archaeology* (editor). University of Illinois Press, Champaign.

1991c   *The Traditional Potter in 19th Century Illinois: Archaeological Investigations at Two Kiln Sites in Upper Alton* (with Bonnie L. Gums and George R. Holley). Reports of Investigations No. 46. Illinois State Museum, Springfield.

1992   *Calumet and Fleur-de-Lys: Archaeology of Indian and French Contact in the Midcontinent* (co-editor with Thomas E. Emerson). Smithsonian Institution Press, Washington, DC.

1993   Chicago Stoneware Bottles. *Antique Bottle & Glass Collector* 10(2):12–13.

2001   French Colonial Material Culture from an Early Eighteenth-Century Outpost in the Illinois Country (with M. Kimball Brown). *Illinois Archaeology* 13:88–126.

2007   Building a Typology for Faience in the Mississippi Valley. In *French Colonial Pottery*, edited by George Avery, pp. 61–82. Northwestern State University Press, Natchitoches, Louisiana.

DEDICATION
TO PAUL WELKO AND EVA MOUNCE

We wish to dedicate this volume to two wonderful people who worked hard to help make *Bottled in Illinois: 1840–1880* a reality, but who passed away before the final book could be assembled and published. The research skills and historical knowledge that Paul Welko and Eva Mounce brought to our study were vital ingredients for making the finished volume a meaningful contribution to Illinois history and the historical-archaeology literature. Their unflagging enthusiasm for the project helped make it fun to navigate collections of historical records and to share with one another our discoveries of elusive information on 1840 to 1880 Illinois bottlers.

PAUL WELKO *(Figures D.1–D.2)* first discovered historic embossed-glass bottles through his work as an electrician at major construction projects in the Chicago area during the 1960s and 1970s. At the time, most of the 100-year-old embossed Chicago bottles

Figure D.1. Photo of Paul Welko and his wife, Lenore, on their 25th wedding anniversary, April 27, 1987.

Figure D.2. Paul Welko at an early 1970s Chicago bottle show.

xxi

that he encountered on construction-project backdirt piles were historically unknown. His curiosity about early local bottlers and their products led him to begin assembling examples of the many different styles and colors of the embossed Chicago bottles he found, and he soon became involved in helping to establish and nurture the *1st Chicago Bottle Club*. He published his historical discoveries about Chicago bottlers in his "*Did You Know...*" column in *Midwest Bottled News*, the Chicago club's newsletter (*see* Welko 1973, 1974, 1977a, 1977b, 1978, 1979, 1982).

We first met and became friends with Paul in the 1980s, so when we talked to him in 2005 about our idea of preparing a summary volume on historic Illinois bottlers and embossed-bottle use, and asked him to let us pick his brain and record his 40-year collection of early Chicago-area bottles, we knew his reaction would be enthusiastic and positive. He jumped into the project with both feet and also put us in touch with other dedicated Chicago-area bottle researchers like Ray Komorowksi and Bob Kott. Paul's offer to help was of pivotal importance: the Chicago section of the book ended up filling almost half the volume.

Before the book could be completed, however, Paul Welko passed away at age 66, on Sunday morning, August 2, 2009. We couldn't have done it without your help, Paul. We hope the finished book measures up to your contributions and to your expectations.

EVA DODGE MOUNCE (*Figure D.3*) began her historical research career as a northern Illinois housewife with a strong urge to learn all she could about the nineteenth-century stoneware industry in the Big Bend area of the Illinois River Valley—an antebellum and Civil War–era pioneer Illinois industry her grandfather had been involved in. For several years, as time allowed after she got her husband off to work in the morning, she (often with her children) visited county and state government archives and libraries in search of information on Illinois potteries, sought out the actual pottery kiln sites themselves, and spent many hours making collections of the kiln-site pottery sherds and kiln furniture they found while walking cultivated fields in northern Illinois.

Eva eventually parlayed her considerable skills at navigating state and local historical archives and libraries into a full-time position as historical archivist for the archaeological survey and excavation studies being conducted in advance of Illinois Department of Transportation road construction projects (working first for the Center for American Archeology in Kampsville and then for the University of Illinois). At that time she also helped form *The Foundation for Historical Research of Illinois Potteries*, subsequently authoring

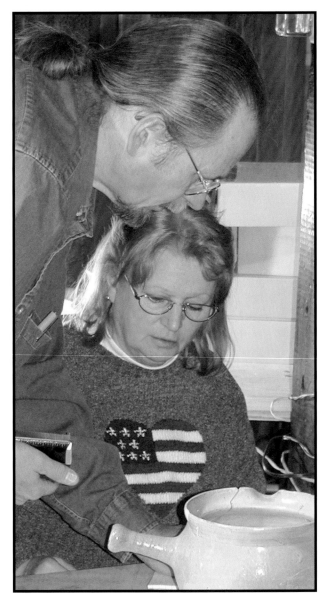

Figure D.3. Eva Mounce studying a collection of Alton, IL stoneware and earthenware in 2002 with Ken Farnsworth.

and coauthoring several historical publications on her favorite subject: Illinois historic stoneware manufacture (see Gums, Mounce, and Mansberger 1997; Mansberger and Mounce 1990, 1993; Mounce 1988, 1989; Mounce, Walthall, and McGuire 1988).

When in 2005 we started to work seriously on our book of historic bottles, we enlisted Eva's services as historical archivist. She embraced the project with her customary abandon, but she was not above chiding us by observing that our wide-ranging research interests (and frequent lack of project focus) suggested to her that we were unlikely to complete such an ambitious volume. When Eva passed away at the age of 61 on September 17, 2007, the bottle book was still far from done, and we were lucky indeed to be able to enlist the services of Curtis Mann (*Sangamon Valley Collection* librarian at the *Lincoln Library* in Springfield) as Eva's worthy successor to help us complete the historical archival research for the volume.

Eva, we sincerely wish you could have lived to see the published book—we would have especially relished the chance to finally be able to respond to your good-natured (and often-expressed) doubts with: *"See. We **can too** finish something!!!"*

Dedication References

Eva Mounce

Gums, Bonnie L., Mounce, Eva Dodge, and Floyd R. Mansberger
    1997   *The Kirkpatricks' Potteries in Illinois: A Family Tradition.* Transportation Archaeological Research Reports 3. Illinois Transportation Archaeological Research Program, University of Illinois, Department of Anthropology, Urbana.

Mansberger, Floyd R., and Eva Dodge Mounce
    1990   *The Potteries of Peoria, Illinois.* The Foundation for Historical Research of Illinois Potteries, Circular Series, Volume II, Number 1. Springfield, IL.
    1993   *The Potteries of McDonough County.* The Foundation for Historical Research of Illinois Potteries, Circular Series, Volume III. Springfield, IL.

Mounce, Eva Dodge
    1988   *The Potteries of La Salle County.* The Foundation for Historical Research of Illinois Potteries, Circular Series, Volume I, Number 1. Springfield, IL.
    1989   *Checklist of Illinois Potters and Potteries.* The Foundation for Historical Research of Illinois Potteries, Circular Series, Volume I, Number 3. Springfield, IL.

Mounce, Eva Dodge, John A. Walthall, and David A. McGuire
    1988   *The Potteries of White Hall.* The Foundation for Historical Research of Illinois Potteries, Circular Series, Volume I, Number 2. Springfield, IL.

Paul Welko

Paul Welko
    1973   J. A. Lomax—Revisited. *Midwest Bottled News* 4(2):1–2.
    1974   Chicago Soda Scene—A. J. Miller. *Midwest Bottled News* 5(1):5.
    1977a   Did You Know About Hutchinsons? *Antique Bottle World* 4(3):24–26.
    1977b   Bottle of the Month. *Midwest Bottled News* 8(4):9–10.
    1978   Did You Know About...Arthur Christin's Patent? *Antique Bottle World* 5(3):12–14.
    1979   Bottle of the Month. *Midwest Bottled News* 9(5):9–10.
    1982   J. A. Lomax—A Giant in the Early Chicago Bottling Industry. *Antique Bottle World* 9(1):6–7.

THE EMBOSSED AND STAMPED BOTTLES USED BY ILLINOIS MERCHANTS FROM 1840 TO 1880

STUDY FOCUS

The first excavations at Euro-American historic archaeological sites dating to the Domestic French Colonial and Transitional Control periods (1730–78), the American Frontier Development and Frontier Expansion periods (1778–1840), and the American Industrial Development period (1840–70) in Illinois (see Mazrim 2009) were conducted by university- and museum-based historical researchers as part of government-sponsored historical reconstruction projects during the early and mid-twentieth century. More recently, the scale of historical archaeology research has been greatly expanded—especially at residential sites dating to the American Industrial Development period—by legally mandated site survey, documentation, and excavation studies conducted in advance of publicly funded building projects (particularly those related to highway construction).

Such studies have produced an ever-increasing Illinois historical archaeology database during the past several decades. Unfortunately, age identification and functional documentation of many of the historical artifact classes recovered by such excavations has lagged woefully behind the pace of artifact recovery and site-interpretation needs. The present study addresses two such poorly documented artifact classes of the American Industrial Development period: the embossed glass bottles and stamped stoneware bottles used by nineteenth-century Illinois merchants from 1840 to 1880. The many embossed-glass bottles, jars, vials, and flasks from this early period reflect a rich Illinois heritage of pioneer, Civil War–era, and early industrial commerce by capitalists and con men alike. In the pages of this book, we describe and illustrate nearly 1,100 different Illinois embossed-bottle varieties produced before, during, and after the Civil War for close to 500 Illinois merchants operating in over 100 small towns and cities across the state, with populations ranging from just a few hundred souls to more than 100,000 people. Several small towns are represented by only a single embossed-bottle style used by one enterprising merchant (see, e.g., Henry Burger's Frogtown, IL, soda listing). At the opposite end of the spectrum, Chicago bottler John A. Lomax created the "Largest Bottling House in the U.S." during the 1860s and 1870s, using over 60 different embossed-bottle styles, in a rich variety of colors, to help sell his beverages. The authors have worked with historical archivists Eva Mounce and Curtis Mann to research the bottlers and bottled products included in our book—and 14 additional historical-research contributors have added their local and regional expertise and knowledge to help make the volume a reality.

Because of the daunting scale of the effort needed to document embossed and stamped bottle styles, user/maker marks, bottle contents, and product histories, the few existing pioneering published studies of such bottles used by early Illinois merchants provide only partial, often regional, thumbnail-outline lists with little associated historical information on the merchants and their products (Graci 1995, 2001; Oertle 1990, 1995, 2008, Paul and Parmalee 1973; Walthall 1993). For example, the most extensive of these studies (Paul and Parmalee 1973)—which focuses only on embossed Illinois soda bottles dating from 1840 to the 1940s—lists just six embossed sodas dating to the 1840–60 period (only three of which were recognized by the authors as dating to pre–Civil War times). Our current study documents, illustrates, and provides historical-context studies of 87 embossed soda/mineral water bottles of this age, used by bottlers in 46 Illinois towns ranging from Chicago to Cairo. For the 1860–80 period, Paul and Parmalee document only 43 blob-top soda styles statewide. Our current study provides comprehensive historical and stylistic information on 247 such bottles. The two Graci volumes list just 17 stamped stoneware bottles from Illinois, several of which were not identified as the product of Illinois bottlers at the time his publications appeared. Our current study provides historical and stylistic information on 37 such bottles *(see also "Late Additions" bottle section, this chapter).*

Because of the scale and scope of the present volume and the limited utility of previously published sources, our focus has been to provide detailed information on bottle styles and their embossed lettering and bottle contents, in addition to the histories of Illinois bottlers who used these particular types of glass and stoneware containers. The product manufacture and use information provided within these pages, combined with information from the archaeological sites where complete and fragmentary examples of the bottles were discarded, will no doubt be of use for overview studies of consumer behavior and patterns of product movement. But our immediate study focus has been to provide archaeologists and historians from Illinois and surrounding areas with clear and comprehensive information on 1840–1880 bottle styles, product contents, product functions (both real and imagined), and merchant histories, to aid in reconstructing the age of archaeological site occupations and in interpreting site functions and occupant activities.

Sources of Information

The best sources of information on the embossed bottles used by early Illinois bottlers are the bottles themselves. Although complete and nearly complete examples of such bottles are often found archaeologically, the vast majority of bottles in university and museum collections are fragmentary. By far the best source for complete examples of early embossed Illinois bottles are those preserved in private collections. Local and regional bottle collectors across the state were contacted at the beginning of our study to help create the initial photographic database for our book, and without exception, everyone we contacted enthusiastically agreed to open their homes to us to help build our photo inventory of pre-1880 Illinois embossed bottles *(see the Research Contributions list and Acknowledgments section, and our "Why Do the Study Now?" section, this chapter).*

Using the embossed-bottle database created in this fashion as the "outline" for our book, several standard categories of historical information were sought for each bottler/bottle entry in the volume. These include:

1. Full names and birthdates or ages of bottlers (to aid in correct identification and tracking of bottlers with common surnames)
2. First known year in bottling-related business (or last known year in unrelated business)
3. Last known year in bottling business (or first known year in subsequent unrelated business)
4. Names of all known or suspected bottled products, and the supposed or declared function of each (e.g., from labels or ads)
5. Colorful related stories and photographs or historical facts about the bottlers and their businesses that humanize individual bottlers or indicate business chronologies [e.g.,

factory settings/locations/sizes, number of employees (if any), months in business each year, setbacks/disasters, bottler national origins and gender, relationships or partnerships with other bottlers, and historical sequences of bottlers in particular towns—who bought out who, and why]
6. Photographs and/or drawings of the bottles themselves, as well as related nonbottle pictures, drawings, and maps whenever possible

Several categories of in-text citations did not rise to the level of bibliographic citations. These include federal, Illinois, and local municipality census information; commercial and industrial census data; city, state, and railroad business-directory listings; U.S. patent listings, and federal and state design-patent and trademark listings; local newspaper advertisements and articles; bottler-product catalogs and advertising almanacs; early credit agency reports (e.g., the early R. G. Dun files on file at the Harvard Business School's Baker Library); county history files; genealogical records; corresponding personal communications of local historical researchers; and online collector listings and bottle sales records.

Full citations (in-text and bibliography) were used for published books, journal and magazine articles, and privately and publicly held unpublished manuscripts containing sole-source information.

Sources of Misinformation

It must be noted that some—especially privately issued—historical data sources contain notoriously unreliable information. For instance, some smaller credit-reporting agencies did not have the resources (or the inclination) to check up on small-business credit-worthiness as often as quarterly or annually, and so continued to list no-longer-active businesses in their credit report volumes long after the businesses had closed their doors. (The worst examples consulted for the present study were the early-1870s volumes *McKillop & Sprague's "Commercial Agency Report"* of the early 1870s).

Statewide and railroad-generated annual and biennial business directories were also occasionally outrageously tardy in removing defunct businesses from their directories. For instance, the 1886 *Illinois State Gazetteer and Business Directory* listed Mrs. William Gerhardt as a mineral-water bottler in Litchfield. Mrs. Gerhardt had only run the business for a single season 10 years earlier to close down the business affairs of her recently deceased husband. The same volume listed both Albert Kershaw and John Ricks as Jacksonville soda-water bottlers. Kershaw sold his bottling business to Ricks in 1875, and then reassumed it for just a few years during the late 1870s and early 1880s when Ricks died in 1878. After 1883 Kershaw operated a Jacksonville saloon.

Finally, genealogically recorded personal-recollection narratives often include a wide range of good and bad information, depending on the reliability of the informant's memory.

It was also often the fate of home-based seasonal businesses (such as those of warm-weather soda and mineral-water bottlers) to be missed by census takers and city-directory compilers if they happened to gather their data at the wrong time of year. The seasonal disappearance and reappearance of not only soda bottlers but also those of ales, beers, and porters, was widely noted during the *American Industrial Development* era:

> The parching heat of summer demands some palatal alleviation; the fast-evaporating juices of the body require some judicious recruiting; the dry and fevered throat must be moistened; the dust, both physical and emotional, that comes to all men with the torrid airs of the dog-days, must be washed away; and ingenuity has provided,—in soda and mineral waters, in ales, beer, and porter,—the very beverages best calculated to afford relief and confer comfort. [*Chicago Tribune*, April 25, 1875, p. 5, col. 4—Article entitled "Malt, Beer, Ale, and Soda. Facts and Statistics Regarding Their Manufacture"]

Study Organization, Format, and Terminology

The embossed-bottle text, photographs, and line drawings that follow this introduction are organized alphabetically by town name. Individual bottlers, bottle styles, and product names can also be tracked by using the master index at the end of the volume. The text listings include 1840–80 descriptive bottle-style details and bottler histories for 489 bottling companies operating from 1840 to 1880 in 101 Illinois towns located in 60 of the state's 102 counties.

Recent bottle discoveries that have come to light since the text section of the volume was completed (*see Study Completeness section*) have added 15 new embossed-bottle styles (for a total of 1,093) and two new stamped stoneware bottle styles (for a total of 39). These late additions add five new bottling companies (for a total of 494) and one new bottling town (now 102) to the bottlers and bottling locales discussed in the text. The new bottling town, De Kalb, is a county seat, increasing the number of county seat towns in the study to 45. Table I.1 incorporates these news additions.

Table I.1. County Distribution of 102 Bottling Towns and 494 Bottlers in 60 of 102 Illinois Counties (*see Figure I.1*).

County	Bottling Towns (# of Bottlers)	County	Bottling Towns (# of Bottlers)
Adams	**Quincy** (19)	Kane	Aurora (4)
Alexander	**Cairo** (6)		Elgin (7)
Bond	**Greenville** (2)	Kankakee	**Kankakee** (1)
Boone	**Belvidere** (1)	Knox	**Galesburg** (6)
	Capron (1)	Lake	**Waukegan** (1)
Bureau	**Princeton** (1)	LaSalle	LaSalle (8)
Cass	Beardstown (2)		Mendota (4)
Champaign	West Urbana/Champaign (2)		**Ottawa** (7)
Christian	Pana (1)		Peru (1)
Clinton	Breese (1)		Streator (1)
	Carlyle (4)	Lee	Dixon (2)
	Frogtown (1)		Franklin Grove (1)
Coles	**Charleston** (2)		Nelson (1)
	Mattoon (1)	Livingston	**Pontiac** (1)
Cook	Blue Island (4)	Logan	**Lincoln** (2)
	Chicago (185)	Macon	**Decatur** (3)
	Colehour (2)	Macoupin	**Carlinville** (3)
	Arlington Heights/Dunton (3)		Staunton (2)
	Evanston (1)		Virden (1)
	Lemont (2)	Madison	Alton (11)
De Kalb	Sandwich (1)		Collinsville (2)
	De Kalb (1)		**Edwardsville** (1)
Du Page	Naperville (3)		Highland (5)
Edwards	**Albion** (1)		Lebanon (2)
Fayette	Louden City (1)	Marion	Centralia (3)
Fulton	Farmington (1)	Marshall	**Lacon** (1)
	Lewistown (1)	McHenry	**Woodstock** (1)
Henderson	**Oquawka** (1)	McLean	**Bloomington** (17)
Henry	Galva (1)	Monroe	Columbia (1)
	Kewanee (1)		**Waterloo** (4)
Greene	**Carrollton** (1)	Montgomery	Litchfield (3)
Grundy	**Morris** (1)		Nokomis (1)
Iroquois	Sheldon (1)	Morgan	**Jacksonville** (9)
Jackson	**Murphysboro** (1)		Prentice (1)
Jefferson	**Mt. Vernon** (1)	Ogle	Forreston (1)
Jersey	**Jerseyville** (3)		Maryland (1)
Jo Daviess	**Galena** (3)	Perry	DuQuoin (2)

Table I.1 (continued). County Distribution of 102 Bottling Towns and 494 Bottlers in 60 of 102 Illinois Counties.

County	Bottling Towns (# of Bottlers)	County	Bottling Towns (# of Bottlers)
Peoria	**Peoria** (29)	Stephenson	**Freeport** (2)
Pike	Perry (1)	Tazewell	**Pekin** (3)
	Pittsfield (1)	Union	**Jonesboro** (1)
Randolph	Chester (1)		Western Saratoga (1)
	Red Bud (1)	Vermillion	**Danville** (4)
Richland	**Olney** (1)	Will	Braidwood (1)
Rock Island	**Rock Island** (5)		**Joliet** (3)
Sangamon	**Springfield** (20)		Monee (3)
Scott	Naples (1)		Wilmington (1)
St. Clair	**Belleville** (9)	Winnebago	**Rockford** (2)
	Centreville (1)	Woodford	Metamora (1)
	Darmstadt (3)		
	East St. Louis (3)		
	Freeburg (1)		
	Mascoutah (3)		
	New Athens (1)		
	O'Fallon (2)		

County seats are **bolded**.

Descriptive Format

In the text, each alphabetically listed town is followed by county location and by census data for the time of our study, when available. Bottlers are then listed alphabetically, followed by a listing of each embossed or stamped bottle style used, and the exact embossing/stamping employed on each bottle. For instance, sample Alton embossed-bottle listings include:

Bitters:

(1) **MAGNOLIA / BITTERS** // *(blank)* // **FRED. INGLIS /**   *[amber, olive-amber]*
    ALTON. ILL // *(blank)* //

    »SB w/ keyed hinge mold; tapered collar; square case bottle; 10" tall. *[Figure 9]*

Ale:

(1) **KEELEY & BRO** // **ALTON ILL**   *[dark olive "black glass"]*

    »IP; quart; ring-neck tapered collar; 3-piece mold; 9 3/4" tall. Embossing confined to the bottle's shoulder. *[Figure 10]*

Medicine:

    ESSENCE / JAMAICAN GINGER / *(embossed "propeller" decoration)* /   *[aqua]*
    QUIGLEY. HOPKINS & LEA / ALTON / ILL

    »SB w/ keyed hinge mold; shouldered oval; squared ring collar; 5 1/2" tall. *[Figure 11a]*

Soda ~ Mineral Water:

(1) **C. WEISBACH** // **ALTON / ILL**   *[aqua]*
    Note: Early 1860s style; flat-lettered front and back embossing. *[Figure 14]*

    »SB w/ keyed hinge mold; shouldered-blob-top soda.

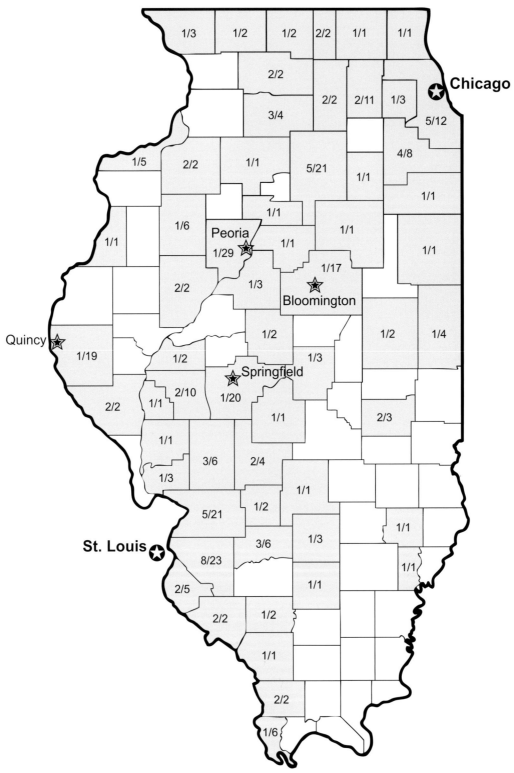

Figure I.1. Illinois county distribution map of 1840–1880 bottlers using embossed or stamped bottles. Numbering shorthand [#/#] = towns per county/bottlers per county. Note the highlighted "empty zones" in a 7-county area of northwestern Illinois (one bottler) and a 29-county area of the southeastern third of the state (five bottlers). This is a reflection of the low immigrant population densities in these areas during our study period (less than 5/mi^2: *see Figure I.39*).

(2a) **C. WEISBACH** *(arched)* / **ALTON.ILL** //   [*aqua, blue teal*]
   (heel): **A. & D. H. C**
   (base): **C.W.**

   »SB w/ keyed hinge mold; oval-blob-top soda. *[Figure 15]*

(2b) **C. WEISBACH** *(arched)* / **ALTON.ILL** //   [*aqua*]
   (heel): **A. & D. H. C**

   »SB w/ keyed hinge mold; oval-blob-top soda. *[see Figure 15a–b]*

(3) *[no body embossing]*   [*aqua*]
   (heel): **A. & D. H. C** *[Figure 15b, #2]*
   (base): **C.W.** *(larger base embossing than #2a above—see Figure 15c, #2)*
   Note: Early hinge-molded, high-shouldered cylinder, "selters-water"–style bottle. *[Figure 16]*

   »SB w/ keyed hinge mold; high-shouldered; oval-blob-top soda.

(4) **C. WEISBACH** / **ALTON** / **ILLS.**   [*amber*]
   Note: Embossed in late-1870s circular slug plate with decorative embossing surrounding **ALTON** and **ILLS.**

   »SB w/ post mold; oval-blob-top quart; 11¼" tall. *[Figure 17]*

The standard symbols and terms used for these listings include those for line breaks of embossed or stamped text:

   / = line break for text on one side of bottle
   // = text break between opposite sides (on oval or round bottles) or between adjacent panels (on square, rectangular, or multiple-paneled bottles)
   // // = blank panels following panels with embossed text. Thus a four-sided bottle with two lines of embossed text on only one side would be listed as: **J. JONES** / **ALTON ILLS.** // // // //

The only other symbolic notation (») is the lead-in key for the descriptive list of bottle shape, style, size and basal mold-line or empontiling attributes. Bottle colors are listed in square brackets following the first line of embossed or stamped text. Figure numbers appear in square brackets following the list of bottle shape, size, mold, and manufacturing-technology details.

Several standard abbreviations are used in the bottle descriptions and the bottler history sections that follow the descriptions. These include shorthand notations for several glass-manufacturing technology terms (**OP**, **IP**, **SB**) and many heavily used sources of commercial history information (**CAR, CD, IIC, ISD, USC, RRD**).

Abbreviated Bottle-Technology Terms

See further discussion under *Manufacturing Technology Terms*, this chapter.

   OP = **O**pen **P**ontiled (also called "open-rod" pontiled and "blowpipe" pontiled). This refers to the scar (an open ring of roughly broken glass) that appears on the bases of smaller bottles manufactured for use in Illinois prior to 1860.

   IP = **I**ron **P**ontiled (terms such as "graphite" pontiled and "sand" pontiled are misleading and have little utility for bottles used in Illinois; see discussion below). "**IP**" refers to the scar comprised of a thin veneer of oxidized iron rust attached to a depressed area on the bottle base left on the bases of larger bottles manufactured for use in Illinois prior to 1860.

SB = **S**mooth **B**ased. This term highlights the shift—at about the midpoint of our study (ca. 1860)—from bottles with pontil-scarred bases to those produced in snap-case molds and having smooth bases (see patent-chronology discussion below).

ABBREVIATED TERMS FOR COMMERCIAL-DATA SOURCES

CAR = The *Commercial Agency Register* was an annual business credit-worthiness report produced by McKillop and Sprague during at least the early-to-mid 1870s (similar in concept to those being produced today by Dun and Bradstreet). The company's lack of ability (or concern) to annually update their listings lead to surprisingly inaccurate bottler listings (*see Sources of Misinformation, this chapter*).

CD = **C**ity **D**irectory. City business directories were produced for midsized Illinois towns as often as biannually. For major metropolitan areas like Chicago, business directories were produced every year, often by two competing companies. They were published irregularly for smaller towns and villages—and often only in conjunction with one-time business-community efforts or as part of the publication of county histories. They are an invaluable source for documenting the business histories of small, lightly advertised, short-lived companies, but depending on publication schedules, seasonal bottling businesses were often missed. Also, few such directories were published during the disruptive era of the Civil War. The number of such directories published during our 40-year study period is so great that they could not be individually listed in the bibliography; instead they are noted only by year (e.g., 1861 CD).

IIC = **I**llinois **I**ndustrial **C**ensus. The state of Illinois gathered extremely valuable industrial census data at 10-year intervals throughout the 1840-to-1880 period of our study. The industrial census was timed to coincide with the U.S. Census, but focused solely on the scale and extent of business operations within the state. Census takers gathered a fascinating array of operational and cost information from the business owners they interviewed (including months in operation; number of male, female, and child employees; salary rates; kinds and costs of raw materials; and annual value of product produced). Businesses recorded by each census were only a small fraction of those in operation at the time, however, and some of the census files indicate that large areas of the state were left uncovered during particular census periods. However, the information that was gathered provides a fascinating window into the operational details of early Illinois bottling businesses, and was searched in some detail by historical archivist Eva Mounce to help inform our study.

ISD = **I**llinois **S**tate **D**irectories were produced during the time of our study for many of the same reasons that city business directories were published, but the ISDs claimed to cover wider regions of the state, or even Illinois as a whole (see also the railroad directory entry in this list). Because of the large area included in such directories, their potential inaccuracy and incompleteness in detail was greatly increased. Vast numbers of state businesses were not included in their pages at all, and they had a tendency to retain listings of long-dead business operations and operators. As the result, the ISDs are relatively poor data sources for developing Illinois business chronologies (*see Sources of Misinformation discussion, this chapter*).

USC = **U**.**S**. **C**ensus data has been extremely valuable for tracking the activities and movements of individual bottlers, glassmakers, and associated wholesalers and product developers (e.g., doctors, pharmacists, and other "concoctors" of bottled products) for the 40-year period of our study. Census data was searched in some detail by historical researcher Curtis Mann to help inform our study.

RRD = RailRoad Directories produced during the period of our study were very similar to the regional and statewide **ISD**s discussed above, in that they provided business directories for wider regions of the state than individual **CD**s did. While they suffer some of the same weaknesses mentioned for **ISD**s, **RRD**s were focused only on the businesses operating in the several towns along each railway route. As a result, they have generally been much more reliable sources of business history and location than the **ISD**s.

Terms Used to Describe Bottle Color, Shape, and Manufacturing Technology

Colors

The bottle color terms used for bottle descriptions in the text have been chosen to reflect terms used by archaeologists and glass collectors to distinguish between the natural colors of aqua glass and the gray-clear appearance of flint glass, and to identify colors, color shades, and color mixtures used for individual glass orders. When colored, the Illinois bottles discussed in the text were characteristically shades of blues, greens, ambers, and various mixtures thereof. Other colors (e.g., pink, red, puce, or purple) occur very rarely, and such instances have been individually described.

The color term "aqua" denotes the pale blue or green shades of glass approximating that of water. This "natural" color of raw "green" glass will vary somewhat depending on the original formula for its manufacture and the kinds and quantities of refuse-bottle "cullet" repurchased by glass manufacturers and added back into the vats of raw glass. More dramatic glass colors intentionally added to specific bottle orders have been divided into the following groups:

1. light blue ("robin's egg") ~ blue ~ cobalt ~ dark cobalt
2. light green ~ yellow green ~ lime green ~ forest green
3. yellow (or yellow-amber) ~ brown amber ~ red amber ~ dark amber ~ amber "black glass"
4. olive-amber ~ olive-green ~ olive "black glass"
5. teal (blue and green varieties).

Some of these colors were chosen for visual impact, and probably sometimes to help distinguish between several similar products offered for sale by one bottler. Others, especially the darker ambers and "black glass" colors, were used to help preserve the quality and flavor of some of the more unstable bottled products (such as ales, porters, and lagers) in an era that largely predates pasteurization or refrigerated transportation and storage.

Embossing and Stamping

To aid in determining the chronology of bottle manufacture for individual Illinois merchants, as well as the manufacturing glass house or pottery works for early Illinois product bottles, the style and line-by-line placement of embossed and stamped lettering on each bottle is described in detail, and whenever possible are illustrated photographically and/or by artist drawings. The locations of lettering on the sides or faces of the vessel body, heel, and base are also noted. Punctuation and plain or embellished (serifed/ornate) lettering style (and superscript style for abbreviated words) are also described and illustrated. This level of detail is also provided as an aid to archaeologists, who often must identify embossed and stamped product bottles of Illinois merchants from very fragmentary excavated examples.

Measurements and Volumes

At least one metric measurement is provided for most described bottles, which when combined with the bottle illustrations will allow close estimations of the scale of each product-bottle described. However, it should be noted that these are handmade products, and there will be some size variation in individual examples. This is especially true of the "long-neck" blob-top soda and mineral water glass bottles recorded in our study. After being blown into their mold, such bottled were trimmed of at the neck for attachment of the blob tops. As a result, individual examples of these bottles—which were manufactured for Illinois bottlers primarily during ca. 1850–80—exhibit a wide range of variation in their final height. Individual soda bottle heights, for bottles of essentially the same capacity, may range from about 63/4 to 77/8 inches (*Figure I.2*). Their height generally decreases during the 30-year period of their manufacture for Illinois merchants (perhaps because of increasing concern for glass economy), but this is by no means a rigorous progression. Thus, for soda and mineral-water bottles in particular, no height measurements are provided in the text.

Bottle capacity terms are also somewhat generic, and thus fairly unreliable as actual measures of content volume. Small (usually medicinal) bottles are often sized as 1-ounce, 2-ounce, or 4-ounce containers (assuming that each larger size of the bottle had about twice the product capacity of the next smaller size), and larger beverage bottles are generally casually referred to as "pints" and "quarts." But these should be considered terms of convenience only, and the present study has not made an effort to more rigorously quantify these generically estimated vessel capacities, since for most of these early bottles few complete examples are known. In the case of early blob-top soda bottles, for instance, depending upon blown-glass thickness, high-shouldered quarts (qt = 32 oz) would have held about 28 oz of liquid, while low-shouldered and squat Chicago-style quarts would have held about 25 oz of liquid. "Large" Lomax-style pints (pt = 16 oz) would have held about 12 oz of liquid, while standard high-shouldered and slope-shouldered "pints" were actually half-pint bottles that would have held about 8 oz and 7 oz of liquid, respectively.

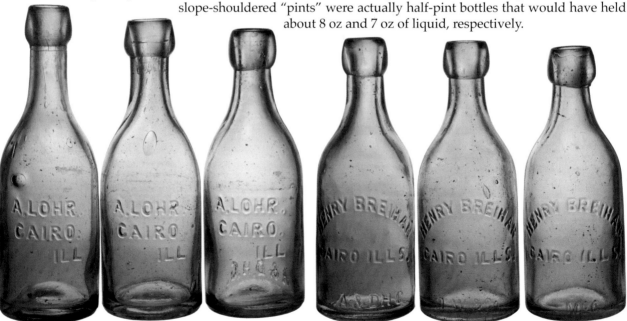

Figure I.2. Variable neck lengths of soda bottles used in Cairo, IL, during the 1860s and 1870s by the Andrew Lohr and Henry Breihan bottling works.

Manufacturing-Technology Terms

Open Pontiled
(Also called "open-rod" pontiled and "blowpipe" pontiled.)

This term refers to the scar left on the bases of smaller bottles manufactured for use in Illinois prior to 1860 (*Figure I.3a*). For small bottles (usually small medicine bottle; rarely bottles as large and heavy as sodas), the punty rod attached to the bottle base when it was removed from the mold into which the embossed body of the bottle was blown was a hollow glass rod—usually about 1/2" to 1" in diameter. The rod was used to hold the bottle while the lip was finished or a separately formed lip was attached and then broken away, leaving a scar comprised of an open ring of glass attached to the base.

a1

a2

Solid glass rods were sometimes used for this purpose on bottles made for midwestern markets, but examples are very rarely seen. Solid-rod empontiling is usually a European or South American manufacturing hallmark. The only such bottle recorded for our Illinois study is a Chicago liquor decanter (*see the Chicago text listing for Thomas & Co.*).

Iron Pontiled
(Terms such as "graphite" pontiled and "sand" pontiled are misleading, and have little utility for bottles used in Illinois.)

a3

This term refers to the scar left on the bases of larger bottles (usually larger medicine bottles, soda-sized bottles, and large, heavy quart bottles) made for use in Illinois prior to 1860 (*Figure I.3b*). In this case, the punty rod tip that was pressed into the bottle base was made of solid iron. The rod was usually round—although occasional iron pontil marks are oval or rectangular—and about 1" to 2" in diameter. As with the hollow glass rods used for smaller bottles, the iron punty was used to hold the bottle when it was removed from the mold into which it was blown, while a finished lip was shaped or a separately formed lip was attached. The rod was then dislodged from the bottle, leaving a scar comprised of a thin veneer of oxidized iron rust attached to the bottle base.

a4

Rods used during the 1840s and early 1850s produced a deeper kick-up in the bottle base: some of the earlier 1840s examples came to a slightly rounded point; the attachment ends of later 1840s and early-to-mid 1850s rods were characteristically hemispherical; by the end of the 1850s ("last-gasp" pontiled times at about the time of the advent of snap-case technology; *see the patent chronology discussion, this chapter*) were cut almost flat at their attachment end.

a5   b1

Terms such as "graphite" and "sand" empontiling seem to have entered the blown-glass literature to describe the early manufacture of more refined tablewares (Jones 1971; Van den Bossche 1999:62–64) and to accommodate the fact that modern-day "traditional" glassblowers often coat the tips of their punty rods with graphite or sand to facilitate detaching them from bottle bases after the lips have been applied. The fact that iron oxide appears to be the pontil-rod material attached to pre-1860 bottles used by Illinois bottlers (graphite

b2

Figure I.3. Examples of the open glass-rod pontil scars (a1–5) seen on smaller embossed bottles and the iron-pontil scars (b1–3) seen on larger embossed bottles used by Illinois bottlers during the 1840s and 1850s.

b3

Smooth Based

This term highlights the shift—at about the midpoint of our study (ca. 1860)—from bottles with pontil-scarred bases to those produced in snap-case molds and having smooth bases (see patent chronology discussion, this chapter). Smooth-based bottles often have chronologically sensitive mold-seam impressions preserved in the glass. When present, these seam patterns are noted to help date the bottles (*see the terminology discussion, this chapter*).

Slug Plate

This somewhat redundant term (Ring 1980:32) refers to sharp-edged, slightly depressed areas surrounding the embossed lettering (or covering a portion of previously embossed lettering that was later removed) beginning with later commercial product bottles.

Later removable slug-plates were characteristically circular or sub-oval in outline (occasionally "tombstone" shaped) on circular and oval bottles, and were usually panel shaped on square or rectangular bottles. They resulted from the development of stamped insert plates, patented in the later 1870s, which allowed commercial bottles to be embossed without having to recut one face of an iron mold to accommodate each new order (*see the patent chronology discussion, this chapter*).

Peened Out

During much of the 1840 to 1880 period, prior to the development of insertable/removable slug plates in the later 1870s, embossed lettering for glass bottles was cut directly into the metal mold forms into which the bottle was blown. When portions or all of the cut-in lettering was removed or changed during these earlier years, "solder or other soft metal [*was*] tapped into the lettering cut into the mold. The marks of the peening hammer seem like a string of round marks. Usually the removed word(s) can be detected when the bottle is viewed under intense light" (Ring 1980:32).

Body Mold Lines

Mold lines preserved on the body and base of blown-glass bottles indicate the closure points for iron (or in some cases, brass) bottle molds. Most bottle molds were simple two-piece molds. The round or oval bottles blown in them have vertical mold lines from heel to neck on opposite sides. Usually bottle bases were incorporated into the two-piece molds by means of a central straight-line or "keyed" hinge-mold closure, but sometimes (especially during the 1870s) a third "cup" mold or "post" mold element was incorporated into the bottle base (*see the Basal Mold Line discussion, this chapter*).

Ale-bottle manufacturing technology, in particular—especially that used for making quart-sized early "black glass" amber and olive ale bottles during the 1850s and early Civil War years—resulted in bottles with mold lines indicating that three-piece molds and four-piece (perhaps five-piece) molds were used. The body of three-piece mold ales was blown into a single-piece dip mold. The second portion of the metal mold, from the shoulder to the neck, was a two-piece hinge mold, resulting in shoulder-to-neck vertical mold lines on opposite sides of the bottle and a horizontal mold line ringing the shoulder. Some of these early ale bottles have vertical mold lines on opposite sides that extend from heel to neck, as well as a horizontal mold line ringing the shoulder, indicating that the upper and lower portions of the metal mold were "stacked" two-piece hinge molds, resulting in a four-piece mold bottle. In some instances keyed hinge-mold lines on the vessel base indicate that the lower two-piece mold included the bottle base as well (*see the Basal Mold Line discussion, this chapter*). However many pre-1860 ales show no evidence of either basal mold lines or pontil scars. This may be evidence that some of the molds into which the earliest embossed ale bottles were blown were actually five-piece molds, in the sense that the metal stand on which the two stacked two-piece molds were mounted included a shaped basal depression for the bottom of the bottle (below the heel). If so, the fifth piece of the mold would essentially be a very shallow dip mold for the bottle base.

Body Shape and Style Terms

Most of the terms used in this volume to describe the body shapes of embossed pre-1880 bottles are self-evident: cylindrical, oval, barrel-shaped, tenpin-shaped, cone-shaped, and sunken-panel or flat-panel rectangles or squares (there are also a few paneled bottles with more than four sides). Some round bottles have decorative panels on the lower body above the heel: these are referred to as mug-based (*see Figure I.4a*). These and other decoratively paneled bottles (e.g., those with shoulder-panels) may have concave rather than flat panels. These are called fluted (*see Figure I.4b*). The panels are so narrow on some multisided cylindrical bottles that such bottles are referred to as "xx-sided cylinders" (*Figure I.4c*).

As with the full beveled-edge sunken panels on square and rectangular bottles, smaller sunken panels on early Illinois bottles were designed as depressed areas to protect embossed lettering or attached paper labels from use-wear. The partial sunken panels are also referred to by their shape (e.g., square, arched, round, oval, tombstone-shaped, shield-shaped). The "shield-shaped" designation has been traditionally applied to square or rectangular sunken panels that also have cut corners (see, e.g., George Lomax sodas in the Chicago text section below), even though such panels do not have a classic shield outline.

From the early days of factory production in the Midwest, bottles were designed as distinctively shaped containers for particular contents, probably so that they would be easily recognizable to consumers. This was especially true for bottled products that did not need attached paper labels to clarify the functions or dosages of their contents. But the style names assigned by collectors and archaeologists to the many distinctive bottle types that were manufactured during the 40-year period of our study often do not particularly evoke the bottle shapes themselves (e.g., case gin, shoofly flask, schoolhouse ink, Saratoga mineral water, citrate of magnesia, ginger ale). For that reason, examples of each of the named types frequently used by early Illinois bottlers are shown in the photographs below.

Glass Bottles

    Liquor Products
      Barrel Whiskey (*Figure I.5*)
      Whiskey Cylinder (*Figure I.5*)
      Case Gin (*Figure I.5*)
      Onion-shaped Whiskey Decanter (*Figure I.5*)

      Handled Chestnut Flask (*Figure I.6*)
      Strap-Sided "Union" Flask (*Figure I.6*)
      Sharp-sided "Coffin" Flask (*Figure I.6*)
      "Hotel Oval" Flask (*Figure I.6*)
      Pumpkin Seed Flask (*Figure I.6*)

Figure I.4. Varieties of paneled and fluted bottle decoration.

Figure I.5. Shape varieties of larger-size whiskey bottles and decanters used by Illinois bottlers during the study period.

Figure I.6. Shape varieties of smaller whiskey flasks used by Illinois bottlers during the study period.

Brewery Products
  4-Piece Mold Quart Ale (*Figure I.7*)
  3-Piece Mold Quart Ale (*Figure I.7*)
  2-Piece Mold Quart Ale (*Figure I.7*)
  3-Piece Mold "Pint" Ale (*Figure I.7*)
  2-piece Mold "Pint" Ale (*Figure I.7*)

  Porter (*Figure I.8*)
  Squat Porter (*Figure I.8*)

  Quart Blob-top Beer (*Figure I.9*) (amber: cf. Quart Soda)
  Beer (*Figure I.9*)

Square Bitters (*Figure I.10*)

Sodas and Mineral Waters
  High-Shouldered Soda (*Figure I.11*)
  Slope-Shouldered Soda (*Figure I.11*)
  Quart Blob-top Soda (*Figure I.11*) (aqua and colored: cf. Quart Beers)
  Squat "Chicago-style" Aqua Quart (*Figure I.11*)

  Cantrall-style Ginger Ale (*Figure I.12*) (round, semi-round, & flat)
  Aerated Medicated Water (*Figure I.12*)
  Cantrall-style "Dumpy" Soda (*Figure I.12*)

  Saratoga-style Mineral Spring Water (*Figure I.13*)
  Selters-style Mineral Spring Water (*Figure I.13*)

Figure I.7. Shape and size varieties of quart- and pint-size ale bottles used by Illinois bottlers during the study period. Bottles shown include (a–b) four-piece and three-piece mold quart ales, (c–e) three-piece and two-piece mold pint ales, and (f) a two-piece mold quart ale.

Figure I.8. Shape and size varieties of half-pint porter bottles used by Illinois bottlers during the study period.

Figure I.10. Examples of 1860s (left) and 1870s (right) beveled-edge square quart bitters bottles used by Illinois bottlers.

Figure I.9. Examples of mold-engraved and slug-plate embossed pint (a) and quart (b) lager beer bottles used by Illinois bottlers during the 1870s.

Figure I.11. Examples of high-shoulder and low-shoulder pint sodas (a–b), slope-shoulder quart sodas (c), and long-neck vs. squat Chicago-style quart sodas (d–e) used by Illinois bottlers.

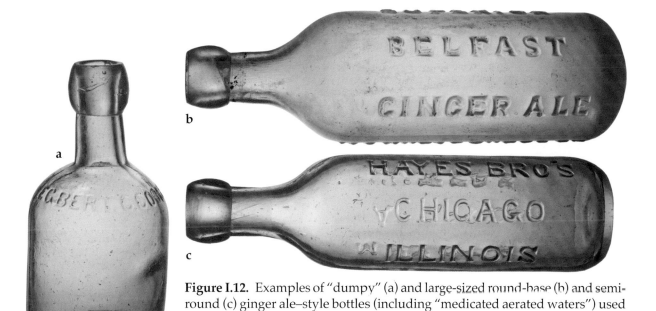

Figure I.12. Examples of "dumpy" (a) and large-sized round-base (b) and semi-round (c) ginger ale–style bottles (including "medicated aerated waters") used by Illinois bottlers during the 1870s.

Figure I.13. Examples of Saratoga-spring style (a) and Selters-spring style (b) mineral-water bottles used by Illinois bottlers in the post–Civil War era.

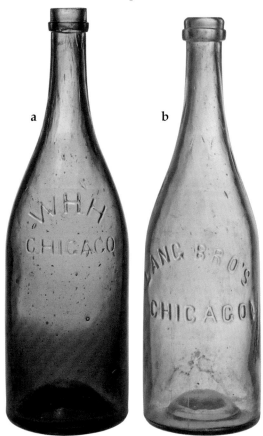

Figure I.14. Examples of early (left) and late-style (right) ring-necked cider quarts used by Illinois bottlers during the 1870s.

Ciders
  Ring-neck or "Champagne" Cider (*Figure I.14*)

Inks
  Master Ink (*Figure I.15*)
  Cone Ink (*Figure I.15*)
  Schoolhouse Ink (*Figure I.15*)

Patent Medicines
  Cylindrical and Oval Medicine Vials (*Figure I.16*)
  Oval Medicines (*Figure I.16*)

  Rectangular and Square Medicines (*Figure I.17*)

  Unguent or Salve Jars (*Figure I.18*)

Citrates (*Figure I.19*)

Pickles and Preserves
  Pickles (*Figure I.20*)
  Perrine Honey Bottles (*Figure I.20*)
  Fruit (Preserve) Jars (*Figure I.20*)

Stoneware Bottles

Tapered Stone Sodas (*Figure I.21*)
Sharp-Shouldered Stone Sodas and "Small" Beers (*Figure I.21*)
Smith and Merrill Patent Paneled Stone Beers (*Figure I.21*)

Figure I.15. Embossed Illinois ink-bottle shapes of the 1860s and 1870s: (a) master ink; (b) schoolhouse ink bottles; and (c) cone ink.

Figures I.16. Round, cylindrical (a) and oval, lens-shaped (b) medicine bottle shapes used by Illinois bottlers during the study period.

Figure I.17. Beveled-edge flat-paneled (a) and sunken-paneled square and rectangle (b) medicine bottle shapes used by Illinois bottlers during the study period.

Figure I.18. Wide-mouthed salve and unguent medicine bottle shapes used by Illinois bottlers during the study period.

Figure I.19. Citrate-style cylinder medicine bottle shape used by Illinois bottlers during the study period.

Figure I.20. Illinois preserve-jar styles of the 1860s and 1870s: (a) pickle bottles; (b) honey jars; and (c) canning jars.

Figure I.21. Stamped stoneware bottle shapes used by Illinois bottlers during the study period: (a) wide-shoulder and slope-shoulder sodas or mineral waters; (b) sharp-shouldered pint and quart "small" or "root" beers; (c) ten-paneled Smith patent quart root beers; (d) twelve-paneled Merrill-patent narrow and wide "small" or "root" beers.

Base Shapes, Kick-ups, "Decorations," and Mold Lines

Most embossed bottles from the 1840 to 1880 period in Illinois are flat based, but some (particularly cylindrical sodas, medicated waters, and ginger ales) are round based and would not stand alone (*Figure I.22*), while other such cylindrical bottles have semi-round bases (*Figure I.22*) and would stand alone only precariously. Tenpin-shaped sodas also had unstable small, flat base areas (*Figure I.22*). Such bottles were likely placed in a pouring-harness for use, and stored on their sides. Received wisdom (and basal embossing on many bottles) suggests that wine-rack-like side storage and upside-down case storage of beverage and aerated-liquid bottled products helped keep corks moist and bottles tightly sealed.

Cup-mold and deeply indented "champagne-style" central basal kick-ups on both small and large flat-based bottles were perhaps used to invoke a more refined, genteel image for champagne-cider-style carbonated beverages.

Some bottle bases also have raised hemispherical glass "dimples" or "bumps" either at the central base (1–3 raised-glass bumps in a line; *see Figure I.23*) or near the outer edge (a triangular array of three raised-glass bumps). These may have been decorative, but they more likely resulted from depressions cut in the basal mold to form blown-glass "anchors" or "feet" to hold the newly formed bottle in place while it initially cooled thus keeping the embossing as crisp as possible. Raised bumps such as these are usually seen on larger (quart-size) ale and bitters bottles.

Basal mold lines in glass bottles reflect the closure of individual two-piece iron molds. Such mold patterns changed through time, largely to accommodate basal embossing of bottles, and can be an important chronological marker for bottle age during the period of our study. Some early embossed bottles (usually small ones) used by Illinois bottlers have a simple straight line mold seam across the mid-base (*Figure I.24*). However, most such bottles have half-circle keyed hinge molds (*Figure I.24*). Square keyed hinge molds are also seen occasionally. During the first half of our study period (1840–60), these mold lines are obscured by iron-rod and open-glass-rod pontil scars (*see pontil discussion, this chapter*). With the advent of smooth-based bottles made in snap-case molds (ca. early 1860 in Illinois), keyed hinge molds are clear and nearly ubiquitous on most bottles. However, some bottles of this age show no mold lines on the base. Some of these may be bottles blown into cold molds; others have clear heel-mold lines, indicating that they were formed in molds with a slight basal dip mold on the mold platform. By ca. 1871, keyed hinge molds are largely replaced by post molds in Illinois (*Figure I.24*; *see the patent chronology discussion, this chapter*). These molds closed around a central post element, upon which small, interchangeable slug plates could be mounted to easily accommodate and alter basal embossing. Although smooth-based and heel-mold bottles still appear, most Illinois bottles manufactured during the 1870s (and for decades thereafter) have post-mold bases.

Shoulder Shape and Style Terms

Shape terms used to indicate how bottles narrowed at their shoulders are again largely self-evident. Bottles are characterized as high-shouldered, slope-shouldered,

Figure I.22. Round (a), semi-round (b), and tenpin-shaped (c) bottle bases.

Figure I.23. Examples of raised-glass embossed "dimples" on the bases of Illinois quart ales of the 1860s and 1870s.

sharp-shouldered, domed, or bell-shaped. Wide-mouthed bottles or vials with very little shoulder constriction are described as weak-shouldered.

The shoulder areas of many bottles also have decorative detailing. These include applied (usually embossed) glass seals (*Figure I.25*), raised shoulder rings (*Figure I.25*), ridged shoulders (*Figure I.25*), decorative shoulder bumps (*Figure I.25*), fluted shoulders (*see Figure I.4*), and petal-paneled (front) shoulders (*Figure I.25*). Petal-paneled soda bottles are pontiled, with five full panels on the reverse side, and they usually date to the later 1850s. Petal-paneled bottles are also sometimes associated with Selters-style mineral waters.

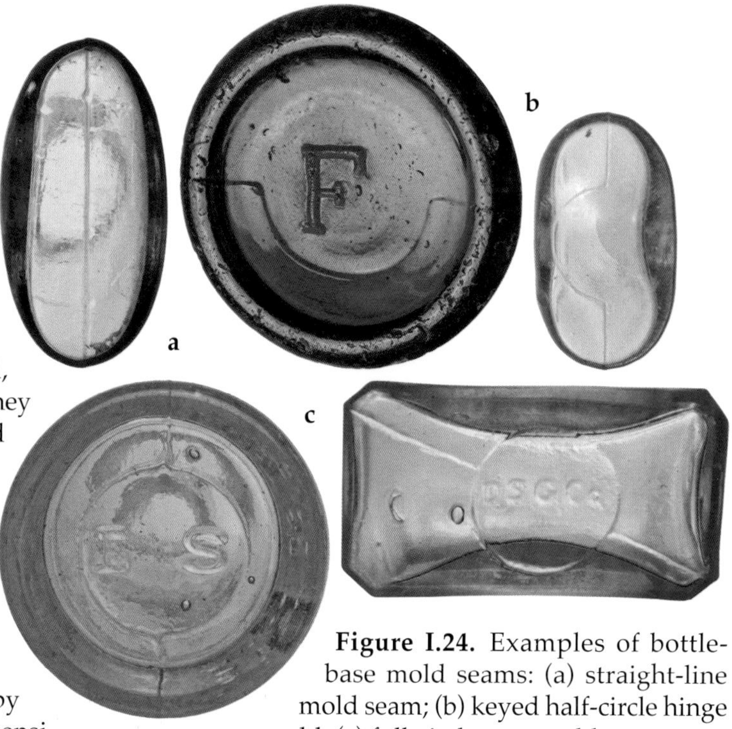

Neck Shape and Style Terms

Simple neck shapes require little descriptive terminology: long-necked, short-necked, stubby necked, collared. But these terms can be time sensitive (in the order listed), since there is a general tendency for neck length (of soda bottles in particular) to become progressively shorter during the 40-year period of our study.

The only two separately named categories of neck styles are the "lady's leg" shape (*Figure I.26*), usually found on alcoholic-beverage bottles, and the "ring neck," "champagne finish," or "cider finish" collar (*Figure I.26*), typically found on wines, ciders, and some whiskey flasks (*see the lip forms discussion, this chapter*).

Figure I.24. Examples of bottle-base mold seams: (a) straight-line mold seam; (b) keyed half-circle hinge mold; (c) full-circle post mold.

Figure I.25. Examples of bottle shoulder decorations: (a) applied seal; (b) raised-ridge decorative design; (c) shoulder rings and raised upper-shoulder ridge; (d) upper shoulder raised-glass "dimples"; (e) petal-paneled front shoulders.

Lip Styles

Figure I.26. Examples of decorative neck styles and finishes: (left) "lady's leg" neck shape; (right) ring-neck "cider" finish (*see Figures I.14 and I.33*).

Many formal shape varieties of bottle lip finishes were applied to glass bottles dating to the 1840–80 period of our study. Some of them are time sensitive; some of them are product-specific; and several of them were used on more than one category of product bottle. These are listed and illustrated below, grouped according to their associated product bottles and chronology of use.

The earliest lip finishes found on embossed Illinois medicine bottles are the paper-thin flared-lip and rolled-lip forms (*Figure I.27*). The thin flared-lip form was particularly short lived (ca. 1840–45), probably because it was very easily broken, leaving razor-sharp fragmentary lip edges. The rolled lip was used in the early 1840s as well, but the style was more popular since it was stronger and less likely to break and was used throughout the 1840s.

The earliest pontiled soda-bottle lip form was the tapered collar (*Figure I.28*). It is ubiquitous on embossed sodas dating to the late 1840s and early 1850s. By 1853, tapered collars had been replaced by a more rounded, softer-edged form: the shouldered blob top (*Figure I.28*). Shortly thereafter, oval blob-top shapes and lozenge-top varieties came into use as well (*Figure I.28*), and both forms continued in use throughout the remainder of the 1850s, 1860s and early 1870s. The corks used in these bottles were held in place with wire swing-stoppers (*Figure I.29*). By the later 1870s, shorter necked blob-top soda forms began to be replaced by collared blob sodas. These collared-blob bottles had a short cylindrical collar beneath their oval blob top, apparently to accommodate a wider-based, easier to replace snap-on swing stopper to hold the cork (*Figure I.29*). But by 1879, a revolution in mass-produced and mechanically filled soda bottles with Hutchinson-style internal stoppers rapidly terminated the corked soda-bottle era (*see discussion about ending our study at ca. 1880, this chapter*).

Many larger and quart-sized mineral-spring bottles, medicine bottles, and alcoholic-beverage (especially bitters) bottles of our study period were finished with ring-neck tapered collars (*Figure I.30*) or tapered collars (*Figure I.30*). Near the end of the 1870s, and during the 1880s and 1890s, a number of different lipping tools were invented that were used to smooth, refine, and remove glass drips from the upper necks of applied-lip finishes. Such tooled lips on larger bottles (*Figure I.30*) are a convenient chronological marker for products marketed at and after the end of our study period.

In addition to tapered collars, any smaller medicine bottles and other similar-sized product bottles were fitted with a variety of thickened-ring collars. These included narrowly thickened square-ring and broad-ring collars, more widely flared square-ring collars, rounded-ring collars, lozenge-top collars, and double-ring collars (*see Figure I.31*).

Whiskey flasks embossed for Illinois bottlers were fitted with distinctive lip finishes as well. These include cider-style ring-neck finishes and ring-neck tapered collar "brandy" finishes.

Figure I.27. Examples of 1840s lip-finish styles used by Illinois medicine bottlers: (top row) the relatively unstable paper-thin flared lip of the early 1840s; (bottom row) rolled-lip varieties.

Figure I.28. Soda and mineral-water bottle lip finishes used by Illinois bottlers during the study period: (a) earlier and later tapered-collar styles (used prior to 1853); (b) earlier and later shouldered-blob styles used from 1853 onwards; and (c) oval blob and lozenge-shaped lip styles used primarily during the 1860s and 1870s.

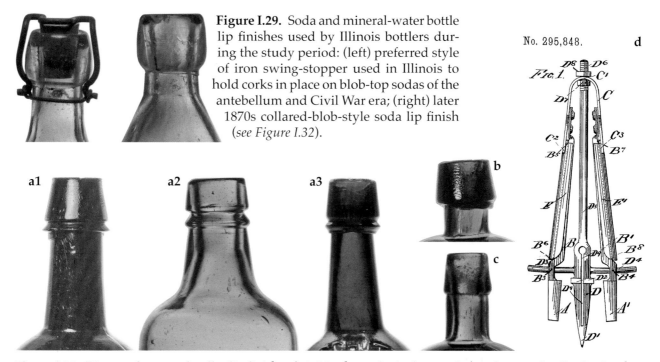

Figure I.29. Soda and mineral-water bottle lip finishes used by Illinois bottlers during the study period: (left) preferred style of iron swing-stopper used in Illinois to hold corks in place on blob-top sodas of the antebellum and Civil War era; (right) later 1870s collared-blob-style soda lip finish (see Figure I.32).

Figure I.30. Ring-neck tapered-collar lip finishes (a1–3 in chronological use order) and tapered-collar lip finishes (b–c in chronological use order) found on embossed ale, whiskey, and bitters bottles used by Illinois bottlers during the study period. (d) is an example of the lipping tools used by glass houses beginning in the later 1870s to form the tooled tapered collars of later whiskey and bitters bottles.

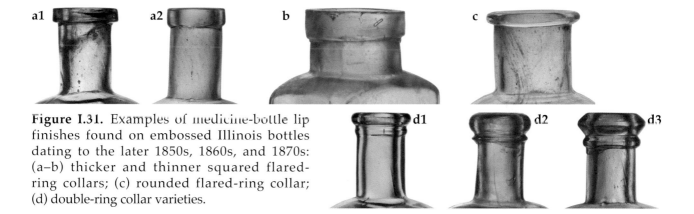

Figure I.31. Examples of medicine-bottle lip finishes found on embossed Illinois bottles dating to the later 1850s, 1860s, and 1870s: (a–b) thicker and thinner squared flared-ring collars; (c) rounded flared-ring collar; (d) double-ring collar varieties.

Closures

Literally hundreds of beverage-bottle lip closures were patented and marketed in the United States from the mid-1800s to the World War I era (see Graci 2003). However, very few of the more creatively elaborate bottle closures patented during the period of our study were used by Illinois bottlers. By far the most popular bottle closure used in Illinois from 1840 to 1880 was a simple cork—held in place by foil or paper wrapping for most nonvolatile beverages and secured by a variety of heavy-gauge wire-bale or metal-sheet swing stoppers in the case of volatile carbonated sodas, ciders, and mineral waters. The most common swing-stopper form seen (with a twisted wire neck band) is similar to that patented (#127,851) by Joseph Connor on June 11, 1872 (*Figure I.32a*). Other patented variations had more easily removable wide neck bands that allowed easier replacement of damaged or broken swing-stopper parts. These included the Cree swing stopper (#106,557) of August 23, 1870 (*Figure I.32b*) and the Weatherbee fastener (#113,603) of April 11, 1871 (*Figure I.32c*). Slightly different varieties of metal swing stoppers were designed to accommodate the ring-neck collars of champagne-cider style bottles. These included the Bates fastener (#110,421) of December 27, 1870, and the Dewey fastener (#122,579) of January 9, 1872 (*Figure I.33a–b*). The security of the swing stoppers on soda bottles was particularly important for containing their carbonated contents (see Harper's New Monthly Magazine 1872:341–346; Priestley 1772; Yates 2006:72–75). Because of the pressures exerted by their carbonated contents, soda and mineral-water bottles were also characteristically thick-walled to help reduce the likelihood of their exploding during refilling (Riley 1958:80–87; Mullikin 1978; *see Figure I.34*: bottler's mask).

In addition to the cork closures used by the vast majority of soda and mineral-water bottlers in Illinois, some local bottlers also experimented with a few types of glass gravitating-stopper bottles. Because of their weight, these types of stoppers would fall into place in the neck of the bottle when bottles were inverted for filling and were later held in place by the carbonated pressure of the beverage. In Illinois, four types of gravitating stoppers were occasionally used as soda bottle closures, but none of them were very widely accepted, and the bottles manufactured for such gravitating stoppers are seldom seen. The stoppers were usually made of glass, and were round (Codd-Rylands "marble" stoppers) or torpedo shaped (Matthews gravitating stoppers, Arthur Christin gravitating stoppers, and Kelly patent gravitating stoppers). They were either cork-lined themselves or they fitted into a cork ring fitted into a recessed area of the inner bottle lip. Their principal perceived advantage may have been that they were reusable during multiple fillings of the bottle (see Riley 1958:94–102).

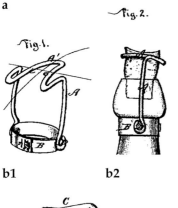
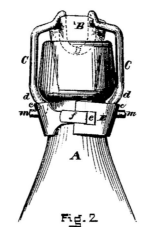
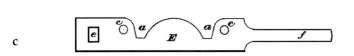

Figure I.32. Patent illustrations for swing stoppers and cork bales used to secure corks in blob-top bottles during the study period.

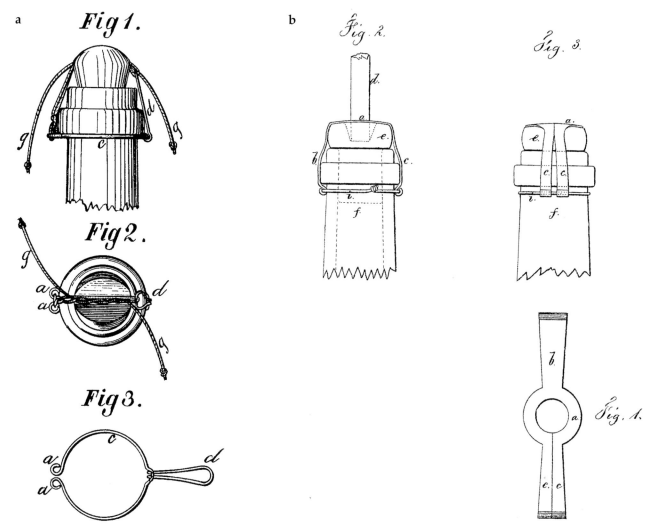

Figure I.33. Patent illustrations for swing stoppers and cork bales used to secure corks in ring-neck bottles during the study period.

The most widely distributed patented gravitating-stopper bottles used in Illinois were Matthews gravitating stopper base-embossed bottles. Base-embossed examples (touting John Matthews of New York as manufacturer) were used by the Buff & Kuhl company in Alton, D. H. Kammann in Kankakee, and William Bower in Olney. Slightly later examples of the bottle (lacking the Matthews base embossing) were used by four Illinois bottlers: Barothy & Cook in Chicago, James Allison in Nelson, Hubert Fellrath in Peoria (two styles), and William Galtermann in Danville. Prototype examples of this gravitating-stopper style were first patented by Albert Albertson in 1864 (#44,686 on October 11 and #44,912 on November 1). The gravitating-stopper configuration and style that was used by Illinois bottlers was patented by John Matthews on August 13, 1867 (#67,781), with a further stopper improvement patented (#137,941) on April 15, 1873 (*see Figure I.35*).

Just a single Illinois bottling company used both recorded styles of Illinois-embossed Codd-marble stoppered bottles: Carse & Ohlweiler of Rock Island (see maker's mark discussion below, and Carse & Ohlweiler text listing and Figures 983–984 in this volume). The two bottle styles they used conform to Hiram Codd's American patents of July 23, 1872 (#129,652) and April 29, 1873 (#138,230).

The final two torpedo-style patented glass-gravitating-stopper closures have only been found on bottles used in the Chicago area in Illinois. Chicago bottler Arthur Christin filed his

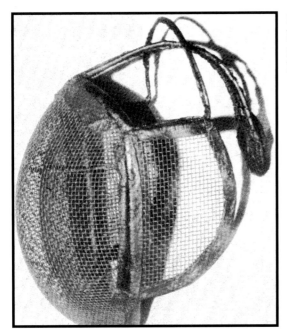

Figure I.34. An example of a protective wire mask worn by bottlers during the study period to protect them against injury from exploding glass sherds while filling carbonated-beverage bottles (see Mullikin 1978).

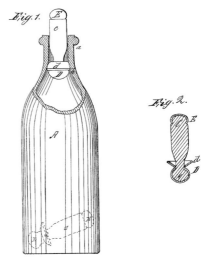

Figure I.35. An 1860s patent illustrations for the Matthews gravitating-stopper bottle closure used by Illinois bottlers.

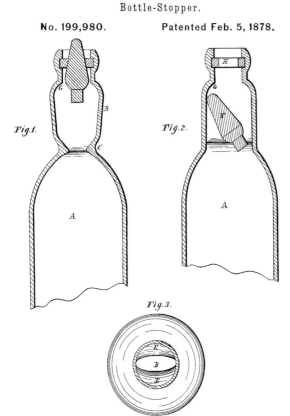

Figure I.37. An 1870s patent illustrations for the Kelley gravitating-stopper bottle closure used by Illinois bottlers

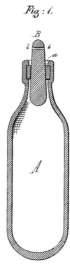

Figure I.36. A patent illustration for the Christin gravitating-stopper bottle closure used by Illinois bottlers.

gravitating-stopper patent application in December 1874, and was awarded a patent for it (#161,863) on April 13, 1875 (Welko 1978; *Figure I.36*). Both Christin and Chicago bottler John Lomax used bottles designed for this stopper, but its only other documented use in Illinois was by J. G. Bolton in Lemont (another Cook County town located southwest of Chicago). William Kelley was awarded a patent for a similar gravitating-stopper closure on February 5, 1878 (#199,980). It had a glass stopper similar to that used in the Arthur Christin bottle, but was also designed with a constricting neck like that in the Hiram Codd patent "marble" bottle, to keep the stopper from falling below the neck area of the bottle (*Figure I.37*). Kelly-patent bottles have also only been found in Chicago, where they were used by three bottlers: Frederick Lang, William Hausburg, and Gottleib Wurster.

Another distinctive bottle-closure style was limited to whiskey flasks and barrel-shaped whiskey bottles marketed in Chicago. These Chicago whiskeys were embossed for the *Chapin & Gore* liquor company, the Tremont House hotel (John B. Drake), the Palmer House hotel (Potter Palmer), and the *Kirchhoff Brothers* liquor dealership. These flasks and slightly larger barrel-shaped bottles were fitted with glass stoppers with threaded stems, designed to screw into internally threaded bottles. A cork cushion with a central hole was fitted against the inner screw cap. Many of the glass (sometimes hard rubber) caps and bottles are embossed **PAT^D AUG6TH1872 / H FRANK**. William Frank & Sons of Pittsburgh, a.k.a. Frankstown Glass Works, made the distinctive barrel-shaped bottle whose threaded internal stopper (patented by son Himan Franks) was used by Chapin & Gore. They produced this bottle for only a brief time. In June 1874, the Frankstown Glass Works where the bottle was blown burned down (McKearin and Wilson 1978:159). Shortly afterward, the Hawley Glass Works, another Pennsylvania manufacturer, started producing the bottles, but with a hard rubber threaded stopper.

The only other distinctive noncork closures noted for Illinois bottles included in the present study were the externally threaded ground lips of two preserve jars embossed for Chicago dealers *Wheeler & Bayless* and *A Leibenstein & Co.* These jars would have been fitted with threaded tin lids with rubber- or cork-rimmed milk glass or white-metal inserts (Leybourne 1993:201).

STUDY LIMITS: THE 40-YEAR PERIOD BRACKETING THE CIVIL WAR

Why Begin the Study at ca. 1840?

No glass or stoneware product bottles have been found predating 1840 that are known to have been embossed or stamped for Illinois merchants. The earliest documented embossed bottles in the state are those made for A. & M. Lindsay's Liniment, in Springfield, IL. The proprietor's newspaper ads indicate they were packaging and selling their liniment in embossed bottles as early as 1841 (*see their Springfield listing*). Evidence gathered for this study suggests that H. G. and W. B. Farrell of Peoria may have begun packaging their Farrell's Fever & Ague Drops and their Farrell's Arabian Liniment in embossed bottles at about this same time (*see text discussion*). The early locally advertised production of such generic pain relieving liniments and generic (probably cholera-related) disease cures is a reflection of the rapid influx and settlement of pioneer European immigrants into the Illinois country during the 1830s and 1840s, along developing early Riverine, Great Lakes, and overland transportation routes (see Mazrim 2002).

In the pre-1840 era, there were no established production or mercantile infrastructures and not enough cash-based consumer support to foster local development of medicinal or beverage-based bottling enterprises, so the first embossed bottled products that came to Illinois were imported (or traveled with immigrants) from established East Coast cities and European locales (Davis 1972; Griffinhagen and Young 1959; Jones 1981, 1983a, 1983b; Larson 1937; Murschell 1998;

Figure I.38. Unembossed, paper-labeled, thin-walled "puff"-style pharmacy bottle used during the 1840s by Geneseo, IL, druggist N. L. Lutz.

Young 1953). Even though a few bottles (usually for medicinal or pharmacy products) were being embossed for merchants during the early 1840s, most were still paper labeled (*see Figure I.38*)—even those for larger merchants who used early embossed bottles as well (*see Farrell Peoria listing, Figures 884–885*).

For the most part, beverages and medicinal/pharmacy products bottled for sale in the Mississippi Valley western frontier during the early nineteenth century were packaged in unembossed glass containers with paper labels. For instance, March and April issues of the 1825 *Illinois Gazette* (Shawneetown, IL) carried a nearly full-column ad by Dr. John Reid and son, announcing their "general and extensive assortment of genuine DRUGS and MEDICINES." They offered literally dozens of medicinal mixtures for wholesale and retail sale, and their Eastern patent medicine stock included "LEE'S PILLS, GODFREY'S CORDIAL, BATEMAN'S DROPS, HARLEM OIL, OPEDELDOC, BRITISH OIL, WORMSEED OIL, SHINN'S PANACEA, &c. &c." A separate ad on the same page for Philadelphia-chemist John Shinn's Panacea proudly announced that "the subscriber having discovered the composition of SWAIM's celebrated PANACEA has now a supply on hand for sale & he has reduced the price…All charitable institutions in the United States and the poor, will be supplied gratis."

> This medicine is celebrated for the cure of the following diseases:—Scrofula or king's evil, ulcerated or putrid soar throat, long standing rheumatick afflictions, cutaneous diseases, white swelling and diseases of the bone, and all cases generally of an ulcerous character, and chronick diseases, generally arising in debilitated constitutions, but more especially from syphilis, or afflictions arising therefrom; ulcers in the layrnx nodes; &c. And that dreadful disease occasioned by a long and excessive use of mercury, &c. It is also useful in diseases of the liver. [*Kaskaskia Republican,* June 1, 1824, p. 4, c. 5]

The growing consumer desire for many of the bottled products later produced and sold locally (e.g., ales, porters, mineral-spring waters, and bitters) was also a reflection of European immigrant origins. A vivid illustration of how and why the upper Great Lakes area of the Northeast became an early focus of consumer settlement and capitalist production can be seen in a series of maps published in the interpretive summary volume of the results of the 9th U.S. Census (1870), showing the developing patterns of population density in the Great Lakes-Riverine area (*Figure I.39*), patterns of German immigrant settlement across the area (*Figure I.40*), and the developing consumer base as reflected in the regional distribution of wealth in the region (*Figure I.41*).

Blunt-force trauma and the ubiquitous cuts, burns, and abrasions associated with pioneer rural lifestyles gave rise to Illinois' numerous pre–Civil War liniment-bottling enterprises, with product names which often invoked secret recipes from exotic locales, and which were often advertised for use "by man or beast." These included *Bristol's Nerve & Bone Liniment* (Belvedere); *Wakefield's Egyptian Liniment* (Bloomington); *Wells German Liniment* and *Dr. Butt's Excelsior Liniment* (Chicago); *Farrell's Arabian Liniment* (Peoria and Chicago); *Floyd's American Liniment* (Greenville); *Dr. Hamilton's Indian Liniment* (Jacksonville); *Dr. Freeman's Eclectic Liniment* (Jerseyville); *Keith's Persian Liniment* (Naperville); *Dr. Thomson's Galvanic Liniment* (Pittsfield); *Bunnel's German Liniment* (Quincy); *Lewis' Everybody's Liniment, Dr. North's Cherokee Liniment,* and *A&M Lindsay's Liniment* (Springfield). The other dramatic contributing reason for the rise of early local patent-medicine bottling industries in Illinois during the 1840s and 1850s was the widespread occurrence of epidemic cholera. We discuss the devastating impact of the late 1840s central-Illinois cholera outbreaks as part of the Wakefield Bottling Works text in the Bloomington section of this volume, and 1849–51 "cholera seasons" in Quincy, IL, were well documented during the later nineteenth century by the last surviving pioneer physician there (see Zeuch 1927:431–435). During the Quincy epidemics, the *Quincy Weekly Whig* ran a regular column titled "Health of the City":

> We regret to be compelled to report that there were an increased number of deaths during this past week *[19]*, as compared with the week previous—from cholera, or a disease that resembles it in many

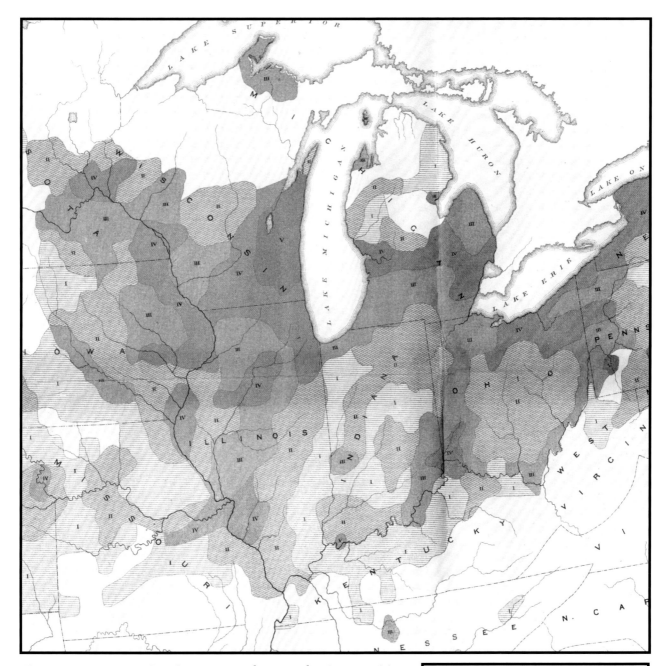

Figure I.39. An 1870 distribution map showing the density of foreign immigrant settlement in the upper Great Lakes-Riverine area. Published in a population-statistics volume (Walker 1872) evaluating the results of the 9th U.S. Census.

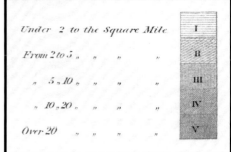

of its symptoms.... These deaths in the most instances were among a class of people but poorly provided with the necessities or comforts of life, and as a general thing can be traced to imprudence in their manner of living. We were informed by the City Sexton on Saturday morning, that in the vicinity of the lower mills there were not well people enough to take care of the sick.... Our citizens feel no alarm and there is no cause for any, in fact. Prudence in diet—a cheerful disposition—will do more towards warding off the disease than all the medicines ever invented or compounded. [*Quincy Weekly Whig,* June 24, 1851, p. 3, Col. 1]

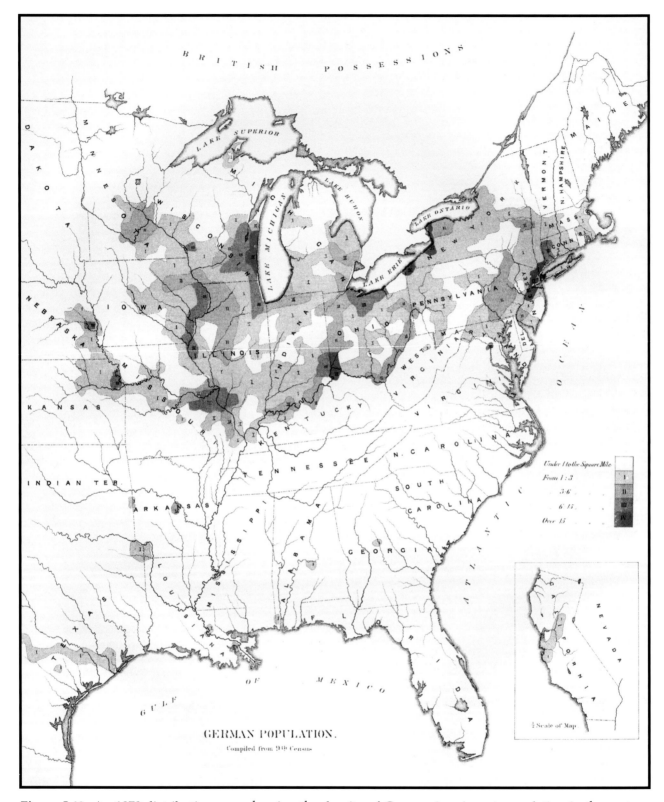

Figure I.40. An 1870 distribution map showing the density of German immigrant population in the eastern United States, concentrated in the upper Great Lakes-Riverine area. Published in a population-statistics volume (Walker 1872) evaluating the results of the 9th U.S. Census.

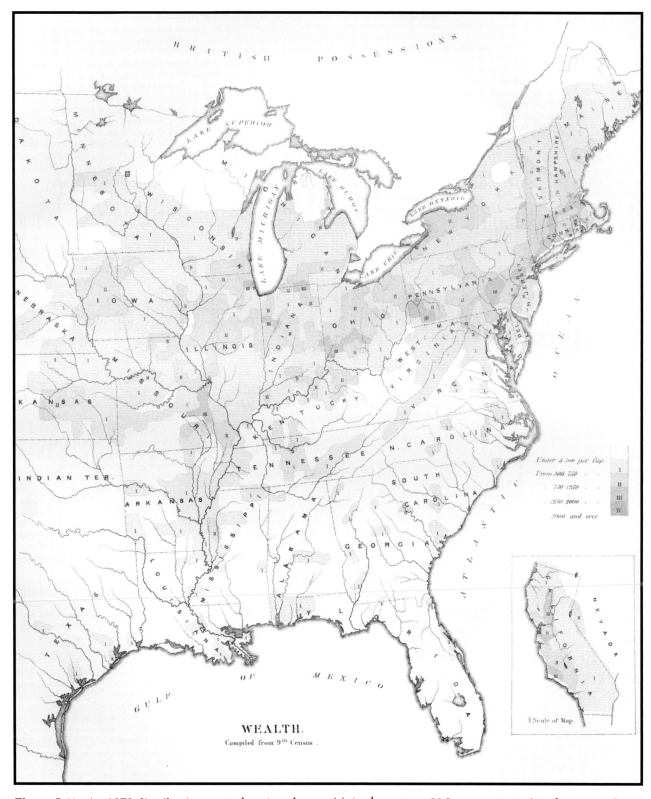

Figure I.41. An 1870 distribution map showing the wealth in the eastern U.S., concentrated in the upper Great Lakes-Riverine area. Published in a population-statistics volume (Walker 1872) evaluating the results of the 9th U.S. Census.

> From the books of the two Sextons, it appears that the number of deaths from Monday morning 23rd up to Monday morning 30th was 30, of which 28 were of cholera. Last week the number of deaths during the same number of days was 19. Several of our oldest and most respectable citizens, who to all appearances, were among the most prudent in their habit and diet—have paid the great debt of nature. [*Quincy Weekly Whig*, July 1, 1851, p. 3, Col. 1]

The medicines compounded and sold by local physicians and capitalists to combat cholera were generally referred to by the sellers, and on their bottles, as "Fever & Ague" cures, or "Fever Specifics." This name arose from a French phrase for the symptoms suffered by cholera patients: *fievre aigue*—which in French simply meant "acute fever." Such cholera cures in pre–Civil War Illinois include: *Wakefield's Fever Specific* and *Wakefield's Fever & Ague Pills* (Bloomington); *Sawyer's Fluid Extract of Bark* (Chicago); *Dr. Cavarly's Ague Bitters* (Ottawa); *Farrell's Fever & Ague Drops* (Peoria); *Dr. Bodley's Fever & Ague Balsam* (Princeton); *Wallace & Diller's Fever & Ague Bitters, Dr. North's Fever & Ague Killer*, and *Dr. Van Deusen's Ague Bitters* (all three from Springfield).

Many of the ensuing decade-by-decade developments in local bottled-product production and distribution in Illinois during our 40-year study period are summarized in the concluding section of this introduction.

Why End the Study at ca. 1880?

The four decades following the first appearance of bottles embossed for local merchants witnessed important technological changes in bottle production and embossing, and an associated rapid expansion and diversification in the number of Illinois bottlers using embossed bottles and in the kinds and quantities of bottled products. Our study can be roughly divided into four decade-long "eras" of embossed-bottle use by Illinois merchants:

Pioneer	1840s
Antebellum	1850s
Civil War	1860s
Industrial Expansion	1870s

During the 1840s and 1850s, Illinois and the Upper Mississippi Valley was still a Western frontier expansion area of growing immigrant settlement. Even the port of Chicago was still in its infancy as a commercial and industrial center.

But in the northeastern United States, the Civil War changed all that: commerce, industry, and industrial innovation of all kinds were thrown into high gear in support of the Civil War effort. Those developments and capitalist energies were turned to civilian/consumer production during the later 1860s and early 1870s after the war ended. The commercial energy of the time is reflected in the cover illustration for the 1868 Chicago City Directory (*Figure I.42*). This resulted in a rapid state-wide expansion of Illinois bottlers and bottling works—and in the rapid expansion in the kinds and quantities of embossed bottles being produced for these businesses (supported by advances in glass-making technology and associated reductions in per-unit product costs).

Thus, the middle and late 1870s saw an explosion in the quantities and types of embossed bottles being produced by glass houses for Illinois wholesale and retail merchants—ranging from individual (often seasonal) home bottlers to the Chicago-based "Largest Bottling House in the U.S." (*see J. A. Lomax "Fire and Soda" under the Chicago listings*). At the same time, important new related developments—like refrigerated railway cars and pasteurization—explosively accelerated the growth of bottled brewery-product operations (*see, e.g., Gipps & Co. Peoria listing and discussion in this volume*). Beer-bottling, even in small Illinois towns, expanded by orders of magnitude. For instance, over 50 different embossed blob-top beer-bottle styles are known from Kewanee in Henry County (Tim Wallace, personal communication 2010).

The potential magnitude of these developments is best illustrated by the meteoric growth at this time of the brewery-bottling industry in St. Louis, on the central western border of Illinois. There, the breweries of William Lemp and E. Anheuser & Co. (the latter managed by Adolphus Busch) were pioneers in developing and expanding the market for bottled brewery products during the later 1870s:

> During the last year Mr. Lemp has added a bottling department to his brewery, with a capacity for putting up twelve thousand bottles daily, which will soon be increased to one hundred thousand daily, as the demand for his bottled beer...is enlarging so rapidly that it is impossible now to fill the orders. [Dacus and Buel 1878:277]

> In the year 1873, the E. Anheuser & Co.'s Brewing Association was incorporated...At this time, a new departure was made which had revolutionized the trade in American beer...the introduction of bottling, a procedure that at once lifted the manufacture of this beer from a local to a national trade.... The first year one million bottles were sold. The increase has also been large and constant, amounting to about one and one-fourth million each year, until during the last year sales have aggregated seven million bottles.
> The manufacture of the glass alone is an important item...the glass companies of the city have been placed upon a foundation so strong as to compete with eastern rivals. The Mississippi and Lindell Glass Companies manufacture almost all of the glass consumed in these bottling operations. [Reavis 1879:24–25]

> The new *[1878]* bottling house is a building about two hundred feet long and thirty broad, provided with apparatus for putting up one hundred thousand bottles of beer daily...Refusing to restrict itself to the ordinary transportation facilities offered by railroads, the Association built and is now running one hundred and ten of its own refrigerating cars over the different roads. [Dacus and Buel 1878:280–281]

In addition to the bottled-brewery-product revolution, the period from the mid-1870s to World War I was the heyday for production of the many embossed patented styles of pharmacy bottles; literally thousands of different embossed varieties and sizes of these bottles were manufactured for Illinois pharmacists statewide. For Chicago alone, more than 900 pharmacy companies used an estimated 3,500 styles and sizes of such bottles (Tom Majewski, personal communication 2010). Downstate, a recent study of pharmacy bottles from Springfield, IL, (Brown 2010) has documented over 300 style/size varieties of such bottles produced for over 50 different pharmacy companies in the capitol city.

Similarly, at the end of the 1870s, Hutchinson-style soda bottles—with their wire-loop internal rubber stoppers and their more mechanized bottling process—(*Figure I.43*; see Paul and Parmalee 1973:12–13, 16), took Illinois soda-bottling businesses by storm. Hutchinson-style bottles and their close cousins, collared-blob-style sodas (*see Figure I.29b*)—which also could be adapted for use with elongated Hutchinson-style stoppers (*Figure I.43*)—essentially eliminated blob-top cork-sealed sodas from the bottling marketplace during the 24- to 36-month period between 1879 and 1882 (*see the patented-bottle-closure discussion, this chapter*). Without a doubt, a volume the size of our present study could be produced to document just the embossed Hutchinson-style bottle varieties used by Illinois soda-bottling companies between ca. 1880 and the turn of the century;

Thus, in order to keep our study from becoming a 20-volume *Encyclopedia "Bottallica"* (and thus maintain our sanity), our study end limits during the later 1870s and early 1880s are somewhat more "fuzzy" and pragmatic than our rock-solid 1840 start date. To make this volume a practical possibility, we established five specific rules for inclusion vs. exclusion of classes of embossed bottles produced at this time:

1. **CONTINUE TO RECORD bottle styles whose production-history was nearing an end during the later 1870s and early 1880s.**
   These bottle styles include blob-top (cork-closure) soda bottles, two-piece-mold (ring-neck tapered collar) ales, porters and mineral waters, and ring-neck-quart champaign ciders.

Figure I.42. Illustration commissioned for the 1868 Chicago City Directory showing Liberty amid the fruits of creativity and capitalism in the foreground of the consumer Mecca of the Midwest.

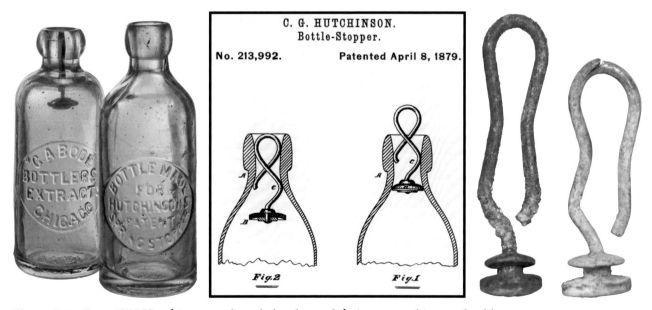

Figure I.43. Post-1878 Hutchinson-style soda bottles and their patented internal rubber stoppers.

2. **RECORD new horizon-marker bottle styles and bottle-manufacturing technologies that first appeared during the mid-to-late 1870s, AND which could be specifically dated to that time.**

    These bottle styles and technological developments include collared blob-top sodas with embossing cut directly into the mold, early removable-slug-plate embossed bottles (particularly beers produced by the *Cunningham & Ihmsen* bottling works in Pittsburgh), and early tooled tapered-collar square and rectangular quart medicines and bitters.

3. **CEASE RECORDING at 1880 all other continuously produced bottle styles (i.e., those with an early-bottler use-history but which were also produced well beyond 1880).**

    These bottle styles include medicines, inks, flavoring extracts, bluing, whiskey flasks, and bitters. The only bottles of these categories that were included in the study are those that could be determined to predate 1880 by their maker's marks, lip treatments, and bottler histories.

4. **EXCLUDE new horizon-marker bottle styles and technologies that first appeared during the mid-to-late 1870s, but which often could not be reliably dated to that time and which were early produced in such massive quantities as to be impractical to study.**

    These include:

    - Patented slug-plate pharmacy bottle styles that were first manufactured in vast quantities during the later 1870s. As discussed further in the section on 1840–1880 *Changes in Manufacturing Technology,* these included at least four early druggist-bottle styles: the "French Square" (*Figure I.44*), the "Millville Round" (*Figure I.45*), the "Philadelphia Oval" (*Figure I.46a–b*), and the "Double Philadelphia" (*Figure I.46c–d*)
    - Most pasteurization/refrigeration-generated quart beer bottles (except those examples that could be dated to the later 1870s: see Rule #2 above)
    - ALL Hutchinson-style sodas; Collared-blob-top sodas (except those included under Rule #2 above); and, of course, all hand-blown crown-top or "bottle cap" sodas (which were not first manufactured for Illinois bottlers until ca. 1898)

5. **EXCLUDE all new bottle styles and product-container categories that did not appear until the 1880s or the later nineteenth century.**

    These include:

    - Weiss beer bottles (*Figure I.47*)
    - Late (usually pint) stoneware beer bottles (*Figure I.48*)
    - Distinctive late blob-top (usually quart) sodas and beers made for patented stoppers like the Lightning stopper and Baltimore Loop stopper (see Graci and Rotille 2003)
    - Embossed milk bottles (*Figure I.49*), which, like most beer bottles, reflect the advent of pasteurization and refrigeration)
    - Target balls and fire grenades (see Bushell 1956; Finch 2008; Ketchum 1975:200–206; Kleinstuber 1989; Munsey 1970:196–197; O'Malley 2008)
    - Cigar jars (*Figure I.50*)
    - Poison bottles (see Ketchum 1975:182–187; Munsey 1970:161–164; *Figure I.51*)
    - Slug-plate flasks, except for those that could be dated to the later 1870s under Rule #2 above (see Thomas 1974; Wilson and Wilson 1968; *Figure I.52*)
    - Slug-plate pharmacy citrate bottles, except for those that could be dated to the later 1870s under Rule #2 above (*see Figure I.53*)
    - Etched seltzer-water dispensers (see Paul and Parmalee 1973:9–11; *Figure I.54*)

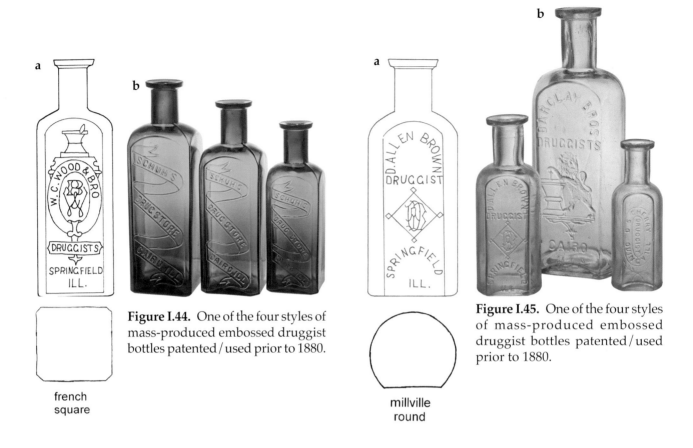

Figure I.44. One of the four styles of mass-produced embossed druggist bottles patented/used prior to 1880.

Figure I.45. One of the four styles of mass-produced embossed druggist bottles patented/used prior to 1880.

Figure I.46. Two of the four styles of mass-produced embossed druggist bottles patented/used prior to 1880.

Figure I.47. Examples of post-1880 embossed, stamped, and etched bottle styles used in Illinois too late for inclusion in the present 1840–1880 study: Weiss beers.

Figure I.48. Example of a post-1880 embossed, stamped, and etched bottle style used in Illinois too late for inclusion in the present 1840–1880 study: half-pint stoneware small beers.

Figure I.49. Examples of post-1880 embossed, stamped, and etched bottle styles used in Illinois too late for inclusion in the present 1840–1880 study: flint-glass milks.

Figure I.50. Examples of post-1880 embossed, stamped, and etched bottle styles used in Illinois too late for inclusion in the present 1840–1880 study: cigar jars.

Figure I.51. Examples of post-1880 embossed, stamped, and etched bottle styles used in Illinois too late for inclusion in the present 1840–1880 study: poisons.

Figure I.52. Examples of post-1880 embossed, stamped, and etched bottle styles used in Illinois too late for inclusion in the present 1840–1880 study: flint-glass whiskey flasks.

Figure I.53. Examples of post-1880 embossed, stamped, and etched bottle styles used in Illinois too late for inclusion in the present 1840–1880 study: drugstore citrates.

Figure I.54. Examples of post-1880 embossed, stamped, and etched bottle styles used in Illinois too late for inclusion in the present 1840–1880 study: colored and flint-glass seltzers.

Products Packaged in Embossed Glass and Stamped Stoneware Bottles

The product categories packaged in embossed glass and stamped stoneware bottles during the 40-year period of our study can generally be grouped into 10 categories: (1) alcoholic distilled beverages; (2) brewed alcoholic beverages; (3) alcoholic root beers (usually bottled in stoneware containers: see Riley 1958:111–112); (4) sparkling ciders; (5) sodas and mineral waters, usually carbonated; (6) pickles and preserves; (7) chemical-company products like ink, laundry bluing, and flavoring extracts; (8) pharmacy medical products; (9) patent medicines; and (10) miscellaneous bottled products like hair and leather treatments, glass-factory give-aways, gun oil, and rat killer.

In 1840–80 Illinois, the association between bottle shapes and styles and their packaged contents was very strong. This was probably done for ease of consumer recognition of generic product categories—especially for generic beverage categories like soda, mineral water, and mineral spring waters, which seem to have never had applied paper labels during this era, even when found or seen in historic photos with their contents intact (*see*

Figure I.55. Examples of blob-top and Hutchinson-style sodas still filled with product contents but lacking paper labels.

Figure I.55). This minimized confusion as to the contents of even sodas bottled at a brewery and embossed with the brewery's name (*see, e.g., Red Bud listing for Excelsior Brewing Co.,*). Unusual or exotic soda and mineral-water products were generally further identified by means of embossed-glass labeling (e.g., Selters Water, Ginger Ale, Medicated Aerated Water, California Pop, and Persian Sherbet).

Similarly, hard liquors were characteristically marketed in decanters, cylindrical bottles, or flasks; softer liquors, like brandy and gin, and "medicinal" alcoholic stomach bitters, "fever & ague" bitters, and wine bitters, were put up in square or rectangular bottles. Brewed ales, porters, and beers were also associated with distinct bottle shapes and container types: for instance, the stoneware root-beer containers in which root beers such as spruce beer, lemon beer, and sarsaparilla were bottled (*see discussion of bottle shape and style terms, this chapter*). Embossed-glass preserves (food syrups, honey, pickles, canning jars) and chemical products were generally associated with similarly distinctive bottle shapes, even though they often carried paper labels as well.

Exceptions to this rule include generic pharmacy medicinal products (although citrates were bottled in distinctive containers) and "patent" medicines (it was usually the product trademark name and or label style that was patented). For these products, the testimonial claims of miraculous cures (the more outrageous the better) and the style, flamboyance, and ostentatious presentation of brand-name advertising was far more important than container identity (Carson 1961; Young 1960, 1961a, 1961b, 1985):

> The faith of mankind in patent medicines has been carefully educated. As a matter of fact, faith is one of those delicate plants which flourish in the shade.... To produce a healthy specimen of faith, the penetrating rays of curious reason must be banished. And to do this requires persistence, shrewdness, and money, especially money. The marvelous power of any particular nostrum must be brought out by a wood-cut, or better, by a brace of wood-cuts, giving a glimpse of what mankind would be, and too often is, without this heaven-sent remedy: the companion picture showing the change which has come over the patient by the use of four or five bottles. In addition to these cuts, there must be a collection of letters from grateful patients pouring out thanks for their restoration to life and health...Then the newspapers must be brought in. Standing notices are ordered at so much a line, from a cent in Calumet to a dollar in Chicago, and 10 cents in religious publications....
>
> Every man has his tune, and the advertisement must be in long before the advertiser is sure of success. Hence, while it is the easiest thing in life to compound a few drugs, call the mixture by a long name, and fill, label, and pack a thousand gallons of it, it is not so easy a matter to dispose of it. Kindly nature has provided humanity with a safeguard against quacks. Of 1,000 who sail out upon a sea of invigorants, alteratives, purgatives, and pain-killers, all but one founder in mid-ocean. The survivor is rewarded....
>
> [Henry T. Helmbold, the compounder of an eastern patent medicine, Helmbold's Extract of Buchu, on a sales visit to Buck & Rayner's drug store in Chicago] remarked to Mr. Rayner, a leading druggist in this city, that no man could hope to succeed who could not make a profit at least 400 per cent on his nostrum. The expense of advertising would gobble 350 per cent, leaving the balance for legitimate expenses and a fair margin of profits....
>
> A well-stocked drug-store contains between 800 and 1,000 varieties of these special remedies. They run from corn plasters to fever and ague medicines, touch lightly on ointments, wallow deep in rheumatic antidotes, pick their way through pills of all kinds, and dash madly among cosmetics and cough mixtures. [Investigative reporting column titled "Patent Medicines" in the *Chicago Daily Tribune*, June 14, 1874, p. 10, col. 3–4]

In Illinois, prior to the advent of patented shapes and styles of removable slug-plate embossed pharmacy bottles during the later 1870s, individual druggists only occasionally used embossed prescription bottles (*see, e.g., Melvin's Springfield listing*) or embossed their names on bottles containing their own brands of citrates (especially Citrate of Magnesia) or Essence of Jamaican Ginger. Occasionally pharmacists produced their own patent medicine concoctions in embossed bottles (*see, e.g., Springfield listing for Diller's "Turner's Lotion"*).

Most of the patent medicines produced and sold in pre-1880 Illinois were either imported from eastern cities or concocted by a wide variety of more or less competent Illinois entrepreneurs (from unemployed laborers to physicians), heavily advertised, and marketed wholesale to drug stores and other retail outlets.

Illinois patent medicines included a bewildering array of products, named using a mystifying hodge-podge of historical terminologies and packaged in a wide variety of bottle shapes and styles (which seldom, if ever, overlapped with the nonmedicinal bottle styles discussed above). Patent medicines were bottled in the form of powders, pills, and liquids; liniments, lotions, and oils; ointments, salves, creams, and unguents; and medicinal cough/pectoral and teething syrups. Their medicinal properties were referred to as alterative, cathartic, aperient, sudorific, and uterine. Irrespective of their claimed curative properties, medicines were marketed as cordials, balsams, balms, anodynes (i.e., stress or pain relievers that usually included opium), tonics, panaceas, embrocations, and chologogues (i.e., medicines promoting discharge of bile from the system).

Many Illinois patent medicines were sold as universal health renewers or universal pain cures, but most were said to be cures for one or more specific ailments or injuries. These include cures/antidotes/remedies for asthma, cholera, diarrhea, catarrh, liver problems, malaria, "fever and ague" (i.e., acute fever), consumption, rheumatism, and neuralgia, as well as vermifuges, "red drops" (a socially acceptable reference to venereal-disease cures), and eye water. The multiple curative powers of individual medicines are almost laughably random, and often seem to have been assembled at the whim of copywriters. From an archaeological standpoint, however, since the claimed health benefits of specific medicines dictated whether or not consumers purchased them, it has been important to record the advertised claims.

Study Completeness: Late Additions to the Text Bottle Listings

With the help of many concerned archaeologists and glass collectors over a period of several years, capped by a focused effort during the past 24 months, we have worked to document and photograph every known type and variety of Illinois pre-1880 embossed-glass and stamped-stoneware bottle variety, and to reconstruct Illinois-bottler histories and associated Eastern and midwestern glass-manufacturer chronologies. But with some outstanding exceptions, relatively few complete or reconstructable examples remain of most of the early Illinois bottles included in our study. As many as a third of all the bottles documented in the present study are known from only one to five surviving examples. Yet based on data collected for the Illinois Industrial Census in 1860, 1870, and 1880, local bottlers obviously purchased and used bottles in vastly greater numbers (from many hundreds to many thousands of embossed bottles) in their bottling businesses than suggested from the few remaining examples. Beverage bottles were probably embossed not just for advertising purposes but also to encourage their return for refilling, and merchants may have offered nominal financial incentives to return bottles as well. But ceramic and glass containers were obviously brittle, and rough treatment during filling, transport, and use would have taken its toll on intact examples. "Bottle attrition" was likely the result of at least two added economic forces as well. First, some bottles would have no doubt been retained by as reusable household containers, not to be discarded until they were too damaged for further use.

Much more important, however, is the substantial evidence that waste glass, or "cullet," was actively accumulated by glass-manufacturing houses throughout the Midwest and used to enhance the glassmaking process. According to the 2008 *Encyclopedia Britannica*, cullet was aggressively sought out by glass makers: any given glass batch of raw glass prepared by manufacturing houses might typically consist of 25 to 60 percent cullet. Such manufactured-glass refuse was added to similar new batches of glass because its early melting in the furnace was thought to help "bring the mineral particles together, resulting in accelerated reactions."

There would obviously have been substantial bottle attrition through damage and loss at nineteenth-century bottling houses. Illinois bottlers likely accumulated damaged bottles and fragments to return to glass manufacturers as cullet in exchange for discounts on future orders. Depending on the pound-for-pound value of glass cullet to refuse dealers, salvagers might actively gather waste glass for resale to glass manufacturers as well. And of course, whenever an Illinois bottler was faced with going out of business, bulk sale of his bottle inventory back to glass houses for cullet would be a quick way to recoup some of his business losses. It seems very likely that these combined recycling efforts would tend to quickly diminish the number of embossed glass bottles available for deposit in archaeological contexts.

Add to this the attrition rate for surviving preserved specimens as the result of subsequent construction and rebuilding cycles in urban contexts and it becomes fairly obvious why discoveries of reasonably intact examples of such bottles 100 to 150 years later would be rare. Given the rarity of these surviving bottles, the "completeness" of our present study is dubious. As a testament to the fact that our study is destined to be a growing database over time, a recent discoveries listing is included below for 17 previously unrecorded pre-1880 Illinois embossed bottles that have come to our attention in the two months since we submitted the text portion of our volume, the editorial production office for page-proofing. Notably, seven of these newly discovered bottle varieties are the products of short-term, small-scale Illinois bottling houses, recently found by glass-insulator collectors in an apparent bottle-refuse dump along a north-central-Illinois railway line above the big bend of the Illinois River southwest of Chicago.

Recently Documented Bottles

Figure I.56

AURORA
(Kane County)

Albert [?] Stege

Soda ~ Mineral Water:

(1) **A. STEGE / AURORA** [aqua]

»SB w/ keyed hinge mold; slope-shouldered soda; oval blob top. *[Figure I.56]*

(2) **A. STEGE / AURORA** [aqua]

»SB w/ keyed hinge mold; slope-shouldered quart; oval blob top; 10" tall x 3 1/2" diameter. *[see Figure 26]*

The quart bottle listed above was originally interpreted in our text as a beer or cider bottle because of Stege's possible association with the Edward Stege brewery in Chicago (*see text discussion*). However, with the new discovery of an A. Stege half-pint Aurora embossed soda, it seems likely that both of his embossed bottles were used in a short-lived Aurora soda-bottling business.

No reference to this bottler or the bottle's contents could be found in any Kane County historical reference or census data. However, during the late 1860s a man named Albert Stege (perhaps a relative of Edward?) worked as a teamster for the Metz & Stege Brewery. He later worked as a laborer in other fields for a few years before becoming an "agent" (for the brewery?) at the time of the break-up of the Metz/Stege partnership in the mid-1870s. By 1882, he is listed as a Chicago brewer.

From the 1860s keyed-hinge-mold technology of the Stege Aurora soda bottles, perhaps Albert attempted (and failed) to develop his own bottling operation in Aurora for a short time during the mid-1860s before moving to Chicago to become involved in Edward Stege's brewery operation.

BELLEVILLE
(St. Clair County)

Joseph Fisher (*aka* Fischer)

Soda ~ Mineral Water:

(1) **J. FISCHER** *(arched)* / **BELLVILLE** *(arched)* / **ILL S** // **MINERAL WATER** *(arched)* [aqua]
  Note: Both **BELLVILLE** and **FISCHER** are misspelled.

  »IP; blob-top soda with top broken away. *[Figure I.57]*

--- ❖ ---

Joseph Fisher is listed in the 1860, 1870, and 1880 USC, and in the 1875 ISD, as a soda-water manufacturer in Belleville. According to Miller (2007a:159–165), Fisher moved to Belleville in 1849 and is listed in the 1850 USC for Belleville as a stone mason. He entered into a partnership with Louis Ab Egg in the mid-1850s in the soda and mineral-water bottling business (*see their text listing*). He also formed a brief business partnership with Ernst Rogger in the later 1850s (*see their text listing*), but he was listed as the sole owner of a soda-bottling business in the 1860, 1870, and 1880 USC.

The fact that this newly discovered "Fischer" soda is deeply iron-pontiled indicates that Fisher began in business on his own by 1858 or 1859.

Figure I.57

August G. Koob

Soda ~ Mineral Water:

(1) **A. KOOB, / BELLEVILLE / ILLS //**   [aqua]
    (heel): **L.G.CO.**
    (base): **A.K**

»SB w/ post mold; short-neck, shouldered-blob-top soda. [*Figure I.58*]

Mr. Koob bottled in Belleville only from 1879 onward, but prior to the discovery of the bottle listed above, his known Belleville embossed bottle styles were manufactured too late to fall within the scope of this study. This newly recorded bottle was his earliest embossed Belleville soda, and should date to ca. 1879–80.

For a discussion of Koob's earlier bottling history, see his Mascoutah text listing.

BLOOMINGTON
(McLean County)

Calvin B. Castle and Co.

Figure I.58

Medicine:

(1) **DR FALOON'S ROSIN / WEED BALSAM // // // //**   [pale & dark aqua]
    (base): **I G W Co** *(embossed in central hinge-mold circle)*

»SB w/ keyed hinge mold; beveled-edge rectangle w/ 4 sunken panels; double-ring collar; 7 1/2" tall. [*Figure I.59*]

From its keyed hinge-mold construction, this bottle is the earliest (ca. 1870–71) produced by C. B. Castle & Co., Bloomington, Illinois *(in the text section of this volume, see discussion of Castle's later-variety "R. W. BALSAM" bottle, its contents, and its advertised curative powers)*.

An 1888 trade card listing the company's products has been found *(see Figure 57)*, indicating that Castle had been producing and selling Faloon's medicines since 1870. The later bottle described and illustrated in the text is actually larger that the bottle listed above, so its abbreviated **DR FALOON'S / R. W. BALSAM** embossing likely indicates that the brand was well known by the time the second bottle-style was produced. Castle's *Rosin Weed Balsam* was marketed as a cure for diseases of the lungs, kidneys, and bladder.

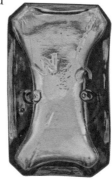

Figure I.59

The "**I G Co**" maker's mark was used by the Indianapolis Glass Works Company ca. 1870–76 (see website database of David Whitten (www.myinsulators.com/glass-factories). The mark has not been seen on any other bottle recorded by this study (*see discussion of the maker's mark and glassworks, this chapter*).

CAIRO
(Alexander County)

Andrew Lohr

Soda ~ Mineral Water:

--- ❖ ---

As discussed in the text section of this volume, Andrew Lohr (in addition to his embossed glass bottles) used a uniquely shaped, Albany-slipped soda-sized stoneware bottle that was stamped **A. LOHR** on its upper body or neck (*see Figure 131a*). The bottle was likely made and used during the 1860s. Some bottles of this style were also heel-stamped by Noah Ailiff, the stoneware potter who manufactured them (see Mounce 1989:19): **N. AILIFF, MANUFACTURER, MOUND CITY, ILLS.** (*see Figure 131b*). Mound City is located on the Ohio River about six miles north of Cairo.

The recent discovery of an **A. LOHR** stamped stoneware soda bottle made by a different potter (*Figure I.60*), suggests a wider use and multiple orders of this bottle by Lohr during the Civil War era. Based on the manufacturing technology and clay body of this second vessel style, it was likely made across the Mississippi River in the Commerce, MO, pottery manufacturing locale (see Shrum 1995:203–218, 1996:124–129), likely by Charles Koch, who had earlier operated a pottery at Mound City (Mounce 1989:20).

Figure I.60

CHICAGO
(Cook County)

Victor Barothy and Egbert C. Cook

Soda ~ Mineral Water:

(1) **BAROTHY & COOK / CHICAGO //** [aqua]
 (*heel*): **C&I**
  Note: No "**B.C**" base embossing; no base-embossed Matthews Company patent information.
   However, the bottle contains a glass Mathews gravitating stopper with embossed patent dates:

compare the Buff & Kuhl Alton text listing below and text Figures 4–5). [John Panek, personal communication 2010: no photo]

»SB w/ post mold; Matthews-style flat-topped-blob soda with short neck and Matthews gravitating internal-glass-stopper closure; 7 1/4" tall.

———————— ❖ ————————

We have recorded only three Illinois embossed Matthews gravitating stopper (MGS) bottles with the Matthews base-embossed patent information, and four such bottles with glass-house maker's marks but lacking the base-embossed patent information. This *Barothy & Cook* embossed soda adds a fifth such MGS bottle to our database. It is the only example of either type of MGS-style soda known from Chicago.

Barothy and Cook are listed in Chicago CDs as bottlers of champagne and soda water from 1872 to 1873 *(see their Chicago entry)*. Sometime in 1873 or early 1874, Barothy retired from the partnership and Cook took over the business *(see his separate Chicago entry)*.

Lang Brothers (August, William, and Frederick)

Soda ~ Mineral Water:

(1) **LANG BROS / CHICAGO.** //   [aqua]
    (heel): **A & D. H. C.**
    (base monogram): **L.B**R
    Note: *There are two raised dots beneath the full-sized* **S** *in* **BROS**.

»SB w/ keyed hinge mold; "full-pint"-size shouldered-blob-top soda; 8 1/2" tall. [Figure I.61]

———————— ❖ ————————

Figure I.61

According to Chicago CDs, the Lang Brothers set up two different, sequential, companies to bottle mineral water and soda water between 1868 and 1878 *(see Chicago listings)*. The first attempt was in 1868, by brothers August and William. They were listed in the 1868 CD as "mineral water manufacturers" at 171 N. Clark. In 1870, August and Frederick began a second *Lang Bros.* soda and mineral water bottling operation at 268 Wells. This business venture was a more successful one, and the two brothers were bottling partners until 1878, after which Frederick C. Lang took over and ran the bottling works on his own *(see his separate listing, this chapter)*.

If embossed bottles were made for the earlier one- or two-season bottling effort by August and William Lang, the only recorded bottle (prior to the discovery of the large-size pint listed above) whose hinge-mold manufacturing technology is early enough to qualify as an 1868–70 product is #1a in the text listing *(see Figure 423)*. This newly discovered large-size soda is a second hinge-mold example that may date to the 1868–70 period of operation.

We have recorded only two other Chicago soda bottles of this size. Both are cobalt examples used by John A. Lomax during the short time (i.e., 1870–72) he operated at 14 & 16 Charles Place *(see the Chicago Lomax discussion)*.

Frederick C. Lang

Soda ~ Mineral Water:

(1) **F. C. LANG / CHICAGO** //   [aqua]
    (heel): **A & D. H. C.**

(*base*): **PATENT. FEBY 5, 1878** (*outer ring*)
Note: *Known from photo only, which suggests that there is also an "F. C. L." monogram embossed at the center of the base.*

»SB; Kelly patent Codd-marble-type gravitating stopper bottle (*see discussion below*); about 7 1/2" tall. *[Figure I.62]*

Frederick C. Lang took over and ran the Lang Brothers bottling works on his own from 1878 to 1884 (*see their text listing*).

The unusual glass-gravitating-stopper bottle listed above is only known to have been used by two other Chicago bottlers, William Hausburg and Gottleib Wurster (*see their Chicago listings*), and is not known to have been used elsewhere in Illinois. The bottle was designed by William Kelly of Cleveland, Ohio. Kelly submitted his patent application on May 17, 1877, and received the patent for his bottle-closure on February 5, 1878 (Patent #199,980). Kelly's patent claimed that "the bottle was able to be opened with no additional tools, because of the high projection of the stopper through the neck, and that once the stopper was pushed in, it was contained in the neck where it would not agitate or disturb the contents. Because the stopper did not have to be held in any particular position, the bottle could be easily cleaned and refilled" (Welko 1978:9).

According to Paul Welko, "It is probable that upon seeing the success of the Arthur Christin company, who used a similar stopper" (*see Christin's Chicago listing*), the Hausburg Co. "decided to compete with their own style bottle. It doesn't seem likely the venture was successful due to the scarcity of the bottle" (Welko 1978:9).

Figure I.62

August and Henry Mette

Ale:

(1) **A. METTE & BRO. / CHICAGO. ILL.** //   [*red-amber*]
    (*base*): **W. M^cC. & Co**

»SB w/ depressed center and post mold; 4-piece mold quart ale; ring-neck tapered-collar; 9 1/4" tall. *[Figure I.63]*

Quart Cider:

(1) **A.METTE & BRO / CHICAGO** //   [*yellow-amber*]

»SB w/ depressed center and post mold; ring-neck cider-finish quart; 11 1/2" tall. *[Figure I.64]*

August Mette, along with his brothers Henry and Louis, first appeared in the Chicago CDs in 1872, as a brewing company manufacturing "tonic beer" at 328 Archer. The three brothers continued as business partners in the firm of *A. Mette & Bros.* between 1873 and 1875 (at 985 S. Halsted in 1873 and at 370 Archer in 1874–75), but during these years they are listed in the CDs only as "soda water manufacturers." Louis was no longer a partner in the business after 1875, and from 1876 onward the firm became *August Mette & Bro.*

Figure I.64

Figure I.63

Prior to the recent documentation of the quart ale and cider bottles listed above, only aqua sodas had been recorded for *Mette & Bro. (see their Chicago listing)*. Perhaps the ale-style bottle contained the tonic beer the brothers manufactured early in their bottling history.

Andrew J. Miller

Beginning with the 1846 CD and continuing through 1848, A. J. Miller is listed as a brewer (one of only three in town) with a "small beer" sales stand opposite the American Temperance House hotel. Small beers (e.g., spruce beer, lemon beer, sarsaparilla/root beer) were characteristically put up in stoneware bottles, and two styles of hand-turned *A. J. Miller* stoneware quart bottles are known: one is stamped **A. J. MILLER** (in small, unadorned letters) on its upper body or sloped shoulder area (*see Figures 514–515*); the second is a sharp-shouldered quart, horizontally stamped **A. J. MILLER** in large seriffed lettering. Kiln-failure sherds of this second Miller bottle style are known from the Kirkpatrick family pottery kiln site in Vermillionville, some 80 miles southwest of Chicago (Gums, et al. 1997:25).

A third style of **A. J. MILLER** pottery bottle has recently been located that is very different in shape and clay body from the two previously known varieties. Both of the earlier known varieties were made of stoneware and were manufactured in northern Illinois (Vermillionville) and/or southern Wisconsin. The newly documented Miller bottle (*Figure I.65*) is a sharp-shouldered, wide-mouth, Ohio yellowware-style container, and is much shorter (7" vs. 8 1/2") than the earlier-known styles.

Figure I.65

The Chicago CDs available to us for the period from 1849 to 1851 are fragmentary, and the sections that would have Miller's business listings for those years are missing. But by 1852, he was listed only as a "mineral water" bottler on Randolph Street, and the 1853 CD listed him as a "soda water manufacturer" from his residence at the corner of Lake and Carpenter streets.

William Morrison

Soda ~ Mineral Water:

(1) **MORRISON / SODA WATER //** [aqua]
    (*heel*): **A & D.H.C.**
    (*base*): **S**

   »SB with post mold; short-neck oval blob-top soda. *[Figure I.66]*

(2) **MORRISON / SODA WATER** [aqua]
    (*heel*): **C&I**

   »SB with post mold; short-neck oval blob-top soda. *[Figure I.67]*

These two varieties of Morrison soda bottles were recently discovered from a railroad-side bottle dump in north-central Illinois above the big bend of the Illinois River. All but a few of the hundreds of embossed bottles in the dump had been made for manufacturers who operated within an approximately 70-mile in diameter area from Chicago (N), Dixon (E), Rockford (N), and LaSalle (S). Since the company was previously unknown, and since the town of Morrison

Figure I.66

Figure I.67

is located near the western edge of distribution of embossed bottles found in the dump, we initially sought the bottling works there. The post-mold manufacturing style of the Morrison bottles dated the company to the 1870s; the fact that that a few shorter-neck Morrison bottles (although they are not Hutchinson-style bottles) had been fitted with unusually long Hutchinson-style internal rubber-ring stoppers (*see Figure I.43*: made after late 1878), more closely dated the bottles to ca. 1879–80.

Historical society contacts and historical-documents research for that period found no soda bottling works operating in Morrison or in surrounding Whiteside County. But Illinois marriage and death records for the late nineteenth century did indicate that numerous families with the Morrison surname resided in the region during the late nineteenth century—particularly in the vicinity of Sterling, LaSalle, Elgin, and especially Chicago. The Morrison families in Sterling and LaSalle were well documented genealogically, and family members were not known to have been involved in soda bottling. However, when the many dozens of Morrisons living in late-nineteenth-century Chicago were checked in the CDs of the 1870s, a one-or-two-season bottling company run by William Morrison was found in the directories at exactly the right time: 1879. Mr. Morrison was listed as a "pop maker" from his residence at 134 W. Harrison Street.

The significance of the large embossed "S" on the base of the A&DHC bottle variety is not known. Base-embossing (usually the bottler's initials) was often used as a "short-hand" way of identifying bottles in wooden carrying cases, since they were often transported upside-down to moisten the corks and help keep the contents sealed. Perhaps in this case the "S" indicated soda water.

DE KALB
(De Kalb County)

1850 census: 500
1860 census: 1,700
1870 census: 2,200
1880 census: 2,400

Richard Dee

Soda ~ Mineral Water:

(1) **R. DEE / DE KALB / ILLS. //** [aqua]

»SB w/ post mold; blob-top soda w/ shouldered-blob and flat-topped oval-blob varieties. *[Figure I.68]*

❖

Richard Dee established his bottling business in De Kalb in 1878. By 1883, he was a manufacturer of "bottled soda-water and ginger ale, bottled beer, ale and porter," as well as a "dealer in ice in any quantities" (Chapman Brothers 1883:715). He was born in England in 1829 and came to America as a young man. He settled first in Wisconsin, coming to Illinois in 1852. Before settling in De Kalb in 1863, he mined coal in LaSalle and then managed saloons in Colchester, Chicago, and De Kalb. After the Civil War years, he moved to Dixon, where he built and operated a brewery. He then returned to De Kalb and operated a meat market for several years before opening his soda-bottling works there.

Later (early to mid-1880s) slug-plate embossed **R. DEE** and **RICHARD DEE** beer bottles from his De Kalb bottling years have also been seen, but they were manufactured and used too late to be included in the present study. Since the only pre-1880 bottles known are sodas, it seems likely that his first bottling efforts were focused on soda and ginger ale.

Figure I.68

DIXON
(Lee County)

Frank J. Finkler and Co.

Soda ~ Mineral Water:

(1) **F. J. FINKLER & Co /** [light blue]
    DIXON / ILL^S //
    (heel): **A & D. H. C**

»SB w/ keyed hinge mold; shouldered-blob-top soda. *[Figure I.69]*

❖

Figure I.69

Frank J. Finkler was the brother of Alex and John Finkler, who bottled soda water in LaSalle, Ottawa, and Streator during the 1860s and 1870s (*see their text listings*). Frank worked for his brother Alex in the La Salle soda water factory in 1860 (USC) and, after the Civil War, moved to Dixon and opened his own soda water business. He is listed in the 1870 and 1880 USC as a soda water manufacturer in Dixon.

The embossed A&DHC soda bottle shown in Figure I.69 was produced and used during the late 1860s or early 1870s. It represents a technologically earlier variety of the bottle (using the same debossed mold) than the **C&I** variety of the bottle recorded in the text section of our volume. The last year the *Cunningham & Ihmsen* (*C&I*) glassworks of Pittsburgh produced embossed bottles was 1878.

LASALLE

(LaSalle County)

Alexander Finkler and Co.

Soda ~ Mineral Water:

(1) **ALEX. FINKLER / LASALLE / ILLS //**   [aqua]
(*left front heel):* **F. A & CO**

»SB w/ keyed hinge mold; shouldered-blob-top soda. [*Figure I.70*]

Alex and John A. Finkler were owners of a LaSalle soda-water bottling company in the 1860 and 1870 USC and in the 1867 ISD I (*See A. Finkler & Co. listing and the company's text section*). Early on, John had also been a silent "& Co." partner with J. B. St. Clair in 1864 (according to the IRS assessment that year) in St. Clair's Peoria bottling business there. A third brother, Frank J. Finkler, also worked at the LaSalle bottling plant according to the 1860 USC, before leaving to serve in the Civil War. After serving in the Civil War, Frank had expanded the brothers' bottling partnership to a new bottling operation in Dixon (*see the F. J. Finkler & Co. listing, this chapter and the company's text section*), where he bottled during the late 1860s and early 1870s.

The significance of the newly discovered FA&Co bottle shown in the figure is that it indicates Alex Finkler was operating his own bottling works in LaSalle by 1863 or earlier, since the *Fahnstock, Albree & Co.* glass-manufacturing business produced embossed soda bottles in Pittsburgh for only four years: 1860–63 (*see Maker's Mark section, this chapter*). This in turn suggests that Finkler was not an actual participant in St. Clair's Peoria bottling operation, since he was operating his LaSalle bottling works at the same time. More likely, he was an investor with a financial interest in the Peoria bottling works.

Figure I.70

MENDOTA
(La Salle County)

Frederick Molln

Soda ~ Mineral Water:

(1) **F. MOLLN** *(arched)* / **MENDOTA. ILL**ˢ /   *[aqua]*
(front heel): **C&I**

»SB w/ keyed hinge mold; shouldered-blob-top soda. *[Figure I.71]*

Frederick Molln was listed in the 1880 USC for Mendota as a 56-year-old retired saloonkeeper. During the post–Civil War era, saloons often provided bottled soda and mineral waters for mixing drinks.

From the stylistic and technological attributes of the newly discovered Molln bottle shown (and the previously known *Molln & Co.* embossed bottle listed in the text section of the volume), his embossed bottles were likely made and used during the 1860s. Molln may have bottled soda water on a part-time basis during the mid-to-late 1860s to sell at his saloon. The "& Co." partner in Molln's sideline bottling business has not been determined *(see Wohlers listing in the text section)*.

Figure I.71

The significance of the newly discovered Molln bottle is that Molln bottled on his own at first, before taking on a partner and adding "& Co." to the bottle mold.

GLASS-FACTORY MAKER'S MARKS ON ILLINOIS EMBOSSED BOTTLES

For the period of our study, we have recorded embossed glass-factory maker's marks used by 38 different glass manufacturers on 408 (37%) of the 1,093 Illinois embossed bottles we recorded including 1 New York glass factory; 1 Maryland factory; 15 Pittsburgh-area factories and 1 other Pennsylvania factory; 4 Ohio factories; 2 Indiana factories; 1 Kentucky factory; 2 Milwaukee factories; 2 St. Louis factories; 5 Illinois factories; and even 2 glass makers from Barnsley and London in England.

Considering the number and regional diversity of maker's marks we documented, an initially surprising result of our overview study of these marks on Illinois bottles was the realization that *there are no embossed glass-factory marks on any Illinois embossed bottles during the entire first half of our study period*. The earliest maker's marks on Illinois bottles date ca. 1859–61 including an AA-marked Chicago soda; two FRL-marked Belleville sodas; an A&DHC-marked Quincy soda; two Chicago ales (Keeley and Hutchinson) and an early Chicago Hutchinson Co. soda with Wm. McCully maker's marks; and perhaps as many as 15 FA&Co-marked sodas *(see discussions under specific maker's marks)*. However closer scrutiny of the years our 35 glass houses were in business *(see Table I.2)* shows that prior to 1859/60 only three of them were in business, all in

Table I.2. Glass Factory Maker's Marks on Pre-1880 Illinois Embossed Bottles

Maker's Mark	Factory	Location	Business Years	No. of IL Marks
AA	Arbogast	Pittsburgh, PA	1858–62	1
A&Co	Agnew	Pittsburgh, PA	1866–ca. 92	3
A&DHC	Chambers	Pittsburgh, PA	1843–ca. 89	167
AGWCo	American	Pittsburgh, PA	1866–79	1
AGWL	American	Pittsburgh, PA	1880–1905	1
ALFRED ALEXANDER		London	1883–ca. 87	1
ALTON GLASS Co		Alton, IL	1871–72	1
BB&Co	Baker Bros.	Baltimore, MD	ca. 1870s–95	1
BFGCo	Beaver Falls	Pennsylvania, PA	1869–79	10
BGCo	Belleville	IL	1882–96	1
BODE	[Not a glassworks—see text]			2
C&I	Cunningham & Ihmsen	Pittsburgh, PA	1865–78	36
C&Co	Cunninghams	Pittsburgh, PA	1878–86	2
C&Co LIM	Cunninghams	Pittsburgh, PA	1886–1907	1
CV or CVCo	Chase Valley	Milwaukee, WI	1880–1881	1
DeSGCo	DeSteiger	LaSalle, IL	ca. 1876–78	14
DSGCo	DeSteiger	LaSalle, IL	ca. 1878–96	9
DOC	Cunningham	Pittsburgh, PA	ca. 1880–1931	10
EHE	Edw. Everett	Newark, OH	ca. 1880–94	2
FA&Co	Fahnestock-Albree	Pittsburgh, PA	ca. 1860–63	18
FBCo	Findlay	Ohio	ca. 1888–93	1
FRL	Frederick Lorenz	Pittsburgh, PA	ca 1854–59/60	2
HAWLEY		Hawley, PA	ca. 1872–85	2
IGCo	Illinois Glass	Alton, IL	1873–1929	14
IGCo L	Chr. Ihmsen	Pittsburgh, PA	ca. 1870s	1
IGWCo	Indianapolis, GW	IND	ca. 1870–76	1
KH&G ZO	Kearns et al.	Zanesville, OH	ca. 1868–85	2
LGCo	Lindell	St. Louis MO	1874–92	6
LG WKS	Lockport	NY	1840–72	1
LG WKS	Lockport	NY	1840–72	1
LOCKPORT NY			1840–72	1
L&W	Lorenz & Wightman	Pittsburgh, PA	1862–74	20
MATTHEWS	[Not a glassworks—see text]			3
MGCo	Mississippi GCo	St. Louis, MO	1873–84	4
N (star)	Star Glass	Newark, OH	1873–1904	2
PUTNAM	[Not a glassworks—see text]			1
RYLANDS & CODD		Barnsley, England	ca. 1877–81	1
SB&GCo	Streator	IL	1881–1905	1
(large star symbol)	Star Glass	New Albany, IN	ca. 1860s & 70s	4
TW&Co	Thos. Wightman	Pittsburgh, PA	ca. 1874–95+	8
WIS G CO MILW		Milwaukee, WI	ca. 1881–85	1
Wm. McCully*		Pittsburgh, PA	ca. 1841–1909	39
WM. F. FRANK & SONS		Pittsburgh, PA	1866–75	4

*Wide variety of marks: see text.

Pittsburgh: A&DH Chambers (from 1843 on); Frederick R. Lorenz (from 1854 on); and Wm. McCully (from 1841 on). Apparently when a fourth Pittsburg company (Fahnestock, Albree & Co., ca. 1860–63) took over the Lorenz factory and began aggressively putting FA&Co marks on their bottles, the maker's mark race was on. During this time, Chambers and McCully, along with a third Pittsburgh glassworks that opened its doors at the end of the Civil War (Cunningham & Ihmsen), continued to dominate the Illinois glass trade—accounting for nearly two-thirds of all maker's marks on Illinois bottles from 1860 to 1880 (A&DHC = 164; C&I = 48; McCully = 38). More generally, the 14 Pittsburgh-area glassworks accounted for 326 of the 397 maker's marks recorded by our study, while there were only 71 maker's marks from the remaining 21 glassworks combined. In fact, 17 of these works are known in pre-1880 Illinois from one (12 glassworks) or two (6 glassworks) maker's marks each.

During the 1860 to 1880 period, many new eastern and midwestern glass factories opened their doors for business and entered the fray. We should note that three of these later companys' "maker's marks" found on 1840–80 Illinos bottles listed and discussed in this section *are not* included in the totaled numbers and percentages summarized here. This is because these three companies (Bode, Matthews, and Putnam) were not actually glassmakers. Rather, they were middleman jobbers or glass patent-holders who took orders from Illinois bottlers for their products and then farmed-out the glass-blowing work to unnamed glass houses. The table also includeds maker's marks used by nine glass houses which began in business at or shortly after 1880. These are manufacturers who, for a time, continued to produce a some bottles using the 1870s molds or in the "old fashioned" 1870s style.

Pre-1860 Illinois Bottle Manufacturers

So who manufactured the embossed bottles *without* maker's marks produced for Illinois bottlers between 1840 and 1860? (*See summary discussion of these Illinois pontiled bottles in the section on 1840–80 changes in manufacturing technology, this chapter.*) During the 1840s, the best-candidate Illinois bottle manufacturers for downstate bottlers are the William McCully companies in Pittsburgh (McCully opened his first glass factory there in 1832 and took on additional partners and expanded the business in 1841), Alexander and David H. Chambers of Pittsburgh (from 1843 on), and the St. Louis Glass Works (across the Mississippi River on the "Western Frontier") that first opened its doors in 1842 (Toulouse 1971:167). From serendipitously discovered 1850–51 sales receipts (preserved in probate records in Quincy, IL), we know that the St. Louis Glass Works (under the proprietorship of Mssrs. Sell & Coe) manufactured bottles for the short-lived pioneer Lundblad soda bottling works in Quincy (*see discussion under Quincy in this volume*). During the 1850s, a new downstate Illinois-trade competitor Pittsburgh glass house was first run by Frederick Lorenz, Sr., and Thomas Wightman (1851–54) and then by Frederick Lorenz, Jr. (from 1854 on).

Early Illinois bottlers in the Chicago area (where the first local bottling works appear in the late 1840s–early 1850s) also had access, through the Great Lakes trade, to glass houses in the Northeast. The best-known eastern bottling works connection is that of the Hutchinson family Chicago bottling works in the early 1850s. William and Joseph Hutchinson had emigrated to Chicago from Williamsville, New York (on the outskirts of Buffalo). Most of their early bottles are cobalt to sapphire blue, and from the bottle shapes (*see the Hutchinson Chicago entry*), they appear to have been produced by New York glass houses. The two glassworks that the Hutchinsons would have been most familiar with would have been the Lockport works (located 12 miles north of Williamsville) and the Lancaster works (located eight miles SE of Williamsville). Family connections there and Buffalo's location on Lake Erie would have facilitated ordering and Great Lakes transport of the bottles.

Post-1860 Illinois Bottle Manufacturers

Studies of early Pittsburgh glass houses and other eastern and midwestern bottle makers is still a work in progress, as H. J. Seymour noted in his early study of "Glass and Glass Makers" published while many of the glass houses discussed here were still in operation:

> It cannot but be regretted that the active and venturesome men who planted at such cost and labor this thriving business in western Pennsylvania, did not leave more detailed records of their trials and experiences, as the story would have been one of absorbing interest. But those were stern,

trying days, and the founders of the industry had more pressing things to think about than the wishes of posterity. [1886:373]

But recent researchers have made great strides. Unless otherwise noted (as our own research), the glassworks summaries below are largely informed by the works of Julian Harrison Toulouse (1971), Jay Hawkins recent 600-page volume on Pittsburgh glass houses and glass manufacturers (Hawkins 2009), the continually growing website database of David Whitten (http://www.myinsulators.com/glass-factories), and a series of important, in-depth articles by Bill Lockhart and his colleagues serially published as "The Dating Game" in the pages of *Bottles and Extras* (Lockhart et al. 2004, 2007, 2008a and b, 2009).

1. **AA** *[KF/JW = 1858–62]*
   Alexander Arbogast, Pittsburgh
   (Whitten: 1860–ca. 1880)

The only bottle included in our study with this maker's mark is a Francis Shonwaltd (aka Franz Schonwald) soda from Chicago. This bottle was made and used ca. 1859–60. The embossed "**A A**" on Schonwald's bottle #1 (*see his Chicago listing*) is the mark of Alexander Arbogast, Pittsburgh. Only two other bottles with this maker's mark are known. Both are Civil War–era black-glass quart whiskey-style 3-piece-mold cylinders—one unembossed and the other shoulder-embossed U.S.A. / MEDICAL SUPPLIES / FROM / PIKE & KELLOGG, ST. LOUIS. Both bottles have a base mark "A. ARBOGAST PITTS." Rex Wilson (1981:113) reports a base of one of these bottles from a Civil War–era western military context. Andrew Arbogast appears in Pittsburgh CDs in the 1850s and early 1860s as a glassblower (ca. 1850–57) and glass manufacturer (ca. 1858–62). From 1863–67, he operated a window-glass factory in New Castle, Pennsylvania (von Mechow 2008). An advertisement in the 1861 Pittsburgh city directory (pg. 23) for the Arbogast glassworks states: "Particular attention paid to Private Molds."

2. **A & Co** *[KF/JW = ca. 1866–92+]*
   Agnew & Co., Pittsburgh
   (Whitten: 1876–92+)

Agnew maker's marks are found only on three *Red Jacket Bitters* bottles produced for and sold by Bennett Pieters in Chicago. The marks are: **A&Co., A&Co. No 4,** and **A&Co No 5**. The bottles are embossed with Pieters 31–33 Michigan address, and so were used ca. 1866–68.

3. **A&DHC** *[1843–ca. 1889]*
   Alexander & David H. Chambers, Pittsburgh
   (Whitten: 1843–ca. 1889)
   (Toulouse: 1843–86+)

From the predominance of their embossed maker's marks, the Chambers glassworks was the single most popular source for embossed bottles used by Illinois bottlers between 1860–80. The A&DHC mark is never seen on embossed Illinois bottles of the 1840–59 period (*see earlier discussion*), although Chambers company glassworks ads have been noted in Illinois and St. Louis city directories as early as 1866.

The A&DHC maker's marks were primarily used on embossed soda bottles statewide (by 90 bottlers, with embossed maker's marks on 136 embossed-bottle styles). The A&DHC mark has also been recorded on 10 cider-finish quarts used by seven different bottlers in the Chicago & Cook County area from the post-CW era through the 1870s. It also appears on 12 ale and porter bottles (11 bottlers) and 9 beer bottles (4 bottlers) statewide. The A&DHC mark is rarely seen on medicinal products (2 medicines and one bitters) or on other bottle styles across the state (1 Chicago whiskey and 1 Chicago chemical works bottle only).

Although the A&DHC maker's marks we have recorded on the heels and bases of Illinois embossed bottles always incorporate the same five characters, the embossed height of the characters ranges from 4 to 14 mm, and there are often noticeable differences in lettering style and thickness. Even on bottles of similar shape and size, A&DHC maker's marks also exhibit subtle, seemingly random variations in the placement of periods between any two characters (including the ampersand):

    A. & D. H. C   A. & D. H. C.   A & D. H. C.
    A. &. D H. C   A & D. H. C   A & D H. C.
    A & D H C

The reasons for these differences are not known, but are not obviously related to time differences or bottle style or size differences. Like the variable use of differently styled William McCully maker's marks *(see discussion later in this section)*, the differences could have included some sort of coded information for factory workers, or they may just reflect a lack of standard company instructions to mold makers.

    4. **AGW Co** *[1866–79]*
        AGWL *[1880–1905]*
        American Glass Works [Limited], Pittsburgh
        (Whitten/Toulouse: AGW = 1866–79; AGWL = 1880–1905)

The AGW Co. mark was only seen on a quart beer used by the Gipps, Cody & Co. bottling works in Peoria. The Gipps Cody company was in business ca. 1871–73.

The AGWL mark was only seen on one post-1880 "old fashioned" Saratoga-style mineral-water bottle from Ypsilanti Springs company in Chicago ca. 1885.

    5. **Alfred Alexander & Co.** *[KF/JW = 1883–ca. 1887]*
        London ENGLAND

The only bottle with this mark documented by our study is an 1870s-style quart blob-top soda used by the John A. Lomax bottling works in Chicago. The Lomax bottle was manufactured sometime during his years at 14–18 Charles Place (ca. 1873–86).

The maker's mark indicates that this particular bottle actually postdates 1880 by a few years (ca. 1883). According to the London trade publication *Ridley & Co.'s Monthly Wine and Spirits Trade Circular (January 24, 1884, p. 27)*, the Alexander & Austin glassworks had been reformed as Alfred Alexander & Co. in 1883, and was now operating from 72 Bishopsgate Street in London. In the later 1880s, the company relocated to Leeds.

Notably, this Lomax quart represents one of only two known examples of bottle used by Illinois bottlers which were manufactured by foreign glassworks *(see also Rylands & Codd maker's mark discussion, this section)*.

    6. **ALTON GLASS Co.** *[KF/JW = 1871–72]*
        Alton, IL (see text listing)

The company name has only been seen embossed on a give-away advertising whiskey flask used a glassworks company that operated for 2 years in Alton. The company became Illinois Glass Co. in 1873 *(see text listing)*.

    7. **BB & Co** *[ca. 1870s–95]*
        Baker Bros. & Co., Baltimore:
        (Lindsay et al. suggest ca. 1870s–95)
        (Toulouse suggested Berney-Bond Co., a Pennsylvania medical-products company of the early 1900s)

This maker's mark has only been recorded on a single John A. Lomax soda bottle with an embossed 14 & 16 Charles Place address. Lomax only operated from this address ca. 1870–72 *(see the Lomax discussion in the Chicago section)*.

Toulouse (1971:70) suggested the mark was associated with the Berney-Bond Company, but this was an early 1900s medical-products company. Several possible alternative glass house identifications have been suggested by internet searches: David Whitten's list of "Glassmaker's Marks found on Bottles" identifies the mark as Baker Bros. & Co. (proprietors of the Baltimore, MD, Federal Hill and Spring Garden Glass Works): "Mark seen on the base of Baltimore-area blob beer bottles from the ca. 1880–1895 period." This company advertised in the 1870s and 80s, with a strong focus on window-glass production. Bill Lindsay's massive Historic Glass Bottle Identification and Information Website (2009) indicates the mark "likely stands for the *Berney-Bond Glass Company* (Pennsylvania), and illustrates it on an 1880s Trommer Extract of Malt (Freemont, Ohio). However, from Pennsylvania historical records, this company was only in business from ca. 1902–30. There was also a British glass company with these initials (perhaps too late for the Lomax context), and a pharmacy bottle stamped "B. B. & Co. / London" is known.

So, from present information, the Baker Bros. & Co. attribution seems most likely, especially since **B. F. G. Co** marks on several Lomax bottles indicate that another Pennsylvania glassworks, the Beaver Falls Glass Co. (1869–79), was also patronized by Lomax at the same time (ca. 1870–72).

8. **BFG Co** *[1869–79]*
   Beaver Falls Glass Co.: Beaver Falls, PA
   (Whitten suggests Beaver Falls, PA [1869–1879], and Toronto ONT [1897–1948])

The possibility was also considered that the mark was a Belleville Glass Co. variant (BFGCo-marked bottles are found in the greater St. Louis area, so the mark has been suggested as a variant for the Belleville, Illinois, glassworks—usually represented by BG Co).

In Illinois, however, the mark has only been found on 9 sodas, ciders, and ales used by 4 Chicago bottlers of the early 1870s, and on one soda bottle made for the Johnson & Peterson Co. in Springfield at about the same time, suggesting glass-company access via the Great Lakes trade. Since one of the Chicago bottlers (Lomax) also used bottles made by another Pennsylvania glassworks at the same time (*see BB&Co in this section*), and since the Belleville glassworks did not begin in production until after 1880, the Beaver Falls Glass Co. of Pennsylvania was the likely producer of the BFGCo bottles.

9. **BG Co** *[1882–1896]*
   Belleville Glass Co., IL
   (Whitten = 1882–96)
   (Toulouse = Burlington Glass Co., Hamilton, ONT: 1887–1909)

The Belleville Glass Co. formally opened for production in early September 1882, with 75–80 employees (*Decatur Morning Review,* Sept. 14, 1882). Only one BGCo bottle is included in the present study: an example of an "old fashioned" quart soda or beer used by Edward Schroeder in East St. Louis (near Belleville) in the early 1880s. There is no good evidence that it was a late-1880s Canadian product from Burlington Glass Co.

10. **BODE** *[KF/JW = at least 1879–1882]*
    Chicago, IL
    (Whitten = Chicago extract company which "seems to have produced some bottles early in their history")

The BODE maker's mark has only been documented on bottles of the ca. 1840–80 era for one Chicago bottler (E. Y. Cronk), who was in the bottling business there for just a short time at the very end of this era. He bottled in Chicago from 1879 to 1882.

Bode was not a glassworks, and there is little doubt the Bode extract and bottler's supply company served (in this case at least) as a middleman supplier who subcontracted E. Y. Cronk bottle production to an unknown glassworks *(see also Matthews gravitating stopper and Putnam listings, this chapter).*

11. **C&I** *[1865–78]*
    Cunningham[s] & Ihmsen, Pittsburgh, PA
    (Whitten/Lockhart = Cunningham & Ihmsen: 1865/1866–1878)
    (Toulouse = Cunninghams & Ihmsen: 1865–79)

This is another example of a Pittsburgh glass-maker (like A&DHC) represented by numerous embossed C&I bottles used by Illinois bottlers in Chicago and many other northern and west central Illinois towns (distribution: 16 towns in 11 counties—*Figure I.72*).

The total number of bottles with this mark are as follows: sodas = 36 embossed bottle styles used by 28 bottlers in 15 towns; 3 quart ales (1 Bloomington; 1 Chicago; 1 Peoria); 6 beers (4 Chicago; 1 Jacksonville; 1 Peoria); 3 quart ciders (2 Chicago; 1 Rock Island); 1 Chicago whiskey; 1 Evanston pickle; and 1 Pekin patent medicine.

Illinois embossed bottles were manufactured throughout the company's business history—including the earliest (ca. 1876–78) examples of embossed bottles produced using a removable slug-plate to change the embossing, rather than creating a new mold. (A C&I innovation?—*see discussion of 1840–80 developments in glass-blowing technology.*)

12. **C&Co** *[ca. 1878–86]*
    C&Co LIM *[ca. 1886–1907]*
    Cunninghams & Co., Pittsburgh, PA
    (Whitten = 1879–85 // 1886–1907)
    (Toulouse = 1879–1907)
    (Lockhart = 1878–91 // 1892–1907)

See the discussion of the earlier C&I glassworks. C&Co continued to produce a few Illinois "old fashioned" blob-top sodas for Illinois bottlers at the end of the 1870s and beginning of the 1880s: three are listed in the present volume—Hausburg sodas from Blue Island and Chicago and a Singer soda from Peoria. Two are C&Co sodas, and one is a later C&Co LIM soda.

13. **CV** *[1880–81]*
    CV Co *[1880–81]*
    Chase Valley Glass Co, Milwaukee, WI
    (Whitten/Toulouse = 1880–81 only)

Only one CV maker's mark soda is listed in the present volume, a W. H. Hutchinson Chicago quart soda embossed C.V. No.2 / MILW. As such, the bottle can be tightly dated to 1880–81.

14. **DeSG Co** *[KF/JW = ca. 1876–78]*
    DSG Co *[KF/JW = ca. 1878–96]*
    DeSteiger Glass Co, LaSalle, IL
    (Whitten/Toulouse = 1879–96; Toulouse: "Information...fragmentary"—operation may have started ca. 1877–79)
    (Lockhart = 1878–96)
    (All agree only known mark = DSGCo)

Three varieties of early quilted "Shoofly"-style presentation flasks were made for advertising purposes by the *DeSteiger Glass Co*. They were similar in function to the advertising flask made by the Alton Glass Company on its opening in 1873–74 (*see Alton Glass Co. listing in this section*). The DeSteiger flasks were likely made about three years later, and date to the later 1870s (ca. 1877–78), when the DeSteiger brothers first began manufacturing glass bottles in LaSalle (*see introduction in this volume and Lockhart et al. 2007:31*).

Although the recognized maker's mark for the glassworks is DSGCo, the majority of bottles documented for this volume that have DeSteiger maker's marks are heel- or base-embossed DeSGCo. This mark appears to be the preferred maker's mark during their first several years of business in the late 1870s (ca. 1876–80). The bottling works seems to have been opened earlier

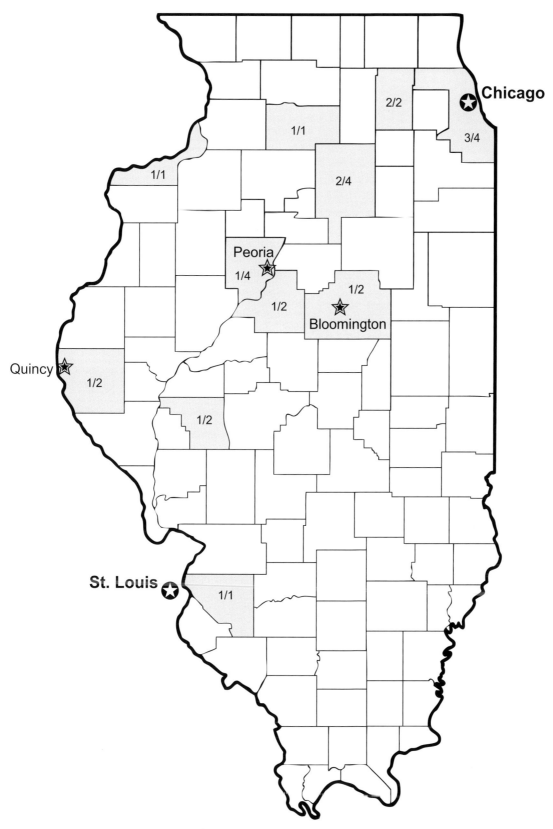

Figure I.72. Illinois county map showing the distribution of embossed bottles documented by our study that have **C&I** maker's marks (*Cunningham & Ihmsen*, Pittsburgh, PA). These bottles were probably imported/distributed via Chicago and Lake Michigan. (Numbering shorthand [#/#] = towns per county/bottlers per county.)

than is generally believed: an 1876 LaSalle Township Directory published in Kett & Co. 1877 (pages 598–620) lists 24 "glass-blowers" as Ottawa Twp. residents. The previously operating factory in town (Phoenix glassworks) made only window glass.

In addition to the three bottling-works flasks, we have documented four quart beers (one from Bloomington, two from Peoria, and one from Rock Island) and two patent medicines (both from Chicago) with the better-known DSGCo mark. By contrast, the DeSGCo mark was used on three beers (two from a single bottler in LaSalle itself and one from Peoria), a Joliet ale, and 10 northwestern Illinois sodas (by eight bottlers in six towns, on bottles dating from the late 1870s to ca. 1881–82).

With the three exceptions noted, beers and medicines are marked DSGCo. All of the sodas are marked DeSGCo. So perhaps the different maker's marks were intentionally applied to different categories of bottles. However, the fact that two quart beers from the same bottler (Weber in Peoria, who bottled there from 1875–1880) have different marks may support the suggestion that the two varieties are earlier and later maker's mark styles. Since the DSGCo mark came into general use during the 1880s, the DeSGCo mark may have been used only during the late 1870s.

15. **DOC** *[ca. 1880–1931]*
    Dominick O. Cunningham, Pittsburgh, PA
    (Whitten = 1882–1931; Toulouse = 1882–1937;
    Lockhart = 1880–1931)

This glassworks began operation at the very end of our study period. But early on it seems to have specialized in accommodating preexisting or newly produced "old fashioned" iron bottle molds from bottlers already in operation during the decade of the 70s. This seems to have only applied to soda bottlers in a few towns: five soda bottle styles from five Chicago bottlers and three styles from three Jacksonville bottlers have been documented, as have single soda-bottle styles from a Kewanee and a Springfield bottler.

16. **EHE** *[ca. 1880–94]*
    Edward H. Everett Co., Newark, OH
    (see also: N (star) and NORTH (star) marks below)
    (Whitten = 1880–1904; Toulouse = ca. 1883–1904)

Like the DOC glassworks above, the EHE glassworks began in operation at the very end of our study period, but seems to have specialized early on in accommodating preexisting or newly produced "old fashioned" iron bottle molds from bottlers already in operation during the decade of the 1870s. Again, only embossed soda bottles have been documented, and in this case, only two of them: one from the F. C. Lang bottling works in Chicago, and one from the J. J. Flynn bottling works in Quincy.

17. **FA & Co** *[KF/JW = ca. 1860–63]*
    Fahnstock, Albree & Co., Pittsburgh, PA
    (Whitten = 1860–69; Toulouse = 1860–62)

Several sources list the Pittsburgh glassmaking firm of B. L. Fahnestock, Robert C. Albree, & Co. as having leased the old Penn Glass Works factory there during the early 1860s (ca. 1860–63). The factory had earlier been operated by Frederick R. Lorenz & Co. from ca. 1854 to 1859 *(see the FRL listing in this section)*. After 1863, Moses Lorenz and partner William Wightman *(see the L&W listing in this section)* resumed glassmaking operations there (Innes 1976:28; Knittle 1927:318–319; Toulouse 1971:195; van Rensselear 1971:178–179; White 1993:6–7).

But in addition to FA&Co marked sodas, numerous fruit jars and whiskey flasks with FA&Co maker's marks are known. Also, several fruit-jar collector's guides published during the 1980s and 1990s have reported that Pittsburgh city directories list Fahnestock & Albree as continuing in business until 1869. This information has led Whitten to conclude on his website that "very probably they were in business during almost the entire decade, since bottles and jars seen with this maker's mark are relatively plentiful.... The actual location of their glass manufactory was the Eclipse Glass Works, Temperanceville (SW Pittsburgh), PA *[after 1863, when L&W reassumed management of the Penn Works — KF/JW]*. Evidently they made glass at both locations at times during their existence." After 1869, FA&Co. also sold the Temperanceville factory to L&W.

From our reading of the Pittsburgh CDs of the 1860s, FA&Co glass production at Temperanceville seems to have been focused on window-glass manufacture. During the early Civil War years at Penn Glass Works (63–65 First Street), FA&Co were listed in the CDs as manufacturers of flint glass, black glass, and vials. But from 1864 to 1869 they were listed in CDs only as manufacturers of window-glass (except in 1866, when "bottles" are also mentioned and in 1868 when "glassware" is also mentioned). Considering other Fahnestock company wholesale and retail drug businesses at the time, these references likely refer to pharmacy glassware production.

Also, our Illinois private mold embossed-bottle data also support the shorter (three-year) time span for FA&Co glassblowing operations. We have documented 18 different FA&Co-marked bottles (all sodas) made for 12 Illinois bottlers in eight towns: Alton, Chicago, Galesburg, Highland, Lacon, La Salle, Peoria, and Springfield. From bottle style and manufacturing technology, combined with our research on local bottler histories, all of these bottles were produced during the 1860–63 period. If bottles were later blown at the Temperanceville plant, perhaps the work there included non-private-mold fruit jars and flasks.

18. **FB Co** *[1888–93]*
    Findlay Bottling Co, Findlay, OH
    (Whitten = 1888–93)

Obviously, this Ohio glassmaker postdates the period of our study. Only one bottle with this mark is included in the text (a Hennessey ale from Chicago), as an illustration of the occasional continuation of an "old-fashioned" bottle style into the 1880s.

19. **FRL** *[ca. 1854–1859/60]*
    Frederick R. Lorenz, Jr., Pittsburgh, PA
    (Whitten = c. 1854–1859/60)

Frederick Lorenz operated the Penn Glass Works in Pittsburgh during the mid- and late-1850s, before leasing the factory to Fahnestock, Albree & Co. during the 1860–63 period *(see the FA&Co discussion in this section)*.

Only two early Illinois FRL-marked sodas have been documented in our text—but as with the FA&Co sodas, they are important "horizon markers" for documenting the end of empontiling and the beginning of smooth-base "snap case" glassmaking technology available to western bottlers.

Both of our Illinois FRL bottles are from Belleville (adjacent to neighboring St. Louis bottling markets just across the Mississippi River). The earliest is an iron-pontiled variety used by the Fisher & Rogger company ca. 1858–59. The later variety is a smooth-based bottle made by Fisher alone, just after he took over the bottling business on his own ca. 1860 *(see text listing)*.

20. **Hawley Glass Co.** *[ca. 1872–85]*
    Hawley, PA
    (Whitten = c. 1872–85)

The beginning and end dates of this Pennsylvania glassworks are uncertain. The sole Hawley-made Illinois bottles we have documented are whiskey-bottle varieties used by Chapin & Gore in Chicago. One is a paneled cylindrical quart and one is a barrel-shaped whiskey bottle. The Hawley-made bottles were produced ca. 1875 *(see Chapin & Gore text discussion)*.

21. **IGCo** *[KF/JW = 1873–1929: see Alton text discussion]*
    Illinois Glass Co., Alton, IL

22. **IGCo L** *[ca. late 1870s]*
    Christian Ihmsen Glass Co., Pittsburgh, PA
    [Ihmsen may start as early as 1855 (McKearin 1941:590)—Toulouse says Ihmsen used IGCo ca. 1870–75. Whitten says the company began using mark IGCo L ("Limited") in 1878]

With one exception, we have encountered no evidence to suggest that our IGCo maker's marks are associated with Christian Ihmsen. All of our bottles with this mark were used during the mid and late 1870s, when both companies were in operation. Only a single 1878/79 Chicago beer marketed by Michael Brand has the IGCo L maker's mark, and likely was produced by the Ihmsen Co. in Pittsburgh.

All of our other IGCo maker's marks have been documented on bottles from west-central Illinois bottlers operating in the Alton vicinity during the mid-to-late 1870s. These include 14 sodas, inks, and medicine-bottle styles used by 11 bottlers operating in 10 western counties (*see Figure I.73*). Curiously, no "hometown" Alton bottles of this age have the IGCo mark (*see Alton Glass Co. and Illinois Glass Co. discussions under Alton, this volume*).

23. **IGW Co** *[ca. 1870–76]*
    Indianapolis Glass Works Co., IN
    (Whitten = "probably" Indianapolis GW Co: ca. 1870–76)

This maker's mark has only been documented on a single pre-1880 Illinois bottle (a *Faloon's Rosin Weed Balsam* from Bloomington), found just as the present study was wrapping up (*see Figure I.59*). From the bottle's hinge-mold technology, it appears to have been marketed by the C. B. Castle bottling company in Bloomington during Castle's first few years in business there (ca. 1870–71).

24. **KH&G ZO**
    Kearns, Herdman & Gorsuch, Zanesville, OH
    (Whitten = 1876–86; Toulouse = 1876–84; Lockhart = 1868–85)

This maker's mark *(see detailed discussion in Lockhart et al. 2008)* has only been found on two very different embossed Illinois bottles used by merchants located half a state apart: a hinge molded shoo-fly whiskey-flask style bitters bottled in Prentice in west-central Illinois during the late 1860s or very early 1870s, and a post-mold blob-top soda bottled in Murphysboro in deep southern Illinois during the early-to-mid 1870s.

A possible third use of this maker's mark in Illinois is on a late-variety *Wallace's Tonic Stomach Bitters* marketed in Chicago by George Powell during the 1870s *(see Powell's Chicago text discussion)*. One variety of the bottle has been recorded as having a base-embossed **ZKH & Co.** mark (Ring 1980:80). Lockhart et al. (2008:54) strongly suggest the recorded mark is a corruption of **KH&GZO**. The authors conclude that the maker's mark "may have been misreported to (or misrecorded by) Ring, or it may have been *[an]* engraving error." If so, the bottle and its maker's mark date to the late 1870s or early 1880s.

25. **LGCo** *[1874–92]*
    Lindell Glass Co., St. Louis, MO
    Lindell = 1875–90 (Whitten); 1874–92 (Lockhart)

26a. **LG WKS** *[1840–72]*
    Lockport Glass Works, NY *[KF/JW, this volume]*

  Best candidate companies for use of this mark are:
    Louisville Glass Co/Glass Works, KY (Toulouse)
    Louisville = 1855–73 (Toulouse, Whitten)
    Lockport Glass Works, NY (KF/JW, this volume)

Since manufacturer histories indicate that all of the LGCo-marked private-mold bottles from Illinois date to the later 1870s, they are here attributed to the Lindell Glass Works, and

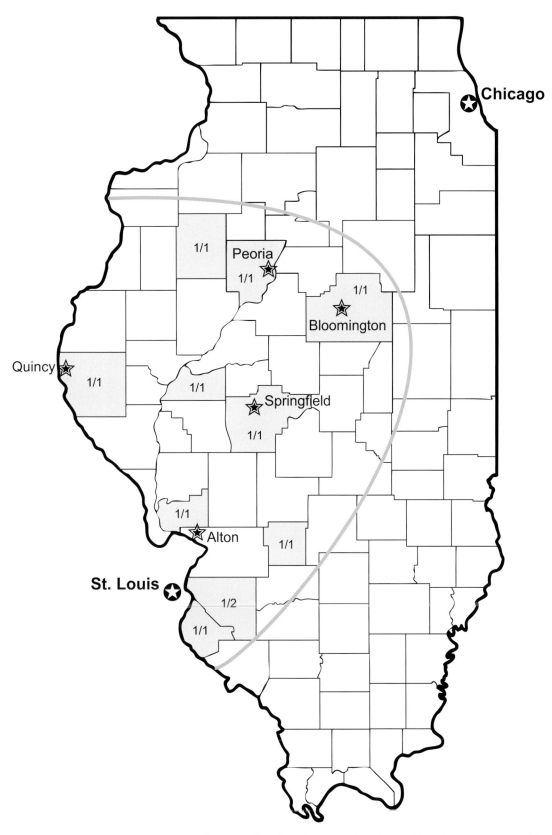

Figure I.73. Illinois county map showing the distribution of embossed bottles documented by our study that have **IGCo** maker's marks (*Illinois Glass Co.*, Alton, IL). These bottles were likely distributed via Illinois River-based and local overland transport. (Numbering shorthand [#/#] = towns per county/bottlers per county.)

the discussion in this section focuses on St. Louis glass manufacturing operations *(see also MGCo maker's mark discussion in this section)*.

The first St. Louis Glass Works opened its doors in 1842, and was in business, under a variety of proprietors, continuously from that time and throughout the 1840–80 period of the present study. So far as is known, the company never used glassmaker marks to identify their bottles, but through a quirk of sales record preservation in probate records we do know that the St. Louis Glass Works (under the proprietorship of Mssrs. Sell & Coe) produced bottles for the Lundblad soda bottling works in Quincy in 1850–51 *(see discussion in the Quincy section of this volume)*. By the mid-1870s, most beverage-bottle manufacture in St. Louis had been taken over by the Lindell Glass Co. (LGCo) and the Mississippi Glass Co. (MGCo: *see discussion in this section*), and the St. Louis Glass Works focused their production efforts on "crystal glass, such as lamp chimneys, bottles for perfumery and prescriptions, fruit jars, and kindred articles" (Dry 1875:57).

The late-nineteenth-century history and maker's marks of the Lindell and Mississippi glass companies in St. Louis has been admirably summarized by Lockhart et al. (2009). The company's origins can be summarized here by quoting from Reavis:

> This is one of the later and more vigorous competitors in the race for supremacy in the glass manufacture of St. Louis, and is doing an immense trade with a large section of the country in the bottle trade...It was organized in 1874...and operations have since been continued, with great energy, and with a remarkable measure of success.... When work was opened only twelve men were employed, now the company has forty-four blowers alone, and an aggregate force of 150 to 160 men.
>
> When work first started merchants and others in this city were shy of placing orders with the new company, for fear that they might be discontinued, and that they would have again to buy from the East under new arrangements. The growth of the bottled beer business in St. Louis, however, introduced a new factor into the trade, and gave a great impulse to the manufacture. The demand for beer bottles by the million gave a new element of stability to the factories, and Pittsburgh has been effectively cut off from the West in the supply of goods in which she once had a monopoly. All of the brewers of St. Louis who bottle beer use largely of the bottles made by the Lindell Glass Company, and some of the largest of them are almost exclusively supplied from this establishment alone. [1879:50–51]

During the 1874 to ca. 1880 period in Illinois, our study has documented only six bottles (all sodas) from six different bottlers with LGCO marks: three in St. Clair County, one in Monroe County, one in Randolph County, and one in Macoupin County. St. Clair, Monroe, and Randolph, of course, are just across the river from St. Louis. The Macoupin County bottler was based at Carlinville, about 40 miles to the northeast.

In addition, a single Chicago bottler (John Lomax) used a soda bottle with an LG WKS maker's mark. This is not a known Lindell mark, and it may possibly be a Louisville Glass Works mark. However, Whitten has noted on his website that no marked Louisville Glass Works bottles have yet been documented. So for now, the mark must be considered to be that of an unknown manufacturer *(but see Lockport Glass Works discussion in this section)*.

26b. LOCKPORT NY *[1840–72]*
Lockport Glass Works, NY
(Whitten = 1840–72)

This mark has been found in Illinois on only a single soda bottle used by John Lomax in Chicago during only the 1870–72 period. Although the mark is not a common one for the Lockport Glass Works, it must surely be theirs. Considering its context, it is intriguing to consider the possibility that the slightly later LG WKS maker's mark on another John Lomax soda bottle *(see Illinois Glass Co. discussion)* is from the Lockport Glass Works as well.

27. L&W *[1862–74]*
Frederick Lorenz, Jr. & Thomas Wightman, Pittsburgh, PA
(Toulouse = 1862–71; Whitten = c. 1862–74)

As discussed in the FA&Co discussion, Moses Lorenz and Thomas Wightman took over the Penn Glass Works glassmaking operations in Pittsburgh from Fahnestock, Albree & Co. in 1863. During the later 1860s and early 1870s the company was widely patronized by Illinois bottlers.

It should be mentioned that an earlier L&W partnership was in the glassmaking business in Pittsburgh from 1851 to 1854 (operated by Frederick Lorenz, Sr., and Thomas Wightman). There are no maker's marks from that era in Illinois, and the company was reformed as FRL in 1854 (Frederick R. Lorenz, Jr.—*see FRL and FA&Co listings in this section*).

Our study has documented 20 different bottles with L&W maker's marks from bottlers in nine towns across the northern 2/3 of Illinois—including seven sodas, six patent medicines, three bitters (two from Chicago and one from Peoria), a Chicago gin, two Peoria beers, a Peoria ale, and a Peoria hair treatment bottle.

28. **MATTHEWS (NY)** *[KF/JW = ca. mid-1860s]*
    No one lists Matthews Co. as bottle makers, but "Matthews gravitating stopper" patented bottles dating to the late Civil Ware–era (ca. 1864 to ??) are often base-embossed as having been "MADE BY" the company.

As with the BODE maker's mark and the PUTNAM maker's mark discussed in this section, Matthews was not an actual glassmaker. The company bottled soda and mineral water and produced bottlers supplies—and patented the Matthews gravitating stopper bottle closure during the late Civil War era. Although the base embossing on the earliest of these bottles that indicates they were made by Matthews, in actuality, as patent holders, the company simply served as a middleman between bottlers and one or more unknown glass companies to produce and sell the bottles.

Base-embossed Matthews gravitating stopper bottles were only used by three Illinois bottlers (Buff & Kuhl in Alton, Kammann in Kankakee, and Bower in Olney). However, bottles of the same style/closure were later produced with a variety of maker's marks for Illinois bottlers *(see text reverences for Galtermann/Danville [no maker's mark], Allison/Nelson [DeSGCo maker's mark], and Fellrath/Peoria [A&DHC and IGCo maker's-mark varieties).*

29. **MG Co** *[1873–84]*
    Mississippi Glass Co., St. Louis, MO
    [Missouri Glass Co. also suggested, also St. Louis, MO]
    (Whitten = MissG Co, 1874–84+; MoG Co ca. 1859–1911)
    (Lockhart = MissG Co.,1873–84)

The exhaustive study by Lockhart et al. (2009) lists and discounts earlier suggestions by Toulouse that this maker's mark may be from the Modes *IND* Glass Company (1895–1900) or the Millgrove *IND* Glass Co. (1898–1911). Ayers et al. (1980) also discussed possible connections to the Milwaukee *WI* Glass Co. (early 1880s) and the Muncie *IND* Glass Co. (1888–1906) but, like the Lockhart study, concluded that the mark was used by the Mississippi Glass Co. in St. Louis.

In light of the possible Modes Glass Company attribution for the MGCo mark, it is interesting to note that William F. Modes, who was manager of the Beaver Falls glassworks in the early 1870s *(see BFGCo maker's mark discussion in this section)*, had many St. Louis and Illinois connections during the late nineteenth century as well. He became manager of the Mississippi Glass Co. during the mid-1870s, and moved on to manage operations at the DeSteiger glassworks in LaSalle in the late 1870s *(see DeSGCo maker's mark discussion in this section)*. During the 1880s, he moved on to the Streator Bottle & Glass Co. as manager there *(see SB&GCO maker's mark discussion in this section)*.

The few Illinois bottlers whose bottles have this heel mark (two Litchfield sodas, one East St. Louis soda, and one Peoria beer—all from the west-central Illinois area near St. Louis) were all in business during the later 1870s, which tends to confirm that their bottles were made by one of the two St. Louis firms.

Although the Missouri Glass Company had been in operation in St. Louis since the Civil War era, they are not known to have included embossed maker's marks on any of their bottles—and with the local ascendancy of the Mississippi and Lindell glassworks as bottle manufacturers in the mid-to-late 1870s, the Missouri Glass Company appears to have ceased local beverage-bottle production altogether. Their catalogs from the 1870s and 1880s (*"Missouri Glass Company's Commercial Solicitor"*) indicate they sold "glassware, lamps and bottles, Queensware, china, &c., &c." and that their bottles were limited to fruit jars and other preserve containers.

By contrast, the Mississippi Glass Company was characterized in a late 1870s publication on St. Louis transportation systems (Reavis 1879:50) as follows: "The class of goods made is of what is known as colored glass, such as colored glass bottles and green glassware. Bottles for proprietary medicines and bitters, made in private moulds, are an important part of the business. The number of hands employed is in all about sixty, and weekly wages about eight hundred dollars." Clearly, the Mississippi Glass Company was the source of the MGCo maker's marks on Illinois bottles during the later 1870s.

30. **N (w/ embossed star)** *[1873–1904]*
    Star Glass Works, Newark, OH
    (Whitten = Most likely marks associated with the Star Glass Works of Newark, Ohio: 1873–1904. *See also EHE maker's mark discussion in this section.*)

The N (star) maker's mark was found on just two late-1870s medicine bottles documented by our study. Both were late-1870s medicines: a *Wakefield Blackberry Balsam* from Bloomington and a *Warner's White Wine & Tar Syrup* from Chicago. Whitten has identified the N (star) mark as being associated with the Newark Star glassworks, which was taken over by Edward H. Everett in 1880 (*see EHE maker's mark discussion in this section*).

31. **PUTNAM** *[KF/JW = ca. late 1870s–early 1880s]*
    Whitten attributes this mark to Lindeborough Glass Works, NH (1866–1886), because of its connection to manufacturing Putnam's Lightning fruit jars. Toulouse attributes the mark to Henry W. Putnam, Bennington, VT, also due to the Lightning fruit jar connection—Putnam bought the patent, and may have had the jars made at Lindeborough. Toulouse also notes a Putnam Glass Works at Putnam, OH (1852–71).

Our study documented only a single Grossenheider lager beer bottle from Chicago with the PUTNAM mark. Bottle technology and bottler history data suggest the bottle was made and used in the late 1870s or early 1880s. Toulouse (1971:330) indicates that Putnam was not a glassmaker, per se, but a middleman jobber who bought glass patents and farmed out orders to other glass factories for production (*see also Bode and Matthews maker's mark discussions in this section*).

32. **RYLANDS & CODD** *[KF/JW = 1877–81]*
    BARNSLEY, Yorkshire, England
    [Only Toulouse notes a RYLANDS BARNSLEY mark. He dates Ben Rylands at 1867–1919, and Dan Rylands (usually associated with Codds) at 1884–1901.]

Only a single Illinois pre-1880 bottle bears this maker's mark: it is an 1872/1873 Hiram Codd patent Codd-marble gravitating-stopper soda bottle used by Carse & Ohlweiler in Rock Island just after the mid-1870s.

The bottle is unique for pre-1880 Illinois in that it was manufactured by the Dan Rylands and Hiram Codd glass factory in Barnsley, England. According to British glass-collector websites, this glassworks partnership had only a short business history: ca. 1877–81. A detailed published study of the Rylands and Codd partnerships (Addams 1987:54–58) agrees with these dates for the Rylands-Codd Company, and puts the end date of a subsequent Codd-Rylands business at 1884 (Hiram Codd's wife passed away that year and his bottle patent ran out at about the same time):

> In 1877 Hiram Codd entered into a partnership with Ben Rylands of Barnsley, the partnership being known as "Rylands and Codd" with Rylands holding four fifths of the capital. These early Codd's usually have pronounced shoulders with a long neck and the embossing of "NO-4."
> Ben Rylands died in 1881 and his son Dan Rylands took over until that stormy partnership known by "Codd and Rylands" ended in 1884. After which, (after settling the partnership with a "large sum of money"), the Codd bottle was just embossed with "Dan Rylands." [Addams 1987:56]

Only one other bottle documented by our study was made in England. It is an 1870s style quart blob-top soda made for John Lomax of Chicago—likely during the early 1880s—by the Alfred Alexander glassworks of London (*see Alfred Alexander maker's mark discussion in this section*).

33. **SB&G Co.** *[1881–1905]*
    Streator Bottle & Glass Co., Streator, IL

Only a single "old-fashioned" (1870s style) quart-size blob-top soda bottle manufactured during the early 1880s by the Streator glassworks was documented by our study. It was made for Hayes Bros. bottling works in Chicago. The Streator glassworks is also known to have produced a similarly "old fashioned" pint blob-top soda for an Iowa company (H. J. Witt) during the early 1880s (Burggraaf and Southard 1998:214).

34. **Large 5-pointed STAR** *[KF/JW = ca. 1865–75]*
    Star Glass Co., New Albany, IN
    (Toulouse = "1860s to 1900s")

We have documented four pre-1880 Illinois bottles with the large embossed-star maker's mark, all of them from Chicago: a Hutchinson ale, a King ale, a Keeley soda, and a Northwestern Chemical Company ink. The three Chicago ale and soda bottles were manufactured and used during the late 1860s and early 1870s. The Chicago ink bottle was used during the late 1870s.

35. **TW & Co.** *[ca. 1874–95]*
    Thomas Wightman & Co., Pittsburgh, PA
    (Whitten = 1874–95+; Toulouse = 1871–95)

Following the death of Moses Lorenz, this company was the successor glassworks to Lorenz & Wightman (*see L&W maker's mark discussion in this section*), and the new TW&Co maker's mark first appeared ca. 1872–74.

The Wightman glassworks was patronized by several Illinois bottlers during the mid-to-late 1870s. Our study has documented eight different bottles with the TW&Co mark: three sodas, an ale, and a spruce beer used by three bottlers in Braidwood (*Blood*), Cairo (*Breihan*) and Joliet (*Paige*), and two Winchell patent-medicine vials from Chicago.

36. **WIS. G. Co. / MILW.** *[ca. 1881–85]*
    Wisconsin Glass Works, Milwaukee, WI
    (Whitten & Toulouse = 1882–86)
    (Reilly 1997 = 1881–85)

Just after this glassworks was formed in 1882, it produced an 1870s-style quart amber soda for a one or two season (ca. 1881–82) bottling company in Colehour (south Chicago).

37. WIDE variety of marks *(see list below)*
    Wm. McCully & Co. Pittsburgh, PA *[ca. 1841–1909]*
    (Whitten = 1841–ca. 1909)
    (Toulouse = 1841–ca. 1886)

William McCully—with 39 marked Illinois sodas, ales, ciders, bitters, whiskeys, and patent medicines used by 22 bottlers in 11 towns statewide—was the third most common glassmaker mark documented on pre-1880 Illinois embossed bottles. Bottles marked by two other Pittsburg manufacturers *(both discussed in this section)* were more commonly found: Cunningham & Ihmsen (48 bottles) and A&DH Chambers (164 bottles).

For the number of bottles documented, the McCully Company used a surprisingly wide range of maker's mark varieties on 1860–1880 Illinois bottles (no pre-1860 McCully-marked bottles were found). Excluding small variations, such as differences in use of larger and smaller-case letters, and variable use of periods and superscripting (either underlined and not underlined), our study recorded 4 different soda heel marks (17 bottles), and 9 different base marks on 15 ales and 4 other bottle styles (21 bottles).

Soda heel marks included:

McC *(1 Chicago soda; 1 Cairo soda)*
W. McC & Co. *(9 Chicago-area sodas; 2 west-central Illinois sodas)*
Wm McC & Co. *(1 Galena soda; 2 west-central Illinois sodas)*
W. McCULLY & Co. *(1 Chicago soda)*

Base marks *(unless noted)* on ales and other bottles included:

McC *(1 Chicago ale heel mark; 1 Rockford whiskey base mark)*
W. McC *(1 Elgin medicine)*
W. McC & Co. *(2 Chicago ales; 1 Chicago bitters; 1 Dixon bitters)*
Wm McC & Co. *(1 Peoria bitters)*
W. McC & Co. PITTS- *(1 Chicago ale)*
W. McCULLY & Co. PITTS PA *(3 Chicago ales)*
W. McCULLY & Co. PITTSBURGH. *(4 Chicago-area ales; 1 Cairo ale)*
W. McCULLY & Co. PITTSBURGH PA *(3 Chicago-area ales; 1 Chicago-area cider)*

The reasons for these differences are not known, but are not obviously related to time differences or bottle style or size differences. Like the variable use of punctuation in the A&DHC maker's mark, the differences may have included coded information for factory workers, or they may just reflect a lack of standard company instructions to mold makers.

38. **Wm Frank & Sons** *[1866–75]*
    Pittsburgh, PA
    (Whitten & Toulouse = 1866–76)
    (Lockhart et al. 2008 = 1866–75)

Just four Illinois bottles have been documented that were likely produced by this glassworks: all from Chicago. One of the three, a Schwab, McQuaid Co. *Red Jacket Bitters* (ca. early 1870s) has a base-embossed **WM. F. & SONS, PITTS. PA.** maker's mark.

The other three—a barrel-shaped whiskey (base-embossed **H. FRANK'S PAT^D / AUG. 1872**) and pint and half-pint sized strap-sided whiskey flasks (base-embossed **PAT^D AUG6TH1872 / H FRANK**)—were used by the Chapin & Gore Co. (ca. 1869–75).

The Himan Franks patent referred to in the embossing is for an internally threaded glass stopper closure. Although the bottles do not have the standard Wm Frank & Sons maker's mark, we assume they were produced by Frank's glass factory *(see H. Frank patent discussion in Lockhart et al. 2008:32–36)*.

Changes in Embossed-Bottle Manufacturing Technology from 1840–1880

Using the chronological information developed in this volume for Illinois bottlers and the technological and stylistic attributes of the embossed bottles they used, several time-diagnostic chronological trends are apparent. With few exceptions, these chronological horizon markers seem to be independent of the specific glass houses that manufactured the bottles—probably because the modernization of glass-house technology moved generally (and fairly slowly during the period of our study) from East to West. The glass houses serving Illinois bottlers during this time were mainly western ones (i.e., located on the "western frontier" rather than the northeast coast): primarily located in the Pittsburgh area during the first two decades of our study, and expanding to include Illinois and St. Louis glassworks during the 1860s and 1870s.

Keeping in mind that historical bottles made before 1880 were sometimes reused, not discarded, after being purchased for their original contents by consumers (Busch 1987), and that many such bottles even survived to the present day, it is best to evaluate the ages of *groups* of embossed bottles found together in good historical contexts rather than individual artifacts. Using the Illinois-bottler business histories assembled for this volume and glass-manufacturing patent data, the most reliable time-diagnostic attributes of changing bottle styles and technologies during our study period include the following:

1. Essentially all of the glass bottles made for Illinois bottlers during the 1840s and 1850s have iron-pontiled or open-pontiled bases. The transition to smooth-base, snap-case held bottles occurred rapidly during a single year (1860), after which all Illinois bottles are smooth-based, snap-case-held varieties. At eastern glass houses, this shift away from pontiled bottles appears to have occurred as early as 1857 (von Mechow 1993), but this early transition had little effect on Illinois bottlers.
2. During the early-to-mid 1840s most small medicinal and pharmacy bottles were made with paper-thin flared-lip finishes. During the later 1840s and early 1850s, most such bottles were made with more sturdy rolled-lip finishes.
3. From the 1840s to ca. 1853, most soda-size and quart beverage bottles were made with applied sharp-shouldered tapered-collar lip finishes. During the later 1850s, most soda-style bottles were made with applied shouldered or oval-blob-top finishes, and most rectangular quart beverage bottles were made with applied ring-neck tapered collars.
4. Hutchinson-style soda bottles, with their internal "figure-8" wire pull closures above rubber-ring stoppers, were patented in Chicago in October 1878. Their more mechanized filling process and reusable rubber stoppers (long varieties of which could even be adapted for use in existing short-neck blob-top sodas), along with aggressive marketing and patent infringement battles by Charles Hutchinson's Chicago-based company, led to this bottle style dominating the soda-bottling world within 24 to 36 months after their initial appearance.
5. Prior to the Civil War, most stoneware beverage bottles used in Illinois were hand-thrown, with smooth sides that expanded to a broad, sharp or somewhat rounded shoulder. During the 1860s, such stoneware bottles were characteristically eight- or ten-sided Merrill or Smith patent containers that were produced in molds. The Merrill and Smith molded stoneware bottle styles were first patented during the late 1840s and 1850s (Graci and Rotilie 2003), but the paneled varieties were not popular among Illinois bottlers until the Civil War years.
6. As discussed in detail in the "Applied Finishes" section of Bill Lindsay's massive "Historic Glass Bottle Identification and Information Website" (Lindsay 2009), these applied-lip finishes continued during the 1860s and early and mid-1870s, after which the invention of more sophisticated lipping tools ushered in a late-nineteenth-century era of more refined

tooled-lip finishes (e.g., J. B. Wilson 3/25/1884 patent #295,848: Figure 30d above, and C. H. Beach 2/7/1893 patent #491,069). These developments are apparent in the refined lips of smaller pharmacy bottles by the mid-1870s, and are more often seen on larger beverage and medicine bottles beginning in the early-to-mid 1880s.

7. The most time-diagnostic aspect of mold-seam change on smooth-based bottles (see Toulouse 1969) occurred on non-dip-mold bottles ca. 1871. Prior to that time, bottles made in two-piece hinge molds characteristically show keyed hinge-mold lines on their bases: these molds were characteristically "keyed" hemispherically, but sometimes the keyed areas are rectangular, and an occasional straight-line hinge is seen on Civil War–era bottles. Pontiled bottles were hinged in this way as well, but the mold lines are usually obscured by the pontil scar. After 1871, a new "post mold" technology was developed that resulted in the mold bases being hinged hemispherically around both sides of a central post. The post area could accommodate an embossed slug plate for easy bottle-base embossing.

8. Prior to the early 1870s, Illinois embossed bottles were embossed primarily by having the embossed lettering cut into the privately made iron molds themselves (a major bottle-mold maker during this era was Charles Yockel: *see Figure I.74*). As early as the late 1860s, however, flat, changeable slug-plate-embossing blocks were patented to change bottle embossing without cutting entirely new iron molds for each new bottler (see J. J. Christie 1867 patent #72,368). Shortly thereafter, flat slug-plate molds for sunken-panel rectangular and square bottles were improved so that individual parallel lines of lettering could be exchanged (see J. J. Christie 1872 patent #132,897 and Gustavus Storm 1875 patent #163,276).

9. Exchangeable slug-plate technology for round and oval bottles was developed at about the same time, and is more easily recognizable on embossed-glass bottles. For two-piece mold slug-plate bottles, the key patent (#19,162) is that of Lancaster Thomas, awarded on June 23, 1868. However this patent development was somehow not well adapted to western bottling-house use, and slug-plate technology is not seen used on Illinois bottles before the later 1870s. From ca. 1876 to 1880, the only apparent Illinois use of the technology was by Cunningham & Ihmsen (C&I) of Pittsburgh, who somehow overcame the difficulties of rapidly and profitably mass-producing curved-sided Illinois bottles using slug-plate embossing (see, e.g., Bradbury/Jacksonville text listing below). At about that

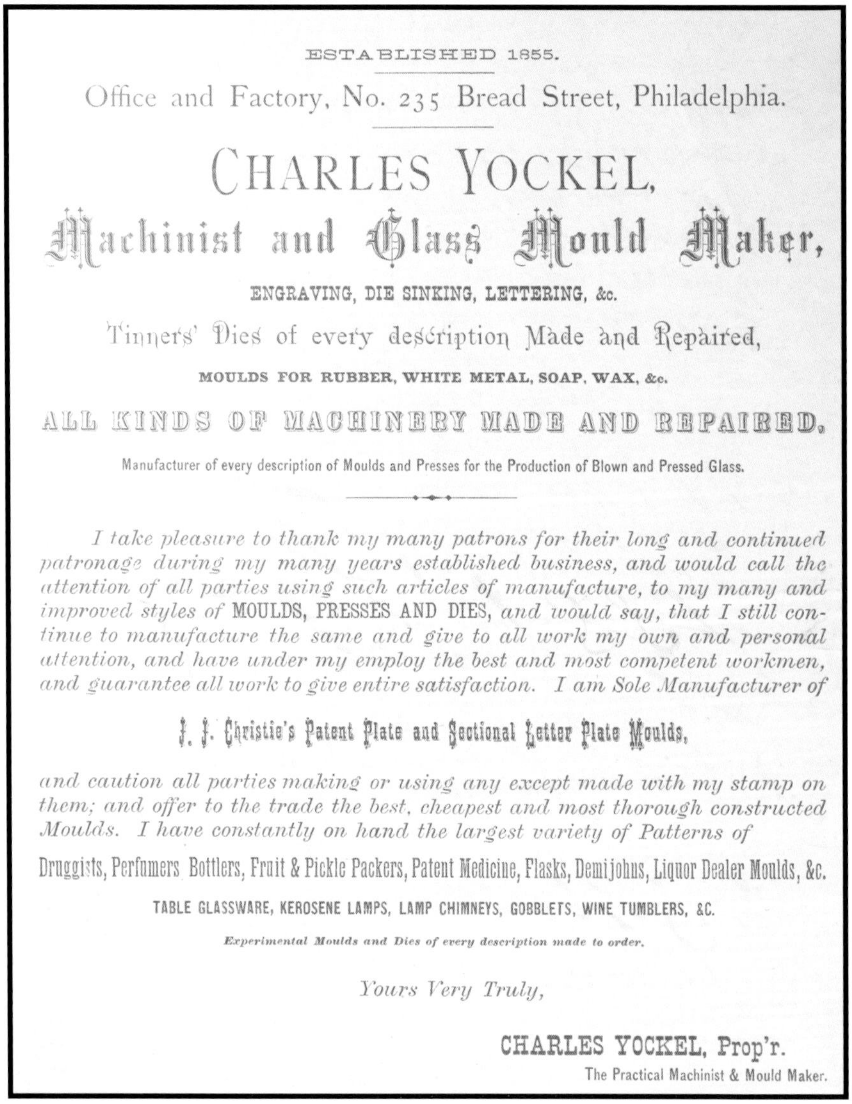

Figure I.74. One of Charles Yockel's early 1870s ads touting his experience in the production of metal "proprietary" embossed-bottle molds for private bottlers.

same time, a similar exchangeable slug-plate patent (#174,514) was awarded to Samuel Garwood for 3-piece dip-mold bottles. After 1880, essentially all embossed-bottle production was done using slug-plate technology.

10. Design-patented, distinctively styled pharmacy bottles were first produced in massive quantities for drug stores during the later 1870s—first in both aqua and clear glass, but soon only in clear or intentionally colored glass (*Figures I.75 and I.76*; see Brown 2010; Burggraaf 2010:xiii–xvii). The earliest known design style was the French square (*see Figure I.44*). No design patent has been found for this bottle, but early graduated French squares were patented in 1866 (Griffenhagen and Bogard 1999:35ff), and bottles of this shape were in use in Illinois by the early 1870s, if not before. The French square druggist-bottle style continued in wide use throughout the late nineteenth century. *Whitall, Tatum & Co.* began producing and embossing their Philadelphia oval style druggist bottle around the mid-1870s, but never patented the shape (Griffenhagen and Bogard 1999:37–38). By the later 1870s, both the Millville round style and double Philadelphia oval pharmacy bottles appear (*see Figures I.45 and I.46 above*). Both are often base-stamped with their 1878 design patent dates. During the later nineteenth century and early twentieth century, pharmacy glass houses designed and produced a dizzying array of similar bottle styles in graduated sizes (e.g., Keystone, Blake, Short Blake, Phoenix, Knickerbocker, Union, Manhattan, Baltimore, Billikin, Buffalo, Prima, Golden Gate, and Metric and Plain ovals). Since these druggist bottles were mass-produced, universally used, and characteristically discarded when empty (often in privies), a careful study of their diagnostic attributes and appearance sequences would create an ideal chronological gauge for dating 1880–1915 archaeological sites.

11. A variety of bottle ownership declarations are first embossed on returnable beverage bottles (primarily sodas and beers) manufactured during the later 1870s and afterwards. These disclaimers, seem to have been added to bottles to avoid some sort of manufactured-goods taxes. They make it clear that only the bottle contents are for sale, not the bottles themselves. Such disclaimers include: THIS BOTTLE NOT TO BE SOLD; THIS BOTTLE MUST BE RETURNED; THIS BOTTLE IS NEVER SOLD; THIS BOTTLE NEVER SOLD [OR ABANDONED]; THIS BOTTLE LOANED, NEVER SOLD; LOANED—NEVER SOLD.

Figure I.75. Grouping of flint-glass pharmacy bottles produced ca. 1875–1915. Several thousand varieties and sizes of such bottles were produced for Illinois druggists statewide.

Figure I.76. Grouping of intentionally colored pharmacy bottles produced for Illinois druggists ca. 1875–1915. Because of the increased cost for colored bottles like these, relatively few varieties of such bottles were produced.

Patterns of Embossed-Bottle Use in Illinois from 1840–1880

Assemblage Patterns

Percentages of embossed vs. unembossed beverage bottles, medicine bottles, and chemical-product bottles found in 1840 to 1880 Illinois archaeological contexts is variable through time and by product category. Illinois soda, mineral water, and cider bottlers almost always embossed their products made during our 40-year study period, perhaps to help ensure return of the bottles to the bottling works for refilling. Impressionistically, well over 90 percent of all such bottles were embossed.

Liquor bottles were seldom embossed before the 1870s, and apparently relied primarily on paper labeling before that. Early brewery products were seldom bottled at breweries prior to the advent of pasteurization in the mid-1870s. However, secondary bottlers bought more-stable brewery products by the barrel, and bottled them for local sale in very darkly tinted embossed ale and porter-style bottles. Ale and porter bottling began around 1850 in Illinois. There were comparatively few such bottlers, but like sodas, most ale and porter bottles were embossed. By the end of our study period, bottled lager beers had replaced these earlier bottled ale products, and local bottlers throughout Illinois were beginning to use embossed beer bottles. During the 1880s and 1890s, the numbers and variety of Illinois embossed bottled beers skyrocketed.

Chemical company products (such as ink, bluing, baking powder, and flavoring extracts), were produced and bottled throughout our study period, but were seldom packaged in embossed bottles before the 1870s. Even then, embossed bottles were used for less than a third of all such products produced statewide.

A diverse array of patent medicines and bitters were put up in embossed and unembossed bottles across the state prior to 1880. During the 1840s, the medicinal products of early Illinois bottlers were almost never embossed, and less than one in four were embossed during the 1850s. By the Civil War era, the patent medicine business in the upper Midwest had become a profit-making powerhouse for savvy bottlers who had the resources to bombard uneducated consumers with outrageous advertising, and embossed patent medicines from Illinois and many surrounding midwestern states began entering the Illinois archaeological record in great numbers. Impressionistically, embossed patent-medicine products produced by large and small Illinois companies during the 1860s and 1870s make up no more than 50 percent of the total. The bulk of these "brand name" medicine products were retailed by community pharmacists and drug stores across Illinois, but Illinois drug stores themselves did not begin extensively embossing their own bottled products before the mid-1870s.

Percentages of Illinois embossed beverage and medicine bottles vs. those produced by *non-Illinois* bottlers during in 1840 to 1880 period varied over time and by product category as well. Non-Illinois embossed soda and mineral water bottles are seldom found within the state, except along Illinois' borderlands near towns large enough to have supported local bottling works in adjacent states (especially the St. Louis area). Scattered embossed soda bottles from more distant areas are also occasionally encountered, and are interpreted as having arrived in the luggage of travelers or immigrants. Like sodas, embossed ale and porter bottles are characteristically the products of Illinois bottlers.

Finally, uncountable numbers of varieties patent medicines and bitters were put up in embossed bottles for distribution across wide regional (often multistate) areas by both Illinois bottlers and those in towns and cities in other eastern and midwestern states. The products of Illinois bottlers likely represent less than half the embossed bottles of these categories found in Illinois archaeological contexts of the 1860s and 1870s, and less than 10 percent of the medicinal bottles found in pre–Civil War Illinois contexts.

Some categories of distinctive 1840–1880 product bottles were seldom if ever embossed by Illinois bottling houses. To a large degree, these product categories were regionally marketed throughout the eastern United States by only a few wholesale bottlers who used distinctive but unembossed glass containers for their products. Such bottles include wine bottles, canning and preserve jars, ornate whiskey flasks, poison bottles, "cathedral" pickle and peppersauce bottles, and cologne, perfume, and snuff bottles (see Ham 1988; Ketchum 1975; McKearin and McKearin 1941, 1950; McKearin and Wilson 1978; Munsey 1970; van Rensselaer 1971). Specialty-use glass containers such as target balls and fire grenades (Bushell 1956; Finch 2008; Ketchum 1975:200–206; Kleinstuber 1989; Munsey 1970:196–197; O'Malley 2008) did not appear on the market until the end of our study period.

Antebellum Product-Use Patterns

The first appearance and early patterns of growth of Illinois beverage and medicine companies using embossed product containers, as documented below in this volume, can be summarized as follows.

The 1840s Pioneer Period

During the early 1840s, the only embossed Illinois bottles we have identified are five patent medicines produced by three central Illinois manufacturers: two sizes of liniment bottles produced by Alexander and Morris Lindsay in Springfield; the *Arabian Liniment* and *Fever & Ague Drops* of the Farrell brothers in Peoria; and two sizes of Dr. Hamilton's *Vegetable Cough Balsam* in Jacksonville.

During the later 1840s, there is a 460 percent increase in the number of embossed medicine bottles documented: 23 bottle varieties produced by 14 manufacturers in 10 towns located in 10 scattered counties across the northern two-thirds of the state (*Figure I.77*). These medicines are primarily liniments, cough balsams, fever and ague cures, and vermifuges. The first Illinois embossed beverage bottles appear during the late 1840s: three used by soda and mineral water manufacturers (one in Chicago and two in St. Clair and Monroe counties across the river from St. Louis), and one was used by a St. Louis "expatriate" ale bottler in Galena in the northwestern corner of the state.

The 1850s Antebellum Period

During the 1850s Illinois was no longer a "Western frontier" characterized by a dramatic inflow of arriving settlers and by the spread of immigrant farming communities. A substantial number of west-central Illinois cities (including Springfield, Peoria, Bloomington, Alton, and Quincy) and county-seat towns statewide were rapidly growing as part of a network of developing commerce. At the northeastern margin of the state, the Great Lakes port of Chicago was becoming a metropolitan area and economic powerhouse to rival St. Louis on Illinois' Riverine southwestern border. Reflecting these developments, local bottling of embossed beverage products thrived—including enterprising residents setting up one-room bottling operations from home and the development of first substantial bottling factories in larger cities and the Chicago metropolitan area.

The three soda and mineral-water bottling businesses recorded during the later 1840s grew exponentially, to include 48 such businesses in 15 cities and towns statewide, using 87 different embossed bottle styles. Thirty of these soda bottlers were clustered in Chicago and in the three Illinois counties closest to St. Louis (*Figure I.78*).

Patent-medicine manufacture experienced a similar vigorous growth. The 14 manufacturers recorded in 10 towns during the late 1840s increased by 270 percent, to include 38 bottlers in 19 cities and towns statewide, using 83 different embossed bottle styles (*Figure I.79*). Most of these embossed patent-medicine bottlers were headquartered in Chicago or along the Illinois River valley in north-central and west-central Illinois. The variety of patent medicines bottled was also much more diverse than earlier. In addition to the liniments, vermifuges, and fever and cough/pectoral cures seen in the late 1840s, the 1850s medicines included at least 10 universal pain and disease cures, two chologogues (a "bile" removing purgative), two cures for "liver" complaints, a Chicago purveyor of *Red Drops* (purportedly a venereal disease cure), and a Quincy

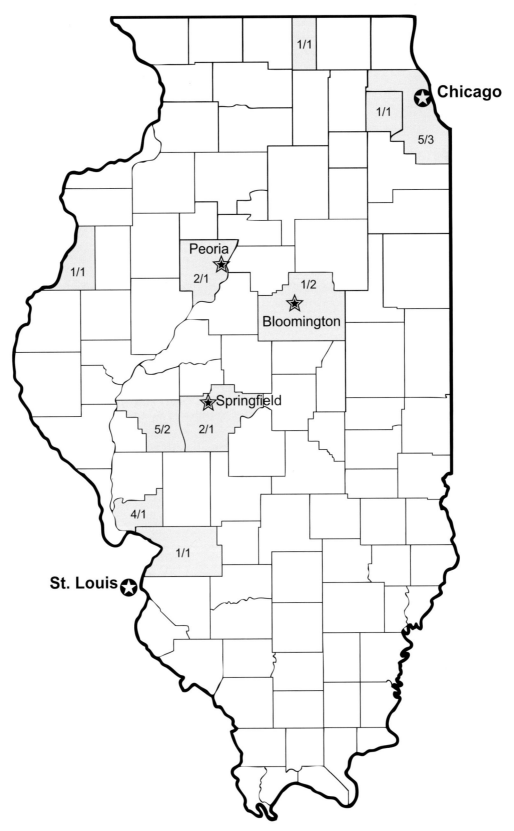

Figure I.77. Illinois county map showing the distribution of embossed medicine bottles documented by our study that were made and used during the late 1840s. (Numbering shorthand [# / #] = towns per county/bottlers per county.)

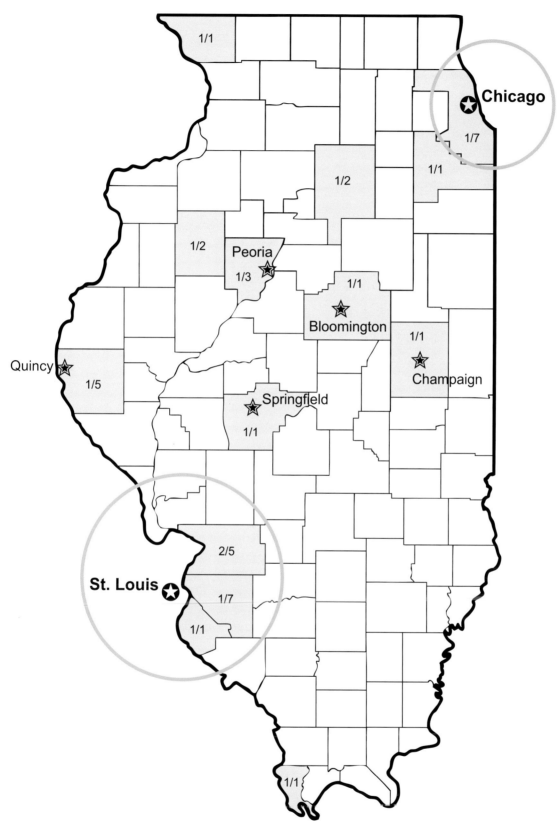

Figure I.78. Illinois county map showing the distribution of embossed soda and mineral water bottles documented by our study that were made and used during the 1850s. Note the concentration of bottlers with embossed bottles at Chicago and near St. Louis. (Numbering shorthand [#/#] = towns per county/bottlers per county.)

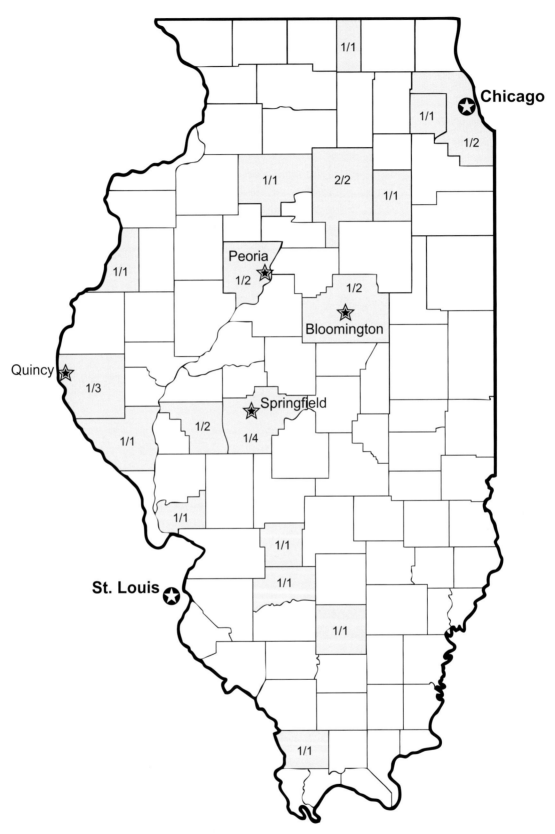

Figure I.79. Illinois county map showing the distribution of embossed medicine bottles documented by our study that were made and used during the 1850s. Note the statewide distribution of scattered bottlers, except at Chicago. (Numbering shorthand [# / #] = towns per county / bottlers per county.)

"eye water." The first generic embossed bottles used by pharmacy businesses appear during this time as well: the bottles were used by three pharmacies in Chicago, one in La Salle, one in Peoria, one in Springfield, and one in Jacksonville.

Bitters, a new kind of distilled liquor-based medicinal beverage introduced into the Midwest by arriving German immigrants, was also first put up in embossed bottles in Illinois during the 1850s. The success and spread of this product in Illinois, as documented in the pages below, was phenomenal. Prior to the Civil War years, however, we have documented just four embossed varieties: three of them produced in the Chicago area and one in Springfield. At this early date, only the Springfield example was put up in the square, amber quart bottles that later came to characterize this "medicinal" liquor product. In the Chicago area, Cory's and Collins' bitters emphasized the medicine-quality of their products by bottling them in patent-medicine-style small sunken-panel rectangular bottles, while *Foerster's Teutonic Bitters* was packaged in a handled "onion"-shaped liquor bottle. A few distilled whiskeys were also first put up in embossed bottles during the 1850s. Both manufacturers were Chicago based: the *Thomas & Co.* product was put up in an onion-shaped bottle like the Foerster's bitters above, and the Bennett Pieters pint handled-jug-shaped bottle was a short-lived precursor to his *Red Jacket Bitters* line of the ensuing 1860s—and may in fact have been an unlabeled bitters itself.

Ale and porter bottling also expanded from its humble Galena beginnings in the late 1840s, especially in Chicago. Ten ale bottlers have been documented in four towns during this time, including Evans and an additional bottler in Galena, six Chicago bottlers, one in Alton (whose brother was a Chicago ale bottler), and one in Belleville. These 10 bottlers used 16 different ale and porter bottle styles during the decade. Most of them bottled in black-glass quart and pint bottles to help preserve their unpasteurized brewed contents, but the short-lived Belleville Heberer bottle and a few of the Chicago ales were made using aqua or teal colored glass.

The 1860–1880 Civil War and Industrial Expansion Periods

During the 1860s and 1870s, events associated with the Civil War and its aftermath brought industrial expansion in Illinois to a new fever pitch. Even though the 1850s saw rapid spread and growth in the embossed bottling of beverages and medicines in Illinois, this growth was eclipsed several hundredfold during the ensuing two decades. As George Hawes, writer and compiler of the 1858/59 *Illinois State Gazetteer and Business Directory*, colorfully observed regarding the development of Carlinville in Macoupin County:

> The population in 1851 amounted to about 500 persons. It now numbers about 2300 souls, and is still making rapid strides in the way of improvements.... Twenty years ago, the spot where Carlinville now stands was the domain of the red man, the country in which it is located bearing the name of "Black Hawk hunting ground." On the shores of the little creek near which the town is situated, their lodges rose in myriads, their dusky inmates thronged the forests, their council-fires crackled, and the old chiefs gathered round in silence, smoking their pipes, and musing on the daring deeds of their fathers, or forming plans of vengeance on some offending foe. They now live only in the memory of those who have followed where the star of civilization pointed the way, and where the soil, teeming with life and richness, invited them to share its blessings. Thus has the hand of the white man made subservient to his uses all the natural resources of a country whose yieldings are sent to supply almost a world with the means of existence. [p. 29]

During and after the Civil War years, continued immigration and rapid commercial growth resulting from Civil War–generated industrial expansion combined to greatly expand the scale, number, and diversity of Illinois bottling operations. The pace of Illinois commercial growth continued to accelerate during the 1870s. Dramatic evidence of expanding product diversity among Illinois bottlers can be seen, for example, in the many new beverage-bottle styles first used in Illinois during the 1870s by the massive John A. Lomax bottling works in Chicago (*see his post-1870 "Fire and Soda" text listing in this volume*). Among the "horizon marker" Illinois embossed-bottle styles first used after 1870 are ring-neck quart ciders, cylindrical medicated aerated waters, and ginger ales. Chicago-style quart sodas first appeared just a few years earlier (ca. 1868).

The 1860–1880 Illinois bottling companies that used embossed and stamped glass and stoneware bottles, the bottled products they distributed, and the shapes, styles, and embossing details of their marked containers are documented in the text pages that follow, but an historical evaluation of the companies themselves and their Civil War era and post–Civil War meteoric growth are far beyond the scope of this preface. Suffice it to say here that the 1840s and 1850s Illinois bottlers and bottled products summarized above represent less than 22 percent of bottlers (107 of 494) and 6 percent (196 of 1,093) of embossed bottle styles used in 102 cities and towns in 60 Illinois counties during our 40-year study period.

During recent decades, historical archaeologists in Illinois have ventured far beyond their early focus on French forts and their mid-twentieth-century tentative first steps toward an understanding of the social and commercial environments of pioneer rural Illinois, and the origins and growth of early Euro-American utopian settlements, villages, and towns across the state. On-the-ground residential and settlement developments associated with this rapid pioneer influx of Illinois immigrants and their capitalist ideals are poorly recorded in the pages of history and are fertile ground for archaeological studies. But our knowledge of nineteenth-century Illinois household and business material needs and material inventories, and our understanding of the history of midwestern commercial and industrial technology, have lagged far behind our government-sponsored fieldwork efforts.

Archaeologically recovered glass, ceramic, and metallic artifacts are usually corroded and often fragmentary. As a result of project time constraints, laboratory analysts can often do little more than count and weigh the recovered fragments in ways that provide minimal information about the lives and activities of the people who lived at the archaeological sites being studied, let alone inform us about the commercial environment in which products were locally and regionally produced and distributed. Broken bottle fragments, for example, are characteristically identified and grouped only by color and the count and weight of each color category (e.g., *xx* kilograms of green container glass)—a method offering precious little insight as to *why* or *when* such containers were originally brought to the site and used there. The more complete vessels and larger embossed or stamped vessel fragments are often illustrated, but because of project completion deadlines and the wide variety of material classes characteristically recovered from excavated historic sites, the opportunity for associated product or business-history research studies is limited (see, e.g., Mansberger 2009a and 2009b for 945-page project completion technical reports on the results of his excavations in downtown Springfield conducted in advance of Lincoln Library and Museum construction).

Bottled in Illinois was undertaken in an effort to create a baseline historical archaeology reference volume to help *date* and *functionally interpret* one commonly recovered artifact class: 1840–1880 embossed glass containers made for and used by Illinois bottling houses.

References

Addams, Cris
   1987   Hiram Codd [And His International Bottle]. *Antique Bottle & Glass Collector* 4(2):54–58.

Ayers, James E., William Liesenbien, Lee Fratt, and Linda Eure
   1980   *Beer Bottles from the Tucson Urban Renewal Project, Tucson AZ.* Unpublished manuscript, Arizona State Museum Archives, RG5, SG3, Series 2, Subseries 1, Folder 220.

Brown, Frederick M.
   2010   *"Good for What Ailed You" in Illinois. Volume #1: Springfield. The Embossed Pharmaceutical Bottles used by Springfield Druggists from the Civil War Era to the Early Twentieth Century.* Bottle identification guide privately distributed by the author.

Burggraaf, Mike, and Tom Southard
   1998   *The Antique Bottles of Iowa: 1846–1915 (Volume 1).* Privately published by the authors. Ohio Wholesale Copy Service, Northfield.

Burggraaf, Mike
   2010   2010 Update to The *Antique Bottles of Iowa: 1846–1915 (Volumes Three & Four).* Privately published by the author. Frontline Printing and Design, Fairfield, IA.

Busch, Jane
   1987   Second Time Around: A Look at Bottle Reuse. *Historical Archaeology* 21(1):67–80.

Bushell, William
   1956   *The Life of Captain Adam Bogardus.* Undated pamphlet privately printed around 1956 at Lincoln, IL.

Carson, Gerald
   1961   *One For a Man, Two For a Horse. A Pictorial History, Grave and Comic, of Patent Medicines.* Bramhall House, New York.

Chapman Brothers
   1883   *Portrait and Biographical Album of DeKalb County, Illinois.* Chapman Brothers, Chicago, IL.

Dacus, J. A., and James W. Buel
   1878   *A Tour of St. Louis; or, the Inside Life of a Great City.* Western Publishing Co., Jones & Griffin, St. Louis, MO.

Davis, Derek C.
   1972   *English Bottles and Decanters: 1650–1900.* The World Publishing Co., New York.

Dry, Camille N.
   1875   *Pictorial St. Louis: The Great Metropolis of the Mississippi Valley. A Topographical Survey Drawn in Perspective A.D. 1875.* Designed and edited by Rich. J. Compton, St. Louis, MO.

Finch, Ralph
   2008   What Are Target Balls? I'm Glad You Asked... *Bottles and Extras* 19(1):47–48.

Graci, David
   1995   *American Stoneware Bottles.* Calem Publishing Co., South Hadley MA.
   2001   *(More) American Stoneware Bottles.* Privately published by the author, South Hadley, MA.
   2003   *Soda and Beer Bottle Closures: 1850–1910.* Privately published by the author.

Graci, David, and David Rotilie
   2003   Smith vs. Merrill. *Antique Bottle & Glass Collector* 20(4):36–39.

Griffenhagen, George B., and Mary Bogard
   1999   *History of Drug Containers and Their Labels.* American Institute of the History of Pharmacy, Madison, WI.

Griffenhagen, George B., and James Harvey Young
   1959   Old English Patent Medicines in America. In *Contributions from the Museum of History and Technology* (Paper 10, pp. 155–183), United States National Museum Bulletin 218. Smithsonian Institution, Washington, D.C.

Gums, Bonnie L., Eva Dodge Mounce, and Floyd R. Mansberger
   1997   *The Kirkpatrick's Potteries in Illinois: A Family Tradition.* Transportation Archaeological Research Reports 3. Illinois Transportation Archaeological Research Program, Department of Anthropology, University of Illinois, Champaign-Urbana, IL.

Ham, Bill
   1988   Sunburst and Similar Scents. *Antique Bottle & Glass Collector* 5(1):14–22.

Harper's New Monthly Magazine
   1872   Soda-Water: What It Is and How It Is Made. *Harper's New Monthly Magazine* 45(267):341–346.

Hawkins, Jay W.
   2009   *Glasshouses & Glass Manufacturers of the Pittsburgh Region 1795–1910.* iUniverse, Inc., Bloomington IN and New York NY.

Innes, Lowell
   1976   *Pittsburgh Glass: 1797–1891. A History and Guide for Collectors.* Houghton Mifflin Co., Boston, MA.

Jones, Olive
   1971   Glass Bottle Push-Ups and Pontil Marks. *Historical Archaeology 1971.* Volume 5:62–73.
   1981   Essence of Peppermint, a History of the Medicine and Its Bottle. *Historical Archaeology* 15(2):1–57.
   1983a   London Mustard Bottles. *Historical Archaeology* 17(1):69–84.
   1983b   The Contribution of the Ricketts' Mold to the Manufacture of the English "Wine" Bottle, 1820–1850. *Journal of Glass Studies* 25:167–177.

Ketchum, William C., Jr.
   1975   *A Treasury of American Bottles.* Bobbs-Merrill Company, New York.

Kett, H. F.
   1877   *Past and Present of La Salle County, Illinois.* H. F. Kett, Chicago, IL.

Kleinstuber, David
   1989   Target Balls Became "Game" of Choice in Late 1800s. *Antique Week* 22(24):1, 36.

Knittle, Rhea Mansfield
   1927   *Early American Glass.* The Century Co., New York, NY.

Larson, Cedric
   1937   Patent Medicine Advertising and the Early American Press. *Journalism Quarterly* 14:333–341.

Leybourne, Douglas M.
   1993   *The Collector's Guide to Old Fruit Jars.* Produced and distributed by the author, North Muskegon, MI.

Lindsay, Bill
   2009   *Historic Glass Bottle Identification and Information Website. Part III: Types of Bottle Closures.* Bill

Lindsay's massive, deeply layered, and highly illustrated labor-of-love website on historic-bottle style and technology. Hosted by the Society for Historical Archaeology. [We used it primarily in 2009, but it is a constantly expanding research resource.]

Lockhart, Bill, David Whitten, Bill Lindsey, Jay Hawkins and Carol Serr
    2004   The Dating Game—Cunningham Family Glass Holdings. *Bottles and Extras* 16(3):2–8.

Lockhart, Bill, Carol Serr, and Bill Lindsey
    2007   The Dating Game—De Steiger Glass Co. *Bottles and Extras* 18(5):31–37.

Lockhart, Bill, Pete Schulz, Carol Serr, Jay Hawkins, and Bill Lindsey
    2008   The Dating Game—William Frank & Sons, Pittsburgh, Pennsylvania (1866–1875). *Bottles and Extras* 19(2):32–36.

Lockhart, Bill, Pete Schulz, Carol Serr, Bill Lindsey, and David Whitten
    2008   The Dating Game—The Kearns Glass Companies. *Bottles and Extras* 19(24): 50–58.

Lockhart, Bill, Pete Schulz, David Whitten, Carol Serr, and Bill Lindsey
    2009   The Dating Game—Marks Used by the Mississippi and Lindell Glass Companies. *Bottles and Extras* 20(1):34–43, 56–57.

Mansberger, Floyd
    2009a   *Block Technical Report I: Archaeological Investigations of the Library Project Area (Block 12, Original Town Plat), Abraham Lincoln Presidential Library and Museum, Springfield, Illinois.* Project completion report submitted to the Illinois Historic Preservation Agency by Fever River Research, Springfield, IL.
    2009b   Block Technical Report II: *Archaeological Investigations of the Museum Project Area (Block 1, Original Town Plat), Abraham Lincoln Presidential Library and Museum, Springfield, Illinois.* Project completion report submitted to the Illinois Historic Preservation Agency by Fever River Research, Springfield, IL.

Mazrim, Robert
    2002   *"Now Quite Out of Society." Archaeology and Frontier Illinois.* Transportation Archaeological Bulletin 1. Illinois Transportation Archaeological Research Program, University of Illinois, Urbana-Champaign.
    2009   A Proposed Fine Grained Chronology of European Presences in the Illinois Country. *Illinois Archaeology* 21:213–219.

McKearin, George S., and Helen McKearin
    1941   *American Glass.* Crown Publishers, New York.

McKearin, Helen, and George S., McKearin
    1950   *Two Hundred Years of American Blown Glass.* Crown Publishers, New York.

McKearin, Helen, and Kenneth M. Wilson
    1978   *American Bottles and Flasks and Their Ancestry.* Crown Publishers, New York.

Miller, Thomas R.
    2007a   *Book I (A–Z). St. Clair County, Illinois, Soda and Related Beverage Bottles (Includes Biographical Sketches and other Information).* Thomas R. Miller, Smithton, IL.

Mounce, Eva Dodge
    1989   *Checklist of Illinois Potters and Potteries.* Foundation for Historical Research of Illinois Potteries, Circular 1(3).

Mullikin, Sean
    1978   Portrait of a Small Town Bottler. *Old Bottle Magazine* 11(3):9–11.

Munsey, Cecil
    1970   *The Illustrated Guide to Collecting Bottles.* Hawthorn Books Inc., New York, NY.

Murschell, Dale
    1998   American Black Glass Seal Bottles. *Antique Bottle & Glass Collector* 14(9):44–45.

Oertle, Ben (ed.)
    1990   *Central Illinois Bottles and Glasses.* Pekin Illinois Bottle Collectors Association.
    1995   *Supplement to Central Illinois Bottles and Glasses: July 15, 1995.* Pekin Illinois Bottle Collectors Association.
    2008   *Central Illinois Bottles and Glasses with Special Listing of Illinois Mini-Jugs. Supplement #2.* Pekin Illinois Bottle Collectors Association.

O'Malley, Mike
    2008   Target Balls. *Bottles and Extras* 19(2):60–61.

Paul, John R., and Paul W. Parmalee
    1973   *Soft Drink Bottling. A History with Special Reference to Illinois.* Illinois State Museum Society, Springfield, IL.

Priestly, Joseph
    1772   *Impregnating Water with Fixed Air; in Order to Communicate to It the Peculiar Spirit and Virtues of Pyrmont Water, and Other Mineral Waters of a Similar Nature.* Printed for J. Johnson, No. 72, in St. Paul's Church-Yard, London.

Reavis, L. U.
    1879   *The Railway and River Systems of the City of St. Louis, with a Brief Statement of Facts Designed to Demonstrate that St. Louis Is Rapidly Becoming the Food Distributing Center of the North American Continent, also, a Presentation of the Great Commercial and*

Manufacturing Establishments of St. Louis. Printed by Woodward, Tiernan & Hale, St. Louis, MO.

Ring, Carlyn
    1980   *For Bitters Only.* The Nimrod Press, Inc., Boston, MA.

Riley, John J.
    1958   *A History of the American Soft Drink Industry. Bottled Carbonated Beverages, 1805–1957.* Published by American Bottlers of Carbonated Beverages, Washington, D.C.

Seymour, Henry James
    1886   Glass and Glass Makers. *Magazine of Western History*, February 1886, pp. 367ff.

Shrum, Edison
    1995   New York and the New York Settlement. In *The History of Scott County, Missouri, Up to the Year 1880.* 2nd Edition, by Edison Schrum, pp. 203–218.
    1996   Commerce Potteries. In *Commerce, MO: 200 Years of History,* by Edison Schrum, pp. 124–129.

Thomas, John L.
    1974   *Picnics, Coffins, Shoo-Flies.* Maverick Publications, Bend, OR.

Toulouse, Julian Harrison
    1969   A Primer on Mold Seams, Parts 1 and 2. *Western Collector* 7(11):526–535 and 7(12):578–587.
    1971   *Bottle Makers and their Marks.* Thomas Nelson Inc., NY.

Van den Bossche, Willy
    1999   *Antique Glass Bottles: Their History and Evolution (1500–1850). A Comprehensive, Illustrated Guide with a World-wide Bibliography of Glass Bottles.* Antique Collectors' Club, Ltd., Suffolk, UK.

van Rensselear, Stephen
    1971   *Early American Bottles and Flasks.* John Edwards, Stratford, CT.

von Mechow, Tod
    1993   The Great Pontil Debate. *Antique Bottle & Glass Collector* 9(10):12–14.
    2008   *Soda and Beer Bottles of North America.* Detailed, database website maintained and frequently updated and expanded by von Mechow at www.sodasandbeers.com/index.html

Walker, Francis A.
    1872   *The Statistics of the Population of the United States, Embracing the Tables of Race, Nationality, Sex, Selected Ages, and Occupations.* Ninth Census—Volume I. Compiled from the original returns of the Ninth Census (June 1, 1870). U.S. Government Printing Office, Washington, D.C.

Walthall, John A.
    1993   Chicago Stoneware Bottles. *Antique Bottle & Glass Collector* 10(2):12–13.

Welko, Paul
    1978   Did You Know About...Arthur Christin's Patent? *Antique Bottle World* 5(3):12–14.

White, Harry Hall
    1993   Early Pittsburgh Glass Houses. *Federation Glass Works,* June, 1993:3–8. Reprinted from *Antiques Magazine*, November 1926.

Wilson, Bill, and Betty Wilson
    1968   *Spirit Bottles of the Old West.* Privately published by William L. Wilson.

Wilson, Rex L.
    1981   *Bottles on the Western Frontier.* University of Arizona Press, Tucson.

Yates, Donald
    2006   John Matthews: Father of the Soda Fountain. *Bottles and Extras* 17(3):72–75.

Young, James Harvey
    1953   Patent Medicines: The Early Post-Frontier Phase. *Journal of the Illinois State Historical Society* 46(3):254–264.
    1960   Patent Medicines: An Early Example of Competitive Marketing. *The Journal of Economic History* 20(4):648–656.
    1961a   *The Toadstool Millionaires. A Social History of Patent Medicines in America before Federal Regulation.* Princeton University Press, Princeton, NJ.
    1961b   American Medical Quackery in the Age of the Common Man. *The Mississippi Valley Historical Review* 47(4):579–593.
    1985   Folk into Fake. *Western Folklore* 44(3):225–239.

Zeuch, Lucius H.
    1927   *History of Medical Practice in Illinois. Volume 1: Preceding 1850.* Printed for The Illinois State Medical Society by The Book Press, Inc., Chicago, IL.

ALBION
(Edwards County)

1850 census: 400
1870 census: 600
1880 census: 900

Alexander Stewart

Veterinary Medicine:

(1) **A. STEWART'S / HORSE OINTMENT / ALBION ILL**   [aqua]

»SB w/ keyed hinge mold; 10-sided cylinder; oval flare-ring collar; 5 1/2" tall. [Figure 1]

Stewart arrived in Albion in 1827, at the age of 21, "where he rented a fully equipped blacksmith shop from George Flower, and sold his horse and saddle to purchase iron. After 16 years of strenuous exercise, he began a successful career as a merchant at Albion [in 1843]" (McDonough & Co. 1883a:223). Stewart was listed in the 1860 Illinois State Business Directory (ISD) and appeared in the 1860 ISC as a "merchant" with real estate valued at $8,600 and $8,000 in personal property.

From the technological and stylistic attributes of the above-listed bottle, his *Horse Ointment* appears to have been manufactured and sold during the Civil War era.

Figure 1

ALTON
(Madison County)

1850 census: 3,600
1860 census: 6,300
1870 census: 8,700
1880 census: 9,000

Figure 2

Alton Glass Co.

Whiskey Flask:

(1) **MADE BY ALTON GLASS CO. / ALTON ILL** [aqua]
»SB w/ keyed hinge mold; "shoofly" flask; ring-neck collar; 5 1/2" tall. *[Figure 2]*

This was an advertising whiskey flask manufactured and given away (with contents) by the Alton Glass Company, a short-lived predecessor to the Illinois Glass Company. The *Alton Glass Co.* began production in 1871, was closed down and sold in 1872, and re-emerged as the *Illinois Glass Co.* the following year (*see Introduction and the I. G. Co. listing, this volume*).

George H. Betts & Co.

Liquor:

(1) **GINGER / BRANDY // GEO. H. BETTS & CO // ALTON ILL //** *(blank)* [dark lime-green]
»SB w/ keyed hinge mold; beveled-edge, slightly tapered, square case bottle; short neck; tapered collar; 9 1/4" tall. *[Figure 3]*

In 1850 George Betts was a 31-year-old merchant from Long Island, NY, living in St. Louis (1850 USC). Betts moved to Alton during the 1850s and died before 1860, when his wife is listed in Alton as a widow in the 1860 USC.

According to an 1858 Alton CD ad and alphabetic entry, the company name changed that year to *Betts & Russell*, and the senior partner had become William J. S. Betts (perhaps George's son or brother):

BETTS & RUSSELL
(LATE GEO. H. BETTS & CO.)
RECTIFIERS
IMPORTERS AND DEALERS IN
FOREIGN AND DOMESTIC
Wines, Brandies, Gins, Whiskies,
SCHNAPPS, CIDER, ALES,
Porters, Cigars, Oysters, Sardines, &c.

DEPOT FOR
BETTS' CELEBRATED GINGER BRANDY.
RECTIFIED WHISKEY,
Constantly on Hand at the Lowest Market Price.
Corner Fourth and State Streets,
ALTON, ILLINOIS.

Figure 3

Since 1857 was the last year in business for *Geo. H. Betts & Co.*, the fact that the listed bottle is smooth-based suggests that the bottle was produced at an East Coast glass house (where the earliest smooth-based bottles appeared ca. 1856) rather than at the Pittsburgh glass houses more commonly used by the river trade (where the earliest smooth-based bottles were not produced until ca. 1860). An alternative possibility is that the bottles were not changed and continued to be marketed under the same name in later years. This is suggested by the fact than an 1858 ad for the product still refers to the earlier company name. From the ad, it is interesting to note that Betts' Ginger Brandy was marketed as a medicine as well as an alcoholic beverage:

> **Betts' Pure Old Ginger Brandy.**—This splendid Beverage has been carefully prepared from a genuine receipt, in the possession of the Proprietors, from an Extract of Ginger, Pure Brandy, &c., &c., and will be found very useful in severe cases of Cholera, Cramps, Colic, Pain in the Stomach, &c., &c. The ingredients of which *Betts' Ginger Brandy* is made have been fully tested in New York city, during the prevalence of the Cholera in 1832, and proved successful in all the above complaints.
>
> Persons Traveling, who are subject to sickness, &c., from change of water, will find that *Betts' Ginger Brandy* is invaluable to them on their travels. By adding a small quantity to the water they drink, it will obviate all danger arising from the same.
>
> Price $1.00 per bottle. Manufactured by GEO. H. BETTS & CO., sole Proprietors. Corner of Fourth and State streets, Alton, Illinois.
>
> For sale by W. A. HOLTON & CO., and all Druggists.

Probably *Geo. H. Betts & Co.* was in operation in Alton only during ca. 1856–57. The earliest ad that has been found for the company was in the *Alton Weekly Courier* for April 30, 1857 (p. 4, col. 5). It announced that the *Geo. H. Betts & Co.* "Ale & Porter Depot" were also agents for the "Alton Soda Water Company." The only soda-bottling company known to have been in business in Alton at this time was that of A. & F. X. Yoerger (*see their listing*).

Jacob Buff and Max Kuhl

Soda ~ Mineral Water:

(1) **BUFF. &. KUHL. / ALTON. / ILLS.**   *[aqua]*
(base): (outer ring) **GRAVITATING STOPPER / MADE BY /**
(inner ring) **JOHN MATTHEWS / NEW YORK /**
(center) **PATD / OCT 11 / 1864** *[Figure 4]*

Note: Patent dates are also embossed on the glass stopper itself:
PATENTED / AUG. 26. 1862 / OCT. 11. 1864 / APRIL 16. 1873 *[Figure 5]*

»SB w/ post mold; Matthews-style flat-topped-blob soda with short neck and Matthews "Gravitating" internal-glass-stopper closure; 6 1/4" tall.

Max Kuhl and Jacob Buff are listed in the 1880 USC as St. Louis mineral manufacturers. However, their partnership is also listed in the 1880 Illinois Industrial Census for Alton. The company was first established in St. Louis near the end of the Civil War. Their earliest business ad (*1865 St. Louis CD*) indicated that *Buff & Kuhl* were "manufacturers of Artificial Mineral Waters" located at Fourth and Elm, and that "Carlsbad, Kissengen, Selters, Vichy, Spaa, Pyrmont, Pullna, Eger, Marienbad, Racoczy, and other MINERAL WATERS *[were]* kept constantly on draught and in bottles." Both men resided in St. Louis throughout the 1860s and 1870s, with their mineral-water factory and business headquarters located first (1865–66) at 48 S. Fourth, then (1867–74) at 121 S. Fourth and 315 Poplar, and finally (1875–82) at 1706 Decatur.

According to an *Alton Daily Telegraph* announcement, Buff & Kuhl expanded their bottling operations into Alton in the spring of 1875:

> SODA AND MINERAL WATER DEPOT.—Messrs. Buff & Kuhl, of St. Louis, are fitting up the building on the corner of Second and Ridge streets, and putting in new machinery for the manufacture of soda and mineral waters. This is a branch of their St. Louis house, and will be an extensive establishment. [*Alton Daily Telegraph,* March 25, 1875]

The 1880 IIC indicated that the Alton *Buff & Kuhl* bottling depot had capital of $6,000 and 5–7 workers seasonally (June 1879 to June 1880). The company paid $500 in wages during that year, their cost of materials was $1000, and $5000 worth of bottled product was produced.

After 1880, however, Buff & Kuhl sold their Alton bottling depot and returned their focus solely to their St. Louis bottling business:

> BUSINESS CHANGE—The firm of Buff, Kuhl & Co., manufacturers of soda and mineral waters, have sold their business in this city to Messrs.

Figure 4

Schmidt & Knecht who will carry it on as before. Both members of the new firm have experience in the business and will make it successful. [*Alton Daily Telegraph,* January 5, 1881]

Figure 5

Max Kuhl left the *Buff & Kuhl* partnership in St. Louis after the 1882 season. Jacob Buff took on a new St. Louis partner, and beginning in 1883 *Buff & Preisig* bottled mineral water at the earlier Buff & Kuhl address. Max Kuhl returned to Alton and reopened a new bottling operation of his own there in 1883 (using Hutchinson-style internal stopper bottles embossed **MAX KUHL** and **MAX KUHL & CO.**). He bottled soda/mineral water in Alton until his death in the early 1890s. That post-1880 operation is too late for inclusion in our present study.

Illinois Glass Co.

Ashtrays:

(1) *(obverse):* **SMITH //** I *(within diamond outline)* **// LEVIS**   [aqua]
    (reverse): **ILLINOIS GLASS CO.** *(arched)* **// 1873 // ALTON, ILL.** *(arched up)* [Figure 6]
    Note: approximately 4,000 made during the fall of 1873.

(2) *(obverse):* I *(within diamond outline) //*   [aqua]
    (reverse): **ILLINOIS GLASS CO.** *(arched)* **// 1873 // ALTON, ILL.** *(arched up)*
    Note: approximately 1,000 made during November of 1873. Mold modified
        after the presser was broken and forge-welded.

(3) *(obverse):* **MERRY // 18** I **73** *(within diamond outline)* **// CHRISTMAS //**   [clear]
    (reverse): **COMPLIMENTS OF ILLINOIS GLASS CO. ALTON, ILL.** *(embossed
        round outer margin of base)* [Figure 7]
    Note: approximately 3,000 made during November/December 1873. In some examples,
        the embossed "S" in ILLINOIS and the embossed "G" in GLASS overlap.

(4) *(obverse):* **SMITH //** I *(within diamond outline)* **// LEVIS**   [pale green]
    (reverse): **MERRY CHRISTMAS** *(arched)* **// ILLINOIS GLASS CO.** *(arched up)*
    Note: This was the final run. This last variety is a larger and heavier than #3 above.
        "Less than five hundred" were made in December 1873.

---------- ❖ ----------

Although not bottles, these ashtrays were the very first embossed-glass containers produced by the fledgling *Illinois Glass Co.* in 1873 and, as such, they deserve mention (*see Alton Glass Co. listing, this chapter and Figure 8*). The ashtrays were produced by the new *IGCo* owners (William Smith and Edward Levis) as advertising items to be sold to hotels (#1 and #2) or given away (#3 and #4) to spread the name of the new glass works, and they seem to have been a modified continuation of the giveaway-flask idea initiated two years earlier by the predecessor *Alton Glass Co* (*see their listing, this chapter*).

Figure 6

Figure 7

Due to the methods used for filling the mold and pressing the ashtray, "the weight of these ash trays varies considerably, anywhere from a pound and an eighth to a pound and a half. The average weight is 21 ounces" (Cundall 1923:6). Ashtrays made in December for the final "Merry Christmas" run are larger, weighing from 28 ounces to two pounds. Ashtrays #1 and #2 above are 4 1/2" in diameter at the base and 3" in diameter within the "cup."

The "Merry Christmas" ashtrays are somewhat larger: about 5 1/2" in diameter at the base and 3 1/2" in diameter within the cup. The top-side measurements can vary as much as 3/4" as the result of sagging glass "caused by too hot or too much glass at the time it was made" (Cundall 1923:11).

Among four ashtrays examined for this study, two examples of IGCo ashtray #1 above weighed 1.11 and 1.42 pounds, respectively. Two examples of the larger IGCo ashtray #3 above weighed 1.69 and 2.02 pounds, respectively.

Figure 8. The 1873 exterior and interior views of the Illinois Glass Company's fledgling five-pot glass-factory building at the original Belle Street location. The drawings are from the *Illustrated Historical Atlas Map of Madison County* (W. R. Brink & Co. 1875:161).

Unexpected obstacles to the early production of a good "dollar ash tray," these heavy glass objects gave rise to the idea of a "moving oven," an important development in glass manufacturing technology:

> On Halloween night in 1873 everything is ready for the fourth try. As each ash tray is made and placed on a brick it advances into the oven. Every three minutes one is placed on a brick and advances six inches. Three hours later the first ash tray is almost at the "cool wheels." It has not broken. It is taken off and is good. At last they have succeeded. The brick is unwired and removed and ready to be carried to the "hot wheels" for another trip through the oven. Six hours work has produced 120 good ash trays. Breakage is practically non-existent and the quality excellent. Mr. Smith does not realize he has fathered the glass annealing lehr (modern-day glass factories use this basic principle but in a much improved version). [Cundall 1923:8]

A smaller, clear-glass version of ashtray #1 above was made by the *Illinois Glass Co.* in 1923 as a 50th anniversary gift to senior employees of the company in Alton.

Fred Inglis

Bitters:

(1) **MAGNOLIA / BITTERS** // *(blank)* // **FRED. INGLIS / ALTON. ILL** // *(blank)* // *[amber; olive-amber]*

»SB w/ keyed hinge mold; tapered collar; square case bottle; 10" tall. *[Figure 9]*

In the 1864 ISD the firm of *Inglis & Lowe* was listed as Alton retailers in foreign and domestic liquors. In the 1864 IRS tax assessment, *Inglis & Lowe* were listed as retail and wholesale liquor dealers, rectifying less than 500 barrels per annum. The company placed newspaper ads for its products in the Alton newspaper as late as June 1865, but *Magnolia Bitters* was never mentioned. A dissolution-of-partnership notice was placed in the *Alton Courier* on July 7, 1865 (p. 2, col. 7)—perhaps precipitated by Fred Inglis' desire to develop his *Magnolia Bitters* (see below).

From July to November 1866, Mr. Inglis (through his Jerseyville agent) placed a series of ads for *Magnolia Bitters* in the *Jersey County Democrat*:

> "John, what gives you such an excellent appetite, and makes you look so well?" "What gives me an appetite? Why, I take MAGNOLIA BITTERS every morning. Its none of your rot gut preparations, but a superior article made here at home, by Fred. Inglis. Try it, and if it don't do you good, I'll pay for a bottle." It is sold by Fred Inglis, at Alton, and by all Druggists. [July 6, 1866 p. 3, col. 1]

> MAGNOLIA BITTERS—For many years it has been an object of interest to the public and consequently to the physicians generally, to find a preparation calculated to build up and renovate the system when run down, or to invigorate and maintain the natural equilibrium when, from exposure or constant fatigue, the general health is in danger. As a tonic and gentle stimulant, those who make a trial will find that the Magnolia Bitters…are bound to strengthen and invigorate when regularly used, and are especially useful in disorders incident to this Western country, such as

Figure 9

affections of the liver, kidneys, etc. The Magnolia Bitters may be depended upon as a preparation from different barks and roots, carefully selected, unmixed with any mineral or poisonous ingredients whatever, and are prepared under the personal inspection of the proprietor. [August 17, 1866 p. 3, col. 1&2]

MAGNOLIA BITTERS—The public generally are beginning to appreciate the great value and importance of these bitters as a medicine. They invigorate the system, cure bilious complaints by their powerful action on the liver, stomach and kidneys, and are withal a safe and healthy tonic.... the youngest child may take them, with no other effect that that of producing tone and cheerfulness. [August 24, 1866 p. 3, col. 1]

MAGNOLIA BITTERS—The ready sale of these bitters show the high esteem in which they are held.... They are manufactured by Fred Inglis at Alton, and no humbug about them. The *Magnolia Bitters* are sold in Jerseyville by Buffington & Bro. [September 28, 1866 p. 3, col. 2]

The 1866 Madison County Gazetteer placed Fred Inglis' business at 2nd Street near State (the 1868 Alton CD gives the address as No. 16 Second) and his residence at 3rd and Market. By 1868, he was living on Grafton Road. His business ads in the *Jersey County Democrat* for 1867 and 1868 only list him generally as a rectifier, wholesale dealer in "liquor, wine & cigars," and agent for four national bitters brands. But the newspaper indicates he sold this business out to Henry Buckmaster in April 1869. After that, he may have gone back to trying to market his *Magnolia Bitters* for a while.

According to Ring (1980:316) *Magnolia Bitters* was advertised in the Baton Rouge *Tri-Weekly Advocate* for June 27, 1870, and Fred Inglis was listed in the 1871 Alton CD as a "Rectifier and Wholesale Dealer. Proprietor of Magnolia Bitters." No further mention of the product has been found after that.

Thomas and Michael Keeley

Ale:

(1) **KEELEY & BRO // ALTON ILL**   *[dark olive "black glass"]*
»IP; quart; ring-neck tapered collar; 3-piece mold; 9¾" tall. Embossing confined to the bottle's shoulder. *[Figure 10]*

———————— ❖ ————————

Thomas Keeley came to America from Ireland in 1847 and arrived in Madison County in 1850. He began bottling ale in Alton in 1852, with the long-distance assistance and advice of his older brother Michael, who followed him to America in May 1851. Michael had "several years" experience in the bottling business, apparently in Ireland, since he had arrived in America only a year earlier, and he helped Thomas set up his Alton ale bottling business (see quote later in the entry). Michael then relocated to Chicago in 1854 to establish a *Keeley & Brother* ale depot in northern Illinois (*see Chicago Keeley listings*).

Although few early Alton CDs have been preserved, Thomas was a surprisingly prolific newspaper advertiser, and his Alton newspaper ads for his three-year *Keeley & Brother* ale-bottling business downstate may represent the most detailed public record of a pre–Civil War bottling enterprise anywhere in Illinois.

Figure 10

An advertisement on page 1 of the *Alton Weekly Courier* for June 11, 1852, announced the opening of T. Keeley & Brother, dealers in port and ale, located under the Franklin House, State Street, Alton.

> Have on hand and to arrive…Ale of the following brand[s], Dayton, Ohio, Cincinnati, Pittsburgh, Wheeling, Va.… All orders promptly attended and will be furnished by the barrel, half barrel, [and] bottle, at the lowest cash prices.
> N.B.—The above advertisement is to inform the citizens of Alton that I have received a part of my stock. I have now commenced bottling, with the assistance of my brother, who is experienced in the business for several years. THOS. KEELEY

By the 1853 bottling season, Thomas Keeley's newspaper ads announce that Keeley & Bro. are producing their own brand of "Claret Wine," the likely contents of the bottle shown in the figure:

Ale And Porter
THANKFUL to my customers for the liberal patronage heretofore extended by them, I would announce…that I have on hand and will continue to keep a full supply of Ale of the different brands, together with Porter of the finest domestic quality, which I offer for sale at the lowest market prices, by the barrel, or bottled in any way to suit customers. **I have also on hand and offer for sale a fine article of Claret Wine, of our own bottling, which we warrant** [*Emphasis ours: "claret" usually refers to a dark red, often spicy, Bordeaux wine*]. [*Alton Weekly Courier,* May 20, 1853:3(7)]

Ale And Porter
Buckeye, Pittsburgh and Wheeling Ale Depot.
T. KEELEY & BRO. are now receiving their Summer supply of stock Ale and Porter, which they offer at the lowest cash rates.
200 barrels Harries Dayton Ale,
300 do Middlewood Cincinnati Ale,
150 do Geo. W. Smyth, Pittsburgh Ale,
150 do Wheeling Ale,
250 do Cincinnati Porter,
100 do Pittsburgh do
Also, a superior article of Brown Stout, and Scotch Ale, pints and quarts. Groceries and private houses supplied with Ale and Porter by the barrel, half barrel, or bottle.
We are bottling a superior article of Claret Wine, which we recommend and keep for sale at reasonable terms. [*Alton Daily Telegraph,* August 4, 1853:1(3)]

As noted in the *Alton Daily Telegraph* for March 19, 1853 (p. 3, col. 1), *Keeley and Bro.* were also ice dealers in Alton: "We observed a large flat boat of choice ice from the Illinois River, at the levee yesterday. It is for Messrs. KEELEY & BRO.'s ice house, and will come into play before long." However the success of such a business must have depended on local winter cold spells and ability to import ice from elsewhere, as shown by Thomas Keeley's announcement in the *Alton Weekly Courier* for September 5, 1854 (p. 2, col. 6):

ICE—SPECIAL NOTICE
After some time spent in negotiating for a supply of ICE from Chicago, I am compelled, although reluctantly, to inform my customers that I shall not be able to…supply them with ice from Chicago. As I have ascertained, through my correspondent in that city *[likely his brother Michael]*, that two houses in Alton have obtained contracts for the exclusive sale of the article, my customers will see that no efforts on my part were wanting to supply them to the end of the season.

In addition to ice, Thomas Keeley added another bottled product to his line in 1854: bottled soda water:

Soda Water, Ice, Ale, and Porter
Determined as I am to meet the wants and suit the tastes of my customers, in some of the greatest luxuries of the season, I am now prepared to fill with promptness and dispatch all orders for ICE, ALE, PORTER, and BOTTLED SODA. Al persons desiring a full, regular, and satisfactory supply of

any of the above named articles will find it [to] their advantage to call before making arrangements elsewhere. [*Alton Weekly Courier*, June 5, 1854:2(7)]

Since Keeley's locally produced claret wine was likely put up in embossed *Keeley & Bro.* black glass bottles (May 20, 1853, *Alton Weekly Courier* ad), it is likely that *Keeley & Bro.* used embossed bottles for the soda water they bottled in1854 as well. However, we have not yet been able to document any embossed examples of such a bottle. For a while, it appeared that 1854 would be the first and only Keeley soda-bottling season. Thomas Keeley placed a notice in the *Alton Daily Morning Courier* on January 25, 1855 (p. 3, col. 2), announcing the sale of his business:

SPECIAL NOTICE
A RARE CHANCE FOR GOING INTO BUSINESS.
I hereby offer for Sale my ALE BUSINESS, horses and wagons, and all that pertains thereto, which I intend to sell at a great bargain, by early application, as I am winding up my business here, and consequently will sell low.

But just two months later, beginning in late March of 1855, Thomas Keeley placed new announcements in the Alton paper, first retracting his earlier published intention to sell the business, and then four weeks later announcing a new plan to manufacture soda water in Alton for a second season with an unnamed St. Louis bottler as his new partner:

SODA. SODA.
ACCORDING TO PROMISE I hereby take pleasure in announcing to my old customers that I have succeeded in making arrangements to manufacture bottled soda in Alton the coming season: consequently I hope to be able to fill all orders with promptness and dispatch, as during the coming season I will not be subject to many disappointments connected with obtaining supplies of soda from St. Louis. Notwithstanding my previous notice to the contrary, this fact will enable me to supply my customers at last year's prices, which are thirty five cents per dozen to retailers. [*Alton Daily Telegraph,* April 3, 1855:4(6)].

SODA WATER, ALE AND ICE:
NOTWITHSTANDING MY PREVIOUS notices of offering my interest at Alton for sale, I am still on hand.
I am now prepared to fill all orders, to any reasonable amount, with any of the above articles, as I have on hand a large and well selected stock of Ale and Porter of different brands. Also a fine stock of Ice, the finest ever offered [in] the market, together with Bottled Soda Water, of our own manufactory, which is said not to be surpassed, *by those who are already using it [emphasis ours]*; and as my partner in this branch of the business is one of the St. Louis Soda Company, who has had long experience in the business [*probably Gerard Timmerman—see discussion below*], I have no fears of a full supply to fill orders from any quarter: so I will say that in case any of my friends in the city or country should want for a cooling beverage on the Fourth of July coming, it will not be on account of not having an abundance of fine Ice and Soda Water, *from the most improved patent bottling machine [emphasis ours]*, subject to their order; which never could be said heretofore at Alton. Thus far for Alton enterprise! [*Alton Daily Telegraph,* May 4, 1855:3(3)]

The "St. Louis Soda Company" bottler that Thomas Keeley mentions taking on as a partner in the above ad was very likely Gerard Timmerman. Prior to this time, the St. Louis Soda Company had three partners (Cairns, Timmerman, and Block). In the later 1850s, Cairns and Block were the only listed partners.

We have not yet been able to document any examples of embossed Keeley & Timmerman Alton soda bottles, and Michael Keeley died soon afterward. So we do not know if this final joint bottling effort was ever realized.

Thomas Keeley's bottling efforts ended for good when he died in Alton just a year later, on May 14, 1856. The reason he died is unknown, but was likely misadventure or disease (cholera perhaps?), since he was just 26 years old. Probate records filed at the Madison County Courthouse in Edwardsville, August 1857, by his brother Michael (his sole heir and executor), indicate that Thomas' estate was not documented in detail but did not exceed $7,000.

Beginning in 1855, Michael Keeley opened a modest *Keeley & Brother* ale-bottling depot in Chicago, and he went on to establish larger and more diversified bottling works there under his own name (*see M. Keeley Chicago listings*).

Quigley, Hopkins, and Lea

Medicine:

(1) **ESSENCE / JAMAICAN GINGER** / *(embossed "propeller" decoration)* / [aqua]
QUIGLEY. HOPKINS & LEA / ALTON / ILL

»SB w/ keyed hinge mold; shouldered oval; squared ring collar; 5 1/2" tall. *[Figure 11a]*

In 1866 (*Jersey County Democrat*), *Quigley Bros. & Co.* operated a wholesale drug house in Alton. According to the *Alton Daily Telegraph* for February 7, 1868 (p. 1, col. 7), one of the Quigley brothers "has withdrawn from the wholesale drug house. The firm will now be known as Quigley, Hopkins, and Lea." Principals were Webb C. Quigley, George K. Hopkins, and Charles G. Lea.

Quigley, Hopkins & Lea were listed in the 1871 ISD as wholesale druggists located on 2nd Street at the corner of 6th Street in Alton. Their wholesale drug business was focused on nationally popular brands (e.g., Figure 11b–d). Their only known Alton-store brand is the Essence of Jamaican Ginger listed above.

By 1875 (ISD), the company's name had been changed to *Quigley, Hopkins & Co.*, and in 1879 (*Alton Daily Telegraph*), the firm became *W. Quigley & Sons*.

From the keyed hinge mold on the listed bottle, it likely dates to the first few years of the company's operation (ca. 1868–69).

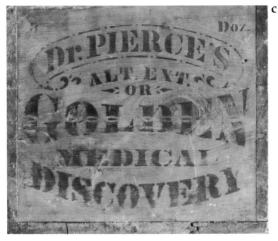

Figure 11. Stenciled shipping crate for national-brand patent medicine (Dr. Pierce's Golden Medical Discovery) used in Alton by Quigley, Hopkins, & Lea during the late 1860s or early 1870s.

Although we have listed the QH&L Essence of Jamaican Ginger as a medicine, it was also often used as a flavoring extract because of its spicy, peppery taste (ginger bread, ginger beer, ginger ale). Medically, it was used externally as an anti-inflammatory (like pepper cream) for arthritis and muscle aches and pains, and was taken internally as a digestive aid and to relieve stomach cramps and nausea.

[Henry] E. Rupert

Soda ~ Mineral Water:

(1) **H. E. RUPERT / ALTON, / ILL'S /** [aqua]
 (left front heel): **L&W**
 (base): **R**

»SB w/ keyed hinge mold; oval-blob-top soda. *[Figure 12]*

--- ❖ ---

H. E. Rupert's first soda-bottling operation was at Lacon, IL, a small town on the Illinois River about 30 miles downstream from Pekin. From the F. A. & Co. glass house mark on the heel of his smooth-based Lacon bottles (*see his Lacon listing*), they were produced and used ca. 1861–62.

Rupert moved further downriver to Alton soon afterward. His Civil War–era Alton bottles were made from the same glass mold and embossed the same as his earlier Lacon bottles, except that the Alton bottles were produced by a different Pittsburgh glass house (*Lorenz, Wightman & Co.*), so the earlier *Fahnstock, Albree & Co.* heel mark was been replaced with **L&W**.

H. E. Rupert was not listed in the only Civil War–era Alton CD (1866) or the 1868 Alton CD, so it appears he did not come to Alton until the later 1860s. An Alton digger active in the 1970s and 1980s has indicated he found (but did not save) broken Rupert bottles embossed **SPRINGFIELD, ILL'S** (not seen by us), so perhaps Rupert made an interim stop there before setting up shop in Alton in the late 1860s.

Rupert was in his late 20s when he was listed in the 1870 USC as an Alton manufacturer of soda water. However the *Alton Telegraph* for November 1, 1872 (p. 3, col. 1), reported that Rupert's mineral water factory "in the eastern part of the city" had been destroyed by fire. There is no evidence to indicate that Rupert attempted to resume bottling in Alton after that.

Figure 12

Jacob Weisbach

Soda ~ Mineral Water:

    (1a) **J. WEISBACH / ALTON / ILL**   *[aqua, light blue]*

        »IP; slope-shouldered soda; shouldered-blob-top. *[Figure 13, left]*

    (1b) **J. WEISBACH / ALTON / ILL**   *[aqua]*

        »SB w/ keyed hinge mold and large centered glass "dimple;" slope-shouldered soda; shouldered-blob-top. *[Figure 13, right]*

Figure 13

Jacob Weisbach was the father and originator of the generation-spanning Weisbach family bottling operations in Alton. He was not listed yet in the 1858 Alton CD, but was listed in the 1860 ISD and the 1860 IIC as operating a soda water bottling plant there. The two bottle varieties listed were identically embossed and made from the same mold; they may even have been blown at the same time—just when Pittsburgh glass houses were undergoing a rapid shift from iron-pontil to snap-case manufacturing technology: i.e., 1860. This appears to have been his only bottle order, and it is likely that he was in charge of the Weisbach bottling operations in Alton for just a few years, from ca. 1859 to 1861.

From his 1860 IIC listing, it seems that Jacob Weisbach's soda factory produced *flavored* soda waters (as suggested by the fact that 20 barrels of "syrups" are listed in his raw-material inventory, in addition to marble dust, water, sugar, and vitriol). The IIC indicates that he had 4 employees, who cost him $80/month, and produced 10,000 "boxes of soda water" annually with an overall value of $6000.

Jacob was the father of Christian Weisbach (*see his listing*), and at some point during the early Civil War years he transferred control of the Alton bottling business he had started in the late 1850s to his son.

Christian Weisbach

Soda ~ Mineral Water:

    (1)   **C. WEISBACH // ALTON / ILL**   *[aqua]*
        Note: *Early 1860s style; flat-lettered front and back embossing. [Figure 14]*

        »SB w/ keyed hinge mold; shouldered-blob-top soda.

    (2a) **C. WEISBACH** (*arched*) **/ ALTON.ILL //**   *[aqua, blue teal]*
        (*heel*): **A. & D. H. C**
        (*base*): **C.W.**

        »SB w/ keyed hinge mold; oval-blob-top soda. *[Figure 15]*

    (2b) **C. WEISBACH** (*arched*) **/ ALTON.ILL //**   *[aqua]*
        (*heel*): **A. & D. H. C**

        »SB w/ keyed hinge mold; oval-blob-top soda. *[embossed as in Figure 15b, but without base-embossed* **C.W.**]

Figure 14

(3) *[no body embossing]* [aqua]
   (heel): **A. & D. H. C** *[Figure 16b]*
   (base): **C.W.** *(larger base embossing than #2a above—see Figure 16)*
   Note: Early hinge-molded, high shouldered cylinder, "selters-water" style bottle. *[Figure 16a]*

   »SB w/ keyed hinge mold; high-shouldered; oval-blob-top soda.

(4) **C. WEISBACH / ALTON / ILLS.** [amber]
   Note: Embossed in late-1870s circular slug plate with decorative embossing surrounding **ALTON** and **ILLS.**

   »SB w/ post mold; oval-blob-top quart; 11 1/4" tall. *[Figure 17]*

Figure 15

Few Alton CDs are available for the Civil War period, but Christian Weisbach (abbreviated "*Christ.*" and occasionally listed as *Christopher*) likely took over the family soda bottling business from his father (Jacob Weisbach—*see his listing*) in the early war years. He is listed as a bottler in Alton IRS tax assessments in 1864–66, and he appears in Alton CDs as a mineral-water manufacturer from 1866 to the mid-1880s. The 1882 CD states that Christian Weisbach was a "Proprietor and Manufacturer of Soda and Mineral Waters," that his family emigrated to the U.S. from Bavaria, and that he and his father came to Madison County in 1856 (when he was about 18 years old).

In the 1868 Alton CD, his ad indicated he was a "manufacturer of Soda Water and Sarsaparilla" (the latter perhaps in the selters-water style bottle listed as #3), working from his

Figure 17

residence at the corner of Second and Walnut streets. He continued to bottle at the same address throughout his career. When he retired, his eldest son (William) took over the Weisbach-family bottling business, and was later joined by his brother Edward, but their operation was too recent for inclusion in the present volume.

In the 1880 IIC, Christian Weisbach's soda-water factory was listed as having 2 to 3 employees seasonally who were paid an average of $2.50 per day. The factory was only in operation six months a year, and Weisbach estimates the value of his annual product at ca. $2,500 (about half the production of the contemporary Buff & Kuhl operation—*see B&K listing*).

Augustin and Francis X. Yoerger

[Note: the Yoerger name was sometimes spelled "Ioerger" and "Joerger" on their embossed bottles and in their advertising]

Soda ~ Mineral Water:

Figure 16

(1) **A. YOERGER & BRo / ALTON. ILL.** *[blue teal, green teal, aqua]*
»IP; paneled (10-sided); shouldered-blob-top soda. *[Figure 18a–c]*

Note: In the mistaken belief that this bottle was one of the Illinois Glass Company's first products when they began operation in 1873, 100 replica aqua presentation-copies of the bottle (Figure 18d, e)

Figure 18

were produced and handed out to senior employees by the company in 1973 (molded from an original example, with a fake open-glass-rod pontil mark: see Figure 18e) to mark the 100th anniversary of the Illinois Glass Company bottle manufacturing in Alton (Dodd 1973; McKearin and Wilson 1978:691–692). Around the margin of its base, the reproduction bottle is debossed: ILLINOIS GLASS COMPANY / 1873 OWENS-ILLINOIS 1973.

(2) **A. IOERGER & BRo / ALTON. ILL** [aqua]

»IP; squat soda with high-shouldered body with long neck; oval blob top. *[Figure 19]*

(3) **A & F X. JOERGER / ALTON / ILL //** [aqua]
  (heel): **F.A & Co**

»IP; shouldered-blob-top (?) soda. *[Figure 20]*

One fragmentary example known. Top inferred from other contemporary F.A & Co soda bottles.

Augustin Yoerger first announced his entry into the soda and mineral water business in Alton in the *Alton Daily Telegraph* beginning March 30, 1855:

SODA, SODA.
ACCORDING to promise, I hereby take pleasure in announcing to my friends that I have made arrangements to manufacture bottled Soda Water in Alton the coming season. Pure stuff all the time.
Saloon Fountains charged with pure Mineral Water on the shortest notice and on liberal terms.

ICE! ICE!! ICE!!!
I have ice enough to supply the citizens of Alton the coming summer, and keep a supply constantly at my dwelling house, for the accommodation of the public.
AUGUSTIN IOERGER, Third St.
Kossuth House

The Alton CD for 1858 lists A. & F. X. "Joerger" as ice dealers (on Second between George and Langdon) and soda water manufacturers (on Belle near Hamilton). August Yoerger also operated a boarding house and saloon (presumably Kossuth House—back flyleaf CD ad lost to rebinding) on Third Street east of State.

No Alton CDs are available from 1859 to 1866, but from the single **A. & F. X. Joerger** smooth-based *F. A. & Co.* soda known (#3 above), their bottling works was in operation until at least 1861–62.

Figure 19

Figure 20

ARLINGTON HEIGHTS
(Cook County)

1880 census: 995
(1874–present: see **Dunton**, 1861–1874)

Frederick W. Müller

Soda ~ Mineral Water:

(1) **F. W. MULLER / ARLINGTON / HEIGHTS // ILL. //**   [aqua]
 (*heel*): **C & I**
 (*base*): embossed outline image of soda bottle (see Figure 21b).
 Note: Large embossed "double apostrophe" (probably an umlaut) over the **U** in **MULLER**. [Figure 21a]

   »SB w/ post mold; oval-blob-top soda.

    Frederick W. ("Pop") Müller emigrated to the United States from Germany, arriving in Chicago in 1868, when he was 19. He soon found employment with the *Sass & Haffner* soda bottling company (*see their listing*), where he worked and learned the bottling business for four years (Daniels 1971:112; Simon 1971:1). He moved from Chicago to Dunton in 1872, where he set up his own bottling business at 116 S. Dunton, just before the town name was changed to Arlington Heights. During 1872–73, when the town name was still Dunton, he initially bottled in partnership with his half-brother—Chicago bottler Louis Sass—as *Sass & Bro* (see their Dunton listing and *Sass & Haffner* Chicago listing).

    The products he bottled in the 1870s included ginger ale, sarsaparilla, lemon soda, and strawberry soda (Daniels 1971).

    Frequent trips to Chicago were necessary to purchase supplies. Traveling over the famous plank road along Milwaukee Avenue was not easy and, of course, it was a toll road. All deliveries of the Muller products had to be made using horses and wagons *[see Figure 22]*. The area covered extended over many miles, and included deliveries to such towns as Wheeling, Half Day, Diamond Lake, Schaumburg, Itasca, Bensenville, Des Plaines and others. In the winter a bobsled was used and straw and blankets were added to prevent freezing. [Daniels 1971:112–113]

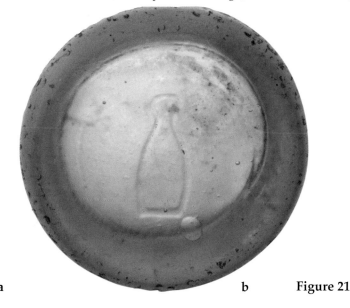

a   b   **Figure 21**

Figure 22. Late nineteenth-century photo of "Pop" Müller and his soda-water delivery wagon in Arlington Heights. (Photo provided by and used courtesy of the Arlington Heights Historical Society.)

In 1882, Müller built a new home and factory at the corner of Vail and Freemont streets, where he continued his bottling operations for 40 more years, two sons taking over after he retired. The bottles listed above are only those that fall within the scope of the present volume—that is, those likely blown and used during the 1870s.

AURORA
(Kane County)

1850 census: 1,900
1860 census: 6,000
1870 census: 11,000
1880 census: 11,900

William C. Budlong

Medicine:

(1) *(side):* **WILLIAM C. BUDLONG // BERNARD'S RADICAL //**   *[clear]*
    (side): **AURORA. ILLS. // CURE FOR / RHEUMATISM //**

»SB; rectangular; 4 sunken embossed panels; double-ring collar; 7 1/2" tall. *[Figure 23]*

---❖---

William Budlong was a druggist in Aurora for over 30 years. He is first mentioned in the 1870 USC as an 18-year old living in his father's household. From the technological attributes

Figure 23

of the above-listed bottle, it was likely produced early in Budlong's pharmacy career in Aurora—probably during the late 1870s. William Budlong was a long-time Aurora druggist, and was still in active practice at the turn of the twentieth century.

James W. and George W. Green

Soda ~ Mineral Water:

(1) **GREEN BRO'S / AURORA ILLS** // [aqua]
 (*heel*): **C & I**
 (*base*): **G** (*large double-outlined letter, embossed backwards*)
 »SB w/ post mold; high-shoulder soda; oval blob top. [Figure 24]

Figure 24

The Green Brothers are listed in the CAR for 1873 and in Aurora CDs for 1874–75. According to the Kane County history (Le Baron 1878a:762), the Green brothers began their soda water business in 1871 and continued as partners until J. W. Green retired in 1877. George (G.W.) continued the business as sole proprietor into the 1880s. He was listed as a "manufacturer of bottled soda water, seltzer water and syrups, champaign cider and Belfast ginger ale."

Their plant was located at 53 N. Broadway in Aurora and could produce up to 600 dozen bottles of soda water per day during the late 1870s.

John Heck

Medicine:

(1) **JOHN HECK'S / CHAMOMILE / TONIC / AURORA / ILL.** [clear]

»SB; oval; strap sides; ring-neck tapered collar; 4 1/2" tall. *[Figure 25]*

❖

John Heck was listed in the 1880 USC for Aurora as a 25-year-old clerk. We have not been able to find him associated with patent-medicine production, but from the bottle's stylistic attributes and production technology it appears to have been manufactured for Heck in the later 1870s or early 1880s.

Often prepared as a health aid, German chamomile (as a tonic or tea) was generally used as an analgesic, anti-inflammatory, and anti-spasmodic medicine. Taken as a general tonic, it was also thought to help control coughing and heartburn.

A. Stege

Figure 25

Beer or Cider:

(1) **A. STEGE / AURORA** [aqua]

»SB w/ keyed hinge mold; slope-shouldered quart; oval blob top; 10" tall x 3 1/2" diameter. *[Figure 26]*

❖

This Stege bottle is another Aurora mystery. No reference to this bottler or the bottle's contents could be found in any Kane County historical reference or census data.

It is tempting to interpret this keyed-hinge-mold bottle as evidence of a late-1860s Aurora brewery because of the nearby famous Edward R. Stege brewery at 538 N. Clark in nearby Chicago. In Chicago, Stege had a partner named Metz from the late 1860s until the time of the Great Chicago Fire in 1871. Then—from the 1870s until its sale in 1923—it was known as the *E. R. Stege Brewery*. However, the Stege brewery is not known to have used embossed bottles during its early years in business, and no family member could be found with the initial "A." (E. R. Stege's three sons were Edward A., Richard, and George.)

Figure 26

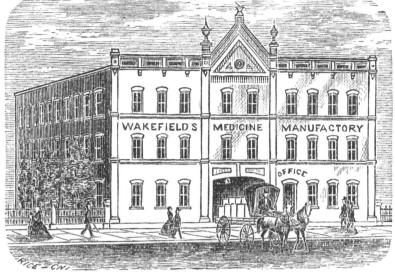

BEARDSTOWN

(Cass County)

1850 census: 1,600
1860 census: 3,800
1870 census: 2,500
1880 census: 3,100

Frederick W. Ehrhardt

Soda ~ Mineral Water:

(1) **F. W. EHRHARDT** *(embossed vertically on one panel)* //   [aqua]
(heel): **IGCo**

» SB w/ post mold; 10 vertical body panels below a ring of 10 oval embossed-glass dots or dimples; tapered body (2 3/4" heel dia.; 2" shoulder dia.); oval-blob-top soda. *[Figure 27]*

(2) **SUPERIOR / BELFAST / GINGER . ALE / MANUFACTURED . BY /**
F. W. EHRHARDT / BEARDSTOWN . ILL   [aqua]

» SB w/ round base and heel mold; round-base cylinder; vertically embossed; oval-blob-top, 9 5/8" tall; 2 3/4" dia.). *[Figure 28]*

(3) *(encircling heel)*: **BELFAST & DUBLIN / AROMATIC GINGER ALE**   [aqua]
(base): **E** *(within an embossed 5-pointed star)*.

» SB w/ heel mold; semi-round-base; 2 1/2" diameter cylinder; oval-blob-top, 8 1/2" tall. *[Figure 29]*

❖

Frederick Ehrhardt began business in Beardstown as a manufacturer of baking powder and extracts in 1873 (Perrin 1882a:238). He is listed in the 1875 Beardstown CD as a maker of ginger ale and white beer, and his round-bottom embossed bottles are ginger-ale style. The E-star base embossed bottles are attributed to Ehrhardt because many of them were

Figure 27

Figure 28

found during 1970 construction work for a Central Illinois Public Service Company office building at the site of his Beardstown soda factory.

F. W. Ehrhardt
Wishes to say to the people of Beardstown and vicinity that he has completed his building and arrangements for manufacturing

    BAKING POWDER,
    TRUE FLAVORING EXTRACTS,
    EAU DE COLOGNE,
    SUPERIOR BELFAST GINGER ALE,

Figure 29

And is now ready to fill orders for the same. He especially solicits the patronage of the community, who should encourage home industry, particularly when the articles can compete with the best, and excel the majority in quality and purity. The BELFAST GINGER ALE is the best, healthiest and most refreshing summer drink ever offered to the market, and is warranted not to contain alcohol. Families desiring to keep it in their homes can have it delivered free to any part of the city. Orders may be addressed to F. W. Ehrhardt, P.O. Box 410, or left at the Factory, Corner Clay and Second streets. The other articles are for sale by all Grocers. Cans and Bottles of my own manufacture will be exchanged. [*The Central Illinoisan,* Aug. 26, 1875, p. 3, col. 4—ad run from June 24th]

In the later 1870s, he added bottled soda and mineral water to his product line (Perrin 1882a:238). His aqua 10-sided soda bottle is an early commercial container made by the fledgling Illinois Glass Company in Alton. After 1880, Ehrhardt continued bottling soda water in a fluted-base Hutchinson-style soda bottle (see Paul and Parmalee 1973:Figure 34)—not listed above because of the 1840–1880 focus of the present volume.

Paul and Parmalee (1973:36) indicate that the cylindrical, round-bottom "Belfast Ginger Ale" style bottles used by Ehrhardt were "of foreign make." This is not necessarily the case. English and Irish bottled ginger ale was shipped to America in distinctive cylindrical round bottles for sideways cartage (in part to keep the cork moisturized and in place), and were very popular with shippers as "Ballast" style cargo. However, when ginger ale popularity in America outgrew the availability of imported products, regional glass houses mimicked their easily recognizable bottle style, producing and embossing them for local bottlers.

George Schneider and Henry W. Krohe

Soda ~ Mineral Water:

(1) **SCHNEIDER & KROHE / BEARDSTOWN / ILLS** [aqua]

»SB w/ post mold; shouldered-blob-top soda. *[Figure 30]*

--- ❖ ---

According to the *Biographical Review of Cass, Schuyler and Brown Counties, Illinois* (1892), Schneider & Krohe, who were brothers-in-law, began their business partnership by opening a saloon in Beardstown in 1866. They began their soda water manufactory in 1869 as an adjunct to their saloon business at the corner of Main and Washington. In 1872–72 they built the Opera House block in Beardstown (undaunted by a tornado that hit the town while construction was underway), and were the first managers of the opera house itself during the 1870s and 1880s. They placed an ad for the new facility, and for their ongoing soda bottling business, in the "Business Notices" section of the 1874 Cass County Atlas (Brink & Co. 1874):

> George Schneider   Henry W. Krohe
> SCHNEIDER & KROHE'S OPERA HOUSE. This building was erected in 1873, – seventy-six by sixty-six feet, two stories high, and basement. The first floor is occupied with three large and commodious storerooms. The Opera House occupies the whole upper floor, which is seated with opera chairs, with seating room for eight hundred. The stage is well arranged and fitted up in the latest style, with fine scenery and an elegant drop curtain; two large dressing rooms, with a dining-room eighteen by fifty feet above them….
>
> Messrs. Schneider & Krohe are also engaged in the manufacture of Bottled Soda and Mineral Waters on a large scale. They are prepared to furnish to the trade, in quantities, at the lowest living rates; they are also prepared to ship on order to all points, either by railroad or river.
>
> The citizens of Beardstown may well feel proud of the persevering energy and public enterprise of these gentlemen, prompt and courteous in their business transactions.

In the section of the 1874 Atlas listing its contributors and sponsors, Krohe is listed as a manufacturer of soda waters and "President" of the Opera House; Schneider is listed

Figure 30

as a soda water manufacturer and "Proprietor of the Opera House. About 1884, Henry Krohe sold his share of the Opera House to his brother Fred, who remained partners in its operation with George Schneider ("now of Omaha": Biographical Review Publishing Co. 1892:282–283).

Henry Krohe continued in the soda and mineral water business, however, until the time of his death in 1889. Only the one style of embossed *Schneider & Krohe* soda bottle listed above is known. From the bottle's shape and technological attributes, it appears to date to the 1860s or very early 1870s: the heyday of Schneider & Krohe business development and expansion in Beardstown. The embossed bottles may have been ordered in 1869, the year the *Schneider & Krohe* soda factory first opened.

BELLEVILLE
(St. Clair County)

1850 census: 2,900
1860 census: 7,500
1870 census: 8,100
1880 census: 10,700

Louis Ab Egg

Soda ~ Mineral Water:

(1a) **L. AB.EGG'S. SODA / MANUFACTORY / BELLEVILLE. ILL**ˢ [blue teal, green]
»IP; shouldered-blob-top soda. *{Figures 31 and 33}*

(1b) **L. AB.EGG'S. SODA / MANUFACTORY / BELLEVILLE. ILL**ˢ [green teal, aqua]
»SB w/ keyed hinge mold; shouldered-blob-top soda. *[Figures 32 and 33]*

(2) **L. ABEGG. / BELLEVILLE / ILL.** [aqua]
Note: *No original photo available, but a photo is given in Miller (2007a:13, see Figure 34), and the bottle embossing is identical to that shown for bottle #3 below, with the* **ILL.** *embossing removed.*
»SB w/ keyed hinge mold; oval-blob-top soda.

(3) **L. ABEGG. / BELLEVILLE** [aqua]
(base): **AB** *(stylized, attached letters)*
»SB w/ keyed hinge mold; oval-blob-top soda. *[Figure 35]*

(4) **L. ABEGG. / BELLEVILLE** *[Figure 36]* [aqua]
»SB w/ post mold; oval-blob-top soda.

Louis Ab Egg is listed in the 1860 ISD, the 1860 Belleville CD, and in the 1863 Belleville CD as a soda water manufacturer located on the east side of South Second, between Spring and Richland. Miller (2007a:9–13) reproduces Ab Egg's obituary, which was published in the *Belleville Weekly Advocate* for January 17, 1879. In part, the obituary reads: "Mr. Ab-Egg was a Swiss by birth, emigrated to America in 1850, since which time he has resided in Belleville—a quiet, diligent, upright and useful citizen. He died at the age of 64 years."

Ab Egg formed a partnership with Joseph Fischer during the mid-to-late 1850s to bottle soda and mineral water (*see Fisher & Ab Egg listing, this chapter*). The partnership had dissolved by

Figure 31

Figure 32

Figure 33

Figure 34

Figure 35

1860 when they are listed in the CD as competitors. Ab Egg established his own soda/mineral water factory at 118–120 South Richland (now Second Street). Fisher's competing factory was at 522 Fulton Street (Miller 2007a:170). Bottles #1a and #1b in the listing would have been manufactured at this time: the bottles are made from the same mold, some pontilled and some smooth-based, just at the 1860 transition period between these two manufacturing technologies.

Ab Egg was 45 years old when he was listed in the 1860 IIC as a soda water manufacturer with three employees, in operation six months a year. Under "Raw Material Used" he listed 12,000 dozen bottles valued at $600. He valued annual sales of the filled bottles at $3,300. He continued in operation in the 1870s, adding his son Charles to the business. The 1877 Belleville CD also lists Ab Egg as operator of an ice house with a partner named Dintelman at his bottling-business address (Miller 1989).

After Ab Egg's death in 1879, August Koob (*see Koob entry, this chapter*), who had recently moved from Mascoutah to a house next door to the Ab Egg's (1880 USC), bought the bottling business (Miller 2007a:9). However, since only one embossed Ab Egg bottle is known from the 1870s period (#3 in the listing—technologically, the other listed bottles are of 1860s manufacture), the bottling works may have sat idle during the later 1870s, or served largely as an ice house.

George N. Clark

Soda ~ Mineral Water:

(1) **J. N. CLARK / BELLEVILLE / ILL** [aqua]
  Note: The embossed "J" is an early glass-manufacturer's error.
  »IP; tapered-collar soda. *[Figure 37, right]*

Figure 36

(2) **G. N. CLARK / BELLEVILLE / ILL** [aqua]
  Note: First initial corrected, otherwise same mold as #1 above.
  »IP; tapered-collar soda. *[Figure 37, left]*

Figure 37

George Clark was listed in the 1850 USC as a 38-year-old Belleville "soda-maker," who lived next door to Francis Stoltz, another Belleville soda water bottler (*see Stoltz listing, this chapter*). At the time of the census, Clark's youngest child, four years old, was listed as having been born in Missouri. So Clark arrived in Belleville sometime after 1846. The only post-1850 documentation for Clark as a soda bottler was in the *Belleville Weekly Advocate* for September 11, 1851, in which an ad for "Pensoneau & Stephen's Saloon" mentions that they receive "a daily supply of bottled soda water from the well-known establishment of G. N. Clark" (see Miller 1989).

Miller (2007a:542–3) believes that Clark purchased the Stoltz bottling works. But there is only one early Clark bottle style (perhaps dating as early as the late 1840s), while several styles of Stoltz bottles are known, reflecting a greater 1850s time range of use (see Stoltz listings below). So, with the information currently at hand, it seems more likely that Stoltz bought out Clark ca. 1852. (It is also possible that the two men pursued their soda-bottling interests separately.)

Clark and his family are not listed in the 1860 Belleville USC, and appear to have left the area sometime before the onset of the Civil War. By contrast, Stoltz was active in real estate and politics later in life, and he is listed in the 1860 census and 1868 directory for Belleville, but without occupation.

Joseph Fisher (*aka:* Fischer)

Soda ~ Mineral Water:

(1) **JOSEPH FISHER / BELLEVILLE / ILL // MINERAL WATER /** [aqua]
   (heel): **FRL**

   »SB w/ keyed hinge mold; shouldered-blob-top soda. *[Figure 38]*

(2) **JOS. FISCHER'S / SELTERS WATER / BELLEVILLE / ILL_ //** [aqua]
   (heel): **A. & D. H. C.**

   »SB w/ keyed hinge mold; 2 "petaled" front shoulder panels; 5 rear body panels; shouldered-blob-top soda. *[Figure 39]*

(3) **J. FISHER / BELLEVILLE / ILL** [aqua]

   »SB w/ keyed hinge mold; selters-style soda bottle w/ long cylindrical neck and cylindrical blob top. *[Figure 40]*

(4) **J. FISHER. / BELLEVILLE / ILL** [aqua]

   »SB w/ keyed hinge mold; long neck; shouldered-blob-top soda. *[Figure 41]*

(5) **J • FISHER • / BELLEVILLE / ILL //** [aqua]
   (base): **F**

   »SB w/ post mold; oval-blob-top soda. *[Figure 42]*

(6) **J. FISHER. / BELLEVILLE / ILL //** [aqua]
   (heel): **A. & D. H. C**
   (base): **F**
   Note: No original photo. Bottle recorded from Paul and Parmalee (1973:78, Figure 71). Miller (1989) lists a variety without the **F** base mark).

   »SB w/ probable post mold; short-neck, oval-blob-top sodas.

(7) **J. FISCHER / BELLEVILLE / ILLS //** [aqua]
   (heel): **L.G. C°**
   (base): **F**

   »SB w/ post mold; short-neck, oval-blob-top soda. *[Figure 43]*

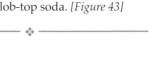

Figure 38

Belleville — Joseph Fisher (aka: Fischer)

Figure 39

Figure 40

Figure 41

Joseph Fisher is listed in the 1860, 1870, and 1880 USC, and in the 1875 ISD, as a soda water manufacturer in Belleville. According to Miller (2007a:159–165), Fisher was born in Austria and had immigrated to the United States by 1848, when he was living in Missouri. He moved to Belleville in 1849 and is listed in the 1850 USC for Belleville as a stone mason. He entered into a partnership with Louis Ab Egg in the mid-1850s in the soda and mineral water bottling business (*see their listing*). He also formed a brief business partnership with Ernst Rogger in the later 1850s (*see their listing*), but was listed as a manufacturer of soda water and the sole owner of the company in the 1860, 1870, and 1880 USC.

Figure 42

He appears in the IIC index for 1880 as operating six months full time, two months three-quarters time, and four months half time, with a seasonal maximum of nine employees (five adults and four children) earning from $1.25 to $2 per day. Subtracting $4,200 for labor and supplies, his annual profit from the operation was about $3,800. The bottling operation was listed as being located at Fisher's residence on Mascoutah, south of Abend, in the early 1870s, but he must have later constructed a separate factory building, because the works were

Figure 43

located at 522 Fulton Street (at the corner of Mascoutah) at the time the business was purchased by Fisher's nephew, John Winkler, in 1882. After retiring from the bottling works, Fisher continued to live in Belleville, where he died ca. 1885.

The Fischer bottle containing "Selters Water" is not a misspelling of "seltzer." Selters Spring, along with Vichy and Kissingen, were popular European spas of the mid-to-late nineteenth century (Waddy 1996; 1997). Many American bottlers used these names on their labels to increase the appeal of their local mineral waters. During this time, mineral waters were usually not flavored and were advertised for their medicinal qualities. Likely, their mineral taste made them seem to be more effective as health drinks.

Joseph Fisher and Louis Ab Egg

Soda ~ Mineral Water:

(1) **FISHER & ABEGG / BELLVILLE / ILL^S // MINERAL WATER /**   [aqua]
   Note: **BELLVILLE** is misspelled; two embossed dots under superscript **S** in **ILL^S**; small-lettered variety of **MINERAL WATER**.

   »IP; shouldered-blob-top soda. *[Figure 44]*

(2) **FISHER & ABEGG / BELLVILLE / ILLS // MINERAL WATER /**   [aqua]
   Note: **BELLVILLE** is misspelled; two embossed dots under full-sized **S** in **ILLS**; large-lettered variety of **MINERAL WATER**.

   »IP; shouldered-blob-top soda. *[Figure 44]*

Little is known about the Fisher and Ab Egg partnership. Based on the single known style of embossing on the iron-pontiled bottle, combined with the dating of the manufacturer (the

Figure 44

Frederick R. Lorenz glass company), the fact that Ab Egg was in business on his own in the late 1850s (*see his listing*), and Fisher's late-1850s partnership with Ernst Roger (*see their listing*), the Fisher & Ab Egg association likely took place during the mid-1850s and was of short duration.

Joseph Fisher and Ernst Rogger

Soda ~ Mineral Water:

   (1) **J. FISHER & ROGGER / BELLEVILLE / ILL //**   [aqua]
       NINERAL WATER
       [Note: "NINERAL" misspelling]

       »IP; shouldered-blob-top soda. [Figure 45]

   (2) **J. FISHER & ROGGER / BELLEVILLE / ILL //**   [aqua]
       MINERAL WATER
       (left rear heel): **FRL**
       [Note: "MINERAL" corrected spelling and added FRL maker's mark]

       »IP; shouldered-blob-top soda. [Figure 46]

Historical documentation concerning the partnership between Fisher and Rogger indicates that it took place in the late 1850s—after he was involved in a partnership with Louis Ab Egg in the mid-1850s (*see listing*)—and

Figure 45   Figure 46

lasted for only a brief period, perhaps just one or two summer seasons. The partnership had been terminated by 1860, when Fisher alone is listed in the USC (*see his listing*). Rogger's first name has also been transcribed "Erna" or "Irne" in local historical records (Miller 2007a:166).

Edward T. Flanagan

Medicine:

(1) (side) **NORMAN'S // MAGIC / LINIMENT //** [aqua]
    (side) **BELLEVILLE, ILL // //**

»SB; rectangular bottle w/ 3 sunken panels; flared square-ring collar; 4 1/16" tall. *[Figure 47]*

(2) (side) **NORMAN'S // CINCHONA AND / PINE KNOT CORDIAL // //** [aqua]
    (side) **BELLEVILLE ILL // //**
    Note: No complete examples of this bottle could be located to photograph. Figure 48
    is a redrawn sketch taken from Miller (2007b:233), who saw and recorded fragments
    of a broken example of the bottle.

»SB w/ keyed hinge mold; beveled-edge rectangle w/ 4 sunken panels; tapered collar; 9" tall. *[Figure 48]*

E. T. Flanagan & Co. of Belleville was founded in the late 1860s to manufacture and distribute a patent-medicine product line supposedly invented by a local medical practitioner, Dr. George Norman, who cannot be be found listed in local historical records (Miller 2007b:227). Norman's *Celebrated Magic Liniment* and Norman's *Pine Knot Cordial* were heavily advertised in the local newspapers (the *Belleville Democrat* and the *Belleville Weekly Advocate*) between 1869 and 1872. One of the advertisements (January 26, 1872), written in article style, states that "Messrs. E. T. Flanagan & Co. received...a large consignment of Pine Knots to be used in the manufacture of their already celebrated Pine Knot Cordial." The same article goes on to say that a local carriage manufactory had built "a new medicine wagon...for E. T. Flanagan & Co. to be used in the sale of their medicines. It is one of the strongest, lightest, and prettiest vehicles we have ever seen in Belleville."

Figure 47

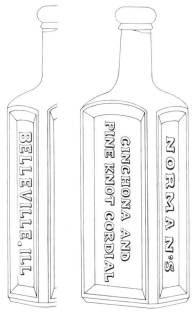

Figure 48

Figure 49

The May 16, 1873, issue of the *Weekly Advocate* contains a notice that E. T. Flanagan & Co. filed for bankruptcy and was going out of business (Miller 2007b:228).

A keyed hinge-mold bottle embossed **NORMAN'S // CHALYBEATE / COUGH SYRUP / S^T LOUIS** is also known *[Figure 49]*. It was distributed there by *Brown, Weber & Graham* (and later by *Meyer Bros. & Co.*), but we have as yet been unable to document a Belleville-embossed variety. But since Flanagan did advertise it heavily in Belleville, perhaps it was paper-labeled only. A large ad he placed in the *Historical Review of Belleville* in 1870 (Hinchcliffe 1870, reproduced in Miller 2007b:229) indicated that "one retail house alone in our city is selling this medicine at the rate of nearly one thousand dollars per annum. The demand has increased so rapidly within the last two years that several men are constantly employed in putting up the medicine and in filling orders for shipments to various parts of the country. While we write, a large order from the state of Texas has just been received." *Norman's Chalybeate Cough Syrup* was advertised as an effective cure for croup, colds, sore throat, hoarseness, enlargement of the tonsils, and would lessen the symptoms of whooping cough and measles: also "teachers, public speakers and singers will find it invaluable."

The Belleville Democrat for February 16, 1871, included an ad for *Norman's Magic Liniment* which indicated that it was a superior remedy "for the cure of Rheumatism, Toothache, Spinal Affections, Cholera Morbus, Bruises, Sore Throat, Neuralgia, Earache, Cuts, Sprains, Colic, Burns, Cramps, Headache, Inflammatory Rheumatism, Pain in Stomach and Bowels, Inflamed Sores,…Corns and Bunions on the Feet."

No detailed ads have yet been found for *Norman's Pine Knot Cordial*, but a testimonial in the *Belleville Weekly Advocate* for March 1, 1873, declared: "I never saw such a medicine in my life. When I commenced using it, I was unable to leave my bed. I did not take quite four bottles, until I was able to be about and do a good day's work on my farm, and I have grown stouter ever since. My disease was of the back and kidneys."

Thomas Heberer and Brothers

Ale ~ Beer:

(1) **T. HEBERER & BROS / BELLEVILLE ILLS** [aqua]

»IP; quart barrel-effigy bottle with a long neck and ring-neck tapered collar, 9 1/2" tall. Embossed on the upper shoulder only. *[Figure 50]*

Thomas Heberer, along with his brothers Adam and Henry, founded the City Park Brewery in Belleville during the 1850s. "In 1859 they completed the construction of the City Park Theater as an adjunct to the brewery. Beer manufacturing continued at the corner of Richland and North First until June of 1865, when Adam Heberer was arrested for evasion of revenue taxes. The brewery was seized by the Federal Government and later sold at auction" (Kious and Roussin 1997:5). After that, the brewery continued in operation for a while, but under new management: "The 1868–69 Belleville City Directory listed HENRY HEBERER and TOBIAS SCHIFFER at the same address as the City Brewery. By 1873 the brewery was closed and sold at auction" (Miller 2008:79).

The Heberer brewery is listed in the 1860 and 1864 ISD and was one of seven local breweries listed in the 1860 Belleville CD. *Heberer & Brothers* were also listed in the 1860 IIC, which indicated they had $10,000 in capital and real estate invested in the business and had four employees and an average monthly salary outlay of $120. Their 1860 production rate was 2500 barrels of ale annually, with an estimated product value of $17,000.

Since their barrel-shaped bottle is pontiled, it probably dates to their business expansion in 1859; the (filed) embossed bottle was likely distributed that year to advertise their product.

August G. Koob

Mr. Koob bottled in Belleville only from 1879 onward, and his embossed bottle styles do not fall within the scope of this study. For his earlier bottling efforts, see his Mascoutah listing.

Dr. George Norman

Dr. Norman's medicines were bottled and marketed by Edward T. Flanagan & Co. in Belleville (*see his listing, this chapter*).

Francis (Franz) Stoltz

Soda ~ Mineral Water:

Figure 50

(1) **F. STOLTZ / BELLVILLE / ILL** [aqua]

»IP; slope-shouldered soda. Three closure varieties—tapered-collar, lozenge-top, and blob-top—are known. [*Figure 51*]

❖

Francis (Franz) Stoltz was born in Barvaria and immigrated to the Unites States in 1837 (USC). Miller (2007a:542–543), citing articles in local newspapers, states that Stoltz came to Belleville in the early 1840s. He established the Napoleon Hotel on the Belleville public square in 1848. Miller suggests that Stoltz only bottled soda water for a short time in 1849–50, and then sold his soda and bottling business to his neighbor, George Clark, who is listed as a Belleville bottler in the 1850 USC.

However, Clark bottles are seldom seen, and the known examples are all of the late-1840s and early-1850s tapered-collar style (*see Clark listing*). By contrast, Stoltz bottles are commonly found, and were widely distributed on both sides of the Mississippi River. Moreover, the Stoltz bottles are known in tapered-collar (a late-1840s to early-1850s closure), lozenge-top (a mid-1850s closure), and blob-top (a late-1850s closure) varieties [*see introduction, this volume*].

Thus, in the absence of closer documentation, Miller's suggestion could be reversed: Clark seems to have bottled for only a short time and then sold out to Stoltz, who continued bottling seasonally, likely as a second line of work, for several years during the 1850s.

Figure 51

BELVIDERE
(Boone County)

1860 census: 1,100
1870 census: 3,200
1880 census: 2,900

Solomon W. Bristol

Medicine:

(1) **S. W. BRISTOLS / NERVE & BONE LINIMENT //**   *[aqua]*
BELVIDERE / ILL^S

»OP; shouldered cylinder; rolled lip; 5 1/8" tall. *[Figure 52]*

Figure 52

S. W. Bristol advertised himself in local 1840s newspapers (from at least March 11, 1846 on) as a druggist and agent for patent medicines. He was recorded in the 1850 USC as a 46-year-old Belvidere merchant. By the time of the 1860 USC, he had become a nurseryman.

Bristol's Nerve & Bone Liniment was produced during his druggist and early merchant years in the late 1840s and early 1850s. It was advertised in the May 1, 1850, *Rockford Forum,* accompanied by local testimonials from 1849 citing its effectiveness on horses and for external human use. In November 1849, it was advertised for sale at five different local drug stores. At the time, it was "manufactured exclusively" at Belvedere by J. Jeffley. It was advertised in the February 2, 1850, *Milwaukee Sentinel and Gazette* as "manufactured exclusively" at Belvedere by Hamilton & Co., "for sale in all portions of Illinois, Wisconsin, and Iowa" (Odell 2007:48).

BLOOMINGTON
(McLean County)

1850 census: 1,600
1860 census: 7,000
1870 census: 14,600
1880 census: 17,200

Bottling House

Soda ~ Mineral Water:

(1) **BOTTLING HOUSE / BLOOMINGTON, ILL'S**   *[aqua]*
(base): **J. E.**

»SB w/ keyed hinge mold; shouldered blob top; 10" tall quart. *[Figure 53]*

There is no known "J. E." glassmaker's mark, so the base embossing on this bottle likely represents the bottler's initials. One other "BOTTLING HOUSE" bottle, also with "J. E." base embossing is known from *Metamora* (*see Metamora Bottling House listing*). That bottle is a pint soda or mineral water, so the Bloomington bottle was likely a quart soda, not a beer. Since both bottles were made with snap-case technology and have keyed hinge-mold seams, the Bloomington company probably operated for a short time during the 1860s.

No direct documentary evidence has surfaced for a bottler with the initials "J. E." in historical records from Bloomington. But a soda factory operator with these initials *has* been documented for Metamora. A soda-factory operator there named **Joseph Erl** was recorded in both the 1870 USC ("Soda Water Maker") and the 1880 USC ("keeps Soda Factory"), and in the 1879–80 Illinois Industrial Census. In the 1879–80 IIC, Erl was listed as operating the factory, with four adult male employees, full-time eight months a year (the factory was idle for the remaining four months). His workers were paid from $1.25 to $1.75 per day. He paid $550 in wages and $2000 for bottling materials and supplies during the census year, and estimated the annual retail value of his bottled beverages at $3100.

There was only one known soda-bottling factory in Bloomington during the 1840–1880 period of our study, at the corner of Madison and Mulberry streets. J. E.'s *Bottling House* likely operated from there. *Mueller & Stein* first established a Bloomington soda factory at the Madison and Mulberry location in the late 1850s (*see their listing, this chapter*). They bottled there ca. 1858–61; S. M. Hickey likely used it ca. 1863–64 (*see his listing, this chapter*); *Petterson & Bro.* used it ca. 1870–74 (*see their listing, this chapter*); and Peter Kreis used it from ca. 1876 until the late 1880s (*see his listing, this chapter*). This suggests the Bloomington soda manufactory was available for use by Joseph Erl during the later 1860s (ca 1865–68), before he left (ca. 1869) to establish his new bottling works in Metamora.

Figure 53

Browen's Rat Killer

See the Cyrenius Wakefield listing, this chapter.

Calvin B. Castle and Co.

Medicine:

(1a) **DR FALOON'S / R. W. BALSAM** [aqua]

» SB; beveled-edge rectangle w/ 4 sunken panels; double-ring collar; 8 1/8" tall. *[Figure 54]*

(1b) **DR FALOON'S ROSIN/ WEED BALSAM** [aqua]
 (base) **I G W CO**

» SB w/ keyed hinge mold; beveled-edge rectangle w/ 4 sunken panels; double-ring collar; 7 3/8" tall. *[No figure, similar to Figure 54; see introduction]*

(2) **FALOON'S INSTANT RELIEF** [aqua]

» SB w/ post mold; beveled-edge rectangle w/ 4 sunken panels; double-ring collar; 3 3/4" tall. *[Figure 55]*

Figure 54

Calvin Castle was a 34-year-old medicine manufacturer in Bloomington in 1880 (USC). Dr. Mathew Faloon came to Bloomington during the Civil War era and was listed in the 1880 USC as a 51-year old Bloomington physician. According to the 1879 McLean County History, "he attended and became a graduate of the Eclectic Medical College, of Cincinnati: he also became a graduate in allopathy in Philadelphia…he is a thorough botanist, frequently having herbs sent to him from the laboratories of different colleges, for the purpose of learning their medical qualities; he is also the propounder of different medicines now before the public, prominent among them is Dr. Faloon's Tonic, Dr. Faloon's Instant Relief and many other well-known remedies. The Doctor has been in constant practice for twenty-six years, and has been practicing in Bloomington for the past nineteen years" (Le Baron 1879:780).

A paper label preserved on one of the R. W. BALSAM bottles (*Figure 56*) identifies the contents as "**DR. FALOON'S / ROSIN / WEED / BALSAM**... / C. B. Castle & Co. / BLOOMINGTON, ILL. / *Price, $1.00.*"

Throughout the entire vegetable kingdom there is not anything which gives such marked specific action upon the lungs as this plant. For all forms of

DISEASES OF THE LUNGS,

Such as Coughs, Colds, Catarrh, Whooping Cough, Croup, Bronchial Affections, Short or Difficult Breathing, Inflammation, Tightness of the Lungs and Chest, Palpitations or Hard Beating of the Heart, Asthma and all forms of Consumption—if not too far advanced, although it will give ease and comfort even in the last stages of the disease,—Diseases of the Kidneys and Bladder, Dropsy of the Heart, Sinking Chills, all Female Irregularities, &c.

Another crowning feature of this Balsam, is its being a great Blood Purifier, correcting the impurities and removing all excretions from the face and surface of the body, giving a healthy complexion, and invigorating the whole system.

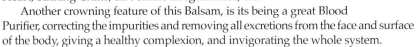

Figure 55

An 1888 trade card listing the company's products has also been found (*Figure 57*), indicating that Castle had been producing and selling Faloon's medicines since 1870. (The significance of the 1838 date on the card is unknown). Of the nine products listed and described, three are pills and two are creams or salves. The other four are medicines, and in the 1870s they were probably put up in embossed bottles like the two examples documented here. The **R.W. BALSAM** was a cure for diseases of the lungs, kidneys, and bladder. The **INSTANT RELIEF** was "the best Painkiller in existence." Also produced were *Dr. Faloon's Tonic*, for general debility, and *Dr. Faloon's Blackberry Diarrhoea Specific*.

Dr. Mathew Faloon

Dr. Faloon was born and raised in Carlisle, Pennsylvania, and graduated from the Cincinnati Eclectic Medical College. He received graduate training in allopathy and "the Eurescopian system" before coming to Bloomington to practice as a botanical physician ca. 1860 (Le Baron 1879:780). Faloon was listed in the 1866 and 1868–69 Bloomington CDs as a "physician" with offices at 111 E. Washington. In the 1870–71 CD, he was listed as selling "medicines" at the same address, in partnership with John Geltmacher. The first two Faloon and Geltmacher newspaper ads in the *Bloomington Pantagraph*—for Faloon's

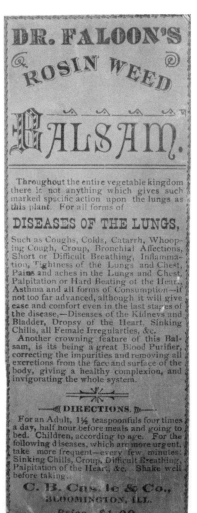

Figure 56

Figure 57

Instant Relief and *Infant Cordial*, respectively—date to November 28 and December 20, 1870 (Anthony Green, personal communication 2009). No embossed bottles are known from the Faloon and Geltmacher years.

Faloon was at the same address, alone, in the 1873 CD, selling "trusses, supporters, and medicines. From 1873 onward, he is listed only as a physician. He does not appear to have embossed his medicine bottles during this time, as no *Faloon* or *Faloon & Geltmacher* embossed bottles have been found.

The Faloon patent medicines compounded and marketed during the later 1870s and 1880s were not sold by Dr. Faloon himself. This was done through arrangement with *Calvin B. Castle & Co.* (*see their listing, this chapter*).

Francis M. Funk and Ira Lackey

Medicine:

(1) *(side):* **BLOOMIMGTON // FUNK & LACKEY //** [aqua]
    (side): **ILLINOIS**

»SB w/ post mold; rectangular; beveled corners; 4 sunken panels; square ring collar; 6 1/2" tall. *[Figure 58]*

The above-listed bottle is in the collection of the McLean County Historical Society. A paper label on the bottle (*Figure 59*) reads: LACKEY'S JUSTLY CELEBRATED BLACKBERRY CORDIAL... PREPARED BY IRA LACKEY & BRO., DRUGGISTS & CHEMISTS, SOUTH SIDE PUBLIC SQUARE, BLOOMINGTON, ILL.

According to listings in Bloomington CDs, the *Funk & Lackey* partnership dates from 1878 to 1891, located at 110 W. Washington. According to the 1879 McLean County History, "in 1877, Mr. Funk entered the drug business with Ira Lackey, which is one of the leading drug firms of Bloomington" (Le Baron 1879:784).

Since the *Ira Lackey & Brother* firm was dissolved in 1874, the paper label was at least three years out of date when it was used on the *Funk & Lackey* bottle described in the listing. [*See Ira Lackey and Bro. entry*]. The *Funk & Lackey* embossed bottles are found in contexts that suggest they date to the early years of the business.

Although the Historical Society example of the bottle contained Lackey's "Justly Celebrated" Blackberry Cordial, these were no doubt generic pharmacy bottles during the early years of the Funk & Lackey partnership, that were likely filled with a wide variety of medicinal products. For instance, a Funk & Lackey trade card is known (*Figure 60*) that advertises their "BRADNER'S HONEY BALSAM For Coughs, Colds, &c." (*also see Ira Lackey & Bro. listing, this chapter*).

Stephen M. Hickey and Co.

Soda ~ Mineral Water:

(1) **S. M. HICKEY & Co / BLOOMINGTON / ILLS**   [aqua]

»SB w/ keyed hinge mold; shouldered-blob-top soda. *[Figure 61]*

Figure 58

Figure 59

Figure 60

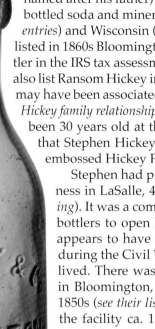

Stephen M. Hickey (*aka:* "Sephremus" or "Sephreness" Marona, named after his father) was a member of the large Hickey family who bottled soda and mineral water in Illinois (*see Joliet, LaSalle, and Peoria entries*) and Wisconsin (Peters 1996). S. M. Hickey has not been found listed in 1860s Bloomington CDs, but he was listed as a Bloomington bottler in the IRS tax assessments for 1863 and 1864. These same documents also list Ransom Hickey in Bloomington in 1864 as an "ale dealer," and he may have been associated in business with Stephen (*see the discussion of Hickey family relationships under their Peoria listing*), who would have been 30 years old at the time. The IRS tax assessments indicate that Stephen Hickey had moved to Freeport, IL, by 1865 (no embossed Hickey Freeport bottles are known).

Stephen had previously been in the soda bottling business in LaSalle, 40 miles to the north (*see his La Salle listing*). It was a common practice at this time for soda water bottlers to open branch facilities in other cites. Stephen appears to have expanded his business to Bloomington during the Civil War, but the branch operation was short lived. There was only one known soda-bottling factory in Bloomington, started by Mueller & Stein in the late 1850s (*see their listing, this chapter*). Mueller & Stein used the facility ca. 1858–61, S. M. Hickey likely used it ca. 1863–64. J. E.'s *Bottling House* may have operated from there during the later 1860s (*see Bottling House listing, this chapter*) Petterson & Bro. used it ca. 1870–74 (*see their listing, this chapter*), and Peter Kreis used it from ca. 1876 until the late 1880s (*see his listing, this chapter*).

An embossed **S. M. HICKEY & Co** soda, unearthed at a construction site 45 miles south of Bloomington in Springfield, IL, has no embossed town name (*Figure 62*). There is no evidence the Hickey family ever bottled in Springfield,

Figure 61

and all the northern Illinois Hickey bottles have embossed town names, so perhaps it is a variant of the Bloomington soda described earlier.

William P. Hill and Sanford K. Vanatta

Figure 62

Medicine:

(1) (side): **HILL & VANATTA // OIL OF LIFE //**   [aqua]
    (side): **BLOOMINGTON ILL**
    (base): **L&W**

»SB; beveled-edge rectangle; 4 sunken panels; short neck; tapered collar; 6" tall. *[Figure 63]*

Hill & Vanatta were only listed in the 1870–71 Bloomington CD as partners in the druggist business at The People's Drug Store at 120 S. Main. They were not listed in the 1868–69 CD, and the 1872–73 CD indicates that R. D. Bradley had taken over operation of *The People's Drug Store*.

William Hill was a physician who had served as a U.S. Army surgeon during the Civil War. At the close of the war (when he was 37 years old), he came to Bloomington and set up a medical practice there, an occupation he was still engaged in 1879 (Le Baron 1879:787). Sanford Vanatta had been a salesman prior to his two-season association with Hill (*see newspaper ads discussion*), and returned to that occupation afterward. Their short partnership may have been a marriage of skills, with Hill developing the medicine and Vanatta marketing it. Given the rarity of the bottles and the shortness of the partnership, their efforts clearly failed.

Hill & Vanatta advertised their drugstore and their products in the *Bloomington Pantagraph* from February 1870 through the end of 1872 (Anthony Green, personal communication 2009). No advertisements or paper-labeled bottles have yet been located for their *Oil of Life* to indicate its specific claimed therapeutic value, but Anthony Green has noted (personal communication, 2008) that "it didn't do Mr. Vanatta any good, as it appeared he was murdered by his wife.... Although his body was exhumed twice (!), the evidence against her was inconclusive" (*see Bloomington Pantagraph* 10/26/1899, p. 7; 1/5/1900, p. 7, col. 3; 1/6/1900, p. 5, col. 3–4; 1/11/1900, p. 5, col. 3).

Bliss S. Howe

Medicine:

(1a) (side): DR B. S. HOWE // //　　　　　　　　　　　　[aqua]
　　　(side): **BLOOMINGTON ILLS** // //

»SB w/ post mold; beveled-edge rectangle; 4 sunken panels; tapered collar; 6¾" tall (x 2½" x 1½"). *Note: larger than #1b.* [Figure 64, right]

Figure 63

(1b) (side): DR B. S. HOWE // //　　　　[aqua]
　　　(side): **BLOOMINGTON ILLS** // //
Note: smaller than #1a, above. The bottle also has a tooled lip, so it may not be pre-1880.

»SB w/ post mold; beveled-edge rectangle; 4 sunken panels; tooled, tapered collar; 6" tall (x 1⅞" x 1"). [Figure 64, left]

❖

According to an 1899 biographical sketch (Clark & Co. 1899:424–427), Bliss S. Howe arrived in Bloomington in 1861. In 1864 he formed a partnership with Dr. Zera Waters (*see the Waters' entry, this chapter*)

manufacturing a general line of family medicines under the firm name of *Waters & Howe*. Our subject traveled with these medicines throughout this state and had other agents in different parts of the country. Dr. Waters later sold his interest to a Mr. Woodard, and the firm became *Woodard & Howe*. They did a good business until the winter of 1876–7 when the partnership was dissolved. Up to that time the medicines had been known as the Waters medicines, but in 1878 Dr. Howe began manufacturing them under his own name, and has since been alone in business.

Figure 64

Among the preparations he puts up are Dr. Howe's vegetable liver pills, Egyptian salve, linament [*sic*], instant pain relief, tonic bitters, honey balsam and blackberry balsam. [Clark & Co. 1899a:427]

Howe was listed in the 1870 USC as a 32-year-old Bloomington physician. In the 1880 USC, he is listed as a patent medicine manufacturer. From ca. 1872 to 1877, he bottled patent medicines as the junior partner in the firm of *Woodard & Howe*. When *Woodard & Howe* dissolved their business partnership in Bloomington and went their separate ways, both continued to produce medicines (*see Woodard listing, this chapter*). Woodard apparently retained the rights from their previous partnership to continue producing Dr. Waters' medicines.

But after the 1878 Woodard/Howe split, Dr. Howe apparently tried to continue to sell Dr. Waters' recipe medicines as well, together with an "Instant Relief" medicine that had been concocted by Dr. Woodard, by omitting Waters' and Woodard's names from the labels:

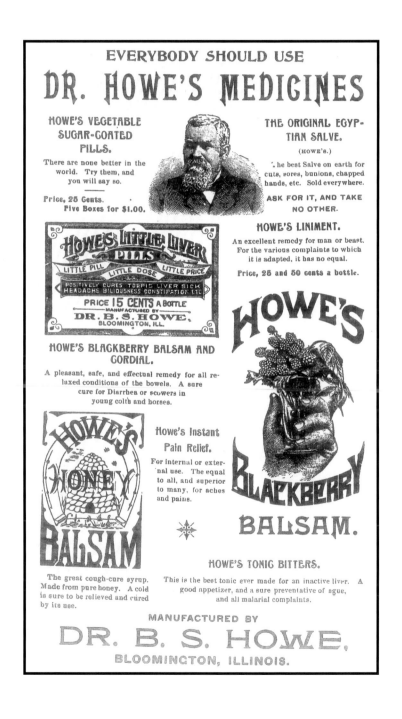

This resulted in Dr. Waters warning Bloomington residents in an 1878 issue of the *Daily Pantagraph* that "imitations" of his medicines were being produced (*see Woodard entry, this chapter*). In the mid-1880s Howe introduced a series of his own medicines, which continued to "bear an uncanny resemblance to Dr. Waters' medicines" (Anthony Green, personal communication 2008). Howe (*Figure 65*) continued to sell his medicines in Bloomington until around the turn of the century.

Howe's embossed bottles are found in contexts that suggest they date to the early years of his business.

Otto Kadgihn, Jr.

Ale:

(1) **O. K. JR** [aqua]
»SB w/ post mold; 2-piece-mold quart; ring-neck tapered collar; 9½" tall. *[Figure 66]*

Figure 65

❖

These 1870s ales have turned up in Bloomington-area contexts, but the business has not been found advertised in local newspapers. The rarity of these bottles (about 3–5 known) suggests that the *O.K., Jr.* ale business was short-lived.

In 1879, 52-year-old German-born Otto Kadgihn, Sr., ran a saloon and restaurant in Bloomington. He arrived in Bloomington just prior to the Civil War from St. Louis, where he had worked as a painter. Soon after he came to Bloomington, he "engaged in the grocery business; after that he entered the restaurant and saloon business, which he has followed ever since" (Le Baron 1879:794). He was listed in the 1868–69 Bloomington CDs as running a saloon at 114 W. Washington. In the 1870–71 CD, he and a partner (Hubert Neuerburg) had moved to 110 N. Centre, and their business was listed as "saloon, restaurant, wines, liquors, etc." Their CD ad that year (p. 334), added: "families supplied with the best bottled beer."

In 1871, Kadgihn was a litigant in an Illinois Supreme Court case *[Otto Kadgihn v. The City of Bloomington]* regarding Kadgihn's selling of spirituous liquors within city limits without a proper permit, his argument being that the license was improperly withheld. The Court ruled in favor of the city.

By 1872–73, O. K., Sr., was operating his saloon and restaurant alone, at two addresses (110 and 118 N. Centre), and his son, Otto Kadgihn, Jr., was employed at one of the locations. In the 1873 CD, *Otto Kadgihn & Son* were listed as business partners running the saloon and restaurant at 110 N. Centre. According to O. K. Senior's obituary (*Bloomington Pantagraph*, Oct. 26, 1899, p. 7), they ran two saloons: one was "a noted resort in the early days, and the second ("The Sebastopol") "became famed all over central Illinois for the skill and delicacy with which wild game was cooked and served."

In the 1874–75 and 1875–76 CDs *Otto Kadgihn & Son* still operated the saloon and restaurant at 110 N. Centre, but their business at 118 N. Centre now sold "boots and shoes." O. K., Jr., apparently attended to this part of the operation. But by 1876 (and during the later 1870s), the footware business was closed, O. K., Jr., worked as a salesman at another shoe store, and O. K., Sr., ran the restaurant/saloon operation on his own.

Figure 66

From the technological (post-mold base) and stylistic attributes of the *O.K., Jr.* bottle, it seems to have been manufactured and used during the early 1870s. This fits with the 1872–73 date of O. K. Jr.'s, involvement with the Kadgihn saloon and bottled-spirits business.

Peter Kreis

Soda ~ Mineral Water:

(1) **PETER KREIS / BLOOMINGTON / ILL //**   [aqua]
 (heel): **A. & D. H. C**

»SB w/ keyed hinge mold; oval-blob-top soda. *[Figure 67]*

Peter Kreis is listed in Bloomington CDs from 1876 to 1887 as a soda water manufacturer. He is also listed as a soda water dealer in the 1879 *History of McLean County* (Le Baron 1879:1069). The bottles listed above date to the later 1870s, and exclude his post-1880 Hutchinson-style internal-rubber-stopper bottles. He was never listed in the CDs in the 1870s or early 1880s as *Peter Kreis & Co.* Peter Kreis beer bottles from the later 1880s were embossed *P. Kreis & Co.*, however. In 1880, when Peter Kreis was 36 years old, he was listed in the IIC as operating a soda water bottling factory full time for six months a year and half time for six months a year. His estimated annual value of bottled products was $3360.

There was only one known soda-bottling factory in Bloomington, started by Mueller & Stein in the late 1850s *(see their listing, this chapter).* Mueller & Stein used the facility ca. 1858–61, and S. M. Hickey likely used it ca. 1863–64. J. E.'s *Bottling House* may have operated from there during the later 1860s *(see Bottling House listing, this chapter),* Petterson & Bro. used it ca. 1870–74 *(see their listing, this chapter),* and Peter Kreis used it from ca. 1976 until the late 1880s.

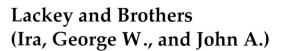

Lackey and Brothers
(Ira, George W., and John A.)

Figure 67

Bitters:

(1) *(side):* **BLOOMINGTON // LACKEYS / IRON BITTERS //** *(side):* **ILLINOIS // //**   [amber]

»SB w/ keyed hinge mold; rectangular; 3 sunken panels; double-ring collar; 9" tall. *[Figure 68]*

(2) *(side):* **BLOOMINGTON // LACKEYS / IRON BITTERS //** *(side):* **ILLINOIS // //**   [aqua]
 Note: The bottle is embossed like #1, but has a tooled tapered collar, and thus may date after 1880. If so, it is a Funk & Lackey product (see F&L listing). No photo.

»SB w/ post mold; rectangular; 3 sunken panels; tooled lip; 6" tall.

Medicine:

(1) *(side):* **IRA. LACKEY & BRO // DR MAJOR / QUININE / SUBSTITUTE //**   [aqua]
 (side): **BLOOMINGTON / ILLINOIS**
 Note: This bottle is known only from a single fragmentary example, curated in the McLean County Historical Society Museum collections.

»SB w/ keyed hinge mold; squat beveled-edge rectangle; 2 sunken side panels; double-ring collar; 4 1/2" tall x 2 1/4" x 1 1/2". *[Figure 69]*

Figure 68

Figure 69

The *Lackey & Brother* druggist partnership began in the early 1860s:

> Mr. Ira Lackey, of the firm of Funk & Lackey, is another of the old residents of Bloomington...he is a native of Wayne Co., Ind.; was born in 1838, and, in 1855, came to Illinois, locating at Bloomington; his first business engagement was in the capacity of a clerk for the drug firm of Paist & Elder, with whom he remained for about three years; the firm changed to Paist, Marmon & Co., he still remained with them for two years; then he and his brother engaged in business in the Ashley-House block, where they carried on the drug business for five years, the firm being known as I. & G. W. Lackey; his health failing, he sold out to his brother, went to Chicago and began traveling for Fuller, Finch & Fuller, wholesale druggists; he traveled for this firm until 1869 *[Ira Lackey is listed as a "salesman" for F., F. & F. in the 1866 and 1867 Chicago CDs only]*; having regained his health, he returned to Bloomington, and again began the retail drug business by buying the drug establishment of J. M. Major *[See Major's listing, this chapter]*, since which time he has been permanently located, gradually increasing his business until now he has one among the finest retail drug houses in this part of Illinois, doing a retail trade of about $33,000 per annum, besides his jobbing trade, which is gradually increasing; his establishment is located at No. 110 W. Washington street [Le Baron 1879:796–797].

According to the Anthony Green (personal communication 2009), there were *two different* druggist firms involving Lackey brothers. The *I. & G.W. Lackey* firm, with Ira and George, operated only during the Civil War years, when George left to become a Chicago pharmaceutical salesman. When George returned and bought out J. M. Major's drugstore in ca. 1869, his was a separate operation from the ongoing *Ira Lackey & Brother* firm. Bloomington CDs for the late 1860s and early 1870s indicate that Ira Lackey's partner during that time was John A. Lackey. John resided in St. Louis, so he was likely only a financial partner in the business.

Lackey & Brother continued to be listed in city directories as patent-medicine manufacturers up to 1874, when Ira was listed alone. Ira Lackey was listed alone as a druggist in the Bloomington CDs until 1877. The following year, he appeared as a partner in the patent-medicine firm of *Funk & Lackey* (see their listing, this chapter).

Because Lackey & Bro. also manufactured "liquid slating for blackboards," the *Alumni Journal* for October 1873, (Vol. III, No. 10) that carried their "liquid slating" ad, also mentioned their patent medicine products (*see Figure 70*):

Figure 70

Lackey's Iron Bitters
TO THE PUBLIC!
The proprietors of **LACKEY'S IRON BITTERS** come before the people with no desire of palming off a mixed up whisky, – advertised humbug, but a remedy that they honestly recommend for Dyspepsia, Liver Complaint, Nervous Debility, and Impurities of the blood. All we ask is a trial. **Price One Dollar per Bottle.**

– PROPRIETORS OF –
Bradner's Honey Balsam!
SAFE FOR CHILDREN OR ADULTS,
FOR THE CURE OF
Coughs, Colds, Croup, Whooping Cough, and all Diseases arising from the Throat or Lungs.

Dr. John M. Major

John Milton Major was a pioneer Bloomington-area physician who first came to the McLean County area with his family as a young man in 1835. Like Dr. Faloon, Dr. Major graduated from the Eclectic Medical College of Cincinnati (in 1849). He returned to Bloomington as a physician in 1855 after practicing medicine for six years in Quincy and then Macomb, IL:

> In the summer of 1849 the Asiatic cholera was very bad at Quincy, and the doctor had much practice with it. He only remained at Quincy one year before he removed to Macomb, where he again met the cholera, which was very wide spread. He remained at Macomb five years, when he again attended lectures in the hospital in the Ohio Medical Institute at Cincinnati. After this he returned to Bloomington, and continued his practice. In 1855, the doctor says, the cholera again broke out among our Irish friends in the forty acres.... In 1857 he bought out the interest of Dr. Wakefield in the drug store of Wakefield & Thompson, and the new firm became R. Thompson & Co. In 1867, he bought out Thompson and gave up the practice of medicine, but soon afterwards sold out the establishment to Ira Lackey & Bro. Since then Dr. Major has been engaged in trading. [Duis 1874:297]

> ...in 1855, he came to this city and began the practice of his profession and engaged in the drug trade; this he followed until [1858], when he was obliged to give it up on account of his health failing; after regaining his health, he engaged in the manufacture of metallic caskets and burial cases; this he also gave up, and, in 1868, again began the practice of his profession; his office at present is at C. Wakefield's drug store, corner Center and Jefferson sts. [Le Baron 1879:805]

So far as is known, Dr. Major never put up any of his medicinal concoctions in embossed bottles himself. This he did through arrangement with *Ira Lackey & Bro.* (see their listing).

James W. Maxwell and Co.

Ink:

(1) **THE / WESTERN / INK CO. / BLOOMINGTON / ILL.** [aqua]
   Note: *Four subtle mold varieties of this bottle have been documented (Figure 71). Three have post-mold bases (one of them is base-marked* **A.& D.H.C**), *and the fourth is a later (post-1880) cup-mold version.*

   »SB w/ post mold; schoolhouse-style ink, 2 1/2" tall.

(2) **WESTERN INK C**^{O.} **/ BLOOMINGTON / ILL.** [aqua]
   »SB w/ post mold; cylinder master ink; raised neck and shoulder rings; no pour spout; squared ring collar. *Three sizes known*: 1/2 pint (6" tall); pint (7 1/2" tall); quart (10" tall). *[Figure 72]*

According to an early letter from John Maxwell in the files of what is now *S. A. Maxwell & Co.* in Mundelein, IL, Maxwell first opened a store selling books, stationery, and wallpaper in Lacon, IL, in 1851. He moved the store to Bloomington during the 1850s. From the early Maxwell letter, the company believes that in the late 1850s "a young Abraham Lincoln...and other lawyers in state government service made the shop their after-hours headquarters to discuss politics and current events" (D&WC magazine 2001: "Paper with a Past and Future").

Maxwell's company was located on 214 N. Centre Street in Bloomington from the mid-1860s until ca. 1879–80. The business was operated by *John W. Maxwell & William Hogg* until 1870; *J. W. Maxwell, George W. Batchelder & Co.* from 1870 to 1875; and *J. W. Maxwell & Co.* (with Samuel A. Maxwell) from 1876 until 1880.

A full-page ad in the 1870–71 Bloomington CD lists Maxwell's company as "Wholesale and Retail Dealers in BOOKS, STATIONERY, Picture and Picture Frames, Blank Books, Wall Paper, Window Shades, CORNICE, &c. Also Manufacturers of **MAXWELL'S CELEBRATED INKS**." A paper label on one of

Figure 71

Figure 72

Figure 73

the company's 2 1/2" post-mold bottles (the center example in Figure 71) reads: MAXWELLS / VIOLET / INK / WESTERN / INK CO. / BLOOMINGTON / ILLS (*Figure 73*). Given the post-mold manufacturing technology of Maxwell's documented *Western Ink* bottles, his inks were probably not bottled for sale much before 1870.

A short business essay on Maxwell's company on page 31 of the Bloomington 1870–71 CD indicates that

> The business...has been established many years, about 15, and now ranks as one of the most extensive of the kind in central Illinois.... they possess facilities equal to any other house in the west,...the store is 22 feet by 100, 3 stories high and basement, each room literally groaning under its load of stock. They carry about one hundred thousand dollars worth of goods.... The trade of this house extends through all of Central and Southern Illinois *[and]* they can successfully compete with Chicago and St. Louis houses, buying as they do from the same manufacturers, while their expenses are considerably less.

The company relocated to Chicago around 1879 (ISD), but no Chicago-embossed *Western Ink* bottles are known.

At about the time of their Chicago move, the 1880 IIC (for June 1879 to June 1880) lists them still in Bloomington with $4,000 in invested capital and four employees (two adult males, one female, and a child). The company was in operation eight months a year (or perhaps closed and moved after the eighth month of the census year). Employees were paid 95¢ per day, and the annual value of product produced was estimated at $6,000.

Gustavus Mueller & Louis Stein

Soda ~ Mineral Water:

(1) **MUELLER & STEIN / BLOOMINGTON / ILL**ˢ [aqua]
»IP; shouldered blob-top-soda. *[Figure 74]*

Gustavus Mueller and Louis Stein were partners in the soda water bottling business in Bloomington from 1858 to 1861 (CDs), when both men were in their mid-20s. Since the listed bottle has an iron-pontiled base, it likely dates to 1858–59. No smooth-based Mueller & Stein bottles are known.

The 1860 USC states that Mueller and Stein were both bachelors at that time, and both lived in the household of William Hilbert, a saloonkeeper. Louis Stein was operating a brewery by 1862, which he continued to run through the 1860s and perhaps into the early 1870s (*Bloomington Pantagraph,* March 1868). Mueller went into the grocery business in Bloomington, which he ran until 1868.

Mueller & Stein are listed in the 1860 IIC as having three employees and producing 5,000 dozen bottles of soda water annually. They estimated their annual wholesale product value at $2,000 (or a little less than 3 1/2 cents per bottle of soda water).

An 1859 advertisement placed in the Bloomington *Pantagraph,* contemporary with the Mueller & Stein bottle described earlier, states:

BLOOMINGTON
SODA WATER MANUFACTORY,
CORNER OF
MADISON AND MULBERRY STs.

SODA WATER IN BOTTLES

Always on hand and delivery wagon constantly running.
FOUNTAINS CHARGED at any hour of the day.
All orders promptly attended to,

MUELLER & STEIN
May 18, 1859

Figure 74

Charles and Alfred Petterson

Soda ~ Mineral Water:

(1) **C. PETTERSON & BRO / BLOOMINGTON / ILL //** [aqua]
(heel): **A. & D. H. C**

»SB; oval-blob-top soda. *[Figure 75]*

(2) **C. PETTERSON & BRO. / BLOOMINGTON / ILL //** [aqua]
(heel): **C & I**

»SB (heel mold?); oval-blob-top soda. *[Figure 76]*

The "Petterson" spelling on these bottles seems to be in error. Peterson and Brother are listed as a soda water manufacturing company in the January 1874, CAR. C. Peterson and Co. is listed in the 1875 ISD. Alfred Peterson (the "brother") is listed in the 1870 USC for Bloomington as a soda water manufacturer. Despite the company name variations, Bloomington CDs indicate that the Peterson soda water business was operated by Alfred during its brief existence there (1870–74). During this time, Charles lived in Springfield 45 miles to the south, where he was also a partner with John Johnson in a soda water bottling business there (*see Johnson & Peterson Springfield entry*).

There was only one known soda bottling factory in Bloomington, started by Mueller & Stein in the late 1850s (*see their listing*). Mueller & Stein used the facility ca. 1858–61, and S. M. Hickey likely used it ca. 1863–64. J. E.'s *Bottling House* may have operated from there during the later 1860s (*see Bottling House listing, this chapter*), Petterson & Bro. used it ca. 1870–74, and Peter Kreis used it from ca. 1976 until the late 1880s (*see his listing, this chapter*).

Figure 75

William Schausten

Ale / Beer:

(1) **W^M SCHAUSTEN /** [aqua]
BLOOMINGTON / ILLS //
(heel): **C & I**

»SB; quart, 2-piece mold ale; ring-neck tapered-collar; 9 1/2" tall. *[Figure 77]*

(2a) **CINCINNATI BEER / WM. SCHAUSTEN /** [amber]
BLOOMINGTON. ILL. //
(heel): **THIS BOTTLE / NOT TO / BE SOLD**
(base, in post-mold circle): **RBB** (flat) / **D. S. G. Co.** (arched)

»SB w/ post mold; tall quart beer; shouldered blob top; 11 3/4" tall. *[Figure 78]*

(2b) **CINCINNATI BEER / WM. SCHAUSTEN /** [yellow-amber, aqua]
BLOOMINGTON. ILL. //
(base, in post-mold circle): **D. S. G. Co.** (arched) *[Figure 79]*
Note: on some examples of this bottle style, the side embossing is upside-down [see Figure 79].

»SB w/ post mold; tall quart beer; shouldered blob top; 11 3/4" tall.

Figure 76

Figure 77

Figure 78

Figure 79

William Schausten was listed as a 37-year-old Bloomington saloon owner in the 1870 USC. Schausten is first listed (as "William Schauffhausten") in the 1870–71 Bloomington CD as running a saloon and retail bottled wine and liquor business at Centre and Washington. The 1872–73 CD also lists his business in some detail: "saloon, liquor, wines, beer, cigars, refreshments, reading room, n.w. cor. Washington and Centre, in basement."

During the mid-1870s, Schausten (with the shortened spelling after the first CD listing) advertised his bottled ale and Cincinnati beer, as well as wines and liquors. A quarter-page ad in the 1875–76 Bloomington CD (p. 487) lists Schausten as a dealer in "Pure Wines & Liquors, Genuine Cincinnati Lager Beer, Wholesale and Retail, 206 North Centre Street."

He was listed in the Bloomington CDs as a saloonkeeper throughout the 1870s, up to 1881. From 1878 onward, his establishment was listed as a "saloon and restaurant."

Cyrenius [and Zera] Wakefield and Co.

Medicine:

(A1) **WAKEFIELD'S / BLACKBERRY / BALSAM // // // //**   [aqua]

»OP; rectangular with 3 sunken panels; rolled lip; 5 1/4" tall. *[Figure 80]*

[Note: Odell 2007:361 indicates there is a larger size (7" tall), but this cannot be confirmed from local collections.]

(A2) **WAKEFIELD'S / BLACKBERRY / BALSAM // // // //**   [aqua]
(*base*): **A. & D. H. C**

»SB w/ keyed hinge mold; beveled-edge rectangle; 3 sunken panels; double-ring collar; 5 1/4" tall. *[Figure 81]*

(A3a) **WAKEFIELD'S / BLACKBERRY / BALSAM // // // //**   [aqua]
(*base*): **N** *with* **5-pointed Newark Star** (*Ohio glass co.*)

»SB w/ keyed hinge mold; beveled-edge rectangle; 3 sunken panels; thick flared-ring collar; 5 1/4" tall. *[Figure 82]*

(A3b) **WAKEFIELD'S / BLACKBERRY / BALSAM // // // //**   [aqua]

»SB w/ keyed hinge mold; beveled-edge rectangle; 3 sunken panels; thin flared-ring collar; 5 1/4" tall. *[Figure 83]*

(A4) **WAKEFIELD'S / BLACK BERRY / BALSAM // // // //**   [aqua]
(*base*): **I.G.CO** (*late 1870s variant*)

»SB; beveled-edge rectangle; 3 sunken panels; narrow thickened-ring collar; 5 1/4" tall. *[Figure 84]*

(A5) **WAKEFIELD'S / BLACK BERRY / BALSAMCOMPOUND // // // //**   [aqua]

»SB; beveled-edge rectangle; 3 sunken panels; thick flared-ring collar; 5 1/4" tall. *[Figure 85]*

(B1) **WAKEFIELD'S / COUGH SYRUP // // // //**   [aqua]

»IP; beveled-edge rectangle; 3 sunken panels; thickened flared lip; 3 sizes?: 4 1/2", 5 1/4", and 6 1/2" tall. *[Figure 86]*

(B2) **WAKEFIELD'S / COUGH SYRUP // // // //**   [aqua]

»OP; beveled-edge rectangle; 3 sunken panels; rolled lip; 3 sizes?: 4 1/2", 5 1/4", and 6 1/2" tall. *[Figure 87]*

(B3) **WAKEFIELD'S / COUGH SYRUP // // // //**   [aqua]

»SB; beveled-edge rectangle; 3 sunken panels; double-ring collar; 3 sizes?: 4 1/2", 5 1/4", and 6 1/2" tall. *[Figure 88]*

Figure 80

Figure 81

Figure 82

Figure 83

Figure 84

Figure 85

Figure 88

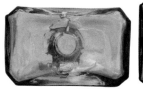

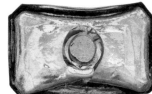

Figure 86 **Figure 87**

(B4) **WAKEFIELD'S / COUGH SYRUP // // // //** [aqua]

»SB; beveled-edge rectangle; 3 sunken panels; broad flared-ring collar; 3 sizes?: 41/2", 51/4", and 61/2" tall. *[Figure 89]*

(C1) **WAKEFIELD'S / EGYPTIAN / LINIMENT** [aqua]

»OP w/ straight-line hinge mold; cylinder; rolled lip; 41/8" tall x 11/4" diameter. *[Figure 90]*

(C2a) **WAKEFIELD'S / EGYPTIAN / LINIMENT** [aqua]

»SB w/ keyed hinge mold; cylinder; thickened-ring lip; 4" tall x 11/4" diameter. *[Figure 91, left]*

(C2b) **WAKEFIELD'S / EGYPTIAN / LINIMENT** [aqua]

»SB w/ heel mold; cylinder; flared-ring lip; 4" tall x 11/4" diameter. *[Figure 91, right]*

(C3) **WAKEFIELD'S / EGYPTIAN / LINIMENT** [aqua]

»SB w/ post mold; wider-bodied cylinder; flared-ring lip; 5" tall x 2" diameter. *[Figure 92]*

Figure 89

Figure 90

Figure 91

Figure 92

(D1) **WAKEFIELDS / FEVER / SPECIFIC // // // //** [aqua]
»OP; beveled-edge rectangle; tapered collar; 4 flat panels; 5 7/8" tall. *[Figure 93]*

(D2) **WAKEFIELDS / FEVER / SPECIFIC // // // //** [aqua]
»IP; beveled-edge rectangle; flared-ring collar; 3 sunken panels; 5 1/2" tall. *[Figure 94]*

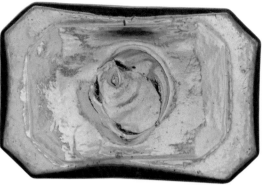

Figure 93

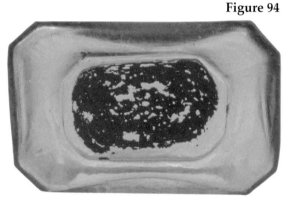

Figure 94

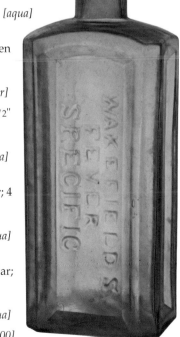

(D3) **WAKEFIELDS / FEVER / SPECIFIC // // // //**   *[aqua]*

»SB w/ keyed hinge mold; beveled-edge rectangle; flared-ring collar; 3 sunken panels; 5 1/2" tall. *[Figure 95]*

(D4) **WAKEFIELD'S / FEVER / SPECIFIC // // // //**   *[aqua]*
(base): **IGCº**

»SB w/ post mold; beveled-edge rectangle; broad, thickened-ring collar; 3 sunken panels; 6" tall. *[Figure 96]*

(E) **WAKEFIELD'S GOLDEN OINTMENT**   *[clear]*

»SB w/ heel mold; shoulder-embossed cylinder; rolled flared collar; 2 3/4" tall x 1 1/2" diameter. *[Figure 97]*

(F1) (side): **WAKEFIELD'S // MAGIC / PAIN CURE //**   *[aqua]*
(side): **BLOOMINGTON ILL // //**

»SB w/ post mold; beveled-edge rectangle; long neck; wide thickened-ring collar; 4 sunken panels; serifed embossed lettering; 6" tall. *[Figure 98]*

(F2) (side): **WAKEFIELD'S // MAGIC / PAIN CURE //**   *[aqua]*
(side): **BLOOMINGTON ILL. // //**

»SB (heel mold?); beveled-edge rectangle; long neck; narrow thickened-ring collar; 4 sunken panels; plain embossed lettering; 6" tall. *[Figure 99]*

(G) **WAKEFIELD'S / NERVE & BONE / LINIMENT**   *[aqua]*

»SB; shouldered cylinder; flared square-ring collar; 3" tall x 1 1/4" diameter. *[Figure 100]*

Figure 95

Figure 96

Figure 97

Figure 98

Figure 99

Figure 100

Figure 101

Figure 102

(H1) **WAKEFIELD'S / STRENGTHENING / BITTERS**　　　　　　　　　　　　　　　*[aqua]*
　　　Note: a.k.a. "Wakefield's Wine Bitters," beginning in the 1860s. (paper label—see Ring 1980:476)

　　»SB w/ keyed hinge mold; beveled-edge rectangle; 4 sunken panels; short neck; wide square-ring collar; 7¼" tall. *[Figure 101, left; Figure 102, top]*

(H2) **WAKEFIELD'S / STRENGTHENING / BITTERS**　　　　　　　　　　　　　　　*[aqua]*
　　　Note: a.k.a. "Wakefield's Wine Bitters," beginning in the 1860s. (paper label—see Ring 1980:476)

　　»SB w/ keyed hinge mold; beveled-edge rectangle; 4 sunken panels; long neck; tapered collar; 8" tall. *[Figure 101, right; Figure 102, bottom]*

(I) **BROWENS RAT KILLER / C.WAKEFIELD & Co**　　　　　　　　　　　　　　　*[aqua]*

　　»SB w/ keyed hinge mold; wide-mouth shouldered cylinder; rounded flared-ring collar; 3" tall x 1⅝" diameter. *[Figure 103]*

(J) **WAKEFIELD & CO / AGUE & FEVER / PILLS**　　　　　　　　　　　　　　　*[clear]*

　　»SB w/ post mold; cylinder with 8 vertical panels; vertically embossed on 3 adjacent panels; weak shoulder and plain neck w/ ground sheared lip; ⅞" diameter x 2¼" tall. *[Figure 104]*

───────────── ❖ ─────────────

　　Cyrenius Wakefield was born in New York in 1815 and came to the Bloomington area in 1837 at the age of 22, where he first taught school and farmed. His older brother, Dr. Zera Wakefield, came up from Arkansas and joined Cyrenius in the Bloomington area in 1846–47. In the summer of 1847 brothers Cyrenius and Zera Wakefield opened up a farmer's store near Marion in DeWitt County. Cyrenius ran the store while Zera established a new medical practice in the surrounding countryside. The recipes for the fever and ague medicines that Zera Wakefield used eventually

Figure 103

became a patent-medicine dynasty in Bloomington throughout the latter half of the nineteenth and early twentieth century.

[Their country store] was quite successful under the management of Cyrenius, while Dr. Z. Wakefield entered immediately into the practice of his profession. After about two months, the miasmatic fevers commenced and were of unusual prevalence and severity. With his accumulated skill in subduing the violent congestive fevers of the South, he was able to break up the most severe cases here in a few hours. His wonderful success created a great sensation, and his fame soon extended fifty miles around. With the aid of a driver and a change of horses, he was quite unable to fill all of the demands upon him. When they could not get him, they wanted some of his medicine, and this necessitated keeping it prepared, with directions, to supply the demand. This soon necessitated printed directions and a uniformity of bottles.... This condition of events changed their country store into a medicine laboratory. In June, 1848, after two years of practice here, Dr. Z. Wakefield took a violent congestion of the lungs, which caused his death in thirty-six hours. This was a heavy stroke on Cyrenius,...*[but]* demand for these medicines was great, and, much of his means being invested in it, he wisely bought out his brothers interest *[from his widow]* and continued to extend their usefulness.

He sold his farm and moved to Bloomington in February, 1850, in order to get better postal and express facilities. He applied himself diligently to the study of medicine and pharmacy to prepare himself for more extensive work, and here gained the title of Doctor. During his first few years in Bloomington, he did quite an extensive drug business in company with Robert Thompson. His medicine laboratory was then in the rear of the drug store. *[In the great fire of October 16, 1855, he lost, in addition to his half of the drug store and building, his medicine factory and a three-story brick building, with small insurance.]* In 1856, he built a brick laboratory near his dwelling-house, to which subsequent additions have given its present extensive proportions *[see Figures 105 and 106]*. In 1857, he retired from the drug trade, and gave his whole attention to his medicine business. [Le Baron 1879:826–827; see also Duis 1874:354–357 and Chapman Brothers 1887a:201–202]

Wakefield's unpublished autobiography (1882 *McLean County Historical Society* manuscript; see also unpublished "Biographical Sketch of Dr. Cyrenius Wakefield" by Homer Wakefield, ca. 1900) indicates he began his drugstore partnership with Thompson in 1852. According to the 1855–56 Bloomington CD, the *Wakefield & Thompson* drug store was located on the south side of the square. A full-page ad in the CD indicates that in addition to patent medicines and standard druggist stock, "**Wines and Liquors for medicinal use**" and the "**Best Alton Stone Ware, &c**" were offered for sale.

Figure 104

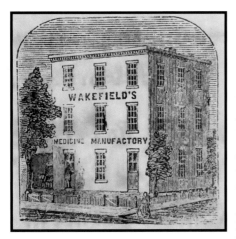

Figure 105. Early image of Wakefield's medicine factory published in his second annual *Western Farmer's Almanac* in 1862.

Although Zera Wakefield died shortly after their patent medicine production business began, Cyrenius went on to grow the company into one of the largest and most renowned patent-medicine manufacturing and distribution firms in the Midwest between the time of his arrival in Bloomington in 1850 and his death in February 1884. Beginning in 1860–61, Wakefield produced an annual advertising publication he called the *Western Farmer's Almanac*. In its heyday during the 1870s, more than 1,500,000 copies of this publication were distributed annually throughout the Midwest:

> They are sold largely in the states of Illinois, Indiana, Missouri, Kansas, Nebraska and Iowa, while there is a good demand in all the Western and Southwestern states. He has a team, with a fine wagon in six of these states, and thus keeps a watch over the territory through salaried agents, while he has over six thousand local agents, mostly druggists and dealers, who sell his medicines on commission. He employs from twenty-five to fifty hands, according to the season. He has four printing-presses, run by steam-power, by which he prepares his advertising matter. In 1860, he got up 100,000 almanacs for his agents to circulate, and he now sends out 1,500,000 annually. He consumes nearly fifty tons of printing paper each year.
>
> "Wakefield's Almanac"…is printed in the English, German, Norwegian, and Swedish languages. [Burnham 1879:88]

According to the 1870–71 CD, Wakefield employed "thirty-five hands" and that year had $100,000 invested in the business. In 1874, Duis (1874:357) noted that "He gives employment to forty persons in his medicine business (one-half of whom are females) and his annual sales amount to $100,000." In the 1880 IIC, the Wakefield company is listed as a manufacturer of "patent medicines & compounds" with $120,000 invested in the business enterprise. The company employed approximately 60 employees who were paid a total of $18,000 annually and who produced approximately $84,000 worth of medicines and compounds that year.

During the mid-1870s, at the height of his company's early success under his leadership, one of Wakefield's biographers attempted to paint a "word picture" of his appearance (*see Figure 107*):

> Dr. Wakefield is about five feet nine inches in height, is well proportioned, and has a wiry, good constitution. His features are regular, and his eyes are small but sharp and penetrating. His hair and beard are dark and full, but now are turning gray. His whole appearance is that of a careful, calculating, straightforward, energetic business man. [Duis 1874:358]

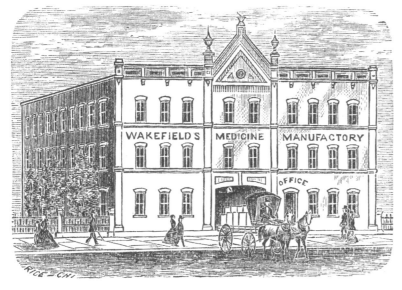

Figure 106. Later image of Wakefield's expanded medicine factory, used on his company's letterhead during the 1870s.

The embossed bottles listed earlier are limited to those varieties that were used between ca. 1850 and 1880, and which fall within the scope of the present study. An early Wakefield printed order form Wakefield sent to Springfield druggists *Birchall & Owens* on July 21, 1851 (curated by the Illinois State Historical Library) lists the following Wakefield medicines for which embossed pontiled bottles are known: *Blackberry Balsam; Cough Syrup; Egyptian Liniment;* and *Fever Specific*. Also listed are *Nerve and Bone Liniment* and *Strengthening Bitters* (for which only post-1860 smooth-base embossed bottles are known) and *Worm Medicine* (for which no embossed bottles have been found).

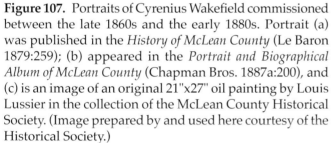

Figure 107. Portraits of Cyrenius Wakefield commissioned between the late 1860s and the early 1880s. Portrait (a) was published in the *History of McLean County* (Le Baron 1879:259); (b) appeared in the *Portrait and Biographical Album of McLean County* (Chapman Bros. 1887a:200), and (c) is an image of an original 21"x27" oil painting by Louis Lussier in the collection of the McLean County Historical Society. (Image prepared by and used here courtesy of the Historical Society.)

Wakefield's earliest and most enduringly popular products, in addition to Zera Wakefield's fever and ague medicine that started the company on its road to success (*Wakefield's Fever Specific*), included *Wakefield's Blackberry Balsam, Wakefield's Cough Syrup,* and *Wakefield's Egyptian Liniment.*

According to the second (1862) issue (1862, page 3) of his *Western Farmer's Almanac*, his bottled product line and per-bottle retail prices that year were as follows:

Wakefield's Fever Specific	$1.00
Wakefield's Cough Syrup (*small*)	.25
Wakefield's Cough Syrup (*large*)	.50
Wakefield's Blackberry Balsam	.25
Wakefield's Egyptian Liniment (*small*)	.25
Wakefield's Egyptian Liniment (*large*)	.50
Wakefield's Nerve & Bone Liniment	.25
Wakefield's Strengthening Bitters	.50

According to the 1870–71 Bloomington CD, p. 34, several new medicines were added during the 1860s ("During the past ten years he has found it necessary to add Worm Lozenges, Eye Salve, Liver Pills, **Wine Bitters**, and **Bowen's Rat Killer**, in the preparation of which he has been remarkably successful").

During 1868, Cyrenius Wakefield took his son Oscar and his son-in-law A. S. Eddy into the company as partners and changed the corporate name to C. *Wakefield & Co.* So the two *Wakefield & Co.* embossed bottles listed above ("H" and "I") would have been produced after that time. Both products—Wakefield's Ague and Fever Pills and Browen's Rat Killer were first advertised in 1869 as new to the Wakefield line "during the past year" (*see summary product listing*).

An approximate chronology of the appearance and use of his embossed medicines is as follows:

Pontiled Bottles

Late 1840s: *Wakefield's Fever Specific*

Early 1850s: *Wakefield's Blackberry Balsam*
Wakefield's Cough Syrup
Wakefield's Egyptian Liniment

Smooth-Base Bottles
All of the Above, plus—

Civil War years (1860–65):
Wakefield's Nerve & Bone Liniment (until 1870)
Wakefield's Strengthening Bitters

1868: *Strengthening Bitters* becomes *Wine Bitters*
(Note: embossed bottle unchanged)
Browen's Rat Killer

1871: *Wakefield's Magic Pain Cure*

1878: *Wakefield's Golden Liniment*

In his early almanacs, Wakefield advertised a wide range of curative powers for his line of bottled medicines:

Blackberry Balsam—"It speedily cures Diarrhea, Cholera Morbus, Flux, Dysentery, Camp Diarrhea, and attacks of Cholera. It regulates the bowels of children while teething, cures Colic in children or adults, and is used successfully for Flatulency and Sour Stomach" (1869:15).

"In more violent cases of this character, make a strong decoction of Oak Bark, and to one gill of it, add one tablespoonful of the Balsam, administer by injection, and continue to give the Balsam freely, and relief will most assuredly soon follow.... In attacks of Cholera, give large doses of the

Balsam with equal quantity of the best Brandy, and repeat often until relief is obtained; keep the feet warm, and wet compresses over the stomach and bowels" (1862:17).

Browen's Rat Killer—First produced by the Wakefield Company in 1868, and first advertised in the 1869 *Almanac*.

Cough Syrup—"...for Coughs, Colds, Pain and tightness in the lungs, Typhoid and Lung Fever, Inflammatory attacks, *and cases where the vital forces of the system are unequally distributed, causing an undue amount of heat in one organ, and coldness in another*" (1862:3).

"...when used freely it will combat Typhoid Fever, sudden Colds, Engorgements of the system, Measles and Whooping Cough more effectually than any remedy before the public. It relieves engorgements by equalizing and harmonizing the action of the circulating fluids and stimulating the torpid organs to healthy action" (1869:3).

Egyptian Liniment—"This is the most stimulating and penetrating external curative agent before the public. It warms up the indolent nerves and muscles to vigorous action, softens the callous formations, relieves pain, cures lameness and strengthens the weaker parts. It has a good reputation for curing Sprains, Stiffness of Joints, Soreness in the Liver, Spleen or Kidneys, Weakness in the Back, Frosted parts, and also for Sweeney, Spavin, windgalls, chafes, and all external diseases about horses (1869:27).

Fever Specific—"... a certain remedy in all cases of Bilious Fever, Ague and Fever, Dumb Ague, Chills and Fever, Tiedoloreaugh and many cases of Nervous Debility (1862:9).

"This old standard Western Remedy has been successfully battling down Ague and Fever, congestive, bilious, and all forms of fevers of this humid country for the past twenty years. It routs the enemy in a few hours and never experienced a defeat.... It is a purely vegetable compound, contains no poisons or deleterious substances, and is safe for infants or females in any condition" (1869:17).

Ague and Fever Pills—"We are also manufacturing *Ague and Fever Pills*, having all the virtue of the Fever Specific and more acceptable to the stomach. They are sugar coated and put up in glass vials (1869:1a).

Golden Ointment—First advertised under the Wakefield name in the 1878 *Almanac*: "We now issue a new remedy under the name of *Wakefield's Golden Ointment*, the effects of which, we have thoroughly tested during the last few years.... It is the same preparation, with some important additions, that has for several years been used in Kansas and Nebraska, under the name *Water's Golden Ointment*.... We confidently recommend this article for Burns, Piles, Felons, Old Putrid Sores and all of the numerous uses recommended in the directions that go with it *[including Corns, Sore Throat, Diptheria, Fever Sores, Chaps, Neuralgia, Scald Head, Lameness, Cold Skin, etc."].*

Golden Ointment in not included in the list of embossed Wakefield bottles because the only documented embossed style appears (technologically) to postdate 1880.

Magic Pain Cure—The *Pain Cure* was first produced by Wakefield in 1871, and was first advertised in the 1872 *Almanac*: "During the past year I have issued a new remedy, (Magic Pain Cure), a volatile and stimulating compound adapted to internal and external use, and very powerful in relieving pains, aches, and local torpid conditions.... I expect it will supercede the use of the **EGYPTIAN LINIMENT** in most cases on human flesh. On horses I expect the **LINIMENT** will continue to be extensively used" (1872:1).

Nerve and Bone Liniment—"Is very penetrating, softening and healing, but less stimulating than the Egyptian Liniment. It is a superior remedy for burns or chilblains, weak joints, &c." (1869:23). *Note: No longer advertised by 1871.*

Strengthening/Wine Bitters—"...demand for it has doubled within the last eighteen months. Since June, 1868, we have added to its former recipe two valuable blood alteratives...Its strengthening and regulative effects on the digestive organs are well known. It is not designed for a beverage. There is no cheap rum or whisky in it. Pure Native Wine and Neutral Spirits are used in extracting the medical properties from the drugs. It is purely vegetable and free from all deleterious substances.

SYMPTOMS REQUIRING THE USE OF Wakefield's Strengthening Wine Bitters: Dyspepsia, Kidney Disease, Loss of appetite, Rheumatism, Constipation, Jaundice, Heartburn, Enlargement of the Spleen, Waterbrash, Declining of Health, Acidity of the Stomach, Pain or heaviness in right side, Dull Headache, Fullness of blood in the head, Swimming in the Head, Nausea of the stomach, Numbness of Limbs, Aching in the bones, Frequent moveable Pains, Slow Fever, Fluttering at Pit of Stomach, Dimness of Vision, Wasting of flesh and strength, General weakness, Scrofula, Constant Imaginary Evils, and Great Depression of Spirits" (1869:10–11).

Consumption—"Persons declining in health and thinking they are in any way threatened with this disease should use the **Cough Syrup** and **Wine Bitters** together, taking the **Bitters** before meals and the **Cough Syrup** between meals. The **Cough Syrup** relaxes the tissues and starts the impurities towards the surface, while the **Bitters** regulate the stomach and liver, cleanse the blood and invigorate the system so as to enable it to overcome the disease" (1869:5).

After Cyrenius died in 1884, "two sons, Oscar and Homer, and a grandson, Louis Eddy, kept the business going until 1969. Black Berry Balsam, their best seller, was produced all 122 years. It was for summer complaints and contained roots of black berry, rhubarb, and culvers, prickly ash bark, catechu gum, potassium carbonate, ginger, cranesbill, erigeron, sodium benzyl succinate, camphor, and alcohol, 12%" (Cannon 1995:46).

Dr. Zera Waters

Like Dr. Faloon and Dr. Major, Dr. Waters was yet another early Bloomington physician who trained at and graduated from the Eclectic Medical Institute of Cincinnati (*see Faloon and Major listings, this chapter*).

Early in life, he decided on the profession of medicine, and read the allopathic system, and, in 1859, commenced practice in Van Buren, Ark., where he continued for three years; in July, 1861, he returned to Bloomington, and, in October of the same year, entered the Eclectic Medical Institute of Cincinnati, where he graduated in the spring of 1862. He then enlisted in the army...and was soon promoted to the position of Assistant Surgeon...he returned with his regiment to Springfield, where they were discharged in September, 1862; in October, 1862, he commenced the practice of Medicine in Bloomington, Ill., where he has continued to the present time. Dr. Waters is the author of the well-known "Waters' Family Medicines;" also the inventor of the "Waters' Abdominal and Uterine Supporter." He is a member of the Illinois State Eclectic Medical Society, and of the U.S. Eclectic Medical Society. [Le Baron 1879:827–828]

According to historical/biographical sketches and Bloomington CDs researched by Anthony Green (personal communications 2009), Zera Waters moved with his family from New York to DeWitt County, IL, in 1844 (when he was 16), settling just east of the Bloomington area. He and his two brothers (Henry and Orin) worked for a time in the mid-1850s as traveling salesmen for the *Wakefield Family Medicine Co.* in Bloomington before Zera Waters embarked on his medical career. Henry had a long career selling Wakefield's patent medicines. By the late 1850s, he moved to Lawrence, Kansas, and took over the "Western Branch" of Wakefield's patent-medicine distribution network. In Kansas, he developed patent medicine varieties for Wakefield, sold in the West as *Waters' Renovator* and *Waters' Golden Ointment.* Wakefield eventually produced the ointment in Illinois as well, as *Wakefield's Golden Ointment* (*see the Wakefield medicine chronology in his listing*).

Although Dr. Zera Waters established his own patent medicine production after the Civil War as well (*see Waters & Howe discussion under B. S. Howe listing, this chapter*), he is not known to have used embossed bottles for his products. Embossed Waters' medicines seem to have first appeared after Henry S. Woodard bought out Waters' interest in the company in 1872, and Woodard & Howe started producing the medicines (*see the Woodard and Woodard & Howe listings, this chapter*). After the *Woodard & Howe* era (together during 1872–1877, and then individually during the later nineteenth century), Waters' popular medicines continued in production through C. B. Castle & Co. As a measure of the enduring popularity of Waters' medicines, an unembossed pharmacy bottle produced by the Illinois Glass Company for O. A. T. Stewart of Normal, IL, retains a paper label that reads:

DR. Z. WATERS'
System Builder and Blood Purifier
SHEA—WA
[image of an Indian on a bluff-top overlook]

Nature's Laboratory, the great treasure and secret to health, that makes new, pre blood and strong nerves and banishes diseases—a true friend to Indian Life. From the cradle to the grave he didn't know what sickness was, the only thing assailing him was accident and death in battle or old age. Instinct taught the Indian to find in their woods and plains, the secret herb eliminant to poison from the blood which made him strong, which came to him like a dream from the Great Spirit. This great remedy contains the Balm, Flowers, Roots, Barks, Herbs and Berries, the healing of nature.
12 1/2 per cent alcohol by volume

PRICE $2.00
SERIAL NO. 14817
MANUFACTURED BY
O. A. T. STEWART
710 S. MAIN ST. NORMAL, ILL.

Edwin M. Wells

Medicine:

(1) **WELLS / GENUINE / LINIMENT // BLOOMINGTON / ILLS**   [aqua]
 »OP; shouldered cylinder; rolled lip; 3 7/8" tall x 1 1/4" diameter. *[Figure 108]*

(2) **WELLS // PECTORAL SYRUP / OF / WILD CHERRY // BLOOMINGTON ILL //**   [aqua]
 GREAT WESTERN FAMILY / MEDICINE DEPOT
 »IP; beveled-edge rectangle; 4 sunken panels; tapered collar; 8 3/4" tall. *[Figure 109]*

Figure 108

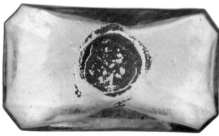

Figure 109

During the 1850s, Edwin M. Wells' patent-medicine company operated in Bloomington, and they later expanded their manufacturing outlets to Chicago and St. Louis. E. M. Wells first opened a drug store in Bloomington, at the corner of Main and Front streets, in 1851 (Deiss 1981:98). When he established his "Great Western Medical Depot" at Bloomington in the 1851–1853 period, he packaged **WELLS GENUINE LINIMENT** in small open-pontilled rolled-lip bottles. In the early 1850s, he appears to have changed the product name to **WELLS GERMAN LINIMENT** (the earliest ad we have located thus far for his "German" Liniment was in the December 1, 1852, *Bloomington Intelligencer*). His December 1852 ad is primarily a listing of national-brand "family medicines"—and famous Illinois brands such as *Sloan's* (Chicago), *Farrell's* (Peoria), and *Hamilton's* (Jacksonville) that he was sales agent for—but he also mentions his *German Liniment* and a *Blackberry Cordial, a German Condition Powder,* and *Great Western Vegetable Pills* that may have been his own products.

By March 1853, E. M. Wells' *Intelligencer* ads were focused on his own medicine products—including his **German Liniment** ("Recipe obtained from a German farrier...Burns, bruises, sprains, rheumatism, stiff joints, swellings, pains of all kinds, frozen limbs, galls, Sweeney, cuts, old sores, chapped hands and injuries, are rapidly cured.... In connection with the German Condition Powders, a cure is absolutely certain in all diseases of horses and cattle, if a persevering course is pursued"), his **Great Western Pills** ("For All Bilious Diseases. Impurity of the blood, inflammations, dropsy, cholera morbus, indigestion, costiveness, liver complaint, measles, scarlet fever, salt rheum, jaundice, dyspepsia, sick headache, fevers of all kinds, pains in the head, back, side, limbs, and breast, erysipelas, all diseases of the skin, colds, scrofula, croup, lowness of spirits, tumors, female complaints, spitting of blood, pleurisy, &c."), and his newly introduced **Pectoral Syrup of Wild Cherry** ("For the cure of Incipient Consumption, Bronchitis, Coughs,

Colds, Hoarseness, Asthma, Whooping Cough, Croup, Spitting of Blood, Difficult Breathing, Pain or Soreness of the Breast, throat or lungs").

He was joined in his Bloomington patent medicine business at about this time (for a short period) by his brother, Alvin T. Wells, who later returned to help set up and run the Chicago branch of the business. Wells continued to run his drug store in Bloomington and sell his medicines there through 1854. But his newspaper ads for February 1855 indicated he had taken on a new junior partner in the business: George W. Lichtenthaler—and by late spring of that year, Wells' ads in the 1855–56 Bloomington CD (p. 76) indicated that Lichtenthaler was now managing the Bloomington drug store, while Wells himself had gone down to St. Louis to establish his patent medicine operation there:

E. M. WELLS,
Proprietor and Manufacturer of
WELLS GREAT WESTERN FAMILY MEDICINES,
No. 38 Second St., St. Louis.

As part his Chicago/St. Louis expansion in the middle and late-1850s, Wells packaged his **WELLS GERMAN LINIMENT** in larger-sized, tapered-collar containers, separately embossed **CHICAGO** and **ST. LOUIS** (*Figure 110*). A. T. Wells managed the Chicago branch of the firm, which was established during 1857 (Burggraaf 2003; *see Wells Chicago entry below*); he may have helped run the short-lived St. Louis branch. Embossed bottles from all three cities are known. All are pontiled and were made between 1850 and 1860.

In the late 1850s, Edwin Wells moved his family and his E. M. Wells patent-medicine operation to Chicago (*see Wells Chicago listing*). His first Chicago listing was in the 1858 CD. But he died on April 11, 1859 (*Bloomington Daily Pantagraph*, April 20, 1859, p. 3), and his brother, Alvin T. Wells, took over the company for its final three or four years of operation in Chicago. Since all known Wells' bottles are pontilled, the company may have just been selling off existing stock after 1859.

Perhaps coincidentally, the June 1, 1859, *Illinois Statesman* announced that the old Wells drug store—now managed by G. W. Lichtenthaler alone—had been purchased by Dr. E. Thomas.

From the last almanac that Edwin Wells produced while he was in Chicago (Wells 1860), his *German Liniment* (embossed "*Genuine*" *Liniment* on the earlier Bloomington bottles) was touted as a cure for "the Worst Old Sores, Wounds, Poll-Evil, Sweneys, Snake-Bite, Spavin, Rheumatism, Frozen Limbs, Burns, Scalds, Colic, &c." His *Pectoral Syrup of Wild Cherry* was sold as a cure for consumption. According to a discussion in the *Chicago Daily Times* that Wells reproduced in his 1860 *Almanac* (a publication he had issued annually since 1854 to advertise his patent medicine products):

> Some six years since, E. M. Wells, of Chicago, thought a remedy could be prepared that would not only cure diseases of the lungs, but restore the general health, and thus remove the cause. He kept this object steadily in view, studied and experimented day and night, visited the Eastern cities several times, and consulted the best physicians, chemists, and pharmaceutists, and when he was satisfied the object was attained, if it could be, **commenced a course of treatment on himself for what had been considered hereditary Consumption**. Every symptom was removed, the general health restored, and full confidence established in the remedy. It was then dispensed as a prescription for about a year, when it became exceedingly popular, and was brought out under the name of **Wells' Pectoral Syrup of Wild Cherry**, and is the most popular cure for lung diseases ever introduced.

Figure 110

Perhaps, considering Wells' death as the 1860 *Almanac* was being prepared, the cure was not as effective as he had hoped.

He also produced and sold *Wells' Great Western Vegetable Pills*, *Wells' German Condition Powder*, and (as wholesale agent) *Denie's Electro-Magnetic Plaster* (in a "tin box").

Western Ink Co.

See the *J. W. Maxwell & Co.* listing, this chapter.

Henry S. Woodard

Medicine:

(1) (side): DR H. S. WOODARD //   [aqua]
(side): **BLOOMINGTON ILLS**

»SB w/ keyed hinge mold; rectangular; 4 sunken panels; slightly tapered collar; 6 1/4" tall. [Figure 111]

Based on the presence of a keyed hinge mold on the base of this bottle, it likely dates to the 1860s. This suggests that Dr. Woodard was bottling medicines on his own prior to the formation of the *Woodard & Howe* patent-medicine partnership in the early 1870s.

Dr. H. S. Woodard was not listed in the 1866 CD, but after that he was listed alone (but perhaps as an employee of the Waters & Howe company—*see 1878 Pantagraph ad later in the entry*) in the Bloomington CDs for at least four years—from 1868 until 1872—at which time the Woodard & Howe partnership was formed. Early *Woodard & Howe* ads in the 1872–73 and 1873–74 Bloomington CDs described the company's origin and listed their principal patent-medicine products:

DRS. WOODARD & HOWE
Successors to Dr. Z. Waters & Co.
Proprs. & Manufs. Of Waters' Family Medicines

NE PLUS ULTRA LINIMENT
The Great Pain Eradicator.

CATHARTIC and LIVER PILLS
The Best Pills in use.

THE RENOVATOR!
For Itch and all Humors.

RASPBERRY BALSAM
For Dysentery, Diarrhoea, &c., &c.

HONEY BALSAM
For Coughs, Colds, and all Lung Diseases.

EGYPTIAN SALVE
For Pains, Burns, Sores, &c., &c.

THE BEST PREPARATIONS IN USE.
Warranted Genuine, and to give Satisfaction in all cases for which they are recommended.
Thousands are using them, and attest their virtues.
COR. OF NORTH AND MADISON STREETS.

Figure 111

The *"Dr. Z. Waters & Co."* mentioned in the ads was an earlier partnership between *Waters & Howe* (*see their separate entries, this chapter*). Woodard bought out senior partner Waters, and he and Howe continued to produce and sell the Waters medicine line. The Woodard & Howe partnership lasted until 1876–77 (*see Woodard & Howe entry, this chapter*), and from ensuing conflicts over medicine-brand production rights (*described later in the entry*), the break-up does not seem to have been a friendly one.

From 1878 on, both men continued to manufacture and bottle patent medicine products on their own. As the purchaser of Waters' earlier business interests, Dr. Woodard continued to sell Waters name-brand medicines. His medicine-bottle paper labels were reproduced on the backs of scenic trade cards distributed by the Woodard Company. He also developed a medicine of his own: *Woodard's Instant Relief.*

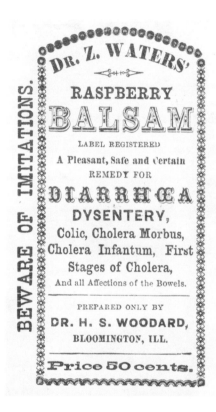
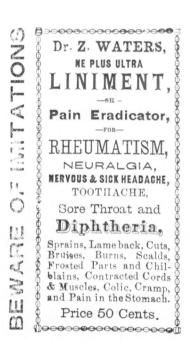

Apparently, Dr. Howe tried to do the same by using the product names but omitting Waters' and Woodard's names on the labels and advertising (*see B. S. Howe entry, this chapter*). This prompted Woodard to run a large ad in the 1878 *Bloomington Pantagraph,* and a series of similar warnings in the Bloomington CDs during the early 1880s:

THE NAME NOT CHANGED.
Beware of a Fraud.
I have been advised that certain parties are seeking to place upon the market imitations of
Dr. Z. Waters' Family Medicines
Known as Z. Waters' Honey Balsam, Raspberry Balsam, Liver and Cathartic Pills, and Ne Plus Ultra Liniment. The public are hereby cautioned against buying any such imitations, as none are genuine unless the name of Dr. Z. Waters is printed on each label and wrapper on or around the bottles and pills, and the words Ne Plus Ultra on the liniment. To prevent fraud, I have secured the registration of these names and trade marks in the patent office at Washington, and am the sole proprietor of these medicines. **I have been connected with their manufacture**

for the last 11 years [*emphasis ours*], and the exclusive compounder of the same for the last nine years.

DR. H. S. WOODARD,
Bloomington, Illinois.

Bloomington, Ill., May 18th, 1877

I hereby certify that Dr. H. S. Woodard is the sole proprietor and manufacturer of the following named medicines known as Dr. Z. Waters' Family Medicines, and bearing my name: Dr. Z. Waters' Ne Plus Ultra Liniment, Dr. Z. Waters' Honey Balsam, Dr. Z. Waters' Raspberry Balsam, Dr. Z. Waters' Liver and Cathartic Pills, Dr. Z. Waters' Itch Renovator. Having perfect confidence in the ability and integrity of Dr. H. S. Woodard, I cheerfully recommend the above medicines to the public as long as he controls the manufacture of the same, he having been the manufacture of the same for the last nine years.

Z. WATERS, M. D.

Dr. Woodard's own medicine brand, *Woodard's Instant Relief*, was definitely produced during the mid-to-late 1870s and was a very successful product during the early 1880s (in the 1885–86 Bloomington CD, Woodard declares, "Over 10,000 Bottles of Woodard's Instant Relief sold by one man within Four Years." No early ads have yet been located, but the earlier embossed bottle in the listing perhaps contained an earlier version of this same medicine. An 1886 Woodard trade card indicated this medicine was for internal and external use by both men and livestock:

DR. H. S. WOODARD'S INSTANT RELIEF.

Directions.—for Rheumatism, Neuralgia, Headache, apply freely and often; also take internally. For pains in back, Soreness or Tenderness of Spine, apply freely along the spine; Inflamed or Sore Breast, Sprains, Bruises, Sores, Burns, Frost Bites, Chilblains, Tumors, if not too large; Felons (when they first commence.) For Tooth and Earache, on cotton in tooth or ear, also apply under the ear and side of face. For Colic, Pain, Cramp, in Stomach and Bowels, Sinking or Congestive Chills, Cholera Morbus, apply over Stomach and Bowels, take 1/3 to 1/2 teaspoonful to 1/2 bottle in 1/2 pint of water every few minutes. For Horses' Colic and Botts, give from a tablespoonful to 1/2 bottle in 1/2 pint of water every few minutes. For Founder, give same dose 2 or 3 times a day. For Sprains, Bruises, Sores, Gall, Grease Heel and Scratches, apply freely twice per day. For Cattle, give and apply the same as for horses.

Price, 50 cents per Bottle.

The 1880 state IIC lists H. S. Woodard & Co. as a patent-medicine manufacturing company with eight employees during the period from June 1879 to June 1880. He remained in the medicine business throughout most of the late nineteenth century, but apparently did not package his medicines in embossed bottles during the later nineteenth century.

Henry S. Woodard and Bliss S. Howe

Medicine:

(1) **WATER'S / NE PLUS ULTRA / LINIMENT //** [aqua]
(*side*): **WOODARD & HOWE //**
(*side*): **BLOOMINGTON. ILLS // //**
(*base*): **L & W**

»SB w/ keyed hinge mold; beveled-edge rectangle; 4 sunken panels; long neck; tapered collar; 6 1/2" tall. [*Figure 112*]

(2) (*side*): **WOODARD & HOWE // //** [aqua]
(*side*): **BLOOMINGTON. ILLS // //**

»SB w/ post mold; beveled-edge rectangle; 4 sunken panels; long neck; slightly tapered collar; 6 1/2" tall. [*Figure 113*]

Figure 112

Curiously, a paper label preserved on one example of bottle #2 indicates that it contained **DR. Z. WATERS' HONEY BALSAM** and that it was prepared and sold only by **Dr. Z. WATERS & ___, Bloomington, Ill.** (*see Figure 114*).

An 1873 advertisement in the *Bloomington Pantagraph* states that Drs. Woodard & Howe were successors to Dr. Z. Waters and Co. (for whom no embossed bottles are known—*see Howe and Waters listings, this chapter*), corner of 9th and Madison Streets, Bloomington. The Waters company had previously been run by Waters and Howe (ca. 1864–71) until Woodard bought out Waters' interest and became senior partner in 1872. According to Kemp (2009), a *Bloomington Pantagraph* article in fall 1874 reported that Woodard & Howe "employed six men who traveled in wagons as far south as Texas to supply some 600 local agents."

Woodard & Howe remained together producing Dr. Waters' medicines and Woodard's *Instant Relief* until 1876–77, after which they split up and pursued separate careers in patent-medicine manufacture in Bloomington (*see their individual listings, this chapter*).

Figure 113

Figure 114

BLUE ISLAND
(Cook County)
1880 census: 1,500

R. Boil and Co.

Soda ~ Mineral Water:

(1) **B. & CO. / BLUE ISLAND //** [aqua]
(heel): **A. & D. H. C**

»SB w/ keyed hinge mold; oval-blob-top soda. *[Figure 115]*

Figure 115

R. *Boil & Co.* could only be found listed in the 1875 ISD as a soda-water manufacturing company in Blue Island. The company was likely short-lived (perhaps just a season or two), as the bottles are very seldom seen.

Early Blue Island bottlers are particularly difficult to research because there are no CDs available before the late 1890s, and the Blue Island newspaper archives were destroyed in a fire in the 1920s.

Gustave H. Hausburg

Soda ~ Mineral Water:

(1) **G H. HAUSBURG / BLUE ISLAND ILL.** [red amber]

»SB w/ heel mold; oval blob top; slug-plate embossed pint; 8 3/4" tall. *[Figure 116]*

(2) **G. H. HAUSBURG / BLUE ISLAND, ILLS.** [amber, yellow]
(base) **C & CO / LIM**

»SB w/ post mold; oval-blob-top quart w/ circular-slug-plate embossing; 12 1/2" tall. An 11 3/4" tall slope-shouldered quart blob-top variety of this bottle is also known. It is heel-embossed with an "E H E24" maker's mark (the Edward H. Everett glassworks of Newark, OH), so the bottle was not made earlier than ca. 1880, and may date somewhat later than that. *[Figure 117]*

The bottles described date to the late 1870s and early 1880s, and reflect a later return to Blue Island by Gustave Hausburg a decade or so after the earlier *Hausburg & Dettmer* and *Hausburg & Ennis* Blue Island soda bottling partnerships had been dissolved (*see their listings, this chapter*).

During the interim, Gustave Hausburg moved to Chicago in the early 1870s. Together with his brothers William A. and Charles, he helped establish a family soda and mineral water bottling works in there during 1872 and 1873 (*see Hausburg Brothers Chicago listings*), and he married Elizabeth Wilhelms there in 1875.

By the later 1870s, Gustave may have been involved with his brother Charles in setting up a soda-bottling operation in Elgin (*see the Hausburg Brothers Elgin listing*). The listed bottles are evidence that he had returned to Blue Island by 1878–79 to reestablish a bottling works there on his own.

Gustave Hausburg is also known to have bottled in Blue Island much later in his career, so his post-1880 bottling works there must have been fairly successful. There are Hutchinson-style *G. H. Hausburg* Blue Island sodas dating to the 1880s and 1890s, and an early 1900s "So. Chicago"

Figure 116

Figure 117

crown-top soda bottle bearing his name is known. Hausburg is listed in the 1900 USC for Blue Island Village as a 51-year-old manufacturer of soda water.

Since Gustave's previous bottling experience had been focused on soda and mineral water, these first G. H. Hausburg Blue Island bottles are listed in the same category, but there are no early Blue Island CDs to confirm this, and from the bottle's shapes and amber color they may have held brewery products.

Gustave H. Hausburg and William Dettmer

Soda ~ Mineral Water:

   (1) **HAUSBURG & DETTMER / BLUE ISLAND / ILLS //**   [aqua]
         (heel): **A. & D. H. C**

     »SB w/ post mold; oval-blob-top soda. *[Figure 118]*

———————— ❖ ————————

Hausburg formed a partnership with Dettmer during a season or two at approximately 1870, just after the dissolution of the earlier Hausburg/Ennis partnership (*see their listing, this chapter*). The partnerships with both Ennis and Dettmer were apparently short-lived, as Hausburg was bottling on his own by the early 1870s (*see his listing, this chapter*). William Dettmer was listed as a soda bottler in Elgin by the time of the 1880 census (*see his Blue Island listing, this chapter*).

Early Blue Island bottlers are particularly difficult to research because there are no CDs available before the late 1890s, and the Blue Island newspaper archives were destroyed in a fire in the 1920s.

Interestingly, in the late 1990s bottle diggers in Leadville, Colorado, recovered examples of both of these Hausburg partnership bottles together in a boomtown privy there.

Figure 118

Gustave H. Hausburg and [August] Ennis

Soda ~ Mineral Water:

(1) **HAUSBURG & ENNIS / BLUE ISLAND / ILLS //**   [aqua]
    (*heel*): **A & D. H. C**

    »SB w/ keyed hinge mold; oval-blob-top soda. [*Figure 119, right*]

(2) **HAUSBURG & ENNIS / BLUE ISLAND / ILL^S //**   [aqua, amber]
    (*heel*): **A. & D. H. C**

    »SB w/ post mold; squat Chicago-style quart soda bottle; 3 3/4" diameter, 9 3/4" tall. [*Figure 119, left*]

Based upon the keyed-hinge-mold bottle described above, the Hausburg and Ennis partnership was the first of the Hausburg bottling works, dating to the later 1860s. August Ennis is listed as a member of a Blue Island fraternal organization in the 1870s, and perhaps he was the Hausburg bottling company partner. By ca. 1870, Hausburg had formed a new partnership, also short-lived, with William Dettmer (*see their listing, this chapter*).

Early Blue Island bottlers are particularly difficult to research because there are no CDs available before the late 1890s, and the Blue Island newspaper archives were destroyed in a fire in the 1920s.

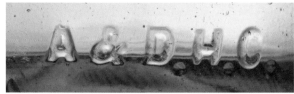

Figure 119

BRAIDWOOD
(Will County)

1880 census: 5,500

Horace W. Blood

Soda ~ Mineral Water:

(1) **H. W. BLOOD. / BRAIDWOOD / ILL'S //** [aqua]
(*heel*): **TW & Co**

» SB w/ hinge mold; oval-blob-top soda. *[Figure 120]*

(2) **BLOOD'S / BOTTLING HOUSE / BRAIDWOOD / ILL'S. //** [aqua]
(*heel*): **DeS. G. Co.**
Note: No photo available. Only a single example of this bottle is known, from a detailed sketch and rubbings of the embossed lettering made in the 1980s by Robert Kott (see Figure 121).

» SB w/ probable post mold; oval-blob-top soda.

Horace W. Blood first came to Illinois from New York in 1863 (at the age of 20) and farmed for two years before taking employment in 1865 at J. D. Paige's bottling house in Joliet. He remained in Joliet for two seasons learning the bottling business, then moved 15 miles south to Wilmington to oversee operations at J. D. Paige's bottling depot there (*see the Paige Joliet listing*), having purchased a half-interest in the operation. He also engaged in the ice business in Wilmington.

In 1870, he purchased a controlling interest in Paige's Wilmington bottling works (Kott 1978:5; Le Baron 1878b:751) and commenced soda and mineral water bottling there as H. W. Blood & Co. (*see his Wilmington listing*). In 1873, he moved his operation five miles further southwest to Braidwood and set up a new bottling works there

which today is one of the most complete in this vicinity. His many ingenious inventions have rendered the employment of a large number of men unnecessary, and with but few men, Mr. Blood can turn out of his establishment an equal quantity of goods, with firms who employ a far greater number of persons. Thus...Mr. Blood is enabled to sell his goods cheaper than any other firm, and at the same time give the purchaser as fine a quality of goods as can be found in the country. The building is 24x50, two stories high; capacity of 200 boxes per day; employs three hands; two tow-horse wagons find sales for goods in Braidwood and towns in this vicinity. Mr. Blood manufactures soda and mineral water, bottled lager beer, ale, porter, cider and spruce beer. [Le Baron 1878b:751–752]

After 1880, Blood's bottling works continued to produce soda and mineral water in at least two varieties of embossed Hutchinson-style internal-stopper bottles. These are beyond the scope of the present study and are not listed above. Blood apparently retired from active management of the bottling house in the early 1890s (a June 1891, ad lists H. W. Blood & Son as "successors" to H. W. Blood), and he apparently died around the turn of the century (Kott 2005b).

Figure 120

Figure 121

BREESE
(Clinton County)

1880 census: 600

Charles and Henry Dorries

Soda ~ Mineral Water:

(1) **C. DORRIES & BROTH / BREESE ILL**^{S} //
(heel): **W^m McC & Co.**

»SB; oval-blob-top soda. *[Figure 122]*

[aqua]

Charles Dorries emigrated to Breese with his family in 1851 from Brunswick, Germany. He first worked in his father's blacksmith shop in Breese before becoming a soda water manufacturer. Dorries appears in both the 1870 and 1880 USC for Breese as a soda water bottler. From its maker's mark and base mold, the company's one pre-1880 embossed bottle order seems to have been made and used during the mid-to-late 1870s.

The 1870 IIC lists Dorries & Bro. as a "soda water factory" operating six months a year with two employees. Their estimated annual product that year was "4000 boxes or 96,000 bottles," or soda water (24 bottles per box) cumulatively valued at $2,800.

Figure 122

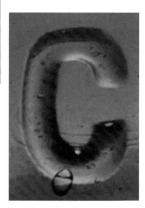

CAIRO

(Alexander County)

1850 census: 240
1860 census: 2,200
1870 census: 6,300
1880 census: 9,000

Henry Breihan

Ale:

(1) **H. BREIHAN / CAIRO ILLS //**   [amber]
    (heel): **A & D. H. C.**

»SB; 2-piece mold quart ale; ring-neck tapered collar; 8 3/4" tall. *[Figure 123]*

Soda ~ Mineral Water:

(1) **HENRY BREIHAN / CAIRO ILLS. /**   [aqua]
    (front heel): **A. &. D H. C**

»SB w/ keyed hinge mold; oval-blob-top soda. *[Figure 124]*

(2) **HENRY BREIHAN / CAIRO ILLS. /**   [aqua]
    (front heel): **M^{c}C** *[Same mold as soda #1]*

»SB w/ keyed hinge mold; oval-blob-top soda. *[Figure 124]*

(3) **HENRY BREIHAN / CAIRO ILLS. /**   [aqua]
    (front heel): **T W & Co** *[Same mold as sodas #1 and #2]*

»SB w/ keyed hinge mold; oval-blob-top soda. *[Figure 124]*

❖

The earliest reference to Henry Breihan as a Cairo bottler is in a biographical sketch of Cairo resident George Becker, who was said to

Figure 123

Figure 124

have left school at the age of 12 (i.e., 1867–68) and "shortly afterwards entered the employ of Mr. Henry Breihan, where he acquired his first knowledge of the soda water business."

> He remained with Mr. Breihan in the capacity of bookkeeper and afterward manager until he accepted a position with Mr. Andrew Lohr *[see Lohr listing]*, at the time conducting a modest bottling plant. Into his new vocation Mr. Becker threw his entire personality, and by force of hard work, not only became vice-president of the newly incorporated company in 1889, but ten years later reorganized the company, Mr. Lohr retiring, and as president and general manager...he has added to, improved and modernized the entire plant, until today there stands on the little bottling shop of 1879 an immense factory covering an entire city block." [Federal Publishing Co. 1905:265]

Mr. Breihan was listed as a teamster in the 1870 USC, so perhaps he bottled only seasonally. But the Becker memoir suggests he left Breihan's employ to go to work at Mr. Lohr's "little bottling shop" in 1879. This comment and the technological and stylistic attributes of the Breihan bottles listed above suggest he was in business from the late 1860s to the late 1870s. A comment published in the *History of Alexander, Union and Pulaski Counties, Illinois* suggests he was still in business in the early 1880s as well: "The beer bottling, soda and seltzer and mineral trade has grown to immense proportions here recently. Mr. A. Lohr and Henry Brenhan [sic] each have extensive concerns, and a wide market to supply in this and adjacent states" (Perrin 1883:167).

The known Breihan bottles themselves are unique in that, although they likely represent a considerable time span of use from ca. mid-1860 through the 1870s, they were all made from the same bottle mold—by three different Pittsburgh glass companies. All of the Breihan bottles even have the same piened-out and filled-in embossed name and town of an earlier bottler (unreadable) on the reverse side. Given the blob-top soda pattern of shorter bottle neck lengths over time, a possible sequence of bottle manufacture is *A. & D. H. Chambers* (7 1/8" tall); *Thomas Wightman & Co.* (7" tall); *William McCully &Co.* (6 7/8" tall).

No later-designed (with post-mold bases) Breihan embossed blob-top soda styles have been documented, and Thomas Wightman did not begin in business until 1874, so apparently the old mold continued in use, and the bottles in the listing were filled seasonally throughout the 1870s.

John H. Kump

Soda ~ Mineral Water:

(1) **J. H. KUMP** / **CAIRO. ILL** /   [aqua]
   (*right front heel*): **A. & D. H. C**

   »SB w/ keyed hinge mold; shouldered-blob-top soda. *[Figure 125]*

John Kump had embossed bottles from many midwestern towns. In addition to the Cairo bottle, four styles of his embossed soda bottles are known from Memphis, TN—including his two earliest bottles, both pre-1860 iron-pontiled varieties (see American Pontiled Soda Database at http://bottleden.com). Three of Mr. Kump's later Memphis bottles were recovered from the wreck of the Union ironclad *Cairo*, which sank in 1863 near Vicksburg. The ship and the bottles are on display at the *Cairo* Museum there. Embossed J. H. Kump sodas are also known from Columbus, OH; Oil City and Petroleum Centre, PA; and Portland, IN.

A biographical essay on Kump's brother-in-law, Philo M. Clark (Blackmar 1912), indicates that Clark and Kump established the Memphis bottling works together in 1858, but closed the works and headed north at the outbreak of the Civil War. According to Blackmar's (1912) account "they had cleared some $20,000; and it was contrary to Confederate orders for any Northerners to take money with them.... Mrs. Kump concealed it in a hand satchel, which she used as a pillow for the train, and it was never discovered."

During the war, Clark went on to establish contemporary bottling works at Louisville, New Albany, Jeffersonville, Lexington, and Indianapolis "and was thus able to supply soda water to the soldiers of the Union Army during all the years of the war" (Blackmar 1912:1509–1511). After the war, he closed these operations and went into the bottling business in Oil City, PA "during the oil excitement."

Kump bottled soda and mineral water at both Columbus and Cairo during the war. (These bottles are nearly identical long-neck blob-top sodas, and both were made at the A. & D. H. Chambers glassworks in Pittsburgh: *Figure 126*.) He also apparently reestablished his Memphis bottling works in 1863/64, after the Battle of Memphis. Apparently he and Clark were in business

Figure 125

Figure 126

(together?) in several cities at once during the Civil War to distribute mineral water to Union Army soldiers. For instance, the 1865 Cincinnati CD lists the Clark & Kump soda business at 47 E. 4th. J. H. Kump was a resident in the city that year. Their continued alliance seems to be further indicated by the fact that both of them ended up at Oil City after the war.

The 1860 USC lists Kump as a 22-year-old mineral water bottler in Memphis. A May 1863 IRS tax assessment lists Kump as a bottler in Cairo. Portland, IN, was apparently his last stop, as his Portland bottle is the only known post-1880 Hutchinson internal-stopper Kump soda.

Andrew Lohr

Ale:

(1) **ANDREW LOHR / CAIRO ILL //** [amber]
    (heel): **A & D. H. C.**

   »SB w/ keyed hinge mold; quart ale; ring-neck tapered collar; 9 1/4" tall. *[Figure 127]*

Beer or Soda:

(1) **ANDREW LOHR / CAIRO ILLS** [aqua]
    (base): **A & D. H. C.**

   »SB w/ post mold; shouldered quart w/ long "lady's leg" style neck; tapered collar; 11 3/4" tall. *[Figure 128]*

Soda ~ Mineral Water:

(1a) **A. LOHR. / CAIRO. / ILL //** [aqua]

   »SB w/ keyed hinge mold; raised embossed lettering; long neck; oval-blob-top soda. *[Figure 129, left]*

(1b) **A. LOHR. / CAIRO. / ILL //** [aqua]
    (heel): **A. & D. H. C** *[Figure 129, right]*

   »SB w/ keyed hinge mold; flattened embossed lettering; short neck; oval-blob-top soda.

(2) **ANDREW LOHR / CAIRO, ILLS.** [aqua; teal]

   »SB; shouldered-blob-top quart soda; embossed in faint slug plate; 11 1/4" tall. *[Figure 130]*

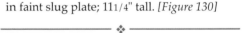

According to the 1883 history of Alexander, Union, and Pulaski counties (Perrin 1883: Pt. 4, pp. 29–30), Andrew Lohr sold his dairy herd in 1861 and purchased a soda water factory in Cairo. The factory mentioned would have to be that recently vacated by Edward Schroeder (*see his listing, this chapter*). Prior to that time, Lohr appears to have been working in the Schroeder soda factory. According to the June 1860 USC, both men were 25-year-old "manufacturers of soda." Lohr apparently took over the operation when Schroeder left.

Lohr is listed in the 1870 IIC, which indicated that he operated his soda factory year-round with two male employees who were paid $1,600 annually. He produced 5,000 boxes of soda (120,000 bottles at 24 bottles per box) with a value of $5,000 (about 4¢ per bottle).

Lohr is also listed in the 1875 ISD, and in Chandler's (1878) Illinois Railway Business Directory, as a Cairo soda water manufacturer. Perrin's 1883 county history (p. 167) notes that in the early 1880s Lohr's soda business

Figure 127

Figure 128

Figure 129

Figure 130

Figure 131

had "a wide market to supply in this and adjoining states." He continued his Cairo bottling business into the early twentieth century; the 1900 USC lists Lohr as a 68-year-old operator of a soda water factory there.

The bottles listed are only those that would predate 1880. Probably during the 1860s, Andrew Lohr also used a uniquely shaped, Albany-slipped soda-style stoneware bottle that was shoulder stamped **A. LOHR** (*Figure 131, left*). Some bottles of this style were also heel-stamped by Noah Ailiff, the stoneware potter who manufactureded them (see Mounce 1989:19): **N. AILIFF, MANUFACTURER, MOUND CITY, ILLS.** (*Figure 131, right*). Mound City is located on the Ohio River about six miles north of Cairo.

Charles Pfifferling

Hair Treatment:

(1) **PFIFFERLINGS // TRIONADRAPHIA / CAIRO. ILL _**   [aqua]

»SB w/ keyed hinge mold; rectangular w/ 3 sunken panels; ring collar; 7" tall. [*Figure 132*]

Figure 132

Pfifferling worked in St. Louis prior to the Civil War as a hotelkeeper (1860 USC). He is listed in IRS tax assessments for Cairo between 1862 and 1865 as a retail liquor dealer and as a hotel operator. He was the proprietor of the Rising Sun Hotel in Cairo in 1864 (ISD). The *Cairo Times* of May 25, 1874, described "Charlie" Pfifferling as "the old time caterer of the St. Charles Hotel, now landlord of the hotel at Anna.... He is as big and fat as ever, and jolly as a sailor."

Pfifferling's Trionadraphia was first mentioned in Cairo newspapers in 1863. It was advertised in issues of the October 1865 *Cairo Daily Democrat* as follows:

Pfifferling's Trionadraphia—An infallible remedy for baldness and diseases of the hair, $2.00 bottle. To be had at the principal drug stores; at Eshback & Miller's Barber Shop, and at Charlie Pfifferling's on 7th opp. Winters Block.

His hair treatment appears to have been manufactured and sold in Cairo only during the mid-1860s.

Edward Schroeder

Soda ~ Mineral Water:

(1) **E. SCHROEDER / CAIRO / ILL**   [aqua]

»IP; shouldered-blob-top soda. [*Figure 133*]

Edward Schroeder is listed in the 1860 USC as a 25-year-old manufacturer of soda water in Cairo. He arrived in southern Illinois from Germany in the late 1850s, when he was in his mid-20s. He apparently bottled in Cairo for just a season or two (ca. 1858–60).

He then moved to St. Louis during the Civil War years and continued his beverage-bottling activities there. During the Civil War years he also became associated with pioneer St. Louis bottler John Cairns (as part of *John Cairns & Co.*) in a grocery and soda bottling operation in both St. Louis, MO, and East St. Louis, IL.

After the war, he relocated across the Mississippi River to East St. Louis and established a long-lived family soda-water business there (*for a more detailed business biography see Miller 2007a and Schroeder's East St. Louis listing*).

Figure 133

William H. Schutter

Ale:

(1) **W H SCHUTTER / CAIRO / ILL'S** [amber]
    (base): **W. McCULLY & Co. PITTSBURGH.**

»SB; 2-piece mold; tall, narrow, long-necked quart ale; ring-neck tapered collar; 9¼" tall. *[Figure 134]*

According to the 1864 ISD, Wm. Schutter was proprietor of a billiard saloon and was a wholesale dealer in wines, liquors, and cigars in Cairo during the late Civil War years. In the 1870 USC, he is listed as a wholesale liquor dealer. The 1871–72 ISD carried an ad for Schutter's liquor business on page 39:

> WILLIAM H. SCHUTTER
>
> Wholesale and Retail Dealer in
> **Wines, Liquors, and Cigars,**
> AGENT FOR
> **SANDS' PALE CREAM ALE,**
> No. 4 Springfield Block,
> CAIRO————ILLINOIS

From their technological and style attributes, his embossed bottles appear to have been in use during the early and mid-1870s. Perhaps they were used to bottle *Sands' Pale Cream Ale*. By 1880 he had become a Cairo deputy sheriff (1880 USC).

According to his obituary in the July 2, 1891, *Cairo Weekly Citizen,* "He was an old citizen of Cairo, coming here before the war. In 1865 he was one of our wealthiest men, but reverses came and swept away his property."

Figure 134

John Sproat

Sproat was a prominent Cairo merchant who transported his goods via the river trade, and apparently only embossed bottles bound for other towns, such as Chicago and Texarkana. (*See a more detailed discussion under his Chicago entry*)

CAPRON
(Boone County)
1880 census: 300

Platus A. Field

Medicine:

(1) **P. A. FIELD'S / TONIC QUEEN / CAPRON. ILL.** [aqua]

»SB; rectangular w/ 4 sunken panels; thick flared-ring collar; 8" tall. *[Figure 135]*

The 1870 USC listed Platus Field as a clergyman in Capron. From the style and technology of the listed bottle, he apparently turned at least some of his attention to patent-medicine manufacturing and distribution during the 1870s. By the time of the 1880 USC, he had left the area.

The village of Capron was so small during Field's residence there that few historic records can be searched for ads about the reputed curative powers of his *Tonic Queen*, but tonics in general were marketed for their ability to improve a person's mental and physical "tone," or sense of well-being.

The city of Rockford was located about 20 miles southwest of Capron, so possibly 1870s ads for the medicine will eventually be located early Rockford newspapers.

Figure 135

CARLINVILLE
(Macoupin County)
1850 census: 400
1860 census: 3,200
1870 census: 5,800
1880 census: 3,100

Charles Labbe and Son

Soda ~ Mineral Water:

At this time no embossed bottles are known for the *Labbe & Son* soda water bottling operation, but unembossed soda bottles are rare in Illinois during the 1840–80 period, and none are known from Carlinville, so *Labbe & Son* embossed bottles may yet turn up. *Labbe & Son* are listed in the 1880 IIC, covering the manufacturing period from June 1879 to May 1880. They had two employees during this period: a skilled laborer making $2.50/day and a laborer making $0.50/day. They worked 12-hour days from May to November and 8-hour days between November and May. Annual costs of material and labor were $1,130 and the annual retail value of product produced was estimated at $1,750.

Labbe & Son would have operated in competition with the August Zaepffel bottling works in Carlinville (*see his Carlinville listing, this chapter*), and the local market may not have supported the competition for long. Alternatively, they may have produced bulk soda water for drugstores or other soda fountain outlets.

John Weber

Soda ~ Mineral Water:

(1) **JOHN WEBER / CARLINVILLE / ILL //**   [aqua]
    (*heel*): **A & D. H. C**

»SB w/ keyed hinge mold; oval-blob-top soda. [*Figure 136, left*]

---- ❖ ----

John Weber is listed in the 1870 USC and in the 1871 ISD as a soda water manufacturer in Carlinville. He first started the bottling works in the late 1860s with a partner named (John?) Mueller (*see their listing below*). After 1873, Weber's Carlinville bottling works was taken over by August Zaepffel (*see his Carlinville listing, this chapter*).

John Weber and [John A.] Mueller

Soda ~ Mineral Water:

(1) **WEBER & MUELLER /**   [aqua]
    CARLINVILLE / ILL //
    (*heel*): **A & D. H. C**

»SB w/ keyed hinge mold; shouldered-blob-top soda. [*Figure 136, right*]

---- ❖ ----

Figure 136

John Weber and (John?) Mueller were partners in the Carlinville bottling business, probably for only one or two seasons, in the late 1860s. Their partnership was dissolved by 1871 when John Weber is listed in the ISD as sole proprietor of the bottling works (*see his listing above*).

Both Weber and Mueller were common names associated with early west-central Illinois bottling enterprises. *Frederick Weber* bottled alone in Litchfield and in partnership with *Jacob Weber* in Pana (*see their Pana listings*). Gustavus Mueller bottled with Louis Stein in Bloomington (*see the Bloomington listing*), and Anton Mueller (a.k.a. "Miller") bottled with Alfred Beck in Highland (*see their Highland listing*).

It has not yet been determined if the Carlinville bottler named Mueller was Gustavus or Anton, or someone else. From genealogical records, there was a John A. (Andrew/Andreas) Mueller living in Carlinville at about the time of the bottling partnership who would have been in his late 20s at the time.

August Zaepffel

Soda ~ Mineral Water:

(1) **AUGUST ZAEPFFEL / CARLINVILLE / ILL'S**   [aqua]

»SB w/ keyed hinge mold; oval-blob-top soda. *[Figure 137]*

A biographical sketch of August (*a.k.a.* Augustus) Zaeppfel appears in the *Biographical Record of Macoupin County, Illinois* (Richmond and Arnold 1904:83). He began "the manufacture of carbonated beverages of all kinds" in 1874. A list of businessmen in the *1875 Atlas of Macoupin County* (Warner and Beers 1875a) lists Zaeppfel as a manufacturer of soda water, Excelsior, Belfast Ginger Ale, and champaign cider. He was born in Alsace, France, the son of a wine merchant. Zaeppfel died just after the turn of the century, and his business was taken over by his son, who was still operating the bottling works in 1910, according to the U.S. census.

Many August Zaepffel bottles are Hutchinson-style internal stopper blob tops and turn-of-the-century-era crown-top bottles, but these bottles were manufactured too late for inclusion in the present study.

CARLYLE
(Clinton County)

1850 census: 300
1870 census: 1,350

John S. Gunn

Medicine:

(1) DR J. S. GUNN'S / GOLDEN / VERMIFUGE   [aqua]

»OP w/ straight-line hinge mold; shouldered cylinder w/ short neck and rolled lip; 4¼" tall. *[Figure 138]*

According to the U.S. Census for 1850, John S. Gunn was a physician living in Carlyle, Clinton County, IL. At the time, he was 33 years old and had a wife and three children. He was also listed in Clinton County in the 1860 census.

Figure 137

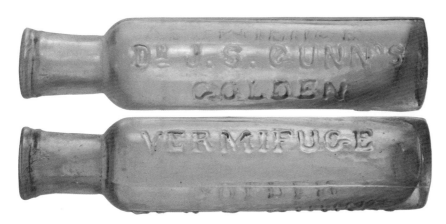

Figure 138

Examples of his cylindrical vermifuge bottles have been recovered from pre–Civil War privy excavations in Alton, about 40 miles northwest of Carlyle. The bottles were likely manufactured and used during the 1850s.

Henry Hess [a.k.a. Heiss] and Co.

Soda ~ Mineral Water:

(1) **H. HESS & C⁰ / CARLYLE / ILL** [aqua]

»SB w/ keyed hinge mold; blob-top soda. *[Figure 139]*

❖

The 1860 USC listed Henry Hess as a 28-year-old barber in Carlyle. The firm of *Hess & Hutter* was first found listed in the August 1866 IRS tax assessment for Carlyle. *Hess & Co.* was listed under Carlyle soda water manufacturers in the 1867 ISD.

There are also "*Heiss*" *& Hutter* embossed soda bottles of the same era from Centralia, ca. 15 miles southeast of Carlyle *[see listing, this chapter]*. H. C. Chandler's *Railway Business Directory* for 1868 lists William Hutter as operating a soda water factory in Centralia. Clearly "Hess" and "Heiss" are the same man, and *Hess & Co.* refers to the bottling association of Henry Hess and William Hutter, who bottled together in both towns during the mid- to-late 1860s, with Hess running the Carlyle bottling, and Hutter running the Centralia bottling operation.

By 1870 their bottling venture had been dissolved, and Hess had become a Carlyle saloonkeeper (USC).

Anthony Hubert

Figure 139

Soda ~ Mineral Water:

(1) **A. HUBERT / CARLYLE / ILL //** [aqua]
    (heel): **A. & D. H. C**

»SB; blob-top soda. *[Figure 140]*

❖

The 1870 USC lists Anthony Hubert as a hotelkeeper in Carlyle. T. Peter, a resident of Hubert's hotel, is listed in the same census as a 19-year-old employee of a soda water factory. Hubert likely operated the former Hess & Co. soda factory seasonally during the

Figure 140

Figure 141

1870s, since Hess & Hutter had ceased operations ca. 1870 (*see Hess listing*), and J. M. Menkhaus (*see the next listing*) did not take over the soda-factory operation until the end of the 1870s. According to Brink, McDonough & Co. (1881a), the factory was located on 11th Street between Fairfax and Franklin.

John M. Menkhaus

Soda ~ Mineral Water:

(1a) **J. M. MENKHAUS / CARLYLE / ILL //** [aqua]
(heel): **A. & D. H. C**
Note: The embossed "J" on the bottle is reversed. [Figure 141, top]

»SB w/ keyed hinge mold; shouldered-blob-top soda.

(1b) **J. M. MENKHAUS / CARLYLE / ILL //** [aqua]
(heel): **A. & D. H. C**
Note: The reversed embossed "J" on the bottle has been corrected. [Figure 141, bottom]

»SB w/ keyed hinge mold; shouldered-blob-top soda.

John Menkhaus bottled soda and mineral water in Carlyle for over 40 years, from the Civil War era (first working for *Hess & Co.*, then *Hubert*, and finally taking over the plant himself ca. 1880) until the turn of the twentieth century (see 1900 USC).

Menkhaus also ran a grocery store, so perhaps he bottled soda water on a part-time, seasonal basis. The 1880 USC for Carlyle states that Menkhaus was 44 years old and was a grocer/merchant. The soda factory was described as follows in the 1881 County History: "It is a frame building, and with apparatus for manufacture, is estimated at $1,200. It contains two fountains of thirty gallons each, and has the capacity of manufacturing 200 boxes of mineral water per day. The annual value of manufactured product is about $4,000, and it gives employment to five men and two teams" (Brink, McDonough & Co. 1881a).

The embossed bottle listed above likely dates to Menkhaus' first year operating the bottling house: ca. 1879–80.

CARROLLTON
(Greene County)

1850 census: 800
1860 census: 2,800
1870 census: 2,800
1880 census: 1,900

Dr. Alexander H. Smith

Veterinary Medicine:

(1) **D^R. A. H. SMITHS / GOLDEN.LINIMENT //** (side) **FOR // //** (side) **ANIMALS** [aqua]
(base): **L & W**

»SB w/ keyed hinge mold; rectangular; 4 sunken panels; tapered collar w/ long ringed neck; 5 1/2" tall. [Figure 142]

Figure 142

Medicine:

(1) **DR A. H. SMITHS / CELEBRATEDE** // *(side):* **AGUE** // **UNION.MEDICINE. CO /**   *[aqua]*
SOLE.PROPRIETORS // *(side):* **CURE**
(base): **L & W**
Note: CELEBRATED is misspelled.

» SB w/ keyed hinge mold; rectangular; 4 sunken panels; tapered collar; ring around lower neck; 7" tall. *[Figure 143]*

Bitters:

(1) **DR A. H. SMITH'S / CELEBRATED / OLD STYLE BITTERS** // //   *[amber]*
(horizontal ~ upper shoulder): **O. S. / 2781** *(vertical ~ lower shoulder):* **THE STANDARD TONIC / AND / BLOOD PURIFIER** // //

» SB; square bitters with beveled corners; 2 opposite embossed sunken panels and 2 opposite unimbossed flat panels; tapered collar; 87/8" tall. *[Figure 144, bottom illustration]*

(2) **DR A. H. SMITHS / OLD STYLE BITTERS** // // **O.S / 2781** // //   *[amber]*

» SB w/ post mold and cupped central base; beveled-edge square bitters; 2 opposite embossed sunken panels and 2 opposite unimbossed flat panels; tapered collar; 83/4" tall. *[Figure 145]*

    The two published histories of Greene County (Donnelley et al. 1879; Miner 1905) list Dr. A. H. Smith as a physician and druggist in Carrollton. Dr. Smith, who was listed as 52 years old in the 1870 U.S. census, had come to Illinois from New Jersey with his family in the late 1830s. His father was also a physician.

    A. H. Smith operated as a merchant for several years in nearby Kane (where his store came to focus on pharmacy products: Miner 1905:128), before moving to Carrollton and establishing his drug store there in 1867. According to Miner (1905:480), he carried on his Carrollton pharmacy business until his death in 1892.

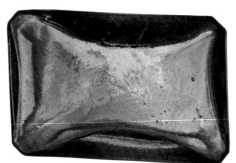

Figure 143

DR. A. H. SMITH'S OLD STYLE BITTERS ★★★

Marked in two horizontal lines: "DR. A. H. SMITHS/OLD STYLE BITTERS." Reverse in domed indented panels: "O. S./2781." Two sides flat and plain.
Shape: Square
Color: Amber
Mouth: LTC
Measurements: 8¾ x 2½
Base: Smooth
Corners: Beveled
Label reads "Union Medicine Co. Sole Proprietor, St. Louis, Mo."
Variant A. Marked in three vertical lines: "DR. A. H. SMITH'S/ CELEBRATED/OLD STYLE BITTERS." Reverse in one horizontal line below shoulder: "O S/2781." Below, vertically in three lines: "THE STANDARD TONIC/ AND/BLOOD PURIFIER."
Shape: Square
Color: Amber
Mouth: LTC

Figure 144. Hand-drawn illustration of Smith's "Celebrated" Old Style Bitters embossed variety (lower illustration) published by Watson 1965:201. No examples of this bottle have yet been located to photograph.

Figure 145

During the 1870s, he brought three of his sons into the pharmacy business in Carrollton (Donnelley et al. 1879:499), which no doubt gave him more time to develop and market his patent-medicine line. His eldest son Edward joined the pharmacy in 1871

> where he learned the business, gaining accurate knowledge of the medicinal qualities of the various goods carried, so he was registered as a pharmacist when the first law passed *[in the late 1870s]*. He continued with his father until the latter's death and has since carried on the business, having a large and well selected stock, which, carefully arranged, makes his a neat and attractive store. [Miner 1905:480]

On March 20, 1873, an advertisement that touted A. H. Smith's *Old Style Bitters* as a general health tonic and blood purifier, was printed in the *Decatur Republican* (IL). The ad indicated that the *Old Style Bitters* sole proprietors were the *Union Medicine Co.* of Carrollton, IL. Smith's *Old Style Bitters* were also advertised in the *Red River Gazette* (MN) on February 6, 1873, and have been found listed in drug catalogs for 1874 and 1878 (see Ring & Ham 1998:505–506). Smith's *Union Medicine Co.* at Carrollton was also listed in the 1875 ISD under "patent medicines."

According to Ring & Ham (1998), a paper label on one of Smith's bitters bottles indicated that the *Union Medicine Co.* of St. Louis, MO, was sole proprietor of the product. A November 13, 1873, ad in the *Decatur Republican* also indicated that sales headquarters for Smith's *Old Style Bitters* had relocated to No. 6 and 8 Second Street, St. Louis.

As the result, it has been assumed that Smith took his company to St. Louis (or sold the manufacturing rights to a company there) in the late 1870s. However, a search of the St. Louis CDs for the 1870s and 1880s failed to reveal the presence of any "Union Medicine" company there, or of anyone marketing bitters from the 2nd Street address. (The closest facsimile name found was for an *American Medical Co.* operated during the late 1870s by G. W. Young & Co. at 921 N. Levee Street.) Apparently any mid-1870s attempt to expand the Old Style Bitters sales market to St. Louis rapidly failed, and the company was never listed in the St. Louis CDs.

Technologically and stylistically, all of the embossed A. H. Smith bottles appear to date to the decade of the 1870s. In fact, the enigmatic "**O.S. 2781**" embossing on his *Old Style Bitters* bottles may date their first appearance. "**O.S.**" is probably an abbreviation for "Old Style," and "**2781**" could be "1872" written backwards.

CENTRALIA
(Marion County)

1870 census: 3,200
1880 census: 3,600

Daniel J. Besant

Soda ~ Mineral Water:

(1) **D. J. BESANT / CENTRALIA / ILL** [aqua]

»SB w/ keyed hinge mold; shouldered-blob-top soda. *[Figure 146]*

❖

Daniel J. Besant was listed in the 1860 USC as a 25-year-old liquor dealer living in Centralia. Besant was a bachelor at the time. He was married in Centralia on September 5, 1861, to Mary Stedlin. From its stylistic and technological attributes, the soda bottle listed above appears to date to the early 1860s. The only known specimen of this bottle is a large fragment unearthed at a Civil War camp in Cairo, IL.

Figure 146. Rubbing of side-embossed lettering taken from restored excavated sherds of a fragmentary Besant blob-top soda bottle. No complete examples of this bottle have yet been located to photograph.

Figure 147. Rubbing made by Ron Ridder of side-embossed lettering on a blob-top soda bottle previously in his possession. No examples of this bottle have yet been located to photograph.

Henry Heiss [*a.k.a. Hess*] and William Hutter

Soda ~ Mineral Water:

(1) **HEISS & HUTTER / CENTRALIA / ILL //**   [aqua]
(*heel*): **A & D. H. C.**

»SB w/ keyed hinge mold; blob-top soda. *[Figure 147]*

The firm of "Hess" & Hutter was first found listed in the August 1866 IRS tax assessment for Carlyle, ca. 15 miles northwest of Carlyle (*see listing, this chapter*). Hess & Co. was listed under Carlyle soda water manufacturers in the 1867 ISD.

Technologically, the *Heiss & Hutter* embossed soda bottles from Centralia were manufactured at about the same time. H. C. Chandler's *Railway Business Directory* for 1868 lists William Hutter as operating a soda water factory in Centralia. Clearly "Hess" and "Heiss" are the same man, and *Hess & Co.* refers to the bottling association of Henry Hess and William Hutter, who bottled together in both Carlyle and Centralia during the mid-to-late 1860s, with Hess running the Carlyle bottling operation, and Hutter running the Centralia operation.

Henry G. Wehrheim

Soda ~ Mineral Water:

(1) **H. G. WEHRHEIM / CENTRALIA / ILL _**   [aqua]
(*heel*): **A & D. H. C.**

»SB w/ keyed hinge mold; oval-blob-top soda. *[Figure 148]*

Wehrheim was listed as a Centralia soda water manufacturer in the 1870 USC. His occupation was listed as a farmer in the 1880 USC, but his son Adam is listed as

Figure 148

a soda water bottler in the same census. The style and technological attributes of the embossed bottle are consistent with early 1870s manufacture.

CENTREVILLE [CENTERVILLE]

[now Millstadt]

(St. Clair County)

1870 census: 1,100

Christian Fischer

Soda ~ Mineral Water:

(1) **CHR FISCHER / CENTREVILLE / ILL //**   [aqua]
 (heel): **A. & D. H. C**

 »SB w/ keyed hinge mold; shouldered-blob-top soda. *[Figure 149]*

(2) **CHR FISCHER / CENTERVILLE / ILL'S**   [aqua]

 »SB w/ post mold; oval-blob-top soda. *[Figure 150]*

Christian Fischer is listed as a Centreville soda water manufacturer in the 1870 USC, the St. Clair Co. Atlas of 1874, the 1875 ISD, the 1881 *History of St. Clair County,* and the 1884 Belleville CD. His residence and soda water factory were located at the corner of La Fayette and Van Buren streets. Miller estimates the time of Fischer's entry into the soda water business during the 1860s, as follows:

Figure 149

> He probably acquired the machinery for soda water manufacturing from Samuel Just who lived in *[Centreville]* and in the 1860 census was listed as an owner of a "soda factory." Samuel was probably related to *[Fischer's wife]* Augusta, possibly an uncle. Since Samuel was not listed in the 1870 census, he probably was dead. If so, the machinery for soda manufacturing was accessible to Christian.... The 1862 military census for *[Centreville]* required all men in the village...to register. Christian Fischer...was listed as 23 years old, born in Germany, and a "stone mason" as his occupation. This means that he did not operate a soda factory before 1862. [Miller 2007a:149]

From the style and technological details of the oldest embossed Fischer bottles, they were not made until the late 1860s, sometime after the end of the Civil War.

The name of the town was changed to Millstadt in 1887. Coincidentally, Christian Fischer died in early 1888 and the last embossed bottle with his name on it has the new town name (Miller 2007a:152). The bottle was manufactured too late for inclusion in the list, which does not include bottles made after 1880. Fischer's son, Charles, took

Figure 150

Figure 151

over the business after 1887, and continued bottling soda water in Centerville into the early twentieth century (1900 USC).

The Fischer bottling works was illustrated in the 1881 St. Clair County history published by Brink, McDonough & Co. (1881b; *Figure 151*).

CHAMPAIGN
(Champaign County)

1860 census: 1,700
1870 census: 4,600
1880 census: 5,100
(see **West Urbana**, 1855–1860)

Nicholas Miller

Ale:

(1) **N. MILLER / CHAMPAIGN / ILL _**   [amber]

»SB w/ keyed hinge mold; quart ale; ring-neck tapered collar; 9 1/2" tall. [*Figure 152*]

---❖---

The 1860 USC lists Nicholas Miller in Champaign as a 27-year-old saloonkeeper who had immigrated to the United States from Prussia. The 1864/65 ISD also lists Miller as a Champaign saloon keeper.

In the 1870 USC, Miller's occupation was listed as spruce-beer manufacturer. Miller was listed in the 1871 ISD as a "bottler" whose business was located at the corner of University Avenue and First Street in Champaign. However, the 1870–71 Champaign CD indicated that Miller was a "manufacturer of mineral water, etc." from his residence at the corner of University Avenue and First Street. He was listed in the 1875 ISD as a "soda water manufacturer."

Figure 152

In light of its hinge-mold construction, the ale-style bottle listed was probably produced and used during the late 1860s or early 1870s, when Miller was primarily selling (and bottling) alcoholic beverages.

There are no known early or mid-1870s soda bottles embossed with his name.

CHARLESTON
(Coles County)

1850 census: 800
1860 census: 2,200
1870 census: 2,800
1880 census: 2,900

Henry C. Cunningham and Silas Barnes

Medicine:

(1) **CARMAN'S / BITTER SWEET // CUNNINGHAM & BARNES / PROPRIETORS // CHARLESTON / ILLINOIS** [yellow-olive]

»SB w/ post mold; bitters-style, beveled-edge square case bottle; 3 sunken panels; tapered collar; 9 7/8" tall. *[Figure 153]*

---❖---

The bottlers of *Carman's Bitter Sweet* have been difficult to identify, but were likely the proprietors of the Crystal Palace Drug Store in Charleston during the late 1860s and early 1870s. Henry C. Cunningham was listed as a druggist in Charleston in 1870 (USC), and was listed as a physician in the 1880 USC. Silas Barnes was not listed in the 1870 census, and was likely a post-1870 associate (since their *Bitter Sweet* bottle was manufactured with 1870s post-mold technology. Barnes was listed as a 33-year-old Charleston druggist in the 1880 USC. Neither man is mentioned in the Coles County historical volumes or biographical albums, but an 1879 Charleston CD included in the *History of Coles County* (Perrin 1879) includes the following Charleston entry:

> Calvert, D. H., Proprietor Crystal Palace Drug Store. Dealer in Pure Drugs, Medicines, Paints, Oils, Dyestuffs, Toilet Articles, and **Manufacturer of Bitter Sweet, the greatest tonic of the age.** West side Public Square. *[emphasis ours]*

Calvert was probably a successor proprietor to *Cunningham & Barnes'* pharmacy operation, who inherited the *Bitter Sweet* bottling rights along with the business. According to the Chapman Brothers (1887b) *Portrait and Biographical Album of Coles*

Figure 153

Figure 154

County, Calvert entered the pharmacy business in Charleston in 1876, so the listed *Cunningham & Barnes* bottle probably dates to the first half of the 1870s. Their *Bitter Sweet* was obviously marketed in Charleston as a general health tonic. The product was likely named for local physician Dr. W. H. Carman, who probably originally concocted it.

Dr. Samuel Van Meter

Medicine:

(1) **DR. S. VAN METER / CHARLESTON. ILL** [aqua]

»SB w/ keyed hinge mold; strap-sided oval; embossed oval sunken panel; thick flared-ring collar; 6 7/8" tall. *[Figure 154]*

Dr. Samuel Van Meter founded the Illinois Infirmary in Charleston in 1867. He practiced medicine there until his retirement in 1877. He was initially exposed to the practice of medicine in Charleston in 1844, when he was 20 years of age. At that time

He was taken into the office of Dr. T. B. Trower, at Charleston, as an office boy and general helper, and while there put in his spare time studying medical works. After spending five years with Dr. Trower he engaged in other pursuits *[until 1851]*...and again went into the office of Dr. Trower, this time to prepare himself in earnest for the practice of medicine. He began his professional career in 1854, and soon built up a wide practice all over the country *[county?]*. About 1867 he founded the Illinois Infirmary, at Charleston, an institution which enjoyed a national reputation, patients coming from all parts of the United States for treatment. About 1877 he retired from active practice. [Wilson 1906:874]

From its smooth-based, keyed-hinge-mold technology, the generic embossed medicine bottle listed dates to the years of Dr. Van Meter's general medical practice during the 1860s. Bottles of this style were likely filled with a variety of medicines prescribed by Dr. Van Meter.

CHESTER
(Randolph County)

1850 census: 1,000
1860 census: 1,200
1870 census: 1,600
1880 census: 2,600

John Wenda

Soda ~ Mineral Water:

(1) **JOHN WENDA / CHESTER / ILL //** [aqua, green]
(heel): **A & D. H. C.**

»SB w/ keyed hinge mold; blob-top sodas w/ shouldered and oval varieties. *[Figure 155]*

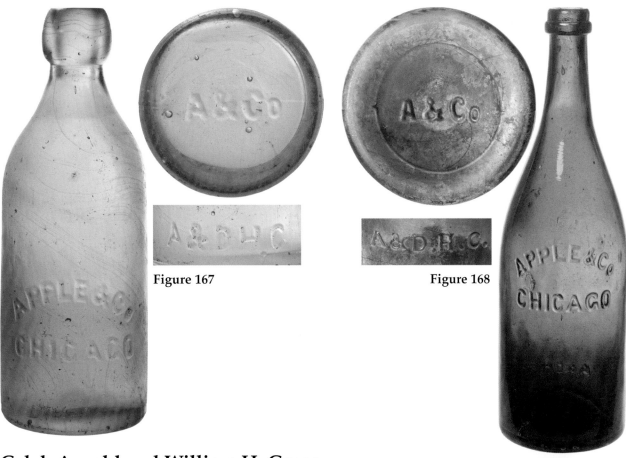

Figure 167

Figure 168

Caleb Arnold and William H. Green

Spruce Beer:

(1) **ARNOLD & GREEN / CHICAGO / ILL**ˢ [amber]
»SB; ale-style 2-piece-mold quart; oval blob top; 7" tall. *[Figure 169]*

Caleb Arnold, Jr., is first listed in the 1862 Chicago CD as a "spruce beer maker" at his home address. He becomes associated with William Green later that year, and Arnold and Green are listed as manufacturers of spruce beer in the 1862 and 1863 IRS assessments. The Chicago CDs for 1863 through 1865 list them together as spruce beer manufacturers, first at Washington near Wells and later at 99 & 101 W. Monroe. In the 1866 Chicago CD, Caleb Arnold was back at his home address making spruce beer, and Adam Amberg and William Green had become partners (*see Amberg & Green listing, this chapter*) producing spruce beer and mineral water at the 99 & 101 W. Monroe factory earlier used by Arnold & Green.

In addition to the glass bottle listed above, Arnold & Green also sold their spruce beer in quart stoneware bottles. Two debossed styles of such bottles have been recorded: one is a 12-sided, shoulder-stamped Merrill-patent style (*Figure 170*); the other is round with a sharp shoulder, and side-stamped cobalt-highlighted lettering within a half-circle border (*Figure 171*).

Aromatic Golden Bitters

See the Parker Mason listing.

Figure 169

Figure 171

Figure 170

William C. Badeau

Medicine:

(1) (side): **BADEAU'S // PURE BLOOD MAKER //** [aqua]
(side): **CHICAGO**

»SB w/ keyed hinge mold; rectangular w/ beveled edges; 3 sunken panels; tapered collar; 9 1/4" tall, 3 7/8" wide. *[Figure 172]*

(2) **BADEAU'S / PURE / BLOOD MAKER** [aqua]

»SB; cylinder w/ rounded shoulders; squared flared-ring collar; 3 1/4" diameter, 8 1/2" tall. *[Figure 173]*

---❖---

William C. Badeau came to Chicago in 1863 from Troy, NY, where he had operated a drug store. His first Chicago business listings (from 1864 to 1866) are as a member of the firm of Post (George W.) & Badeau at 192 E. Lake. They are listed variously as wholesale druggists, proprietors of Badeau's Pure Blood Maker and Liver Cure, and importers of California wines and liquors. In 1867, Badeau is listed alone at the same address, as proprietor of his Pure Blood Maker "&c." In 1868, he joined Garrison & Co. (with Herod Garrison and Allen Murray), "botanic and electric druggists" at 171 Lake. But in 1869 he is listed alone as a patent-medicine manufacturer at 66 Monroe. After that—from 1870 until his last Chicago CD listing

Figure 172

Figure 173

in 1881—he is listed alone as a druggist and drug/patent medicine dealer from his residence and a variety of small offices.

From the technological attributes of his *Blood Maker* bottles, they were produced for at least a decade after his arrival in Chicago, where he was involved primarily in the wholesale drug trade. The earliest ad he published in the Chicago Tribune for *"Badeau's Pure Blood Maker"* appeared on March 15, 1866. The ads announce only the medicine's name and manufacturer, with no information as to its curative properties or where it was sold retail.

According to a nineteenth-century British traveler (Marshall 1882), Badeau's blood maker was still prominently advertised in Chicago when he passed through in June 1878. Badeau conducted a wholesale drug business in Chicago until 1882 when he moved to Los Angeles (see *History of Los Angeles County*: Lewis 1889).

In patent-medicine terms, a "blood maker" was an iron or other tonic used as a strength-builder to combat anemia and to cure diseases caused by "impure blood."

William Bailey and A. M. Eaton

Gun Oil:

(1) *(side):* **BAILEY & EATON // GUN OIL //**   *[aqua]*
   (side): **PARAFFINE**

   »SB w/ straight hinge mold; rectangular; square collar; 3 sunken panels; squared flared-ring collar; 3¾" tall. *[Figure 174]*

Figure 174

William Bailey and A. M. Eaton were business partners in Boston, MA, during the early 1860s as dealers in paraffine lubricating oils produced by the Union Coal and Oil Company of Maysville, KY. Union Coal & Oil mined and made oil from cannel coal and shipped the crude oil in barrels and by oil boat to their Maysville refinery. The products they produced—which were packaged and marketed during the Civil War years by Bailey & Eaton from facilities on Canal Street in Boston and Water Street in Providence, RI—included paraffine lubricating oils (e.g., gun oil, engine oil, machine and sewing-machine oil); packing oil; axle grease; and paraffine / candle wax (1864 Boston CD, ad. page 53; Wright and Potter 1865:127). According to the 1864 Boston CD ad, Bailey and Eaton's products were "Put up in gross packages, carefully prepared. For sale wholesale."

Also, Bailey & Eaton's ad in the December 3, 1864, issue of the *Army and Navy Journal* indicates the company marketed its gun oil directly to the military: "Bailey and Eaton's Paraffine Gun Oil. Prepared expressly for Army use. It prevents rust on locks, barrels, swords, scabbards, etc. Carefully put up in one-gross packages. Wholesale Depot, 28 and 30 Canal St., Boston, Mass." Their bottles embossed specifically for Chicago, and the company's absence from CDs and advertising outlets there, may be a reflection of the fact that Chicago was the site of Camp Douglas, the largest Civil War training camp and Civil War prison in Illinois. If Bailey & Eaton marketed directly to the military, a sutler intermediary might be helpful but advertising would be irrelevant.

But even the local sutler may not have been needed. Amasa Eaton possibly had Chicago relatives in the arms business. According to Chicago CDs, Daniel and Charles Eaton (father and son?) sold guns, pistols, and sporting goods in Chicago from 1859–62, Charles Eaton then worked for the Enoch Woods arms and sporting goods company in the late Civil War years. In the late 1860s, Charles was a partner in the Eaton & Abbey firearms business, and following his death in 1870, his widow Emma took over the business. Her E. E. Eaton Company grew to be a major firearms and sporting goods business in Chicago in the late nineteenth century (*see Figure 175*).

With the discovery of rock oil just after the Civil War years, costs of producing cannel-coal paraffine oils became prohibitive. And from court records, Amasa Eaton declared bankruptcy in March 1867. The shape and mold attributes of the B & E Chicago gun-oil bottle suggest it dates to the early 1860s. So, from historical and technological evidence, their Chicago bottle likely dates between 1863 and 1866.

Victor Barothy (& Co.)

Soda ~ Mineral Water:

(1) **VICTOR BAROTHY**   [amber]
 »SB; slope-shouldered, "champagne-style" bottle with deep basal kick-up and shouldered-blob top; 7 3/4" tall. [*Figure 176*]

(2) **V. BAROTHY / CHICAGO //**   [aqua]
 (heel): **A & D. H. C.**
 (base): **B**

Figure 175

Figure 176

»SB w/ keyed hinge mold; high-shouldered, cylindrical shouldered blob-top soda. [Figure 177]

Victor Barothy "& Co." were listed as a bottling works that distributed champagne and mineral-spring waters at 120 Dearborn from 1869 to 1870 (Chicago CDs). Barothy was listed alone as a manufacturer of mineral spring water in the 1871 Chicago CD, at the time of the great fire. The following year, he took on partner E. C. Cook (*see their listing, this chapter*).

Victor Barothy and Egbert C. Cook

Soda ~ Mineral Water:

(1) **BAROTHY & COOK / CHICAGO //** [aqua]
 (*heel*): **A & D. H. C.**
 (*base*): **B**

 »SB w/ keyed hinge mold; high-shouldered, cylindrical shouldered blob-top soda. [*Figure 178*]

(2) **BAROTHY & COOK / CHICAGO //** [aqua]
 (*right front heel*): **C & I**
 (*base*): **B.C**

 »SB w/ post mold; high-shouldered, cylindrical shouldered blob-top soda. [*Figure 179*]

(3) **BAROTHY / & / COOK'S / AERATED / WATERS / CHICAGO** [pale blue]
 (*heel*): **C & I**

 »SB w/ post mold; "tenpin"-shaped oval-blob-top soda; small-lettered maker's mark lightly embossed on heel. [*Figure 180*]

(4) **BAROTHY / & COOK / CHICAGO / ILLINOIS / MEDICATED / AERATED / WATERS** [aqua]

 »SB w/ rounded base; slightly tapered blob-top soda; 9" tall. Lettering embossed vertically around entire body of bottle. [*Figure 181*]

(5) **BAROTHY & COOK / CHICAGO //** [aqua]
 (*heel*): **C & I**

 »SB w/ post mold; Chicago-style quart oval-blob-top soda; 8¾" tall; small-lettered maker's mark lightly embossed on heel. [*Figure 182*]

Barothy and Cook are listed in Chicago CDs as bottlers of champagne and soda water from 1872 to 1873. Sometime in 1873 or early 1874, Barothy retired from the partnership and Cook took over the business (*see the Cook entry, this chapter*).

Figure 177

Figure 178

Figure 179

Figure 180

Figure 181

Figure 182

William M. and William H. Barrett

Cider:

(1) **TRADE MARK** *(within apple-shaped outline)* / [amber]
**BARRETT & BARRETT / YORK STATE /
CLARIFIED / CIDER / 55 MARKET ST. /
CHICAGO. // THIS BOTTLE / IS / NEVER SOLD**
(base): **B&B**

»SB w/ post mold; slightly tapered shouldered-blob-top quart; 11 1/2" tall. *[Figure 183]*

William M. and William H. Barrett were partners in the firm of Barrett & Barrett, listed as apple cider manufacturers with offices at 45 Market in the 1877 Chicago CD.

An 1870s trade card for the Chicago branch (*Figure 184*) indicates that the Barrett cider company of Holley, NY, expanded its bottling operations to include Chicago for a time in the late 1870s. The reverse side of the Chicago trade card makes it clear that Barrett's cider was marketed as more than just a refreshing drink:

THE CIDER CURE.

Its devotees pronounce it an infallible remedy for Rheumatism. The Blue Grass Cure and the Boiling Water Remedy for Dyspepsia, have been succeeded by the Cider Cure for Gout, and has received the sanction of some of the best physicians in Europe and America.... A correspondent in the London "Times" claims that he has been cured from rheumatism of long standing by the use of Crab Cider, and it will be well to try the experiment by ordering Crab Apple Cider made by B. & B.... By using this cider, invalids can pursue their routine of treatment at their homes to the best advantage.

Figure 183

Figure 184. Both sides of a printed trade card distributed in the 1870s by the Chicago branch of Barrett & Barrett cider company.

Edward K. Bebbington and Co.

Soda ~ Mineral Water:

- *(1a)* **E. K. B // AERATED / SODA WATER** [cobalt]
  Note: *This bottle has crudely modified embossing made by piening-out the lettering on the earlier* **E&B** *and* **E&L** *tapered-collar soda mold (see Entwistle listings below). [Figure 185, left]*
  »IP; shouldered-blob-top soda.

- *(1b)* **E. K. B // AERATED / SODA WATER** [cobalt]
  Note: *This bottle has had new larger lettering recut into a new, unaltered bottle mold. [Figure 185, right]*
  »IP; shouldered-blob-top soda.

- *(2a)* **OERATED / SODA.WATER / EKB** [cobalt]
  Note: *"***AERATED***" is misspelled. [Figure 186]*
  »IP; shouldered-blob-top soda.

- *(2b)* **AERATED / SODA.WATER / EKB** [cobalt]
  Note: *"***AERATED***" misspelling corrected. [Figure 187]*
  »IP; shouldered-blob-top soda.

- *(3)* **BEBBINGTON.&.Co** [cobalt]
  »IP; shouldered-blob-top soda. *[Figure 188]*

Figure 185

Edward Bebbington was a junior partner with Joseph Entwistle and George Lomax from 1852–55, on Canal Street between Lake and Fulton *(see J. Entwistle and G. Lomax listings, this chapter)*. However, all of the listed bottles date to the period of the late 1850s (ca. 1856–59), after George Lomax had left the group to form a new bottling business of his own (with John Meagher) and Joseph Entwistle had retired.

Bebbington continued to bottle soda water on his own, at the address of the original partnership (Canal between Lake and Fulton) until 1858, when he moved to new quarters at 110 Carroll for his final year in the bottling business (Wagner 1980; Welko 1973b:9). The Bebbington "**& CO**" embossing on the final bottle listed above likely refers to a new junior partner Bebbington took into the business ca. 1857–58: John A. Lomax, another member of the Chicago bottling "Lomax clan" *(see the Lomax listing, this chapter)*.

Bebbington's embossed bottles are listed above in their probable 1856–59 production sequence.

Bennett Pieters and Co.

This bottler is listed alphabetically under Pieters.

Thomas Bligh

Medicine:

(1) **BLIGH'S TONIC** // // **CHICAGO ILLS** // //   [aqua]

»SB w/ keyed hinge mold; rectangular; 3 sunken panels; tapered collar; 10 1/2" tall. *[Figure 189]*

Figure 186

Figure 188

Figure 187

No specific ads have been found for this tonic bottle, but there were four different (perhaps related?) Blighs involved in the liquor trade in Chicago between 1864 and 1872. Two of them were only associated with saloons: Andrew, from 1865–69; and George in 1871. The third, James, was a grocery and liquor merchant from 1869–72.

Perhaps the most likely candidate is Thomas Bligh. Thomas is listed as a 33-year-old liquor merchant in Chicago in the 1870 USC. He was involved in the retail and wholesale wine and liquor trade at various Chicago business locations from 1864 through 1869. But until a specific product advertisement can be found, this must be considered a tentative attribution.

Ebenezer A. Bower and Wesley Bryan

Soda ~ Mineral Water:

(1) **BOWER & BRYAN / CHICAGO / ILLS** [aqua]
(base): **B & B**

»SB; shouldered-blob-top soda. *[Figure 190]*

(2) **BOWER & BRYAN / CHICAGO //** [aqua]
(heel): **A & D H C**
(base): **B & B**

»SB w/ keyed hinge mold; oval-blob-top soda. *[Figure 191]*

(3) **BOWER & BRYAN //** [aqua]
(heel): **A & D H C**
(base): **B & B**

»SB w/ keyed hinge mold; oval-blob-top soda. *[Figure 192]*

Figure 189

Ebenezer A. Bower and Wesley Bryan were Chicago mineral water manufacturers in the late 1860s and/or early 1870s. They were either a very short-term company, or very shy of being included in the commercial directories. They are only listed as mineral water bottlers in the 1873 Chicago CD, and they are only listed separately that year (at their residential addresses) with no indication that they were working together. They are not included in the *Soda and Mineral Water Bottlers* summary listing at the end of the volume, and they do not appear in any prior or subsequent CD years.

Michael Brand and Co.

Beer:

(1) **M. BRAND & Co / CHICAGO //** [amber]
(heel): **THIS / BOTTLE / NOT TO BE / SOLD**
(base): **I G Co L**

Figure 190

Figure 191   **Figure 192**

»SB w/ post mold; slope-shouldered quart; slightly tapered shouldered-blob-top beer; 9 1/2" tall. *[Figure 193]*

Michael Brand is listed in the 1870–75 Chicago CDs as President of Busch and Brand Brewing Company. In 1878 he is listed as head of the Brand Brewery and the 1879 Chicago CD lists Brand as the president of M. Brand and Co. In the 1860–67 CDs, the brewery was located at 25–31 Cedar.

The bottle listed above probably dates to 1878–79, marking the inauguration of the Brand Brewing Co. A similar inaugural embossed brewery bottle was produced in the 1870s by John E. Bradbury & Co. of Jacksonville (*see his Jacksonville listing*) to mark his business start-up there.

William P. Brazelton and Co.

Soda ~ Mineral Water:

(1) **W. P. BRAZELTON / CHICAGO //**   [aqua]
    (heel): **A. & D. H. C**

»SB; shouldered-blob-top soda. *[Figure 194]*

(2) **W. P. BRAZELTON'S / SELTZER / CHICAGO //**   [aqua]
    (heel): **A. & D. H. C**

»SB w/ keyed hinge mold; shouldered-blob-top soda. *[Figure 195]*

Figure 193

Figure 194

William Brazelton began in the confectionary supply business in Chicago in 1869 at 100 W. Madison. In the 1869 CD he listed soda water manufacturing first and ice cream manufacturing second under his business entry.

But that focus changed rapidly. W. P. Brazelton and Co. (F. A. Benson was the new partner) is listed in the 1870 Chicago CD as a manufacturer of ice cream (first) and soda water (second) at 100 Madison. The 1871 listing is the same, with new partner F. A. Brazelton.

By 1872 the company is no longer listed as manufacturing soda water. During the mid- and later 1870s, with a variety of partners, Brazelton focused his confectionary-supply business on the distribution of ice cream and ice cream machinery, coffee, confections, and fruit.

Brown's Oriental Hair Renewer

See the Walker & Taylor listing in this chapter.

Figure 195

Figure 196

Frederick A. Bryan

Medicine:

(1) **CLARKE'S // GENUINE // RED DROPS // PREPARED ONLY //**   [aqua]
    BY. F. A. BRYAN // CHICAGO. ILL

»OP; vertically embossed; octagonal sided cylinder; w/ short neck and rolled lip; 4 3/8" tall. [Figure 196]

---❖---

Frederick A. Bryan is listed as a druggist in Chicago CDs beginning in 1852 and appears in the USC as a Chicago druggist for 1860 and 1870. In the 1875 edition he has retired from the pharmacy business and is listed as a "capitalist." He was retired by 1880 (USC). His embossed bottle is open-pontiled, so it dates from the 1850s.

We have not yet been able to locate any advertising discussion of the purported medical benefit of Clarke's Red Drops. However, three of four Red Drop medicines from elsewhere in the eastern United States described in Odell (2007) had advertisements indicating they were cures for venereal disease. Advertising for Levison's Red Drops in New York pointed out that "Great Sins Require Great Repentance" (Odell 2007:224).

George Buck

Medicine:

(1) **BUCK APOTHECARY / CHICAGO**   [aqua]
    Note: This is a citrate-of-magnesia style bottle. The embossing forms a circle but is not in a recessed slug-plate.

»SB; citrate-style cylinder; rounded shoulders; wide double-ring collar; 2 1/4" diameter, 7" tall [Figure 197]

Figure 197

Figure 198

(2) **BUCK // CHICAGO** [clear]

»SB; sharp-shoulder cylinder; wide "unguent'-style neck and flared lip; 1 3/4" diameter, 4 1/2" tall. [Figure 198]

❖

George B. Buck was a pharmacist in Chicago from the 1850s through the 1880s. For most of his career, he was associated with the prominent Chicago druggist firm of *Buck & Rayner* with James B. Rayner (*see the Buck & Rayner listing, this chapter*). Before he became a partner in the *Buck & Rayner* firm in 1858, he had been in charge of the prescription department for J. H. Reed & Co. (*see the Reed listing, this chapter*).

For unknown reasons, James Rayner left Chicago for three years just after the Civil War. The *Buck & Rayner* firm was discontinued between 1866 and 1870, during which time the George B. Buck pharmacy operated from the *B&R* 93 Clark Street address (*see the Buck listing, this chapter*). James Rayner reappeared in the Chicago CDs in 1869 as a "druggist" at the 93 Clark Street address; the next year, in 1870, the Buck & Rayner partnership was reestablished. In addition to his successful pharmacy partnership with Rayner, George Buck was a faculty member at the Chicago College of Pharmacy during the 1870s. During this time, he served as the College's President and on its Board of Trustees. The College of Pharmacy was formed at the outset of the Civil War and the Class of 1880 was its 17th graduating class.

Both George Buck and James Rayner retired or passed away ca. 1890, and ownership of the *Buck & Rayner* firm was taken over by George's son (pharmacist Charles G. Buck) and Martin Schnitzler. The firm of *Buck & Rayner* continued as a successful drugstore operation well into the twentieth century.

The two embossed pharmacy bottles listed above could only have been manufactured and used during the four-year period when George Buck was sole operator of the pharmacy at 93 Clark Street: 1866–1869.

George Buck and James B. Rayner

Soda ~ Mineral Water:

(1) **BUCK & RAYNER'S / MINERAL / WATER / CHICAGO.** [cobalt]

»SB; Saratoga-style mineral-water bottle; ring-neck tapered collar; 6 3/4" tall. [Figure 199]

Hair Dressing:

(1) (*above an embossed circle*): **BUCK & RAYNOR /** [clear]
(*beneath the circle*): **CLARK ST. CHICAGO /**
(*within the circle*): **GENUINE OX MARROW / FOR THE HAIR**

»SB; shouldered cylinder; wide-mouth "unguent"-style bottle w/ flared lip; 4 3/4" tall. [Figure 200]

❖

Buck and Rayner became partners in the Chicago pharmacy business beginning in 1858 (Ebert 1905:248). Buck had previously been in charge of the prescription department for J. H. Reed & Co. (*see the Reed listing, this chapter*). The *Buck & Rayner* firm is listed in the 1860 ISD and in the 1862 IRS assessment. For unknown reasons, James Rayner left Chicago for three years just after the Civil War. The *Buck & Rayner* firm was discontinued between

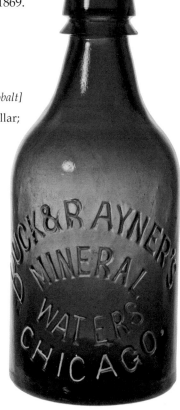

Figure 199

1866 and 1870, during which time the George B. Buck pharmacy operated from the *B&R* 93 Clark Street address (*see Buck listing, this chapter*). James Rayner reappears in the Chicago CDs in 1869 as a "druggist" at the 93 Clark Street address; the next year, in 1870, the Buck & Rayner partnership is reestablished.

Buck & Rayner are not known to have produced many medicinal products themselves, and seem to have been cautious about endorsing the patent-medicine products they sold. In an 1874 *Chicago Daily Tribune* article about the Chicago patent-medicine business (*CDT*, June 14, 1874, p. 10, col. 3–4), Mr. Rayner responded to a customer's question about the curative qualities of *Plantation Bitters* by saying: "We never express an opinion, sir, upon these medicines. We do not know what they are made of, and will not be made responsible for their effects."

Beginning with the 1870 Chicago CD, Buck and Rayner had multiple business locations. In the 1875 CD, they are listed as having drug stores at 132 State and 117 Clark. Both George Buck and James Rayner retired or passed away ca. 1890, and ownership of the Buck & Rayner firm was taken over by Buck's son (pharmacist Charles G. Buck) and new partner Martin Schnitzler. Theirs was a very prominent and successful pharmacy business, and the firm remained in business well into the twentieth century.

Figure 200

They began their in-store soda and mineral water "Spa" soon after they began their partnership in the late 1850s, and from numerous *Chicago Daily Tribune* ads, it became a renowned and resplendent feature of their business during the 1870s. At this time, ticket books were sold for glasses of their mineral waters and soda-fountain beverages (including "hot soda water"), redeemable at either store. A ticket for a single glass was 10¢, but 16 tickets could be bought for $1 (see especially "A Peep Below Stairs, Buck & Rayner's Soda Water," *CDT* June 10, 1877, p. 8, col. 6):

THE "SPA"
IF YOU COULD CATCH THE GAS as it goes from a foaming Seidlitz Powder and imprison it in a goblet of Pure Spring Water, you would have the secret of the Sparkling Mineral Waters of the
"SPA"
The waters are fresh from Nature's fountains, and touch nothing but glass, Banca tin, and pure silver. BUCK & RAYNER. (*CDT* 3/1/74, p. 1, col. 2).
THE SPA
Snow-flakes from the revolving Ice-Plane, Fruit Juices, Sweet Cream, and Sparkling Soda Water announce the 17th opening of the "Spa" at both stores of BUCK & RAYNER (*CDT* 5/10/74, p. 1, col. 3).

A HEAVY TAX ON DAIRYMEN
Between 2,700 and 2,800 glasses of soda water and mineral waters went from the "Spa" on Friday. To furnish sweet cream for such insatiable consumers is a tax on dairymen. Half a barrel of cream has been used in a single day by Buck & Rayner (*CDT* 5/20/77, p. 5, col. 3).

CARD FROM BUCK & RAYNER
A Midsummer Reverie
There is a difference you little dream of between "Soda Water" and "Soda Water." The regulation article that greets you at every street corner in city or town, mixed up by the hogsheadful in Soda Water Factories, and sold to the trade at a few pennies a gallon, is one thing: Buck & Rayner's superior Soda

Water, made by themselves in all its details, of pure and genuine materials, is another and very different beverage.... Hence the familiar adage, "Buck & Rayner draw the best Soda Water in the United States, and their Mineral Waters are excellent" (*CDT* 7/16/79, p. 1, col. 2).

Popular bottled mineral waters from famous eastern and midwestern springs were imported and served at Buck & Rayner's Spa (including Sheboygan Mineral Water, Gettysburg Katalysine Water, Kentucky "Blue Lick" Water, Maine "Poland" Water, Oswego "Deep Rock," Dunbar's Bethesda, and the Saratoga "Vichy," Congress, Empire, Hathorn, Geyser, Triton, "&c., &c., &c."

However, *Buck & Rayner* seldom bottled druggist products or mineral waters under their own name. Their first, and longest advertised signature bottled product, which appeared in their *CDT* newspaper ads throughout the 1870s, was their Mars cologne. No embossed bottles for this product have yet been seen. They also bottled a few other products during the mid- and late-1870s, but with little commercial success.

Buck & Rayner bottled a local mineral water in the cobalt Saratoga-style bottles listed— probably as a hoped-for Midwest competitor for the eastern Saratoga-type mineral waters they imported for sale at their in-store spas. Their "Genuine Glen Flora Water," a bottled mineral water "for families," was advertised in the *CDT* only during fall 1874 and spring 1875. The Glen Flora springs were located in Waukegan, along the Lake Michigan shore about 30 miles north of Chicago.

> [I]n the mid-19th century, Waukegan was one of the busiest harbors on the Great Lakes.... As merchants reliant on the lake were turning the town into a bustling commercial center, another aqueous feature was drawing hundreds of visitors annually: the alluring natural mineral springs that flowed through Waukegan's ravines.
>
> Various of these springs were commercialized as bottlers shipped the ostensibly healing waters to eager consumers around the United States. One group of five springs was located near the center of Waukegan and another, the Glen Flora Springs, eventually lent its name to what is today the Glen Flora Country Club just west of Sheridan Road. [Bielski 1998:174]

Buck & Rayner likely hoped Glen Flora Water would become popular enough for them to bottle and export to other northeastern and midwestern cities where Saratoga waters were popular. But the rarity of their bottles indicates that the effort was not successful. Their newspaper ads were terse, and gave no details as to any possible health-renewing properties of the water.

The two embossed *Buck & Rayner* bottles listed above are seldom seen and, from their *CDT* newspaper ads, were only marketed for a very limited time. The ads for their "Ozonized Ox Marrow" for the hair appear only from May 1876 to August 1877:

Mars' New Satellite
It was named "Ozonized Ox Marrow," and has been seen revolving about the "Mars Cologne" for some months past.
It is a rich Pomade for the Hair, elegantly perfumed, and NEVER BECOMES RANCID.

In 1878 and 1879, they advertised their own brands of bottled "malt cough mixture," "moth powder," and cold cream, but no embossed bottles for these products have been located, and they may have been bottled with paper labels only.

Herbert E. Bucklen and Co.

Medicine:

(1) **DR. KING'S / NEW DISCOVERY / FOR CONSUMPTION //** [aqua]
(side): **H. E. BUCKLEN & Co //**
(side): **CHICAGO. ILL.**
(base): **D S G Co.**

»SB w/ post mold ; rectangular; 4 sunken panels; double-ring collar; 6 1/2" tall. *[Figure 201]*

Figure 201

Figure 202

H. E. Bucklen obtained proprietorship of the patent-medicine formulas of one "Dr. Z. L. King" of Elkhart, IN, in 1878 (Fike 1987:109). Bucklin almost immediately moved his bottling headquarters for Dr. King's medicines to Chicago, but his earliest embossed version of the Chicago bottle listed above is side-embossed "ELKHART IND_" instead of "CHICAGO, ILL" (*Figure 202*).

H. E. Bucklin & Co. is listed in the 1880 Illinois Industrial Census as having $5,000 capital invested in the business, having a maximum of 17 employees (12 males and 5 females) who were paid from $1.00 to $1.25 per day, with an annual payroll cost of $5,304. He estimated the annual value of material used to produce his medicines at $50,000 and the annual value of his products at $60,000.

Bucklin's patent medicine company remained in Chicago until 1913, and then moved its headquarters to St. Louis. They produced several of the Dr. King brands (and others) but the only medicine they bottled in the late 1870s was the *New Discovery for Consumption*, which by 1885 had become one of the best-selling brands of medicine in the world. The company claimed on a later paper label that the medicine had been originated (by Dr. King, presumably) in 1869 (Fike 1987:109). The "New Discovery" was a medicine for tuberculosis, which at the time was killing millions of people. Dr. King's *New Discovery for Consumption* was targeted by Samuel Hopkins Adams in a series of articles he published in *Colliers Weekly* magazine from 1905 to 1907 attacking the patent medicine industry (Adams 1907). In his January 13, 1906, installment (*"Preying on the Incurables"*), he noted:

> It is proclaimed to be the "only sure cure for consumption." Further announcement is made that "it strikes terror to the doctors." As it is a morphine and chloroform mixture, "Dr. King's New

Discovery for Consumption" is well calculated to strike terror to the doctors or to any other class or profession, except, perhaps, the undertakers. It is a pretty diabolical concoction to give anyone, and particularly to a consumptive. The chloroform temporarily allays the cough, thereby checking Nature's effort to throw off the dead matter from the lungs. The opium drugs the patient into a deceived cheerfulness. The combination is admirably designed to shorten the life of any consumptive who takes it steadily. [Adams 1907:45]

Charles Butt

Medicine:

(1) *(side):* **DR BUTT'S // EXCELSIOR / LINIMENT //**   [aqua]
(side): **CHICAGO.ILL.**

»IP; Beveled-edge rectangle; 3 sunken panels; double-ring collar; 8" tall. *[Figure 203]*

(2) *(side):* **DR BUTT'S // EXCELSIOR / LINIMENT //**   [aqua]
(side): **CHICAGO.ILL**

»SB w/ keyed hinge mold; rectangular w/ beveled corners; 3 sunken panels; tapered collar; 5" tall. *[Figure 204]*

(3) *(side):* **DR BUTT'S // EXCELSIOR / LINIMENT //**   [clear]
(side): **CHICAGO.ILL**

Note: There are two embossed dots or periods beneath the superscript **R** in **D^R**.

»SB w/ keyed hinge mold; rectangular w/ beveled corners; 3 sunken panels; flared lip; 4 1/2" tall. *[Figure 205]*

(4) *(side):* **DR BUTT'S // EXCELSIOR / MEDICINE /**   [aqua]
(side): **CHICAGO ILL**

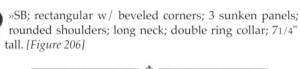

»SB; rectangular w/ beveled corners; 3 sunken panels; rounded shoulders; long neck; double ring collar; 7 1/4" tall. *[Figure 206]*

Figure 203   **Figure 204**

From their technological attributes, the three Dr. Butt *Excelsior Liniments* bottles listed above date from the late 1850s through the late 1860s. The *Excelsior Medicine* bottle dates from the early 1870s. Although this was obviously a Chicago medicine of long-standing, Charles Butt is close to invisible in the Chicago CDs during the period he made his liniment—especially if his name is misinterpreted as "Butts." This is because he made and bottled the medicine from his residence, which changed several times during the 1860s, and so was sometimes missed by the directories.

Charles Butt first appears in the Chicago CDs in 1861, as a "liniment manufacturer" residing at 391 Clark. Since his earliest Chicago bottle is pontilled, he likely arrived a little before his first listing—ca. 1859–60. He is listed in the 1870 USC as a 50-year old physician from England, so he was about 40 when he came to Chicago. In 1862, he is listed as making "Butt's liniment;" in 1863, as "manufacturer of excelsior liniment;" in 1864 and 1866, as a "liniment manufacturer;" in 1869, as "manufacturer, excelsior liniment and pills;" and in 1870 as "physician and manufacturer of Excelsior Ointment." In 1871 and 1872 he is listed only as a "physician." In 1874 and 1875, his wife Margaret is listed separately, as a "clairvoyant," and apparently Margaret is selling the *Excelsior Medicine* represented by the latest bottle we have recorded.

We have been able to locate just one (early) ad for his liniment, which ran in the *Chicago Tribune* the week of September 5, 1862 (Sept. 7, p. 3, col. 6). It is quite detailed:

Figure 205

DR. BUTT'S
EXCELSIOR LINIMENT

Dr. CHARLES BUTT wishes to be distinctly understood that his Great Remedy cures the worst cases of Rheumatism, Neuralgia, Cramps, Sprains, Bruises, Chilblains, Cuts, Contracted Sinews, Scalds, Toothache, Nervous Headache, Pains in the Back, Pains in the Side, Mosquito Bites, Ring Worm, Catarrh, Earache, Deafness, Sore Eyes, Sore Throat, Stiff Neck, Fever and Ague, Stiff Joints, Burns, Boils, Frost Bites, Warts, Corns, Milk Leg, Dysentery, Mumps, Cholera Morbus, and Diphtheria.

DIRECTIONS FOR USE ACCOMPANY
EACH BOTTLE

No ship should proceed to sea, nor House remain without this valuable liniment. Manufactured and sold wholesale and retail by

DR. CHARLES BUTT,
P.O. Box 4455, 393 South Clark Street, Chicago.
And sold by Druggists generally, and country stores.

Ask for Dr. BUTT'S EXCELSIOR LINIMENT. If they have not got it, ask them to get it. All orders promptly attended to. Retail prices, 25 cents and $1 per bottle. Try one bottle, and if not found all it is represented, return the bottle empty and demand your money where you purchased the bottle.

[*pointing hand*] All that is required is, rub it on till the pain is gone, in from three to five minutes.

After 1872, the Chicago CDs list Charles Butt and his wife Margaret at separate addresses. Charles may have become ill or incapacitated in some way (physically or legally) during this time, because Margaret seems to have taken over sales of the medicine and repackaged it in new bottles embossed *Excelsior Medicine*.

Figure 206

This appears to have been the case at least during 1872 and 1873, when she is listed in the Chicago CDs as a "clairvoyant." An ad from that time, located in the *Chicago Tribune* for February 23, 1873 (p. 3, col. 6) reads as follows:

MRS. BUTT, NATURAL CLAIRVOYANT, TEST, and business medium; also physical examinations; price, $1. Dr. Butt, Magnetic and Eclectic Physician, 21 South Desplaines-st. Dr. Butt's Excelsior Medicine and never-failing pills. Sold by all Druggists.

No further reference has been found to Excelsior Medicine after 1873, and Charles and Margaret Butt are no longer listed in the Chicago CDs after 1875.

Andrew Carpenter

Ale:

(1) **A. CARPENTER** / *AC (monogram)* / **TRADE MARK** / **CHICAGO** / **ILLS.**   [amber]
(base): **TRADE** / *AC (monogram)* / **MARK**

»SB w/ post mold; pint ale; 2-piece mold; ring-neck tapered collar; 6 1/2" tall. [Figure 207]

———— ❖ ————

Andrew Carpenter worked for Amberg & Stenson in the late 1860s and early 1870s as a teamster and soda peddler (1869–1871 Chicago CDs). He briefly went into the ale-bottling business on his own in the mid-1870s.

Figure 207

William B. Cassilly and Co.

Bitters:

(1) **PROFESSOR** / **LENNORDS** // **CELEBRATED** / **NECTAR BITTERS** //   [amber, olive green]
CASSILLY & Co / **CHICAGO** / **AGENTS** // *(blank)* //

»SB w/ keyed hinge mold; square case bottle w/ beveled edges; tapered collar; 9 1/2" tall. [Figure 208]

(2) **CASSILLY & CO.**ˢ // **NECTAR** // **COCKTAIL** // *(blank)* //   [aqua]

»SB w/ keyed hinge mold; rectangular w/ beveled edges; tapered collar; 9 3/4" tall. [Figure 209]

———— ❖ ————

Cassilly & Co. first appear in the 1864 Chicago CD as "rectifiers and wholesale dealers in liquors" located at 33 S. Water Street. Principles were William B. Cassilly and James Mix. The firm was apparently not very successful. They were listed again in the 1865 Chicago directory, but not after that. Their only Chicago newspaper ad first appears in February 1864 (*Chicago Tribune*, February 25, 1864, p. 4, col. 4):

CASSILLY & CO.,
Distillers, Rectifiers,
AND WHOLESALE DEALERS IN
FOREIGN & DOMESTIC
LIQUORS,
33 South Water Street, Chicago.

———

Manufacturers of the celebrated Nectar and XXX Magnolia Whiskey, V. A. Brown's Monongahela, Rose Gin, and all domestic liquors.

Agent's for Professor Leonard's celebrated Nectar Bitters.

Ring and Ham (2004:69) reproduce another newspaper ad for this product from the September 3, 1864, *St. Paul Pioneer* that lists the reputed curative powers of **PROFESSOR LENNORD'S** *(spelled "Leonnard's")* **CELEBRATED NECTAR BITTERS**:

N. B.
Prof. Leonnard's
CELEBRATED NECTAR BITTERS

A CERTAIN PREVENTATIVE OF **FEVER AND AGUE**, AND A SURE CURE FOR **DYSPEPSIA**, **CONSTIPATION**, CHOLIC, CHOLERA MORBUS, Cramps, Nervous Headache, Debility, Depression of Spirits, Loss of Appetite and Liver Complaint.
FOR SALE BY
C. Ragnet & Co., P. F. McQuillan, J. I. Beaumont,

CASSILLY & CO.

Just who "Professor Lennord" might have been, if he was a real person, is unknown. But it is interesting to note that both years *Cassilly & Co.* operated in Chicago one of their employees was Richard H. "Leonard." In 1864 he was listed as a "rectifier" there, and in 1865 he was listed as a "clerk." Also, since the *Cassilly & Co. Chicago Tribune* ad in 1864 indicates they were "agents for," rather than "manufacturers of" Professor Leonard's Bitters (note that the professor's name is spelled the same way as their employee's name in the newspaper ad), perhaps the bitters originated and was distributed to sales agents from somewhere other than Chicago.

Figure 208

Note also that, in addition to the "celebrated Nectar" (i.e., the **NECTAR COCKTAIL** listed above), Cassilly & Co. also manufactured three other name-brand products which may have been put up in embossed bottles. These were: **XXX MAGNOLIA WHISKY, V. A. BROWN'S MONONGAHELA**, and **ROSE GIN**.

Figure 209

Gardner S. Chapin and James J. Gore

Whiskey:

(1) (shoulder): **CHAPIN & GORE / CHICAGO //**   [amber, olive green]
    (body): **SOUR MASH / 1867**
    (base): **H. FRANK'S PAT^D / AUG. 1872**

»SB w/ post mold; ringed barrel-shape; ring-neck cylindrical blob top w/ internal screw threads; 8 1/2" tall. Top of glass screw-cap embossed: **PAT. AUG6 72** around central raised ring and dimple. *[Figure 210]*

(2) (shoulder): **CHAPIN & GORE / CHICAGO //**   [amber]
    (body): **SOUR MASH / 1867**
    (base): **HAWLEY GLASS CO / HAWLEY. PA.**

»SB w/ post mold; ringed barrel-shape; ring-neck cylindrical blob top w/ internal screw threads; 8 3/4" tall; 4/5 quart. Screw cap made of hard black rubber. *[Figure 211]*

(3a) (outer ring): **CHAPIN & GORE / CHICAGO**   [amber]
    (inner ring): **SOUR MASH / 1867**
    (base): **PAT^D AUG6^TH 1872 / H FRANK**
    Note: Bottle manufactured in two sizes, half-pint and pint, and in light and dark shades of amber. [Figures 212–213]

»SB w/ post mold; strap-sided flask; ring-neck cylindrical blob top w/ internal screw threads; about 6" and 7 1/2" tall. Top of glass screw-cap embossed: **PAT. AUG6 72** around central raised ring and dimple. *[Figure 210]*

(3b) (outer ring): **CHAPIN & GORE / CHICAGO**   [amber]
    (inner ring): **SOUR MASH / 1867**
    Note: Unimbossed base and screw cap. Bottle manufactured in two sizes, half-pint and pint, and in light and dark shades of amber. [Figures 212–213]

»SB w/ post mold; strap-sided flask; ring-neck cylindrical blob top w/ internal screw threads; about 6" and 7 1/2" tall.

(4a) (shoulder): **CHAPIN & GORE /**   [yellow-amber]
    CHICAGO //
    (base): **HAWLEY GLASS CO / HAWLEY PA**

»SB w/ post mold; cylindrical whiskey-style quart with 4 fluted panels on upper body below shoulder-embossing; internally-threaded ring-neck tapered collar with embossed 6-pointed star on glass screw cap; 10" tall. *[Figure 214]*

Figure 210

Figure 211

Figure 212

(4b) (shoulder): **CHAPIN & GORE / CHICAGO //**   *[amber]*
(base): **CHAPIN & GORE / CHICAGO. ILLS.**

»SB w/ post mold; cylindrical whiskey-style quart with 4 fluted panels on upper body below shoulder-embossing; unthreaded ring-neck tapered collar; 10 1/2" tall. *[Figure 214]*

--- ❖ ---

Gardner Chapin and James Gore began in business in 1865, with their Chapin & Co. grocery store located at the corner of State and Monroe:

> Gore did not like the grocery business and in 1867 he finally convinced Chapin to add a liquor department to the store. The department had a separate entrance on Monroe Street and the business was transacted on a wholesale basis only, which meant in those days, a jug trade. Soon the grocery business ended altogether and thus, the partners were listed in 1868 at 162 State Street as liquor wholesalers.
> Business was very good and they decided in 1869 to open a retail establishment at 65 Monroe Street. Dame Fortune continued to smile on them for in early 1871 they opened another store at 152, 22nd St., putting out a brand of their own whiskey...called "1867." Indeed, they were fortunate to have this location because the Great Fire later that year destroyed their other establishments to the extent of a $75,000.00 loss.
> Undaunted by this loss and encouraged by the successful marketing of their "1867" brand, on May 13, 1872, they opened what became the most famous liquor establishment in the world at 73 and 75 Monroe Street, the former site of the old Majestic Theatre. Gore originated the words "Sour Mash" and I believe... the barrel whiskey pictured *[#1 in this listing]* dates very near this period in time. [Panek 1977:21]

William Frank & Sons of Pittsburgh, *a.k.a.* Frankstown Glass Works, made the distinctive barrel-shaped bottle whose threaded internal stopper (patented by son Himan Franks) was

Figure 213

Figure 214

used by Chapin & Gore. They produced this bottle for only a brief time. In June 1874, the Frankstown Glass Works where the bottle was blown burned down (McKearin and Wilson 1978:159). It was rebuilt within 10 weeks, but the following year Frank & Sons shuttered their business for good (Lockhart et al. 2008:32). Immediately afterward, the Hawley Glass Works, another Pennsylvania manufacturer, started producing the bottles, but with a hard rubber threaded stopper (*#2 in this listing*). At the time, Chapin & Gore were operating three retail liquor stores in addition to their wholesale liquor business (1875 Chicago CD).

[A] third Chapin & Gore bottle next appeared, most likely in the mid-1880s. Its overall side embossing remained unchanged, but the base was made smooth.... Additionally, the bung hole on the reverse side of the bottle was eliminated, and the barrel's eight rings were made to appear more wavy.... Finally, unlike either bottle which came before, the collar was neatly tooled. The bottle was made both with and without inside threads.

The last Chapin & Gore barrel went into production near the turn of the century. This bottle still included the words "Chapin & Gore" and "Chicago" around the top, but reference to "Sour Mash 1867" was eliminated altogether. In its place, a flat diamond-shaped surface was provided for a paper label. The label identified this bottle's contents as "Old Jim Gore Burbon Whiskey." [Denzin 1993:47]

The Chapin & Gore liquor business is listed in the Chicago CDs well into the early 1900s.

Abel H. Christie

Medicine:

(1) **CHRISTIE'S // AGUE BALSAM // CHICAGO / ILL** [aqua]

»SB w/ keyed hinge mold; beveled-edge rectangle; 3 sunken panels; 6" tall. *[No photo. Similar to Figure 215]*

(2) **CHRISTIE'S // AGUE / BALSAM // CHICAGO** [aqua]

»SB w/ keyed hinge mold; beveled-edge rectangle; 3 sunken panels; double-ring collar; 7 1/8" tall. *[Figure 215]*

❖

Dr. A. H. Christie was a New York patent-medicine manufacturer from 1846 until his death in 1852 or 1853. He first gained fame for his galvanic and magnetic electrotherapy aids and medicines—including his galvanic belts, necklaces, and bracelets, and his bottled magnetic fluid "for cure of paralysis and palsy, epileptic fits and palpitations of the heart" (Odell 207:75;

Baldwin 1973:111). *Christie's Ague Balsam* was later added to his New York medicine line. According to copyright designs for its paper-box label in the Library of Congress [*Figure 216*], it was "unequaled in its power of curing AGUE and FEVER, CHILLS and FEVER, and all the varied forms of BILIOUS DESEASES...and all other Fevers that originate from Malaria."

After Dr. Christie's death, sole proprietorship of his medicines was acquired by J. & J. F. Trippe & Co. of New York. The earliest New York Christie's medicines have pontil-rod base scars, which predate the smooth-based Chicago Christie's bottles. By the late 1850s, a few regional sole-distributorships were also created. The regional proprietor for the Ohio area was H. M. Wooster & Co. of Norwalk (Blasi 1974:30; Odell 2007:75; Wilson and Wilson 1971:110). Another distributor for the upper Midwest, whose Christie's bottles were embossed "CHICAGO," was Bolles, Smith & Co. (*Quincy Whig and Republican*, May 27, 1859, p2/c2). During 1858 and 1859, this company distributed Christie's, Farrell's, Sloan's, Wells', and other patent medicines from their Chicago offices at 14 S. Water (1858) and 124 Lake (1859). A large February 1859 ad for the company in the *Chicago Press & Tribune* referred to their Lake Street facility as the "Great Western Wholesale and Retail Patent Medicine Depot" and listed their product categories, including cough and pulmonic remedies, blood purifiers, fever and ague medications, hair dressings and restoratives, rheumatic pills and liniments, pile remedies, pain killers, purgatives and cathartic pills, "female" pills and "private" disease remedies, liver and dyspeptic remedies, vermifuges, tonic bitters and schnapps, and a range of trusses, strengthening plasters, soaps, tooth powders, toilet articles, perfumery, and "handkerchief extracts."

Figure 215

In 1860 and 1861, patent-medicine distributor John D. Parks occupied the 124 Lake address. Bolles, Smith & Co. were no longer listed in CDs after 1859, so Parks likely took over their medicine distributorship. After 1861, Parks moved his business to New York.

It is probable that Chicago-embossed *Christie's Ague Balsam* was distributed in the area for only four years: 1858–61. The only local newspaper ad we have found for it was run in the *Chicago Press and Tribune* by Bolles, Smith & Co. in fall 1858 (*CP&T*, Sept. 10, 1858, p. 2, col. 9):

THE GREATEST MEDICAL DISCOVERY OF THE AGE
FACTS FOR THE PEOPLE
RESPECTING
Dr. Christie's Ague Balsam

DR. CHRISTIE'S AGUE BALSAM is the best [relief] for the permanent cure of Chills and Fever, Fever and Ague, Dumb Ague, and all malarial diseases incident to the climate.

DR. CHRISTIE'S AGUE BALSAM has never been known to fail where the directions have been strictly followed.

DR. CHRISTIE'S AGUE BALSAM is a
PURELY VEGETABLE COMPOUND,
Containing neither Arsenic, Quinine, Strychnine, or anything of a poisonous or deleterious nature.

DR. CHRISTIE'S AGUE BALSAM has had for years the largest sales of any remedy for specific disease ever offered to the public.

DR. CHRISTIE'S AGUE BALSAM has attained its immense popularity through its own merits, its great power in curing disease, its singular harmlessness and freedom from hurtful ingredients.

Figure 216. Printed design of a pasteboard-box container for *Christie's Ague Balsam* bottles.

It can be taken by the [...] infant, vigorous youth, and feeble age[d], ever without injury, and always with benefit:

[*Four 1857–58 testimonials follow, from regional sales agents in Matoon, IL, Vinton, IA, Metamora, IND, and Delphos, O.*]

Price $1 per bottle, or 6 bottles for $5

BOLLES, SMITH & CO., 14 South Water-st., Chicago, Proprietors for the Northwestern States, also, Wholesale Dealers in Patent Medicines generally.

In general, smooth-based bottles used by Illinois manufacturers do not appear until 1860. But smooth-base bottle technology first appears two to three years earlier on the East Coast (Von Mechow 2008: *"Antique Soda and Beer Bottles: Bottle Bases"* at http://mysite.verizon.net/vonmechow/bases.htm). Since all known CHICAGO-embossed Christie's bottles are smooth-based, they were likely not bottled locally, but rather manufactured in, and shipped from, New York for distribution in the Chicago area during this four-year period.

Arthur Christin

Soda ~ Mineral Water:

(1) **BELFAST GINGER ALE / ARTHUR CHRISTIN / CHICAGO ILL. /**   [aqua]
(*front heel*): **PAT. APR. 13TH 1875** [*2 raised dots below* ᵀᴴ] //
(*reverse heel*): **A & D. H. C / ARTHUR CHRISTIN**
(*base*): **AC** *monogram.* [Figure 217]
Note: There are three shape variations of the base of this embossed bottle: one variety has a round base; a second variety has a semi-rounded base; and a third variety has a flat base and square heel (see Figure 218).

»SB w/ post mold; cylinder w/ vertical embossing; blob top w/ recessed rubber-ring seal for AC-patent stopper; 8" tall.

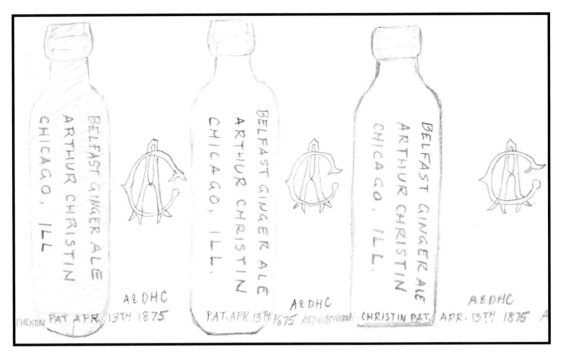

Figure 218. Arthur Christin embossed *Belfast Ginger Ale* varieties drawn by Robert Kott (1988) from fragments recovered by Paul Welko (Welko 1978 and personal communication 2009). Just the flat-based bottle variety (*Figure 215*, and right-hand drawing above) included a base-embossed monogram in addition to the monogram embossed on the reverse side of the body. Fragments of the round-based variety (shown at left above) are known in both aqua and cobalt glass colors. This variety of the bottle was also produced with a glass (rather than hard-rubber) Christin-patent stopper.

(2) **ARTHUR CHRISTIN / CHICAGO / ILLS //**   [cobalt]
(large monogram): **AC**
(heel): **A. & D. H. C**

»SB w/ post mold; tall slope-shouldered quart; tapered collar; 12 1/2" tall. *[Figure 219]*

(3) **CHRISTIN & Cº / CHICAGO**   [amber]

»SB w/ post mold; tall slope-shouldered quart; ring-neck cider finish; 12 1/2" tall. *[Figure 220]*

The earliest Chicago CD listing found for Arthur Christin & Co. (with Alfred Baby) is as a bottler of "champagne cider" at 64 S. Halsted in 1872. In 1873, Christian & Co. (with Edwin H. Horsey) relocated to 862 & 864 S. Halsted. In 1874, Christin was at the same address without a partner. During all three years he is listed as a bottler of champagne cider. Bottle #3 in this listing is likely associated with the 1872–73 business, while bottle #2 likely dates to 1874. According to the 1880 USC, Arthur Christin, who was 49, came to Chicago from Canada.

Arthur Christin's bottling company was located at 362 S. Halsted in 1875, the year he patented an internal hard-rubber gravitating-stopper soda-

Figure 219

bottle closure (April 13th, 1875, see Welko 1978). Beginning at this time, his bottling efforts shift to a focus on soda water and Belfast Ginger Ale manufacture (using varieties of bottle #1). The new Christin patent bottle-closure was widely adopted by a number of bottlers in the United States (and at least one in Canada). Bottles with his patented gravitating-stopper closure were manufactured in Pittsburgh by the *A. & D. H. Chambers* glassworks. The 1880 IIC indicates that the total value of his ginger ale products that year (produced full-time during two months of the year) was $44,000.

In 1883, Christin moved his business to 425 W. Harrison. The latest bottles he uses are Hutchinson-style internal-stopper bottles (Harms 1978; Wright 1992), so perhaps his move was associated with a decision to abandon bottling using his own patented Christin-style hard rubber gravitating-stopper bottles, and adopt the Hutchinson style.

His bottling works is last listed in the 1888 Chicago CD. Beginning in 1889, he is listed as Superintendent and Treasurer of the *Chicago Fountain Soda Water Company.*

In a near-tragedy late in his life, a carriage transporting Arthur Christin's wife and three visiting family members from Ottawa, Canada (Mrs. C. Christin and her two young daughters) was hit by a Rock Island train at the Harrison Street crossing on December 30, 1894. At the time, his wife was not expected to live ("Come Nigh to Death," *Chicago Tribune*, Dec. 31, 1894). However, six years later (September 25, 1900, p. 7, col. 1), the *Tribune* reported that the couple had renewed their marriage vows on the occasion of their 50th golden wedding anniversary, so Mrs. Christin was able to recover from the crash.

Figure 220

Thomas Clowry and James Fitzgerald

Soda ~ Mineral Water:

(1) C & F / CHICAGO.   [light green, light blue]

»SB w/ keyed hinge mold; shouldered-blob-top soda. *[Figure 221]*

Clowry & Fitzgerald are listed in the 1864 Chicago CDs as soda water manufacturers with bottling facilities at 298 Clark. C & F seem to be a one-year company. They do not appear in either the 1863 or 1865 Chicago CDs.

Silas F. Collins

Bitters:

(1) DR S. F. COLLINS / JAUNDICE BITTERS / CHICAGO, ILL   [aqua]
Note: *2 raised dots beneath superscript* ^{R} *in* D^R

»IP; patent-medicine-style rectangular bottle w/ beveled edges; 3 sunken panels; double-ring collar; 8 3/8" tall. *[Figure 222]*

(2) DR S. F. COLLINS / JAUNDICE BITTERS / CHICAGO. ILL.   [aqua]
Note: *2 raised dots beneath superscript* ^{R} *in* D^R

»SB w/ keyed hinge mold; patent-medicine-style rectangular bottle w/ beveled edges; 3 sunken panels; 9 3/4" tall. *[Figure 223]*

Dr. Collins is listed in Chicago CDs from 1853 through 1858 as a physician and druggist at 44 Franklin Street. The 1853 CD also includes a separate advertisement for Collins' "Indian Medical Depot," which lists seven of his prepared medicines for treatment

Figure 221

of chronic diseases (*Ague Killer, Anti-Bilious Physic, Blood Purifier, Cholera Tincture, Diarrhea and Dysentery Cordial, Life Ointment,* and *Lung Syrup*). Note that his *Jaundice Bitters* is not included in this early list. His 1853 add also announces **CASH PAID FOR ROOTS AND HERBS,** and his 1854 Chicago CD listing specifically characterizes him as a "botanic physician."

During spring and summer 1855, *Chicago Tribune* advertisements for Dr. Collins *Jaundice Bitters* also indicate that he still made and marketed the 1853-listed bottled products at his "Indian Medical Depot" in Chicago. However, a new one had been added to the seven listed in 1853: *Rheumatic Mixture*. No embossed bottles of any kind for any of the earlier products have yet been located; the only embossed bottle we have documented (*above*) is the *Jaundice Bitters*—which he likely produced from the time of the 1855 ad until 1858, his last year as a Chicago physician.

Unless his bottles were being manufactured for him by an East Coast glass house (where empontiling was replaced by snap-case technology ca. 1856–57), the fact that the later of his two bottle styles is smooth-based (#2 in the listing) indicates that he marketed and sold his *Jaundice Bitters* until at least 1860.

Collins' ads described his *Jaundice Bitters* as a "universal remedy for the Jaundice, it gives life and action to the liver, regulates and restores the billiary organs, to their habitual functions, and for habitual costiveness its equal is not known" (*Chicago Tribune*, May 25, 1855, pg. 1).

Figure 222

Figure 223

As an interesting side note, Collins announced in a June 19, 1855, *Chicago Tribune* ad that he was now "preparing" and distributing Dr. William Beach's Wine Bitters, which was originated and earlier distributed in New York City (Odell 2007:27). This product had been advertised in the late 1840s in Chicago (*Weekly Chicago Democrat*, June 6, 1848) and bitters-guide listings have given the impression that it might have been a Chicago product (Ring 1980:B39; Ring and Ham 1998:88).

J. M. Connell

Medicine:

(1) **CONNELL'S BRAHMINICAL** / (*small paper label here*) / **MOONPLANT** /   [olive amber]
(*front heel*): **EAST INDIAN** / **REMEDIES** //
(*ring of 10 stars around a pair of bare footprints*) / **TRADE MARK**
(*base; ornate script*): J m Connell / Proprietor
Note: Small rectangular paper label on one example of the bottle reads: **APERIENT.** /
Shake, (upper **M** *monogram***) Pour out** / **Recork (***lower* **M** *monogram***) Speedily**

»SB; oval; double-ring collar; 9" tall. [*Figure 224*]

───────── ❖ ─────────

Although this bottle is not embossed "Chicago," several examples of it have been found in the Chicago area. During the Civil War years (1860 and 1864–65), a merchant named Jeremiah M. Connell is listed in Chicago CDs as a retail dealer in groceries and provisions. From the bottle's technological attributes, however, it appears to have been produced in the mid-1870s. Initial indications are that the name similarity is just coincidence.

J. M. Connell's patent medicines first show up during the late 1860s, in Australia. In 1869, the Melbourne distributor (William Witt & Co.) published a 68-page advertising booklet titled "*Connell's East Indian Remedies.*" It contained numerous user-testimonials "from some of the best-known residents of the Australasian Colonies," as to the effectiveness of his medicines in combating liver problems, gout, rheumatism, and the ague. In 1869 and 1870, the *Brisbane Courier* also carried ads for Connell's medicines, announcing that W. Steele would be the sole agent there. Mr. Steele's ads highlighted the fact that Connell had also developed "A Mild and Active Aperient for BILIOUS, CONSTIPATED, and FLATULENT SUFFERERS" (*Brisbane Courier*, Oct. 6. 1969, p. 4, col. 5).

After 1870, Connell's products seem to disappear from Australia. But in 1873, an advertising booklet similar to the earlier Australian version was published in San Francisco. It for the first time highlighted Connell's Moonplant remedies: "*Connell's East Indian Brahminical Moonplant Remedies. Their origin denoted and usefulness declared in testimonies of many well-known persons.*" After that— probably in the mid-1870s—Connell's bottles show up in the Chicago area (local distributor unknown, advertising booklets unknown). The last "footprint" we have found for Connell's medicines is an advertising booklet published in 1881 in Boston by the local distributor there (Smith & Videto: "*Dr. J. M. Connell's Brahminical Moonplant. East Indian remedies*").

From the fragmentary data we have been able to gather so far, the San Francisco, Chicago, and Boston distribution nodes for Connell's medicines appear to reflect the efforts of either sequential or contemporary regional American sales agents during the decade of the 1870s. The Chicago distributor is yet to be identified, as is J. M. Connell himself. A local "copy" of Dr. Connell's medicine was also produced in Chicago near the end of the 1870s by Dr. George Wolgamott (*see his listing*).

Figure 224

As to the medicine's content, "Brahminical moonplant" was associated with the Vedic "Soma" ritual in India. It is thought to have been a mountain plant with hallucinogenic energizing and intoxicating qualities, but the species has not been positively identified. It was gathered by moonlight and made into an alcoholic fluid, distilled through a goat-hair sieve with barley and clarified butter. It is said to have been "a nasty drink" (Balfour 1885:703).

Since Connell's medicine is called an "aperient" (i.e., a laxative), and he distributed it in America, it may be interesting to note that "Moonplant" is another name for Jimson weed here.

Egbert C. Cook and Co.

Soda ~ Mineral Water:

(1) EGBERT. C. COOK & CO / CHICGO [aqua]
Note: "CHICGO" misspelling. [Figure 225]

»SB; shoulder-embossed only; British "dumpy" soda style; rounded base; round shoulders; long neck; blob top; 6 1/2" tall.

Cook was a junior partner with Victor Barothy in a Chicago soda water bottling company during 1872 and 1873 (*see their listing*). In the 1874 Chicago CD, Egbert C. Cook is listed alone in the manufacturers index (p. 1327) as a

> "MANUFACTURER OF
> *Pure Soda Water,*
> AND DEALER IN
> Soda Water Apparatus and Materials,
> 50 & 52 Fourth Av., near VanBuren St.
> Importer and bottler of English, Scotch, and American
> Ales and Brown Stout, Belfast Ginger Ale,
> Champagne Cider, etc.
> Sole manufacturer of A. W. Caverly's
> Patent Tilting Stand for emptying Carboys."

In an 1875 CD advertisement (1875:1309), E. C. Cook & Co. (Frank H. Peabody, junior partner) state that they are wholesale dealers in bottled goods and soda water apparatus and materials at 10 & 12 Madison Street. However, they are also listed under the "Soda Water Manufacturers" section of the business index. An April 25, 1875, newspaper article about the company in the *Chicago Daily Tribune* (p. 5, col. 6) indicates that *Cook & Co.* were indeed still bottling that year:

> The work of charging fountains is the sole employment of large establishments. The firm whose name heads this sketch are at present doing a more extensive business in this direction than any other house west of New York. They have in their establishment every facility for executing the work in the most speedy and complete manner possible, and... they are, continually, during the summer months, in receipt of orders from all parts of the adjacent country. Besides being **general manufacturers of soda water, mineral spring water, Belfast ginger ale, New Orleans mead, and other celebrated beverages, they are extensive bottlers of pale ale, stout and porter** [*emphasis ours*], and what is of still more interest and importance to the public, they have the exclusive right to use the celebrated Mathews' block tin-lined steel fountains, the construction of which entirely prevents the waters becoming contaminated with the metal, which is often the case when copper and iron fountains are used. The office and salesroom of the firm are at Nos. 50 and 52 Fourth avenue, Chicago.

Figure 225

The embossed bottle listed above likely dates from this 1874–75 bottling effort. Although this bottle style is unique for Illinois, its shape is that of a distinctive late-nineteenth-century British ale or ginger-ale used by the Cantrell companies in Brighton (*Cantrell & Co.*; *Cantrell & Cochrane*: see Schmeiser 1970:387–388), and may well have been copied and used for its "ambiance."

In the 1876 and 1877 CDs, Cook & Peabody are listed only as "bottled goods" manufacturers at 12 Madison. By 1878, the company is no longer listed in the Chicago directories.

Since we have recorded just one embossed bottle for the company, it seems likely that most of their bottled products were paper-labeled only.

Dr. Benjamin S. Cory and Son

Bitters:

(1) **DR. CORY'S** (*in small arched sunken panel*) // **STOMACHIC** //   [aqua]
    (*unimbossed flat panel*) // **BITTERS**
    Note: Dr. Cory's embossed bottle is only known from a few broken examples. The bottle's shape and style of embossing emulate that of the popular, contemporary, Ayer's Cherry Pectoral medicine bottle (Figure 226). Dr. Ayer's Medicine products were regionally marketed in the Midwest from the company's base of operations at Lowell, Massachusetts (see Baldwin 1973:46).

    »IP; beveled-edge rectangle; 3 sunken panels; double-ring collar; ca. 6" to 7" tall.

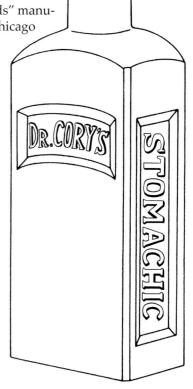

Benjamin Sayre Cory was born in 1805 and was educated as a physician and surgeon in New York State. After finishing his education, he established a medical practice in Canada. During the later 1830s, "he was a regularly commissioned surgeon for the Prince Edward Co. Regiment, raised for active service

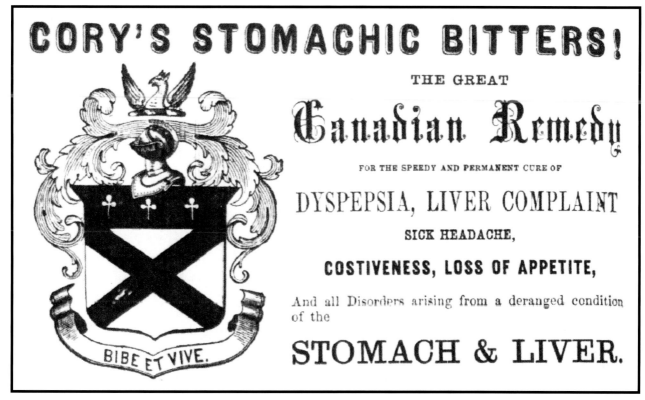

Figure 226. Cory's coat-of-arms logo and "BIBE ET VIVE" banner slogan as it appeared in the company's 1859 *Stomachic Bitters* ad (published in that year's edition of the Chicago City Directory, p. 174).

against the Rebels" (Dickinson 1914:31). In 1854, B. S. Cory moved his family from Wellington, Ontario, to Waukegan, IL, where a number of members of the "Cory Clan" already resided.

B. S. Cory & Son (likely his eldest son, James) list themselves as "druggists and apothecaries" in their ads, and appear to have carried on a two-pronged attack to market their *Stomachic Bitters* in northeastern Illinois in 1859. According to their 1859 newspaper and CD ads, they simultaneously opened wholesale and retail distribution outlets in both their Lake County home-town area of Waukegan (at 47 Washington Street) and in downtown Chicago (at 174 South Clark Street). Except for the business addresses, the large ads were fundamentally the same. They included a very ornate coat of arms (Figure 226) with the inscription "**BIBE ET VIVE**" ("Drink and Live"), and advertised:

CORY'S STOMACHIC BITTERS
The Great ***Canadian Remedy***
for the speedy and permanent cure of
DISPEPSIA, LIVER COMPLAINT,
SICK HEADACHE, COSTIVENESS, LOSS OF APPETITE
And all disorders arising from a deranged condition of the
STOMACH & LIVER.

PRICE ONE DOLLAR PER BOTTLE—SOLD BY ALL RESPECTABLE DRUGGISTS.

An advertisement placed by the company in the *Woodstock Sentinel* beginning in mid-January, 1859, provided more detail on the claimed curative powers of *Dr. Cory's Stomachic Bitters*:

> These bitters are compounded from Gums long known to the Profession as possessing peculiar curative powers in cases of Dyspepsia, Liver Complaint, &c., which operate powerfully upon the Stomach, Bowels, &c., and the Digestive Organs, restoring them to a healthy and vigorous action, and thus by the simple process of nature, enable the system triumph over Disease.
>
> Its adaptation to the cure of all afflictions of the Billiary or Digestive Organs is peculiar to itself.... **For this disease, every physician will recommend Bitters of some kind, then why not use an article known to be infallible? Every country has its Bitters as a preventative of disease and for strengthening of the system in general, and among them all there is not to be found a more healthy people than the Canadians**, from whom this preparation emanated, based upon scientific experiments which have tended to advance the destiny of this great preparation in the grand medical scale of science.... And as it neither creates nausea nor offends the palate, and rendering unnecessary any change of diet, or interruption to the normal pursuits of life, but promotes sound sleep and healthy digestion, the complaint is thus removed as speedily as is consistent with the production of a thorough and permanent cure.

The Cory & Sons firm was also listed in Civil War–era IRS assessments (in 1862 as apothecaries & physicians and in 1863 as an apothecaries and retail liquor dealers). However, the ads appear to have been a one-shot effort, and in Chicago CDs they are only listed (with their ad) in 1859. They are not listed at all, even as Chicago druggists, in 1858 or 1860.

In the IRS assessment for June 1866, B. S. Cory is listed alone as a physician in Waukegan.

Lewis Cox Chemical Company

Ink:

(1) **COX'S / CARMINE / INK** [aqua]

»SB; "schoolhouse" ink bottle; pharmacy-style flared-ring collar with rounded lower edge; 2½" tall. *[Figure 227]*

Figure 227

In the 1860 USC, Lewis Cox is listed as a 44-year-old ink manufacturer in St. Louis, MO. In the 1870 USC, he is listed as a manufacturer of ink in Chicago.

Cox first appears in Chicago in the 1864 CD, as an inkmaker at 172 N. Water Street (with William Brandt). Cox and Brandt remain together as manufacturers of writing fluids and inks through the 1868 CD listing. But in the 1869 through 1871 Chicago CDs, the business becomes Cox and Co. (Lewis Cox and 29-year-old son Lewis H. Cox, proprietors). The company still manufactures writing-fluid, and it is still located at 172 N. Water Street. In the 1872 and 1873 CDs, Lewis Sr. has taken on a new partner in the ink business (Benjamin Wright) and has moved to a new address (143 Kedzie). In the 1874 CD, Lewis Cox is still listed as an ink manufacturer, but alone and at his residence.

By 1875, the only Lewis Cox in the Chicago CD is a bookkeeper (probably the son, who worked as a bookkeeper when he first came to Chicago in 1867). There is no separate listing for Cox and Co. ink manufacturers.

John S. Cram and Samuel H. Melcher

Medicine:

(1) *(side):* **DR CRAMS // FLUID LIGHTNING //** [aqua]
    (side): **CHICAGO**

   »SB w/ keyed hinge mold; rectangular; 4 sunken panels; [lip form?]; only example seen was broken at the neck: bottle would be ca. 3" tall. *[Figure 228]*

(2) *(side):* **DR CRAMS // FLUID / LIGHTNING //** [aqua]
    (side): **CHICAGO**

   »SB; rectangular; 4 sunken panels; thick flared-ring collar; 2 3/4" tall. *[Figure 229]*

———————————— ❖ ————————————

Dr. John Cram is first listed as a Chicago physician in the 1866 CD. By 1867 (CD) he is associated with an "Inventor's Exchange," and by that time may be formulating and beginning to sell his *Fluid Lightning* to patients through his practice (since the earliest embossed bottles documented have a keyed hinge mold, and likely date to the late 1860s).

However, there is no indication of wider marketing of *Fluid Lightning* until November and December 1873, when its first newspaper ads appear in the *Chicago Daily Tribune*. In the 1873 CD, Cram is still listed alone as a physician at 673 W. Lake Street. The earliest

Figure 228　　　　　　　　　　　　　　　　　　　　　　　　　　　　　　　　　**Figure 229**

CDT ad (Nov. 23, p. 3, col. 5) indicates that the product might have first been "introduced" (free of charge) to a wider audience at the 1873 Chicago Exposition:

"THE THOUSANDS WHO HAD THEIR HEADACHE and neuralgia pain cured at the Exposition can find DR. CRAM'S Celebrated Fluid Lightning at 673 West Lake-st."

And during November and December, anyone who took the trouble to come to the "Howard Medical Association" at 673 W. Lake could still have it applied free of charge:

A SURE THING!
DR. CRAM'S FLUID LIGHTNING instantly cures Neuralgia, Headache, Toothache, and all nervous pains. Applies FREE OF CHARGE, daily, at Howard Medical Association, 673 West Lake-st., Chicago Ill., or sent to any address for $1.00 per bottle. For sale by druggists.

In the 1874 CD, Dr. Samuel H. Melcher first appears as a physician at 8 Kingsbury. The following year, Drs. Cram and Melcher are listed as partners in the patent-medicine business at 115 Randolph Street (1875 Chicago CD). The listing identifies their business as simply "Fluid Lightning."

Dr. Cram's *Fluid Lighting* was advertised in the April 3, 1876, Rock Island *Daily Argus*, which identifies the manufacturers as Cram & Melcher, 115 E. Randolph Street, Chicago. The product was sold in large and small sizes. A second advertisement in the same newspaper dated April 2, 1877, declares that *Fluid Lightning*, "Will positively afford relief by external application. It cures on the instant Neuralgia, Nervous Headache, Rheumatism, Toothache, Earache, and all nervous pains, SO AS BY MAGIC. Sold by all Druggists at 50 cts and $1.00 per bottle."

According to Baldwin (1973:133), Dr. Cram had moved to Philadelphia by 1878, where his *Fluid Lightning* is advertised in the *Public Ledger* on July 12th of that year.

Edward Y. Cronk

Root Beer:

(1) **E. Y. CRONK / ROOT BEER / CHICAGO //**   *[cobalt, light blue]*
   (monogram): **E. Y. C. /**
   (right rear heel): **BODE**

   »SB w/ post mold; high-shoulder, cylindrical blob-top soda; 7 3/8" tall. *[Figure 230]*

(2) **E. Y. CRONK / ROOT BEER / CHICAGO // REGISTERED /**   *[cobalt, light blue, lime green]*
   (monogram): **E. Y. C. / TRADE MARK /**
   (right rear heel): **BODE**

   »SB w/ post mold; high-shoulder, cylindrical blob-top soda; 7 3/8" tall. *[Figure 231]*

———————— ❖ ————————

Edward Cronk was a younger brother of Albert Cronk of Detroit, where Albert was in charge of the main bottling plant producing "Dr. Cronk's" beverages, a business started in the early 1850s by their father, "Dr. Cronk." Their famous entrepreneurial father was Warren Cronk, an early bottler and "medicine man." He was born Warren "Cronkhite" in New York in 1804. He had shortened his family name to Cronk by the time of the 1830 census, and had given himself the title of "Dr." by the time of the 1850 census, in which he is listed as a "compounder of medicine" (Cronkite 1998:7.2).

Dr. Cronk began as a brewer in Albany, NY, during the 1840s (Graci 1994:2), and his earliest glass-bottle bottling enterprise appears to have been in Buffalo, NY, at about the time of the 1850 census—where pontiled bottles embossed CRONK & SULLIVAN / BUFFALO, N.Y. are known (Spoelstra 1994:20).

Also, many of Dr. Cronk's earlier bottles are hand-turned and 12-sided Merrill- and Smith-patent stoneware containers stamped DR. CRONK'S ROOT BEER and DR. CRONK'S SARSAPARILLA BEER (Graci 1995; Graci and Rotilie 2003):

> True sarsaparilla was obtained from various tropical American species of the genus *Smilax*.... A couple of substitute plants used for sarsaparilla were the wild sarsaparilla and spikenard, both North American plants of the ginseng family. Their extracts were used mostly for medicinal purposes, hence <u>Doctor</u> Cronk. This fellow was no G. P., however, that much I **do** know. [Spoelstra 1994:23]

Dr. Cronk's root beer and sarsaparilla were among the company's most popular products, and stoneware bottles produced by several Great Lakes area potteries continued in use for several years after the move to Detroit. Dr. Cronk also produced mineral water, sodas, and ginger beer. (Canon 1993:14). By 1877, Cronk's son Albert had taken over the Detroit operation (in partnership with Charles Palmer, who had joined the firm in 1873). The firm ceased bottling operations and sold out to Schafberg & Jarrait in 1883 (Spoelstra 1994:21).

In 1870, Warren and Albert Cronk had also established a Windsor, Ontario, branch of the firm. Albert's younger brother Edward Y. Cronk joined the operation there in 1876. He had previously operated one of the ginger beer and soda water plants in Detroit

Figure 230

Figure 231

(1868–72). Edward left the Ontario branch of the firm in 1879 to attempt to start up a branch of the family bottling operations in Chicago (see Spoelstra 1994).

He is first listed in the 1879 Chicago CD as a root beer manufacturer at 49 W. Van Buren. The listings continue during 1880–1882, but by 1883 he and his Chicago branch of the Cronk-family bottling operations are gone. This is the same year that the Dr. Cronk bottling operations in Detroit were closed, so his Chicago departure appears to correspond with a general downfall of the family's Great Lakes area regional bottling empire, rather than a failure of business in Chicago specifically. Edward Cronk was listed in the 1880 IL Industrial Census as a Chicago "sasparilla" (root beer) bottler who was active full-time for four months, half time for four months, and idle for four months each year. His total 1880 annual product value was only $2,400.

During Edward Cronk's four-year Chicago bottling effort, he not only bottled in glass (see listings), but also packaged his root beer in a large (quart-size or larger) hand-turned, salt-glazed stoneware bottle shoulder-stamped **E. Y. CRONK** with cobalt-highlighted lettering (*see Figure 232*). This is a very late instance of stoneware bottling in the Midwest, and it likely indicates that the Dr. Cronk family of root beer and sarsaparilla beer products had become so identified in consumer's minds with the Cronk root beer brand that the company's regional branches continued to bottle in stone containers, as a sideline, throughout their nearly 40-year business history.

Abraham F. Croskey

Gin:

(1) **CROSKEY'S** / **COLUMBIAN GIN** / [amber]
(embossed shield logo*) / **CHICAGO ILL'S**
*Note: The logo consists of a banner reading **CROSKEY'S** / (image of a shield containing **COLUMBIAN GIN** / (sheaf of wheat) / **CHICAGO ILL.**

»SB w/ keyed hinge mold; square w/ beveled corners; 2 opposing sunken panels; 9 1/2" tall. [Figure 233]

Figure 232

Abraham Croskey first appears in Chicago CDs as a distiller and liquor merchant in 1859. Located at 51 (and then 51 & 53) S. Water Street, he advertises boldly in each year's Chicago CD between 1859 and 1863 (but always on the CD's back cover, which is usually not preserved in the microfilm files). There is an advertisement for A. F. Croskey's whiskies and liquors in the *Chicago Tribune* for November 9, 1859. The 1862 IRS assessment listed A. F. Croskey as a dealer and manufacturer in the liquor trade. Croskey appears as a rectifier in the May 1863 assessment.

In the March 4, 1863, *Chicago Tribune*, Croskey, located at 51 and 53 Water Street, advertised alcohol, pure spirits, whiskies, gins, imported wines, brandy, and liquors. He also offered for sale an assortment of Pittsburgh glass bottles including:

> Brandy and wine bottles, 3 sizes
> Quart & pint squares
> Porter bottles, quarts & pints
> Hocks & Clarets
> Flasks—Quarts, pints, and half-pints
> Demijohns "of all sizes"

However, his independent business effort is only successful for a short time. By 1864, he is listed as a distiller working for Wm. Phelps & Co. Since the bottle listed above is not pontiled, Croskey's Columbian Gin was bottled during the early Civil War years, between ca. 1860 and 1863.

Figure 233

George F. Davies

Beer:

- (1) **DAVIES / LAGER BEER / PEORIA & MADISON ST / CHICAGO // THIS / BOTTLE / NOT TO BE / SOLD** *(base):* **C & I** [aqua]

    »SB w/ post mold; sloped-shoulder cylinder; narrow blob top; 9¼" tall. *[Figure 234]*

Figure 234

George F. Davies is listed in Chicago CDs as the proprietor of a restaurant and a saloon from 1872 to 1875. He likely bottled beer to sell at his establishments. Glassmakers Cunningham and Ihmsen (*C&I*) went out of business in 1878, and the Davies bottle is a post mold type, so the bottle can be closely dated to the time Davies was in business.

See the Conrad Seipp Chicago entry for a nearly identical bottle.

Dr. Amos B. DeMor and Co.

Medicine:

(1) (shoulder): **D**[R] **DEMOR / CHICAGO. ILLS -** // [amber]
(shoulder): **IMPROVED MEDICATED / BLACKBERRY BRANDY**
Note: *Two embossed dots beneath superscript* **R** *in* **D**[R]

»SB w/ post mold; two-piece mold quart whiskey cylinder; shoulder-embossed only; long "lady's-leg" style neck; ring-neck tapered collar; 11" tall. [*Figure 235*]

❖

DeMor & Co. (Amos and Edgar DeMor) were first listed in the 1874 Chicago CD as manufacturer's agents. The next year (1875 Chicago CD), they were listed as "bitters" manufacturers at 233 N. Wells.

In the 1876 and 1877 CDs, Amos is listed as a physician (so he is the "**DR DEMOR**" on the bottle listed above), and the company is listed as "blackberry brandy" manufacturers located at 424 W. Chicago. Finally, in the 1878 Chicago CD, DeMor & Co. is listed as "dealers in DeMor's stomach bitters" at 146 N. LaSalle. Amos, Edgar, and their company were no longer listed in Chicago CDs after 1878. Thus it appears that the embossed bottle in this listing may date to just the two-year period from 1876 to 1877.

It should be noted, however, that A. B. Demor was listed as still involved in "patent medicine compound" production in the 1880 IL Industrial Census (perhaps as a wholesaler only?). The 1880 IIC indicated his investment in the business was about $3,000, with four employees and an annual labor cost of $1,500. His cost of production materials was listed as $1,000 with an annual product sales value of $5,000.

Daniel Dempsey

Ale:

(1) **D. DEMPSEY / CHICAGO. / ILL.** [olive-amber, red-amber]

»SB w/ post mold; two-piece mold quart ale; ring-neck tapered collar; 8¾" tall. [*Figure 236*]

Porter:

(1) **D. DEMPSEY / CHICAGO. / ILL_** [aqua]

»SB; two-piece mold pint porter; lip broken away, but probably ring-neck tapered collar; about 6" tall. [*Figure 237*]

Figure 235

Figure 236

Figure 237

Soda ~ Mineral Water:

(1) **D. DEMPSEY / CHICAGO / ILL^S //**   *[aqua]*
    (*rear-shoulder monogram:*) **D & H**
    (*heel*): **W. M^{c}C & Co**
    (*base*): **D. D.**
    Note: The rear-shoulder monogram is a mold hold-over from predecessor company Dempsey & Hennessey (see their listing), and was removed in the second Dempsey soda bottle produced by the company (see below). [Figure 238]

    »SB w/ hinge mold; shouldered-blob-top soda.

(2) **D. DEMPSEY / CHICAGO / ILL^S //**   *[aqua, blue]*
    (*heel*): **W. M^{c}C & Co**
    (*base*): **D. D.**

    »SB w/ post mold; short-neck, oval-blob-top soda. *[Figure 239]*

Daniel Dempsey is listed in the 1880 USC as a 50-year-old soda water manufacturer. He was in a partnership with Thomas Hennessey from 1872 to 1881 (*see Hennessey listing, this chapter*), but he *never* appears in Chicago CDs as sole owner of a bottling company during the post–Civil War period of this study. (City directories were checked through 1886.)

Although Dempsey and Hennessey were partners for nine years, they only produced a single **D & H** embossed-bottle style before Hennessey bought out Dempsey's interest and continued in business alone for a few years after 1882 (*see the Hennessey listing, this chapter*).

Figure 238

Figure 239

It seems highly probable that the four Dempsey-only bottle styles listed above date to the earlier years of the partnership, a time during which Hennessey may have been too junior a partner in their bottling business for his name to appear on the bottles (*Note*, e.g., soda/mineral water bottle style #2 above, which has **DEMPSEY** embossed on the obverse but also has a **D & H** monogram embossed on the reverse shoulder).

Daniel Dempsey again appeared in city directories as a Chicago bottler in 1884, when he formed a soda and (stoneware) small-beer bottling partnership with Cornelius Ryan. Since Dempsey & Ryan operated in the Hutchinson-style bottling era only, their bottles date too late for inclusion in the present volume.

Daniel Dempsey and Thomas Hennessy

Soda ~ Mineral Water:

 (1a) **D & H** / **CHICAGO** / **ILLINOIS** / **GINGER** / **ALE**   *[aqua]*
  (base monogram): ***D&H***

  » SB with round base; shouldered-blob-top cylindrical soda; 8 1/4" tall. *[Figure 240]*

 (1b) **D & H** / **CHICAGO** / **ILLINOIS** / **GINGER** / **ALE**   *[aqua]*
  (base monogram): ***D&H***

  » SB with flat base; shouldered-blob-top cylindrical soda; 7 3/4" tall. *[Figure 241]*

The partnership of Daniel Dempsey & Thomas Hennessy is listed under the soda water manufacturers heading in the 1875 ISD. The partnership was listed in the Chicago CDs from

Figure 240

Figure 241

1872 to 1881. Hennessey is listed alone in the directories from 1882 to 1886 (*see his listing, this chapter*). Daniel Dempsey reappears in city directories in1884 when he forms a partnership with Cornelius Ryan (*see the Dempsey discussion*).

The Dempsey & Hennessey soda water bottling partnership was listed in the 1880 IL Industrial Census as having been a full-time, 12-months-per-year business, with a total annual product value of $9,600.

George Dickinson

Soda ~ Mineral Water:

(1) **GEORGE / DICKINSON / CHICAGO //** [aqua]
(*heel*): **A & D. H. C.**

»SB w/ keyed hinge mold; shouldered-blob-top soda. *[Figure 242]*

(2) **GEO. DICKINSON / CHICAGO ILL //** [aqua]
(*heel*): **C&I**
(*base*): **D**

»SB w/ hinge mold; oval-blob-top soda. *[Figure 243]*

George Dickinson appears to have been associated with beer-bottler Julius Grossenheider in the early 1870s (*see Grossenheider listings, this chapter*). He is listed alone as a soda manufacturer at 122 Archer in the 1873 Chicago CD. In both 1872 and 1874 he is listed in the CDs as a brewing-company bookkeeper.

It is surprising that he made two embossed-bottle orders from Pittsburgh glass manufacturers in so short a time. The hinge-mold manufacturing technology on his *A&DHC* bottle suggests he may have started bottling under his own name as early as 1871.

Figure 242

Figure 243

Joseph Dinet

Soda ~ Mineral Water:

(1) **J. DINETS // SUPERIOR // // SODA WATER // // CHICAGO** *[cobalt]*
   »IP; six-sided "ten-pin" shape; long neck and sloped shoulders; tapered collar; 8 1/4" tall. *[Figure 244, top left]*

(2) **SUPERIOR // // SODA WATER // // CHICAGO** *[aqua]*
   Note: This bottle was made from the same mold as bottle #1, but with Dinet's name removed. As discussed in the entry, this variety of the bottle was likely made for and used by Joseph Entwistle.
   »IP; six-sided;" ten-pin" shape; long neck and sloped shoulders; 8 1/4" tall. *[Figure 244, bottom right]*

Figure 244

Both of the listed bottles are identical, except that #2 is aqua and "J. DINET'S" has been peened-out of the first embossed panel. Dinet was a Chicago confectioner between 1845 and 1851, and is historically referred to as a "prominent candy manufacturer" (Clark 1912:468). He was also Chicago's earliest known soda water bottler. Technologically the bottles look like they were a late 1840s product. Dinet also put up some of his confectionary drinks in early hand-turned broad-shouldered stoneware bottles (*Figure 245*). These 8" tall mineral-water-style bottles were characteristically albany slipped and are shoulder-stamped **J. DINET**. At the height of Dinet's confectionary business he had four employees.

According to a Chicago newspaper article published on May 26, 1886, two years after his death, Joseph and his wife Elizabeth were married there in 1838. They "opened a small candy store with $5 capital and prospered in business, and retired in 1851 to invest their earnings in real estate." In fact, one City of Chicago subdivision is known as "Joseph Dinet's Subdivision." When he died, he left an estate estimated at $300,000.

Joseph Entwistle (*see his listing, this chapter*) was listed in the Chicago CD for 1849–50 as a barkeeper; in the 1851 Chicago CD he was listed as a soda water manufacturer. This would make him only the second known Chicago soda water bottler, following Joseph Dinet, to have embossed bottles. But his **J. ENTWISTLE** embossed bottle has the manufacturing characteristics of a mid-1850s style and is definitely not an artifact of his first years in business. In this light, an alternative interpretation of the aqua-variety Dinet bottles—which no longer have his name embossed on them—is that they were adapted and used by Entwistle in 1851 by piening out Dinet's name and leaving the rest of the bottle unaltered.

Oddly, most of Dinet's embossed bottles (and both of the varieties listed) have been found in New Orleans. Members of the Dinet family are long-time residents in the New Orleans area, so he may have come to Chicago from there, and his bottles may have been sent down to family members in New Orleans.

Figure 245

James A. Dorman and Co.

California Pop Beer:

(1) **H. C. DORMAN & CO / CELEBRATED / CALIFORNIA / POP BEER /**   *[amber]*
TRADE MARK / PAT. OCT 29. 1872 // CHICAGO / ILL^S.

»SB w/ post mold; shouldered-blob-top quart; 11 1/2" tall. *[Figure 246, right]*

❖

The 1878 Chicago city directory lists the firm of Geer and Dorman (Shubael Geer, Jr. and James A. Dorman, Jr.) as "pop beer" bottlers at 24 N. Green. The following year, 1879, the firm's name had changed to Dorman & Company, with Geer now the junior partner. By 1880 the partnership had been dissolved, Shubael Geer is listed alone as a bottler at 24 N. Green (*see the Geer listing, this chapter*).

The fact that the Dorman bottle is mis-embossed "H. C." DORMAN is strange. There was no Dorman in Chicago with the initials "H. C." during the years Dorman was a bottler, so perhaps the "H. C." initials are just a glass-manufacturer's error.

The Dorman and the Geer quart bottles appear to have been made from the same mold. The mold for the Geer bottle appears to have been modified by peening-out DORMAN & CO. and substituting S. GEER. Based on their business chronology, these bottles date to 1879–1880.

Paul Rouze, and John Lomax were also "California Pop" distributors in Chicago (*see their listings, this chapter*).

Harvey C. Doty

Ale ~ Porter:

(1) **H. C. DOTY / PITSBURGH. ALE** [aqua]
  Note: that "PITSBURGH" is misspelled on the bottle.

  »IP; squat-porter shape; squat "lozenge" style blob-top; 6 1/2" tall. *[Figure 247]*

Figure 246

In 1852, Harvey Doty is listed as a boarder at Erastus Doty's hotel in Chicago. In 1853, the first advertisement appears in the Chicago CD for:

DOTY & WAUGH
BOTTLERS PITTSBURGH ALE
*REAR OF DOTY'S HOTEL, RANDOLPH ST.
ALL ORDERS PROMPTLY ATTENDED TO.*

By 1854, R. W. Waugh (who also boarded at the hotel) is no longer involved, and Harvey Doty is listed as sole proprietor of the "Pittsburgh Ale" bottling business. While Doty continues to be listed for years as a clerk at his father's hotel, or in the billiard hall business, 1854 is the last year he appears in the CDs as an ale bottler.

Figure 247

John B. Drake

Whiskey:

> *(1)* **JOHN B. DRAKE / TREMONT HOUSE / CHICAGO / ILLS** [amber]
>
> »SB; broad-shouldered oval whiskey flask; ring-neck tapered collar; internal-screw-thread closure; threaded amber glass cap with two glass "thumb knobs" at the top; 7" tall full-pint size. *[Figure 248, left]*
>
> *(2a)* **JOHN B. DRAKE / TREMONT HOUSE / CHICAGO / ILLS** [yellow]
>
> »SB; broad-shouldered oval whiskey flask; tapered collar without ring-neck; internal-screw-thread closure; threaded amber glass cap with two glass "thumb knobs" at the top; 7" tall full-pint size. *[Figure 248, right]*
>
> *(2b)* **JOHN B. DRAKE / TREMONT HOUSE / CHICAGO / ILLS** [amber]
>
> »SB; broad-shouldered oval whiskey flask; tapered collar without ring-neck; internal-screw-thread closure; (cap not present); 6" tall half-pint size. *[Figure 249]*

    According to Clark (1912:567–568), John Drake was first associated with the Tremont House in 1855 as a steward. Malik (1980:5) indicates that Drake purchased an interest in the hotel at that time and became comanager with owners Ira and James Couch:

> Being thrifty and saving, he was soon enabled to purchase an interest in this hotel, and before many years became an associate proprietor, and afterwards sole proprietor. This hotel was burned in the great fire in 1871, and while the fire was raging, Mr. Drake made a deal by which he

Figure 248

Figure 249

took possession of the Michigan Avenue Hotel...which he kept for two years afterwards. In 1874 he purchased the lease of the famous old Grand Pacific Hotel of Chicago, and was its leading proprietor until his death in 1895. [Clark 1912:568]

According to Malik (1980:5), Drake had become sole owner of the hotel by 1868. From information in Chicago city directories, Drake was proprietor of the Tremont House Hotel for only three years: from 1871 to 1873. In 1874, he became a vice-president of the Illinois Trust & Savings Bank.

According to the 1875 Chicago CD, Mr. Drake returned to the hotel business in Chicago that year as proprietor of the Grand Pacific. He retained his vice-presidency at the bank, and managed the Grand Pacific with two partners (Tyler Gaskill and Samuel Turner) as *John B. Drake & Co.* from 1875 to 1879. That year, the Grand Pacific management group took on an additional partner, Samuel W. Parker, but retained the *Drake & Co.* corporate name.

In 1885, the corporate name was changed to *Drake, Parker & Co.* Soon afterward, in the tradition of the Tremont House flasks listed, the company issued a distinctive clear-glass, bulbous-necked, embossed shoofly-style whiskey flask for distribution (full, of course) to guests of the Grand Pacific Hotel (*Figure 250*). This full-pint bottle is base-embossed: **C. NEWMAN'S / BAKER BROS. 7 CO. / PAT. OCT 17, 1876**, suggesting it might be a late-1870s product. However, Newman's-patent bottles produced in the 1870s are made from amber glass (Lindsay 2009: SHA website), and the clear-glass Grand Pacific flask was produced five years too late for formal listing in the present volume.

Figure 250

James E. Eaton

Soda ~ Mineral Water:

(1) **J. E. EATON & Co / CHICAGO. ILL'S /**   [aqua]
    (heel): **FA & CO**

»SB w/ keyed hinge mold; shouldered-blob-top soda. *[Figure 251]*

James Eaton first bottled soda and mineral water in the Galesburg/Knoxville area of Knox County from 1857 to 1861 (*see his Galesburg listing*), before moving his bottling operation to Chicago with relative George Eaton. The 1862 IRS assessment lists James Eaton as a soda water manufacturer in Chicago. In the 1862 Chicago CD he is listed (his only year in the directories) as the principal in J. E. Eaton & Co., soda and mineral water bottler, at 261 Kenzie. The "& Co." partner is a relative (son?) George E. Eaton. James apparently bottled in Chicago for only a season, since both he and George are gone from the main body of the directory after that first year, but George is also listed alone in the bottler index

Figure 251

at the back of the CD in both the 1863 and 1864 issues (as a "soda and spruce beer" manufacturer).

Since only three (identical) bottles are known from this partnership, it is likely that his Chicago bottling effort was limited to just 1862. By the mid-1860s he is listed back in the Galesburg/Knoxville area as a resident (not a bottler), but shortly afterward he moved on to become a soda bottler in several towns in Wisconsin. He bottled soda water at Beloit, Burlington, and Waukesha from the mid-1860s to the late 1970s (Peters 1996:24, 28, 183), before moving back to Illinois in the late 1870s and setting up shop in the old Hickey Bottling Works factory in Peoria from the late 1870s into the 1880s (*see his Peoria listing*).

FA & Co. only produced bottles of the kind listed here during the early Civil War years, supporting this early 1860s date for Eaton's use of the Chicago embossed bottle.

"E." Edwards

Soda ~ Mineral Water:

(1) **E. EDWARD°S / CHICAGO / ILL //**   [aqua]
(*heel*): **A. & D. H. C**
(*base*): **E**

»SB w/ post mold; oval-blob-top soda. [*Figure 252*]

❖

From the shape and technology attributes of the listed bottle, Mr. Edwards would have been a soda water or mineral water manufacturer in Chicago during the early to mid-1870s.

A Clark Street physician named Edward W. Edwards was found listed in the Chicago CDs beginning in 1873, but he was never mentioned as having bottled or distributed mineral waters as part of his practice. Other than Dr. Edwards, no "E." Edward/Edwards was found listed as a bottler in the CDs during the 1870s, so the listed bottle was very likely a "misprint" variety of the closely similar "F." EDWARD'S bottle, blown in the mid-1870s for Chicago bottler Francis C. Edwards (*see his listing*).

Figure 252

Francis C. Edwards

Soda ~ Mineral Water:

(1) **F. EDWARD'S / CHICAGO / ILL**   [aqua]
(*base*): **E**

»SB w/ post mold; oval-blob-top soda. [*Figure 253*]

❖

Francis Edwards appears in the 1875 and 1876 Chicago CDs as a neighborhood soda water manufacturer and distributor from his residences on Vernon Av. @ 38th (1875) and 145 Stanton (1876). In the 1877 CD he is listed only as a Chicago resident.

Figure 253

Jacob K. Eilert

Medicine:

(1) *(side):* **EILERT'S** // // *(side):* **EXTRACT OF / TAR & WILD CHERRY**   [aqua]

»IP; beveled-edge rectangle; 4 sunken panels; square shoulders; double-ring collar; 5 3/4" tall. *[Figure 254]*

(2) *(side):* **EILERT'S** //   [aqua]
(side): **EXTRACT OF / TAR & WILD CHERRY**

»SB w/ keyed hinge mold; beveled-edge rectangle; 4 sunken panels; steeply sloped shoulders between neck and rectangular body; double-ring collar; 7" tall. *[Figure 255]*

Eilert (probably Jacob K., see later discussion) is only known from the embossed glass bottles in this listing, which from their manufacturing characteristics likely date to ca. 1859–61. Mr. Eilert never appears in nineteenth century Chicago CDs. But his product is tentatively listed as a Chicago medicine because of its later connection, through the firm of Snyder & Eilert *(see listing, this chapter)* to the hugely successful Chicago patent-medicine distributorship, Emmert Proprietary Company *(see Emmert discussion, this chapter)*.

In addition to other later medicine products, the Emmert Company, which set up shop in Chicago in 1872, acquired the distribution rights to Eilert's *Extract of Tar & Wild Cherry* (and Eilert's *Day-Light Liver Pills*) as well as to Snyder & Eilert's *Dr. Winchell's Teething Syrup* *(see Emmert listing and 27"x37" advertising sign: Figure 256)*.

Figure 254

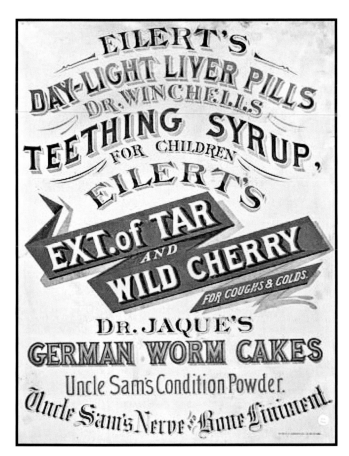

Figure 256

Figure 255

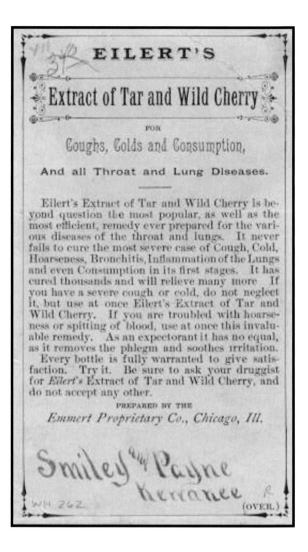

Figure 257

Prior to moving to Chicago, Emmert had been in the drug business for some 20 years (initially as Emmert & Burrell) in Freeport, IL (Hill, ed. 1910:478). In the 1845/1860-era, a druggist named Jacob K. Eilert lived in the Freeport, IL/Monroe, WI area (the two towns are about 20 miles apart), so it is possible that Eilert's and Snyder & Eilert's medicines were first sold by Emmert & Burrell at Freeport.

Eilert's *Extract* was marketed as a cure for coughs, colds, consumption, and all throat and lung diseases: "If you are troubled with hoarseness or spitting of blood, use at once this invaluable remedy" (Emmert trade card—*see Figure 257*). The extract continued to be sold throughout the nineteenth century.

John S. Emmert Proprietary Co.

Medicine:

(1a) **DR WINCHELLS / TEETHING SYRUP / EMMERT /**   *[aqua]*
    PROPRIETARY Co / CHICAGO ILLS
    (base): **L&W**

    »SB w/ keyed hinge mold; tall, narrow cylinder; sloped shoulders; flared-ring collar w/ tapered base; 1" diameter, 5" tall. *[Figure 258]*

Figure 258

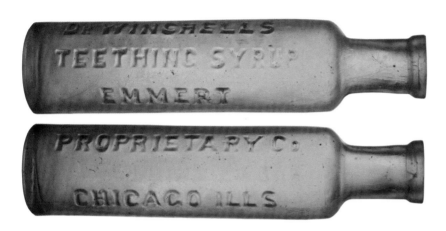

Figure 259

(1b) **DR WINCHELLS / TEETHING SYRUP / EMMERT / PROPRIETARY Co / CHICAGO ILLS** *[aqua]*
(base): **T.W & Co**

»SB w/ keyed hinge mold; tall, narrow cylinder; sloped shoulders; flared-ring collar w/ tapered base; 1" diameter, 5" tall. *[Figure 259, Figure 260]*

John S., Charles F., and Harris L. Emmert are first listed in the 1872 Chicago CD as having formed the Emmert Proprietary Company, located at 296 W. Lake Street. John Emmert was company president. As discussed in the Jacob K. Eilert listing, Emmert had been in the drug business for some 20 years (initially as Emmert & Burrell) in Freeport, IL (Hill, ed. 1910:478). In the 1845/1860-era, a druggist named Jacob K. Eilert lived in the Freeport, IL/Monroe, WI area (the two towns are about 20 miles apart), so it is possible that Eilert's and Snyder & Eilert's *(see their listing, this chapter)* medicines were first sold by Emmert & Burrell at Freeport.

During their early eras in Chicago in the 1870s, it is likely that the Emmert Company also produced an embossed 1870s version of the *Eilert's Extract of Tar & Wild Cherry* medicine. But we have not yet seen an example of the bottle, so it is not listed. From their 1870s/1880s trade cards, they were definitely producing both Eilert's Extract and Eilert's Liver Pills *(see Figure 256)* soon after their Chicago arrival, and soon after 1880 (too late to be included in the present volume) they added several more patent medicine products to their line—most prominently including

Figure 260

 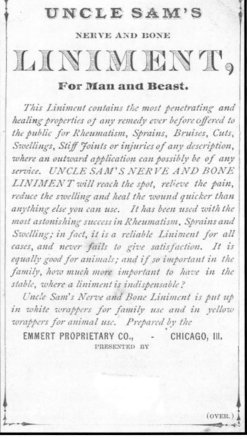

Figure 261

Uncle Sam's Nerve & Bone Liniment and *Condition Powder, Happy Home Blood Purifier,* and *Dr. Jacques German Worm Cakes* (Figure 261).

The *Emmert Proprietary Co.* is listed in the 1880 IL Industrial Census, which estimated the company's investment in property and equipment at $10,000, with a maximum of 40 annual employees (with 13 male and 20 female employees on average) and an annual payroll of $11,700. Annual supplies costs were estimated at $75,000, with an annual product value of $100,000.

Figure 261 (continued)

The Emmert Company marketed *Dr. Winchell's Teething Syrup* as the "best medicine for diseases incident to infancy." Their ads and trade cards (*see Figure 260*) declared "it will positively cure every case if given in time"; that it "quiets and soothes all pain"; that it "cures diarrhea and dysentery in the worst forms"; and that it was "a certain preventive of [and cure for] diphtheria" (Knudson 2006:415).

This product appears to have first been a Civil War era copy of *Mrs. Winslow's Soothing Syrup*, a popular patent medicine first introduced in 1849 by the Curtis & Perkins company of Bangor, ME. The bottle forms and sizes were nearly identical (Mrs. Winslow's syrup came in a cylindrical bottle 11/8" diameter and 5" tall).

Mrs. Winslow's Syrup was used to reduce pain in teething infants (and no doubt to generally keep them quiet). Since one of its primary ingredients was morphine, it was taken off the market soon after the passage of the 1906 Pure Food and Drug Act.

The two embossed bottles are listed in chronological order: Thomas Wightman & Co. was a successor Pittsburgh glass factory to Lorenz & Wightman.

Figure 261 (continued)

Joseph Entwistle

Soda ~ Mineral Water:

   (1a) **J. ENTWISTLE / CHICAGO / ILL.**   *[cobalt]*
      Note: *The embossed "***S***" in* **ENTWISTLE** *is embossed backwards.*

     »IP; shouldered-blob-top soda. *[Figure 262]*

   (1b) **J. ENTWISTLE / CHICAGO / ILL.**   *[cobalt]*
      Note: *The embossed "***S***" is reversed in* **ENTWISTLE**. *This is the same bottle shape and the same style of embossing as #1a, but* **J. ENTWISTLE** *has broader and flatter embossed lettering.*

     »IP; shouldered-blob-top soda. *[Figure 263]*

Joseph Entwistle is listed in the Chicago CD for 1849–50 as a barkeeper and in the 1851 Chicago CD as a soda water manufacturer. This makes him only the second known Chicago soda water bottler following Joseph Dinet (*see Dinet listing, this chapter*) to have embossed bottles. The bottle listed above has the manufacturing characteristics of a mid-1850s style, and is definitely not an artifact of his first years in business. In this light, an alternative interpretation of the aqua-variety Dinet bottles—which no longer have his name embossed on them—is that they were adapted and used by Entwistle in 1851 by piening out Dinet's name and leaving the rest of the bottle unaltered (*see Figure 244, bottom right, in the Dinet listing*).

Figure 262

In the 1852–53 CD, Entwistle had become senior partner with Edward Bebbington (to produce soda water—*see listing*) and with George Lomax (to produce mineral water—*see listing*) on Canal Street between Lake and Fulton. The Bebbington/Entwistle/G. Lomax partnership can only be documented to have lasted through 1854, after which George Lomax left to form a new bottling partnership the next year with John Meagher (*see G. Lomax listing*), and Edward Bebbington continued bottling on his own at the Canal Street address. All of the Bebbington-only embossed bottles are late 1850s styles (ca. 1856–59: *see Bebbington listing, this chapter*).

From the existence of the embossed *J. Entwistle* bottle, it appears that Joseph Entwistle made an unsuccessful effort to start his own Chicago bottling business during at least part of the 1855 season, before retiring from Chicago business and leaving town. Perhaps he died, since only his wife is listed in the 1855 Chicago CD. However, his embossed bottle dates from this time, so he must have made at least a short-lived attempt to continue bottling on his own.

Joseph Entwistle and Edward K. Bebbington

Soda ~ Mineral Water:

(1) **E & B // AERATED / SODA WATER**   *[cobalt]*

»IP; tapered-collar soda. *[Figure 264]*

Joseph Entwistle and Edward Bebbington were partners from 1852 through 1854 (*see discussion under Entwistle*). Their soda water manufactory was located on Canal between Lake and Fulton.

Note that although this bottle and the **E & L** bottle listed below are the earliest embossed soda bottles from Chicago except for the cobalt and aqua Dinet-style bottles (*see Dinet and Entwistle listings, this chapter*),

Figure 263

both bottle styles have had their molds heavily modified by piening-out prior embossing. The only unaltered lettering areas on both the obverse and reverse of these bottles is the "**E &**" portion of the front. Since there are no earlier Chicago bottles or bottling companies that would fit this part of the mold, the glass company that made the bottle must have offered the Entwistle partnerships a discount to accept a modified earlier bottle mold from a defunct company located elsewhere in the eastern United States.

Joseph Entwistle and George Lomax

Soda ~ Mineral Water:

(1) **E & L** // **AERATED** / **SODAWATER**   *[aqua; cobalt; teal]*
»IP; tapered-collar soda. *[Figure 265]*

George Lomax joined Entwistle and Bebbington at their factory on Canal Street between Lake and Fulton in 1852. This bottle was made during the three years of their partnership, 1852–54. The Chicago CDs indicates that George Lomax left the group in 1855, dissolving the Entwistle—Bebbington—Lomax partnership, and formed a new company with John Meagher at 138 W. Madison from 1855 to Lomax's death in

Figure 264

Figure 265

1863 (*see G. Lomax listing, this chapter*). No embossed Lomax/Meagher bottles are known, but Meagher was likely a junior partner.

Note that although this bottle and the **E & B** bottle listed are the earliest embossed soda bottles from Chicago except for the cobalt and aqua Joseph-Dinet-style bottles (*see Dinet and Entwistle listings*), both bottle styles have had their molds heavily modified by piening-out prior embossing. The only unaltered lettering areas on both the obverse and reverse of these bottles is the "**E &**" portion of the front. Since there are no earlier Chicago bottles or bottling companies that would fit this part of the mold, the glass company that made the bottle must have offered the Entwistle partnerships a discount to accept a modified earlier bottle mold from a defunct company located elsewhere in the eastern United States.

Peter Fahrney

Medicine:

(1) **DR FAHRNEY'S** // // (side) **UTERINE** // //   [aqua]
   *Note: There are two embossed dots beneath the **R** in **D^R**.*

   »SB w/ keyed hinge mold; beveled-edge rectangle; 3 sunken panels; double-ring collar; 8 3/4" tall. [*Figure 266*]

(2) **DR P. FAHRNEY'S** // // (side) **PANACEA** // //   [aqua]
   *Note: There are two embossed dots beneath the **R** in **D^R**.*

   »SB w/ keyed hinge mold; beveled-edge rectangle; 3 sunken panels; squared, flared-ring collar; 8 1/2" tall. [*Figure 267*]

Figure 266

Figure 267

Figure 268

(3) **DR P. FAHRNEY'S BLOOD // // VITALIZER CHICAGO ILLS // //** [aqua]

»SB; tall oval with a round base ending in a small flattened area 1 1/2" in diameter; embossing on opposite edges of bottle face; 11 1/2" high. This bottle will stand, although its base resembles that of a round-bottom soda. *[Figure 268]*

Peter Fahrney' company was a long-term—and successful—patent-medicine manufacturer during the second half of the nineteenth century and first third of the twentieth century. The business operated out of Chicago for most of that time. According to the 1875 Chicago CD, Fahrney was a manufacturer of patent medicines located at 690 W. Indiana Street. Over such a long business history, the Fahrney Co. produced many named medicines in numerous glass container styles. The bottles listed are only those that clearly date to the pre-1880 era of our present study.

In the 1865 and 1866 IRS assessments for northern Illinois, Fahrney is listed as a physician in the small Ogle County town of Buffalo, some 90 miles west of Chicago. Chicago CDs indicate that he did not move his business to Chicago until 1875. However, a 1934 letterhead for Dr. Peter Fahrney & Sons patent-medicine laboratory indicates that they first established their Chicago business in 1869—and that their business was founded in 1780 in Washington County, Maryland.

Technological attributes of the first two bottles listed above indicate that they were likely made and used while Fahrney still conducted business in Buffalo, IL.

William B. Farrell

Medicine:

(1) **D^R / W. B. FARRELL'S // INDIAN / OINTMENT / CHICAGO. ILL** [aqua]
Note: Single embossed dot beneath raised **R** *in* **D^R**.

»OP; short, wide-mouthed shouldered cylinder; flared rolled lip; 2 1/2" high; basal diameter 1 1/2". *[Figure 269]*

(2a) **W. B. FARRELL'S // ARABIAN / LINIMENT** [aqua]

»OP; sharp-shouldered cylinder; rolled lip; small size: 3 1/2" tall. *[Figure 270]*

(2b) **W. B. FARRELLS // ARABIAN / LINIMENT** [aqua]

»OP; sharp-shouldered cylinder; rolled lip; small size: 3 1/2" tall. *[Figure 271]*

Figure 269

Figure 270

Figure 271

(3a) **W. B. FARRELLS // ARABIAN / LINIMENT** [aqua]

»OP; sharp-shouldered cylinder; rolled lip; large size: 5" tall. *[Figure 272]*

(3b) **W. B. FARRELLS // ARABIAN / LINIMENT** *[aqua]*

»OP; sharp-shouldered cylinder; rolled lip; large size: 5" tall. Differently styled embossing from 3a, and small S in FARRELLS gone.

(3) **DR W. B. / FARRELL,S / ARABIAN . LINIMENT / CHICAGO . ILL°S** [aqua]

»OP; sharp-shouldered cylinder; rolled lip; 3 1/2" tall. *[Figure 273]*

(4) **DR W. B. FARRELL'S / ARABIAN LINIMENT / CHICAGO . ILL'S** [green, aqua]
*Note: Two embossed dots beneath raised **R** in **D^R***

»IP; slope-shouldered tall cylinder; short neck; tapered collar; 7 1/2" tall. *[Figure 274]*

Figure 272

Figure 273

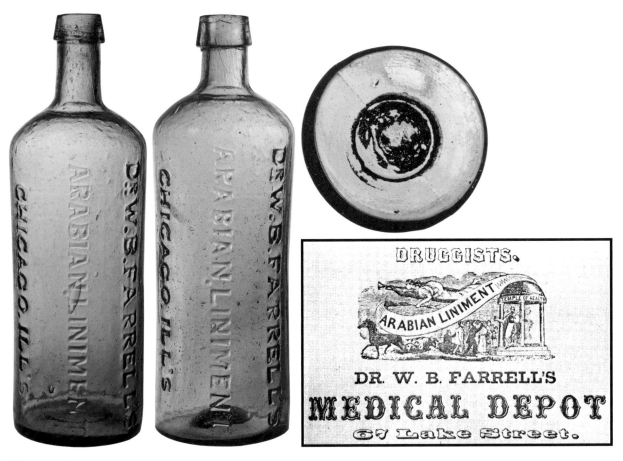

Figure 274

    W. B. Farrell was the brother of Hiram G. Farrell. The two men were very early partners in an extremely successful patent-medicine business in Peoria that primarily promoted *Farrell's Arabian Liniment*, a widely popular liniment in the mid-to-late 1840s (*for a full discussion, see Farrell's Peoria listing*). They had an acrimonious business falling-out, and William moved to Chicago in 1849—where he continued to sell his version of the Arabian Liniment in competition with his brother's Peoria version (with each calling the other a "base counterfeit").

    In 1850, William advertised in the newspaper that "the great Arabian Remedy, W. B. Farrell's Celebrated Arabian Liniment" was now being manufactured at 218 Lake Street in Chicago (Bogard 1982:102). W. B. Farrell operated his medical depot at various Chicago locations in the 1850s (see Odell 2007:25). In the 1851 Chicago CD, he offered "the genuine original Arabian Liniment, Indian black ointment, anti-bilious family pills and rheumatic specific." A single example of his *Indian Ointment* bottle has been located, but no embossed examples of his *Rheumatic Specific* have been found. The 1852 CD carried an illustrated advertisement (*Figure 274*) showing "a temple of health where both humans and animals came for treatment" (Bogard 1982:102).

    William is listed in the Chicago CDs as a patent-medicine dealer until 1858. After that, he became a practicing physician (he is listed in the USC for 1860 as a 45-year-old physician), and later a board-of-trade operator (Bogard 1982:103).

> In the 1870s he was a well-known florist with twelve greenhouses in Hyde Park. In 1887, however, William was again a liniment manufacturer, while his son-in-law, Frank Backus, was president of the Congo Chemical Company. Congo Liniment was listed in the catalogs of wholesale druggists in Chicago.... William B. Farrell died suddenly in 1889. [Bogard 1982:103]

    The advertised uses of *Farrell's Arabian Liniment* are summarized in the Peoria discussion. But *Farrell's Indian Black Ointment* (bottle #4) was produced only by William in Chicago. An

1852 Woodstock, IL, newspaper ad, with a testimonial from an 1850 Chicago user ("Your Ointment...in less that three weeks restored the surface of my body to a sound condition, which previously was perforated almost to the appearance of a sieve"), declared—

Scrofula, Salt Rheum,
AND THE MOST DREADFUL SORES
ARE CURED BY
W. B. Farrell's
Indian Black Ointment!
Grand Depot, 52 State-street,
Chicago, Illinois.

James F. Farwell and James F. Brace

Leather Treatment:

(1) **FARWELL'S / GERMAN LIQUID / POLISH / CHICAGO** [aqua]

»SB w/ heel mold; high-shouldered cylinder w/ vertical embossing; flared, pharmacy-style lip; 1 3/4" diameter, 5" tall. *[Figure 275]*

Figure 275

―――――――――― ❖ ――――――――――

James Farwell was first listed as a druggist at 132 Clark in the Chicago CDs in 1872. In 1873 and 1874, he was listed as a druggist at 108 Clark. By 1875, he had moved his pharmacy operation next door to 106 Clark, where he was listed as a druggist until his final CD business entry in 1877.

In 1874 (CD), Farwell had formed a one-year partnership with James F. Brace at the 108 Clark address to manufacture "German liquid polish." Before their 1874 partnership, Brace had been a bookkeeper with the "saddlery hardware" firm of *Smith & Duncan*. He was no longer listed in the CDs after his one-year partnership with Farwell.

According to a contemporary "Ready Reference and Reservoir of Useful Information" (Blakelee 1884:223), *German Liquid Leather Polish* could be made at home by dissolving "three and one-half ounces of shellac in half a pint of alcohol. Rub smooth twenty-five grains of lampblack with six drams of cod liver oil, and mix. A few drops are to be applied to the leather with a sponge."

Farwell's leather polish bottles were likely filled and sold in his drug store in 1874, and perhaps for a short time afterwards.

Henry Filer

Ale:

(1) *(shoulder):* **H. F & Co** // *(shoulder):* **CHICAGO** [amber]
(base, in circle): **W Mc CULLY & Co PITTS PA**

»SB w/ post mold; 2-piece mold quart ale; *ring-neck tapered collar;* 8 1/4" tall. *[Figure 276]*

Soda ~ Mineral Water:

(1) **H. F & Co. / CHICAGO** // [aqua; blue]
(heel): **W McC & Co**

»SB w/ post mold; blob-top soda. *[Figure 277]*

(2) **H F & Co / CHICAGO** // [aqua]
(heel): **C&I**

»SB w/ post mold; oval-blob-top soda. *[Figure 278]*

―――――――――― ❖ ――――――――――

Figure 276

Figure 277

Henry and Charles Filer are first listed in the Chicago CD for 1872 and 1873 as soda manufacturers at 500 W. Halsted. In the 1873 CD, Charles is listed alone as a soda water maker. During the next three years, he is listed as a tanner and currier.

During the 1860s and 1870s, Henry and Charles are never listed as ale/porter/cider bottlers. But it looks like they must have given it a try, at least during their first year in business as H F & Co., since a search of the 1860s and 1870s CDs turned up no one else with their initials listed as ale or porter bottlers.

Edward H. Flagg

Medicine:

(1) **E. H. FLAGG'S / COUGH KILLER / CHICAGO. ILL** [aqua]

»SB w/ keyed hinge mold; beveled-edge rectangle; ring neck; squared flared-ring lip; 5 1/4" tall. *[no photo]*

(2a) *(side):* **E. H. FLAGG'S // INSTANTANEOUS / RELIEF //** [aqua]
(side): **CHICAGO**

»SB w/ keyed hinge mold; beveled-edge rectangle; 3 sunken panels; double-ring collar; 2-ounce size: 4 1/2" tall. *[Figure 279]*

(2b) *(side):* **E. H. FLAGG'S // INSTANTANEOUS / RELIEF //** [aqua]
(side): **CHICAGO**

»SB w/ keyed hinge mold; beveled-edge rectangle; 3 sunken panels; double-ring collar; 4-ounce size: 6" tall. *[Figure 279]*

Figure 278

Figure 279

 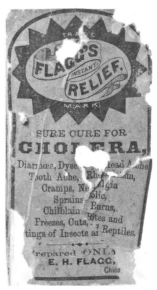

Figure 280

E. H. Flagg's *Instant Relief* pontil bottles that are known are embossed PHILADELPHIA, and likely date to the late 1850s (Greer Sale 1988: #593).

Edward Flagg moved to Chicago during the early Civil War years. He first appears in the 1864 Chicago CD, and is listed as a patent-medicine manufacturer at 181 1/2 Clark in 1865. From 1866 through 1868, he is listed as a manufacturer of medicines from his residential address at 57 S. Paulina. It is probably during this time that he formed an association with long-time Chicago druggist Henry Sweet to be a marketing outlet for his medicines. At about this same time, Sweet was also selling his own patent-medicine brands (*see the Sweet listing, this chapter*).

In 1867–68 Sweet and Schroeder formed a pharmacy partnership, and E. H. Flagg opened a retail outlet at 150 State Street to sell his medicines. In 1868 and 1870, he is listed as a medicine manufacturer located at 54 and 56 Michigan Avenue. By 1871, Flagg is no longer listed in the directories as either a medicine manufacturer or a Chicago resident.

Flagg and Sweet are never listed in the Chicago directories as partners. But for a time, likely in the late 1860s, they formed a retail-sales association—as indicated by the paper label (*see Figure 280*) on a Flagg's *Instantaneous Relief* embossed bottle, which reads as follows:

**SURE CURE FOR
CHOLERA,**
Fever and Ague, Diarrhoea,
Dysentery, Colic, Burns,
Cramps, Sprains,
Rheumatism,
Stomach Ache, Chills, and
Bites of Insects and Reptiles

Prepared **ONLY** by
FLAGG & SWEET,
153 STATE STREET, Chicago

Theobald Foerster and Co.

Bitters:

(1) **FOERSTER'S / TEUTONIC / BITTERS / CHICAGO.**   [*yellow amber*]
»IP; chestnut flask with applied handle on left; ring-neck plain finish collar; 7" tall. [*Figure 281*]

❖

Theobald Foerster is listed in Chicago city directories from 1858 through the Civil War era and the late 1860s. T. Foerster & Co. began as a retail liquor company with two stores—located at 89 W. Lake and 15 Canal. In 1859 and 1860, Foerster is listed as a rectifier at 54 S. Wells. He was also listed in the 1860 ISD and in Civil War–era Chicago CDs as a liquor dealer and rectifier at 54 S. Wells and then 246 Randolph.

The bottle listed above is pontilled, so it dates to 1858 or 1859, when Foerster first opened his retail liquor establishment.

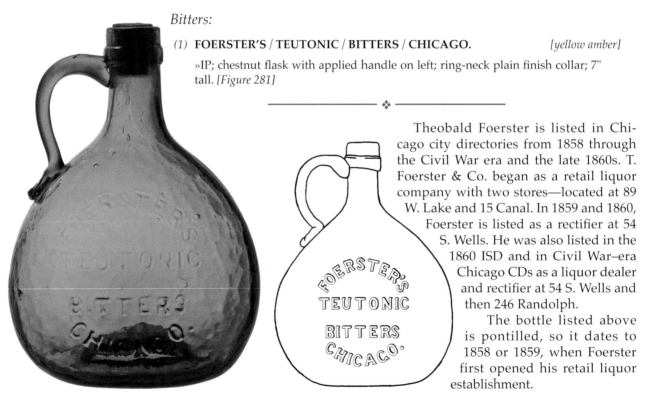

Figure 281

Charles M. Foster

Medicine:

(1) *(side):* **FOSTERS // INDIAN / HEALTH RENEWER //** [aqua]
    (side): **CHICAGO ILL'S**

»SB w/ keyed hinge mold; beveled-edge rectangle; 3 sunken panels; tapered collar; 9" tall.
[Figure 282]

"Dr." Charles Foster's name appears in the 1874 *Atlas of Sangamon County* (Brink, McCormick, and Co. 1874a:103), which illustrates a drawing of his home in Springfield (*Figure 283*). The picture caption states that Dr. C. M. Foster is the manufacturer of *Foster's Indian Health Renewer* and another drug named *Foster's Child's Relief* (no embossed bottles known). Since no Springfield-embossed bottles for these medicines have been found, they may have been put up in unembossed bottles with paper labels only. More likely, the Chicago-embossed bottles were used in the Springfield area (*see later discussion*). Prior to 1871–72, when he is first listed as a "physician" in Springfield, Foster worked there as a butcher and in the hide and leather trade.

From Chicago CD evidence, Foster's first Chicago connection was during 1869 and 1870. In 1869, he moved up to Chicago (boarding at the Keystone House) and joined the "general commission [forwarding] merchant" firm of J. C. Barnes & Co. By 1870, he was a partner in the firm (Barnes, Foster and Stone), but his address was listed as Springfield, IL. It is important to note that the third new member of the firm was R. R. Stone: the manufacturer of *Manzanita Bitters* (*see Stone listing, this chapter*). This shows that the Barnes firm was involved in patent-medicine wholesaling, and suggests that the reason Foster joined the firm was to market his newly concocted *Indian Health Renewer*, and perhaps his *Child's Relief*.

This effort seems not to have been particularly successful, since the firm is no longer listed in the CDs after 1870, at which time Foster seems to have returned to Springfield as a self-styled "doctor" and focused his medicine-sales efforts downstate (many of his Chicago-embossed bottles have turned up in the Springfield area). At the same time, a new commission-merchant company continued to try to market Foster's *Health Renewer* in Chicago:

> WE CAN RECOMMEND FOSTER'S INDIAN
> Health Renewer for strengthening the system against the blasts of winter. It tones the stomach, cleanses the liver, and regulates the kidney and bowels. Sold by REILAND & RANNEY, 143 West Madison-st. (*Chicago Daily Tribune*, 11/1/1874, p. 15, col. 7)

However, Foster did not prosper in Springfield either. He lost his Springfield house (illustrated in the 1874 *Atlas*) to a sheriff's sale in 1876 and moved to back Chicago, where he was then listed as a physician in the 1876 through 1880 Chicago CDs, and in the 1880 U. S. census. In the 1877 Chicago CD, he is listed as "mnfr. Foster's Indian health renewer" at 528 Michigan Ave. This is his only post-Springfield listing offering the *Health Renewer* for sale in Chicago, but apparently the embossed bottles were still available for sale as late as 1877:

> **THE CHILDREN __**
> A Torpid State of the Liver and Kidneys is the great cause of Nearsightedness so general among our children. FOSTER'S Indian Health Renewer is known to be a safe and sure cure for this evil (*Chicago Daily Tribune*, 5/6/1877, p. 15, col. 7)

Figure 282

Figure 283

The 1874 Sangamon Co. Atlas indicated that Foster's *Indian Health Renewer* would cure "all diseases of the Liver, Lungs, Stomach, and Kidneys. It has never failed to cure Heart-Disease, Asthma, Dyspepsia, Costiveness, and Nervousness." He called his *Child's Relief* a "vegetable preparation" and declared that "Where it has been used, a child has never been known to die with Cholera Infantum or Summer-Complaint. Will cure Chronic Diarrhoea."

John Frechette

Bitters:

(1) *(in barred oval):* **J. FRECHETTE / SPANISH / TONIC BITTERS /** [amber]
(front heel): **PATD MAY / 21ST 1872**

»SB; square case bottle; beveled corners; slightly tapered collar; 9 1/2" tall. *[Figure 284]*

Frechette's bitters was poorly marketed and/or unsuccessful, and is seldom seen. Ring's (1980) master inventory of U.S. bitters bottles did not include it, although it was added to the Ring and Ham (1998:232) update volume, but with no information on its maker or city of origin. The bottle we illustrate may be the same one reported to Ring & Ham.

Fortunately, however, Frechette patented his recipe and the patent date is given. A patent search (eventually) produced the following information for Patent #126,948, awarded to John Frechette:

> Specification describing a new and useful Improvement in Medical Compound, invented by John Frechette, of Chicago, in the County of Cook and the State of Illinois.
>
> My invention has for its object to furnish an improved medical compound or tonic-bitters for invigorating the system when reduced or weakened by sickness or other causes; and it consists in the compound prepared of the ingredients, in the proportions, and manner hereinafter set forth.
>
> In preparing this compound, take one pound of orange peel, three-quarters of a pound of calamus, one-quarter of a pound of ginger, one-quarter of a pound of bayberry-bark, and four lemons. These ingredients are ground and put into two gallons of pure spirits, diluted to not less than seventy-five per cent of spirits.
>
> The compound is then allowed to stand for thirty-six hours in a warm place, and is then reduced to forty per cent of spirits is sweetened to taste with crushed-sugar sirup, and is filtered. It is then ready for bottling for use or market.

Figure 284

Edwin O. and William H. Gale

Medicine:

(1) **GALE BROTHERS / CHICAGO** [aqua]

»IP; slope-shouldered oval bottle with vertical embossing; double-ring collar; 8 1/4" tall. *[Figure 285]*

———❖———

E. O. & W. H. Gale arrived in Chicago from New York in 1855; they were partners in the drugstore business in Chicago from 1856 to 1864. They began at 184 Randolph and a few years later moved to their long-time location at 202 Randolph. In 1865, they added Wm. Blocki to the business, and by 1866 they had split up: E. O. Gale and Wm. Blocki were partners, and Edwin's brother William was associated with the Ehrman & Co. druggist firm.

In an April 3, 1861, newspaper ad they published in the *Chicago Tribune*, the Gale Brothers state that they are "Wholesale and Retail Dealers in Choice Drugs and Chemicals of Every Description." The bottle listed above is pontilled and likely dates to soon after their arrival in Chicago (ca. 1856–59).

Garden City Chemical Works

Flavoring Extracts:

(1) **GARDEN CITY / CHEMICAL WORKS** [aqua]

»SB w/ keyed hinge mold; rectangular; 4 sunken panels; double-ring collar; 5 1/4" tall. *[Figure 286]*

———❖———

Figure 285

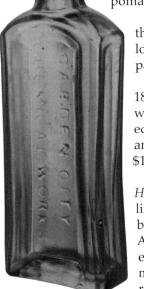

During the late Civil War years (1863–65), John H. Huyck was a clerk at F. A. Bryan's drugstore. He started the Garden City Chemical Works in 1866 at 111 S. Water Street as *Huyck Bros. & Esmay* in 1866 to produce "flavoring and fluid extracts." During the three-year period from 1867 through 1869, the proprietorship became *Huyck & Randall*, first at the same address, and later at 34 S. Water (*see their separate listings, this chapter*). A business card in the 1867 Chicago CD (p. 53) announced that Huyck & Randall were "Manufacturers of Huyck's Standard Flavoring Extracts, Fluid Extract Buchu, Azzaieline for the Hair, pomades, hair oils, perfumeries, etc., etc."

In the period from 1870 to 1876, Amos Randall was replaced by Charles Knox, and the firm became *Huyck & Knox* (*see their separate listings, this chapter*). They were first located at 34 S. Water, then 57&59 W. Lake, then (during the final three years of the partnership) at 33 Michigan Avenue.

After 1876, the firm became *J. H. Huyck & Co.*, with various junior partners, during 1877 and 1878, and then just *J. H. Huyck* alone between 1879 and 1882. Huyck's company was listed in the 1880 IL Industrial Census as having $25,000 invested in property and equipment, with approximately 30 employees at any given time (10 men, 10 women, and 10 children). His annual payroll cost was $11,250 with an annual materials cost of $125,000 and an annual product sales value of $250,000.

A fancifully illustrated trade card produced during this period (Figure 287) for *John H. Huyck's Garden City Chemical Works* at 33 Michigan Avenue lists his ca.-1880 product line as "Huyck's Full Measure Triple Flavoring Extracts, perfumes, German Cologne, backing [baking?] powder, inks, bluing, lemon sugar, chewing gums, Jamaica Ginger, Azzaieline, Seidlitz Powders, washing crystal, soap powder, fancy soaps, hair oils, etc., etc." During the 1860s and 1870s, the success of John Huyck's chemical company may be due in part to the fact that his family's retail drugstore business served as a retail outlet for his flavoring extracts and grooming products.

Figure 286

After 1882, the company became *J. H. Huyck & Co.* again, moving to 109 N. Water, and later to 73 S. Water, with a variety of junior partners—including the temporary reappearance of Charles Knox (in 1891–92) and a partnership with his son, J. H. Huyck, Jr., from 1893 until the business closed down in 1901.

Technologically, the bottle style listed was produced in the 1860s—characteristically as a container for flavoring extracts. Since Huyck & Randall included their name on embossed bottles (*see Huyck and Randall listing, this chapter*), the listed example may date to 1866, the first year Huyck's *Garden City Chemical Works* was formed.

Charles H. Gardiner

Medicine:

(1) (side): **GARDINER'S // RHEUMATIC & /**   [aqua]
    NEURALGIA / COMPOUND //
   (side): **CHICAGO**

»SB w/ keyed hinge mold; beveled-edge rectangle; 3 sunken panels; sharp, slightly rounded shoulders; long neck; tapered collar; 9" tall. [*Figure 288*]

Charles H. Gardiner was a long-time druggist in Chicago beginning in 1863. Chicago CD entries indicate that his drug store was located at 28 N. Clark throughout the 1860s, and at 156 N. Wells in 1875. From its technological attributes, and Gardiner's history, the listed bottle appears to date between

Figure 287

1863 and 1869. No advertising has been found for this medicine, but its function as a cure for rheumatism and neuralgia is evident from its name.

John Garrick and Solon L. Cather

Whiskey:

(1) **GARRICK & CATHER. // CHICAGO. ILL //** [amber]
(*shoulder-embossed above embossed image of a palm tree*)

»SB w/ post mold; quart whiskey cylinder; ring-neck tapered collar; 11½" tall. [Figure 289]

According to the CDs, Garrick and Cather business partners for just two years in Chicago. In 1872, they established a cigar and liquor retail and wholesale business at 36 W. Randolph. In 1873, they maintained their W. Randolph address, probably for the wholesale business, and were also the proprietors of a liquor "sample" room and billiard hall at 71 Monroe Street called the *Royal Palm*. According to a book on Monroe Street in Chicago's early days (Mack 1914:53), the *Royal Palm* was built late in 1872. "This building was four stories high and was occupied, on the ground floor, by a sample room, and upstairs by a billiard room. It was owned by John Garrick and S. L. Cather." A writer in *The Landowner* (January 1873) described the *Royal Palm* as follows: "Paris in the palmy and gorgeous days of the empire never boasted of such apartments."

In late-nineteenth-century parlance, the term "sample room" usually meant a combined saloon and retail liquor store. From the above listing, Garrick and Cather sold some of their liquor in proprietary bottles embossed with a palm tree—i.e., a "*Royal Palm*."

For reasons we have not been able to discover, Cather is gone from the Chicago directories after 1873, and Garrick is involved in unrelated mercantile endeavors. But in any case, the bottle listed above clearly dates to the time of the Royal Palm's grand opening in late 1872 and early 1873.

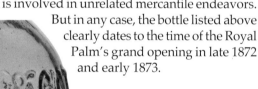

Figure 288

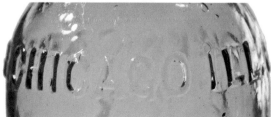

Figure 289

Schubael Geer

California Pop Beer:

(1) **S. GEER / CELEBRATED / CALIFORNIA / POP BEER / TRADE MARK /**   *[amber]*
PAT OCT 29, 1872 // CHICAGO / ILL^S

»SB w/ post mold; slope-shoulder quart; straight-sided shouldered-blob top; 11 1/2" tall. *[Figure 290]*

❖

Shubael Geer was a partner in the pop-beer bottling business with James Dorman from 1878 to 1879. Dorman was also a California Pop Beer bottler (*see his listing, this chapter*). Geer is listed on his own in Chicago CDs from 1880 through 1885 as a bottler of pop-beer at 24 N. Green. He is also listed in the Chicago USC for 1880 as the proprietor of a soda water business.

The Dorman and the Geer quart bottles appear to have been made from the same mold. The mold for the Geer bottle appears to have been modified by peening-out DORMAN & CO. and substituting S. GEER. Based on their business chronology (Geer took over Dorman's business in 1880), these two bottles date to 1879 and 1880 respectively.

According to the 1880 IL Industrial Census, Schubael Geer's *California Pop Beer* bottling business operated full-time, 12 months per year and annually produced bottled products valued at $7,500. Geer was 44 years old in the 1880 USC.

See also John Lomax, and Paul Rouze listings, both of whom were also Chicago "California Pop" (aka: "California Pop Beer") distributors.

Figure 290

Paul W. Gillett Chemical Works

Ink:

(1) **GILLETT'S / CHEMICAL / WORKS / CHICAGO**   [aqua]
(base, arched in central post-mold circle): **A & D H C**
Note: two T's in GILLETT.

»SB w/ post mold; schoolhouse-type ink; thickened, squared, flare-ring lip; 2 1/2" tall. *[Figure 291]*

Bluing:

(1) **GILLET'S / CHINESE / LIQUID / BLUEING / TRIPPLE STRENGTH**   [aqua]
Note: "TRIPPLE" embossing is misspelled.

»SB w/ post mold; high-shouldered oval; tapered collar; 5 1/2" tall. *[Figure 292]*

(2) **GILLET'S / CHINESE / LIQUID / BLUEING / TRIPPLE STRENGTH**   [aqua]
Note: "TRIPPLE" embossing is misspelled.

»SB w/ post mold; slope-shouldered cylinder; flared-ring collar with tapered base; 5 3/4" tall, 2" diameter. *[Figure 293]*

Flavoring Extract:

(1) (side): **GILLETS** // (side): **CHEMICAL WORKS**   [clear]

»SB; rectangular base; semi-cathedral shape; roped corners; steep decorative shoulders; long neck; squared, flared-ring collar; 6" tall. *[Figure 294]*

Figure 291

Figure 292

Figure 293

(2) (side): **GILLET'S // CHEMICAL WORKS //** [aqua]
(side): **CHEMICAL WORKS // //**

»SB; beveled-edge rectangle with 4 sunken panels; long neck; squared, flared-ring collar; 5 1/4" tall. *[Figure 295]*

Product Jar:

(1) **GILLETS CHEMICAL WORKS / CHICAGO ILL** [aqua]
Note: This jar is similar to the one used by C. O. Perrine, Chicago, for honey (see his listing).

»SB; cylindrical wax-seal jar; sloping shoulders; ring neck; flared, rounded lip; 7" tall; 3 1/4" diameter. *[Figure 296]*

Paul W. Gillett was an ink manufacturer in Wheaton, IL, in 1860 (USC). He first came to Chicago in 1863, and is listed that year in the CD as an "agent, flavoring extracts" at 189 Madison—perhaps marketing his earlier Wheaton chemical company products.

In the 1864 CD, he is listed as running his own factory at 189 Madison, manufacturing flavoring extracts (1864 CD) and ink (1864 ISD).

Then in 1865 and 1866, he brought his son Egbert into the business as a bookkeeper (*see the Gillet & Son listing, this chapter*) and moved his operation to 57 W. Lake, manufacturing "flavoring extract, hair oils, etc."

Beginning in 1867, he created an expanded partnership to run the company (including his son Egbert W., and Carlton G. McCulloch) and changed its corporate name to *Gillet, McCulloch & Co.* (*see listing, this chapter*). The Gillet, McCulloch & Co. chemical works was in business from 1867 through 1878, variously highlighting their production of flavoring extracts, baking powder, yeast, and perfumes in their CD listings. Their factory address changed frequently during this time, which could be helpful in dating embossed bottles that retain their paper labels: 85 Market (1867–68); 56 & 60 S. Water (1869); 61 Michigan Av. (1870–71); 51 W. Lake (1872); 34 & 36 S. Water (1873–75); 42 & 44 Michigan (1876–78). In 1879, McCulloch left the company (now expanded to 38–44 Michigan), and in 1880 Egbert "Gillett" took over the company on his own—reliably adding a second "t" to his last name in the CDs, which might be a help in dating post-1880 bottles.

Egbert ran the company until his death in 1903, and his estate officers managed its closure in 1904.

Figure 294

Figure 295

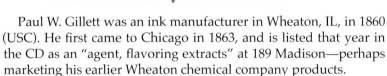

Figure 296

Paul W. Gillet and Son

Bluing:

(1) **P. W. GILLET & SON / CHICAGO** [aqua]

»SB w/ keyed hinge mold; rectangular; 4 sunken panels; long neck; double-ring collar; 5 1/4" tall. *[Figure 297]*

As noted in the *Gillet Chemical Works* discussion, in 1865 and 1866, Paul Gillet brought his son Egbert into the business as a bookkeeper and moved his operation to 57 W. Lake, manufacturing "flavoring extract, hair oils, etc." according to the 1865 Chicago CD.

Beginning in 1867, he created an expanded partnership to run the company (including both his son Egbert and Carlton G. McCulloch) and changed its corporate name to *Gillet, McCulloch & Co.* (*see their listing, this chapter*).

Gillet, McCulloch and Co.

Medicine:

(1) **GILLET'S / ESS JAMAICA / GINGER /** [aqua]
(*shield monogram*): **GCW / GILLET McCULLOCH & CO. / CHICAGO**
(*base*): embossed **five-pointed star**.

»SB w/ post mold; oval; flared, square-ring collar; 6" tall *[Figure 298]*. A second variant has no embossed star on the base.

Figure 297

The firm of *Gillet, McCulloch & Co.* included principles Paul W. Gillett, Carlton G. McCulloch, and Egbert W. Gillett. Their chemical works was in business from 1867 through 1878. Their CD listings variously highlight production of flavoring extracts, baking powder, yeast, and perfumes. From the listed bottle, "Essence of Jamaican Ginger" was another among their stable of products. From a paper label (with the G M & Co. name and a design-patent date of 1874) on one of their unembossed bottles, another of their products was *Gillet's Chinese Liquid Bluing* (*Figure 299*).

Their factory address changed frequently during this time, which could be helpful in dating embossed bottles that retain their paper labels. Their various locations include 85 Market (1867–68); 56 & 60 S. Water (1869); 61 Michigan Av. (1870–71); 51 W. Lake (1872); 34 & 36 S. Water (1873–75); and 42 & 44 Michigan (1876–78). In 1879, McCulloch left the company (now expanded to 38–44 Michigan),

Figure 298

Figure 299

and in 1880 Egbert "Gillett" took over the company on his own—reliably adding a second "t" to his last name in the CDs, which might be a help in dating post-1880 bottles.

Jamaica Ginger was sold as a patent medicine in the nineteenth century to treat nausea and to serve as an antispasmodic. This liquid ginger compound contained 70–80 percent ethyl alcohol by weight and was also used to flavor alcoholic beverages.

Julius Grossenheider and Co.

Soda ~ Mineral Water:

(1) **J. GROSSENHEIDER & Co / CHICAGO. //** [aqua, teal]
(rear shoulder): embossed spread-wing eagle

»SB w/ keyed hinge mold and post mold variants; shouldered-blob-top soda. [Figure 300]

(2) **J. GROSSENHEIDER & CO. / CHICAGO //** [teal]
(rear shoulder): embossed spread-wing eagle)

»SB w/ post mold; Chicago-style blob-top quart soda; 10 1/2" tall. [Figure 301]

Lager Beer:

(3) (in circular slug plate): **GROSSENHEIDER'S / CELEBRATED /** [amber] **BOTTLED / LAGER BEER // THIS BOTTLE / NOT TO / BE SOLD**
(base): **PUTNAM**

»SB w/ post mold; slope-shouldered cylindrical beer; long neck and shouldered-blob top; 9" tall. [Figure 302]

❖

Julius Grossenheider first appears in the Chicago CDs in 1868 as a saloon owner and a wholesale dealer in bottled lager beer. By 1869 and 1870, he is listed as a wholesaler beer bottler at 73 Market Street.

Figure 300

The Grossenheider "& Co." associations begin in 1871, and mark his expansion (largely as a facilitator and agent) into the soda and mineral water realms. In 1871, he is associated with Henry Kruse in the "mineral springs water" bottling business at 66 Clark (*see their listing, this chapter*). In 1873, he is associated with George Dickinson (who had been a barkeeper and brewing-company bookkeeper in 1871 and 1872) in the soda bottling business at 122 Archer (*see Dickinson listing, this chapter; see Grossenheider & Dickinson listing, this chapter*). In 1874, he is associated with M. Grossenheider in his bottling businesses. Given uncertainties of short-term CD listings, Grossenheider's "& Co." association with soda and mineral water bottling seems to span ca. 1870 to 1875.

From 1875 onward, he is listed alone, usually as an agent, and only in the wholesale beer-bottling business, just as he was in 1868 and 1869 prior to his soda and mineral water ventures. He continued alone in this business until his retirement in 1885. Toward the end of his career, he is also listed as a bottler of "ale and porter" (in 1883) and as a "bottle manufacturer" (in 1884).

From the style and manufacturing technology of the embossed bottles used by these various Grossenheider & Co. incarnations, they all seem to be soda and mineral water bottles that date from the early 1870s. But we should note that two of the Grossenheider & Co. bottles (the only teal-colored ones) have an embossed eagle emblem on the rear shoulder, an image often associated with brewery symbolism. If they were used as beer bottles, they represent a rare example of the use of soda-style containers for brewery products.

It is also intriguing to note that during the short four-to-five-year span of the soda and mineral water ventures sponsored by Grossenheider, at least eight differently embossed bottles and three color variations were used, an unusual diversity for such a short time.

Figure 301

Figure 302

Julius Grossenheider and George Dickinson

Soda ~ Mineral Water:

(1) **GROSSENHEIDER / & / DICKINSON / CHICAGO //** [amber, aqua]
(heel): **A. & D. H. C**

»SB w/ post mold; shouldered-blob-top soda. *[Figure 303]*

Figure 303

Julius Grossenheider and George Dickenson were briefly associated in the soda bottling business during the early 1870s (*see discussion in the previous Grossenheider listing*).

Julius Grossenheider and Henry Kruse

Soda ~ Mineral Water:

(1) **J. GROSSENHEIDER & KRUSE / CHICAGO** [aqua]
(base): **G&K**

»SB w/ post mold; shouldered-blob-top soda. *[Figure 304]*

Julius Grossenheider and Henry Kruse were briefly associated in the "mineral spring waters" bottling business in the early 1870s (*see discussion in the Grossenheider listing, this chapter*). The only Chicago CD listing for Henry Kruse is for 1871. He is not listed in the 1870 or 1872 directories at all.

Figure 304

Zebulon M. Hall

Unguent / Spice:

   (1) **Z. M. HALL / CHICAGO**   [aqua]

      »SB w/ keyed hinge mold; shouldered squat cylinder; wide mouth; ring collar; 2 1/2" diameter, 5 1/2" tall. *[Figure 305]*

Figure 305

During the Civil War years, Zebulon Hall was a junior partner with a wholesale grocery business and a commission hide-merchant business, before beginning a new wholesale and retail grocery business with his brother in 1865 and 1866 at 293 Kedzie. Hall was also listed in the 1863 IRS assessment for Chicago merchants, but the listing does not indicate his type of business.

Beginning in 1867, he was in business by himself at 259–261 Randolph. He still dealt in groceries, but with an unusual twist: in 1867 he was listed as a dealer in "teas, groceries, and cordage"; in 1868 as a "grocer and ship-chandler"; and in 1869 as a dealer in "groceries and naval stores." He was still listed in this same business in the 1870 and 1875 Chicago CDs, but he went bankrupt in 1876, and his entire retail stock was sold at auction (*Chicago Tribune*, 1/31/77, p. 1, col. 1). His son, who had been involved in business with his father, resumed in the wholesale and retail grocery trade in 1881 as *Edgar A. Hall & Co.* (*Chicago Tribune*, 9/25/81).

From the technological attributes of the bottle listed, it dates from Hall's early business efforts in the late 1860s, during which time he was providing supplies and provisions to the marine trade. The bottle's shape suggests only that it could have contained any of a variety unguents, creams, or spices. However, a *Chicago Tribune* ad has been found (7/19/74, p. 8, col. 1) which suggests it was a spice container:

PURITY IS PRICELESS — WHY BUY THE
Cheap adulterated trash that is vended throughout the city and country for Teas, Ground Coffee, and Ground Spices? We roast and grind our own Coffee daily; also, grind our own Spices, selecting the choicest and best goods for this purpose, which is the only way sure to get Coffee and Spices that are really pure.   **Z. M. HALL**
Wholesale and retail grocer, Lind Block, cor. Randolph and Market-sts.

John A. and Lysander B. Hamlin

Medicine ~ Early Oval Wizard Oil Bottles:

   (1) **HAMLIN'S / WIZARD / OIL // CHICAGO / ILL_**   [aqua]

      »OP; shouldered oval; long neck; ring-neck collar; 5 3/4" tall. *[Figure 306]*

   (2) **HAMLIN'S / WIZARD / OIL // CHICAGO / ILL_**   [aqua]

      »OP, shouldered oval; short neck; rolled lip; 4" tall. *[Figure 307]*

   (3) **HAMLIN'S / WIZARD / OIL // CHICAGO / ILL_**   [aqua]

      »SB w/ straight hinge mold, shouldered oval; short neck; thin flared-ring collar; 4" tall. *[Figure 308]*

   (4) **HAMLIN'S / WIZARD / OIL // CHICAGO / ILL_**   [aqua]

      »SB w/ keyed hinge mold, shouldered oval; short neck; double-ring collar; 4" tall. *[Figure 309]*

Figure 306

Figure 307

Figure 308

Figure 309

Medicine ~ Later Rectangular Wizard Oil Bottles:

(5) *(side):* **HAMLIN'S // WIZARD OIL //** *(side):* **CHICAGO** [aqua]

»SB w/ keyed hinge mold; beveled-edge rectangle with four sunken panels; short neck; double-ring collar; 5 1/2" tall. *[Figure 310]*

(6) *(side):* **HAMLINS // // ** *(side):* **WIZARD OIL // //** [aqua]

»SB w/ keyed hinge mold; beveled-edge rectangle with four sunken panels; long neck; double-ring collar; 8" tall. *[Figure 311]*

(7a) **HAMLINS / WIZARD OIL // // //**

Note: Smaller, narrower embossed lettering than variety 7b.

»SB w/ keyed hinge mold; beveled-edge rectangle with four sunken panels; short neck; double-ring collar; 6 1/4" tall. *[Figure 312, left]*

[aqua]

Figure 310

(7b) **HAMLINS / WIZARD OIL // // /** [aqua]

Note: Larger, wider embossed lettering than variety 7a.

»SB; beveled-edge rectangle with 4 sunken panels; short neck; double-ring collar; 6" tall. *[Figure 312, right]*

(8a) **J. A. HAMLIN & BRO / CHICAGO // // WIZARD OIL // //** [aqua]

»SB w/ keyed hinge mold; beveled-edge rectangle with four sunken-panels; double-ring collar; 6 3/4" tall. *[Figure 313]*

(8b) **J. A. HAMLIN & BRO / CHICAGO // // WIZARD OIL // //** [aqua]

»SB w/ post mold; large embossed X in central circular depression on base; beveled-edge rectangle with four sunken panels; double-ring collar; 6" tall. *[Figure 314]*

274 Chicago — John A. and Lysander B. Hamlin

Figure 311

Figure 312

Salve Jar

(9) **J. A. HAMLIN / & BRO / CHICAGO** // // // //　　　　　　　　　　　　　　　　　　　　　　　　　　[clear]

»SB; beveled-edge square wide-mouth salve jar with 4 flat panels; narrow flared-ring collar; 2 1/2" tall. *[Figure 315]*

"Cough Balsam:"

> Note: This later Hamlin's product was first advertised during the early 1870s. However, no embossed **HAMLIN'S / COUGH / BALSAM** bottles (shaped like the rectangular Wizard Oil bottles) have been documented predating the 1880s. Thus the product appears to have been separately bottled too late for formal inclusion in the present volume.

In 1859, while still in his early 20s, John Austin Hamlin first developed and began to sell his *Hamlin's Wizard Oil* in Cincinnati, OH. His Chicago biographer indicates that the formula for the medicine originated with John's father, William "a pioneer circuit rider, who, while faithfully administering to the souls of men and women, did not forget the bruises, sprains and other hurts of their bodies. When he died he left his son little except the formula for the oil which he dispensed with his kind and Christian words" (Waterman 1908:976). Another publication, without attribution, indicates that

> Hamlin bought the *Wizard Oil* formula from Dr. C. M. Townsend, of Lima, Ohio (for whom John had been a traveling agent) and moved to Cincinnati to begin manufacture of the brand in 1859. He later got caught up in the Civil War before he could really get it widely sold. While still in his

Figure 313 **Figure 314**

early 20s, he located to a new and larger facility in Chicago, Ill., and within a few years the brand caught on. [Wilson and Wilson 1971:118]

Regardless of the medicine's origin, John Hamlin and his brother Lysander moved from Cincinnati to Chicago in the early 1860s to produce their *Wizard Oil* there.

Whether Hamlin developed his sense of traveling showmanship riding the circuit with his father, or as a traveling-show agent in Ohio for Dr. Townsend, or because he had been a traveling magician in his youth (Carson 1961:37)—perhaps with Dr. Townsend's show, as Townsend's products included a *Magic Oil* (Young 1961:194)—Hamlin's showmanship acumen became the hallmark of his career.

> In the first year of the war he came to Chicago to develop the business, and about this time originated the "medicine show" as his star advertising medium. He employed comedians, ventriloquists and fakirs to draw the crowds throughout the country, after which the lecturers came up on the stage and sold the oil as fast as they could hand out the bottles. It was one of the most successful advertising schemes of the day and placed the business on a splendid foundation. [Waterman 1908:977]

Figure 315

Figure 316

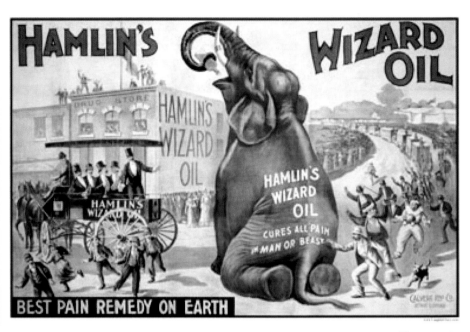

Figure 317

Touring the highways and byways of the country were numerous troupes, each made up of a lecturer, a driver, and a male quartet. The group traveled in a special wagon, pulled by a four- or six-horse team, into which was built a parlor organ. The wagon, in the torch-lit evening, became a stage, from which the quartet sang and played. A stylish sight they were, clad in silk top hats, frock coats, pin-striped trousers, and patent leather shoes—with spats. At times the assembled audience sang with them. One of Hamlin's stunts was the lavish distribution of pamphlets [see Figure 316] in which the words of such songs as "I'se Getting' Up a Watermelon Party" and "Is Life Worth Living?" were interspersed with promises as to how Wizard Oil could grapple with asthma and neuralgia. [Young 1961:193–194]

Figure 318

Hamlin's traveling shows were hugely popular (see Figure 317 and 318), and with the help of their circus-like atmosphere well over a million bottles of Wizard Oil were sold during the late nineteenth century.

It would be a daunting task indeed to list all of the sizes and mold variations of embossed bottles made for this product during the time it was produced by the Hamlin brothers in Chicago in the late nineteenth and early twentieth century. Even the above listing of embossed bottles from the first 20 years of their business is conservative. For instance, their embossed post-mold bottles, first made in the late 1870s and early 1880s, have been omitted if there was any question as to their pre-1880 age.

The first *Hamlin's* bottles (all of them are oval and pontiled) were made only during 1859–60. The earliest variety is vertically embossed **HAMLIN'S / WIZARD OIL // CINCINNATI, / OHIO** and is rarely seen (as noted in *American Bottle and Glass Collector* 11(10):39, see also Holst 1994:125; *see Figure 319*). But within the year, two sizes of very similar pontiled bottles were also produced (#1, misspelled, and #2), embossed **CHICAGO / ILL** on the reverse side. All of the rectangular, smooth-based, hinge-mold bottles in the list would have been produced during the 1860s and early 1870s.

The **CHICAGO**-embossed pontiled bottles were puzzling, at first. The Chicago CDs have no listing for John Hamlin, his brother Lysander, or their *Wizard Oil*, until 1864, when John appears

Figure 319

Figure 320

as a "manufacturer of patent medicines" at 102 Washington Street, along with a large ad for *Hamlin's Wizard Oil*. The ad is accompanied by a strange illustration of a man painting a Wizard Oil ad on the side of a massive cliff-side rock (*Figure 320*): "This article is used both Internally and Externally, and cannot be surpassed, or even equalled, for curing all kinds of lameness and pain." The ad declared that Wizard Oil would cure toothache, headache, earache, neuralgia, and cramp colic in a matter of minutes; sore throat, lame back, sprains, and diphtheria in a matter of hours. Rheumatism would be cured "in a few days." The two bottle sizes of *Wizard Oil* sold for 25¢ and 50¢, and the ad pointed out that "the large bottles contain nearly three times as much as the small ones."

But the 1864 ad itself contained the answer to how pontiled and early smooth-based bottles of *Hamlin's Wizard Oil* had been distributed in Chicago while the Hamlin brothers were "caught up in the Civil War." Two Chicago wholesale distribution agents are listed at the bottom of the ad: Lord & Smith; and Fuller, Finch & Fuller. In 1859, according to Chicago CD data, Thomas Lord and Lafayette Smith sold drugs and patent medicines wholesale at 43 Lake Street. Oliver Fuller and Edward Finch sold drugs and patent medicines wholesale at 244 Lake & 265 S. Water. They no doubt acted as Chicago sales agents for the Hamlin brothers from 1859 to 1863, and heavily advertised the *Wizard Oil* there (*Figures 321–324*).

In the 1865 CD, Lysander Hamlin is listed as having joined John in the Chicago patent-medicine business. But they are absent from the directories during the next three years (1866–68). During this time, they may have been living "on the road," developing their patent-medicine traveling-circus shows. The success of these efforts seems to have made it possible for them to settle down in the Chicago area to develop the bottling business itself after 1868, leaving the road shows to agents. They are

Figure 321

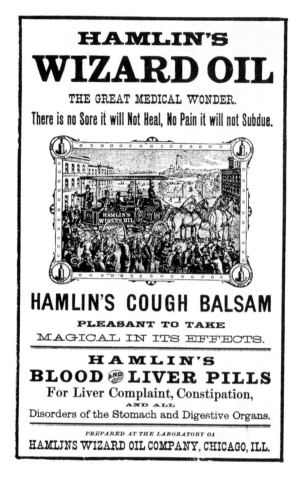

Figure 322

Figure 323

listed in the 1869 CD and in an 1869 design patent (#3435) for their bottle's paper labels (*Figure 325*) as manufacturers of *Hamlin's Wizard Oil* at 49 Franklin in Chicago, residing in Elgin, IL.

John and Lysander Hamlin are then listed in the CDs as partners in the Chicago patent-medicine business from 1869 through 1878, during which time they expanded their line to include a *Hamlin's Cough Balsam*. The firm's name during this time was always "J. A. Hamlin & Bro." Then, in 1879, Lysander Hamlin took over as sole manager of the "Wizard Oil Company," and John Hamlin retired from the business to pursue another of his show-business passions:

> In 1872, a few months after the great fire, he built the Hamlin Theater in Chicago, which afterward became the Grand Opera House. Of this he had been the sole proprietor for many years, and around it were long clustered his most earnest work and best thoughts for the advancement of legitimate and high-class theatricals. [Waterman 1908:977]

The company, run by a succession of family members, remained a massive and healthy ongoing concern through the first third of the twentieth century, and to some extent, even beyond that.

However, it suffered some serious image set-backs in the late nineteenth and early twentieth century after medical analysis of their Wizard Oil's content showed it to be 65 percent alcohol, with a little capsicum and chloroform to help numb pain, along with ammonia and a few pungent/aromatic vegetable oils (camphor, sassafras, myrrh) to enhance its medicinal "signature" (Douglas 1886:427). Subsequent attention from the Food and Drug Administration forced the company to scale back their "Cures-Anything" style of advertising and to limit their claims to the product's benefit in the temporary relief of "minor aches and pains." Their grow-

Figure 324

Figure 325

ing image problem is perhaps best exemplified by a popular American folk ditty collected and published in 1927 by Carl Sandburg for his booklet *The American Songbag*:

> Oh! I love to travel far and near throughout my native land;
> I love to sell as I go 'long, and take the cash in hand.
> I love to cure all in distress that happen in my way,
> And you better believe I feel quite fine when fools rush up and say:
> [Chorus:]
> I'll take another bottle of Wizard Oil,
> I'll take another bottle or two;
> I'll take another bottle of Wizard Oil,
> I'll take another bottle or two.

William A. Hausburg and Brothers

(William A., Gustave H., and Charles)

Soda ~ Mineral Water:

(1) **H & BRO'S / CHICAGO** [aqua]
   (base): **H&BRO'S**
   Note: This variety is the smallest of the 3 sizes of the bottle (ca. 6 1/2" tall), and has large-lettered embossing around the margin of its base.

   »SB w/ keyed hinge mold; squat, slope-shouldered, oval-blob-top soda. *[Figure 326]*

(2) **H & BRO'S / CHICAGO** [aqua]
(base): **H&BRO'S**
Note: *This variety is the larger of the two "pint" soda bottle styles (about 7 3/4" tall) and has much smaller-lettered embossing around the margin of its base.*

»SB w/ keyed hinge mold; tall, high-shouldered, oval-blob-top soda. [Figure 327]

(3) **H & BRO'S / CHICAGO** [aqua]
(base): **H&BROS**
Note: *This variety is a much larger "Chicago-style" quart size of the bottle (about 9 1/2" tall) and has small-lettered embossing (no apostrophe) around the margin of a central depressed post-mold circle on its base.*

»SB w/ post mold; squat, high-shouldered Chicago-style quart soda; oval-blob-top. [Figure 328]

Ale:

(1) **H & BRO'S / CHICAGO** [amber]
(base): **H&BRO'S**

»SB w/ post mold; squat, 2-piece mold ale with ring-neck tapered collar; 9 1/2" tall. [Figure 329]

From the technological attributes and post-mold construction of the *H & BRO'S* bottles listed above, they likely date to the early and mid-1870s. The Chicago CDs, list only one soda or mineral water bottling operation active in Chicago during those years involving three brothers: William, Gustave, and Charles Hausburg were listed as bottlers of mineral water from their residence at 112 Claybourn Avenue in the CDs for 1872 and 1873.

Prior to this time, one of the Hausburg brothers (Gustave) had been senior partner in two soda or mineral water bottling operations in the sleepy South Chicago "suburb" of Blue Island (*see Hausburg & Innes and Hausburg & Dettmer listings, this chapter*). From the technological attributes of the Hausburg Blue Island bottles, they appear to date from the mid-to-late 1860s.

The first Chicago Hausburg-brothers partnership was short-lived, however. By the time the 1874 Chicago CD was published, Gustave had left the company, probably to help set up a family bottling works at Elgin (*see the Elgin Hausburg Bros. listing*). By the late 1870s/early 1880s, he had established the *G. H. Hausburg* bottling works in Blue Island (*see his Blue Island listing*).

During the mid-to-late 1870s, the remaining two brothers (William and Charles) became the "Hausburg Brothers" bottling works in Chicago

Figure 326

Figure 327

Figure 328

(*see their listing, this chapter*). They continued to bottle together until 1878, after which William took over the Chicago bottling operation on his own (*see his listing, this chapter*), while Charles presumably moved to the bottling operation in Elgin.

Hausburg Brothers
(William and Charles)

Soda ~ Mineral Water:

(1a) **HAUSBURG BROS / CHICAGO //** [teal; aqua]
(*heel*): **A & D. H. C.**

»SB w/ keyed hinge mold; oval-blob-top soda. [*Figure 330*]

(1b) **HAUSBURG BROS / CHICAGO //** [aqua]
(*heel*): **C&I**

»SB w/ post mold; oval-blob-top soda. [*no photo; see Figure 330*]

(2) **HAUSBURG BROS / GINGER ALE / CHICAGO / ILLINOIS /** [aqua]
(*heel*): **A & D H C**

»SB w/ round base; shouldered-blob-top soda; 8 1/4" tall. [*Figure 331*]

❖

According to the Chicago CDs and the 1875 ISD, William and Charles Hausburg continued operating their Chicago soda water manufactory together for five years (1874–1878) after the departure of Gustave (*see the Hausburg & Bros. listing, this chapter*). They continued bottling at their home address, 112 Claybourn Ave., during 1874–75 and then moved to new quarters at 143 Larabee from 1876 to 1878.

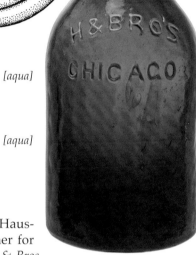

Figure 329

Figure 330

Figure 331

After William and his brother Charles dissolved the Chicago family bottling works following the 1878 season to pursue separate bottling interests, Charles continued to bottle on his own in Chicago for awhile (*see Figure 332*) before leaving town—perhaps to join Gustave in Elgin. But there are later (1880s–90s) narrow, cylindrical, amber Baltimore Loop and Hutchinson-closure bottles from Blue Island with Charles Hausburg embossing, as well as much later (crown-top) Charles Hausburg "South Chicago" bottles, so at some point he moved (returned to?) Blue Island.

William Hausburg remained as the sole proprietor of the Chicago bottling plant for its final 11 years of operation (*see the Wm. Hausburg listing, this chapter*).

William A. Hausburg

Ale:

(1) **W**^M **A. HAUSBURG / CHICAGO**   [amber]
(base): **W. A. H**

»SB w/ post mold; two-piece mold ale; ring-neck tapered collar; shoulder embossed; 9 1/2" tall. [*Figure 333*]

Soda ~ Mineral Water:

(1) **W**^M **HAUSBURG / CHICAGO / ILLS //**   [aqua]
(heel): **A & D. H. C.**
(base): **PATENT. FEBY 5, 1878**
Note: There are two embossed dots beneath the raised **in W**^M

»SB; Kelly patent Codd-marble-type gravitating stopper bottle (*see discussion in the listing*); 7 1/2" tall. [*Figure 334*]

(2) **W**^m **A. HAUSBURG / CHICAGO**   [amber]
(heel): **C & Co**
(base): **W. A. H**

»SB w/ post mold; narrow shouldered-blob-top quart; 11 1/2" tall. [*Figure 335*]

Figure 332

After William and his brother Charles dissolved the Chicago family bottling works following the 1878 season, William Hausburg formed his own soda bottling company and continued bottling soda (and ale) under the Hausburg name until he finally closed the factory for good following the 1889 season. He first moved the bottling operations next door to the residence he and Charles shared (to 145 Larabee) until 1885. He then moved to a location at 297 Claybourn for a few years, and finally bottled from a new residential location at 96 Dayton in 1889 before closing the operation down permanently.

The bottles listed above are only those that are likely to have been used during William Hausburg's early years of bottling alone in Chicago (ca. 1878–early 1880s).

Figure 333

Figure 334. The 1 3/4" glass portion of the internal gravitating stopper from an 1871 Kelly-patent soda bottle produced for Wm. A. Hausburg. The rubber portion of the stopper was not preserved.

Figure 335

In the late 1870s or early 1880s, he filed a handwritten trademark registration pursuant to an 1873 State of Illinois act "to protect manufacturers, bottlers and dealers in ale, porter, lager beer, soda, mineral water, and other beverages, from the loss of their casks, barrels, kegs, bottles and boxes:"

On all bottles the words "Wm A. Hausburg Chicago Ills"
On all kegs barrels and boxes the letters "W. A. H."
Wm A. Hausburg [signature]

The unusual glass-gravitating-stopper bottle above (S&MW #1) was used by only a few Chicago bottlers, and has not been recorded elsewhere in Illinois. The bottle was designed by William H. Kelly of Cleveland, OH. Kelly submitted his patent application on May 17, 1877, and received the patent for his bottle-closure on February 5, 1878 (Patent # 199,980). Kelly's patent claimed that "the bottle was able to be opened with no additional tools, because of the high projection of the stopper through the neck, and that once the stopper was pushed in, it was contained in the neck where it would not agitate or disturb the contents. Because the stopper did not have to be held in any particular position, the bottle could be easily cleaned and refilled" (Welko 1978:9).

According to Paul Welko, "It is probable that upon seeing the success of the Arthur Christin company, who used a similar stopper" *[see the listing, this chapter]*, the Hausburg Co. "decided to compete with their own style bottle. It doesn't seem likely the venture was successful due to the scarcity of the bottle" (Welko 1978:9).

Michael and Patrick Hayes

Medicated Beverages:

(1) **HAYES BRO'S / CHICAGO / ILLINOIS / MEDICATED / AERATED / WATERS** [aqua]

»SB w/ semi-round base (will stand); oval-blob-top soda style; 8 1/2" tall. *[Figure 336]*

(2) **NERVE TONIC / HAYES BROS / TRADE MARK /** [aqua]
(embossed horseshoe—inside horseshoe): **ESTABLISHED / 1871 // REGISTERED / CHICAGO, ILL.**
(heel): **D. O. C.**

»SB; high-shouldered cylindrical soda-style bottle; short neck; squat shouldered-blob top; 10" tall. *[Figure 337]*

Soda ~ Mineral Water:

(1) **HAYS BROˢ / CHICAGO. ILL /** [aqua]
(heel): **A & D. H. C**
Note: "HAYS" is misspelled. There are two raised dots under the superscript **S** in **BROˢ**

»SB; shouldered-blob-top soda. *[Figure 338]*

(2) **HAYE'S BROˢ / CHICAGO ILL //** [aqua]
(heel): **L&W**
(base): **H**

»SB w/ post mold; oval-blob-top soda. *[Figure 339]*

(3) **HAYS BROS / TRADE MARK / HB / REGISTERED / CHICAGO. ILL /** [yellow, amber]
(heel): **D. O. C**

»SB w/ post mold; shouldered-blob-top quart soda; 10 1/2" tall. *[Figure 340]*

(4) **HAYS BROS / TRADE MARK / HB / REGISTERED / CHICAGO. ILL. /** [yellow]
(base): **SB&GCO**
Note: Side-embossed logo and ornate trademark monogram are within a circular slug plate.

»SB w/ post mold; oval-blob-top quart soda; 10 3/4" tall. *[Figure 341]*

Figure 336 **Figure 337**

Figure 338

Figure 339

Figure 340

(5) **HAYS BROS. / TRADE MARK /**   [aqua]
(embossed horseshoe—inside which): **ESTABLISHED / 1871) // REGISTERED / CHICAGO, ILL.**
(heel): **THIS BOTTLE / IS NEVER SOLD**

   »SB w/ post mold; oval-blob-top quart soda; 11 1/2" tall. [Figure 342]

Ale and Porter:

(1) (shoulder embossed): **HAYES BROS / CHICAGO ILLS //**   [amber]
(heel): **A & D. H. C.**

   »SB w/ central glass "nipple"; 4-piece-mold quart ale; ring-neck straight collar; 9 1/2" tall. [Figure 343]

(2) (shoulder embossed): **HAYES BROS / CHICAGO ILLS**   [amber]

   »SB; 2-piece mold quart ale; ring-neck tapered collar; 9 1/2" tall. [Figure 344]

(3) **HAYES BROS. / TRADE MARK / HB / REGISTERED / CHICAGO ILLS**   [olive amber]
Note: Ornate monogram **HB** trademark.

   »SB w/ post mold; 2-piece-mold pint porter; applied double-ring collar; 7 3/4" tall. [Figure 345]

Figure 341

Figure 342

Figure 343

Some *Hayes Brothers* bottles listed indicate that the bottling company was established in 1871. But the brothers must have established their bottling business late that year. The 1871 Chicago CD lists Michael Hayes as a tinsmith living at 812 Washington, while Patrick Hayes (same home address) is listed as working for Fuller & Fuller. Patrick's Fuller & Fuller connection was very likely an important one for the Hayes brothers starting their own bottling business. F&F were prominent Chicago patent medicine wholesalers and the Hayes Brothers early product line had an unusual focus on "medicated" waters, "nerve tonic," and light "therapeutic" alcoholic beverages.

The Hayes Brothers bottling works entrance was listed as the rear of their residence for the first decade of their company's operation (at 812 Washington until 1875, and then at 908 Carroll until 1882). So the 1870s bottles listed were produced for a relatively small-scale household bottling operation.

But the Hayes Brothers home-based business was becoming more and more successful by the end of the 1870s. The Illinois Manufacturer's Census for the period from June 1879 to June 1880 indicates that they were producing "soda water and seltzer" with 5 to 8 employees (working full time from May to November and half to 2/3-time for November to May). Their material and labor costs that year were $8,000 and $2,800, respectively, and annual product sales amounted to $16,000.

The brothers made a large business leap into the soda and mineral water realm after the 1882 season. They bought the large, established Hutchinson & Sons bottling works at 241 Randolph in 1883 when that company decided to turn its full attention to manufacture and sales of bottler's supplies, syrups, and apparatus (*see their listing, this chapter*). They operated out of the Hutchinson factory from 1883 through 1886.

But by 1887, they were no longer bottling in the old Hutchinson factory. Michael had left the business entirely, and Patrick had returned to producing soda water from his current residence at 893 Carroll (probably still using "Hayes Bros." bottles). The next year he took on a new partner, John J. Sullivan, and they began producing soda water at 906 Carroll under the corporate name of "Hayes & Sullivan." At that time, Michael Hayes is listed as a "pop manufacturer" at 904 Carroll. But this was the final gasp of the business. After 1888, the Hayes family was no longer bottling soda and mineral water in Chicago.

Figure 344

Although the company's last bottles were used too late for consideration in this study, it is interesting to note that two post-1880 circular-slug-plate blob-top amber quart sodas are known that date to the final years of the Hayes Brothers bottling operation and the one-year Hayes & Sullivan bottling company. At this time, and probably from the time the Hayes Brothers took over the Hutchinson bottling plant in the early 1880s, their sodas were bottled in Hutchinson-style internal stopper bottles. Their Hutchinson-style bottles were used too late for inclusion in the present volume.

While the Hayes Brothers' trademark HP monogram appears to have been in use on bottles by the final years of the 1870s, the "horseshoe" monogram device may have been a later development that was only used from the early 1880s onward.

John M. Hedlund and Co.

Soda ~ Mineral Water:

(1a) **HEDLUND & CO / CHICAGO / ILL** [cobalt blue]
   »IP and OP varieties; shouldered-blob-top soda. *[Figure 346]*

(1b) All ditto #1a above, except larger embossed lettering [cobalt blue]
   and "**CO.**" *[Figure 347]*

Figure 345

Figure 346

Figure 347. The small-lettered version of the Hedlund & Co. bottle includes both a cobalt variety (with an iron-pontil scar on the base) and a teal variety (with an open-glass-rod scar on the base).

    (2) **HEDLUND & Co. / CHICAGO. ILL**   *[cobalt blue]*
        »IP; shouldered-blob-top soda. *[Figure 348]*

    (3a) **HEDLUND / CHICAGO / ILL**   *[cobalt blue]*
        Note: "& CO." *peened out of mold (see #1).*
        »IP; shouldered-blob-top soda. *[Figure 349]*

    (3b) **HEDLUND / CHICAGO. ILL**   *[cobalt blue]*
        Note: "& CO." *peened out of mold (see #2).*
        »IP; shouldered-blob-top soda. *[Figure 350]*

    Soon after they emigrated to American from Sweden, John Hedlund and Isaac Johnson settled in Chicago and formed the Hedlund & Co. soda water bottling company on Payton between Michigan and Kinzie. They are listed as soda water bottlers for only three or four seasons (in the 1856 through 1858 Chicago CDs).

    In their last bottling season, Isaac Johnson withdrew from (or played a much-reduced role in) the company, since Hedlund's later bottles have the "& CO." embossing peened out.

    Notably, at least five embossed-lettering mold variations are known for the Hedlund company bottles. These mold varieties are an interesting illustration of patterns of glass production and consumption by early bottlers, and the possibilities for chronological sequencing of bottles used

Figure 348 **Figure 349** **Figure 350**

by a single company. In this case, bottle style #3 was created in the final year of the company's production, after Johnson had left the business. Since bottle styles #1b and #2 were modified to reflect the change, both were in contemporary use during Hedlund's second/third bottling seasons, and bottle style #1a was no longer in use—perhaps due to mold damage or loss.

This number of glass changes over a three-to-four-season company history indicates that they were more successful than expected and that orders for #1b and #2-style bottles quickly followed the initial order (perhaps from two separate glass companies to fill the need as fast as possible, since two different molds were created). Finally, both of these second-order molds were modified for the last bottle order. This gives us almost a year-by-year chronology of glass consumption by the Hedlund Company.

Dr. William A. Henley's California IXL Bitters

Note: Dr. Henley's California IXL Bitters ("prepared from the root of the Oregon wild grape") is not formally listed or illustrated in the present volume because Dr. Henley's company was primarily a San Francisco business, despite a short abortive attempt to establish their national production operation in Chicago. In the late 1860s Dr. Henley, who was an Oregon physician, reportedly developed the bitters as part of his medical practice in Portland, and then brought it to San Francisco to try to market it more widely. He and a business partner named Henry Epstein decided to come to Chicago to organize national production and distribution for his IXL Bitters (Wichmann 1999:63).

From 1869 to 1871, Epstein, Goodman & Co. worked to establish production of the brand at 184 Kinzie. In 1870, William Henley himself took up residence in Chicago to develop marketing, advertising, and sales for his IXL Bitters under the banner of Epstein, Henley & Co. at 7 LaSalle.

Their efforts were not successful: "Apparently financial arrangements could not be agreed to and in 1873 Henley decided to start his own company, at which time he relinquished control of the IXL brand" (Wichmann 1999:63). Epstein was given California and Nevada distribution rights, and both men returned to the far West. During the later 1870s, Midwest distribution of the IXL brand was handled in Chicago by the wholesale drug and patent-medicine firm of Van Schaack, Stevenson & Reid from their Old Salamander Wholesale Drug Warehouse at 92 & 94 Lake Street (Ring and Ham 1998:274–275).

Thomas Hennessy

Ale:

   (1) **THOMAS HENNESSEY / CHICAGO / ILL //** [amber]
   (heel): **M$^{\underline{c}}$ C**

   »SB w/ post mold; two-piece mold quart ale; applied ring-neck tapered collar; 9 1/2" tall.
   [Figure 351]

   (2) **THOMAS HENNESSEY / CHICAGO / ILL //** [amber]
   (base): **F. B. Co. / 1**
   Note: The base mark on this bottle refers to the Findlay Bottle Co., an Ohio glassmaker during the later 1880s and early 1890s.

   »SB w/ post mold; two-piece mold quart ale; applied ring-neck tapered collar; 9 1/2" tall.
   [Figure 352]

Figure 351

Figure 352

According to the Chicago CDs, Thomas Hennessy and Daniel Dempsey were partners in the Chicago bottling business from 1872 until 1881 (*see Dempsey & Hennessy listing, this chapter*). They are listed as partners in the 1880 manufacturers census.

Thomas Hennessy is listed as the sole proprietor of the former D & H bottling company at 322 S. Des Plaines in Chicago in the CDs from 1882–1886. Hutchinson-style Hennessy soda bottles (post-blob-top era) were characteristically used by the company for their soda products. So, strictly speaking, Hennessy's company and its bottles are too late for inclusion in the present volume.

Interestingly, however, the two ale bottles listed are styles that were usually associated with the late-1870s. Hennessy's use of them, during his short solo business venture, is evidence that their use continued for a few years into the early and mid-1880s as well. An indication that these were some of the first bottles he ordered is the fact that his name is misspelled on both of them.

Sometime in the early 1880s Hennessy filed a handwritten trademark registration with the State of Illinois pursuant to an 1873 State act "to protect manufacturers, bottlers and dealers in ale, porter, lager beer, soda, mineral water, and other beverages, from the loss of their casks, barrels, kegs, bottles and boxes": On all bottles, barrels, kegs, and boxes, the words "Thomas Hennessy Chicago Ills" and the letters "**T. H.**" appear on the bottom of said bottles.

Ernst Hess

Soda ~ Mineral Water:

(1) *(front heel):* **E.H / CHICAGO**   *[teal, aqua]*
 (base): **E.H**

 »SB w/ keyed hinge mold; shouldered blob-top soda. *[Figure 353]*

(2) **E. HESS / MINERAL.SPRING / WATERS / CHICAGO. ILL**   *[teal, aqua]*

 »SB; squat "Chicago-style" quart soda; long neck; oval blob top; 9" tall. *[Figure 354]*

Figure 353

Figure 354

Ernst Hess is listed in the Chicago CDs for just two years as a mineral water manufacturer: 1868 and 1869 (in the 1867 CD, he was a broom manufacturer). In 1868, his partner in the "*E. Hess & Co.*" business at 47 N. Clark was Fritz Sarkander. In 1869, the company moved to 499 N. State, and his new business partner there was Peter J. Atkinson. Then in 1870, 1871, and 1872, Hess is not listed in the Chicago CDs at all, and his company is absent from the mineral water manufacturer summary listings.

However, *E. Hess & Co.* reappears in the CAR for 1873 as bottlers of mineral water. This is a strange holdover listing, because by then the company had been gone from the Chicago CDs for three years. However, 1873 was the first year that Ernst (and his brother Charles) were again listed in the Chicago CDs—now as "commission forwarding merchants" (wholesalers) at 236 Water ("Room 8"). During Hess' forwarding-merchant career in the 1870s, he and his brother are never listed as soda or mineral water bottlers.

During 1874–75, Charles was listed as running the business (which moved to 275 Kinzie). But during the late 1870s, Ernst returned to run the company, and Charles was relegated to a bookkeeper role. This might possibly suggest that after the brothers set up their new company, Ernst turned over the wholesaling activities to his brother and focused his own attention on trying to restart his earlier bottling business for a few years. However, there is no direct evidence for this, and it is more likely that the 1873 CAR listing is a clerical error.

George Hofmann and Brothers

Beer:

> (1) **GEO. HOFMANN & BROS / PREMIUM / BOTTLED BEER / CHICAGO, ILL.** // [amber]
> **THIS BOTTLE / NOT / TO BE SOLD**
> (*base monogram*): **CBCo**
>
> »SB w/ post mold; slope shouldered; narrow shouldered-blob top; no slug plate; 9 3/4" tall. [*Figure 355*]

The Hoffman brothers were Chicago bottlers at the very end of the 1840-to-1880-period that is our focus in this volume, but they were a very short-term company and represent one of the few such early Illinois beer-bottling efforts outside the Peoria area (see e.g., *John E. Bradbury & Co. Jacksonville listing*). As such, the embossed bottle style they used during their two seasons as Chicago bottlers is instructive for developing embossed brewery bottle chronologies.

According to the 1880 USC and the 1881 Chicago CD, George Hofmann Jr. and his brothers Valentine and Alois bottled beer at the corner of Lake and N. Sangamon in Chicago. The USC indicates that at that time they were living in the

Figure 355

household of their Father (George, Sr.), a retired hotelkeeper. They apparently moved to Chicago specifically to start their bottling business, since they are not even listed as Chicago residents before or after their two-season bottling effort.

The significance of the embossed monogram on the base of the bottle is unknown. Perhaps the company was known as the Chicago Brewing Co.

Samuel A. Hogeboom and Robert Wolf

Bitters:

(1) **WOLF'S / STOMACH BITTERS // HOGEBOOM WOLF & CO / CHICAGO.**   *[amber]*
»SB, beveled-edge 2¾" square; flat side panels; short tapered collar; 9" tall. *[example not found for photography]*

--- ❖ ---

Hogeboom, Wolf, & Co. was a short-lived company whose origin is elucidated by the backgrounds of Hogeboom and Wolf. In 1868, S. A. Hogeboom was a Chicago "rectifier" operating from a facility at 185 Kinzie and Robert Wolf was a "traveling agent" for *Bennett Pieters & Co.* (*see listing for Bennett Pieters and his Red Jacket Bitters, this chapter*).

According to the Chicago CDs, Hogeboom and Wolf got together to produce and market *Wolf's Stomach Bitters* at Hogeboom's 185 Kinzie facility during 1869 and 1870. They are listed boldly and in the wholesale liquor dealer summary in 1869, and have just a small listing in the alphabetic directory in 1870. By 1871, Robert Wolf is in the wholesale wines and liquors business at 11 and 13 W. Madison, and Samuel Hogeboom is retired and listed only as a Chicago resident.

Clearly, the *Wolf's Stomach Bitters* bottle listed was manufactured in 1869, but the product enjoyed very little sales success. The bottle is very rarely found (Ring 1984:110; Ring and Ham 1998:584).

Humboldt's German Bitters

See the Charles H. Plautz listing in this chapter.

Hunki Dori Stomach Bitters

See the Henry Matthews listing in this chapter.

Robert L. Hunt

Soda ~ Mineral Water:

(1) **R. L. HUNT / CHICAGO / ILLS /**   *[aqua]*
(L front heel): **F.A & Co**
»SB w/ keyed hinge mold; shouldered-blob-top soda. *[Figure 356]*

--- ❖ ---

Robert L. Hunt first appears in the Chicago CDs in 1863 and 1864 as a fireman. His first listing as a soda-water manufacturer is in 1865 in partnership with David Kenyon (*see Hunt & Kenyon listing, this chapter*). This partnership continues to be listed through the 1868 directory, after which his partner in the soda-bottling business is David Saulpaugh in the directories for 1869 and 1870 (*see Hunt & Salpaugh listing, this chapter*). The last, possibly related listing for Hunt's bottling works is in the 1871 CD, which indicates that Robert "M." Hunt is a "pop maker."

During the first year of the Hunt & Kenyon partnership in 1865, the bottling works was at 293 Clark Street; thereafter it was always located at his 138 W. Madison

Figure 356

residence. From the bottle listed, Hunt obviously began his soda-bottling operation sometime in 1863 or 1864 before he took on Kenyon as a partner—either as a part-time endeavor while he was a fireman, or (more likely) too late in 1864 to be listed as a soda maker in that year's directory.

Either way, the bottle is unusual. Among the dozen or so "F.A&Co" marked soda bottles recorded for Illinois, this is the only slope-shouldered example, and Hunt was the latest Illinois soda bottler ever to make a glass order from the company (*see introduction, this volume*). The Fahnstock, Albree & Co. Pittsburgh Glass Works was in operation only from 1860 to 1866, and all of the other heel-marked soda bottles they produced for Illinois bottlers were likely made during the first half of this period.

In the 1870 USC, Robert Hunt was recorded as a 32-year old soda water manufacturer in Chicago, so his bottling operations were early business efforts on his part. At the time, he was living in the household of Ann Lomax (perhaps soda-maker George Lomax's widow, or a relative of soda-maker John Lomax?), and his in-law connections may have been the reason he went into the business.

Robert Hunt and David Kenyon

Ale:

(1) (shoulder embossed): **H & K** // **CHICAGO** / **ILL.**   [amber]
    (heel): **A & D. H. C.**

» SB; two-piece mold quart ale; ring-neck tapered collar; 9 1/2" tall. *[Figure 357]*

(2) (shoulder embossed): **H. &. K.** // **CHICAGO** / **ILL.**   [olive green]
    (base): **W. MCCULLY & Co. PITTSBURGH.**

» SB; two-piece mold quart ale; ring-neck tapered collar; 9 1/2" tall. *[Figure 358]*

Soda ~ Mineral Water:

(1) **H & K / CHICAGO / ILL**   [aqua]

» SB w/ keyed hinge mold; oval-blob-top soda. *[Figure 359]*

(2) **H & K / CHICAGO / ILL**   [blue teal]

» SB w/ keyed hinge mold; oval-blob-top soda. *[Figure 360]*

(3) **H & K / CHICAGO. ILL.**   [aqua]

» SB w/ keyed hinge mold; oval-blob-top soda. *[Figure 361]*

Figure 357

Figure 358

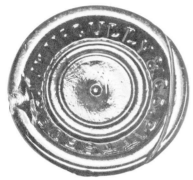

Figure 359

According to the Chicago CDs, Robert Hunt and David Kenyon were partners in the soda water bottling business from 1865 to 1868. This was the first major Hunt soda-bottling partnership following what was likely a single-season solo effort on his part in 1864 (*see Hunt listing, this chapter*). They began business in the basement at 292 Clark and moved to Hunt's new residence at 138 W. Madison in 1866.

Each year of their partnership, Hunt and Kenyon are listed in the directories and associated business summaries only as "soda water manufacturers." Ale bottling is never mentioned, and they are never listed in the "Ale and Porter Bottler" directory summaries (although the 1865 and 1868 directories do not include Ale/Porter summary lists). In this light, it is perplexing that the ale-bottle varieties outnumber the soda varieties during their partnership.

Robert Hunt and David Saulpaugh

Soda ~ Mineral Water:

(1) **HUNT & SALPAUGH / CHICAGO** [aqua]

»SB; blob-top soda. [*not located for study; no photos*]

———◆———

This partnership, which consisted of Robert L. Hunt and David Saulpaugh, was listed in the 1870 IIC and in the 1869 and 1870 Chicago CDs. Hunt & Saulpaugh continued bottling at the Hunt & Kenyon location, 138 W. Madison, which was also Hunt's residence (*see Hunt & Kenyon listing, this chapter*). Embossed examples of bottles from this partnership are said to have been seen, perhaps with Saulpaugh's name misspelled, but none could be located for photography.

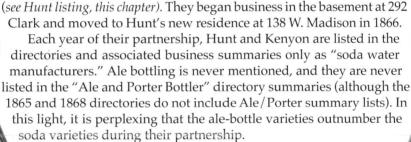

Figure 360

Figure 361

Hunt may have continued bottling for awhile after Saulpaugh left the operation in 1871 (as a Robert "M." Hunt is listed as a "pop maker," but from a different residence in the 1871 Chicago CD, although Saulpaugh is listed as a "painter").

But the listing for the partnership two years later in the regional CAR for 1873 is clearly a holdover clerical error. Large-scale regional directories are notorious for this kind of mistake (*see, e.g., Ernst Hess listing, this chapter*).

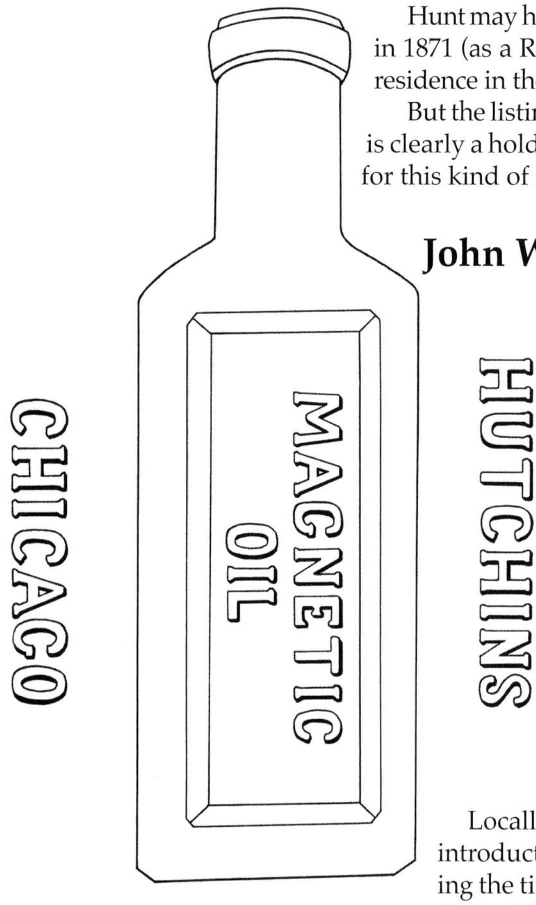

John W. Hutchins

Medicine:

(1) **HUTCHINS // MAGNETIC / OIL // CHICAGO**   [aqua]

»SB; beveled-edge rectangle; 3 sunken panels; thickened-ring lip; ca. 5" tall. *[Figure 362]*

———— ❖ ————

We have not yet found any newspaper advertising for this medicine to definitively tie it to its manufacturer or to specifically indicate its use. However, it does seem to be a local Chicago product, rather than one that originated elsewhere and was marketed in a wide regional area, including Chicago, by local commission agents. (C. S. Hutchins & Co. were contemporary Chicago commission merchants, but their CD listings indicate that they focused on produce, dried fruits, and starch.)

Locally, we have been able to find only one strong candidate for the introduction and sale of *Hutchins Magnetic Oil* in Chicago during the time the bottle's technological attributes suggest it was in use (the early-to-mid 1870s). John W. Hutchins appears as a Chicago physician in the 1871 CD, at 28 Aberdeen Street. Throughout the remainder of the 1870s, he ran his practice from his residence at 386 West Adams. By the end of the 1870s, he had established a separate office for his practice at 125 State, where he carried on his practice well into the 1880s (and perhaps beyond).

Thusfar, however, we have no direct evidence that he is the source for *Hutchins Magnetic Oil*.

Figure 362

William A. Hutchings

Syrups and Extracts:

(1) **W. HUTCHINGS / SYRUP / CHICAGO.**   [aqua]

»SB w/ hinge mold; tall cylinder; long neck; tapered collar; 2 1/4" diameter x 10" tall. *[Figure 363]*

———— ❖ ————

William Hutchings was a highly successful manufacturer and marketer of high-end confectionary syrups and flavoring extracts throughout much of the 1860s. According to the Chicago CDs, he was a liquor dealer in the early 1860s (and a "preserves and liquors agent" in 1859). He turned his attention to the manufacture of flavoring syrups at his Chicago Syrup Factory at 204 West Madison in 1864. His bottling works remained at the same location throughout the 1860s, but in 1868 and 1869 he added a partner (Thomas W. Osborn) and the company became *W. Hutchings & Co.* From 1870 onward, his business no longer appears in the city directories.

Figure 363

His characterization of his products in the city directories over the years is informative as to the contents of his bottles: "syrup factory" (1864); "syrup manufacture" (1865); "syrups and flavoring extracts" (1866); "fancy syrup manufacture" (1867); "fruit syrup manufacture" (1869). The bottle listed likely dates prior to his adding a business partner ca. 1864–1867.

The William H. Hutchinson Companies

Note: The Hutchinson family bottling works, with its various partnerships and different bottling focuses over the years, was one of two generation-spanning corporate giants among nineteenth century Chicago bottling works. (The other was the John A. Lomax Company. Lomax ads in the 1870s and 1880s declared the company to be the largest bottling house in the United States: *see Lomax listings, this chapter*). From pre–Civil War times to the early 1880s, these two companies competed aggressively for the lion's share of the Chicago soda, small-beer, and ale bottling market. After the early 1880s, however, the Hutchinson Company changed course, ceased beverage bottling, and focused its corporate effort on manufacturing and providing bottler's equipment and supplies to the industry.

Because of the complex bottling history of the two companies, the listings and historical discussions of their embossed bottles and bottled products are grouped in historical sequence, rather than listed all together.

The Hutchinson Company listings are organized as follows:

(1) 1849–51: *Hutchinson Family* small beers
(2) 1852–54: *Hutchinson & Co.* soda/mineral water
(3) 1855–58: *W. H. Hutchinson* soda works and ale/porter depot
(4) 1859–64: *Hutchinson & Dunn*
(5) 1865–75: *W. H. H.* soda and ale bottling works
(6) 1876–78: *Hutchinson & Son*
(7) 1879–81: *Hutchinson & Sons*
(8) 1882 onward: *Hutchinson & Son* bottlers supplies only

(1) HUTCHINSON FAMILY BOTTLING WORKS: 1849–51.

There are no embossed glass bottles from this first Hutchinson family bottling operation in Chicago, although several stamped stoneware bottles are known. William Hutchinson and his wife, Jane, moved to Chicago from Williamsville (on the outskirts of Buffalo on Lake Erie) in northwestern New York in the spring of 1849 with his wife and two sons: William A., then seven years old, and Charles G., who had been born in New York just over two years earlier. They were accompanied by his father, Joseph M., and his (uncle?), Andrew J. Hutchinson. Joseph, who was 51 when they relocated to Chicago, had for many years prior to the move been the landlord of the Mansion House Inn and tavern in Williamsville (Calumet Press 1895:234).

Likely because of their earlier experience as inn/tavern keepers, William and Joseph were able to quickly set up a Chicago bottling operation. At first they focused their efforts on stoneware-bottled "small beers." In fact, Joseph and Andrew may have arrived in Chicago a year before William: according to Wagner, the first Chicago CD listing for Hutchinson & Co. is in 1848, and characterizes them as "root beer manufacturers" located on Randolph Street between Clinton and Jefferson (Wagner 1980:23). The company's 1849 CD listing, after William's arrival, characterizes the operation as "Beer Market and Dealers in Cider" (Hoste 1978:7). The company's CD listing as "Beer and Cider" dealers remains the same during 1850 and 1851. Several Hutchinson shoulder-stamped, hand-turned quart stoneware bottles have been documented that date to these first three or four seasons of the Hutchinson family small-beer bottling business (Walthall 1993). They include: **HUTCHINSON'S / LEMON BEER**; **HUTCHINSON'S / SPRUCE BEER**; and **HUTCHINSON'S / SARASAPARILLA** (*see Figure 364*). A hand-turned pint-size **HUTCHINSON'S / LEMON BEER** is also known (*Figure 364*).

(2) HUTCHINSON & CO. BOTTLING WORKS: 1852–54.

In the Chicago CDs for 1852 through 1854, the focus of the Hutchinson bottling business changed. The company, located at the corner of Randolph and Peoria, likely operated only

Figure 364

seasonally, and it is not listed in 1853. But in both 1852 and 1854, William is the junior partner in a "mineral water manufacturing" business that was headed (financed?) in 1852 by Andrew J. Hutchinson and in 1854 by Horace Hopkins. At the time of the 1854 partnership, Hopkins was proprietor of the Bull's Head Hotel in Chicago, and as an innkeeper he may have had an interest in developing the local mineral water trade. The first three glass bottles produced for the *Hutchinson & Co.* business date to this period (*see listing*).

Most of the bottles are cobalt to sapphire blue, and from their styles (particularly of #1 and #2) they appear to have been produced by New York glass houses. Considering the fact that William and Joseph had recently emigrated to Chicago from Williamsville, NY (on the outskirts of Buffalo), the two glassworks the Hutchinson family would have been most familiar with would be the Lockport works (located 12 miles north of Williamsville) and the Lancaster works (located eight miles SE of Williamsville). Family connections there and Buffalo's location on Lake Erie would have facilitated ordering and Great Lakes transport of the bottles.

At this time as well, and during the Civil War era, Hutchinson-stamped small beers were put up in Merrill-patent sided quart stoneware bottles that were Albany-slipped above the shoulder (Walthall 1993). These include **HUTCHINSON & CO.**; **HUTCHINSON & CO. NO. 1**; **HUTCHINSON & CO. NO. 1 / LEMON BEER**; and **HUCHINSON** [sic] **& CO. NO.1** (*see Figure 365*).

Soda ~ Mineral Water:

(1) **HUTCHINSON & C$^\underline{O}$** *(flat)* **/ CELEBRATED / MINERAL /WATERS / CHICAGO**   [cobalt]
»IP; tapered-collar soda. *[Figure 366]*

(2) **HUTCHINSON & C$^\underline{O}$** *(flat)* **/ CELEBRATED / MINERAL /WATER / CHICAGO**   [cobalt]
»IP; shouldered-blob-top soda. *[Figure 367]*

Figure 365

Figure 366

Figure 367

(3) **HUTCHINSON & Cº** *(flat)* / **CELEBRATED** / **MINERAL** /**WATER** / **CHICAGO**   *[cobalt]*

»IP; shouldered-blob-top soda. *[Figure 368]*

(4) **HUTCHINSON & CO** *(arched)* / **CELEBRATED** / **MINERAL WATER** /   *[cobalt, aqua]*
CHICAGO

»IP soda; short-tapered-collar and shouldered-blob-top lip forms. *[Figure 369]*

---

(3) W. H. HUTCHINSON *SODA WORKS AND ALE/PORTER DEPOT: 1855–58.*

Beginning in 1855, William Hutchinson is listed as the sole proprietor of the bottling works, now characterized as located on Peoria between Lake and Fulton (probably the same address as above but stated differently). His CD listing in 1855 and 1858 indicated he operated a "soda manufactory." In the 1856 and 1857 CDs, he is also listed as an "ale and porter" bottler. Several embossed ale and soda bottles are known from this period (*see below*).

During this period as well, and perhaps continuing into the Civil War years, the last of the Hutchinson stoneware "small beer" bottles were being filled at the bottling works. As with the Period #2 stoneware bottles listed, these were also sided Merrill-patent quart bottles Albany-slipped above the shoulder (Walthall 1993). Two stamped varieties are known: ***W. H. HUTCHINSON***; and ***W. H. H.*** (the latter have been found both with and without a cobalt "X" above the shoulder stamping; *see Figure 370*).

Ale:

(1) *(shoulder):* **W. H. H.** // **CHICAGO.**   *[olive-amber]*
(encircling base): **W. McCULLY & Co.** / **PITTS PA**

»SB w/ keyed hinge mold; 3-piece mold quart ale; ring-neck tapered collar; 10 1/4" tall. *[Figure 371]*

Figure 368

Figure 369

CHICAGO — THE WILLIAM H. HUTCHINSON COMPANIES

Figure 370. Late-variety W. H. WUTCHINSON and W. H. H. 12-sided Merrill-patent stoneware bottles. Note that some of these W. H. H.-stamped stoneware bottles have an overlapping ("double-V" style) crossed-bar **W**. This was also done on one post–Civil War variety of Hutchinson soda (*see Figure 384*).

(2) *(shoulder)*: **W. H. H. // CHICAGO.**   *[olive-green "black" glass]*
(encircling base): **W. McCULLY & Co. PITTSBURGH.**

»SB w/ keyed hinge mold; quart; 3-piece mold ale; ring-neck tapered collar; 10 1/4" tall. *[Figure 372]*

Soda ~ Mineral Water:

(1a) **W. H. H. / CHICAGO** *(short blocky embossed initials)*   *[cobalt]*

»IP; sloped-shoulder soda; shouldered blob top; lower body embossed. *[Figure 373]*

(1b) **W. H. H / CHICAGO** *(tall thin embossed*   *[cobalt]*
initials, no final period)
Note: one variant has a large embossed X and two circular raised glass bumps on its base.

»IP; sloped-shoulder soda; shouldered blob top; lower body embossed. *[Figure 374]*

(1c) **W. H. H. / CHICAGO** *(tall blocky*   *[aqua, cobalt]*
initials, w/ final period)

»IP; sloped-shoulder soda; shouldered blob top; lower body embossed. *[Figure 375]*

(2) **W. H. H / CHICAGO / ILL //**   *[cobalt]*
(heel): **W. McCULLY & CO.** *(tall thin initials, no final period;* **ILL** *added)*

»IP; sloped-shoulder soda; shouldered blob top; lower body embossed. *[Figure 376]*

Figure 371

301

Figure 372

Figure 373   **Figure 374**   **Figure 375**   **Figure 376**

(4) W. H. HUTCHINSON and THOMAS O. DUNN BOTTLING WORKS: 1859–64.

From 1859 to 1864, Hutchinson is associated with a new (this time, junior) partner in the bottling business: Thomas O. Dunn. The address of the bottling works (and residence) during this time is given as 8 & 10 N. Peoria, but this is likely just a new way of stating the earlier address locale with building numbers. According to the 1860 Illinois industrial census, the Hutchinson company had five full-time workers at a combined monthly salary of $125. During the year, 15,600 dozen bottles of mineral water and soda water were produced (187,200 bottles filled/refilled!), with a total value of $6,000 and $3,000, respectively.

Dunn was a fairly junior partner. His name is never embossed on the bottles, and in this light it is surprising that the company name is listed as Hutchinson & Dunn in the Chicago directories. Perhaps Mr. Dunn had a connection with Civil War military suppliers and was helping Hutchinson acquire an outlet at nearby Camp Douglas for his bottled soda and ale products.

With one exception (*see later discussion*), the W. H. H. bottles listed above (for the 1855–58 period) seem to have continued in use by the company during the Civil War years. The Chicago CDs during this time tend to highlight the company's soda water production, but they are also listed under the summaries for ale, porter, and cider bottlers. The 245 W. Randolph branch seems to still be operating too, so perhaps ale and soda bottling efforts were run from separate facilities.

The one new bottle added by the company during the war years (*see listing*) is an aqua soda that is smooth-based, so it was introduced after 1860, and it is embossed **H & Co.** rather than **W. H. H**, acknowledging the new partnership in a low-key sort of way. The only other "H & Co." in operation in Chicago during the 1860s was E. Hess & Co., for two years during 1868–69. But from its range of colors and its technical attributes, this bottle was clearly made for the Hutchinson Company during the Civil War era—likely by the St. Louis Glass Works.

Soda ~ Mineral Water:

(1) **H & Co. / CHICAGO. ILL.** [green, pale blue, aqua]

»SB w/ keyed hinge mold; shouldered-blob-top soda. [*Figure 377*]

Figure 377

---

(5) WILLIAM HUTCHINSON & CO. FAMILY BOTTLING WORKS: 1865–75.

As the Civil War was ending, the Hutchinson Company was entering its most successful bottling period, and company partnership entered a new era as well, with the departure of Thomas Dunn and the entry of two of William's sons into the business for the first time. This first short appearance by the sons (William A. and George C.) was for four years only (1865–68) and did not generate any changes in "W. H. H." bottle embossing.

Interestingly, however, the 1866 Chicago CD documents the fact that one of the sons (William A.) was in charge of an ill-fated one-year attempt just after the end of the Civil War to expand the W. H. H. bottling business into Pittsburgh, PA (George C. stayed with the Chicago operation). Although it was not listed in the 1866 Pittsburgh CD, the Hutchinson bottling business there was actually in operation for at least part of the season: an aqua blob-top soda bottle embossed **W. H. H. / PITTSBURGH** is known; and a 7" tall, **W. H. H.**-embossed, amber, pint, long-neck porter-style bottle with a keyed hinge mold base—though not embossed **PITTSBURGH**—is found frequently in the Pittsburgh area (*Figure 378*), but does not occur in Chicago contexts. By 1867, both sons were back with the Chicago operation, and there is no further mention of the Pittsburgh expansion attempt.

During this period, ale and porter bottling was only mentioned during the first two years. After that, the focus is on soda water manufacture—first at 241 Randolph and then at 241/243 Randolph. Also, the 1869 Chicago CD listing for the first time

Figure 378

mentions the manufacture of soda-related products in addition to the bottling operation itself: "manufacturers of soda water, syrup, etc."

In October 1871, the great Chicago fire destroyed the family residence at Erie and N. State, but the bottling works at Randolph and Peoria escaped destruction. William Hutchinson's "prompt loan of a quantity of soda-water boxes, which afforded admirable pigeon-holes at the time, enabled the post office to resume the distribution of the mails with little delay after the fire" (Calumet Press 1895:234).

An Industrial Interests Directory from 1873 "lists the firm as having 75 employees and doing business in excess of $100,000 per year" (Welko 1977a:25).

Ale:

 (1) *(shoulder):* **W. H. H.** / *(heel):* **NEVER SOLD //**   *[green, amber]*
   (shoulder): **CHICAGO.**
   (encircling base): **W. McC & Co**

   »SB w/ post mold; 2-piece mold quart ale; ring-neck tapered collar; 8 1/2 to 9" tall. *[Figure 379a–d]*

 (2) *(shoulder):* **W. H. H.** / *(shoulder):* **NEVER SOLD //**   *[olive "black" glass, amber]*
   (shoulder): **CHICAGO.**
   (encircling base): **W McCULLY & Co PITTS PA**

   »SB w/ post mold; 2-piece mold quart ale; ring-neck tapered collar; 8 1/2 to 9" tall. *[Figure 379a–c, e]*

 (3) *(shoulder):* **W. H. H.** / *(shoulder):* **NEVER SOLD //**   *[olive "black" glass]*
   (shoulder): **CHICAGO.**
   (unimbossed base)

   »SB; 2-piece mold quart ale; ring-neck tapered collar; 8 1/2 to 9" tall. *[Figure 379b–c]*

 (4) **W. H. H.** *(large letters)* // **CHICAGO.** / **ILL**^{S} *[amber]*
   (base): **large 5-pointed star**
   Note: The large, boldly embossed star on the vessel base is a glassmaker's mark used during the 1870s by the Star Glass Works of New Albany, IN (Lindsay, S.H.A. website, 2007).

   »SB; 2-piece mold quart ale; probably with ring-neck tapered collar (top broken off studied example); ca. 8 1/2" tall. *[Figure 380]*

Soda ~ Mineral Water:

 (1) **W. H. H / CHICAGO / ILL**   *[blue, cobalt]*
   (heel): **W. M^{c}C & Co** *(above faceted "mug" base)*

   »SB w/ keyed hinge mold; mug base (10-sided); slope-shoulder soda; shouldered blob top. *[Figure 381]*

 (2) **W. H. H / CHICAGO / ILL _**   *[cobalt]*
   (heel): **W. McCully & Co** *(".." under raised ^{c})*

   »SB w/ keyed hinge mold; slope-shoulder soda; oval blob top. *[Figure 382]*

 (3) **W. H. H / CHICAGO //**   *[dark cobalt]*
   (rear heel): **NEVER SOLD**
   (base): **H** *(surrounded by)* **4 embossed stars**
   Note: a variant has a differently styled H with an adjacent large raised-glass bump on its base.

   »SB w/ post mold; sloped-shoulder soda; oval blob top; lower body embossed. *[Figure 383]*

 (4) **W. H. H / CHICAGO**   *[cobalt]*
   (base): **H**
   Note: Distinctively styled **W** in **W. H. H** has overlapping ("double-V" style) crossed-bar shape.

   »SB w/ post; slope-shoulder soda w/ shouldered-blob top. *[Figure 384]*

Figure 379

Figure 380

Figure 381

Figure 382

Figure 383

Figure 384

Figure 385

(5) **W. H. H / CHICAGO**   *[cobalt]*
(base): *Large embossed* **X** *centered on central raise-glass "dimple."*

»SB w/ keyed hinge mold; slope-shoulder soda w/ shouldered blob top. *[Figure 385]*

(6) **W. H. H / CHICAGO**   *[cobalt]*
(base): *(unimbossed)*

»SB w/ keyed hinge mold; slope-shoulder soda w/ oval blob top. *[Figure 386]*

Quart Soda ~ Cider:

(1) *(large block initials):* **W. H. H / CHICAGO / ILL'S**   *[amber]*
(base): **D.O.C**

»SB w/ post mold; wide-mouth blob-top finish, slope-shoulder quart; 11 1/4" tall. *[Figure 387]*

(2) *(large block initials):* **W. H. H. / CHICAGO**   *[yellow-amber]*
(base): **C.V. No. 2 / MILW**
Note: Side-embossed logo in depressed circular slug plate. Embossed base mark used by the Chase Valley Glass Co. of Milwaukee, ca. 1880–81.

»SB w/ hinge mold; slope-shouldered, shouldered-blob-top quart; 11" tall. *[Figure 388]*

(3) *(large block initials):* **W. H. H / CHICAGO / ILL**   *[aqua]*
(heel): **W M^{c}C & Co**

»SB w/ post mold; high-shoulder, "Chicago-style" quart soda; oval-blob-top finish; 9" tall. *[Figure 389]*

Figure 386

Figure 387

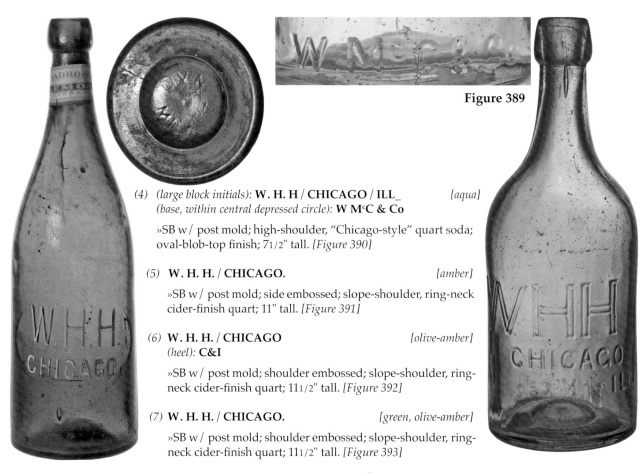

Figure 389

(4) *(large block initials)*: **W. H. H / CHICAGO / ILL_**   *[aqua]*
(base, within central depressed circle): **W MᶜC & Co**

»SB w/ post mold; high-shoulder, "Chicago-style" quart soda; oval-blob-top finish; 7½" tall. *[Figure 390]*

(5) **W. H. H. / CHICAGO.**   *[amber]*

»SB w/ post mold; side embossed; slope-shoulder, ring-neck cider-finish quart; 11" tall. *[Figure 391]*

(6) **W. H. H. / CHICAGO**   *[olive-amber]*
(heel): **C&I**

»SB w/ post mold; shoulder embossed; slope-shoulder, ring-neck cider-finish quart; 11½" tall. *[Figure 392]*

(7) **W. H. H. / CHICAGO.**   *[green, olive-amber]*

»SB w/ post mold; shoulder embossed; slope-shoulder, ring-neck cider-finish quart; 11½" tall. *[Figure 393]*

Figure 388

Figure 390

Figure 391

Figure 392

Figure 393　　　　　　　　　　　　　　　　　　**Figure 394**

(6) WILLIAM HUTCHINSON & SON BOTTLING WORKS: 1876–78.

In 1876, son George C. Hutchinson rejoined the bottling business, and from 1876 through 1878 the business name and the embossed bottles were restyled "**W. H. HUTCHINSON & SON**." In the 1877 Chicago CD, William H. Hutchinson himself was listed as a member of the bottling business for the last time, as he passed away later that year.

Soda ~ Mineral Water:

(1)　**W. H. H. / W H HUTCHINSON & SON / CHICAGO ILL'S**　　　　　　　　　　*[aqua]*
　　　(base): **H**

　　» SB semi-round base; tall shouldered cylinder; vertically embossed; oval blob top; 8 3/4" tall. *[Figure 394]*

---　❖　---

(7) WILLIAM HUTCHINSON & SONS BOTTLING WORKS: 1879–82.

Beginning in 1879, the Chicago CDs list a new partner in the Hutchinson bottling company. William's son George was joined by his brother, Charles G. Hutchinson, and the company name became *Hutchinson & Sons*. Although another son, William A. Hutchinson, had previously been a company partner, this was the first and only appearance of Charles G. as a Hutchinson Co. manager. He was only listed for the year 1879 (although the "& Sons" corporate name was used until 1882).

In 1880 and 1881, the Chicago CDs list Charles as an "artist" and "photographer," still residing in the family home at the bottling works. During that time, however, he patented a new bottle closure that radically changed the company's focus and ended its long history as a bottling works.

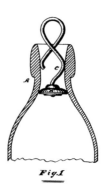

The blob-top cork-closure embossed bottles listed are likely those used by Hutchinson & Sons during the 1879–80 period that ends the bottling era we have focused on in this volume. The 1880 industrial census lists the Hutchinson-company products that year (June 1879–June 1880) as soda water and cider. A total of 48 people were employed full time at the soda works (46 adult males and 2 children), at a total annual labor cost of $18,000. The estimated value of materials used in the production of soda water and cider that year was $40,000, and the total value of the product was estimated at $65,000.

After 1880, for their final two years in the bottling business, *Hutchinson & Sons* started using the new "Hutchinson style" bottles that came into vogue as the result of Charles' invention of his new reusable internal rubber stopper (*see discussion below*). Hutchinson-style bottles are not documented here, but the two varieties used by George and Charles Hutchinson before they closed their bottling works in 1882 are illustrated in Hoste (1978:8) and Welko (1977a:26).

The Hutchinson bottling works was acquired by the Hayes Brothers in 1882. They moved their home-based bottling operation there until they retired from business in 1886 (*see Hayes Bros. discussion, this chapter*).

Ale and Porter:

(1a) **W. H. H. / HUTCHINSON & SONS //**   [amber]
    CHICAGO / ILL^S

    »SB w/ post mold; 2-piece-mold squat quart ale, 9¼" tall. *[Figure 395]*

Soda ~ Mineral Water:

(1) **W. H. H. / HUTCHINSON & SONS / CHICAGO. ILL^S**   [aqua]

    »SB w/ post mold; squat "Chicago-style" quart soda; long neck; oval-blob top; 8" tall. *[Figure 396]*

---

(8) W. H. HUTCHINSON & SON BOTTLER'S SUPPLIES, SYRUPS, AND APPARATUS: 1883–1929.

Although beyond the scope of the present study, the new reusable bottle closure first patented by Charles G. ("Doc") Hutchinson on April 8, 1879 (#213,992), reissued as a design patent on June 17, 1879 (#8755), and expanded on September 16, 1879 (#219,729), rapidly changed and mechanized the soda and mineral water bottling business. To provide the supplies, syrups, and bottling apparatus for their rapidly accepted Hutchinson closure bottles, George and Charles moved to new factory headquarters they built on Desplaines Street, where they employed 40 workers by 1895 (Calumet Press 1895:235).

Figure 395

Figure 396

For quicker acceptance by bottlers using older cork closures and bottles with different neck lengths and diameters, early Hutchinson stoppers were manufactured in three wire lengths and five washer sizes. But glass manufacturers also quickly redesigned their bottles to better accommodate the popular cheap new stoppers, and machines were designed to rapidly fill and close the "Hutchinson-style" bottles. Charles Hutchinson's own design for such a machine was patented March 16, 1880 (#225,475). Interested readers should refer to Fowler (1984), Hoste (1978), Riley (1957), and Welko (1977a).

As bottling-industry technology changed, the Hutchinson bottler's supply company changed with it—for instance, they shifted from Hutchinson internal-stopper closures to crown-top (bottle cap) closures in 1913. They were bought out by G. J. Arnold Bottler Supplies in 1929, but have retained the Hutchinson company name, "and are still in business today, being the fourth oldest company in Chicago" (Welko 1977a:26).

John Huttenlocher

Soda ~ Mineral Water:

(1) **J. HUTTENLOCHER / CHICAGO. / ILLS_** [aqua]
 (*base*): **JH**

 »SB w/ keyed hinge mold; oval-blob-top soda. *[Figure 397]*

Figure 397

John Huttenlocher is listed in only the 1871 Chicago CD as a manufacturer of soda water at 499 N. State. Prior to this time (in the 1870 CD), he is listed at the same address as an ice dealer. The following year (1871 CD), his soda business is no longer listed, and in the 1872 CD listing he has become the junior partner in the *Gottleib Wurster & Co.* soda bottling business (*see the Wurster listing, this chapter*).

It should be noted that the Huttenlocher soda-bottling business is also listed in the McKillop & Sprague 1873 CAR, but this source is notorious for holdover listings of defunct companies, and should not be considered reliable.

Huttenlocher does not appear in Chicago CDs after 1874.

John H. Huyck and Amos S. Randall

Hair Treatment:

(1) (side): **HUYCK & RANDALL** // // (side): **AZZAIELINE** // // [aqua]
Note: Both "**Z**"s are backwards in the word **AZZAILINE**.

»SB w/ keyed hinge mold; beveled-edge rectangle; double-ring collar; 4 sunken panels; 7 1/2" tall *[Figure 398]*. A tapered-collar variety of this bottle has also been reported

Figure 398

Huyck & Randall were proprietors of the Garden City Chemical Works at 111 South Water Street and later at 34 South Water in Chicago from 1867 through 1869. A business card in the 1867 Chicago CD (p. 53) announced that Huyck & Randall were "Manufacturers of Huyck's Standard Flavoring Extracts, Fluid Extract Buchu, Azzaieline for the Hair, pomades, hair oils, perfumeries, etc., etc." The company is also separately listed in the 1868 Chicago CD as a manufacturer of "liquor essences." One of their first-year ads also noted that they manufactured chewing gum.

For a detailed company history, see the entry for *Garden City Chemical Works*.

John H. Huyck and Charles M. Knox

Hair Treatment:

(1) (side): **HUYCK & KNOX** // // (side): **AZZAIELINE** // // [aqua]

»SB w/ keyed hinge mold; beveled-edge rectangle; double-ring collar; 4 sunken panels; 7 1/2" tall. *[No photo, cf. Figure 398 above]*

Huyck & Knox, successors to the Huyck & Randall partnership in 1870, were proprietors of the Garden City Chemical Works from 1870 through 1876. They were first located at 34 S. Water, then 57 & 59 W. Lake, then (during the final three years of the partnership) at 33 Michigan Avenue.

The partnership was dissolved in 1877, after which John H. Huyck became sole principal proprietor of the firm. For a detailed company history, see the entry for *Garden City Chemical Works*.

Stephen Israel

Saratoga Mineral Water:

(1) **ISRAEL'S / MINERAL. WATERS / CHICAGO** [amber]

»SB w/ post mold; broad-shoulder Saratoga-style quart; 12-sided "mug based" lower body; ring-neck tapered collar; 7" tall. *[Figure 399]*

Stephen G. Israel appears in the 1870–74 Chicago CDs as a druggist at 152 Dearborn (1870–71) and later at 522 Wabash Avenue (1872–74). In the 1869 directory, Israel is shown as a partner with John T. Christian in a drug store business at 152 Dearborn. Stephen Israel is listed as a salesman in the 1875 directory and is not listed in 1876. He likely sold bottled mineral water at his drug store during the early 1870s.

Forester F. (aka: Frank) Jaques

Jaques Chemical Co. products—bluing, flavoring extracts, etc.—are not formally included in this volume because the genesis of the company bearing his name just postdates 1880. During the late 1870s, Jaques came to Chicago from Austin, IL, to become a salesman for Gillet, McCullough & Co. (*see their listing, this chapter*). During 1879 and 1880, he served as a junior partner in the Chicago flavoring-extracts firm of Durant, Jaques & Atwood (no embossed bottles known). Beginning in 1881, he became senior partner in the flavoring-extracts company of Jaques, Atwood & Co., and the embossed bottles bearing his name likely date after this time.

Figure 399

Dr. [Hosmer A.] Johnson

Medicine:

(1) **DR JOHNSON'S / VIGOR OF LIFE // CHICAGO ILLS**   [aqua]
   (base): **L & W**

   »SB w/ keyed hinge mold; beveled-edge rectangle; 4 sunken panels; double-ring collar; 5 1/2" tall. *[Figure 400]*

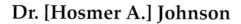

This bottle's **L&W** maker's mark (Lorenz & Wightman bottling works, Pittsburgh) and keyed hinge mold construction date it from 1862 to ca. the late 1860s. No newspaper ads for the medicine or paper-labeled examples of the bottle have yet been located, but if it was a bottle produced for a Chicago medical doctor—rather than a regionally or nationally distributed patent medicine marketed in Chicago by a commission agent or a fictional doctor's name used as a marketing tactic by a local or regional patent-medicine company—the only Dr. Johnson practicing in Chicago in the 1860s was pioneer physician Hosmer A. Johnson.

Hosmer Johnson earned medical degrees in the late 1840s and early 1850s from the

Figure 400

University of Michigan and Rush Medical College, and first appears in the Chicago CDs as a physician at 53 Clark Street in 1852 and 1853. Early in his career, he was frequently associated with medical academies and medical education (Davis 1955:390), but he was also a practicing Chicago physician throughout the 1850s (moving from 53 Clark, to 113 Lake, to 96 Randolph) and the Civil War years (with offices at 101 Washington Street). After the Civil War, he left active practice to become a professor and president of the Chicago Medical College and a commissioner with the Chicago Board of Health in the late 1860s. In the early 1870s, he was an attending physician at the Cook County Hospital, but he seems to have been largely retired after 1873, after which he is listed as a physician at his residence. So the most likely time for him to have produced and sold the *Vigor of Life* medicine, if it was his recipe, would have been during the Civil War years.

During his early years in practice, he was an active speaker at medical conferences, and he is recorded to have spoken at an early conference about possible medicinal remedies for disease:

> Dr. Hosmer A. Johnson spoke of cancerous disease as caused by "morbid accumulation and distribution of calcareous and other salts." He suggested that cure could be obtained by exclusion of any excess of salines or their removal from the system. However, he advocated the use of large doses of common salt in typhoid and in "intermittents" because of its quality of preserving the blood globules and acting as an alterative and tonic.
>
> There was considerable discussion of the use of "extract of beef's blood" or "extractum sanguines" in the treatment of anemias of whatever cause, as such extract provided "globuline hematine salts mixed with fibrin" which are "precisely the ingredients that are most deficient in anemia. [Davis 1955:112]

Either of these suggested courses of treatment could have resulted in a *Vigor of Life* medicine. No national studies of early patent medicines have recorded *Dr. Johnson's Vigor of Life*, although in pre–Civil War times a *Dr. P. Johnson's Golden Eye Salve* (Rochester, NY) and *Dr. Johnson's Liniment Mixture* (Albany, NY) are known (Odell 2007:204). Also, during the Civil War era, *Dr. Johnson's Balsam for the Lungs* was advertised in Bath, NY, and *Dr. Clark Johnson's Indian Blood Syrup* was a widely distributed and very popular patent medicine product (Baldwin 1973:274–5; Wilson and Wilson 1971:122).

John Joss and Charles Hartmann

Soda ~ Mineral Water:

(1) **JOSS & HARTMANN** *(arched)*   [aqua]
(base): **J&H**

»SB w/ keyed hinge mold; oval-blob-top soda. *[Figure 401]*

The various Joss partnerships are fleeting ones, and are relatively difficult to track in detail (*see Joss & Taylor and Schriber & Joss listings, this chapter*). Both Joss and Schriber are first associated with the soda and mineral water business as salesmen for other companies. John Joss is first listed in the 1869 Chicago CD as a "soda water peddler" (company not named) boarding at 160 Milwaukee Avenue. Charles Schriber (sometimes spelled Schrieber) first appears in the 1866 and 1867 Chicago CDs as a salesman for Hutchinson & Sons (*see listing, this chapter*), residing at their factory complex. In the 1868 CD, he is a "peddler" for the Sass & Hafner bottling company (*see listing, this chapter*) at 128 W. Lake. Joss seems to have been associated with Charles Schriber in the soda-bottling business for a short time in 1871

Figure 401

(as a junior partner—*see their listing, this chapter*). At the time, Sass & Hafner were bottling at the same address.

At times when Joss and Charles Hartmann were not involved in bottling, their occupations are variously listed as carpenter (JJ) and box maker or trunk maker (CH).

The partnership of John Joss & Charles Hartmann, soda water manufacturers, is listed only in the 1873 Chicago CD, at 467 W. Chicago Avenue. According to the 1874 CD, Joss is a "foreman" for a "soda water manufacturer" named Henry Moeller (no embossed bottles known) at the same address. Charles Hartmann (name spelled with one "n") may have moved on to produce patent medicines from an address at the Humboldt House hotel. Joss is listed as a "driver" in 1875 and does not appear in Chicago CDs thereafter.

Thomas Kane

Ale:

(1) **T. KANE. / CHICAGO** [teal]

»IP; cylindrical; 2-piece-mold squat quart ale; tapered collar? (Study specimen broken at neck); 7 1/2" tall at break. [*Figure 402*]

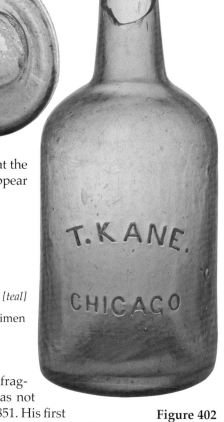

Figure 402

———————— ❖ ————————

Although some of the earliest Chicago CDs used for our study were fragmentary (1849 and 1851) or unavailable (1848, 1850), Thomas Kane was not listed in the 1847 CD and he had no business advertisements in 1849 or 1851. His first listing, in the 1852 Chicago CD, characterized him as a brewer and liquor dealer, with a liquor business and "rectifying distillery" on Canal Street near Randolph. In the mid-1850s, his listings clarify that his retail liquor dealership was at No. 2 S. Canal Street and his rectifying business was at 291 S. Canal.

Most important for the present study, however, is the fact that during his final two years in business in Chicago (1859 and 1860) his focus shifted from liquor products to brewery products. During these last two years, he is listed as a "brewer" with business headquarters at the corner of Polk and Beach. Since the bottle listed is a quart-sized ale, it likely dates to the 1859 inception of his brewery business.

As luck would have it, Thomas Kane's two-year brewery venture was documented in the 1860 Illinois industrial census. According to the 1860 IIC, he had eight male employees at an average labor cost of $200/month. From June 1859 to June 1860, he produced 2,000 barrels of ale (his sole brewery product) with an overall value of $10,000.

Michael and Thomas Keeley

Ale~Porter:

(1) **KEELEY & BROTHER^S / CHICAGO** [olive "black" glass]
Note: Shoulder embossed only.

»SB, pontiled era; 3-piece mold pint; ring-neck tapered collar; 7" tall. [*Figure 403*]

Figure 403

Chicago — Michael and Thomas Keeley

(2) **KEELEY & BROTHER / CHICAGO** [olive "black" glass]
Note: Shoulder embossed only.

»SB, pontiled era; 3-piece mold pint; ring-neck tapered collar; 7 1/4" tall. [Figure 403]

(3) **KEELEY & BRO // CHICAGO, ILL.** [olive "black" glass]
Note: shoulder-embossed only.

»SB, pontiled era; 3-piece mold quart; ring-neck tapered collar; 8 3/4" tall. [Figure 404]

Michael Keeley and his brother Thomas lived in Chicago and Alton, respectively, during the mid-1850s, where they were long-distance partners in a family ale-bottling business in the two cities (*see Alton entry for Thomas Keeley*). Thomas Keeley came to America from Ireland in 1847, and arrived in Madison County in 1850. He began bottling ale in Alton in 1852, with the assistance and advice of his older brother Michael, who followed him to America in May 1851 when he was 24 years old. Michael had had "several years" experience in the bottling business—apparently in Ireland, since he had arrived in America only a year earlier—and helped

Figure 404

Thomas set up his Alton ale bottling business during 1852–53. Michael then relocated to Chicago in 1854 to establish a *Keeley & Brother* ale depot in northern Illinois.

In Alton, Thomas began importing and bottling ale and porter in 1852, and by 1853 he was using *Keeley and Bro.* Alton-embossed black-glass bottles to put up "a fine article of Claret Wine, of our own bottling, which we warrant" (*Alton Weekly Courier*, May 20, 1853). During the 1854 and 1855 seasons, he also bottled soda water in Alton. He died in the spring of 1856 when he was 26 years old (perhaps of cholera?), and the Alton branch of the *Keeley & Brother* business was closed down.

In Chicago, Michael Keeley was first found listed in the annual city business directory as a bottler at 84 Dearborn Street in 1854—two years after he and his brother Thomas began their Alton ale business. This was probably the beginning of the *Keeley & Brother* Chicago Ale Depot. The 1854 Chicago CD has no separate "Ale and Porter" bottling category, but the only other non-soda/mineral water bottler it lists is H. C. Doty, another Chicago ale bottler (*see his listing*).

By 1855, Michael had moved his Chicago branch of the *Keeley & Brother* Ale and Porter Depot in basement/cellar quarters at 43 S. Water Street. The Chicago CDs indicate that he continued to operate the *Keeley & Brother* ale and porter bottling operation at the Water Street location during 1856 and 1857 as well.

But 1857 was likely the last year the *Keeley & Brother* bottling works was in operation. In 1858 and 1859, Thomas expanded and diversified has business and moved to new quarters, revising the name of the business to the *Keeley Bottling Works* in the process (*see Michael Keeley listing*). In 1858, he moved his ale and porter bottling operation to larger quarters at 60 W. Lake Street. Then in 1859 he began a soda bottling operation in addition to his ale business, which he established in separate quarters at 291 S. Canal Street.

The Chicago *Keeley & Brother* black-glass pint bottles listed could have been used for porter bottling, but they also may have been used as single-serving ales or as containers for the "claret" the Keeley brothers produced themselves (*see Thomas Keeley's Alton listing*). In the late 1850s, olive black-glass bottle manufacturing often included a refiring technology that smoothed and largely obliterated pontil scars from the bottle bases. The "SB" base categorization for the bottles listed above likely indicates the use of this glass-house refinishing technology. The *Keeley & Brother*[s] embossing on bottle #3 above is in error. The requested embossing was probably either "*Keeley Brother*[s]" or "*Keeley & Brother.*"

All three of the Chicago *Keeley & Brothers* black-glass ale bottle varieties listed were likely produced and used between 1855 and 1857.

Michael Keeley

Ale~Porter~Cider:

(1) **M. KEELEY / CHICAGO. ILL**   [olive "black" glass]
Note: *2 raised dots under full-sized final* **L** *in* **ILL**

»IP; 3-piece-mold quart; ring-neck tapered collar; 9 1/2" tall. *[Figure 405]*

(2) **M. KEELEY // CHICAGO. ILL**   [green; olive "black" glass]
(base, in circle): **W McCULLY & CO—PITTSBURGH PA**
Note: *2 raised dots under full-sized final* **L** *in* **ILL**

»SB; 3-piece-mold and 4-piece mold quarts; ring-neck tapered collar; 9 1/2" tall. *[Figure 406]*

(3) **M. KEELEY // CHICAGO. ILL**   [amber]
(base, in circle): **W McCULLY & CO. PITTSBURGII.**

»SB; 2-piece-mold quart; long neck; ring-neck tapered collar; 10" tall. *[Figure 407]*

(4) **M. KEELEY // CHICAGO. ILL _**   [olive-amber]
(base, in circle): **W McC & CO. / PITTS-**

»SB; 3-piece mold quart; ring-neck tapered collar; 10" tall. *[Figure 408]*

Figure 405

Figure 406

Figure 407

(5) **M. KEELEY. / CHICAGO ILL. / Nº 1** [olive-amber]

»SB; ring-neck champagne finish quart; 11 1/2" tall. [Figure 409]

Soda ~ Mineral Water:

(1a) **M. KEELEY. / CHICAGO. / ILL** [aqua, green]

»IP; oval-blob-top soda. [Figure 410]

(1b) **M. KEELEY. / CHICAGO. / ILL** [aqua, green]

»SB w/ keyed hinge mold; oval-blob-top soda. [Figure 411]

(3) **M. KEELEY/ CHICAGO. ILL** [aqua]
(base): large embossed X across central glass "dimple."

»SB w/ keyed hinge mold; high-shouldered stubby-necked soda. [Figure 412]

(4) **M. KEELEY. / CHICAGO. / ILL_** [aqua, light blue]
(base): heavily embossed, faceted **5-pointed star**.
Note: The large, boldly embossed star on the vessel base is a glass-maker's mark used during the 1870s by the Star Glass Works of New Albany, IN (Lindsay, S.H.A. website, 2007).

» SB w/ keyed hinge mold; high-shouldered stubby-necked soda. [Figure 413]

Figure 408

Michael Keeley moved out of his cellar Ale & Porter Depot at 43 S. Water Street and relocated his ale bottling operation to 60 W. Lake in 1858. The following year, he established a soda-water bottling operation at 291 S. Canal (see Keeley & Brother listing, this chapter). The 1860 industrial census for Chicago lists Michael Keeley as a soda and ale manufacturer with an annual "Porter, Ale & Soda" production of 3,000 barrels. Since the soda would have been bottled, it is unclear how this was figured into the production estimate.

Keeley bottled at both locations until 1867, when he consolidated his bottling operations into a series of structures at 289–295 S. Canal and ceased ale bottling activities to focus his attention on soda water manufacture. From 1862 to 1864, he also served as Cook County Treasurer.

During the period from 1868 to 1871, he continued to manufacture and bottle soda water, but his bottling operation was burned to the ground during the Great Chicago Fire in October 1871 and did not reopen.

Beginning in 1872, he and Michael Kerwin opened a retail wine and liquor business at 299 and 301 S. Canal:

> The enterprise was so profitable that in 1876 Mr. Keeley withdrew from it and, with his share of the proceeds, in 1876 purchased the F. Binz brewery, at the foot of Twenty-eighth Street.

The name of the plant was changed to the Keeley Brewery and in 1878 to the Keeley Brewing Company. Mr. Keeley was its largest stock-holder and president, the output of the plant the first year being six thousand barrels, which was increased to nine thousand barrels in 1879. The year ending May 1, 1888, the last year of his management of the company's affairs, the sales reached nearly ninety thousand barrels (Hall 1976).

Figure 409

Figure 410

Figure 411

Figure 412

There are no Keeley-embossed Chicago bottles that date after the time of the Chicago Fire. All of the embossed ale, porter, and cider bottles listed were manufactured and used between 1858 (after the *Keeley & Brother* operation closed) and 1867 (when Michael Keeley stopped bottling ale and focused his efforts solely on soda). The black-glass ale bottles (#1, #2a, and #2b) were in use during the late 1850s and early 1860s. The remaining ale/porter bottles were in use during the Civil War era. An early, round, "**M. K.**" stamped, pint stoneware bottle (*Figure 414*) and a later, 10-sided, "**M. K.**" stamped Merrill-patent quart stoneware bottle (*Figure 415*) are also known (Walthall 1993:13). They should date from this 1858–1867 period as well. Stoneware bottles like these generally contained brewed products like "small beers" (e.g., spruce beer) and ciders.

The soda water bottles listed earlier are in probable chronological order of use during the late 1850s and 1960s.

Thomas King

Ale:

(1) **T. KING / CHICAGO. ILLS.** [amber]
   (base): heavily embossed, faceted **5-pointed star**.
   Note: The large, boldly embossed star on the vessel base is a glass-maker's mark used during the 1870s by the Star Glass Works of New Albany, IN (Lindsay, S.H.A. website, 2007).

»SB w/ mold lines obscured; squat 2-piece-mold quart ale; broken at neck; height at break = ca. 8". Lip form unknown, but likely ring-neck tapered collar. *[Figure 416]*

Figure 413

Figure 416

Figure 414   **Figure 415**

Soda ~ Mineral Water:

(1) **T. KING / CHICAGO. / ILLS_**   [aqua]

»SB w/ keyed hinge mold; shouldered-blob-top soda. *[Figure 417]*

───────── ❖ ─────────

Thomas King was a fairly common name in Chicago during the 1860s, when keyed hinge-mold bottles such as these would have been used. Three to eight Thomas Kings were listed each year in the CDs during this time, but only one of them worked in a bottling-related business, and only during two years: 1867 and 1868.

This particular Thomas King was listed in the 1867 Chicago CD as a worker at the Sands *[Cream Ale]* Brewery. The following year, in the 1868 CD, King appeared as a home "brewer" working from his residence at 56 E. Hinsdale. He was not listed in the directory the following year.

The ale bottle listed above is a good fit for the time and the brewing activity listed for King. But soda-style bottles were almost never used (anywhere in the state) for non-soda/mineral-water products. Since this was a one-man home business, perhaps King "broke the rules" and put up pint ales in soda bottles during the one year he tried to start up his own business. On the other hand, soda bottling may have been a last-minute addition to his business to diversify products in an attempt to improve sales.

Kirchhoff Brothers

(Julius, John Diedrich, H. August, and F. Gustavus)

Whiskey:

Figure 417

(1) **KIRCHHOFF BROS / CHICAGO / ILL^S**   [amber]

Note: All embossing on applied circular glass seal. (Company name embossed around edge of seal; town name horizontally embossed within company name: see Figure 418).

»SB w/ straight-line hinge mold; sealed oval whiskey flask w/ applied glass seal; seal-embossed only; tapered ring-neck collar w/ internal screw-thread neck; 8" tall.

───────── ❖ ─────────

The four Kirchhoff brothers were grocers in Chicago during the Civil War years. Then, in 1865, Julius started a vinegar manufactory and distillery at Rush and Pearson, and the three remaining brothers opened a wine, brandy, liquor, and "foreign products" importing business at No. 8 and 9 Custom House Place. Chicago CDs indicate that by 1867 the three brothers had expanded their wine and liquor import business to a second address as well (96 Randolph). During the late 1860s, they consolidated their wine and liquor import business at 103 W. Lake.

Figure 418

Then, in 1870 and 1871, Julius returned to their liquor business, and the firm of J. Kirchhoff Brothers & Co. was formed as "importers and dealers in Kentucky whiskey." This was their one foray into the retail liquor business as *Kirchhoff Brothers,* and the flask listed above likely dates to this two-year period. At that time, the *Kirchhoff Brothers* were competitors of *Chapin & Gore (see their listing, this chapter)* in the Chicago retail whiskey market, and both companies used an embossed, internal-screw-threaded flask during the late 1860s and early 1870s.

The four brothers broke up their partnership in 1872 and went their separate ways in Chicago business. Only H. August Kirchhoff remained in the liquor trade (and only as a wholesaler). From 1872 through the remainder of the 1870s, his only partners in the wholesale liquor business were Robert J. Taylor and Theo Neubarth.

William Knackstead and Co.

Soda ~ Mineral Water:

(1) **W.K & Co / CHICAGO //**   [aqua]
   (heel): **W. McC & Co**
   (base): **WK&Co**

   »SB; shouldered-blob-top soda. *[Figure 419]*

———————— ❖ ————————

In the 1872 Chicago CD, William Knackstead is listed as a coppersmith. The following year (1873), William Knackstead and a partner named Adam "Appel" (also spelled "Apple") are listed as "soda water manufacturers" at Knackstead's residence (W side May between Front and Milwaukee Avenue). In the 1874 Chicago CD, Adam Apple is listed alone as a "soda water maker" from his residence at 10 Smart Street, and Wm. Knackstead is once again listed as a coppersmith. By 1875, Adam Appel/Apple is no longer listed in the Chicago CDs.

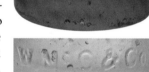

Figure 419

We have been unable to document any Adam Apple embossed sodas that would date to the one year he was bottling on his own. However, embossed *Apple & Co.* bottles are known from the Joseph Apple brewery in Chicago, which date to the 1870s *(see Apple & Co. listing, this chapter).* Perhaps Adam was a relative.

Moses G. Landsberg

Bitters:

(1) *(large shoulder logo):* **1876**   [amber]
   M. G. LANDSBERG / CHICAGO
   (base): **PAT.**

   »SB w/ heel mold; cut-corner square; vertically embossed; very ornately decorated; long neck w/ bell-shaped shoulder ; ring-neck tapered collar; 11" tall. *[Figure 420]*

———————— ❖ ————————

This vertically embossed bottle was not patented by Moses Landsberg until April 11, 1882, and therefore postdates the 1880 end of our study period. But it is mentioned here because it has been strongly associated with (and attributed to) the mid-1870s and the celebrations of the centennial of American independence because of the bottle design, the large **1876** shoulder embossing, and the name of the bitters it contained *(Landsberg's Century Bitters).*

According to New York City CDs of the time, the Landsberg family ran a grocery store in the city in 1874, and in 1875 Moses G. Landsberg opened a liquor store at 169 E. Broadway. It was likely then (and only in New York) that his *Century Bitters* was likely first bottled and sold.

Figure 420

By 1878, the Landsberg family (including Moses) had moved to Chicago, where according to the 1878 CD they manufactured and produced flour. Moses may have still worked from home to bottle and sell his bitters (and when he designed the new ornate bottle in 1882 he was a Chicago resident), but these efforts did not rise to the level of listing in the city business directories.

Landsberg's Century Bitters was undoubtedly first produced as a tie-in to the 1876 U.S. centennial celebrations. But the ornate, old-fashioned-looking embossed bottle, with **1876** prominently embossed on one shoulder opposite a rayed rising sun with **1776** embossed in small characters on its orb, was patented several years afterward, probably as an advertising mechanism to boost regional sales and help the bitters become more widely popular. No evidence has yet been found to suggest that the ornately embossed bottle was manufactured and used in the late 1870s, prior to the patent date.

As described in the 1882 design patent given to "Moses G. Landsberg of Chicago, Illinois" (and patented for 3 1/2 years; Design No. 12,861), the bottle has four rectangular panels, each

> having arched tops, two of the alternating faces or facets being left smooth; or all four of the sides may be left plain; and in the arch spaces over the rectangular faces are represented respectively a shield, the figures 1876, spread-eagle, and a rayed sun. The edges of the sides of the bottle are corrugated in lozenges, while the base is surrounded by a series of hexagons. The neck of the bottle represents the handle, and the shoulder the body of a bell, the bell being encircled midway by a ring of stars.

During the early-to-mid 1880s, Moses Landsberg's patented embossed bottle containing his *Century Bitters* was marketed and sold by the Simon Adler Distilling Company in St. Louis (19 & 21 S. 2nd Street) and Abraham and Adolph Heller retail wine and liquor outlets at 35 & 37 Broad Street in New York (see Ring 1980:291–292). From an embossed aqua bottle of the same Landsberg patent design, the Heller Company also sold a product called **LANDSBERG'S / PURE**

Figure 421

Figure 422

BLACKBERRY / BRANDY (*Figure 421*). Heller's first year in business was 1879/80. The Adler Company was formed in 1878. A search of city newspapers might discover their earliest ads for *Landsberg's Century Bitters*.

Ironically, the final appearance of the bottle was the Chicago variation listed, embossed for Moses Landsberg himself (*Figure 420*). It is the same patented bottle design, embossed **M. G. LANDSBERG / CHICAGO** vertically on one face (replacing the Adler Co. and Heller Co. embossing of the St. Louis and New York varieties discussed earlier), but the **LANDSBERG'S / CENTURY / BITTERS** lettering that was embossed vertically on the opposite side of the Adler and Heller bottles has been largely (but not completely) pieued out (Figure 420). This alteration suggests a subsequent reuse of the bottle in the mid-1880s or later—with different contents—by Landsberg in Chicago.

Landsberg also designed at least one other (similarly ornate) sunken-paneled, pillared-corner, ornately shouldered, rectangular liquor or bitters bottle. No patent description has yet been located for this bottle (see Ketcham 2009:4), which dates to approximately the same late 1870s/early 1880s era. Two of the bottle's opposing panels are plain. The third panel has a large MGL monogram within an encircled shield, surrounded by druggist-style wreathed scales and a mortar-and-pestle. The opposite fourth panel has three large embossed images: two views of the Sphinx and the profile of a Gryphon (*Figure 422*). The use of the druggist symbols implies "medicinal" contents—perhaps a bitters or medicated brandy.

Lang Brothers

(August, William, and Frederick)

Soda ~ Mineral Water:

   (1a) **LANG BRO**ˢ **/ CHICAGO.** //   [aqua]
      Note: *2 raised-glass dots below superscript* **S** *in* **BRO**ˢ
      (heel): **A & D. H. C**
      (base): **L.BR** *(2 raised-glass dots below* **R***)*

       »SB w/ keyed hinge mold; oval-blob-top soda. *[Figure 423]*

(1b) **LANG BROˢ / CHICAGO.** // [aqua]
   Note: 2 raised-glass dots below superscript **S** in **BROˢ**
   (heel): **C & I**
   (base): **L.BR** (*much smaller embossing than base of 1a*)
   »SB w/ post mold; shouldered-blob-top soda. *[Figure 424]*

(2) **LANG BROS. / GINGER ALE / CHICAGO / ILLINOIS** [aqua]
   »SB w/ round base; shouldered cylindrical soda; 9 1/4" tall. *[Figure 425]*

Quart Soda ~ Cider:

(1) **LANG BROˢ / CHICAGO.** [aqua]
   Note: 2 raised-glass dots below superscript **S** in **BROˢ**
   (base): **A. & D. H. C.**
   »SB; Chicago-style quart; oval-blob-top soda; 9 1/4" tall. *[Figure 426]*

(2a) **LANG BROˢ / CHICAGO** // [amber]
   Note: lower-body embossing; 2 raised-glass dots below superscript **S** in **BROˢ**
   (heel): **A & D. H. C**
   »SB; cider-finish ring-neck quart; 11 1/4" tall. *[Figure 427]*

Figure 423

Figure 424

Figure 425

Figure 426

Figure 427

Figure 428

(2b) **LANG BRO'S / CHICAGO //** [green, amber]
  Note: lower-body embossing.
  (heel): **C & I**
  »SB; cider-finish ring-neck quart; 11 1/4" tall. *[Figure 428]*

    According to Chicago CDs, the Lang Brothers set up two different, sequential, companies to bottle mineral water and soda water between 1868 and 1878. The first attempt was in 1868, by brothers August and William. They were listed in the 1868 CD as "mineral water manufacturers," at 171 N. Clark. By 1869, William had joined John C. Meier as a vinegar maker and August is not listed in the CDs (unless he is "Augustus" Lang, a "teacher of gymnastics"). Also in 1869, Frederick Charles Lang first appeared in the Chicago CDs as a vinegar maker in the firm of Lang Bros. & Miller (with John A. Lang and Andrew Miller).

    In 1870, August and Frederick began a second *Lang Bros.* soda and mineral water bottling operation at 268 Wells. This business venture was a more successful one, and the two brothers were bottling partners until 1878, after which Frederick C. Lang took over and ran the bottling works on his own (*see his listing below*). During the *Lang Bros.* 1870s partnership, they were variously (and apparently randomly) characterized as "soda water" or "mineral water" manufacturers. In 1875, they moved their bottling operation from 268 Wells to 58–60 Pearson.

If embossed bottles were made for the earlier single-year mineral-water bottling effort by August & William Lang, the only recorded bottle whose hinge-mold manufacturing technology is early enough to qualify as an 1868 product is #1a in this listing.

Frederick C. Lang

Ale:

(1) **F. C. LANG / CHICAGO //** [amber]
(*heel*): **A. & D. H. C**
(*base monogram*): **F.C. L.**
Note: A variant is known without the A&DHC heel mark.

»SB; quart; 4-piece mold; ring-neck tapered collar; 10" tall. *[Figure 429]*

(2) **F. C. LANG / CHICAGO //** [amber]
Note: circular slug-plate outline around embossing.
(*heel*): **D.O.C**

»SB w/ post mold; quart; 2-piece mold; ring-neck tapered collar; 9¾" tall. *[Figure 430]*

Soda ~ Mineral Water:

(1) **F. C. LANG / CHICAGO //** [aqua]
(*heel*): **A & D. H. C.**
(*base monogram*): **F.C. L.**

»SB w/ post mold; shouldered-blob-top soda. *[Figure 431]*

Figure 429

Figure 430

Figure 431

Figure 432

Figure 433

Quart Soda ~ Cider:

(1) **F. C. LANG / CHICAGO. //** [aqua]
(heel): **M^{c}C**
(base monogram): **F.C. L**

»SB; Chicago-style quart; shouldered blob top; 8 1/2" tall. *[Figure 432]*

(2) (shoulder): **THE F. C. LANG BOTTLING CO. / CHICAGO.** [yellow-amber]
(base monogram): **F.C. L.**

»SB w/ post mold; shoulder-embossed slope-shoulder quart; shouldered blob top; 12" tall. *[Figure 433]*

(3) **F. C. LANG / CHICAGO** [yellow-amber]
(heel): **E.H.E. 37**

»SB w/ post mold; tall slope-shoulder quart; blob top; 11 3/4" tall. Embossed in circular slug plate with EHE maker's mark (the Edward H. Everett glass works of Newark, OH), so the bottle was not made earlier than ca. 1880, and may date somewhat later than that. *[Figure 434]*

———————— ❖ ————————

Frederick Lang formed his owned bottling company in 1878 upon the departure of his brother, August, from their partnership. During that first season, Frederick "Z." Lang is listed alone as a soda water bottler at the 60 W. Pearson address, but the bottling works is still referred to as *Lang Bros.* Beginning in 1879, it is listed in the CDs as the "Frederick Z. Lang" soda works, and from 1880 onward the CDs get the middle initial right and it is the Frederick C. Lang bottling works. The *F. C. Lang* bottling works was listed in the Chicago directories for only six years: 1879–1884. The *F. C. Lang* bottles listed above are only those that, from their technological and style attributes, are likely to have been manufactured prior to 1880 or shortly thereafter. The 1880 Illinois Industrial Census indicates that Lang's bottling operation ran full-time, 12 months a year, and produced an annual value of bottled product estimated at $25,000. In the 1880 USC, Lang was listed as being 30 years old.

Figure 434

After that time, Lang took on a series of bottling partners for his last three years in the soda and mineral water business (including J. Henry Zitt and William Boppart), and from 1885 to 1887 the company name and the embossing on its bottles was changed to *F. C. Lang & Co.* The Lang "*& Co.*" bottles are too late for listing in the present study.

Frederick Lang was never listed in the Chicago CDs as an ale, porter, or cider manufacturer; he only appeared under the soda and mineral water listings. This was usually the case for smaller bottling operations that manufactured their own soda and mineral water, but bought ale or cider (brewed by other companies) by the barrel, and then decanted it into their own bottles for retail sale.

Professor Lennord's Celebrated Nectar Bitters

See the Cassilly & Co. listing.

Abraham L. Liebenstein and Co.

Fruit Jar:

(1) **A. LIEBENSTEIN & C^O** (arched) **/ DEALERS IN CROCKERY / & GLASS WARE** [aqua]
(embossed in circle around address) **/ 177 RANDOLPH S^T. / CHICAGO, ILL. //**
STAR / (above a 6-pointed embossed star)

»SB; slope-shouldered quart fruit jar; ground lip w/ outside screw threads. *[Figure 435]*

———————— ❖ ————————

The screw-top lid for this jar is a threaded tin band with a milk-glass or white-metal insert. Five insert varieties are known (Leybourne 1993:201; see Figure 436):

(1) Glass insert embossed with 6-pointed star in center. In a circle around the star is: HALLERS PATENT / FEB 5TH '67;

(2) Glass insert with three embossed concentric-circle ridges. Embossed in a circle within the outer two ridges is: HALLERS PATENT / FEB 5TH '67.

(3) White-metal insert. Stamped star in center;

(4) White-metal insert, stamped HALLER'S PATENT / FEB 5 67;

(5) White-metal insert. Stamped star in center with HALLER'S PATENT FEB 5 67 stamped around the star.

Figure 435

According to the Chicago CDs, Abraham and Joseph Liebenstein began in business there in 1860 and 1861 as furniture dealers. In 1862, they moved to new quarters at 177 Randolph and expanded into wholesale crockery sales. They were listed as dealers in crockery and glassware in the 1862 IRS assessment of Chicago. According to their ad in the 1864 Chicago CD, they further expanded their wholesale inventory that year to include "upholstery goods, cabinet hardware, undertakers materials, varnishes, and looking-glass plates." In 1865, they also advertised retail-outlet sales of crockery and glassware.

In 1866, Joseph left the business, and Abraham ran it alone during 1866 and 1867. During that time he turned his attention to the import and wholesale and retail distribution of earthenware, queensware, and glassware. In 1868, Abraham took on a new partner (John Ranken), and changed the name of the business to *A. Leibenstein & Co.* The company's china and glassware import business flourished and expanded during the next four years, and by 1872 they were operating from three locations.

However, by 1873 they were operating from just a single location again, and in 1874 the company was not listed in the main business directory—although they were still listed in the china/glass/queensware summary. By 1875 *A. Leibenstein & Co.* was just a "merchandise broker," and by 1876 the company had turned its attention to real estate dealings.

Thus, the Haller's Patent preserve jars with the *A. Leibenstein & Co.* embossed logo were most likely imported and sold by the company in the period from 1868 to 1872.

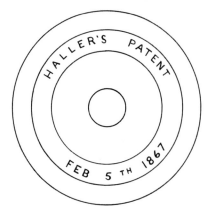 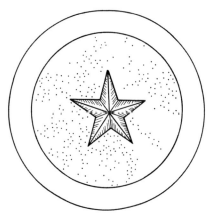 

Figure 436

George Lomax

Soda ~ Mineral Water:

(1) **G. LOMAX. / CHICAGO** [cobalt]
    Note: Paneled (8-sided) below shoulder. Vertical embossing on two adjacent flat panels. Tapered-collar and shouldered-blob lip varieties known. *[Figure 437]*
    »IP; high-shoulder soda w/ 8 flat panels forming lower body.

(2) **G. LOMAX** *(arched)* **/ CHICAGO / ILL** [aqua, green, cobalt]
    »IP; slope-shouldered soda w/ oval blob top. *[Figure 438]*

(3) **G. LOMAX** *(arched)* **/ CHICAGO / ILL** [cobalt]
    »SB w/ hinge mold; slope-shouldered soda w/ oval blob top. *[Figure 439]*

Note: The following six bottle styles listed (#4–9) were all embossed within rectangular sunken panels with beveled corners. The 5 different embossed lettering varieties used were (a) gracile lettering above and below a large central glass "dimple"; (b) small lettering; (c) medium lettering—this is the only variety known to have been made using both iron-pontil and smooth-base technology; (d) large lettering; (e) large faceted lettering with panel rivet depressions visible (*see Figures 440–445 to compare varieties*).

(4) **G. LOMAX** *(arched)* **/ CHICAGO** [light blue, cobalt]
    Note: Gracile letters embossed within shield-shaped sunken panel with central glass "dimple."
    »IP; slope-shouldered, shouldered-blob-top soda. *[Figure 440]*

(5) **G. LOMAX** *(arched)* **/ CHICAGO** [cobalt]
    Note: small letters embossed within shield-shaped sunken panel.
    »IP slope-shouldered, shouldered-blob-top soda. *[Figure 441]*

(6) **G. LOMAX** *(arched)* **/ CHICAGO** [aqua]
    Note: medium-size letters embossed within shield-shaped sunken panel.
    »IP; slope-shouldered, shouldered-blob-top soda. *[Figure 442]*

Figure 437   **Figure 438**

Figure 439

Figure 440

Figure 441

Figure 442

Figure 443

Figure 444

Figure 445

(7) **G. LOMAX** *(arched)* / **CHICAGO** [cobalt]
   Note: medium-size letters embossed within shield-shaped sunken panel.
   »SB w/ keyed hinge mold; slope-shouldered, shouldered-blob-top soda. *[Figure 443]*

(8) **G. LOMAX** *(arched)* / **CHICAGO**  [aqua, cobalt]
   Note: Large letters embossed within shield-shaped sunken panel.
   »IP; slope-shouldered, shouldered-blob-top soda. *[Figure 444]*

(9) **G. LOMAX** *(arched)* / **CHICAGO** [cobalt]
   Note: Very large faceted letters embossed within shield-shaped sunken panel w/ visible rivet depressions.
   »IP; slope-shouldered, shouldered-blob-top soda. *[Figure 445]*

The lives and business histories of early Chicago soda makers George and John Lomax (*see John A. Lomax listing, this chapter*) are closely intertwined, but more than a little clouded. Both men emigrated to the United States from the same town in England (Bury in Lancashire), arriving in Chicago separately early in the city's history (ca. early 1850s). George died there in 1863 while he was still relatively young. At this early time, historical documentary resources are limited to British and American census data, Chicago CDs, and a published biographical sketch (1886) and obituary (1899) for John Lomax, who lived long and prospered in Chicago.

Many details of their early lives and business careers are frustratingly obscured by the clerical and journalistic sloppiness of these early sources, and by the fact that many Chicago CDs and newspapers from the 1848–1852 era are fragmentary or lost. As discussed in the John Lomax section, his two biographical sketches are in almost laughable disagreement about the details of his first years in Chicago. But by combining and cross-checking the primary literature on George and John Lomax, several details of their business histories and relationship are fairly clear.

George Lomax is identified as a 41-year-old Chicago "master soda maker" in the 1860 U.S. Census. George, Jr., born in New York, is listed as seven years old. James Lomax, a 58-year-old "manufacturer" from England (George's father?) was also living in the household. John Lomax, about six years younger than George, is listed as head of a separate Chicago household in a different City Ward. Ten years earlier, the 1850 British Census for Bury, England, lists George Lomax, at about the right age, as still living with his family there. Interestingly, John Lomax is listed as a member of George's household at this time, but under "relationship" the census-taker has written "visitor." This suggests John was not likely George's younger brother; so perhaps he was a nephew or cousin.

From the Census data, George Lomax came from England to Chicago via New York in the early 1850s, when he was about 33 years old. He is first listed in Chicago CDs from 1852 through 1854 as a "mineral water" bottler at Canal Street between Lake and Fulton (he does not appear in the 1851 CD). During these three years, Edward K. Bebbington and Joseph Entwistle are also listed in the CDs as bottlers at the same Canal Street address. Chicago bottles embossed *E & L* (Entwistle & Lomax) and *E & B* (Entwistle & Bebbington) are known dating from this time (*see Entwistle listings, this chapter*).

Beginning with the 1855 Chicago CD, and continuing until his death in 1863, George Lomax is listed as the proprietor of his own soda water manufacturing business—*Lomax & Meagher*, later *G. Lomax & Co.*—in partnership with John Meagher. Meagher was a next-door neighbor in the 1860 Chicago Census, who had also emigrated from England. Their bottling business was located at George's home address: 138 W. Madison (also listed as School between Madison and Monroe). Since Meagher's name was never embossed on the *George Lomax* bottles, he must have been a relatively junior partner in the bottling works. In fact, when George Lomax's business affairs were being put in order and shut down in 1864 after his death, his widow Anne Lomax was listed in the CD as head of the company.

With George's incapacity and death in 1863, there were likely no new glass-company bottle orders after ca. 1861–62. In this light, it is interesting to note that there are only two smooth-based (post-1860) bottle varieties in the list. This suggests that his bottling business was quite successful in the late 1850s, with bottles blown in seven different molds, but was winding down in the early 1860s, with only two (seldom seen) mold varieties reordered.

John A. Lomax

Note: The John A. Lomax bottling works, with its changing partnerships and bottling focuses over the years, was one of two generation-spanning corporate giants among nineteenth century Chicago bottling works. The other was the William H. Hutchinson family bottling works (*see Hutchinson listing above*). From pre–Civil War times to the early 1880s, these two companies competed aggressively for the Chicago soda, small-beer, and ale bottling market. After the early 1880s, the Hutchinson Company changed course, ceased beverage bottling, and focused its corporate effort on manufacturing and providing bottler's equipment and supplies to the industry.

Because of the complex bottling history of the Lomax and Hutchinson companies, the listings and historical discussions of their embossed bottles and bottled products are grouped in historical sequence, rather than listed together by product.

The Lomax listings are organized as follows:

1. Ainsworth & Lomax at 38 W. Lake St. .... ca. 1855/56
2. Bebbington & Co.(??): ....ca. 1856–57
3. J.A.L. at 38 W. Lake St. 1858–59
4. J.A.L. at 4 Charles Place 1860–66
5. J.A.L. at 16 Charles Place 1867–69
6. J.A.L. at 14 & 16 Charles Place 1870–72
7. J.A.L. at 14, 16, & 18 Charles Place 1873–86

The narrative history of the John A. Lomax bottling works is presented following the chronological bottle listings (*see also George Lomax listing, this chapter*).

(1–2) Ainsworth & Lomax; Bebbington & Co.

Prior to establishing his own company at 38 W. Lake Street in 1858, John A. Lomax served a period of apprenticeship as a junior partner in the soda and mineral water bottling operations of Robert Ainsworth and (very probably) Edward K. Bebbington. The historical documentation for these early Lomax bottling years is discussed in the narrative history that follows and in the George Lomax listing above. The embossed bottles used by these two companies during the mid-1850s are listed in the *Ainsworth & Lomax* and *Bebbington & Co.* sections in this chapter.

(3) John A. Lomax at 38 W. Lake Street: 1858–59

John Lomax established his own bottling business at 38 West Lake Street in 1858, and bottled there for just two years. Thus of his pontiled bottles (all sodas) date to these two seasons at the Lake Street address.

Soda ~ Mineral Water:

(1) **J A L / CHICAGO / ILL** [cobalt]

»IP; shouldered-blob-top soda; large, thick, embossed lettering. *[Figure 446]*

(4) John A. Lomax at 4 Charles Place: 1860–66

The sodas and ales used by Lomax during the Civil War years are listed in their likely production sequence. Soda #1 dates to ca. 1860–61; soda #2 dates to ca. 1862–63; sodas #3 and #4 date to ca. 1864–65; and sodas #5 and #6 date to ca. 1865–66. The cider-finish quart bottle and ale #1 date to the early Civil War years, while ale #2 dates to the later part of the Civil War period.

Soda ~ Mineral Water:

(1) **J A L / CHICAGO / ILL** [aqua]

»SB w/ keyed hinge mold; shouldered-blob-top soda; large, thick embossed lettering (same mold as IP version listed above). *[Figure 447]*

Figure 446 **Figure 447** **Figure 448** **Figure 449**

   (2) **J A L / CHICAGO / ILL** *[light blue]*

     »SB w/ keyed hinge mold; shouldered-blob-top soda; smaller, thinner embossed lettering than #1 (new mold). *[Figure 448]*

   (3) **J. A. LOMAX / CHICAGO. / ILL _** *[aqua]*

     »SB w/ keyed hinge mold; shouldered-blob-top soda. *[Figure 449]*

   (4) **J. A. LOMAX / CHICAGO. / ILL _** *[aqua]*
     (heel): **A & D. H. C**

     »SB w/ keyed hinge mold; shouldered-blob-top soda. *[Figure 450]*

Quart Soda~Cider:

   (1) **J. A. LOMAX // CHICAGO** *[amber "black" glass]*

     »SB w/ keyed hinge mold; cider-finish quart; 11 3/4" tall. *[Figure 451]*

Ale ~ Porter:

   (1) **J. A. L. // CHICAGO ILLS** *[shades of amber, olive green]*
     Note: Olive, yellow, brown and red-amber color shades known.

     »SB w/ central glass "dimple" and no mold lines; shoulder-embossed 3-piece-mold quart ale; ring-neck tapered-collar; 9 3/4" tall. *[Figure 452]*

Figure 450

Figure 451

Figure 452

(5) John A. Lomax at 16 Charles Place: 1867–69

Just three styles of soda/mineral water bottles and a single quart soda/cider can be specifically attributed to the three-year bottling period at 16 Charles Place:

Soda ~ Mineral Water:

(1) **J. A. LOMAX / 16 CHARLES PLACE / CHICAGO. // THIS BOTTLE / IS NEVER SOLD /**   [aqua]
(heel): **A & D. H. C**
(base): **J.L**

»SB w/ keyed hinge mold; shouldered-blob soda; large, thick embossed lettering (top two lines). *[Figure 453]*

(2) **J. A. L. / CHICAGO / ILL // Nº 16 / CHARLESPLACE** *(no space)* /   [aqua]
(heel): **C&I**
(base): **J.L**

»SB w/ keyed hinge mold; oval-blob soda. *[Figure 454]*

(3) **J. A. LOMAX / 16 CHARLES PLACE / CHICAGO // THIS BOTTLE / IS NEVER SOLD /**   [teal blue]
(base): **L**

»SB w/ keyed hinge mold; oval-blob soda; smaller, thinner embossed lettering lower on body (different mold than #1). *[Figure 455]*

Quart Soda~Cider:

(1) **J. A. LOMAX / 16 CHARLES PLACE / CHICAGO. // THIS BOTTLE / IS NEVER SOLD /**   [amber]

»SB w/ central-base kick-up; slope-shoulder quart w/ cider-finish ring-neck collar; 11 1/4" tall. *[Figure 456]*

---

(6) John A. Lomax at 14 & 16 Charles Place: 1870–72

During the short time (1870–72) when Lomax operated from 14 & 16 Charles Place, two new embossed additions to his bottles show up for the first time—but only on a few of the later (post mold) bottles likely dating to the end of this period (ca. 1872). The first is the distinctive Lomax 4-bottle trademark logo, seen only on quart sodas/ciders at first (*see Figures 471–473*). The logo was also used on pint porter-style bottles that may date to this time (*see Figures 476–477*). But since the porter bottles are small enough that there was little space on them to emboss street numbers, they may actually date to the subsequent (14, 16, & 18 Charles Place) period. The second development is an added embossed disclaimer ["**THIS BOTTLE / MUST BE RETURNED**"] on several

Figure 453

Figure 454

Figure 456

Figure 455

styles of Lomax bottles with the "**14 & 16**" embossed address. Both of these embossed additions become nearly ubiquitous on Lomax bottles of the subsequent (1873–80) period—occurring on most examples of a wide range of Lomax bottle styles with the "**14, 16 & 18**" embossed address.

Soda ~ Mineral Water:

    (1) **J. A. LOMAX** *(arched)* / **14 & 16** / **CHARLES PLACE** / **CHICAGO.** //   *[cobalt]*
        (heel): **A & D. H. C.**
        (base): **J. L.** *(large faceted letters)*

      »SB w/ keyed hinge mold; oval-blob soda. *[Figure 457]*

    (2a) **J. A. LOMAX** *(arched)* / **14 & 16** *(script)* / **CHARLES PLACE** / **CHICAGO** //   *[cobalt]*
        (heel): **B. F. G. Co**
        (base): **J. L.** *(large faceted letters)*

      »SB w/ post mold; oval-blob soda.

    (2b) **J. A. LOMAX** *(arched)* / **14 & 16** *(script)* / **CHARLES PLACE** / **CHICAGO** //   *[cobalt]*
        (heel): **B. F. G. Co**
        (base): **JAL** *(thick faceted letters)*

      »SB w/ post mold; oval-blob soda. *[Figure 458]*

    (3) **J A LOMAX** *(flat)* / **CHICAGO** / **ILL** // **NOS—14–16** / **S' CHARLESPLACE** *(no space)* /   *[aqua]*
        (heel): **B. F. G. Co**
        (base): **J.A.L** *(small embossed lettering)*

      »SB w/ post mold; oval-blob soda. *[Figure 459]*

    (4) **J. A. LOMAX** *(flat)* / **14 & 16** / **CHARLES PLACE CHICAGO.** //   *[aqua, green teal]*
        THIS BOTTLE / **MUST BE RETURNED**
        (heel): **A & D. H. C**
        (base): **J. L** *(large faceted letters)*

      »SB w/ post mold; oval-blob soda. *[Figure 460]*

    (5) **J. A. LOMAX** *(arched)* / **14 & 16** / **CHARLES PLACE** / **CHICAGO.** //   *[teal blue]*
        THIS BOTTLE / **MUST BE RETURNED**
        (heel): **A & D. H. C**
        (base): **J. L.** *(large faceted letters)*

      »SB w/ keyed hinge mold; shouldered-blob soda. *[Figure 461]*

    (6) **J. A. LOMAX** *(arched)* / **14 & 16** / **CHARLES PLACE** / **CHICAGO** //   *[blue]*
        THIS BOTTLE / **MUST BE RETURNED**
        (heel): **A & D. H. C.**
        (base): **J. L.** *(tall narrow plain letters)*

      »SB w/ keyed hinge mold; shouldered-blob soda. *[Figure 462]*

    (7) **J. A. LOMAX** *(arched)* / **14 & 16** / **CHARLES PLACE** / **CHICAGO** //   *[cobalt]*
        THIS BOTTLE / **MUST BE RETURNED**
        (heel): **A & D. H. C.**
        (base): **J. L.** *(tall narrow faceted letters)*

      »SB w/ keyed hinge mold; short cylindrical neck; shouldered-blob soda. *[Figure 463]*

    (8) **J. A. LOMAX** / **14 & 16** / **CHARLES PLACE CHICAGO** //   *[teal blue]*
        THIS BOTTLE / **MUST.BE.RETURNED** /
        (heel): **LOCKPORT, N.Y.**
        (base): X

      »SB w/ post mold; oval-blob soda with sloping shoulders. *[Figure 464]*

Figure 457

Figure 458

Figure 459

Figure 460

Figure 461

Figure 462

Figure 463

Figure 464

(9) **J. A. LOMAX / 14 & 16 / CHARLES PLACE / CHICAGO. //**   [cobalt]
  (heel): **B. B. &. Cº**
  (base): two varieties: raised central glass "dimple" and **I** *(large raised line).*
  Note: The heel mark is an otherwise unknown glassmaker mark for Illinois. Several possible glass-house identifications have been suggested by internet searches: David Whitten's list of "Glassmaker's Marks Found on Bottles" (2009) identifies the mark as Baker Bros. & Co., proprietors of the Baltimore, MD, Glass Works: "Mark seen on the base of Baltimore-area blob beer bottles from the ca. 1880–1895 period." This company advertised in the 1870s and 1880s, with a strong focus on window-glass production. Bill Lindsay's massive labor-of-love, the "Historic Glass Bottle Identification and Information Website" (2009) indicates the mark "likely stands for the Berney-Bond Glass Company" (Pennsylvania), and illustrates it on an 1880s Trommer Extract of Malt (Freemont, OH). However, from Pennsylvania historical records, this company was only in business from ca. 1902–30. There was also a British glass company with these initials (perhaps too late for the Lomax context), and a pharmacy bottle stamped "B. B. & Co./ London" is known. So from present information, the Baker Bros. & Co. attribution seems most likely, especially since **B. F. G. Co** *marks on several Lomax bottles indicate that another Pennsylvania glass works, the Beaver Falls Glass Co. (1869–79), was also patronized by Lomax.*

  »SB w/ post mold; no neck; slope-shouldered oval-blob soda. *[Figure 465]*

Tall Full-Pint Soda:

(1) **J. A. LOMAX /** *14 & 16 (script)* **/ CHARLES PLACE / CHICAGO //**   [cobalt]
  (heel): **B. F. G. Co**
  (base): **JAL** *(medium-sized thin letters)*

  »SB w/ post mold; oval-blob soda. *[Figure 466]*

(2) **J. A. LOMAX /** Nᵒˢ **14 & 16 / CHARLES PLACE / CHICAGO ILL //
  THIS BOTTLE / NEVER SOLD**
  (heel): **B. F. G. Co**   [cobalt]
  (base): **JAL** *(small thin letters)*

  »SB w/ post mold; oval-blob soda. *[Figure 467]*

Quart Soda~Cider:

(1) **J. A. LOMAX / 14 & 16 / CHARLES PLACE / // CHICAGO.**   [lime green]
  (heel): **A & D. H. C.**
  Note: 4-bottle logo NOT used on this variety. Same embossed lettering as #2, except no TB/MBR on reverse.

  »SB w/ depressed center and post mold; ring-neck cider-finish quart; 11 1/2" tall. *[Figure 468]*

(2) **J. A. LOMAX / 14 & 16 /**   [amber, olive amber]
  **CHARLES PLACE / CHICAGO. // THIS BOTTLE /
  MUST BE RETURNED /**
  (heel): **A & D. H. C.**
  Note: 4-bottle logo NOT used on this variety.

  »SB w/ depressed center and post mold; large, thick embossed lettering; ring-neck cider-finish quart; 11 1/2" tall. *[Figure 469]*

(3) **J. A. LOMAX / 14 & 16** *(script)* **/**   [amber]
  CHARLES PLACE / CHICAGO //
  (heel): **B. F. G. Co**
  (base): **JAL** *(small, plain embossed letters)*
  Note: 4-bottle logo NOT used on this variety. Smaller, thinner embossed lettering than #1 and #2 (different mold).

  »SB w/ depressed center and post mold; ring-neck cider-finish quart; 11 1/4" tall. *[Figure 470]*

Figure 465

Figure 466. Comparative view of standard-size (left) and large-size (right) cobalt soda bottles.

Figure 467

Figure 468

Figure 469

Figure 470

(4) (RAISED 4-bottle logo) **J. A. LOMAX / 14 & 16 / CHARLES PLACE / CHICAGO //**   [amber]
THIS BOTTLE / MUST BE RETURNED

»SB w/ raised ring and post mold; very large, thick embossed lettering; shouldered-blob-top quart; 11 1/4" tall. [Figure 471]

(5) (RAISED 4-bottle logo) / **J. A. LOMAX / 14 & 16 / CHARLES PLACE / CHICAGO //**   [aqua]
THIS BOTTLE / MUST BE RETURNED
Note: This bottle's unique early-1870s shape (with its short, cylindrical neck and narrow blob top) seems to have been designed for an early closure type rarely used by Illinois bottlers. Likely stopper candidates (designed for bottles with similar short cylindrical constricted necks) are: (1) the Matthews or Albertson internal gravitating stoppers, patented in the mid-1860s; and (2) the Weatherbee or Beal cork-closure swing stoppers with broad-metal-band neck fasteners, patented in 1871 (see introduction, this volume).

»SB w/ heel mold; uniquely shaped quart (sloped shoulders, short cylindrical neck, shouldered-blob top); 10" tall. [Figure 472]

(6) (RAISED 4-bottle logo) / **J. A. LOMAX / 14 & 16 / CHARLES PLACE / CHICAGO //**   [aqua]
THIS BOTTLE / MUST BE RETURNED

»SB w/ post mold; Chicago-style quart (low shoulders, long neck, blob top); 9 3/4" tall. [Figure 473]

Ale:

(1a) (embossed horizontal shoulder band) /
J. A. LOMAX / 14 & 16 / CHARLES PLACE / CHICAGO //   [shades of amber, olive green,
THIS BOTTLE / MUST BE RETURNED /   olive "black" glass]
(heel): **A & D H. C.**
Note: Side-embossed below raised embossed horizontal shoulder band. Two amber varieties = brown and red.

»SB w/ post mold; 2-piece mold quart ale; ring-neck tapered collar; 9 1/2" tall. [Figure 474, left and center]

(1b) (embossed horizontal shoulder band) /
J. A. LOMAX / 14 & 16 / CHARLES PLACE / CHICAGO //   [yellow amber]
(heel): **A & D H. C.**
Note: Side-embossed below raised embossed horizontal shoulder band.

»SB w/ post mold; 2-piece mold quart ale; ring-neck tapered collar; 9 1/2" tall. [Figure 474, right]

(2) **J. A. LOMAX / 14 & 16** (script) **/ CHARLES PLACE //**   [amber]
/ CHICAGO
(heel): **B. F. G. Co**
(base): **JAL**

»SB w/ post mold; quart; 2-piece mold quart; ring-neck tapered collar; 9 1/2" tall. [Figure 475]

Figure 471   **Figure 472**

Figure 473

Figure 475

Figure 474

Figure 476

Porter:

> *Note: The bottles in this group lack address numbers (probably due to the small size of the bottles). But they all have the characteristic Lomax 4-bottle embossed logo of the later 1870s, so the bottles likely date to the **14, 16 & 18 CHARLES PLACE** address era.*

(1) (OUTLINE 4-bottle logo) / **J. A. LOMAX / CHARLES PLACE / CHICAGO //**   [blue teal]
    THIS / BOTTLE MUST / BE RETURNED
    (base): **JAL**

» SB w/ post mold; long-neck porter-style pint with ring-neck tapered collar; 7 1/4" tall. *[Figure 476]*

(2) (RAISED 4-bottle logo) / **J. A. LOMAX / CHARLES PLACE /**   [amber, yellow green, green teal,
    CHICAGO // THIS / BOTTLE MUST / BE RETURNED   olive green, olive "black" glass]

» SB w/ post mold; long-neck porter-style pint with ring-neck tapered collar; 7 1/4" to 7 1/2" tall. *[Figure 477]*

---

(7) John A. Lomax at 14, 16, & 18 Charles Place: 1873–80

For the first time, many of the Lomax bottles styles dating to this post-1872 period have distinctive product brand names embossed on them. These include:

    "Medicated Aerated Waters"
    "California Pop"
    "Famous Brown Stout"
    "Famous Weiss Beer"
    "Superior Bottled Lager Beer"
    "Diamond Lager Beer"

Figure 477

For some reason, the bottles with embossed brand-names are embossed with "**14 to 18**" addresses rather than "**14, 16 & 18**" addresses. This must have been a conscious style decision either by the glassworks or Lomax Company, because it looks like space was available on the bottle to emboss them either way. Another notable curiosity is that *only four styles* of Lomax small sodas are documented from this period: both are mold varieties of Arthur Christin-patent gravitating-stopper bottles. By contrast, at least 14 mold varieties of Hutchinson-style and collared-blob small sodas with Lomax embossing are known, which date to the final Lomax Company era (1880–86). Lomax court case records from the late 1870s (vs. Putnam—see *Chicago Tribune* 11/12/1879, p. 11, col. 3 and 4/1/1880, p. 1, col. 4), suggest that Lomax may have been using the 1874 Joel Miller patent internal rubber stoppers as soda-bottle closures during the latter half of the 1870s. Putnam sued, probably for royalties, because the Miller patent had been transferred to him. So it is possible that some of the Lomax soda bottles that look like post-1880 collared-blob styles are actually late-1870s bottles designed for the Miller closures (*see Figures 463, 472, and 478–479*).

Spruce Beer

Also, at this unusually late date, quart and pint stoneware "small beer" bottles were being filled at the Lomax bottling works. A Lomax ad in the 1875 Chicago Tribune (*see later discussion*) indicated that Spruce Beer was one of the many Lomax bottled products. During the Civil War years, this was a very popular Chicago drink (with its own separate listing category in the CDs), and was almost invariably bottled in stoneware containers. So perhaps these late Lomax stoneware bottles were spruce beer containers. The quarts were paneled Merrill-patent salt-glazed stoneware quarts, with *J A L* or *J. A. LOMAX* stamped above the shoulder on each bottle (Walthall 1993). The bottles have hand-painted vertical cobalt stripes from the neck to shoulder on opposite sides. A third paneled quart stoneware bottle variety was shoulder-stamped *J. A LOMAX / 14 to 18 CHARLES PLACE / CHICAGO.*, and heel-stamped ***ESTABLISHED 1851***. The hand-turned salt-glazed pints were round rather than paneled, and were elaborately stamped with the outline 4-bottle trademark and *J. A. LOMAX, / 14, 16 & 18 / CHARLES PLACE / CHICAGO*. The front heel was stamped ***THIS BOTTLE / MUST BE RETURNED***. Two varieties of the pint bottle are known: one with the 4-bottle logo stamped at the shoulder (above the Lomax name and address) and another with the 4-bottle logo stamped at mid-body (below the Lomax name and address). All of these stoneware-bottle varieties were distinctively decorated with two vertical cobalt stripes extending from the lip to the shoulder on opposite sides of the bottle (*see Figures 501 and 503*).

Soda ~ Mineral Water:

(1) **J. A. LOMAX** (arched) / **14. 16 & 18** / **CHARLES PLACE / CHICAGO //**   [cobalt]
   THIS BOTTLE / MUST BE RETURNED
   (heel): **L. G. WKS.**
   (base): **J. L.** (tall narrow letters)
   Note: *The **L. G. WKS.** maker's mark on this bottle is something of a mystery. Lindsay's massive SHA-website "Historic Glass Bottle Identification and Information" guide (2009) does not list the mark, but he does list 3 glassworks that might potentially have used it: Lockport (NY) Glass Works, 1840–72; Louisville (KY) Glass Works, 1855–73; and Lancaster (NY) Glass Works, 1849–90. Since Lomax had previously used at least one soda-bottle style manufactured at the Lockport works (see Figure 464), perhaps the* **L. G. WKS.** *bottle was produced at Lockport as well.*

   »SB w/ post mold; short cylindrical neck; shouldered-blob-top soda. *[Figure 478]*

(2) (Raised 4-bottle logo) / **J. A. LOMAX** (arched) / **14. 16 & 18** / **CHARLES PLACE /**   [cobalt]
   CHICAGO. // THIS BOTTLE / MUST BE RETURNED
   (heel): **A & D. H. C.**
   (base): **J. L.** (tall faceted letters)

   »SB w/ post mold; short cylindrical neck; sloped shoulders; shouldered-blob-top soda. *[Figure 479]*

Figure 478  Figure 479

(3a) *(RAISED 4-bottle logo) / J. A. LOMAX / 14. 16 & 18 / CHARLES PLACE / CHICAGO. //*   [aqua]
THIS BOTTLE / MUST BE RETURNED /
(heel): **A & D. H. C.**
(base): **J. L**
(encircled by): **ARTHUR CHRISTIN. PAT. APR. 13**[TH] **1875.**

»SB w/ post mold; high-shoulder soda w/ large embossed lettering; hard-rubber gravitating-stopper blob-top closure; 6 3/4" tall. *[Figure 480–481]*

(3b) *(RAISED 4-bottle logo) / J. A. LOMAX / 14. 16 & 18 / CHARLES PLACE / CHICAGO //*   [aqua]
THIS BOTTLE / MUST BE RETURNED /
(heel): **A & D. H. C.**
(base): **J. L** *[large raised-glass oval in depressed rectangle below period]*
(encircled by): **ARTHUR CHRISTIN. PAT. APR. 13**[TH] **1875.**

»SB w/ post mold; sloped-shoulder soda w/ small embossed lettering (different mold than #1a); hard-rubber gravitating-stopper blob-top closure; 7" tall. *[Figure 480–481]*

Quart Soda~Cider:

(1) **J. A. LOMAX / 14 16 & 18 / CHARLES PLACE / CHICAGO. // THIS BOTTLE /**   [amber]
MUST BE RETURNED /
(heel): **A. & D. H. C**
Note: This is the only bottle in the quart soda~cider group that does not have the raised 4-bottle logo above the embossing (Ray Komorowski, personal communication 2009). It is otherwise similar to bottle #2 below (cf. Figure 482).

»SB w/ depressed center and post mold; ring-neck cider-finish quart; 11 1/2" tall.

Figure 480

Figure 481

(2) *(RAISED 4-bottle logo)* / **J. A. LOMAX** / **14 16 & 18** / **CHARLES PLACE** / [amber]
CHICAGO. // **THIS BOTTLE** / **MUST BE RETURNED** /
(heel): **A & D. H. C.**

»SB w/ depressed center and post mold; ring-neck cider-finish quart; 11 1/2" tall. *[Figure 482]*

(3) *(RAISED 4-bottle logo)* / **J. A. LOMAX** / **14 16 & 18** / **CHARLES PLACE** / [olive green]
CHICAGO // **THIS BOTTLE** / **MUST BE RETURNED** /
(front heel): **ALFRED ALEXANDER & Cº**
(rear heel): **MAKERS LONDON**

Note: The maker's mark indicates that this bottle may postdate 1880 by a few years, but it represents one of only two known examples of bottle styles used by Illinois bottlers that were manufactured by foreign glassworks (see Figure 483, heel-mark roll-out). The other known example was also a British product (see Carse & Ohlweiler Rock Island listing). Note, however, that an early 1880s Lomax product list includes "Alfred's Ale and Porter (London)," so the heel embossing might refer to the bottle's product-content rather than a glassmaker.

»SB w/ depressed center and heel mold; narrow blob-top quart; 11 1/2" tall. *[Figure 483]*

(4) *(RAISED 4-bottle logo)* / **J. A. LOMAX** / **14 16 & 18** / **CHARLES PLACE** / [amber]
CHICAGO. // **THIS BOTTLE** / **MUST BE RETURNED** /
(heel): **De S. G. Co.**

»SB w/ depressed center and post mold; distinctive early DeSteiger Glass Works mold lettering; blob-top quart; 11 1/4" tall. *[Figure 484]*

(5) *(RAISED 4-bottle logo)* / **J. A. LOMAX** / [olive green, blue teal,
14 16 & 18 / **CHARLES PLACE** / **CHICAGO** //   olive amber]
THIS BOTTLE / **MUST BE RETURNED** /

»SB w/ depressed center and post mold; oval-blob-top quart; 11 1/4" tall. *[Figure 485]*

Figure 482

Figure 483

Figure 484

Figure 485

Medicated Aerated Waters:

(1) (RAISED 4-bottle logo) / **J. A. LOMAX** / **14 TO 18** / **CHARLES PLACE** / **CHICAGO** /   [aqua]
THIS BOTTLE // **MUST BE RETURNED** / **MEDICATED** / **AERATED** / **WATERS** /
(front heel): **ARTHUR** / **CHRISTIAN**
(rear heel): **PAT. APR. 13**TH (2 raised dots under TH) / **1875**

»SB w/ semi-round base (will stand) and heel mold; high slope-shouldered cylindrical pint; vertical embossed lettering encircling body below 4-bottle shoulder logo; horizontal heel lettering; 8" tall. *[Figure 486]*

(2) (RAISED 4-bottle logo) / **J. A. LOMAX** / **14 TO 18** / **CHARLES PLACE** /   [yellow-green, amber]
CHICAGO / **THIS BOTTLE** / **MUST BE RETURNED** //
MEDICATED / **AERATED** / **WATERS**

»SB w/ semi-round base (will stand) and heel mold; high slope-shouldered cylindrical pint; vertical embossed lettering encircling body below 4-bottle shoulder logo; 8 3/4" tall. *[Figure 487]*

(3a) **J. A. LOMAX** / **14 TO 18 CHARLES PLACE** / **CHICAGO** / **THIS BOTTLE** /   [aqua, greens*]
MUST BE RETURNED // **MEDICATED** / **AERATED** / **WATERS**

»SB w/ round base (will not stand) and heel mold; high slope-shouldered cylinder; vertical embossed lettering encircling body; 9 1/4" tall. *[Figure 488]* [*olive, lime, and forest-green shades seen]

(3b) **J. A. LOMAX** / **14 TO 18 CHARLES PLACE** / **CHICAGO** / **THIS BOTTLE** //   [olive green]
MUST BE RETURNED / **MEDICATED** / **AERATED** / **WATERS**
Note: mold recut from #3a above with different line spacing and smaller And less ornate lettering. *[Figure 489]*

»SB w/ round base (will not stand) and heel mold; high slope-shouldered cylinder; vertical embossed lettering encircling body; 9 1/4" tall.

Figure 486

Figure 487

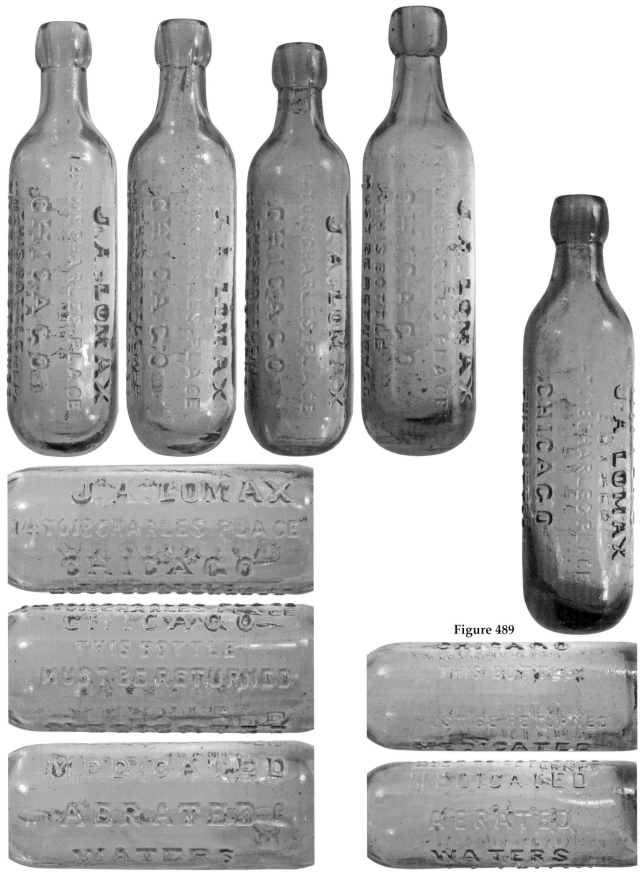

Figure 488

Figure 489

Figure 490

Figure 491

Figure 492

Figure 493

California Pop:

(1a) *(RAISED 4-bottle logo) /* **JOHN A. LOMAX / ORIGINAL / CALIFORNIA POP /**   [amber]
14 TO 18 CHARLES PLACE / CHICAGO / ILLS //
(left front heel): **A. & D. H. C** *(small embossed letters)*
(reverse heel): **THIS BOTTLE / MUST BE RETURNED**
(base): **J.L**

»SB w/ depressed center and post mold; quart soda/cider bottle shape; narrow blob top; 11 1/2" tall. *[Figure 490, left]*

(1b) *(RAISED 4-bottle logo) /* **JOHN A. LOMAX / ORIGINAL / CALIFORNIA POP /**   [olive green,
14 TO 18 CHARLES PLACE / CHICAGO / ILLS //   lime green]
(heel): **THIS BOTTLE / MUST BE RETURNED**

»SB w/ depressed center and heel mold; quart soda/cider bottle shape; narrow blob top; 11 1/2" tall. *[Figure 490, center and right]*

Ale:

(1) *(RAISED 4-bottle logo above raised horizontal shoulder band) /*   [amber]
(body): **J. A. LOMAX / 14 16 & 18 / CHARLES PLACE / CHICAGO**

»SB w/ depressed center and post mold; 2-piece mold quart ale; ring-neck tapered-collar; 10" tall. *[Figure 491]*

Famous Brown Stout:

(1) *(RAISED 4-bottle logo) /* **JOHN A LOMAX / FAMOUS / BROWN STOUT /**   [olive "black" glass]
14 TO 18 CHARLES PLACE / CHICAGO / ILLS //
(heel): **THIS BOTTLE / MUST BE RETURNED**
(base): [undecipherable mark: see Figure 492d]

»SB w/ heel mold; raised oval at center base; sloped shoulders; tapered collar; 9 1/2" tall. *[Figure 492]*

Lager & Weiss Beer:

(1) *(RAISED 4-bottle logo) /* **J. A. LOMAX / 14 16 & 18 / CHARLES PLACE / CHICAGO //**   [amber]
THIS BOTTLE / MUST BE RETURNED /
(heel): **A & D. H. C.**
(base): **J.L**

»SB w/ heel mold; low shoulder and long, bulbous, "lady's-leg" neck; tapered collar; 9 1/2" tall. *[Figure 493]*

(2) *(RAISED 4-bottle logo) /* **J. A. LOMAX / 14 TO 18 / CHARLES PLACE /**   [lime green]
(decorative embossed band) / **CHICAGO / SUPERIOR / BOTTLED / LAGER BEER //**
THIS BOTTLE / MUST BE RETURNED /
(heel): **A. & D. H. C**
Note: Different embossed lettering style and spacing than bottle #3 below.

»SB w/ post mold; slope-shoulder blob-top beer; 9 1/2" tall. *[Figure 494, left]*

(3) *(RAISED 4-bottle logo) /* **J. A. LOMAX / 14 TO 18 / CHARLES PLACE /**   [amber]
(decorative embossed band) / **CHICAGO / SUPERIOR / BOTTLED / LAGER BEER //**
THIS BOTTLE / MUST BE RETURNED /
(heel): **A. & D. H. C**
Note: Different embossed lettering style and spacing than bottle #2 (same lettering mold as bottle #4 below, but including embossed 4-bottle logo.

»SB w/ post mold; slope-shoulder blob-top beer; 9 1/2" tall. *[Figure 494]*

(4) **J. A. LOMAX / 14 TO 18 / CHARLES PLACE /**   [aqua]
(decorative embossed band) / **CHICAGO / SUPERIOR / BOTTLED / LAGER BEER //**
THIS BOTTLE / MUST BE RETURNED /
(heel): **A. & D. H. C**

»SB w/ post mold; slope-shoulder blob-top beer; 9 1/2" tall. *[Figure 495]*

Figure 494

Figure 495

Figure 496

(5) *(RAISED 4-bottle logo)* / **J. A. LOMAX / 14 TO 18 / CHARLES PLACE / CHICAGO /**   *[aqua]*
SUPERIOR / BOTTLED / LAGER BEER // THIS BOTTLE / MUST BE RETURNED /
(heel): **A. & D. H. C**

» SB w/ post mold; slope-shoulder blob-top beer; 9 1/2" tall.

(6) *(RAISED 4-bottle logo)* / **J. A. LOMAX / 14 TO 18 / CHARLES PLACE /**   *[amber, olive green]*
CHICAGO // *(embossed within raised diamond outline):* **J. A. LOMAX /**
DIAMOND / LAGER / BEER

» SB w/ post mold; slope-shoulder beer w/ narrow blob top; 8 1/4" to 8 1/2" tall. *[Figure 496]*

(7) *(RAISED 4-bottle logo)* / **JOHN A. LOMAX / FAMOUS / WEISS BEER /**   *[olive green and*
14 TO 18 CHARLES PLACE / CHICAGO ILLS. // THIS BOTTLE /   *olive "black" glass]*
MUST BE RETURNED
(base): **JAL**

» SB w/ post mold; straight-sided sharp-shoulder cylinder (*like stoneware bottles*); narrow blob top; 8 1/4" tall. *[Figure 497]*

(8) *(OUTLINE 4-bottle logo)* / **JOHN A. LOMAX / FAMOUS / WEISS BEER /**   *[blue teal]*
14 TO 18 CHARLES PLACE / CHICAGO. ILLS // THIS BOTTLE /
MUST BE RETURNED
(base): **JAL**

» SB w/ post mold; straight-sided sharp-shoulder cylinder (*like stoneware bottles*); narrow blob top; 8 1/4" tall. *[Figure 498]*

(9) *(4-bottle logo style unknown)* / **JOHN A. LOMAX / FAMOUS / WEISS BEER /**   *[cobalt]*
14 TO 18 CHARLES PLACE / CHICAGO //
(rear heel): heel mark, if any, unknown
(base): **J.L**

Note: Identified from fragment only (see Figure 499).

» SB w/ post mold; straight-sided sharp-shoulder cylinder (*like stoneware bottles*); narrow blob top; 8 1/4" tall.

Figure 497

Figure 498

Figure 499

SODA & FIRE.
John A. Lomax and the Early History of "The Largest Bottling House in the U.S."

As we discussed earlier in the *George Lomax* bottling-works summary, contradictory and fragmentary historical records have combined to blur the family ties between George and John A. Lomax (both men came to America from Bury, England, ca. 1850) and the connections between their earliest Chicago bottling histories. Since John Lomax eventually became a prominent businessman and political figure in Chicago, he was the subject of a detailed biographical essay in A. T. Andreas' massive *History of Chicago* (Andreas 1886:573–574) and an extensive obituary was published in the *Chicago Tribune* (May 14, 1899, p. 2, col. 7) when he died. But the two sources contradict one another in many details—and the *Tribune* article in particular blatantly misstates many verifiable facts. So the following summary is necessarily tentative on some details of business history. The sources used to gather and corroborate historical information include census records, Chicago CDs, Chicago newspaper advertisements and articles, and the historical essays mentioned above. We have also benefited from reading summary articles on Lomax bottling history by Byron Hughes (1970), Bob Kott (1981), Paul Welko (1973a, 1982), and an unpublished manuscript by Ray Komorowski titled "The Lomax Story" (April 2009).

As John Lomax himself advertised (*see Figure 500*), he had built his Chicago bottling works into "The Largest Bottling House in the U.S." by the late 1870s, so both his business history and the chronology of his company's pre-1880 use of embossed bottles are complex. As a result, our historical summary of Lomax bottling works is divided into seven "eras" of company development, to correspond with the chronology of embossed-bottle groupings listed above:

1. Ainsworth & Lomax at 38 W. Lake St. ...... ca. 1855/56
2. Bebbington & Co.(??): ....ca. 1856–57
3. J.A.L. at 38 W. Lake St. 1858–59
4. J.A.L. at 4 Charles Place 1860–66
5. J.A.L. at 16 Charles Place 1867–69
6. J.A.L. at 14 & 16 Charles Place 1870–72
7. J.A.L. at 14 to 18 Charles Place 1873–86

Figure 500

(1) John and George Lomax in Chicago: The Early Days

George Lomax (*see listing*) first arrived in Chicago in 1851–52 and is listed in the Chicago CDs from 1852 through 1854 as a soda and mineral water manufacturer along with Joseph Entwistle and Edward Bebbington at their bottling works on Canal Street between Lake and Fulton. He left the Canal Street operation after the 1854 season, and from 1855 until his death in 1863 he ran his own bottling works with junior partner John Meagher at 138 W. Madison.

John Lomax's two biographical essays indicate that he first arrived and entered business in Chicago either in 1848 (*Tribune* 1899) or in 1852 (Andreas 1886). Since British Census data indicates that both George and John Lomax were still in Bury, England, in 1850 (not "Berry" as indicated by the *Tribune* article), John probably arrived in 1852. Both sources indicate that he then worked at a "planing mill" at 12th and Clark Streets in Chicago for three years to save up enough money to buy into a soda/mineral water manufacturing operation (in 1855). One of these (Hall & Winch) actually was a planing/saw mill, but its principals did not arrive in Chicago and begin in business until 1858; the other (Abbott & Kingman) made windows and doors at 12th and Clark beginning in late 1854, but their predecessor company (Goss & Abbott) was in the same business at the same address from 1852 onward. So it looks like John Lomax was an engineer at 12th and Clark from 1852 through 1854, while George Lomax was bottling with Entwistle and Bebbington:

> During this period his earnings were only one dollar per day, but out of that he, by close economy, managed to save enough to purchase a small house and lot, valued at $450. In 1855, he traded this property for one-half interest in the soda water business of Joseph Entwistle, which was

then located at No. 39 West Lake Street. The firm of Entwistle & Lomax continued for about three years, when Mr. Lomax purchased his partner's interest and has since been the sole proprietor of the business. [Andreas 1886:573]

The 1899 *Tribune* article essentially agrees, except to indicate that John Lomax uses his earnings directly to purchase the entire business of "John" (*sic*) Entwistle at "26" (*sic*) West Lake Street (actually Entwistle and Bebbington bottled on Canal Street between Lake and Fulton, about seven blocks away). In this case, it looks like *both* biographical sketches were wrong. The Chicago CD indicates that two bottlers left the G. Lomax ~ J. Entwistle ~ E.K. Bebbington Canal Street bottling works in 1855. One was George Lomax (*see his listing*), who left to start his own bottling works at his home at 138 W. Madison (with new partner John Meagher). Perhaps he started his new company using the cash, traded property, or property cash-value he received from John A. Lomax to buy his share of the Entwistle & Bebbington operation (*see their listings, this chapter*). But Joseph Entwistle also disappears from the CD lists at this same time. The 1855 CD shows only his wife, Mrs. Ann Entwistle, as still living in Chicago. Intriguingly, her residence is 38 W. Lake Street: the address from which *Ainsworth & Lomax* were bottling in 1856 (*see A & L listing*) and the address where John Lomax first began his solo bottling operation in 1858. Ray Komorowski (personal communication) has suggested the possibility that after Entwistle left or died, John Lomax may have bought out Entwistle's Canal Street soda-factory interest through his wife Ann, who he then let stay at his house for a time while she settled her affairs.

From 1855 to 1857 the company remaining at the Canal Street address was called *Bebbington & Co.* (*see Bebbington listing, this chapter*).

(2) The Early Lomax and Bebbington Bottling Years: 1855–1857

Just who John A. Lomax bought out to begin his soda-bottling career remains something of a question, but he did not join the profession early in the 1855 season—the 1855 CD listed him (for the first and only time) as an engineer with the Abbott & Kingman sash-and-door company. Also, Bebbington's only partner listed in the 1855 CD was a soda maker named Edward Low. But directory name-takers were sometimes sloppy in their work (see next paragraph), so it should be noted that Bebbington's first name was Edward, and "Low" could be a corruption of "Lomax."

Either later in the 1855 season or in 1856, John A. Lomax may (or may not) have joined *Bebbington & Co.* The 1856 Chicago CD is an indecipherable muddle, the worst of any we have searched. *Bebbington & Co.* was not listed at all (the company reappears in the 1857 CD), but two men were listed individually as soda makers at the Bebbington address: "Robert Lenox" and "George Lomad." These might be twisted versions of the Lomax name—but since George Lomax had left a year earlier to start his own business, perhaps the entries are a hint that John Lomax *was* working there in 1856.

But the waters get muddier still. The soda-bottler summary list in the 1856 CD includes "Ainsworth & Lorax" at 38 W. Lake Street. Since John Lomax started his own bottling business at that address two years later, "Lorax" may be him. And an "**A & L**" embossed bottle of this age actually is known (*see Ainsworth & Lomax listing above*). If this was John, as seems most likely, it appears he was a junior partner at a different start-up firm when he "should have been" working for Bebbington. Neither John Lomax (nor Ainsworth) are listed in the 1857 CD, but 1857 was the last year that Bebbington & Co. were listed in the CDs as bottlers on Canal Street, and in 1858 the John A. Lomax company began bottling at 38 W. Lake.

Early Days Overview

Sorting all this out, our best-guess summary of John Lomax' earliest Chicago business history is as follows:

- ca. 1852–55: carpenter or designer with Goss & Abbott (later Abbott & Kingman) sash and door company at 12th & Clark Streets
- ca. 1855/56: Ainsworth & Lomax bottling works at 38 W. Lake Street
- ca. 1856–57: junior partner with Bebbington & Co bottling works on Canal Street between Lake and Fulton

(3) J.A.L. Bottling Works at 38 W. Lake: 1858–59

According to Alfred Andreas' *History of Chicago*, John Lomax then began his own soda and mineral water bottling business in 1858 at 38 W. Lake:

> The firm of Entwistle & Lomax continued for about three years [*actually Ainsworth & Lomax, then Bebbington & Co.*], when Mr. Lomax purchased his partner's interest and has since been the sole proprietor of the business. He remained at No. 39 [*actually 38*] West Lake Street until 1859, when occurred the big fire, known as the "big fire on West Lake Street"; in this conflagration he lost all he had accumulated by his years of hard work. Not disheartened, however, by his misfortune he at once returned to business at his present location" [*at Charles Place, see below*]. [Andreas 1886:573]

The Andreas summary is partly correct, but incomplete. As reported in the September 16, 1859, issue of the *Chicago Press and Tribune* (p. 4, col. 3–4), the "Great Fire in the West Division" was indeed "one of the most disastrous conflagrations that ever visited our city." Starting near the corner of Canal and Lake, the fire burned its way along both streets. The burned-out area contained primarily one- and two-story structures ("wooden buildings, sheds, outhouses, workshops,...saloons, small stores, &c."), but a large lumberyard and two hotels—the Cleveland House, and the Clinton House—were also destroyed. One of the smaller buildings destroyed was the "Soda-water establishment of John Lummix (*sic*) on West Lake Street."

However, Lomax did not immediately move to the Charles Place location. He first set up temporary bottling operations in a frame structure on Harrison Street, just a stone's throw away from his eventual Charles Place address. Less than two months later (on December 6, 1859) this structure also burned, in a fire that an article in the *Chicago Press and Tribune* of that date reported was apparently set by arsonists in a building near the corner of Harrison and Wells:

> This was a very large structure filled with blinds, doors and lumber, and was soon...giving out an immense volume of flame which from the peculiar character of the clouds overhanging the city, lit up the sky most brilliantly, the reflection being beautiful beyond description....
>
> A two story frame structure next west on Harrison street was partially consumed. It was occupied by John Lomax, soda manufacturer, who was burned out at the great fire on the West Side. He loses $600, no insurance, and a barn, which, with the stock, was worth $300; also uninsured. This building was owned by S. J. Hobbs. [*Chicago Press and Tribune*, "Serious Fire on South Wells St.," 12/6/59, p. 1, col. 3]

(4) J.A.L. Bottling Works at 4 Charles Place: 1860–66

The Chicago CDs indicate that John Lomax, still undaunted, next moved to new quarters at No. 4 Charles Place in another attempt to re-establish his bottling operation. At this time, he appears to have begun to diversify his soda-bottling operation to include spruce beer, and perhaps other root beers, put up in quart stoneware bottles (*Figure 501*). Andreas (1886:573) describes the first Charles Place building as

> a small house 20 x 30 feet and only one story high. This house he rented, but two years later purchased on four years' time. Scarcely had he got fairly started, however, when he was visited a second time by a fire which burned all his property to the ground, and of course entailed on him a loss which he could ill afford to bear.

But from Andreas' fire chronology, this quote is likely referring to the house on Harrison Street discussed earlier (or partly confuses the two). However, Andreas' next comment (1886:573) seems to refer to the subsequent Lomax residence and bottling works at 4 Charles Place:

> He immediately re-built, but one year later, by the explosion of a kerosene lamp, it was again laid in ashes; this time his losses were not quite so heavy, as he was partially insured.

The Chicago CDs indicate that Lomax remained at the 4 Charles Place address (at the SW corner of Wells) for another five years, so if the kerosene lamp blaze occurred just a year after the

Harrison Street fire he must have rebuilt and continued to operate there. However, another source (Hughes 1970), using files from the Chicago Historical Society (*see section 5*), dates the kerosene-lamp fire to early 1870.

(5) J.A.L. Bottling Works at 16 Charles Place: 1867–69

No information has been found on John Lomax's move to 16 Charles Place following the 1866 bottling season or describing his residence and bottling works there. But up to this time, his residence and bottling works had been in one building, and 16 Charles Place was probably no exception. In the 1867 Chicago CD, two Lomax family members are listed as workers at the soda factory (James L. and James H. L.), one boarding at the soda works and one living elsewhere. The 1868 CD lists five additional employees of the Lomax bottling operation:

Figure 501

two "peddlers" of J.A.L. products (John McCabe and John Mayfield) living away from the bottling works, and a "laborer" (Amihien Kaspar) and two "sodawater makers" Frank Dunn and James Horr) boarding at the plant itself. The soda factory jobs seem to have had relatively high turnover—in 1869 only Dunn and Mayfield remained. The other three had been replaced by Frank Damron, Thaddeus Brown, and Matthew Heinskill, and all five 1869 employees resided at the soda works.

The Lomax bottling enterprise was becoming very successful during the later 1860s, and by 1870 it was time to enlarge and diversify the bottling works. The decision to do so was likely precipitated by yet another fire:

> In the spring of 1870 the explosion of a kerosene lamp triggered another fire which totally destroyed the building and equipment. This time with the insurance money, he bought the lots adjacent to his on Charles Place and built an imposing five story plant on the location.
> In 1870 the J. A. Lomax company was the largest independent bottling company in the world. Business volume was about $175,000 per year. His payroll was 80 people and he had 15 wagons delivering his goods. The products were porter, ale, stout, and lager, in addition to many flavored soft drinks. [Hughes 1970:2]

Beginning in 1870, only the soda manufactory would remain at No. 16 Charles Place, and his residence would be at No. 14.

(6) J.A.L. Bottling Works at 14 & 16 Charles Place: 1870–72

But once again, John Lomax's business-expansion efforts were to be thwarted (temporarily) by fire—this time the world-famous Great Chicago Fire of October 1871:

> In 1870, his business having wonderfully increased, he set about building new works on an enlarged scale, his main factory being a brick structure 110 by 40 feet and five stories in height. He

moved into his new quarters just three months before the great fire of October, 1871. This was the heaviest blow of all, and, for a time, so deeply did he feel his losses that he almost gave up in despair. In a short time, however, he recovered himself, and with his old-time energy set about once more to retrieve the fortune which the fire had destroyed. He rented a small place on Cottage Grove Avenue, near Douglas Place, and within twenty-four hours was manufacturing and delivering goods to such of his customers as had not, like himself, been burned out. As soon as was practicable he also re-built, on the site of his former premises, a three-story brick structure, where he has since succeeded in building up his trade to its present important proportions. [Andreas 1886:573]

[T]hat fire wiped out his buildings, plant, and stock, causing a loss of almost $250,000. Undismayed by misfortune, he reestablished his business in the midst of the ashes and debris of the fire, and from that time it increased in volume to the point where he was admitted to be the foremost man in that business in the country. [*Chicago Sunday Tribune*, J.A. Lomax obituary, 5/14/99, p. 2, col. 7]

The 1872 Chicago CD lists only full-time professional staff associated with the Lomax soda works (not hourly workers): they include four delivery drivers, a plant foreman, and a bookkeeper. The earliest Lomax ad found in the *Chicago Tribune* dates to early 1872 (2/2/72, p. 5, col. 6), and shows how quickly he was able to reestablish the Charles Place bottling operation following the Great Chicago Fire:

JOHN A. LOMAX,

14 and 16 Charles-place, between Fifth av. And Franklin st., manufacturer of **SODA AND MINERAL WATERS,** dealer in Bottled Ale, Porter, Lager, and Pure Cider. Sole Agents for the Celebrated Wheeling, West Virginia, Draught Ale.

Thanking my old friends and customers for past favors and respectfully solicit a continuance of the same. I would inform them and all others that want a good and pure article that I am now ready at my old place (14 and 16 Charles place) to fill all orders with promptness and dispatch.

(7) J.A.L. Bottling Works at 14 to 18 Charles Place: 1873–1886

This was the "heyday" era of the Charles Place Lomax bottling operation, which continued to expand during the 1870s. According to Andreas (1886:573):

In 1874 *[Actually 1872, based on the Chicago CD address expansion, and the fact that the building façade itself is emblazoned "Erected 1872"—see Figure 501 above]*, finding his works too small, he erected an additional building on the adjoining lot, 20 by 105 feet; and in 1879, needing still more room, he purchased from the city 200 by 120 feet of ground in the rear of his old factory, and on this he has erected a new building, five stories high and 120 by 100 feet in size *[Figure 502]*, his works now being the largest of their kind in the world. [Andreas 1886:573]

In this capacious building *[i.e., the 1872 structure]* he has facilities for turning out 12,000 dozen (144,000) bottles of soda per day, besides mineral waters and other grateful beverages, and in hot weather his vast resources are taxed to their utmost to supply the demand his reputation and the quality of his wares have created.... Mr. Lomax keeps twenty wagons running constantly during the season...In *[the vaults]* he keeps consistently about 2,000 dozen of bottled ale, lager, and porter, for family use; also, about 1,000 barrels of stock ale and porter.

The line...includes all the milder and most refreshing summer drinks and the various forms or varieties of malt liquor; in the former class are included **the celebrated Ottawa mineral waters, bottled exclusively by Mr. Lomax direct from the springs**; and of the latter he has perfected arrangements for the manufacture of a new and improved quality of Weiss beer. Mr. Lomax is sole agent in this district for the favorite Wheeling (West Virginia) ale and porter and deals largely in Pittsburgh draught ale, bottled seltzer, Kissengen waters, and Spruce beer. Genuine Belfast Ginger Ale...is sold as a specialty by this house. [*Chicago Sunday Tribune*, April 25, 1875, p. 5, col. 5]

By the later 1870s, Lomax' regional sales were increasing as well, and he had begun advertising his bottled products far beyond the limits of Chicago. For instance, during those years the only ads for bottled soda or brewery products that appeared in the *Rock Island Daily Argus* (on the Mississippi River about 150 miles SE of Chicago) were Lomax ads:

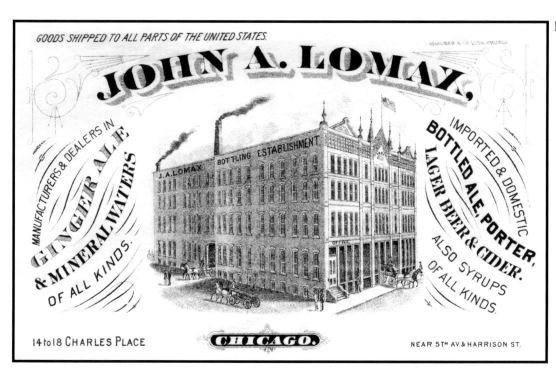

Figure 502

JOHN A. LOMAX
Manufacturer of
Ginger Ale, Seltzer & Mineral WATERS,
Of all kinds, in Jugs and Glass Bottles; also
DEALER IN
Bottled Ales, Porter, Lager Beer & Cider
And Syrups of all kinds.
14 to 18 Charles Place, ...CHICAGO.
Bet. Harrison & Van Buren Sts.
Send for Price List. Goods shipped to all parts
of the United States.

Also, the first major Chicago CD illustrated ad for the John A. Lomax bottling works appeared in 1876. According to the ad, Lomax bottled "Soda, Ginger Ale, and All Kinds of Mineral Waters...Ale, Porter, Cider and Lager Beer. Syrups and Extracts of All Kinds." The 1878 Lomax CD ad was the first to claim that the Lomax factory was "The Largest Bottling House in the U.S." During the early 1880s, the CD ads clarified the "All Kinds of Mineral Waters" comment of the earlier ads by indicating that J.A.L. was the "Sole Agent for the Celebrated Deep Rock Mineral Water, or Oswego, N.Y." (1880 CD) and the "Waukesha Mineral Rock Water (1883 CD).

According to the 1880 Chicago Industrial Census, John A. Lomax ran the bottling works full-time between June, 1879 and June, 1880, and had a maximum of 108 employees during the year, of whom 68 were adult males. Average daily wages for a skilled mechanic were $1.50, and a day's wage for the average laborer was $1.12. Total annual wages paid amounted to $28,000. He estimated the value of materials and fuel used at $60,000 and the value of the bottled products he produced at $125,000.

During the later 1870s in particular, there seems to have been a conscious advertising effort to give a distinctly "British Isles" character to many of the beverages in the Lomax line (*see Figure 502 letterhead*). Among the brewed products, this included distinctively bottled lager beers and famous brown stout, to accompany the more traditional bottled ales and porters. Lomax also reintroduced a new line of spruce and other root beers, put up in pint stoneware bottles (*Figure 503*). Among non-alcoholic beverages, bottling in many colors and styles of quart bottles also added to the Eu-

Figure 503

ropean look—and London glassmakers Alfred Alexander & Co. were commissioned to produce one of the quart bottle styles (see Komorowski 1977:10). The Lomax "Genuine Belfast Ginger Ale," bottled in Cantrell & Cochrane look-alike round-bottom bottles became a local favorite. Another popular British-Isles Cantrell & Cochrane product was "Medicated Aerated Waters," essentially an Irish take on the Saratoga-type minerals so popular in the eastern United States Lomax adopted the name (and the English-looking differently colored cylindrical round-bottom bottles) in the late 1870s, but bottled his own local mineral waters under the J.A.L. Medicated Aerated Waters banner. The 4/25/1875 *Chicago Tribune* article quoted above mentions "the celebrated Ottawa [IL] mineral waters, bottled exclusively by Mr. Lomax direct from the springs." This local mineral water may have been an early Lomax "medicated aerated water" product. But a more likely local candidate was heavily advertised by the Lomax Company during the 1879 bottling season:

THE NEW DISCOVERY.
THE IODINE SPRING
Of Mineral Waters at Montrose, Containing,
Among Other Valuable Ingredients,
Iodine, Chlorate of Sodium, Sulphate of Potassa,
Bicarbonate of Magnesia, Etc., Etc.

Indorsement by the Leading Scientists,
Chemists, and Physicians of the Land.

The discovery of the "Iodine Mineral Spring of Montrose" is a great event for Chicago. The Springs are eight miles from the Court House, at the Crossing of the Northwestern and Milwaukee & St. Paul Railways, and near Milwaukee-av. Profs. Haines and Siebel *[W. S. Haines and J. E. Siebel of the Rush Medical College]* pronounce the water as most valuable for diseases of the liver and kidneys, while for the cure of nervousness and for regulating the bowels and general system it is unquestionably of great worth. The water at the Springs is as cold as ice water, is as clear as crystal, and has no unpleasant taste. **Mr. John A. Lomax, well-known to all Chicago people will deliver this water in jugs, bottles, or siphons**, as may be desired, and families or others wishing to try it can order of him by postal, at 14 Charles-st., and it will receive prompt attention. It is also for sale at the drug stores. [*Chicago Tribune* 6/1/1879, p. 10, col. 7]

The Lomax Company bottles of the early and mid-1880s (e.g., collared blob and Hutchinson soda styles) fall beyond the limits of the present study, and are not included in the embossed bottle list (*see volume introduction*). At least three different embossed styles and seven mold variations of each of these two bottle forms are known (dating ca. 1880–86). After 1886, the Lomax bottling works became the *Chicago Consolidated Bottling Company (CCBC)* by combining the Lomax factory operations with 15 other bottling plants, most of which John Lomax purchased. But John Lomax' intertwined fate with fire disasters was not yet laid to rest:

DAMAGED BY A LIVELY BLAZE.
The Consolidated Bottling Company's Buildings
Injured $25,000 by Fire.

The Chicago Consolidated Bottling Company, formerly the John A. Lomax Bottling company, Nos. 14 to 18 Charles place and Nos. 184 to 204 Congress street, was damaged by fire at 11:30 o'clock last night....

The Congress street front is a three-story and basement building and consists of three buildings divided by light walls. The basement is filled with heavy machinery, the first floor is the shipping room, and the second and third floors are devoted to bottling and storage. The third floor of the west building was used as a storage room for chemicals used in bottling and the fire is supposed to have started in this room from spontaneous combustion. When first discovered smoke was pouring from the west windows and roof, and before the engines arrived the flames had spread to the top floors of the two buildings adjoining. The fire had broken through to the second floor before it could be got under control, and some damage was done to the upper floors of the building.

The company's stables front on Franklin street and run back to the main building. Two hundred horses and 100 wagons with harness and feed were in the stables when the fire broke out. [*Chicago Daily Tribune*, 11/12/1891, p. 3, col. 1]

It is safe to say that John Lomax took a deep breath, squared his shoulders, and got back into production. He served as President and chief owner of the new CCBC corporation until his death in 1899 (see *J.A.L. obituary*).

Dr. Lycurgus L. Lurton

Medicine:

(1) (side): DR LURTONS // CHOLAGOGUE // [aqua]
    (side): CHICAGO // //

»IP w/ rectangular pontil scar; rectangular w/ flat panels; short tapered collar; 5 1/2" tall. [*Figure 504*]

Figure 504

———————— ❖ ————————

The 1860 U.S. Census listed Dr. Lycurgus Lurton as a 39-year old physician in Chicago. He first appeared in Chicago CDs in 1855, with a residential address only. By 1856, Dr. Lurton had opened offices on Water Street, and he was listed in the 1856 to 1858 CDs as a druggist and physician. He expanded his wholesale and retail pharmacy and medicine business interests greatly in 1859, and the Chicago CD of that year lists him as variously partnered with three other pharmacists (Harris, LeGuere, and Johnson) at no less than four wholesale and retail addresses in the city. The 1860 ISD lists Lurton as a partner in three different druggist and medicine businesses in Chicago. But the 1860 CD lists only the druggist firm of Lurton and Stewart. Lurton disappears from the Chicago drug business after 1860 and is no longer listed in Civil War era CDs there.

No ads have been located for Dr. Lurton's Cholagogue, so he could have produced and marketed it in the city at any time in the late 1850s.

Manzanita Bitters

See the Reuben Stone listing, this chapter.

Parker R. Mason

Bitters:

(1) **AROMATIC / GOLDEN BITTERS // PARKER R. MASON / CHICAGO** [amber]

»SB; flat, square bitters w/ beveled edges; tapered collar; 9" tall. [*Figure 505*]

———————— ❖ ————————

Mason was a 28-year-old distiller in Chicago in 1870 (USC). By the 1880 USC, he was a wholesale liquor dealer in Lakeview Township in Cook County. Mason was indicted

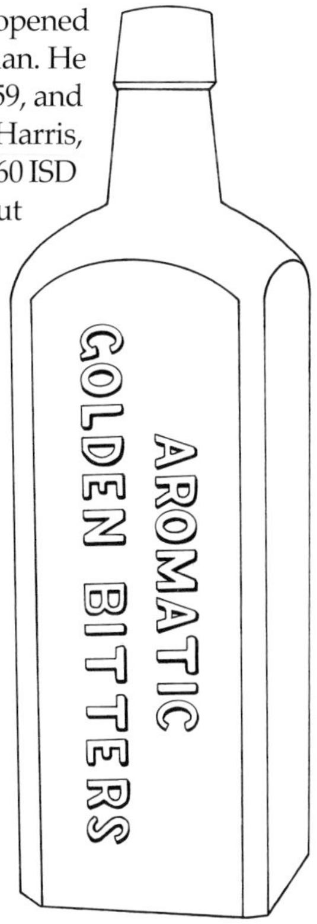

Figure 505

as one of the conspirators in the notorious Whiskey Ring scandal of 1875–76, and he was one of 12 Chicago distillers to plead guilty in January 1876 to the charge of having put 80,000 gallons of "crooked" distilled spirits on the market the previous May without having paid the U.S. government liquor-stamp tax (*New York Times*, 1/15/76). Mason admitted he had been acquiring blank stamp books, "which he filled out to suit himself," evading the tax since May or June of 1872. The disclosures of Mason and others implicated several prominent political figures and government agents.

According to the Chicago CDs, Mason was a home distiller from his residence at 49 E. Kinzie from 1867 to 1871. After that he moved to 38–40 Michigan Avenue and became a large-scale liquor wholesaler (and racehorse owner, according to the August 6, 1873, *Chicago Tribune*). Although no ads have yet been located to document the advertised therapeutic benefits of the product, Mason likely marketed his *Aromatic Golden Bitters* in Chicago for a short time during the late 1860s or very early 1870s when he was a home distiller.

Henry B. Matthews

Bitters:

(1) **HUNKI DORI / BITTERS // H. B. MATTHEWS / CHICAGO** [amber]

»SB w/ post mold; beveled-edge square bitters; flat panels; tapered collar; 9" tall. *[Figure 506]*

(2) **HUNKI DORI / STOMACH BITTERS // H. B. MATTHEWS / CHICAGO, ILL.** [amber]

»SB w/ keyed hinge mold; beveled-edge square bitters; flat panels; tapered collar; 9 1/2" tall. *[Figure 507]*

❖

Henry B. Mathews [*His name is misspelled on the bitters bottles listed*] and two of his three sons (H. B., Jr. and Edwin B.) first appear in the Chicago CDs from 1867 through 1869 as general job printers at 140 S. Water Street. Beginning with the 1870 CD, Henry Mathews is listed as a wood and coal dealer—and "propr. Hunki Dori Bitters"—at 148 State. H. B., Jr. is a clerk at the store. Interestingly, probably because his name is misspelled on the earliest bottles (*see bottle #1 above*), he is also listed separately under the "Matthews" spelling. In 1871, Mathews is listed as a "bitters manufacturer" at

Figure 506

Figure 507

30 N. Dearborn. In 1872, he is listed as a "bitters and bluing manufacturer" at 431 W. Lake and Henry, Jr., has become a "traveling agent" for the company. The only change in 1873 is that the listing is changed to "stomach bitters" manufacturers (*see bottle #2*) and son Edwin B. has also become a traveling agent for the company.

In the 1874 and 1875 CDs, the listings change dramatically: Henry and Henry, Jr. are "patent medicine manufacturers" at the 431 Lake address. The father is no longer listed in the CDs after 1875, and he may have passed away.

In 1876, and continuing through 1879, the company becomes *H. B. Mathews Sons* at 220 Lake Street. Third son David joins the group, and they are once again listed in the CDs as "bitters" manufacturers. At this time, their focus seems to be selling dad's Hunki Dori Bitters wholesale (probably only in paper-labeled bottles considering the rarity of the embossed-bottle examples and the fact that there is no longer an accommodation for the misspelled embossed name). Their bitters is advertised in the Colburn, Birks & Co. (Peoria) wholesale-drugs catalog in 1878 (Ring 1980: 258).

The H. B. Mathews Sons company was listed in the 1880 IL Industrial Census as having a $3,000 investment in equipment and property, with five employees (three men, one woman, one child) and an annual payroll cost of $1,000. Annual cost of production materials was listed at $1,000 with an annual product sales value of $5,000.

During the early 1880s, the brothers continue in business separately and in shifting partnerships. They are variously listed under "drugs," "patent medicines," and "electric belts" early on, and then as "druggists" and chemists." By 1885, two of the brothers have moved on to other trades and only Henry, Jr., remains in the pharmacy business.

As with Mason's *Aromatic Golden Bitters*, no newspaper ads have yet been located to document the advertised therapeutic benefits of Henry Mathews bitters. But based on the Chicago CD listings, bottle #1 was likely used ca. 1870 to 1872, and bottle #2 was used in 1873 and perhaps for a year or two afterward.

August and Henry Mette

Soda ~ Mineral Water:

(1a) **A. METTE & BRO / CHICAGO //** [aqua]
(heel): **C&I**

»SB with post mold; short-neck oval blob-top soda. *[Figure 508]*

(1b) **A. METTE & BRO / CHICAGO //** [aqua]
(heel): **A. & D. H. C**

»SB w/ post mold; short-neck oval blob-top soda. *[No photo. See Figure 508.]*

(2) **A. METTE & BRO / CHICAGO** [aqua]
(base): **MB**

»SB w/ post mold; squat, short-neck oval blob-top soda. *[Figure 509]*

(3) **A. METTE & BRO / CHICAGO** [aqua]

» SB round-base, shouldered cylinder; oval blob top soda; 8 1/4" tall. *[Figure 510]*

(3) **A. METTE & BRO / CHICAGO** [aqua]

» SB, squat, short-necked Chicago-style quart soda; shouldered blob top soda; 8 1/4" tall. *[Figure 511]*

---- ❖ ----

August Mette, along with his brothers Henry and Louis, first appeared in the Chicago CDs in 1872, as a brewing company manufacturing "tonic beer" at 328 Archer. The three brothers continued as business partners in the firm of *A. Mette & Bros.* between 1873 and 1875 (at 985 S. Halsted in 1873 and at 370 Archer in 1874–75), but during these years they are listed in the CDs

Figure 508

Figure 509

Figure 510

Figure 511

only as "soda water manufacturers." We have not yet documented any embossed bottles for the three-brother partnership (i.e., Mette & Bros.).

Louis was no longer a partner in the business after 1875, and from 1876 onward the firm became *August Mette & Bro*. Louis continued to bottle soda water with his brothers, but only as an employee. So after 1876, the company bottles were embossed *A. Mette & Brother*.

August and Henry Mette's soda-bottling operation remained at 370 Archer during 1876–78. They moved their bottling operation to 5 Buena Vista Place in 1879. Their bottling works remained in the 2 to 5 Buena Vista locale until 1886, after which they were bought-out by John A. Lomax and incorporated into his newly formed *Chicago Consolidated Bottling Co.* group (along with several other bottling operations he purchased at this time (*see John A. Lomax listing, this chapter*). August Mette remained with the newly formed conglomerate as a vice president, and continued to run the Buena Vista works. The 1880 IL Industrial Census indicated that the Mette & Bro. bottling works was in business full-time year-round and that they produced an estimated $24,000 annual value of bottled products. According to the 1880 USC, August Mette was 36 years old at the time.

Since our current study is focused on the pre-1880 period, the *Mette & Bro.* collared-blob and Hutchinson-style bottles dating to the early and mid-1880s are not included in the bottle list (see Harms 1978; Wright 1992). The Mette company's "MB" monogram embossed bottles likely date to the post-1880 period, so they are not included in the list.

Louis Mette rejoined the firm as a partner during its final two years of independent operation, so during 1885 and 1886 it is once again listed in the Chicago CDs as *Mette Bros*.

Dr. [Adam] Miller

Medicine:

(1) (side): **DR MILLER'S // INFALLIBLE / ASTHMA REMEDY //** [amber]
    (side): **CHICAGO ILL.**

   »SB w/ keyed hinge mold; beveled-edge rectangle w/ 3 sunken panels; tapered collar; 9 1/2" tall. *[Figure 512]*

(2) (side): **DR. MILLER // //**
    (side): **CHICAGO. ILL. // //** [amber]

   »SB w/ keyed hinge mold; rectangular w/ 3 sunken panels; tapered collar; 8 3/4" tall. *[No photo]*

From their shape and manufacturing technology, Dr. Miller's embossed *Asthma Relief* bottles appear to have been produced and used in the ca. 1865–1875 era. Dr. Miller's product bottles are generally found in the Chicago area, and his medicine(s) do not appear to have had wider regional distribution. No ads for Dr. Miller's medicine have yet been found in Chicago or elsewhere, so we are uncertain of his identity. Chicago CDs for these years do not list any druggists, pharmacists, or patent-medicine manufacturers named Miller. However, four long-term Chicago physicians active during these years could have marketed this asthma-relief medicine through their practices: Dr. Adam Miller (1863–1880 ++); Dr. Benjamin C. Miller (1868–1878); Dr. DeLaskie Miller (1859–1800 ++); and Dr. Truman W. Miller (1866–1880 ++).

Dr. DeLaskie Miller was an academic physician throughout his career (Professor of Obstetrics and Diseases of Women and Children at Rush Medical College), and is unlikely to have been involved in patent-medicine marketing (see Sperry 1904:46–49). Dr. Truman Miller was a Civil War surgeon who after the war became a consulting hospital surgeon, "Surgeon-in-Chief to many of the leading lines of railroads, and Medical Referee and Consulting Surgeon to a number of life and

Figure 512

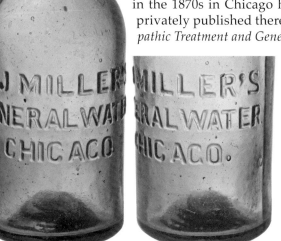

Figure 513

accident insurance companies (Sperry 1904:106). So he too was unlikely to have been a patent-medicine man. Dr. Benjamin Miller was a general physician during part of his 10-year Chicago career, but during the early-to-mid 1870s he served as sanitary superintendent for the Chicago Board of Health. He, again, is somewhat unlikely to have been involved in patent-medicine manufacture.

The most likely Chicago physician to have developed and marketed a patent-medicine product is Dr. Adam Miller. Adam Miller first came to Chicago in late 1862, and remained there the rest of his professional life. He had previously been involved in general medical practice in Springfield and Quincy, IL. He had practiced in Quincy in 1851 during a fearful cholera outbreak there (Bradford 1905). He had wide-ranging intellectual interests, one of which was to seek out and understand effective medications, and in the 1870s in Chicago he wrote a book about homeopathic medicines that he privately published there in 1879: *Plain Talk to the Sick, With Directions for Homeopathic Treatment and General Rules for the Preservation of Health*. He also published a small volume on *Life in Other Worlds*.

So, from the data at hand, Dr. Adam Miller seems to be the most likely developer of *Dr. Miller's Infallible Asthma Remedy* in Chicago.

Andrew J. Miller

Soda ~ Mineral Water:

(1) **A J MILLER'S / MINERAL WATER / CHICAGO** [aqua, blue teal]

»IP; high-shoulder tapered-collar soda; shouldered-blob-top variety also known. *[Figure 513]*

(2) **A J MILLER'S / MINERAL WATER / CHICAGO** [aqua]

»SB w/ keyed hinge mold; high-shoulder soda; shouldered blob top. *[No photo. Identical mold to Figure 513, but with smoothed rather than iron-pontiled]*

❖

According to the 1858 Chicago CD, Andrew J. Miller had emigrated from Denmark and come to Chicago 15 years earlier (ca. 1843). This is supported by the first two Chicago CDS from 1843 and 1844, both of which list him as a barber in the city. An ad for his barbering business appears on page 98 of the 1844 Chicago CD:

A. J. Miller,
Barber, & Fashionable Hair Dresser,
Market Street, near the Sauganash,
CHICAGO, ILL.

French Pomatum, an article superior to every
Other kind of Hair Oil, for beautifying and
Preserving the Hair.
CIGARS of the best quality constantly on hand.

His barber business was no longer listed in the 1845 CD, so he appears to have left the tonsorial trade that year and taken steps to set up his earliest bottling business. Welko (1974:5) has indicated that for a while, soon after A. J. Miller arrived in Chicago, he worked "as a partner of Miller and Clements...grocers and dealers in crockery." Perhaps Miller was involved in this grocery enterprise in 1845.

Beginning with the 1846 CD, and continuing through 1848, A. J. Miller is listed as a brewer (one of only three in town) and "small beer" bottler at a stand opposite the American Temperance House hotel. Small beers (e.g., spruce beer, lemon beer, sarsaparilla/root beer) were characteristically put up in stoneware bottles, and two styles of hand-turned *A. J. Miller* stoneware quart bottles are known: one is stamped **A. J. MILLER** (in small unadorned letters) on its upper body or sloped shoulder area (*Figures 514–515*); the second is a sharp-shouldered quart, horizontally stamped **A. J. MILLER** in large seriffed lettering. Kiln-failure sherds of this second Miller bottle style are known from the Kirkpatrick family pottery kiln site in Vermillionville, some 80 miles southwest of Chicago (Gums, et al. 1997:25).

The Chicago CDs available to us for the period from 1849 to 1851 are fragmentary, and the sections that would have Miller's business listings for those years are missing. But by 1852 he was listed as a mineral water bottler on Randolph Street, and the 1853 CD listed him as a soda water manufacturer from his residence at the corner of Lake and Carpenter streets. The 1854 CD lists him as a syrup manufacturer at the same home address. So, probably ca. 1850, he left the small-beer brewery trade and went into the soda/mineral-water bottling business and the production of flavored syrups for soda fountains.

During 1855–1859, he appears to have expanded his efforts in the area of soda-fountain supplies and apparatus, while continuing his bottling efforts. The Chicago CDs for 1855 and 1856 not only listed him under soda water makers, but also especially indicated he was a "Soda Fountain erector, and dealer in Soda Water Apparatus." The 1857 and 1858 CDs listed him as a soda and syrup manufacturer. The two CDs available for 1859 listed him in business summaries as a soda manufacturer and soda fountain manufacturer.

Figure 514

Andrew Miller's final Chicago business listing was in the 1860 CD. Here he was listed alphabetically as a soda manufacturer from his Carpenter Street residence, but was not listed in the soda-manufacturer summary. So perhaps in this final year he was only filling standing orders for wholesale bottled soda in preparation for closing down the business.

Interestingly, a few of his known later (shouldered blob top) embossed bottles are smooth-based rather than iron pontilled, indicating that his glass suppliers were making the transition to the new post-pontil glassblowing technology ca. 1860.

Figure 515

Charles Moench and William Reinhold

Mineral Water:

(1) **MOENCH & REINHOLD / CHICAGO. ILL** [aqua]

»SB w/ hinge mold; high-shouldered cylinder with vertical embossing; elongated cylindrical "Selters Water"-style neck and shouldered blob top; 2 7/8" diameter, 8" tall. *[Figure 516]*

Charles Moench was first listed as a druggist in the Chicago CD for 1859. Julius "Roemheld" showed up in the Chicago CDs at about the same time as a druggist and apothecary and proprietor of "Roemheld's Chemical Laboratorium." During the Civil War years, Moench became associated with Roemheld as a druggist clerk. We should note, however, that Roemheld and the "Reinhold" partner embossed on the bottle listed above were NOT the same man (*see later discussion*).

Beginning in 1866, a new Chicago pharmacy partnership appeared in the CDs: Charles Moench and physician William Reinhold listed together at 103 N. Clark as dealers in "drugs and medicines." Their partnership lasted for 25 years (until 1890), but their short use of the embossed bottle listed above was closely tied to the Great Chicago Fire in October 1871. The bottle is of somewhat unusual design, but from its stylistic and technological attributes it appears to have been manufactured and used during the late 1860s. No contemporary ads have been located to suggest its use or contents, but the recollections of Chicago pharmacist Wilhelm Bodemann, published in the July 2, 1903, issue of a journal called *The Pharmaceutical Era* (Vol. 30), indicate its likely function:

> I got a job with Moench & Reinhold October, 1867, then at Clark and Indiana streets. Mr. Moench's name was Charles, and that was the reason they called him Henry. When he worked for Louis Wahrlich he found a Charley at work there. Two Charleys in one store would not do, so Henry was his name up to his last day. Lots of the first families of the North side called him Henry, never having known his family name. **Reinhold, who was a pharmacist, a good chemist and a physician, built up quite a mineral water trade, but the fire wiped out the firm's money, trade and ambitions.** Both of the partners died poor. Other Travelers on the Highway.

The Chicago fire no doubt also destroyed nearly all of their stock of embossed bottles—only one example of the bottle has been located to document the short-lived *Moench & Reinhold* mineral water bottling business.

Figure 516

Mountain Root Bitters

See listings for Sylvester J. Smith [Chicago], and Abel, Humiston &Co. [Sandwich].

Isadore Monheimer and Co.

Bitters:

(1) **RED JACKET / BITTERS // // MONHEIMER & Co. // //** [amber]
   Note: MONHEIMER & Co. was cut into the mold after SCHWAB, McQUAID Co. had been piened-out. [No photos—see text]
   »SB w/ heel mold; beveled-edge square bitters; tooled blob-top lip; 9 1/2" tall.

Leonard Monheimer is listed in the 1870 USC as a 23-year-old Chicago liquor distiller. According to Cannon (2000:40–41), Monheimer became proprietor and distributor of the *Red Jacket* bitters brand in 1874. However, from Chicago CD information, this conclusion appears to be in error.

Schwab, McQuaid & Co. (*see their listing, this chapter*)—successors to the Bennett Pieters Company, which first produced and sold *Red Jacket Bitters* (*see Pieters listing, this chapter*), and the one-year (1869) *Schwab, Pieters & Co.* partnership—was in the wholesale and retail liquor business in Chicago from 1870 through 1875. According to Chicago CDs, the company then became *McQuaid & Monheimer Bros.* from 1876 through 1879. The *Monheimer Bros.* (Leonard and Levi) then ran the liquor business during 1880, after which Isadore Monheimer and John S. Cook were proprietors of the *Monheimer & Cook* liquor business from 1881 through 1886. Isadore (not Leonard) Monheimer took over the business after 1886, calling it *"Monheimer & Co."* for the first time.

Strangely enough, after a 10-year hiatus, *Monheimer & Co.* once again produced an embossed *Red Jacket Bitters* bottle. Either they reintroduced the brand for sale, or the 1876–1886 company partnerships bottled *Red Jacket Bitters* in paper-labeled bottles only (*see Bennett Pieters & Co. listing, this chapter*).

The Monheimer & Co. embossed *Red Jacket Bitters* bottle was produced much too late for inclusion or illustration in the present volume, and is discussed here for clarification purposes only.

Samuel Myers and Co.

Whiskey:

(1) **S. MYERS & CO / CHICAGO**   [amber]
(base): **C & I**

»SB w/ post mold; "shoo-fly"-style pint whiskey flask; ring-neck collar; 7 1/2" tall. *[Figure 517]*

❖

Samuel Myers & Company (Samuel Myers, Simon G. Meyers, and M. E. Wilkinson) appears in the Chicago city directories from 1866 to 1917. From the mid-1860s to the late 1870s, the company was located at 268 and 270 East Madison. The company began as whiskey distillers only, but by 1875 they had branched out into wholesale liquor sales.

From the technological attributes of the bottle listed, it appears to date to the mid-1870s. It definitely predates 1878/79, when the Pittsburgh glass company of Cunningham & Ihmsen (*C & I*) was dissolved. It is interesting to note that after 1877 the company shifted to the retail wine and liquor business, with their retail outlet at 270 E. Madison only. Perhaps the embossed bottle was made for the 1878 "grand opening" of their retail establishment, since after that year the glassmaker (*C & I*) was no longer in business.

Figure 517

Northwestern Chemical Manufacturing Co.

Ink:

(1) **N.W. CHEMICAL / M.F.G. Co. / CHICAGO**   [aqua]
(base): *Embossed image of a 5-pointed STAR.*

»SB w/ post mold; bell-shaped cone ink; thickened flared-ring collar; 2 1/2" tall. *[Figure 518]*

❖

The Northwestern Chemical Manufacturing Company was in business in Chicago for five years in the late 1870s. But according to the Chicago CDs, the company manufactured ink for only three years: 1875–1877. During these years only, they were listed in the directories as manufacturers of "Imperial Ink."

The company was located at 155 5th Avenue in 1875 and 1876, and at 200 Jackson in 1877.

Figure 518

Palmer House Hotel [Potter Palmer]

Whiskey:

(1) (outer ring): **PALMER HOUSE / CHICAGO** [amber]
(inner ring): unimbossed
(base): **PAT^D AUG6^TH 1872 / H FRANK**
Note: Bottle seen only in half-pint size, but a full-pint variety may also have been produced (compare Chapin & Gore #3a listing).

»SB w/ post mold; strap-sided flask; ring-neck "straight brandy finish" lip w/ internal screw threads; ca. 6" tall. Top of glass screw-cap embossed: **PAT. AUG672**. *[Figure 519]*

(2) **PALMER HOUSE / CHICAGO** [clear glass]
Note: Side-embossed in circular slug plate. This bottle may have been produced during the early 1880s.

»SB; flat-faced "shoofly"-style flask; ring-neck blob top w/ internal screw threads; ca. 6 1/2" tall. *[Figure 520]*

Figure 519

According to the Chicago Historical Society website, Potter Palmer made his first fortune in the dry goods business. In 1865, when he was 39, Palmer sold the business to a young Marshall Field. During the late 1860s and early 1870s, he bought up commercial property along State Street, turning it into a Parisian-style boulevard with an eye to making it the retail hub of Chicago.

In 1870, Potter Palmer announced his marriage to Bertha Honoré. As a wedding gift, he presented her with a hotel. The Palmer House was, at eight stories, the tallest building in the city. Its 225 rooms were decorated with Italian marble and French Chandeliers....

The Great Fire *[of October 1871, which also destroyed 94 other Palmer-owned buildings along a 3/4-mile stretch of State Street]* destroyed the Palmer House Hotel before it was officially opened, and it was rebuilt quickly. Palmer claimed the new hotel was completely fireproof....

In advertisements, he dared anyone to light a hotel room on fire: "If at the expiration of [one hour], the fire does not spread beyond the room, the person accepting this invitation is to pay for all damages done and

Figure 520

for the use of the room. If the fire does extend beyond the room (I claim it will not), there shall be no charge for the damage done. [CHS "People & Events" webpage]

The Palmer House Hotel was first listed by Chicago CDs in 1874 as being open for business. Interestingly, a similar (but not strap-sided), amber, internally threaded screw-top whiskey flask (like #1) was also distributed during the mid-to-late 1870s by Palmer-competitor John B. Drake at his Drake Hotel in Chicago (*see Palmer House listing, this chapter*).

Dr. Paoli's Sparkling Persian Sherbet

See the Simonds & Paoli listing, this chapter.

Penton, Fisher and Co.

Medicine:

(1) (*vertical*): **PENTON. FISHER & Co / CHICAGO.** [aqua]

»OP; shouldered cylinder; double-ring collar; 8 1/2" tall. (*Note: bottle recorded by Scott Garrow. No photo*).

———————— ❖ ————————

The firm of *Penton, Fisher & Co.* was a partnership between druggists Thomas. B. Penton and Williamson P. Fisher, and Archibald M. Robinson. They were listed together in Chicago CDs for only three years: 1856–1858. The company sold drugs and medicines on a retail and wholesale basis from a store at 94 Lake. Their ad in the 1858 Chicago CD indicated they were dealers in "foreign and domestic drugs and medicines, pharmaceutical preparations, paints, oils, varnishes, etc., etc." The ad goes on to note:

PHYSICIANS PRESCRIPTIONS CAREFULLY PREPARED.
Perfumery, Toilette and Fancy Articles, Pure Cod Liver Oil, prepared from pure livers and warranted. Proprietary Agents for all Popular Patent Medicines. Constantly on hand, a choice stock of Pure Wines, Brandies, etc., vintage from 1830 to 1851. Ales, Porters, etc.

By the publication of the 1859 city directory, the partnership had changed. *Penton & Co.* included earlier partners Penton and Robinson, together with new partners David Smith and John Brinckerhoff, and the business had relocated to 39 S. Water.

Charles O. Perrine

Honey:

(1) **C. O. PERRINE / CHICAGO** [aqua]

»SB w/ post mold; wide cylindrical shouldered jar; ring neck flared-ring collar; 8" tall and 3 1/2" diameter. [*Figure 521, right*]

(2) **C. O. PERRINE / CHICAGO** [aqua]

»SB; cylindrical wax-seal jar; sloping shoulders; ground lip; flared ring below lip; 7" tall; 3 1/4" diameter [*Figure 521, left*]

———————— ❖ ————————

C. O. Perrine is always listed as Charles or by his initials in the Chicago CDs, although a somewhat bizarre early newspaper article about one of his unusual creative ideas

Figure 521

for the honey business (*see later discussion*) refers to him as "Caligula Orpheus" Perrine. Perrine began in the honey business in Iowa (*Chicago Tribune* 2/26/1874, p. 3, col. 2), and in the late 1860s relocated to Cincinnati (*American Bee Journal* (1879), *see below*; Wiley 1892:747, *see below*).

> I began the honey business at Cincinnati, in 1865, and removed to Chicago in 1869, where I continued the business, keeping my hands at work, peddling direct to the consumer. [*American Bee Journal* 15(1) 1879: no pp#]

He first appears in the Chicago directories in 1869 as *C. O. Perrine & Co.*, "honey dealer" (business partner left blank). The company missed being listed in the 1870 CD, but it is listed every year thereafter, from 1871 to 1883. Perrine continues with his "*& Co.*" listing throughout his Chicago tenure, but no partner is ever specifically named until his last year in business (1883), when Alfred R. Bremer is the junior partner. In 1884 Bremer took over the company, which then became *A. R. Bremer & Co.*

Throughout the 1870s, Perrine's business operations moved frequently: 237 W. Madison (1869); 51 Blue Island "late 109 Dearborn" (1870–71); 51 Blue Island (1872); Lake SE corner of Market (1873–74); No. 7 Market Street (1875–77); 54 & 56 Michigan Avenue (1878–1880); 20 & 22 Michigan Avenue (1881–83). Early on, the business was solely focused on honey: "honey dealer" (1869); "honey, wholesale and retail" (1871–72); "honey dealer" (1873). But by the mid-1870s Perrine had diversified somewhat: "honey and maple syrup" (1874–75); "honey and syrups" (1876–77); in 1878, Parker's Dry Hop Yeast is added to honey and syrups; in 1879–80 baking powder is added; and from 1881 to 1883 his product list includes "jellies, honey & syrups, and fruit butter."

During the early 1870s, when Perrine was selling primarily honey and maple syrup, his Chicago business grew fairly quickly. The 1874 Chicago CD lists three Perrine employees that year. But he was selling honey in eastern cities early on as well. In 1890, several years after he had retired from the honey business, he summarized his early business history for Professor H. W. Wiley (in response to Wiley's inquiry for a USDA study of early food "adulterations"):

> In regard to the adulteration of honey as practiced in the United States, all I know is what I did myself.
>
> During the year 1865 I received the idea from a friend that the common dark honey then on the market could be much improved by the addition of a large per cent (sometimes 75 per cent) of good white sugar. I took the idea up, and after making some experiments I worked up quite a large trade among families by selling from house to house; in fact I bought all the cheap honey I could find in the neighborhood (Cincinnati, Ohio), and finally I had to send east and south for supplies of honey. Where 1 pound was used before I commenced, I afterwards sold 100 pounds, as it was much more palatable.
>
> After a few years I introduced my goods into several of the larger western cities [*e.g., Chicago in 1869—K.F./J.W.*], and still later on I opened business in Philadelphia and Brooklyn, N.Y. During all this time I handled large quantities of comb honey, giving customers their choice.
>
> This peddling business was done in wagons, two men to each wagon; each man selling 50 to 100 pounds per day, 90 per cent of which was the mixed article generally.
>
> I have met hundreds of persons who could eat the sugared article, but to whom pure honey was almost a rank poison. I suppose it is like strong black coffee, compared with a milder concoction, with plenty of cream and sugar.
>
> About the year 1870 I tried some French and German glucose, using it as a part substitute for sugar, and when good glucose was made in this country I became patriotic and used only goods of home manufacture.
> C. O. Perrine
> Riverside, Cal., *August 26, 1890*. [Wiley 1892:746–747]

One example of the "large quantities of comb honey" Perrine handled is documented in Lee Watkins' biography of pioneer California beekeeper John Harbison: "As early as 1873 California produced so much honey that the Pacific Coast markets could not absorb it all.... [Harbison] loaded on a steamship in San Diego 21,000 pounds of his comb honey destined for Chicago. In San Francisco the honey was transferred to a freight car, filing it to capacity. After the honey arrived in Chicago, Harbison sold it to C. O. Perrine, a wholesale honey dealer, for twenty-seven cents a pound" (Watkins 1969:17–26).

Perrine's maple syrup business was also conducted on a large scale in the later 1870s. From Chicago Customs House duty records of foreign (Canadian) importations, he imported 80 barrels of maple sugar at one time on 12/30/1876. And three separate import duties he paid during 1880 on Canadian maple sugar were for 150 barrels, 238 barrels, and 64 barrels respectively.

Perrine also had a lively inventor's spirit. A *Chicago Tribune* article titled "Utilizing the Blue-Bottle" published on February 26, 1874 (p. 3, col. 2)—which for some reason records his first and middle names as "Caligula Orpheus" Perrine—reports on a court case that pitted Perrine against his ex-wife:

> At the time of their entering upon the honey business, not a cloud obscured the brightness of their united lives.... Perrine became quite an accomplished finder of the haunts of the Iowa wild bee, and with his wife's aid in the first year of married life, they succeeded in securing several tons of honey, which sold at a high figure in the Chicago market.... But after two years of uninterrupted wedded bliss and uniformly successful bee-hunting...Caligula Orpheus conceived the somewhat novel idea of training the ordinary blue-bottle to the manufacture of honey, an idea which his wife scoffed at from the beginning. It was in vain that Mrs. P. endeavored to explain the futility of her husband's idiotic conception. He was bound to produce a honey-bearing blue-bottle, and to this end squandered the savings of the two successful years of married life, and a large portion of the original capital, and yet...though it was his boast that he had succeeded in securing a hybrid that possessed all the honey-gathering qualities of the wild bee and the domestic affection for fresh meat of the pestiferous fly, *[the blue bottle]* failed in the slightest degree to fall into his way of thinking on the honey subject. At last the patience of Mrs. Perrine gave out and she procured a divorce from her husband on the ground of his strange monomania, and married a man with the unromantic name of Chauncey Spade, who was engaged in the honey business but was entirely devoid of anything like the intellectuality of her late lord.

But C. O. Perrine remained in the honey business, presumably abandoning the blue-bottle line of inquiry, and continued to devise unique, creative ways to make the business prosper. On April 12, 1876, he secured U.S. Patent #176,347 (as "Charles" O. Perrine) on a cylindrical or rectangular glass tumbler to contain comb honey. The prefabricated comb was to be filled by the bees in the hive, already in its glass package, and ready without further processing for transport, labeling, and sale:

> Bees, in forming one or more honey combs in a hollow space, make them nearly straight horizontally, fastening the vertical edges of the comb on two sides of the walls enclosing the space, so that when the package is of glass the two unfinished vertical edges of the comb may be seen through the glass, unlike the finished capped sides of the comb; and the honey in the unfinished sides against the glass is exposed to the action of the light, which causes it to granulate and become opaque; and the glass, in expanding and contracting with heat and cold, breaks the cells, and causes the honey to leak out of them; and where a vessel is placed on a hive with the opening downward, the bees, in building and finishing the comb in the vessel, will fasten the comb to that part of the hive underlying the open space of the vessel; and in removing the vessel or package with the honey from the hive the comb attachment must be cut or broken, allowing the honey in the lower cells to run out, and leaving an unfinished broken part to view. [C. O. Perrine patent-application description, Feb. 18, 1876]

But undoubtedly Perrine's most ambitious creative honey-making venture, with a plot worthy of a made-for-TV movie, was his *Floating Bee Palace*. The plan was first reported by the *Chicago Tribune* on April 8, 1878 (p. 7, col. 5):

A Floating Bee Palace.
The Mississippi as a Basis for the Migratory System of Bee-Feeding and Honey-Making
New Orleans Picayune.

> Mr. C. O. Perrine, of Chicago, has fitted up two barges as a floating apiary. Each barge has a capacity and conveniences for 1,000 hives of bees. These will be towed up to Kennerville next week. They will start up the river with about 1,000 colonies on the two boats.

Mr. Perrine has been in Louisiana eighteen months studying up on the bee business and preparing for the grand onward movement for which he will be ready in a few days. His plan is to start with his bee palaces and his 1,000 colonies from Southern Louisiana when the honey flowers are in full bloom, to remain but a day or two at a landing, and move up each time to another landing and a fresh field. He thinks the bees of from 1,000 to 2,000 colonies will take the cream from the country around the landing from one to two miles distant in one or two days. In this manner he expects to move up the Mississippi to St. Paul, a distance of nearly 2,000 miles, where he will arrive about the last of July.

Returning, he will halt about two months somewhere above St. Louis, and will reach Louisiana with his palaces and bees in October. It will be his object to take the autumnal flowers at each point in their prime precisely as he takes the spring flowers in his advance up the river. He expects his early swarms on his boats to increase his colonies to 2,000 in April and May....

The space below deck is ten feet in width and about seven feet high, and is to be used for sleeping apartments, making and repairing hives, handling and extracting honey, and putting it in marketable shape. The dining-room and cooking will be on the steamer that tows the bee fleet.

To run the steamer and manage the barges and bees from fifteen to twenty hands will be needed. The cost of the whole establishment—barges, bees, steamer, and the complete outfit, will be not much short of $15,000.

Mr. Perrine has been engaged in the honey business in Chicago twelve or thirteen years, and has lately made it a special study. He has dealt largely in California honey. He expects to find the best market for his honey in Europe.

The *Chicago Tribune* followed up on the progress of Perrine's "Floating Apiary" on July 23, 1878 (p. 1, col. 5 & 6):

The floating bee-hive has now passed St. Louis on its way up.... Capt. Perrine purchased two barges of ordinary length and the little stern-wheel steamer James A. Fraser. The two barges were provided with shelving and 400 hives placed upon each. All varieties of honey-makers were introduced, as the enterprising Captain desired to make his experiment as broad and thorough as possible. The hives were painted in contrasting colors in order that the little workers could return to the proper hive, the colors aiding each in distinguishing his home by comparative location. The start was made on May 13, although April 1 was the day set. Capt. Perrine ran up to a point some forty miles south of Vicksburg and released the busy buzzing inmates of the hives.... Capt. Perrine waited patiently and was rewarded. The little ones began to come back, each with his sweet load, and by sundown the success of the experiment was assured. As was expected, a small loss of bees occurred every day, but this was more than made up by the numbers of vagrant bees, a sort of bee tramps, as it were, that joined the moving colony, and the captain has more hives now than when he first set out. The bees are making honey and the Captain is making money, and so all parties are satisfied.

That fall, at the fourth annual meeting of the *Western Illinois and Eastern Iowa Bee-Keepers' Society* in New Boston, IL, Perrine was asked to give a "word-picture" of the results of his floating-apiary experiment. His comments were published in the *American Bee Journal* in 1879:

What first induced me to go into it, was the want of white comb honey. I can get all I want of colored honey, but want hundreds of tons of white honey for my house.... I wanted to extend my business and did so in the Eastern States, afterward in Europe. I received some lots of very nice honey from California, and depended on them greatly for my supply of white honey, but it did not come. I had a big order from Europe, and had a great deal of trouble to fill it....

Then I resolved to try the floating apiary, and began to build two barges. I was kept from starting as early as I wanted to, fully six weeks, by a variety of causes beyond my control.... Owing to the lateness of the season, I concluded not to go far North and put my bees on shore about 60 miles above St. Louis. My bees are in good condition for wintering. The floating apiary is an experiment yet. I put about $12,000 in the venture, and I shall keep trying till I know whether it will succeed or not.... I would like to ask someone who knows, if bees notice color more than form. A great many bees get into the river, possibly 25 per cent....

I propose to put the bees on the boats this time in cold weather, then they will come out and fly a few at a time. I think bees return to their hives more by form than by color. I have tried different colors. We got but few new swarms, the honey did not come in fast enough. I had a little steamer that cost me $2,900, and sold it at a loss of $900. I thought of going up as far as St. Paul, but owing to difficulties could not do it . I propose to tow my bees only by night next year. [*American Bee Journal* 15(1) 1879: no pp#]

The minutes of this same 1879 beekeepers' society meeting include an enlightening discussion by Perrine of how his honey was packaged for shipment, sale to dealers, and "peddling direct to the consumer:"

> I prefer the common [*railway*] car for shipping honey. I would suggest that four to six shipping crates be packed in one bundle, flat; one alone is too light; they get jammed worse, and being handled by hackmen and others are not kept right side up.... I advise the shipper to load his honey into the car himself. Producer to dealer, and dealer to small dealer; we all want to do it right....
>
> Dealers prefer extracted honey in barrels. **I would recommend small sales at home in one and two pound bottles** [*our emphasis*].

The listed Perrine bottles appear to be examples of the one- and two-pound home-sales bottles. Perrine focused his attention on "home sales" of honey early during his business career in Chicago (ca. 1869–1875), and his embossed bottles may date to this time. However, he appears to have continued local retail-outlet sales at some level throughout his Chicago business tenure.

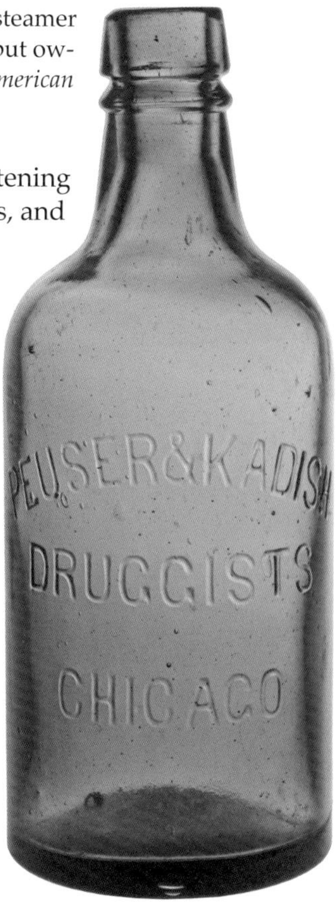

Otto Peuser and Charles J. Kadish

Mineral Water:

(1) **PEUSER & KADISH / DRUGGISTS / CHICAGO**   [blue teal]

»SB; Saratoga-style mineral water bottle; 7¾" tall. [*Figure 522*]

Figure 522

Otto Peuser and Charles Kadish were independent druggists in Chicago throughout the early and mid-1870s. According to the Chicago CDs, they first formed a partnership as the firm of *Peuser & Kadish*, druggists, at 193 Madison, in the 1874. Their partnership continued for just three years (1874–1876), after which Peuser is again a druggist on his own and Kadish is no longer listed in the Chicago directories.

Like other contemporary Chicago druggists, druggists (*see Buck & Rayner and Stephen Israel listings, this chapter*), *Peuser & Kadish* sold bottled mineral water at their retail pharmacy.

Bennett Pieters and Co.

Bitters:

(1) (arched): **BENNETT PIETERS & Co. / 149 S. WATER ST / CHICAGO**   (lavender "black" glass)
   Note: Sole example has base broken away, but was likely iron-pontiled. Bottle dates ca. 1859–60 (see later discussion).

»Probably IP; short-neck whiskey-style cylinder; double-ring collar; applied loop handle; 3¼" diameter, 7½" tall [*Figure 523*]

(2) **BENNETT PIETERS & CO. // 21 RIVER STREET // CHICAGO // //**   (amber)
   (base): **A & Co.**
   Note: Likely an early bottle for Red Jacket Bitters. The bottle dates 1864–66 (see later discussion).

»SB w/ keyed hinge mold; beveled-edge square bitters w/ flat panels; 9½" tall. [*Figure 524*]

Figure 523

(3) **BENNETT PIETERS & Co // 31 & 33 /**   (amber)
    MICHIGAN AVENUE // CHICAGO // //
    (base): **A & Co.**
    Note: *Likely an early bottle for Red Jacket Bitters. The bottle dates ca.1866.*

    »SB w/ keyed hinge mold; beveled-edge square bitters w/ flat panels; 9 1/2" tall. *[Figure 525]*

(4a) **RED JACKET / BITTERS // //**   (amber)
    BENNETT . PIETERS & Co. // //
    (base): **A & Co No 4**
    Note: *This bottle dates ca. 1866–68 (No photo. Example illustrated in 1993 Glass Works Auction catalog #29, Bottle #619).*

(4b) **RED JACKET / BITTERS // //**   (olive-green)
    BENNETT . PIETERS & Co. // //
    (base): **A & Co. No 5**
    Note: *This bottle dates ca. 1866–68.*

    »Both 4a & 4b = SB w/ keyed hinge mold; beveled-edge square bitters w/ flat panels; 9 1/2" tall. *[Figure 526]*

❖

Richard Cannon (2000:40–41) notes that on an example he has seen of bottle #4a the words "CELEBRATED STOMACH" appear to have been peened out preceding the word "BITTERS." If so, a style of this bottle likely exists (not documented in this listing) that slightly predates the embossed versions listed in #4a and #4b. Cannon's article also provides the following chronology for the Bennett Pieters company (Bennett Pieters was the first and last name of a single individual) and

Figure 524

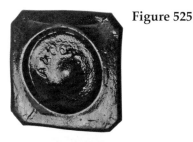

Figure 525

its related successor wholesale and retail liquor-dealer partnerships that continued to bottle and distribute *Red Jacket Bitters*.

1860: *Bennett Pieters & Co.*, 149 Water Street

1864: *Bennett Pieters & Co.*, 21 River Street

1866: *Bennett Pieters & Co.*, 31–33 Michigan Ave.

1869: *Schwab, Pieters & Co.* (*see later discussion*)

1870: *Schwab, McQuaid & Co.* replace BP & Co. and the one-year transitional Schwab, Pieters & Co. partnership and take over its *Red Jacket Bitters* brand.

The first owner and probable originator of Red Jacket Bitters was the Bennett Pieters Company.... They were listed in the 1861 city directory as Bennett Pieters, John F. Stafford and John D. Smedley at 149 Water St. Their bitters was patented in 1864. They apparently soon moved to 21 River Street.

The company moved to 31 and 33 Michigan Ave. in 1866, and was listed as distillers and wholesale liquor dealers.... Back on Water Street in 1866, Edward McQuaid, Charles H. Schwab and John B. Smith were operating a wine and wholesale liquor business of their own. They added partners Leon Monheimer and P. Cavanaugh in 1868, and merged with Bennett Pieters and Co. in 1869. The company was called Schwab, Pieters and Co. [Cannon 2000:41]

[KF/JW Note: Smedley was not a partner, though he may have worked there. BP & Co at 149 Water, from their first year in business in 1861 through 1863, was just Pieters and Stafford. The company moved to 21 River Street during 1864 and 1865.
The 1864 "patent" was likely a design-trademark patent for the paper label applied to the bottle. The annotation at the bottom of the label reads: "Entered according to act of Congress in the year 1864...in the Clerk's Office of the District Court of the United States in the Northern District of Illinois"—see Figure 527]

A history of the *Bennett Pieters & Co.* and its production and distribution of the Red Jacket Bitters brand is also given in an 1866 publication titled *Chicago As It Is. A Strangers' and Tourists' Guide to the City of Chicago. Containing Reminiscences of Chicago in the Early Days; an Account of the Rise and Progress of the City; Description of Public Buildings, Churches, Schools, and Objects of Interest; Etc.* (Religious Philosophy Publishing Association 1866):

> This firm have long and successfully pursued the manufacture of the Red Jacket Bitters, and their business expanding beyond the limits of their former building, they have erected and now occupy a store of palatial proportions.... The building of solid stone and brick, five stories high with basement, and one hundred feet in depth by fifty feet front. The front is of white marble, finely sculptured and embellished. This is used by the store. The upper stories are devoted to the manufacture of the renowned Red Jacket Bitters.

The 1867 Chicago CD (p. 7) highlighted *Bennett Pieters & Co.* as one of 52 companies that personified "the progress which Chicago has made to be deemed a first class manufacturing city. The amount of enterprise and wealth cannot better be recorded...than to insert a list of...the leading firms in their departments of trade, as made to the United States Assessor for this district." *Bennett Pieters & Co.* was valued in 1867 at $1,450,806.

Figure 526

But soon afterwards, the *Red Jacket Bitters* partnerships began to change. A possible reason for the shift in management was suggested by an article in the *Chicago Tribune* in June 1872:

REDUCED TO THE RANKS.
**The Inventor and Proprietor of the Once Famous
"Red Jacket Bitters" Enlists as a Private Soldier
in the United States Army.**

From 1858 to 1867 the name of Bennett Pieters was very well known in this city. It was connected with the Red Jacket Bitters, which were then a popular beverage; advertised largely in the papers, and posted on fences and blank walls *[KF/JW Note: Pieters included a portrait of himself in an advertising strip used on the liquor tax stamp applied to the corked lip and neck of the Red Jacket bottles: see Figure 528]*. Pieters himself was a popular man among his set; he kept good saddle and carriage horses; had rooms at the Sherman House; gave excellent suppers, where the guests were not called on to drink his bitters, but were regaled with champaigne, which was, perhaps, no better, but was certainly much more expensive; and did not openly offend any of the proprieties. His income was a large one, but he spent it easily, and did not, like a good Chicagoan, put it in real estate. He failed to lay up treasure in this orthodox quarter, and hence when calamity came upon him he had nothing to fall back upon. **In 1867 some individual, thinking that he also might make a little by going into the bitter business, began the manufacture of the Red Cloud Bitters. Pieters resented this, as an infringement of his rights, and brought suit to have his rival enjoined. The matter was tried in the United States court. An expert chemist analyzed Pieters' productions, and found they consisted of poor whiskey, flavored with tanay, dogfennel, jimson weed, or some other substance, which possessed no special medicinal virtues. Ultimately the Court decided that his rights had not been invaded, and that Red Cloud and Red Jacket might have separately and independently originated bitters.** This hurt Pieters, but not half as much as the analysis. People had supposed that his bitters owed their rare virtues to simples unknown to the white man, gathered at midnight in primeval forests, by grim Indian chiefs; or dusky Indian girls, dark, but comely, like the tents of Kedar, or the curtains of King Solomon, the secret having been specially communicated to Pieters under circumstances of so private a nature that they never became public. **When the illusion was dispelled, the sale of the Red Jacket stopped**, and Pieters' income diminished. The less money he had the more he drank. He got involved in trouble with his partner *[Charles H. Schwab, see later discussion]*, and finally, in 1869, after the defeat of General Solomon, and his departure for Washington Territory, he accompanied him, along with O'Brien, Hayden, Church, etc. He returned from there a few months ago, and, after loitering around the city doing nothing, last Monday he enlisted as a cavalry recruit in the regular army, and embraced this last desperate resort of a reduced gentleman. [*Chicago Tribune*, June 15, 1872, p. 6, col. 6]

If the details of this newspaper account are accurate, the Red Jacket vs. Red Cloud court case *(not yet located)* may be the earliest recorded chemical test of the hogswallop contents of many early patent-medicine products—later (at the turn of the century) pursued by the government to such effect through the newly formed Food & Drug Administration.

Following the one-year (1869) transitional partnership called *Schwab, Pieters & Co.*, mentioned in the Cannon quote (*see Figure 529*), *Schwab, McQuaid & Co.* was in business from 1870 through 1875. The company then became *McQuaid & Monheimer Bros.* from 1876 through 1879. The Monheimer Bros. (Leonard and Levi) ran the business during 1880, after which Isadore Monheimer and John S. Cook were proprietors of the liquor business from 1881 through 1886. Isadore Monheimer took over the wholesale liquor business after 1886, as *Monheimer & Co.*

Strangely enough, Monheimer & Co. once again produced an embossed *Red Jacket Bitters* bottle. Either they reintroduced the brand for sale or the intervening (1876–1886) company partnerships bottled *Red Jacket Bitters* in paper-labeled bottles only. The Monheimer & Co. incarnation of the embossed *Red Jacket Bitters* bottle was produced much too late for formal inclusion in the present volume.

See the Schwab McQuaid & Co. listing for details of their embossed *Red Jacket Bitters* bottles. See also the "T. Pieters & Co." listing. This bottle/company appears to be a partially altered misprint (or bootleg bottle) using the mold for Bennett Pieters bottle #3 in the listing.

The "**A & Co.**" base mark on bottles #4a and #4b is the mark of John Agnew & Co. (1854–1870), a glass bottle manufacturer in Pittsburgh, PA.

Figure 527

Figure 528

Figure 529

As was usually the case with alcoholic bitters products, Pieters *Red Jacket Bitters* was marketed essentially as a patent medicine, or health drink. This is well-illustrated by a detailed early ad for the product that appeared in May, 1864, in the *Woodstock Sentinel* (Woodstock is about 40 miles NW of downtown Chicago):

These Bitters are prepared, in pure Bourbon Whisky, from a combination of over twenty different kinds of roots, barks and herbs, which act in perfect concert one with the other, prepared from the original formula given by the great chief, Red Jacket, to Dr. Chapin, who used them successfully in his practice for many years, and by their use gained so great a popularity in the4 treatment and cure of Dyspepsia, Liver Complaint, Constipation, Sick and Nervous Headache, Fever and Ague, and all diseases arising from torpid liver or indigestion. Persons suffering from either of these loathsome diseases will find a sure cure by the use of these Bitters, which are perfectly pure and free from all those drugs and poisons usually put up in such preparations and palmed off on an unsuspecting public. A single trial will convince the most skeptical that in the RED JACKET there is virtue which no other Bitters possess.

They strengthen and invigorate the system.
They are unequaled for general Debility.
They are a sure cure for dyspepsia.
They give a good and healthy appetite.
They assist digestion.
They are the best stimulant inexistence.
They are a preventative of Fever and Ague.
They relieve constipation.
They cure Nervous Headache.
They are perfectly pure and palatable.

Aged persons and delicate females will find they can save large doctor's bills by the use of these Bitters.

Beware of counterfeits. The Red Jacket Bitters are only sold in bottles with our names blown on the side.

"T. Pieters & Co."

Bitters:

(1) **PATENTED / BITTERS // T. PIETERS & CO // 31 & 33 / MICHIGAN AVENUE // CHICAGO**   *(amber)*

»SB; beveled-edge square bitters; tooled tapered collar; 9 1/2" tall *[No photo]*.

There is no "T. Pieters" company. The only known bottle with the embossing listed above is in the collections of the Nebraska Historical Society. The embossed name on their bottle is likely a flawed strike. The name should probably read: BENNE**TT**. **PIETERS & CO.**, proprietors of the *Red Jacket Bitters* brand from ca 1859 to 1869 (*see Bennett Pieters listing, this chapter*).

This firm sold their rights to the brand and dissolved the partnership in 1869. Their last business address was 31 & 33 Michigan Avenue. The *Patented Bitters* appears to have been a short-lived brand name or perhaps was part of a different mold referring to *Red Jacket Patented Bitters*.

Charles H. Plautz

Bitters:

(1) **HUMBOLDT'S / GERMAN BITTERS //**   [amber]
   C. H. PLAUTZ / PROPRIETOR / CHICAGO
   (base): **L & W**

   »SB w/ post mold; beveled-edge square bitters; tapered-collar lip; 8 3/4" tall. [*Figure 530*]

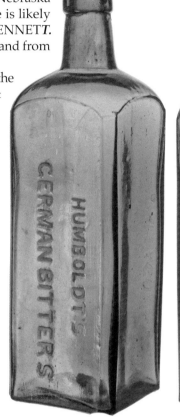

Figure 530

———❖———

Charles Plautz first appears in the Chicago CDs as a clerk for druggist Charles Wuensche at 165 Milwaukee during 1865 and 1866. By 1867 he was listed as a "druggist" at the pharmacy, while Wuensche himself was listed as a "chemist." During 1868 and 1869, Plautz ran the pharmacy, and Wuensche was listed as a clerk. After that, the pharmacy seems to have closed, and neither man is listed in the 1870 Chicago CD.

Charles Plautz reappears as a "druggist" operating from his residence at 513 Milwaukee during 1871–73. Then, beginning in 1874 and 1875, he was a junior partner with Charles W. Grassly in a pharmacy business at 287 12th Street. Plautz again disappears from the directories in 1876 before reappearing finally as an employee with the *August Burgwedel & Co.* druggist firm in 1877. See Wagner (1980:9) for a discussion of Plautz' subsequent business career in Chicago.

The *Lorenz & Wightman* Pittsburgh glassmaker's mark on the bottle listed dates its manufacture to the years *L&W* was in business: 1862–74. Since Charles Plautz operated as a druggist on his own in Chicago from 1868 to 1873, the bottle was likely produced sometime during this six-year period. Ring (1980), indicates that the earliest wholesale drug catalog in which she found the bitters listed was distributed in 1872.

No advertising for his *Humboldt's German Bitters* has yet been located, so we do not know what ills Mr. Plautz may have claimed it cured (*but see Pieters Red Jacket Bitters discussion, this chapter*).

George Powell and Co.

Bitters:

(1) **WALLACE'S TONIC / STOMACH BITTERS // // GEO. POWELL & C**O **/ CHICAGO ILL. // //**   [amber]

   »SB w/ post mold; beveled-edge square bitters; tapered-collar lip; 9" tall. [*Figure 531*]

(2) **WALLACE'S. TONIC / STOMACH. BITTERS // // GEO POWELL & C**O **/ CHICAGO ILL_ // //**   [amber]
   (base): **L & W**

   »SB w/ post mold; beveled-edge square bitters; tapered-collar lip; 9" tall. [*Figure 532*]

Figure 531

(3) **WALLACE'S. TONIC / STOMACH. BITTERS** [amber]
(base): **L & W**
Note: The Powell & Co. name and address has been piened out on the reverse of this bottle.

»SB w/ post mold; beveled-edge square bitters; tapered-collar lip; 9 1/8" tall. [Figure 533]

Figure 532

George Powell & Co. was a patent medicine company located at 141 Kinzie in Chicago for just two years: 1869 and 1870. The partners in the company were listed as George Powell and John C. West (both Chicago residents), and J. H. Wallace, probably the bitters' namesake formulator, a resident of Valparaiso, IN. The *Geo. Powell & Co* letterhead noted that they were proprietors and wholesale dealers in "West's and Wallace's Family Medicines" and manufacturers of "Wallace's Tonic Stomach Bitters" and "West's Liquid Bluing, Hair Oils, Colognes, &c."

The company was not listed in Chicago CD in 1871, but from 1872 to 1875 George Powell is listed in Chicago as president of the *Co-Operative Medicine Company* (no other principals listed), from his residence in Valparaiso, IN. He apparently moved to Chicago in 1875 and after that was a general commission forwarding merchant there.

Probably bottles #1 was made and used during the two *Powell & Co.* years (1869–70), and bottles #2 and #3 were made and used during the first few years of the *Co-Operative Medicine Company* presence in Chicago (ca. 1872–74). Pittsburgh glassmaker's *Lorenz & Wightman* (**L&W**) closed their doors in 1874.

Anther (probably later) variety of bottle #3 has been recorded that is base-embossed **ZKH & Co.** (Ring 1980:80). This may be a post-Powell/Chicago for a subsequent proprietor of the brand. However, Lindsay et al. (2009:54) strongly suggest the mark is a corruption of KH&GZO (Kearns, Herdman & Gorsuch, Zanesville, Ohio)—see, e.g., the Kuehle soda listing in the Murphysboro listing. The authors conclude that the maker's mark "may have been misreported to (or mis-recorded by) Ring, or it may have been [an] engraving error." If so, the bottle dates to the late 1870s or early 1880s.

Wallace's Bitters was still being advertised in the Keokuk *Daily Gate City* for March 6, 1875 (Ring 1980:480). That same year, in the Madison (WIS) *Democrat* and *State Journal*, *Wallace's Tonic Stomach Bitters* was described as providing a "perfect eradication of all Bilious Diseases arising from a foul stomach, strengthening the system...operating on the Liver and Kidneys; purifying the Blood, strengthening the Nerves, and a sure cure for Dyspepsia and Indigestion."

Pratt & Butcher Magic Oil

See the Walker & Taylor listing, this chapter.

Red Cloud Bitters

See the Taylor & Wright listing, this chapter.

Red Jacket Bitters

See Bennett Pieters & Co., Schwab, McQuaid & Co., and Monheimer & Co. listings, this chapter.

Henry O. Redlich

Ink:

(1) **H. REDLICH // CHICAGO**   [aqua]

»SB w/ keyed hinge mold; short shouldered cylinder; flared lip; eight panels; 2 1/2" tall. [Figure 534]

Figure 533

❖

Henry Redlich first appears in the Chicago CDs in 1861 and 1862 as a retail druggist and chemist from his residence at 220 Wolcott. There is no mention of his having produced ink during this time. But he must have shifted professions later in 1862. The 1862 IRS assessment of Chicago lists Redlich as a manufacturer of faucets and ink. The 1863 Chicago CD also reflects this career change. It lists Redlich as a "manufacturer of India rubber saturated faucets, and writing fluid" at his 45 Greenbay residence. (In 1863 the IRS assessment misses the ink-manufacture part of his efforts and lists him only as a manufacturer of faucets.) The 1864 Chicago CD lists him as a manufacturer of "writing fluid" and "cork-lined faucets" at the same address. In 1865, the CD gives the same address and indicates he is an "apothecary and manufacturer of writing fluid" there. In 1866, he is listed simply as an "ink manufacturer" from his residence at 371 N. Clark.

In 1867, Henry Redlich makes another unexpected career shift and becomes a "glass manufacturer" from a glassworks at 373, 375, and 377 N. Clark (No. 375 was his residence). However, his glassworks is only listed for that one year, and he does not appear the Chicago CDs at all from 1868 on.

Redlich appears to have manufactured his "writing fluid" only during 1862–66. The listed bottle would have been manufactured and used during this five-year period.

Figure 534

Josiah H. Reed and Co.

Liquor:

   (1) **J. H. REED & CO. / CHICAGO** [aqua]

   »SB w/ keyed hinge mold; shouldered oval; tapered collar; 11" tall. *[Figure 535]*

--- ❖ ---

The druggist firm of *Stebbins & Reed* came to Chicago from Albany, NY, and opened their first drugstore in Chicago in 1845: "In 1848, having a fine opportunity to go into business in New York, Mr. Stebbins sold his interest in the Chicago firm to Thomas M. Hunt and John Olmsted.... The new firm began business April 1, 1848, under the firm name of J. H. Reed & Co.... In 1853 Messrs. Hunt and Olmsted sold their interest in the firm to Mr. Horace A. Hurlbut who became the company part of the concern" (Ebert 1905:246). *J. H. Reed & Co.*, druggists, advertised that they sold medicines and liquors in the *Chicago Daily Journal* during 1849. The firm was then located at 159 Lake Street. They moved to 144 Lake Street late in 1851:

Splendid Store

Messrs. J. H. Reed & Co., druggists, have removed their business from their old stand at 159 Lake street to the spacious and handsome block at 144 Lake street, erected this season by Mr. J. Price. This store is fitted up in the most magnificent style with marble Mosaic floors, Italian marble counters, etc., while tastefully arranged around are statuary, vases, urns, etc. In fact the fitting up is not excelled by that of any similar establishment in the country.

Messrs. J. H. Reed & Co. intend doing an exclusively prescription business at their new store, the back and upper portions of the building alone being used for the wholesale trade. [*Chicago Daily Democrat*, 10/29/1851]

Chicago Tribune ads for the company in 1856 listed *J. H. Reed & Co.* as forwarding and wholesale agents for national-brand patent medicines located at 144 and 146 Lake Street. An 1859 *Chicago Tribune* ad lists the company as "Wholesale Druggists" and "Jobbers in all descriptions in all descriptions of Manufacturers' and Soap Makers' Goods," still at the 144 and 146 Lake.

Figure 535

Figure 536

During the Civil War, Mr. Reed moved back to New York and became the resident buyer there for *J. H. Reed & Co*. "The firm had large government contracts during the war for drugs and other medicinal supplies.... At one time over sixty men were at work on government orders in their store. Orders for 60,000 or 70,000 pills at short notice were nothing unusual" (Ebert 1905:247, 248). The company is listed in the 1862 IRS assessment of Chicago as wholesale liquor dealers. In an advertisement in the May 1, 1863, *Chicago Tribune* characterized the *J. H. Reed & Co*. as "Importers and Jobbers of Drugs and Chemicals" at 146 Lake Street. In 1868, Josiah Reed sold his interest in the company and the firm name was changed to *Hurlbutt & Edsall*.

From the style and technological attributes of the listed embossed bottle, it should date to the 1860–68 period. During this period, Chicago CD ads for J. H. Reed & Co.—including a full-page 1864 ad with a woodcut illustration of their business headquarters (*Figure 536*)—never mentioned their selling alcoholic beverages of any kind. They were characteristically listed as wholesale and retail dealers in drugs, druggists' glassware, chemicals, paints, oils, dyestuffs, fancy and toilet goods, perfumes, dental and surgical instruments, dry goods, and manufacturers' goods. Their building frieze in the 1864 full-page ad lists "Paints, Oils, Glass, Brushes, Dyes, Soda Ash, Caustic Soda, Rosin, Varnish &c."

Only the 1862 IRS assessment lists them as wholesale liquor dealers. Under the circumstances, it as possible that the embossed bottle listed above contained drugs, oils, or chemicals rather than an alcoholic beverage.

Louis Rodemeyer and Co.

Soda ~ Mineral Water:

(1) **L. RODEMEYER & Co / PREMIUM / MINERAL WATER / CHICAGO ILL**   [green]
   Note: This bottle mold, likely made by a Pittsburgh glasshouse, has on the reverse, ghost letters from a previous merchant's name, incompletely peened-out: M. MONJU & CO. / MOBILE.

   »IP; [rarely:] and SB w/ keyed hinge mold; shouldered-blob-top soda. [Figure 537]

Figure 537

Louis Rodemeyer was listed in the 1858 Chicago CD as a junior partner with his brother Henry (*H. F. L. Rodemeyer & Bro.*) as brewers from their residence at 368 Ohio. He was also listed in the 1860 and 1870 USC as a Chicago saloonkeeper. The soda bottles listed were made from the same mold as Francis Schonwald's bottles (*see listing, this chapter*).

Schonwald's bottles are pontilled, and according to the Chicago CDs were used ca. 1859–60. Nearly all Rodemeyer bottles are smooth-based, although a few pontilled examples are known, so Rodemeyer likely took over the Schonwald business briefly in 1860. His bottling effort was short-lived enough that it was not recorded in the Chicago CDs (although he was listed briefly as a confectioner at about this period, so perhaps his mineral water was bottled as part of that abortive business effort).

Bottles used by both Rodemeyer and Schonwald are the same style, have nearly identical embossing, and share the same reverse-side mold with Monju's name peened-out.

Paul Rouze

Soda ~ Mineral Water:

(1) **PAUL ROUZE** / (*5-pointed-star emblem*) / **CALIFORNIA. POP** //   [aqua]
   (heel): **C & I**
   (base): **P.R.**
   Note: The "Z" in Rouze is backwards.

   »SB w/ post mold; fluted lower body below embossing (10-sided); sloped shoulders w/ long tapered neck; blob top; 7 1/2" tall. [Figure 538]

(2) **PAUL ROUZE** / **CHICAGO** / [aqua]
(heel): **C&I**
(base): **P.R.**

»SB w/ post mold; tall, high-shouldered cylinder; shouldered-blob-top soda; 7" tall. [*Figure 539*]

(3) **PAUL ROUZE** / (5 diamonds in a star shape) / **CHICAGO. ILL** [aqua]
Note: Recorded from R.K. drawing only. Bottle not seen.

»SB; tapered heel; cone shaped body below neck,; oval-blob-top soda; ca. 7" tall. [*Figure 540*]

According to Chicago CDs, *Paul Rouze & Co.* were in the wholesale and retail fruit business between 1873 and 1875, in the first two years at 24 N. Canal and in the final year at 149 S. Water (the two addresses were also Rouze's residence). Rouze does not appear in the Chicago CDs prior to 1873. The "*& Co.*" business partners in 1873 were John Bell and Philip Dazet. In 1874, Bell had retired from the business and Angelo "Matheux" was hired as a clerk. In 1875, Angelo "Mattei" had become the new third partner. In 1876, the company became Angelo Mattei & Co. (with Philip Dazet), and Rouze had left the business. Although there is no hint from the directories that Rouze & Co. bottled fruit-flavored soda drinks as part of their business, it is interesting to note that according to an 1873 newspaper sales ad, soda water apparatus had been of interest to Rouze:

FOR SALE—A COMPLETE SODA WATER APPARATUS; 3 fountains of 14 gallons each, and a No. 1 copper generator, nearly new, to be sold for less than half cost price. Address at No. 24 North Canal-st., Chicago.
P. ROUZE & CO. [*Chicago Tribune 4-27-1873, p. 14, col. 5*]

During 1876 and 1877, Paul Rouze started and ran a new wholesale wine and liquor business from his residence at 157 (later 161) Kinzie. He placed an ad for his new business in the June 17, 1877 *Chicago Tribune* (p. 1, col. 4):

Figure 538

**CALIFORNIA WINES
AND BRANDIES.**
Port, Angelica, Clarets, Muscatel, etc., for wholesale and jobbing trade, just received at warehouse 157 Kinzie –st. Vinyard and distillery Paulinville, Cal. PAUL ROUZE, Grower and Dealer.

But his wine business lasted only for two years, and Rouze no longer appears in the Chicago CDs after 1877.

The embossed glass bottles used by Rouze were likely containers for carbonated (perhaps fermented) fruit drinks sold by the firm, including "California Pop." H. C. Dorman, S. Geer, and John Lomax, were also Chicago "California Pop" and "California Pop Beer" distributors (*see their listings*).

Figure 539

Figure 540

California Pop was patented by Charles C. Haley of Troy, New York, on October 29, 1872 (U.S. Patent #132,574). It was similar to spruce beer and other popular "small beers" of the nineteenth century, but combined oil of spruce, sassafras, and wintergreen with ginger root, hops, malt, and "pure spirits." Most of the California Pop marketed in Chicago was produced toward the end of the 1870s. If Rouze bottled his California Pop in the 1873–1875 period (perhaps 1876–77), it may have been the first appearance of this beverage in Chicago.

Dr. F. A. Sabine's Harvest Bitters

Note: This bottle is not included in the present volume as it was produced just after 1880. F. A. Sabine and Albert J. Lemke formed a short-lived patent-medicine company—the Sabine Medicine Co.—at 1549 Wabash Avenue for a short time in 1883–84. Prior to relocating to Chicago, F. A. Sabine operated a patent-medicine business in St. Louis during the late 1870s and early 1880s. His St. Louis products included **DR. F. A. SABINE'S WORLD'S REMEDY** and **DR. HOFFMAN'S RED DROPS (F. A. SABINE, PRo.)**. The company also sold Dr. F. A. Sabine's Carbolic Ointment in small, embossed tins.

Figure 541

Sanford Manufacturing Company

Ink:

(1) **SANFORD'S INK**   *[aqua]*
(base): **7**

»SB w/ post mold; round; flared lip (angled up); 2 1/2" tall. *[Figure 541]*

❖

The Sanford Manufacturing Co., makers of ink and mucilage, was first formed in Massachusetts in 1857 and relocated to Chicago just after the close of the Civil War:

> In 1857, two men named Frederick Redington and William Sanford Jr., created in Massachusetts the **Sanford Manufacturing Company**. In 1866 the two of them decided to move base to Chicago, Illinois, but were then hit by the "Great Chicago Fire" *[in October, 1871]*. This fire destroyed the **Sanford Manufacturing Company** and the location was burned to bits. Redington and

Sanford Jr. stuck to their guns though and rebuilt, except in 1947 they were asked to move for a new expressway needed to be built. Since then, the **Sanford Ink Corporation**, as it is known nowadays, has remained in Bellwood, Illinois.... Redington and Sanford Jr. probably had no idea that today their sales were almost a billion dollars last year. ["*A Little History on the Company*," from the Sanford Ink website]

According to Chicago CDs, Redington and Sanford did come to Chicago in 1866 to establish the company there, but only Redington stayed beyond that first year to run the new business. According to a 1977 article in *Midwest Bottled News* [MBN 8(10):19–20], Redington had purchased the business from the Sanford brothers several years earlier:

> It was a long time ago—in 1857—when a professor of Latin and Greek, looking for an occupation that would provide a better income with which to support his wife and five children, called on two men he knew in Worcester, Massachusetts. They were the Sanford brothers. They had been making a superior writing ink for some years and had enjoyed a successful sale of this product in the Atlantic States. Frederick W. Redington, the professor, was well acquainted with their ink for it had been used in the Academy at Fredonia, New York, where he taught.
> Mr. Redington bought the business from the Sanford brothers in 1857. His purchase included the ink formula and one for mucilage, which they had developed under the trade name of *Royal Crown*. For nine years, the professor-turned-businessman continued to work out of Worcester then the lure of greater markets moved him to change his base of operations to CHICAGO. [MBN 1977:19]

According to the 1867 Chicago CD, the Sanford Company produced "writing fluid, inks, carmine, mucilage, etc." Redington began operations at 178 and 180 S. Water Street, by 1869 had expanded to 178–182 S. Water, and by 1870 (just before the fire) he further expanded to 178–184 S. Water. The company's first small ad in the *Chicago Tribune* appeared June 8 of that year (p. 1, col. 6):

> **INK.**
> **SANFORD MANUFACTURING cO**
> Sanford's Mucilage is made of Pure Gum.
> His Carmine Ink is brilliant.

After the tragedy of the Great Fire in October, 1871, Redington temporarily relocated to #12 5th Avenue and announced in the CD he was producing "writing and copying ink and sealing wax." By 1872, he had removed to their new headquarters at 223 Fulton. This was to be the new *Sanford Manufacturing Co.* headquarters throughout the 1870s and beyond, but the new start-up was not without its glitches:

> OTHER FIRES.
> The alarm from Box 274, at a quarter past 8 o'clock yesterday morning, was for a fire in the two-story and basement frame house, No. 223 Fulton street, owned by F. A. Eddington [*sic*: Redington], and occupied by the Sanford Ink Manufacturing Company. The fire was caused by the boiling-over of a kettle of borax on the third floor. The building was damaged to the extent of about $1,500; the loss on stock amounts to about $3,500. [*Chicago Tribune*, October 22, 1872, p. 8, col. 3]

The Sanford Company has had a long and prosperous history in Chicago, but during the 1860s and 1870s they apparently almost never embossed their glass ink and mucilage containers, preferring paper labels instead. The listed bottle is the only one we have been able to locate which, from its technological and stylistic attributes, may predate 1880.

Abraham W. Sargent and Co.

Medicine:

(1) (*side*): **A. W. SARGENT & Co** //   [aqua]
    (*side*): **CHICAGO**

»OP; rectangular; 4 sunken panels; rolled lip; 4 1/2" tall. [*Figure 542*]

(2) (side): [A. W. SA]**RGENT & Co** // **PAIN** / [E]**XTERMINA**[TOR] //   [aqua]
(side): [CHICAGO]
Note: No photo. Recorded from bottle fragments collected by John Wilson (bold-face portions of lettering present on sherds).

»OP; rectangular; 4 sunken panels; rolled lip; ca. 4 1/4" tall.

Figure 542

Abraham W. Sargent appears in the Chicago CDs from 1859 through 1863 as a "patent medicine" man—from residences that change every year: first 168 Madison; then 48 Quincy; then 215 State; then 94 Washington; finally 7 Clark. The open-pontil bottles listed stylistically date to the late 1850s. A large Chicago newspaper ad for Sargent's *Pain Exterminator* appeared in the May 4, 1862, *Chicago Tribune* (two-week run):

SARGENT'S
PAIN EXTERMINATOR.

It has become a necessity, a POSITIVE NECESSITY for the Proprietors of the above Invaluable Remedy to extend their facilities for preparing it.

THE DEMAND FOR IT

Has increased so rapidly for the past few months that we have been entirely unable to supply orders with our present means of manufacturing.

For this reason as well as for the humane object of giving all afflicted ones an opportunity of availing themselves of the benefit of its HEALING VIRTUES, we shall, on the FIRST DAY OF MAY, open an extensive Laboratory and Manufactory at
No. 94 Washington Street.

Our facilities will be such that we can manufacture a superior quality of our Remedy, and be able to fill all our orders promptly.

THIS EXTERMINATOR is recommended for only what IT CAN EFFECTIVELY CURE, namely:

Diphtheria, Rheumatism, Ague in the Face, Neuralgia, Toothache, Headache, Sore Throat, Fresh Wounds, Scalds, Burns, Sprains, Chilblains, Dysentery, Diarrhea, Earache, Corns, &c., &c.

All these it will POSITIVELY and RADICALLY cure. We shall hereafter present some few of the numerous testimonials we have received relating the action of its
Healing Properties.

At this time we invite all our former patrons, and all who are afflicted with either of the above complaints, to call at our office and Manufactory.

THE EXTERMINATOR can be had by application to us, or at all the leading Druggists and Apothecaries in the country.

We respectfully ask all to give it at least one trial, and ascertain for themselves its sterling value as a pain reliever.

A. W. SARGENT & CO.,
94 Washington Street, Chicago

Post Office Box 3746. Price 25 cents, 50 cents, and one dollar per bottle. Liberal Discount made to the Trade.

Note that according to this 1862 ad, *Pain Exterminator* could be purchased from *A. W. Sargent & Co.* at 94 Washington Street in three different bottle sizes. The pontilled bottles listed should date to the 1859–60 period. If the company was not particularly successful, these bottles may have continued in use during the early 1860s, but more likely smooth-based bottles were in use during the last few years. Embossed examples may show up, or they may have been unembossed (paper label only).

It should be noted that there was another prominent Sargent druggist firm in Chicago during these years, not related to the production of this patent medicine. Ezekiel H. Sargent and John C. Ilsley were partners in the late 1850s and 1860. Then, during the early 1860s, E. H. Sargent began his own druggist firm (Ebert 1905:249).

Lewis Sass and William Hafner

Ale:

(1) **SASS & HAFNER // S & H / CHICAGO /** [amber, olive-amber]
(*heel*): **A. & D. H. C.**

»SB w/ keyed hinge mold; 2-piece mold squat quart ale; ring-neck tapered collar; 9" tall. [*Figure 543*]

(2) **SASS & HAFNER // S & H / CHICAGO /** [amber]
(*heel*): **C & I**

»SB w/ post mold; 2-piece mold squat quart ale; ring-neck tapered collar; 9 1/2" tall. [*Figure 544*]

Soda ~ Mineral Water:

(1a) **SASS & HAFNER / CHICAGO //** [aqua]
(*heel*): **C & I**
(*base*): **lg. double-embossed oval "O"**

»SB; short neck oval-blob-top soda. [*Figure 545*]

(1b) **SASS & HAFNER / CHICAGO //** [aqua]
(*heel*): **A. & D. H. C**
(*base*): **lg. double-embossed oval "O"**

»SB; short neck oval-blob-top soda. [*No photo; closely similar to C&I variety above: see Figure 545*]

(2) **SASS & HAFNER / TRADE MARK / S & H / CHICAGO ILL /** [aqua]
(*heel*): **C&I**
(*base*): **lg. double-embossed oval "O"**

»SB w/ post mold; short neck oval-blob-top soda. [*Figure 546*]

(3) **S & H / CHICAGO //** [amber]
(*heel*): **A & D. H. C.**

»SB; ring-neck cider finish quart; 11 1/4" tall. [*Figure 547*]

(4) **SASS & HAFNER / CHICAGO** [aqua]
(*base*): **A. & D. H. C**

»SB; high-shoulder Chicago-style quart soda; shouldered-blob top; 9 1/4" tall. [*Figure 548*]

(5) **S & H / CHICAGO //** [amber]
(*heel*): **C & I**

»SB; slope-shoulder oval-blob-top quart; 11" tall. [*Figure 549*]

❖

According to the Chicago CDs, Lewis Sass and William Hafner first formed their bottling partnership in 1866. That first year they are listed in the summary for mineral water bottlers, and their business is characterized as "manufacturing

Figure 543

Figure 544

Figure 545

Figure 546

Figure 547

Figure 548 **Figure 549** **Figure 550**

and mineral water" at 112 W. Lake. Interestingly, although Sass was not included in the 1865 CD, William Hafner was listed in the directory as a "spruce beer maker." It is interesting to note that Civil War–era, sided, Merrill-patent quart stoneware bottles are known that are shoulder-stamped **SASS & HAFNER** (*Figure 550*). This may indicate that very early in their partnership (ca. 1865–66) Sass & Hafner's "manufacturing" activities included stoneware-bottled spruce beer.

Beginning with the 1867 Chicago CD, *Sass & Haffner* are listed as "soda water manufacturers" at 108 W. Lake. Charles Schriber, a soda water "peddler" for them in 1868, took over the Sass & Hafner bottling operation as part of *Schriber & Co.* during the 1869 and 70 seasons, but his focus was on soda fountain supplies so Sass & Hafner continued to bottle soda water under their own names during those years (*see Schriber & Joss listing, this chapter*). By 1871, *Schrieber & Co.* was gone (Schriber was bottling with John Joss at another address), but *Sass & Hafner* were still bottling at 108 W. Lake. They remained at the same address through the 1873 season, and were always listed in the CDs just as soda water manufacturers during this period.

Perhaps notably, their CD listings change during 1874 and 1875. During these two years they are at the same address, but they are listed only as bottlers of ale and porter. So perhaps the two embossed ale bottle styles listed date from this time.

Beginning in 1876, they relocated their bottling works to 41 N. Peoria, and from then until the two partners split up after the 1881 season, they were again listed as soda water manufacturers. The 1880 IL Industrial Census indicated that *Sass & Haffner* were in business full-time 12 months a year and listed their estimated annual value of bottled products at $52,000.

In 1882, Hafner moved on to join the *Hartt Manufacturing Co.* (makers of soda and mineral water apparatus) and during 1882 Louis Sass bottled soda water alone at 41 N. Peoria. An amber tall quart bottle is known from this season that is embossed **L.H. SASS / CHICAGO, ILLS.** (*Figure 551*). During the Hutchinson-style bottling era (and beyond the scope of the present study), Sass took on a new partner, and *Sass & Pomy* were bottlers of soda water at 41 N. Peoria from 1883 onward.

Sidney Sawyer

Medicine:

(1) **SAWYER'S / FLUID EXTRACT / OF BARK //// //// ////** [teal, aqua]

»OP; beveled-edge rectangle; double-ring collar; 5¾" tall. Embossed on one side only. [*Figure 552*]

❖

Sidney Sawyer was one of the earliest pioneer Chicago businessmen. An 1856 Chicago CD mentions that he and his son Nathanial had come to Chicago from New York 15 years previously (ca. 1841). Sawyer was listed in the first three (1843–45) Chicago CDs as a merchant selling "drugs, medicines and groceries" at 124 Lake Street. He placed an ad for his merchandise in the 1844 Chicago CD (p. 87):

**S. Sawyer,
Wholesale and Retail Dealer in
Drugs, Paints, Oils, Dyestuffs, Glass,
Medicines, Chemicals, Perfumery, & Groceries,
124 LAKE ST. Two Doors from Clark St.
CHICAGO, ILL.**

The Chicago CD ads he placed in 1847 and 1849 listed a similarly diversified stock, at the same address, but by that time "DRUGS & MEDICINES" were listed first, in much larger and bolder lettering than his other products. His earliest newspaper ads, in the April 1849 *Chicago Daily Journal* (4/18/49, p. 4, col., 3), listed the same array of products and, for the first time, carried separate ads for *Sawyer's Fluid Extract of Bark*:

Figure 551

**SAWYER'S
FLUID EXTRACT OF BARK**
AN INFALLIBLE REMEDY FOR THE
FEVER AND AGUE,
And the various forms of Billious Diseases, Intermittent, Remittent, Chill Fever and Dumb Ague, having their origin in *malaria* of very fertile or marshy districts, prepared and sold by
Dr. S. SAWYER
Druggist and Chemist,
No. 124 Lake st., Chicago.

Figure 552

Sawyer did not publish newspaper ads in 1850 or 1851 for his *Extract of Bark,* but in 1852 and 1853, new local newspaper ads were placed for his business in general, emphasizing "large additions to his stock," his "fine assortment of FAMILY MEDICINES," and his ability to prepare PHYSICIANS PRESCRIPTIONS using *"Schaplin's Select Powders."* However, the long ads do not mention Sawyer's *Fluid Extract of Bark*.

In 1855, there was a change in business partnership, and two new partners were brought into the business: Sawyer's son Nathanial, and Nathanial Paige. The new business became *Sawyer, Paige & Co.* The following year, Sidney retired from the business, except as a consulting physician, and from that time on, the business was run by Nathanial Sawyer and Mr. Paige.

At the time of this partnership transition, there appeared to be a "second generation" attempt by the company to market *Sawyer's Fluid Extract of Bark.* After a four-year advertising hiatus, a new large ad for Sawyer's medicine appeared in February 1856:

> **Sawyer's Fluid Extract of Bark**
> AN INFALLIBLE REMEDY FOR THE
> **FEVER AND AGUE,**
> And The Various Forms Of
> **BILLIOUS DISEASES,**
> Intermittent, Remittent, Chill Fever and
> **DUMB AGUE.**
> Having their origin in Malaria of very fertile or marshy districts.
> **PREPARED BY DR. S. SAWYER,** and sold by **SAWYER, PAIGE & CO.,** Druggists and Chemists, 124 Lake street, Chicago.
>
> This remedy is prepared entirely from vegetable substances and contains no mercury, arsenic, or anything that can prove injurious to the system. Its effects are those of a deobstruest and tonic, well adapted to restore the healthy action of the stomach whose functions are so generally impaired in this class of diseases.
>
> Quinine and other tonics, serve only in many cases to break the chills, but not remove the diseases of the liver and other organs implicated, on which the chills and fever depend; hence the usefulness of the "Fluid Extract of Bark," (particularly in regions where physicians cannot be readily consulted), which contains all the qualities suited to the indications of these maladies.
>
> Hundreds of testimonials in regard to the efficacy of the medicine from persons of respectability, have been given to the proprietor—but he does not make use of them, as he prefers to have the medicine introduce itself solely by its merits.
> Price $1 per bottle, 6 bottles for $5.

In 1860, the company partners went their separate ways, and Sidney Sawyer retired. From the technological and stylistic attributes of the embossed *Sawyer's Fluid Extract* bottle listed, it was likely produced in the mid-1850s for this short-lived second advertising effort.

"Mrs. Schafer"

Bluing:

(1) **M**ʳˢ **SCHAFER'S // UNEQUALED // BLUING // CHICAGO** [aqua]

»SB w/ keyed hinge mold; beveled-edge square; four embossed sunken panels; 5⁷⁄₈" tall.
[Figure 553]

———————— ❖ ————————

"Mrs. Schafer" was likely a brand name used by one of the early chemical companies in Chicago (see, *e.g.,* Garden City, Huyck & Randall, Gillet, and Gillet, McCulloch & Co.), since no one named Schafer has been found listed in the city directories as a manufacturer of bluing or similar laundry products. From its technological and stylistic attributes, the bottle listed appears to date to the mid-to-late 1860s.

Figure 553

Francis (Franz) Schonwald

Soda ~ Mineral Water:

(1) **FRANCIS SHONWALTD / CHICAGO /**   *[cobalt]*
(front heel): **A A**

»IP; long-neck, shouldered-blob-top soda. *[Figure 554]*

(2) **FRANZ SCHONWALD / PREMIUM / MINERAL**   *[aqua, green]*
WATER / CHICAGO ILL
Note: *This mold was apparently reused. Faintly visible on the reverse side of the bottle is the piened-out embossing:* "**M. MONJU / MOBILE ALA**" (see also Rodemeyer *Chicago entry*).

»IP; long-neck soda; shouldered-blob-top soda. *[Figure 555]*

Francis Schonwald (also "Schoenwald") is first listed in Chicago CDs as a druggist at 354½ State (in 1856) and 352½ State (in 1857). The 1865 CD notes that he arrived in Chicago from Prussia in 1851. The 1857 CD records him only as a "clerk."

In 1859 and 1860, the CDs list Schonwald as a "mineral water manufacturer" (1859) and a "sodawater manufacturer" (1860) from his residence at 16 S. Halsted. The "**SCHONWALTD**" embossing on his earlier bottle, and the "**FRANZ**" Schonwald embossing on his later bottle style appear to have been glass-house errors.

Francis Schonwald does not appear in the city directories after 1860, and sometime during that season he likely sold his bottling business to Louis Rodemeyer, a Chicago brewer and saloonkeeper (*Figure 556: see Rodemeyer listing, this chapter*). Soda bottles (i.e., Schonwald's bottle #2) used by both share a similar, unusual bottle style with nearly identical embossing. Both company's bottles were blown in a two-piece mold and on the reverse-side both have "**M. MONJU & CO. / MOBILE.**" peened-out. One variety of the Rodemeyer bottle has a later Civil War–era smooth base, with a keyed hinge mold.

The embossed "**A A**" on Schonwald's bottle #1 is the mark of Alexander Arbogast of Pittsburgh. Only two other bottles with this maker's mark are known, both Civil War–era black-glass quart whiskey-style 3-piece-mold cylinders—one unembossed and the other

Figure 554

Figure 555 **Figure 556**

shoulder-embossed U.S.A. / MEDICAL SUPPLIES / FROM / PIKE & KELLOGG, ST. LOUIS. Both bottles have a base mark "A. ARBOGAST PITTS". Rex Wilson (1981:113) reports a base of one of these bottles from a Civil War–era western military context. Andrew Arbogast appears in Pittsburgh CDs in the 1850s and early 1860s as a glassblower (ca. 1850–57) and glass manufacturer (ca. 1858–62). From 1863–67, he operated a window-glass factory in New Castle, PA (von Mechow 2008). An advertisement in the 1861 Pittsburgh city directory (pg. 23) for the Arbogast glass works states, "Particular attention paid to Private Molds."

Charles Schriber and John Joss

Soda ~ Mineral Water:

(1) **SCHRIBER & JOSS** *(arched)* [aqua]
   (base): **S&J**

   »SB w/ keyed hinge mold; slope-shouldered oval-blob-top soda. *[Figure 557]*

(2) **SCHRIBER & JOSS** *(arched)* / **CHICAGO** / **ILL** [aqua]
   (base): **S&J**

   »SB w/ keyed hinge mold; slope-shouldered oval-blob-top soda. *[Figure 558]*

---❖---

Charles Schriber and John Joss both began their Chicago careers in the soda and mineral water bottling business as salesmen for other established companies. John Joss is first listed in the 1869 Chicago CD as a "soda water peddler" (company not named) boarding at 160 Milwaukee Avenue. Charles Schriber (sometimes spelled "Schrieber") first appears in the 1866 and 1867 Chicago CDs

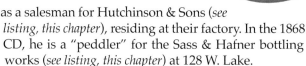

as a salesman for Hutchinson & Sons (*see listing, this chapter*), residing at their factory. In the 1868 CD, he is a "peddler" for the Sass & Hafner bottling works (*see listing, this chapter*) at 128 W. Lake.

Charles Schriber & Co. is first listed in the 1869 and 1870 Chicago CDs as being located at 108 W. Lake. The partners in the Schriber Company (which manufactured soda water fountains and syrups) included Charles Schriber, Louis Sass, and William Hafner. At the time, *Sass & Hafner* were also bottling at the same address. Schriber seems to have focused on fountains and syrups, while *Sass & Hafner* continued bottling under their own names at the same *Schriber & Co.* address.

John Joss missed being listed in the 1870 CD, but in 1871 he is listed as a "soda manufacturer" at 127 N. Green, again with no company association given. That same year, Charles Schriber (no longer part of Schriber & Co.) is listed as a "soda manufacturer," also at 127 N. Green. The *Schriber & Joss* embossed bottles obviously date to their unlisted copartnership during the 1871 season. The year *Schriber & Co.* was formed, John Joss was listed as a "soda water peddler," again without company association, but possibly he was associated with Charles Schriber as early as 1869.

In 1872, Joss is listed in the Chicago CD as a carpenter, and by 1873 he had formed a soda-bottling partnership with Charles Hartmann (*see their listing, this chapter*).

Figure 557

Figure 558

[Otto] Schulz and Ernst Hess

Mineral-Spring Water:

    (1) **SCHULZ & HESS / MINERAL SPRING / WATERS / CHICAGO ILL**   [aqua]

    »SB w/ post mold; shouldered tall pint mineral water; oval blob top. [*Figure 559*]

    (2) **SCHULZ & HESS / MINERAL SPRING / WATERS / CHICAGO ILL**   [aqua]

    »SB w/ post mold; squat shouldered long-neck quart; oval blob top; 9" tall. [*Figure 560*]

―――――――― ❖ ――――――――

This is a mystery company that we have been unable to locate in the Chicago CDs. However, Ernst Hess was a Chicago bottler during 1868–69 (*see his listing, this chapter*), and one of his embossed bottles was an aqua quart "**MINERAL SPRING WATERS**" closely similar in size and style to the *Schulz & Hess* bottle listed as #2. From the technological attributes of the bottles, the *Schulz & Hess* mineral-spring-water bottles appear to have been produced after the *Ernst Hess* mineral-spring quart, so our best guess at present is that the *Schulz & Hess* bottles date to a ca. 1870 partnership just after the Hess years. Mineral spring waters were often marketed by drug stores at the time,

Figure 559

Figure 560

and from 1870 to 1872 Otto W. Schulz was listed in the CDs as a Chicago druggist operating from his residence at 184 Cottage Grove Avenue. Perhaps the two of them formed a short partnership to continue marketing the 1868–69 Hess **MINERAL SPRING WATERS** product.

Schwab, Pieters & Co.

See the Bennett Pieters listing.

Schwab, McQuaid and Co.

Bitters:

(1) **RED JACKET / BITTERS // // SCHWAB. MC.QUAID & CO // //**   *[amber]*
(base): **SCHWAB PIETERS / W McC & CO.**

»SB; beveled-edge square bitters; tapered collar; 9¾" tall. *[Not seen, no photo: see Figure 561 Ring (1980) drawing of embossed side panels]*

(2) **RED JACKET / BITTERS // // SCHWAB. MC.QUAID & CO // //**   *[amber]*
(base): **WM. F. & SONS, PITTS. PA.**

»SB; beveled-edge square bitters; tapered collar; 9¾" tall. *[Not seen, no photo: see Figure 561 Ring (1980) drawing of embossed side panels]*

❖

In the mid-1860s, Edward McQuaid, Charles H. Schwab, and John B. Smith, were in business as wholesale liquor merchants on South Water Street in Chicago. In 1868, on the retirement of Smith, they added partners Leon Monheimer and P. Cavanaugh, and merged with Bennett

Figure 561

Pieters and Co. in 1869 (*see Bennett Pieters listing, this chapter*).

Following their one-year (1869) transitional partnership with Pieters—called *Schwab, Pieters & Co.*—they formed *Schwab, McQuaid & Co.*, and this company was in business from 1870 through 1875. The company then became *McQuaid & Monheimer Bros.* (with Leonard and Levi M.) from 1876 through 1879. The *McQuaid & Monheimer Co.* is listed in the 1880 IL Industrial Census, right at the end of their partnership. The IIC indicated the company had a $25,000 investment in property and equipment, two male employees, and an annual payroll of $1,500. Their annual cost of production materials was estimated at $65,000 with an annual retail product value of $75,000.

The two bottle varieties listed above thus date to ca. 1870–1875 (*see Bennett, Pieters & Co. entry for a general description of the history of Red Jacket Bitters bottling in Chicago*).

John Sears and Charles G. Smith

Medicine:

(1) **SEARS & SMITH** (*in narrow central sunken panel*) // [aqua]
(side): **WESTERN** // **CHICAGO / ILL** //
(side): **CHOLOGOGUE**

»IP; beveled-edge rectangle w/ 4 sunken panels; short neck; double-ring collar; rectangular panel in base with inset rectangular iron-pontil scar; 5 1/4" tall. [*Figure 562*]

(2) **SEARS. SMITH & C°** / **CHICAGO / ILL** // [aqua]
(side): **WESTERN** // //
(side): **CHOLOGOGUE**

»IP; beveled-edge rectangle w/ 4 sunken panels; broken at shoulder, but probable double-ring collar and about 5 1/4" tall, as with bottle #1; sunken rectangular panel in base with inset rectangular iron-pontil scar; broken at shoulder, but probable double-ring collar and about 5 1/4" tall, as with bottle #1. [*Figure 563*]

❖

An 1852 Chicago newspaper advertisement for John Sears, Jr., Wholesale Drug House lists the address at that time as 113 Lake Street. The advertisement states that John Sears' business was a successor to the firm of *Sears & Bay*.

On the retirement of Bay, Charles G. Smith "who had served an apprenticeship with the old firm" was advanced to head clerk, and on January 1, 1854, was made a partner by Sears (Ebert 1905:258–259; Wilson and St. Clair 1868:243). In the 1854 Chicago CD, the firm of *Sears & Smith* was listed at 113 Lake. Sears' partner, Charles Gilman Smith, also had a separate listing in the same volume as a physician at 122 Lake. Edward Burnham was taken into the business in 1855, after which the firm name became *Sears, Smith & Co*. Upon Mr. Sears' retirement in 1857, junior partner Edward Burnham was elevated, and the new firm of *Burnham & Smith* set up retail operations at 23 Lake. In the 1859 Chicago CD, Sears is listed as a partner in the firm of Wells and Smith. Thus, the *Sears & Smith* druggist firm operated only from 1854 to 1855, and their **Chologogue** was likely produced and marketed during that two-year period.

A Chologogue, in early pharmaceutical terms, was a botanical substance that encouraged or increased the flow of bile, acting as a purgative.

Figure 562

Frederick Seibt and Henry Haunschild

Ale:

(1) **SEIBT & HAUNSCHILD / CHICAGO //**   [amber]
 (heel): **A & D. H. C.**

»SB; 4-piece mold quart ale; embossed on lower vessel body; ring-neck tapered collar; 9 1/2" tall. [Figure 564]

Soda ~ Mineral Water:

(1) **SEIBT & HAUNSCHILD /**   [aqua]
 CHICAGO / ILL^S //
 (heel): **A & DHC**

»SB w/ keyed hinge mold; oval-blob-top soda. [Figure 565]

Frederick Seibt (also listed as "Fred" and "Fritz") and Henry Haunschild were first listed in the Chicago CDs in 1873, identified as partners in a "mineral and seltzer water" bottling business located at 303 Division. They continued as bottling partners in 1874 and 1875, but are listed in the CDs for those two years

Figure 563

Figure 564   Figure 565

as "soda" and "sodawater" manufacturers at the same address (Division at Clybourn). There is no directory indication that they bottled ale in addition to seltzer, mineral water, and soda, so perhaps they put up their seltzer in ale-style bottles during their first year in business.

After the 1875 season, Henry Haunschild left the partnership to open a saloon, and Seibt continued soda water bottling on his own at the 303 Division address (*see the Seibt listing*).

Frederick (Fritz) Seibt

Soda ~ Mineral Water:

(1) **FRITZ SEIBT / CHICAGO / ILL. /** [aqua]
    (*front heel*): **De S. G. Co.**

    »SB w/ post mold; oval-blob-top soda. [*Figure 566*]

(2) **FRITZ SEIBT / CHICAGO / ILL^S //** [aqua]
    (*heel*): **A & D. H. C.**

    »SB w/ post mold; oval-blob-top soda. [*Figure 567*]

According to the Chicago CDs, Frederick Seibt was a soda water manufacturer and bottler at 303 Division in 1876. In 1877 and 1878, he was listed in the occupation at 60 N. Clybourn. From

Figure 566

Figure 567

1879 through 1881, he was listed as a soda bottler at 783 N. Halsted. He no longer appeared in the Chicago CDs after 1881.

In addition to the blob-top soda listed, three or four Seibt-embossed Hutchinson-style soda bottles are known (Harms 1978:15; Kott 2005a). The Hutchinson-style closure was patented just prior to the 1879 bottling season. It became the hallmark soda bottle of the 1880s and 1890s, and is therefore not included in the present volume. But it is notable that at least three differently embossed Hutchinson bottles are known for the Seibt Company, since their last bottling season was 1881, and indicates a rapid, immediate transition to the Hutchinson bottling system.

Conrad Seipp

Beer:

(1) **CONRAD. SEIPP. / BREWER. / CHICAGO.** [aqua]
    (base): **C & I**

»SB w/ post mold; tall beer w/ sloped shoulders; straight-sided blob-top; 9 1/4" tall. *[Figure 568]*

❖

The Conrad Seipp Brewing Company was one of the massive historic generational businesses in Chicago. It was located for most of its early history on Johnson Avenue, between 26th and 27th.

Conrad Seipp, an immigrant from Germany, started making beer in Chicago in 1854, after buying a small brewery from Mathias Best. By 1856, Seipp had six employees, who helped him produce about 1,100 barrels of beer each year. In 1858, Frederick Lehman joined the company, which became Seipp & Lehman *[aka Seipp & Lehmann in city directories]*. By the end of the 1860s, when Seipp &

Lehman was one of Chicago's leading brewers, about 50 employees made more than 50,000 barrels of beer (worth close to $500,000) per year. After Lehman died in 1872, Seipp organized the Conrad Seipp Brewing Co. Dominating the Chicago beer market by the late 1870s, Seipp was among the largest breweries in the United States, producing over 100,000 barrels a year. After Seipp died in 1890, the company merged with several smaller Chicago breweries....

At the turn of the century, the Seipp brewery was still active; annual output had reached about 250,000 barrels.... The company limped along through the Prohibition years by producing low-alcohol "near beer" and distributing soda pop.... Ironically Seipp operations ceased in 1933, just before Prohibition was lifted. [Chicago Historical Society 2005]

Despite the scale of the brewery's operations during the late nineteenth century, Seipp embossed-bottle varieties are surprisingly scarce. The example listed no doubt dates to the mid-1870s—that is, between 1872 (when Conrad Seipp took over operation of the brewery on his own following Lehmann's death) and 1878 (when Pittsburgh glassmaker Cunningham & Ihmsen [C&I] ceased operations). In fact, the bottle was likely produced as part of a 1873 advertising effort at the time of the transition to the *Seipp Brewing Co.*

Henry H. Shufeldt and Co.

Gin:

(1) **H. H. S. & Co** // // **IMPERIAL. GIN** // // [yellow amber]
(base): **L&W**

»SB w/ keyed hinge mold; case-gin bottle shape (tapered rectangular body, broader at shoulders than at base); tapered collar; 9 1/2" tall. *[Figure 569]*

Figure 568

Henry H. Shufeldt and his brothers (William T. and George A., Jr.) were listed as distillers, rectifiers, and wholesale liquor distributors in the Chicago CDs from the late 1850s until nearly the turn of the twentieth century. Their first listing was for *W. T. Shufeldt & Co.* (William and George only) at 194 and 196 S. Water Street in 1858. By 1859 and 1860, all three brothers were involved in *W. T. S. & Co.* at 51 S. Water. From 1861 through 1863, William went to work for another local distiller (*A. F. Croskey, see his listing, this chapter*), and Henry and George formed the *Henry H. Shufeldt* company (distillers of "alcohol, whiskies, etc.")—first at 33 S. Water, then 69 S. Water, and then 51 and 53 S. Water. In 1864, William returned to reestablish his own distillery and wholesale liquor business at 51 S. Water, George left the business, and Henry moved to expanded quarters at 56/58/60 S. Water and began his meteoric rise in the distillery business. By 1866, the company became *Henry H. Shufeldt & Co.* (H. H. S., George Taylor, and Abel Smith). Both Shufeldt distilleries continued side-by-side for many years, but it was Henry who particularly prospered. The 1867 Chicago CD (p. 7) highlighted *H. H. Shufeldt & Co.* as one of 52 companies that personified "the progress which Chicago has made to be deemed a first class manufacturing city. The amount of enterprise and wealth cannot better be recorded...than to insert a list of...the leading firms in their departments of trade, as made to the United States Assessor for this district." *H. S. S. & Co.* was valued in 1867 at $1,543,600.

After that, the company experienced some less-prosperous times, and by 1868 Henry Shufeldt was a junior partner in the *Burton M. Ford & Co.* distillery firm at 54 and 56 S. Water. But Henry was back in charge of the company by 1869–70, and the H. H. S. & Co. fortunes continued on an even keel until Henry's death in 1891. After that the company continued in operation for several years under a series of operating managers, but it was no longer listed in the Chicago business directories after 1898.

The Shufeldt Company was listed in the 1880 IL Industrial Census as having a $200,000 investment in property and equipment in Chicago and 14 male employees, with an annual payroll cost of $12,000. Their annual cost of production materials was $750,000, and their annual retail product value was estimated at $810,000. Henry Shufeldt expanded his company's rectifying and distillery operations into the Peoria area in the early 1880s (*too late for inclusion in this volume in the Peoria listings*).

In light of the size and longevity of the Shufeldt Company in Chicago, it is interesting to note that only a few examples of just one embossed-bottle style are known. Clearly, almost all Shufeldt bottles were made for paper labels only. From the style and technological attributes of the embossed H. H. S. & Co. *Imperial Gin* bottle, it appears to date to the 1860s. In this period, "H. H. S. & Co." dates only from 1866 onward, so the bottle was likely a late 1860s product (perhaps dating to 1866, the first year the new Shufeldt/Taylor/Smith corporation was formed).

Later Shufeldt & Co. embossed "signature" bottles (e.g., *Figures 570–571*) likely postdate 1880.

Figure 570

Figure 569

Figure 571

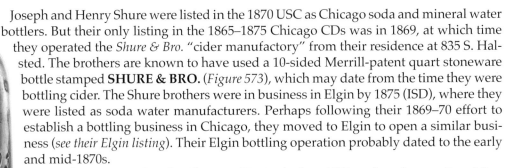

Joseph and Henry Shure

Soda ~ Mineral Water:

(1) **SHURE BRO'S / CHICAGO** [aqua]

»SB w/ keyed hinge mold; oval-blob-top soda. *[Figure 572]*

Joseph and Henry Shure were listed in the 1870 USC as Chicago soda and mineral water bottlers. But their only listing in the 1865–1875 Chicago CDs was in 1869, at which time they operated the *Shure & Bro.* "cider manufactory" from their residence at 835 S. Halsted. The brothers are known to have used a 10-sided Merrill-patent quart stoneware bottle stamped **SHURE & BRO.** (*Figure 573*), which may date from the time they were bottling cider. The Shure brothers were in business in Elgin by 1875 (ISD), where they were listed as soda water manufacturers. Perhaps following their 1869–70 effort to establish a bottling business in Chicago, they moved to Elgin to open a similar business (*see their Elgin listing*). Their Elgin bottling operation probably dated to the early and mid-1870s.

Joseph reappeared in the Chicago CDs in the late 1870s, where he was listed first as a feed store operator, then as a salesman, then as a saloon operator. The Shure brothers were no longer in the soda-bottling business by the time of the 1880 USC.

Figure 572

F. W. Simonds and Gerhard Paoli

Soda ~ Mineral Water:

(1) **DR. PAOLIS. / SPARKLING PERSIAN / SHERBET. //** [cobalt]
F. W. SIMONDS / CHICAGO. / ILL

»SB; "tenpin"-shaped soda with small-diameter base (1"); shouldered-blob top; 8" tall. *[Figure 574]*

Dr. Gerhard C. Paoli was a long-time Chicago physician who first arrived in the city in 1854 and set up a general medical practice there the following year at 142 Lake Street. In 1856, he relocated his practice to 18 Dearborn Street and also formed an association with F. W. Simonds (listed as T. W. Simonds in 1856) to market his "sparkling Persian sherbet" at 75 Huron Street. Their short confectionary business venture at 75 Huron continued for only three seasons (1856–58). Most entries in the regular listings section of nineteenth century city directories are brief, but F. W. Simonds' listing in the 1858 Chicago CD takes up two full lines: **F. W. Simonds, agent & proprietor, Dr. Paoli's Sparkling Persian Sherbet, 75 Huron Avenue**.

Simonds produced one of the most distinctively shaped commercial bottles used in mid-nineteenth-century Chicago (not to mention its unusual contents), but his business venture was not particularly successful. After 1858, Simonds was no longer listed in the Chicago CDs, and *Dr. Paoli's Sparkling Persian Sherbet* was apparently no longer marketed. Dr. Paoli, however, stayed on in Chicago and had a long and successful career there. He was listed as a physician in the Chicago CDs every year from 1855 to the early 1890s. And for a short time in the early 1860s, after the *Persian Sherbet* venture, he and a partner (Martin Hale) also established a

Figure 573

Figure 574

separate business (*G. C. Paoli & Co.*) selling alcohol and spirit gas at 61 Illinois Street.

Besides being distinctively designed, Simonds' *Persian Sherbet* bottle was remarkable for another reason as well: though it predated 1860, it had no iron-pontil scar on the base. It was therefore one of the few early Illinois bottles manufactured in New York (or perhaps elsewhere in the Northeast), where smooth-base, snap-case technology first made its appearance in the late 1850s. Such bottles were available to Chicagoans through the Great Lakes trade, while downstate bottlers had to rely principally on river trade with Pittsburgh glass houses, who continued to use pontil technology until 1860 (*see introduction, this volume*).

According to recipe books and discussions of the era, *Persian Sherbets* were distinctively tart, sweet/sour fruit drinks:

> The varieties...may be divided into those made from the fresh juice of fruit, which are mixed with water and sweetened to the taste; and those made from sirup, in which the juice of fruit has been boiled.
>
> ...the effervescing qualities of royal Persian sherbet only exist in the imagination of the English confectioner. But there is one all-important point that the English vendor would do well to imitate: Persian sherbet is served very cool or iced. Blocks of snow or lumps of ice are always dissolved in the sherbet drunk in Persia, unless the water has been previously artificially cooled. Fresh sherbets are usually lemon, orange, or pomegranate. [Chambers 1884:439]

Walter B. Sloan

Pre–Civil War Pontiled Medicines:

(1) **W. B. SLOAN //**   [aqua]
   (side): **CHICAGO. ILL // //**
   (side): **INSTANT. RELIEF**

   »OP; beveled-edge rectangle; short neck rolled lip; 4 3/4" tall. *[Figure 575]*

(2) **HAIR OIL / WALTER B. SLOAN / CHICAGO. ILL**   [aqua]

   »OP; squat, wide-mouth shouldered cylindrical salve jar; short neck; rolled lip; 3 1/4" tall. *[Figure 576]*

Figure 575

Figure 576

(3) *(side):* [LIFE] **SYRUP // WALTER B. / SLOAN //**   [aqua]
(side): **CHICAGO. ILL**
Note: Fragmentary bottle record. Bracketed embossing not present. [Figure 577]

»OP; beveled-edge rectangle; 3 sunken panels; height unknown. Only available specimen broken at shoulder [neck and collar missing]; 5" tall from base to shoulder.

Post-Civil War Medicines (Marketed by Walker & Taylor):

(4a) **SLOAN'S / OINTMENT // WALKER & TAYLOR /**   [aqua]
PROPRIETORS // CHICAGO. ILL // //

»SB w/ keyed hinge mold; beveled-edge square (1/2 x 3/4"); squat, wide-mouth salve jar; flared-ring collar; 1 3/4" tall. [Figure 578, top and bottom left]

(4b) **SLOAN'S / OINTMENT // WALKER & TAYLOR /**   [aqua]
PROPRIETORS // CHICAGO. ILL // //

»SB w/ keyed hinge mold; beveled-edge square (1 x 1"); squat, wide-mouth salve jar; flared-ring collar; 2 3/4" tall. [Figure 578, center and right]

    Walter B. Sloan established a large patent-medicine business in Chicago during the 1840s. One of the company's later ads indicated that *Sloan's Veterinary and Family Medicines* were "established in Chicago in 1842." But from the Chicago CDs, this seems not to have been the case, and an 1856 Chicago CD indicates he had arrived in the city 10 years previously. Sloan's "*One-for-a-Man & Two-for-a-Horse*" patent-medicine business (apologies to Carson 1961) was first listed in the Chicago directories in 1847: "W. B. Sloan, General Agency Depot, American Temperance House Building." Sloan also published an ad for his medicines in the 1847 CD, indicating he was a

> GENERAL AGENCY FOR THE SALE OF
>   Dr. Beecher's Books and Medicines, McAlisters All-Healing Ointment, &c., Dr. Beekman's Pulmonic Syrup and Pills, Dr. Bragg's Indian Queen Sugar Coated Pills; and also—Many other Medical Books and Medicines, at wholesale or retail, at the manufacturers' lowest prices.

His 1847 focus on these Eastern and St. Louis medicine brands suggests he was not yet producing and selling his own medicine brands. Sloan-entry pages and CDs for 1848 through 1850 were missing in the directories used for this study, but a long-running 1849 *Chicago Daily Journal* illustrated newspaper ad ("*Sloan's Column*"—e.g., June 29, 1849, p. 4, col. 5: *see Figure 579*) and

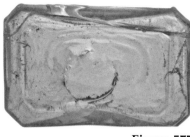

Figure 577

Figure 578

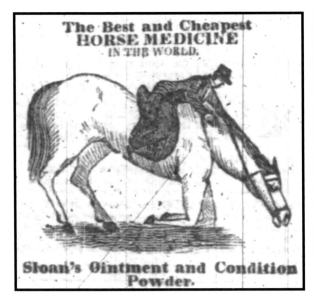

Figure 579

the 1851 and 1852 Chicago CDs indicate that Walter B. Sloan's Medical Depot had relocated to 40 Lake Street, and that by 1849 he had added the manufacture and sale of his own medicine brands and conditioning pastes to his patent-medicine stock. These included:

(1) *Sloan's Family Ointment.* "... now universally acknowledged to be an infallible remedy, in every case where it has been faithfully applied on the human system, for promoting Insensible Perspiration, drawing out the inflammation from a wound, relieving pain of every kind, and in its healing qualities the world over does not produce its equal.... All diseases of the Flesh, Obstinate Ulcers, Old Sores, Chilblains, Sore Throat, Burns, Cuts, Cutaneous Eruptions, Sore Nipples, Sore Breasts, Diseases of the Eye, Ague in the face, side, back, and the other parts of the system, Boils, Ulcers, Scald Head, Bruises, Fresh Wounds, and every kind of sore containing the least particle of inflammation, are permanently cured by this great remedy.

(2) *Sloan's Ointment [Horse Medicine].* "For purity, mildness, safety, certainty, and thoroughness, Sloan's Ointment excels, and is rapidly superceding all other Ointments and Liniments for the cure of the following diseases: Fresh Wounds, Galls of all kinds, Sprains, Bruises, Cracked Heels, Ringbone, Wind Galls, Poll Evil, Callus, Spavins, Sweeny, Fistula, Strains, Lameness, Sand Cricks, Foundered Feet, Scratches or [*Groase*], Mange and Horn Distemper.

(3) *Sloan's Condition Powder [Horse Medicine].* "The Powder will remove all inflammation and fever, purify the blood, loosen the skin, cleanse the water, and strengthen every part of the body, and has proved a sovereign remedy for the following diseases: Distemper, Hide-bound Loss of Appetite, Inward Strains, Yellow Water, Inflammation of the Eyes, Fatigue...also Rheumatism, (commonly called Stiff Complaint), which proves fatal to many valuable horses in this country. It is also a safe and certain remedy for Coughs and Colds, which generate so many fatal diseases.

(4) *Sloan's Tannin Paste [Leather Restorer].* **CHEMICAL BUTTER OIL** Penetrates the stiffest and hardest Leather, softens it at once, removes the crust or blister and imparts a strength that is incredible until seen. This Oil and other life-giving articles form SLOAN'S CELEBRATED TANNIN PASTE, which saturates the pores of leather and renders it **WATER PROOF**.

Aggressive salesmanship and regional advertising (e.g., a large ad for his Ointment is also contained in the *Quincy Weekly Herald* for November 23, 1849) were an early Sloan hallmark. For instance, his 1849 *Chicago Daily Journal* column contained the following "Want Ads":

Great Chance.

100 ENTERPRISING Young Men of good habits wanted to operate in different parts of the Western States. He best assurances will be given that each one can with a cash capital of from $50 to $20, make in one year from $250 to $1,000 clear profit. We don't want you to take our word for it, but call and see for yourselves.

In order to obtain a full description of the business and mode of operation, personal application is necessary. **W. B. SLOAN**

Liberal Offer.

ANY person that furnishes me before the first day of next August, the facts in reference to the most *Extraordinary Cure* effected by the use of my Ointment, shall receive a premium of *Ten Dollars*, and for the second most extraordinary cure, Five Dollars; third, Two Dollars; fourth, One Dollar; and fifth, a 50c box of the Ointment. **W. B. SLOAN**

During the 1850s, he published a Chicago literary journal, *Sloan's Garden City*, which contained advertising for his "Sloan's Remedies," including *Sloan's Family Ointment, Instant Relief, Life Syrup,* and *Sloan's Horse Liniment*. By 1856, he was also publisher of the *Chicago Daily News*, from offices at 42 Lake, also housing his patent-medicine establishment. During the mid-1850s (according to an 1853 ad in the *Woodstock Sentinel*), two additional (internal) Sloan's medicines were introduced:

(1) *Sloan's Life Syrup*. "... **IS** altogether superior to any Medicine heretofore discovered for the rapid cure of *Colds, Coughs, Influenza, Hoarseness, Catarrhs, Whooping Cough, Bronchitis, Asthma, Dyspepsia, Liver-Complaint, Incipient Consumption,* and for the relief of *Consumptive Patients* in advanced stages of the disease; also for *all Diseases* which originate from impurity of the *Blood,* to wit: *Scrofula, Salt-Rheum, Mercurial Syphilis,* or *Venereal and Eruptive Diseases; for Jaundice, Erysipelas, Boils, Ulcers, Rheumatism, Cutaneous Affections,* or *Diseases* of the Skin [such] as Botches, Pimples, or Pustules, and all *Itching Eruptions, Chronic Inflammation of the Liver, White Swellings, Necrosis or Fever Sores, Rickets, Fistula,* and *King's Evil;* in short, in every taint of the *System* or *Blood...One Bottle of Sloan's Life Syrup* will produce more real benefit than *ten times* the same quantity of any other preparation offered to the public."

(2) *Sloan's Instant Relief*. "**FOR** the immediate cure of Colic, Heartburn, Sore Throat, Diarrhoea, Dysentery, Cholera-morbus, all Summer Complaints, Sick Headache, Ague-Cake, Rheumatism, Soreness in the Bones, Pain in the Limbs, Cramps and Spasms, [*Golus*], Spinal complaints, Swelled Joints, &c."

In the 1850 USC, Sloan was listed as 48 years old. In the 1850 Industrial Census for South Chicago, Sloan's patent-medicine business was listed as being worth $10,000 in real estate and invested capital. His annual investment in fuel and raw materials (given as "sundries") to produce his medicines was $20,000. He operated with eight employees at an average annual labor cost of $100, and the annual value of medicine products produced was estimated at $50,000.

Walter Sloan's patent-medicine operation continued at the Lake Street headquarters until 1858–59, when he largely retired to focus his efforts on medicine manufacture only, from his residence, and turned the operation of the business over to his son, Oscar B. Sloan, who moved the operation to 24 S. Water Street. Oscar Sloan placed an ad in the 1859 CD for "Sloan's Medicines, Oscar B. Sloan, successor to Sloan Manufacturing Company," and indicated that he had on hand for sale Sloan's *Family Ointment, Horse Ointment, Condition Powders,* and *Instant Relief* medicines.

After 1860, Oscar Sloan moved the retail and wholesale sales operation to his father's residence, where it remained until the end of the Civil War. In 1866, Walter and Oscar sold out their Sloan medicine line to the Chicago patent medicine firm *of Walker & Taylor* (see their entry). Walker & Taylor placed an ad in the Patent Medicine summary of the 1866 Chicago CD announcing that they had become sole proprietors of Sloan's medicines, including Sloan's *Condition Powders, Horse Ointment, Family Ointment,* and *Instant Relief*. To date, only the family *Ointment* has been

Figure 580

found in a post–Civil War Walker & Taylor embossed bottle. No earlier Sloan Co. embossed version of this ointment has been found.

Charles R. Smith and Co.

Bitters:

(1) **GRAPE / BITTERS // // C. R. SMITH & CO. / CHICAGO. // //**   [amber]

»SB w/ keyed hinge mold; beveled-edge square; flat panels; two opposing sides embossed; tapered collar; 9 1/2" tall. *[Figure 580]*

---❖---

Charles R. Smith is first listed in the Chicago CDs at the end of the Civil War (1865) as a salesman. The following year he was a partner in *Smith & Plows* (with Wm. J. Plows—*see also S. J. Smith & Co. bitters "Mountain Root Bitters" discussion*) in a wine and liquor wholesale business at 103 S. Water Street. The business continued as *C. R. Smith & Co.* in 1867, but by the following year Charles Smith had taken on a new partner and entered the dried fruit business at the same address. Thus his *Grape Bitters* marketing venture must have taken place during 1866–67, his two years as a wine and liquor wholesaler. No advertisements have been discovered as yet for *Grape Bitters*, so its advertised health benefits are unknown.

Samuel Smith

Soda ~ Mineral Water:

(1) **KNICKERBOCKER // MINERAL & / SODA WATER //**   [cobalt]
MANUFACTURED // BY. S. SMITH // CHICAGO

»IP; 10-sided bottle soda bottle with alternate raised and sunken panels (sunken panels embossed); tapered collar. *[Figure 581]*

---❖---

No soda water manufacturer with the name "S. Smith" has been found in pre–Civil War Chicago city directories or newspaper ads. From the tapered-collar lip form of the Chicago Knickerbocker bottles, they were likely manufactured during the late 1840s or early 1850s. It should be noted that between 1841 and 1849, a bottler named Samuel Smith bottled "Knickerbocker Mineral and Soda Water" at 164 W. 18th Street in New York City (Guest 2004). In 1850, he moved his business to Auburn, NY, in cobalt, sided bottles there until 1857 (Jordan 1981), when he encountered hard times and his business was sold at auction:

A special note of interest today about this sale is obviously the 150–300 GROSS of Smith bottles, most likely all iron pontiled, cobalt blue. At today's prices they would be worth $2,700,000 to $5,400,000.

Figure 581

I can't help wishing that my great-grand-father had purchased them. [Jordan 1981:5]

It is possible that Smith opened a branch soda business in Chicago for a short time in the early 1850s. A heavy commercial trade was conducted in the mid-nineteenth century between New York and Chicago *via* the Eire Canal and Great Lakes shipping.

Sylvester J. Smith and Co.

Bitters:

(1) **S. J. SMITH & Co / CHICAGO. ILL** // //   [amber]
     "MOUNTAIN / ROOT BITTERS" // //

»SB w/keyed hinge mold; beveled-edge square; tapered collar; 9 1/2" tall. *[Figure 582]*

Figure 582

Sylvester J. Smith and his father Orrin were listed in the Chicago CDs from 1865 through 1867 as commission forwarding merchants (i.e., product marketing and distribution middlemen). The Smith Company's wholesale sales effort for their **"MOUNTAIN ROOT BITTERS"** seems to have been limited to 1866 and 1867, after which they were no longer listed in the Chicago CDs. Possibly they bought the Chicago distribution rights for *Mountain Root Bitters* from John Abel and Lenson (a.k.a. "Lanson") S. Humiston, who were marketing it on their own in Sandwich just after the Civil War as well (The company is listed in the 1866 IRS records for Sandwich: *see Abel, Humiston & Co. Sandwich listing*). Of course, the opposite could be true as well: Abel & Humiston may have purchased the Sandwich distribution rights for *Mountain Root Bitters* from Smith and Co. Both Smith and Abel/Humiston used the same embossed bottle mold for the product, changing only the embossed name of the distributor on its opposite face.

In 1867, another group of Chicago commission merchants (*Plows, Harris & Upham*) first appeared in the Chicago CDs. They also focused their wholesale distribution business on "alcohols, wines, liquors, &c." By 1868, W. J. Plows had moved the business to 154 Dearborn and had taken on two new partners: John Abel and Lenson (or "Lanson") S. Humiston. In 1869, Abel and Humiston became associated with a large Chicago distiller and rectifier (*J. A. Montgomery & Co.*), and may have continued a short-lived bottling and marketing effort for their **"MOUNTAIN ROOT BITTERS"** at that time (perhaps still labeled *Abel, Humiston & Co.*, Sandwich, Illinois, since Sandwich was located just 40 miles southwest of Chicago).

No newspaper ads or paper labels for **"MOUNTAIN ROOT BITTERS"** have yet been located to indicate the specific supposed health benefits claimed for the bitters.

Snyder and Jacob K. Eilert

Medicine:

(1a-b) **DR WINCHELL'S / TEETHING SYRUP / SNYDER & EILERT / PROPRIETORS**   [aqua]
     Note: One variety of this bottle has all of the "Ns" embossed backwards, while a second (listed above) has the "Ns" corrected.

»SB w/ keyed hinge mold; 1 1/4" diameter cylinder; short neck; flared-ring collar; 5" tall. *[Figure 583]*

Figure 583

From its technological and stylistic attributes, this bottle appears to date to the later 1860s. This predates the time that this product was being produced in Chicago by the Emmert Proprietary Company (They arrived in Chicago in 1872: *see Emmert listing, this chapter*). As discussed in the *Eilert's Extract of Tar & Wild Cherry* listing above, Emmert had been in the drug business for some 20 years (initially as Emmert & Burrell) at Freeport, IL (Hill, ed. 1910:478). In the 1845/1860-era, a druggist named Jacob K. Eilert lived in the Freeport, IL/Monroe, WI area (the two towns are about 20 miles apart), so it is possible that both Eilert's (*see listing, this chapter*) and Snyder & Eilert's medicines were first sold by Emmert & Burrell at Freeport.

The Emmert Company marketed *Dr. Winchell's Teething Syrup* as the "best medicine for diseases incident to infancy." Their ads and trade cards declared "it will positively cure every case if given in time"; that it "quiets and soothes all pain"; that it "cures diarrhea and dysentery in the worst forms"; and that it was "a certain preventive of [and cure for] diphtheria" (Knudson 2006:415). This product appears to been have first been a Civil War era copy of *Mrs. Winslow's Soothing Syrup*, a popular patent medicine first introduced in 1849 by the Curtis & Perkins company of Bangor, ME (*See Emmert listing, this chapter*).

John Sproat

Ale~Porter:

(1) **JOHN SPROAT / CHICAGO //** [amber]
(heel): **A & D.H.C.**

»SB w/ post mold; two-piece mold body; squat pint porter shape; about 7" tall. Not yet seen. Reported by finder Gregory Watt to Robert Kott (2005a). [*Figure 584*] Figure 584b is an identical bottle, but without the "CHICAGO" embossing below the bottler's name.

❖

Figure 584a

From its manufacturing technology and style attributes, the listed bottle likely dates to ca. 1875–80. However, we are only aware of one example of the bottle ever having been found. We have not been able to locate or photograph it, but Chicago historical researcher Robert Kott was able to make a careful drawing of the bottle's shape and embossing (*Figure 584*).

Chicago CDs from 1859 to 1880 were searched for John Sproat's ale-bottling business, but no John Sproat listings were found. Also, he apparently never advertised in the *Chicago Tribune* during the late 1870s. John Sproat did live in Chicago, however. The *Chicago Tribune* for September 5th, 1890, contains a notice of his mortgage foreclosure there.

Notably, A few aqua blob-top soda bottles, dating to ca. the early 1870s and embossed "**JOHN SPROAT / TEXARKANA**" have been unearthed in Cairo, IL, as have several small, squat, long-necked, amber porter-style bottles of the same age. Some of them are embossed "**JOHN SPROAT**" [*Figure 584b*] and some are embossed "**JOHN SPROAT / TEXARKANA**." A brass padlock stamped "**JOHN SPROAT / CAIRO**" is also known. John Sproat is listed in the Cairo CDs in the early 1870s as proprietor of an icehouse (D. Beeler personal communication, 2006).

Perhaps John Sproat came to Chicago from Texarkana (on the southwestern Arkansas/Texas border) by way of Cairo, IL. But is also possible that Sproat shipped specially embossed bottled products from Cairo to Chicago, Texarkana, and elsewhere:

> The beer bottling, soda and seltzer and mineral trade has grown to immense proportions here *[i.e., Cairo]* recently. Mr. A. Lohr and Henry Breihan each have extensive concerns, and a wide market to supply in this and adjoining states. Mr. John Sproat carries on the same, and he adds to this the trade in fresh butter, eggs and vegetables. He loads his own cars and sends them to New Orleans, Mobile and other Southern cities, the *[padlocked?]* seal of the car only broken when it arrives at its final destination. [Baskin & Co. 1883;167]

His ale/porter-bottling business was likely an unsuccessful single-season or single-event attempt that never reached to the level of directory listing, and which may in fact never have been successfully realized as a business after the bottles were ordered (e.g., if the bottles were returned to the glass house for cullet, it would explain their extreme rarity).

[William] Steel and Co.

Whiskey~Ale:

(1) **STEEL & CO // CHICAGO ILL** *(wrap-around embossing on shoulder)*   *[green-teal]*

»SB w/ post/cup mold; cylindrical body w/ "ladies leg" neck; tapered ring-neck collar; 10" tall. *[Figure 585]*

━━━━━━━━━━ ❖ ━━━━━━━━━━

Although this bottle is a whiskey-style shape, *Steel & Co.* was listed in the Chicago CDs during only a single season (1873) as "ale bottlers," at 101 Madison Street. Company principals were not named, but the only Steel listed in the 1873 directory, not in some other business, was William B. Steel (no occupation

Figure 584b

Figure 585

given). In the 1872 CD, Wm. B. Steel and Charles C. Godfrey were grocers at 853 W. Lake; the year after the Steel & Co. ale listing, Wm. B. was listed as a "clerk."

As luck would have it, *Steel & Co.* was also listed in an 1873 *Chicago Tribune* newspaper ad (June 29, 1873, p. 16, col. 3):

ALE.
SCOTCH ALE.

We have just received, in prime condition, a large consignment of Messrs. Robin, McMillan & Co.'s Celebrated Pale India Ale, in stone pints and quarts, brewed especially for export, matured for twelve months, and carefully bottled at their Summer Hall Brewery, Edinburgh. Early orders solicited.

Apply to the importers

P. McINTYRE & CO.,
113 E. Kinzie-st., Chicago.

Or to STEEL & CO., cor. Madison and Dearborn-sts., and WILLIAM DEWAR, 155 East Washington-st.

James Stenson

Ale:

(1) **JAMES STENSON / CHICAGO / ILL^S**   [ambers and greens]
   Note: There are two embossed dots below the superscript **S** in **ILL**^{S} and several amber and green color varieties are known, including amber "black glass, reddish amber, olive-amber, and olive green [see Figure 586].

   »SB w/ keyed hinge mold; two-piece-mold quart ale; ring-neck tapered collar; 9 1/4" tall.

(2) **JAMES STENSON // CHICAGO ILL //**   [amber]
   (heel): **B. F. G Co**
   (base): **J S**

   »SB w/ post mold; two-piece-mold quart ale; embossed around shoulder; ring-neck tapered collar; 9 1/4" tall. [Figure 587]

Figure 586

Figure 587

(3) **JAMES STENSON / CHICAGO ILL //**   [amber]
    (base): **J. S.**

    »SB w/ post mold; two-piece-mold quart ale; ring-neck tapered collar; 9 1/2" tall. [Figure 588]

(4) **JAMES STENSON / CHICAGO / ILL^S //**   [amber]

    »SB w/ post mold; two-piece-mold quart ale; ring-neck tapered collar; 9 1/2" tall. [Figure 589]

Soda ~ Mineral Water:

(1) **JAMES STENSON / CHICAGO / ILL //**   [aqua]
    (heel): **A & D. H. C.**

    »SB w/ keyed hinge mold; oval-blob-top soda. [Figure 590]

(2) **JAMES STENSON / CHICAGO / ILL //**   [aqua]
    (heel): **B F G Co**
    (base): **J S**

    » SB w/ post mold; oval-blob-top soda. [Figure 591]

(3) **JAMES STENSON / CHICAGO ILL'S //**   [aqua]
    (heel): **DE S. G.Co.**
    (base): **S**

    » SB w/ post mold; oval-blob-top soda. [Figure 592]

Figure 588

Figure 589

Figure 590

Figure 591

Figure 592

(4) **JAMES STENSON / CHICAGO. ILL.** //   [aqua]
(heel): **A & D. H. C.**

» SB w/ post mold; oval-blob-top soda. *[Figure 593]*

(5) **JAMES STENSON / CHICAGO / ILLINOIS**   [aqua]

»SB; round-based cylindrical soda; 3 lines of vertical embossing, equally spaced around bottle; shouldered blob top; 9" tall. *[Figure 594]*

Quart Soda ~ Cider:

(1) **JAMES STENSON / CHICAGO ILL**   [aqua]
(front heel): **D. O. C.**

»SB w/ post mold; slope-shouldered quart soda; embossed in circular slug plate; oval blob top; 9 1/2" tall. *[Figure 595]*

(4) **JAMES STENSON. / CHICAGO. ILL^S** //   [amber, olive]
(heel): **A & D. H. C.**

»SB; ring-neck cider-finish quart; 11 1/2" tall. *[Figure 596]*

In 1869 and 1870, James Stenson became the junior partner with John Amberg at a spruce beer, mineral water, and soda water bottling plant that had been in operation since 1866 at 99 &101 Monroe Street in Chicago *(see Amberg and Green and Amberg & Stenson listings,*

Figure 593

Figure 594   **Figure 595**

Figure 596

this chapter). In 1868, a year before his association with Amberg, Stenson had opened a wholesale and retail wine and liquor business at 132 State Street. It is likely that he took the opportunity to join a bottling operation to ensure that he would have an outlet for bottling his own alcoholic beverages (see his *Ale* listings).

When he bought out Amberg and took over the business entirely in 1871, he decided to continue the ongoing soda and mineral water bottling operation as well, under his own name. According to the Chicago CDs, Stenson's soda water and ale bottling business at his Monroe Street factory spanned the entire decade of the 1870s and continued until 1882. At that point, he decided to close the Monroe Street bottling plant and focus his efforts entirely on generic retail-liquor sales at his 132 State Street address.

Toward the end of the 1870s, he switched over to using the new-style Hutchinson internal stopper bottles, four styles of which are known that are embossed with his company name (Harms 1978; Wright 1992). Since these bottles postdate the end of the bottling era studied in this volume, they are not listed above.

The Stenson Company is listed in the 1880 IL Industrial Census as having operated full-time 12 months a year, producing an annual value of bottled products estimated at $16,000. According to the 1880 USC, Stenson was 40 years of age at the time.

Hedding A. Stone and Co.

Flavoring Extracts:

(1) **STONE'S / PURE /** *(monogram):* **HAS / FLAVORS //** [aqua]
    (side): **CHICAGO // //**
    (side): **H. A. STONE & Co**

   »SB w/ post mold; strap-sided beveled-edge rectangle; flared square-ring collar; 5" tall. *[Figure 597]*

Hedding Stone first appeared in the Chicago CDs in 1867 as a real estate agent. In 1868, he was listed as a grocer. The following year, *H. A. Stone & Co.* (with Nathaniel F. Stone) first listed themselves as "manufacturing chemists" at 244 W. Madison. Such companies characteristically produced chemical products like ink, bluing, baking powder, flavoring extracts, and perfumes. In 1870, their business was listed as "manufacturing chemists and perfumers." In 1871, the Stones were listed as "extract manufacturers" at the same address.

Hedding Stone's business partners and business focus changed in 1872. In that year's Chicago CD, H. A. Stone, G. H. Stone, and J. H. Byrne are listed as druggists at 233 W. Randolph.

From 1873 through 1875, H. A. Stone and a new partner (E. J. Harkness) reestablished Stone's "chemists and perfumers" business—first at 143 Kinzie and later at 159 S. Water. Stone and his company are no longer listed in the Chicago directories after 1875.

The only year the company specifically advertised themselves as flavoring extract manufacturers was 1871, so perhaps the listed bottle dates from that time.

Dr. Reuben R. Stone

Bitters:

(1) DR R. R. STONE / CHICAGO ILLS // // MANZANITA / BITTERS // //   [amber]
*Note: Two embossed dots beneath superscript **R** in **D**^R.*

»SB; beveled-edge square case bottle; flat panels; tapered collar; 8 3/4" tall. *[Figure 598]*

Figure 597

❖

Dr. Reuben R. Stone was born in 1824 in New York and came to Illinois in 1845. He studied medicine and graduated from Rush Medical College in 1849. Stone practiced for 15 years in Richmond, IL, and then began full-time mercantile pursuits in 1864. The first newspaper ad we have found for *Manzanita Bitters* is from an 1869 Chicago-area northeastern IL paper (*Woodstock Sentinel,* Jan. 6, 1869). Note that the spelling for the name of the bitters in the ad differs from the spelling on the bottle itself:

A GRAND SUCCESS

Life, Beauty & Happiness
depend upon health. It is better to anticipate
and prevent a disease than cure it with active remedies

DR. R. R. STONE'S
MENZANETA
AND WILD CHERRY STOMACH BITTERS
by their cleanzing taste and alterative properties
accomplish those objects in the most perfect and
surprising manner. They give
NEW LIFE & ACTION
to the whole system. If properly administered they
will not injure the most delicate Constitution, yet
they are
Powerful in their Operation.

The ad ends by listing McHenry County druggist sales outlets for Manzanita Bitters, including Dr. Stone himself at Richmond, IL, and drug stores in Woodstock, Benton, Hebron, Ridgefield, Alden, Greenwood, McHenry, Nunda, Crystal Lake, and Algonquin. These early bottles were probably not embossed "Chicago," and may have just had paper labels.

Figure 598

In 1870, the year after Stone first produced and advertised his bitters, he was involved in an unsuccessful one-year effort to move from the hinterlands and market his bitters in Chicago, as a member of the "commission [forwarding] merchant" firm of Barnes, Foster & Stone (*see Charles Foster listing*). The embossed *Manzanita Bitters* bottles probably date from the time of this effort.

In 1871, he moved to Elgin (*Past and Present of Kane County:* Le Baron 1878a), after which time there is no further evidence he was involved in marketing his bitters.

Alfred and William Strickland

Medicine:

   (1) STRICKLAND'S // WINE OF LIFE   [amber, aqua]

      »SB; 3-piece-mold whiskey-style bottle; high-shoulder w/ long-neck; ring-neck tapered collar; shoulder-embossed only; 11 1/4" tall. *[Figure 599]*

Alfred Strickland was first listed in the Chicago CDs in 1868, in partnership with Isaac Bennett, Jr., as general distribution agents for the "Orange Nectar Co.," with offices at 10 La Salle Street. This was apparently a one-year bottled-product effort, and no embossed bottles are known. The next year, only his brother William was listed in Chicago, and only as a boarder at 133 29th Street.

In the 1870 Chicago CD, Alfred is once again listed, residing with William at 133 29th. The 1870 CD lists him as a "physician" and also lists both brothers as partners with F. W. Herrmann in the "patent-medicine" firm of *Strickland, Herrmann, and Strickland* at 619 State. During their first year in the patent-medicine business, this company began carrying the line of patent medicines marketed by Michigan physician C. D. Warner (*see his Chicago listing*) and ran a six-month series of ads in the 1870 *Chicago Tribune* (from May to November) describing the benefits of several of Warner's medicines. These included *Warner's Pile Remedy, Warner's Dyspepsia Tonic, Warner's Cough Balsam, Warner's Wine of Life,* and *Warner's Emmenagogue*.

The ads were no longer run after this initial six-month blitz, but they must have been very successful because from 1871 to 1873, the Strickland Co. changed its name to *Warner's Proprietary Medicine Co.* (at the same address), and sold only the Warner products. Alfred Strickland was company president, William was superintendent, and F. W. Herrmann was secretary/treasurer.

This relationship was terminated during 1873. In the Chicago CD that year, W. H. Orr, Charles A. Folsom, and T. C. Orr are listed as the local managers of *Warner's Proprietary Medicine Co.*, but this was the last year the company had a Chicago outlet. The *Alfred & William Strickland Co.* was established in 1873 at 174–178 Adams. That first season the company was listed under "patent medicines" and

Figure 599

advertised themselves as manufacturers of *Strickland's Wine of Life*. This name is obviously derived from *Warner's Wine of Life* and suggests the Stricklands may have "modified" the Warner Company recipe to take advantage of the "Wine of Life" name recognition in Chicago. The firm of *A. & W. W. Strickland* was listed in the 1873 CAR as bitters manufacturers, so the brothers may have also marketed *Strickland's Wine of Life* as a bitters in the Chicago area.

Whatever the case, it does not appear to have been a successful front-line product; no mention of it could be found after 1873.

Beginning in 1874, the Strickland Company was listed rectifiers and wholesale liquor distributors. An amber "pumpkin-seed" flask is known, embossed **STRICKLAND'S** (*Figure 600*), which may have been used by the company at this time. By 1877, the brothers had returned to 619 State and changed the name of their wholesale-liquor company to *Strickland's International Cased Goods Co.* After that, from 1878 to 1887, the Strickland brothers ran a more locally grounded liquor and wholesale liquor operation from an almost-annually-changing series of Chicago addresses. Then from 1888 to the mid-1890s, they went back to their patent-medicine roots and became the manufacturers of *Liquid Tonic*. The product must have been successful, as it was taken over by new managers after the brothers retired from the business themselves.

Figure 600

Charles Sundell and Co.

Soda ~ Mineral Water:

(1) **C. SUNDELL & Co. / CHICAGO / ILL**   [cobalt, teal]
Note: normal-size embossing: all one side. Probably the earliest Sundell embossed-bottle style.

»IP; slope-shouldered soda; shouldered blob-top soda. [Figure 601]

(2a) **C. SUNDELL & Co. / CHICAGO. / ILL**   [cobalt]
Note: **C. SUNDELL & Co.** in tall, thin, wrap-around letters encircling the bottle.

»IP; slope-shouldered soda; shouldered blob top. [Figure 602, left full bottle and top row]

(2b) **C. SUNDELL & CO. / CHICAGO. / ILL.**   [cobalt]
Note: **C. SUNDELL & Co.** in tall, thick, wrap-around letters encircling the bottle. Ridges or ripples around heel on base. **ILL** has period.

»IP; slope-shouldered soda; shouldered-blob top. [Figure 602, right full bottle and bottom row]

Figure 601

Figure 602

Charles Sundell and Andrew Lawson (the "& Co." junior partner in the Sundell company) first appeared in the Chicago CDs in 1854 and 1855. In 1854, only Sundell was listed—at a boarding house, with no listed occupation. Lawson first appeared in 1855, as the senior partner in *Lawson & Sundell*, soda water manufacturers at 137 Illinois Street. No embossed bottles are known from this one-year partnership, but the next year the partnership was listed as *Sundell & Co.*, at 155, 157, and 158 Illinois (157 and 158 were Sundell and Lawson's residences), with Sundell listed as the senior partner of the two. Since two differently embossed *Sundell & Co.* bottles are known (perhaps reflecting their two seasons together), the 1855 CD may have accidentally reversed the partnership order. A list published in the *Chicago Democratic Press* in January 1857—headed "Miscellaneous Manufactures of Chicago, January 1, 1857"—includes *Sundell & Co.* and indicates the company had six employees, $3,000 capital invested, and valued their manufactures at $10,000 (see Andreas 1884:571).

From 1857 onward, Lawson was no longer listed in the CDs, and in 1857 Sundell himself is listed as an "upholsterer" at Indiana and Franklin. The 1858 CD indicated that Charles Sundell was back at the "soda water manufactory" business on his own at the 137 Illinois address (*see bottle #3, with "& Co." piened out*).

But this was the end of the soda water bottling business for Sundell, who underwent a radical career change. In 1859, he was appointed Acting Vice Consul of Sweden and Norway at the consulate in Chicago. In 1860, he became the official foreign consul for the two countries there. In 1861, he was appointed to the consulate at Stettin and moved to Prussia.

Calvin H. Swain

Bitters:

(1) **C. H. SWAINS // // BOURBON / BITTERS // //**   [amber, olive amber, olive green]

»SB w/ keyed hinge mold; beveled-edge square bitters; tapers slightly to base; short neck; tapered collar; 9" tall. *[Figure 603]*

(2a) **C. H. SWAINS // BURBON // BITTERS // //**   [amber]

»SB w/ keyed hinge mold; beveled-edge square bitters; short neck; tapered collar; 9" tall. *[Figure 604]*

(2b) **C. H. SWAIN'S // BURBON // BITTERS // //** [amber]
Note: added apostrophe in SWAIN'S.

»SB w/ keyed hinge mold; beveled-edge square bitters; short neck; tapered collar; 9" tall. *[No photo; similar to #2a above]*

❖

Calvin Swain first appeared in the Chicago CDs in 1862 and 1863 as the junior partner of *Hoyt, Pierce & Co.* "wine merchants and agents for A. M. Bininger & Co." at 147 Water Street. Bininger, of course, was a prominent New York City bitters company (supposedly established in 1776) whose flagship brand was *Bininger's Wine Bitters*, endorsed by "physicians, invalids, &c., as a delightful stimulant, healthy TONIC &c., promotive of digestion and unsurpassed as an appetitive" (Ring 1980:82).

In 1864, Swain left *Hoyt, Pierce & Co.* and advertised himself in the Chicago CD for a season as a "physician" at 34 Clark. Then, beginning in 1865, he started marketing his own bitters brand, placing a colored-page ad in that year's CD:

C. H. SWAIN,
Proprietor of
SWAIN'S BURBON BITTERS

OFFICE AND MANUFACTORY,
**34 SOUTH WATER ST.,
CHICAGO.**

Figure 603

The following year, in 1866, he was listed in the CD as the *Swain Manufacturing Co.* "manufacturer of bourbon bitters" (C. H. Swain, president, H. Comstock, vice president, and G. W. Barnard, secretary). An undated (ca. 1866) Keokuk, IA, newspaper ad for "Dr." Swain's Bourbon Bitters reproduced by Ring (1980:448–449) indicates it was marketed as "A healthy tonic, gentle stimulant, and unequaled morning appetizer, prepared in ripe Old Bourbon Whisky. Free from grain oil, with flowers, buds, and barks of the highest medicinal virtue. By increasing the appetite, assisting digestion, regulating the bowels, and giving tone to the system, they impart strength to the body and cheerfulness to the mind."

During the next eight years (1867 to 1874), the company diversified into the *Swain Manufacturing and Distilling Co.* "importers and jobbers of wines, brandies, gins, and whiskies." They no longer focused their wholesale efforts on Swain's Bitters alone, but no doubt continued to sell it. Over the years, their warehouse and distilling operation moved from 34 Water to 86 and 88 Jackson, and then to 43 and 45 State in Chicago.

From 1875 to 1877, his final three years in business, Calvin Swain was listed in the Chicago CDs as just a commission and forwarding agent—perhaps for his Bourbon Bitters—at 14 State Street.

Figure 604

Henry J. Sweet

Medicine:

(1) **SWEET'S / BLOOD RENEWER //** [aqua]
    (side): **CHICAGO, ILL. // //**
    (side): //

»SB; beveled-edge rectangle; 3 sunken panels; thick flared-ring collar; 8" tall. *[Figure 605]*

(2) **SWEET'S / CHOLERA. DROPS** [aqua]

»SB w/ keyed hinge mold; beveled-edge rectangle; 3 sunken panels; double ring collar; 5¼" tall. *[Figure 606]*

--- ❖ ---

Henry Sweet was a long-time Chicago pharmacist who first appeared in Chicago CDs in 1858 in partnership with William J. Jauncey at Canal and Wilson Streets. By 1859, he had removed to his long-time commercial location at Des Plaines and Kinzie, where he conducted business until 1872, often in association with Jauncey and with another druggist named Noble Schroeder. During the 1860s, when the above medicines were likely produced, Sweet was variously characterized in the CDs as a drug and medicine retailer, a druggist, and an apothecary. He moved his pharmacy operation to Milwaukee Avenue in the later 1870s and 1880s, and retired from business in 1889. He was last listed in the directories, as a Chicago resident, in 1891.

No newspaper advertisings have been found for his medicines. The **CHOLERA DROPS** are included as a possible Chicago Sweet's product because the recorded bottle was excavated from a Civil War–era privy in Springfield, IL, and because just after Sweet's probable death an ad appeared in the *Chicago Tribune* (Sept. 1, 1892, p. 5, col. 7) that may have been placed by a commission agent who had acquired his remaining stock (probably in unembossed bottles):

Figure 605

Prepare for Cholera.

In 1866, when the cholera raged in this country, the "New York Sun" printed the recipe for **SWEET'S CHOLERA DROPS,** that the people might know the best remedy for that terrible disease. It saved thousands of lives then and will save them now if kept on hand and taken on the first symptom of diarrhoea or pain in the bowels. Price 25c at all drugstores.

Sweet was associated at one time with E. H. Flagg, probably as a marketing outlet for the patent-medicines that Flagg developed from his residence in Chicago (*see Flagg listing, this chapter*).

Charles R. Sykes

Medicine:

(1) **DR SYKES' / SURE CURE / FOR / CATARRH** [aqua]

»SB w/ post mold; high-shouldered cylinder; squared flared-ring collar; 7" tall. *[Figure 607]*

--- ❖ ---

According to a trade card from the 1890s, the Dr. Sykes catarrh-cure company was founded in Chicago in 1870. He operated in Chicago for 18 years, and then moved to 330 Race Street, Cincinnati, OH (330 Race Street), in 1888. Dr. Charles Sykes is listed in the 1880 USC as a 50-year-old Chicago physician.

Figure 606

Chicago directories do not precisely agree. According to the CDs, Charles Sykes first appeared in Chicago in 1869 as a junior partner in *Parshall & Co.*, an employment agency. In 1870, he was junior partner with "physician" C. A. Miner in the company of *Miner & Sykes* at 114 Dearborn. *Miner & Sykes* produced a product called *"Miner's Erasive Solution,"* which was probably some sort of patent medicine (although it sounds more like office supplies), since Miner was listed as a physician. The 1871 CD listed C. R. Sykes alone at the same address as "manufacturer, Miner's Erasive Solution." No embossed bottles for this product are known.

Beginning with the 1872 CD, Sykes was listed as a doctor, and proprietor of his new catarrh cure:

SYKES C. R. Dr. Propr. of the Sure Cure for Catarrh, 109 W. Randolph.

By 1873, he was listed under "Physicians" and under "Patent Medicines" as proprietor of his "Sure Cure," and advertised a "Free Trial" of his catarrh remedy that year in the *Chicago Tribune*. The next year he was listed under physicians as a "catarrh specialist" and moved to offices at 169 Madison, where he conducted his catarrh-cure business throughout the 1870s. By 1874, a September 20th *Chicago Tribune* ad (p. 2, col. 6) indicated that "About fifty persons daily...are trying 'Dr. Sykes' Sure Cure for Catarrh' free. Will be open this afternoon, and continue giving free trials every day this week." His 1875 ads were headed **A.B.C.**, and promised his medicine would cure asthma, bronchitis, and catarrh (a chronic condition of inflamed mucus membranes accompanied by a runny nose). His advertising from the late 1870s was somewhat more creative:

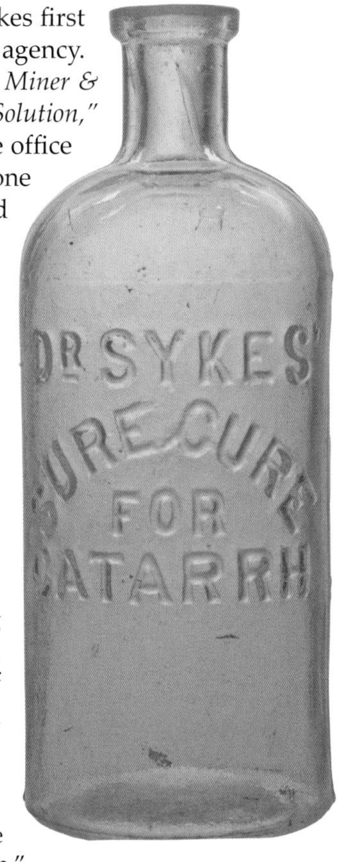

Figure 607

> A gentleman sent by a committee from Colorado to investigate the treatment and remedies for Catarrh, spent three days last week in our city in this work. Finding the preponderance of reliable testimony greatly in favor of "Dr. Sykes' Sure Cure for Catarrh," he invested in that remedy, and left a long list of names of interested parties to be supplied.
>
> The Doctor has recently published a list of cures which, in point of numbers, variety, and severity of cases, stands unparalleled. This, with his invaluable book and a trial of the remedy, he gives free to all who apply at his office, 169 East Madison Street. [*Chicago Tribune*, Dec. 16, 1877, p. 8, col. 6]

> WE SHOULD LIKE TO KNOW OF ANY ONE who has ever enjoyed the holidays when in a poor state of health.... How would "Habby New Yerars" of "Berry Grisman" sound forced trough the nose because the head was clogged by catarrh? All of which is a reminder that Dr. C. R. Sykes, of 169 East Madison Street, who has been so remarkably successful in the treatment of these troubles... has placed his "Sure Cure" for catarrh in the hands of the druggists so that all can procure it without inconvenience. There is nothing which can afford more comfort...and no gift which one friend can bestow upon another more appropriate or acceptable than his "Sure Cure" and Insuffalator." [*Chicago Tribune*, Dec. 19, 1879, p. 3, col. 6]

> THE REASON HE GIVES FOR not advertising is, that from ten years' successful experience he has all he can do to attend to his callers and to fill his orders, with five assistants to help him. This is the best evidence of *merit* and refers to Dr. C. R. Sykes and his "Sure Cure for Catarrh," offices 169 East Madison Street, which are thronged daily. [*Chicago Tribune*, Apr. 24, 1881, p. 9, col. 3]

Then, in 1883, Dr. Sykes business professional reputation hit a serious snag from which he probably never fully recovered, and which may have eventually resulted in his move from Chicago to Cincinnati. The event was detailed in the *Chicago Tribune*:

DIED IN THE SHOP.
THE SAD EXPERIENCE OF MRS. NICHOLS
WHILE CALLING ON HER PHYSICIAN.

A story was in circulation yesterday to the effect that a lady had died in a doctor's office Sunday while undergoing the process of development into a healing medium.... Her name was Mrs. Riley

L. Nichols, and that of the doctor C. R. Sykes [*described as "a tall and rather spare man, apparently about 60 years of age" with "flowing gray hair and beard"*]. He is not a general practitioner, but a catarrh specialist. According to the story Mrs. Nichols was a Spiritualist, and her associates convinced her that she possessed the healing power, and recommended her to go to Dr. Sykes for "development." While there Sunday he mesmerized her, and she died in that state, the only witnesses being the doctor and his assistant. The body, it was said, was removed under cover of night, and every effort made to conceal the actual cause of death.

Dr. Sykes said he was sitting in his office about 7 o'clock Saturday evening when Mrs. Nichols entered, and they chatted for nearly an hour. Suddenly the woman appeared to go into a trance, uttering a number of gibberish sentences.... The doctor conducted her to a seat and began "manipulating" on her head, but it was about half-past 10 o'clock before she recovered herself...and finally accepted the Doctor's invitation to lie down in the rear office. She partially disrobed, and the Doctor chafed her limbs...and then her body, fully restoring the circulation of the blood.... About 7 o'clock [*Sunday morning*] Mrs. Nichols began to froth slightly at the mouth and an electric battery was applied, but the woman suffered another relapse in spite of it all. [*Chicago Tribune*, Sept. 12, 1883, p. 8, col. 4]

In early 1886, Dr. Sykes *Sure Cure for Catarrh* offices were removed "to more congenial quarters" at the corner of Clark and Adams, and by 1888 he had relocated his catarrh clinic to Cincinnati.

John W. Sykes and Calvin M. Fitch

Medicine:

(1) (side): **DR C. M. FITCH** // *(paper label)* // [aqua]
    (side): **DR J. W. SYKES** // **CHICAGO** / **ILL** *(in small shield-shaped sunken panel)* //

»SB w/ keyed hinge mold; beveled-edge rectangle; 2 sunken side panels and small shield-shaped sunken panel on rear face; tapered collar; 6" tall. *[Figure 608]*

Figure 608

Drs. John Sykes and Calvin Fitch were associated physicians in Buffalo, NY in 1856, when Dr. Fitch made an early and well-publicized professional visit to Chicago. As the *Chicago Tribune* noted on September 8th (p. 3, col. 1):

> In Compliance with the expressed wish of many of his Western Patients, **Dr. C. M. FITCH** (Late of New York, author of "The Invalid's Guide and Consumptive's Manual," &c. &c.) will make a brief visit to Chicago, remaining at the **SHERMAN HOUSE** During the month of SEPTEMBER, where he may be consulted by those desiring his advice. He will, he trusts, have assistance sufficient to enable him to avoid detaining patients so long as was often necessary on the occasion of his last visit....
>
> DR. FITCH'S success is due, not only to his ample expertise, but also to the fact that regarding Consumption as both a local and general disease, he treats it as such, and by appropriate **MEDICINAL AND STENOTHROPIC INHALATIONS,** reaches directly the Inflamed or Ulcerated surfaces, while by the employment of appropriate INTERNAL AND EXTERNAL REMEDIES he removes the constitutional taint, which gives rise to the disease, treating, also, at the same time, any other afflictions which may be complicated with Consumption. **DYSPEPSIA, CATARRH, FEMALE DISEASES, ETC.,** claim his special attention, and in the treatment of the latter particularly, his success has been greatly enhanced by the use of proper **MECHANICAL APPLIANCES**, as well as medicines.
>
> In **ASTHMA AND CHRONIC BRONCHITIS** the combination of remedial measures is attended with the most marked and happy results, oftentimes affording speedy and permanent relief in cases which under any partial system of treatment would either have resulted fatally or have been but slightly palliated.
>
> Since his return from Europe, Dr. FITCH has opened in connection with his associate, J. W. SYKES, A.M.M.D., a permanent office at
>
> **458 MAIN STREET, BUFFALO, N. Y.**

According to the Chicago CDs, just three years later Drs. Fitch and Sykes moved their practice to Chicago, establishing offices at 104 Monroe Street in 1959 and 1860. However, these were the only two years they were in business together in Chicago. During 1861 and 1862, Dr. Sykes was listed alone as a Chicago physician, and after that he too was gone.

The fact that the listed bottle is smooth-based rather than pontiled indicates that it dates to 1860 rather than 1859. As luck would have it, the bottle we recorded still retains its paper label (*Figure 608, right*), indicating that it contained inhaling fluid, and reinforcing Fitch and Sykes' focus on lung diseases such as consumption, asthma, and bronchitis (as expressed in the 1856 *Tribune* article above). The paper label on the bottle reads:

DR. C. M. FIT[CH]

INHALING FLUID
No. 2
FOR
CONSUMPTION,
ASTHMA
AND ALL
PULMONARY DISEASES

PREPARED BY
DRS. C. M. FITCH & J. W. SYKES
104 Monroe Street, Chicago

Fragments of another Dr. Sykes bottle have been recovered in Chicago that may have been produced during the Civil War years by J. W. Sykes, but no corroborating documentation has yet been found to confirm this possibility. It is an amber, 6 1/2" tall, beveled-edge square, tapered case-gin-style bottle, with three of four sides embossed within horizontal and pendant vertical ribbon banners: **DR SYKES / NEW ENGLAND // LIVER / TONIC AND // BILLIOUS / ANNIHILATOR.**

The bottle is mentioned here because we are unaware of any widely distributed Eastern patent-medicine brands produced under a "Dr. Sykes" name, and "Dr." Charles Sykes (*see his Chicago listing*) was a catarrh specialist only. It should be noted, though, that Chicago CDs indicate another physician named Sykes (Dr. Charles J. Sykes, boarding at 278 State Street) may have set up shop for a short time in Chicago in 1863.

John and Joseph Taylor

Ale:

(1) **TAYLOR & BRO**. / **CHICAGO** [olive-amber "black" glass]

»IP; 3-piece mold pint; ring-neck tapered collar; 7 1/2" tall; shoulder-embossed. *[Figure 609]*

(2) **TAYLOR & BROTHER** [olive "black" glass]

»IP; 2-piece mold pint; ring-neck tapered collar; 7 1/2" tall; body-embossed. *[Figure 610]*

(3) **TAYLOR & BR**º (*arched*) / **CHICAGO** (*arched*) [olive "black" glass]
Note: *Single embossed dot beneath superscript* **O** *in* **BR**º.

»IP; 3-piece mold quart; ring-neck tapered collar; 9 3/4" tall; shoulder-embossed. *[Figure 611]*

Porter:

(1) **TAYLOR & BRO..** (*arched*) / **CHICAGO.** [aqua, teal green]
Note: *Embossed dash above double embossed dots after* **BRO..**

»IP; squat porter-style bottle; ring-neck tapered collar; 6 1/4" tall. *[Figure 612]*

--- ❖ ---

"Taylor" was a common name in Chicago during the early years, and there were multiple "J." Taylor, "John" Taylor, and "Joseph" Taylor listings in the CDs every year from 1844 to 1860. During much of this time, for instance, there was a John Taylor in the grocery business and a carpenter named John Taylor. However, they never resided at the same address and no Taylor was ever listed as a bottler of any kind until 1856, when the company of *Taylor & Brother* first appeared in the Chicago CDs.

In the 1856 CD, *Taylor & Bro.* were first listed as "ale and porter bottlers" at 18 La Salle Street, and they resided together at 22 LaSalle. That first year John Taylor's brother (later, Joseph) was listed as "James." They were not listed in the 1857 CD, but in 1858 *Taylor & Bro.* were bottling ale and porter and residing at the same two addresses. From 1858 on, they were listed as John and Joseph. In 1859, John and Joseph Taylor moved to new quarters, and during 1859–60, the brothers bottled ale and porter at a home they shared on Wolcott between Scott and Grand Haven Slip (*aka* Wolcott near Division). Paul Welko (1977b:9) has pointed out that "It is interesting to note that one of the city's largest breweries, that of John Huck, was also listed as being located on Wolcott near Division for the same year. It is almost certain that Taylor & Bro. bought and bottled the products of this popular brewer."

Joseph Taylor is listed alone in the 1861 Chicago CD (and in the 1860 ISD) as an ale and porter bottler at the Wolcott address. Joseph was also gone from the CD listings by 1862 (unless he was the Joseph R. Taylor listed as a clerk at the druggist firm of J. H. Reed & Co.).

According to information from Peter Maas (personal communication, 1995), the Taylor brothers may have bottled in

Figure 609

Figure 610

Figure 611

Figure 612

Figure 613

the Milwaukee area before coming to Chicago. *Taylor & Brother* are listed in the 1850–51 Milwaukee CD as bottlers of "soda and small beer." Paneled stoneware bottles stamped "**TAYLOR & BROTHER**" and pontiled soda bottles embossed "**TAYLOR & BROTHERS**" have been found in and around the Milwaukee area (*see Figure 613*).

Newton S. Taylor

Equine Medicine:

   (1) **N. S. TAYLOR / CHICAGO / ILL.** *[aqua]*

   »SB w/ post mold; beveled-edge rectangle; 3 sunken panels; sloped shoulders; flared-ring collar; 6" tall. *[Figure 614]*

---❖---

Newton Taylor first appeared in the Chicago CDs in 1864 as a clerk for patent-medicine wholesale dealer Harvey (also listed as "Harry") Scovill at 76 Randolph, boarding at Scovill's residence. In 1865, Taylor was listed as a salesman for the Scovill Company. Scovill was listed as a drug and patent medicine dealer in Chicago from 1862 through 1866, but no embossed bottles bearing his company name are known. In an 1865 CD ad, Scovill listed himself as a "Wholesale Dealer in FAMILY MEDICINES, perfumery, fancy soaps, flavoring extracts, etc."

Figure 614

After the end of the Civil War, Newton S. Taylor became a partner in the *Walker & Taylor* patent medicine bottling and distribution firm from late 1865 through 1876 (*see their entry*).

From the style and post-mold construction of the listed bottle, it likely dates to the 1870s. After the Walker and Taylor partnership split up in 1876, Taylor remained in the patent-medicine trade on his own. He operated a patent-medicine business at 126 Quincy Street during 1877–78, and moved the operation to 233 5th Avenue in 1879. He closed the business address in 1880, and after that operated from his residence 2336 Prairie Avenue until 1887, during which time he was variously listed in the directories as a "chemist," "manufacturer," and "patent medicine" maker.

Taylor's focus was veterinary medicine, and he was listed in the 1880 USC for Chicago as a 38-year-old manufacturer of horse medicine. So the listed bottle probably contained equine medicine bottled and sold during the early years of Taylor's post-*Walker & Taylor* patent medicine business: ca. 1877–80.

Frank C. Taylor and John M. Wright

Bitters:

(1) **RED CLOUD / BITTERS // // TAYLOR & WRIGHT / CHICAGO // //**   [amber]

»SB; beveled-edge square bitters; 4 flat panels; tapered collar; 9 1/2" tall. *[Figure 615]*

———————————— ❖ ————————————

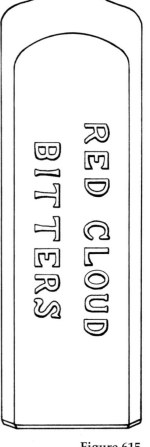

Figure 615

According to the Chicago CDs, Frank C. Taylor and John M. Wright were wholesale grocers in Chicago at 6 and 8 River Street from 1864 through 1869 (successors to *C. G. Wicker & Co.*). In June 1869, they opened the Mammoth Tea House at their 6 & 8 River address, and began a new wholesale tea business at 48 and 50 South Water in 1870. The Taylor & Wright tea business lasted through 1872, after which John Wright continued in the tea business on his own.

Their foray into Red Cloud Bitters manufacture during the late 1860s was highlighted by a highly publicized court battle with Bennett Pieters, of *Red Jacket Bitters* fame in 1867 (*see his listing, this chapter*). Pieters was convinced that Taylor & Wright had closely copied his bitters name to take advantage of its notoriety using an inferior product. But the outcome of the court case, highlighted by the testimony of an expert chemist, was that although the two products were different, both were little more than "poor whiskey" with a variety of added botanical ingredients with little or no medical medicinal value. The result was a marketing collapse for both bitters brands in Chicago.

Red Cloud Bitters resurfaced in 1868 in New York City, where it was distributed for a few years by *Arnold Theller & Co.* with little success. The later *Red Cloud Bitters* bottles are embossed **VOWINKLE & THELLER** instead of **TAYLOR & WRIGHT**, as on the listed example.

No period ads for Red Cloud Bitters have yet been located to indicate what their purported health value was when marketed by Taylor & Wright in Chicago, or to what extent they may have taken advantage of *Red Jacket Bitters* fame to sell the product.

Levi H. Thomas

[Ink, Bluing, Mucilage, Etc.]

Although there are several embossed varieties of Thomas inks, they are not included in this volume because the company was not formed until 1880. Levi Thomas apparently brought his ink company to Chicago from Reading, MI; at least one early-1870's paper-labeled Thomas ink is known from the Reading, MI, years that is base-embossed **L&W** (Lindsay 2009).

The Chicago company was listed under Levi Thomas' direct supervision in the Chicago CDs for only six years: at 51 Wabash Avenue from 1880–85 and at 59 Michigan Avenue in 1886. After 1885, Thomas' personal attention was focused on developing his patent (#331843 awarded December 8th of that year) for a paper ink bottle (Faulkner and Faulkner 2006):

> The manufacture of paper bottles is said to be becoming an important industry at Chicago, and the process adopted is that invented by Mr. L. H. Thomas. These paper bottles, which can be made of all shapes and sizes, are cheaper than those made of glass.... A sheet of paper cemented on one side is rolled on a mandrel, after which the neck is fashioned, and a bottom of paper or wood inserted into the paper vessel. An outer glazed-paper cover is next added; and the interior of the bottle is lined with a fluid composition, which speedily becomes hard, and resists alkalis, acids, spirits, and everything else. The bottles are unbreakable, and require no packing in transit. For various purposes, such as the carriage of ink, blacking, varnishes, and paints, these bottles will doubtless be found useful; but for wines, spirits, medicines, &c., glass, which has the advantage of transparency and great cleanliness, is likely to hold its own. [Chambers 1887:415]

Many of these paper ink bottles are known to collectors —including several varieties of Thomas inks (see Faulkner and Faulkner 2006)—which may suggest that the embossed hand-blown Chicago Thomas inks date primarily to the early 1880s. After the Thomas Company was incorporated into the Sanford group (ca. 1894), glass bottles were again produced and at least one machine-made Thomas ink is known (Lindsay 2009).

Thomas inks are noted here because a glass L. H. Thomas master-ink bottle is base-embossed **PATENTED APRIL 13 1875**. The T. W. Synnott patent referred to (#162117) was for a finishing tool to form the bottle's pour spout, and does not indicate that Thomas inks were being manufactured in Chicago during the late 1870s.

[R. T.] Thomas and Co.

Liquor:

(1) **THOMAS. & Co / CHICAGO** [aqua, clear]

»OP (solid rod pontil); long-neck "onion"-shaped bottle; flat base; flared-ring collar; 7¾" tall. *[Figure 616]* Recently, a smooth-base variety of this bottle has been located, with a long tapered collar. Both varieties were made in the mold, suggesting they were used ca. 1859–1861.

❖

Thomas was a common surname in Chicago in the early days, and several *Thomas & Co.* businesses were located in the CDs from 1850 to the mid-1860s. Since the above bottle is pontiled, it should predate 1860. 1850s *Thomas & Co.* businesses included a coal company, lumber dealer, hardware dealer, railroad-furnishing company, house-furnishing company, real estate company, roofer, and a fruit and produce wholesaler.

The most likely 1850s company to have produced embossed bottled liquor was *R. T. Thomas & Co.*, a commission/forwarding merchant with offices and warehouse at 29 N. Dearborn Street. The company was in business only during 1858 and advertised heavily in the *Chicago Daily Press* that fall and winter: "attention given to the sale and purchase of Flour, Grain, LiveStock, Provisions, &c., on Commission." A common advertising ploy of middleman companies such as this, on entering the trade, was to distribute bottles of ale,

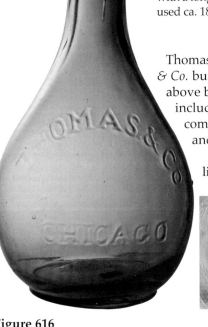

Figure 616

wine, or liquor that advertised the business to prospective customers. Perhaps that was the genesis of the listed bottle.

Beyond that, the only other strong candidate, *Thomas & Co.*, was a wholesale wine and liquor dealer in business at 89 LaSalle Street, but only in 1864. Company partners were Frederick A. Thomas and W. A. Dickerman. They listed their residential address as Rockford, IL.

Thompson

Medicine:

(1) *(side)*: **THOMPSON** // // *(side)*: **CHICAGO**   [aqua]

»SB; beveled-edge rectangle; 3 sunken panels; thickened flared lip; 5" tall. *[Figure 617]*

Figure 617

❖

Thompson is another very common surname in Chicago, and the problem of identifying the bottler is compounded by the fact that the bottler did not use their given-name initials in the embossing. From its technological attributes and style, the bottle likely was made and used during the 1870s. That narrows down the list considerably, but final identification of the bottler will not be possible unless an example of the bottle can be located that still retains its paper label.

Omitting grocers, baking-powder manufacturers, and dentists named Thompson, there were five physicians named Thompson operating in Chicago during the 1870s: Daniel D. (1870–77), Mary H. (1870–81), Eliza A. (1875 only), Merritt W. (1975–88), and Robert C. (1878–80). There was also a druggist to consider: James Thompson at 177 Clark (1876–78), who had worked for Buck & Rayner during the early 1870s.

Further, the bottle might have contained spices rather than medicine. From 1869 to 1872, Alexander M. Thompson was the proprietor of the Western Coffee and Spice Mills Co. at 29 South Canal (in the later 1870s, the company became Thompson & Taylor), and during 1871 Hugh B. Thompson was listed as a Chicago spice and tea dealer.

Tivoli Bottling Company

Ale ~ Beer:

(1) **TIVOLI BOTTLING Co / AD. CLASEN & C[o] / PROP'S / CHICAGO. ILL //**   [aqua]
THIS BOTTLE // NOT TO // BE SOLD
(base): **C & I**

»SB w/ post mold; slope-shouldered quart beer; embossed within a circular slug plate; sharp-shouldered collar; 9½" tall. *[Figure 618]*

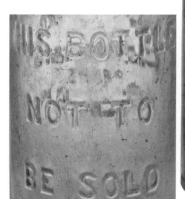

❖

The proprietors of the *Tivoli Bottling Co.* were Adolph D. Clasen and Peter Grube. Both men first appeared in the Chicago CDs in 1878 and 1879, as proprietors of the bottling company, located at 422 State Street. In 1880, only Clasen was still listed as a Chicago resident, as proprietor of the *Adolph Clasen & Co.* bottling company at 152 Dearborn

Figure 618

(Grube was still listed as junior partner). Neither Clasen nor his bottling company were listed in the directories after 1880. During all three years in operation, Clasen and Grube were listed in the CD manufacturer's summaries as bottlers of "ale, porter, and beer." According to the March 24, 1878, *Chicago Tribune* (p. 8, col. 4), the Cook County "Committee on Supplies" awarded the *Tivoli Bottling Co.* that year's contract to supply the county's beer needs.

Since the *Cunningham & Ihmsen* glassworks (*C&I*) ceased operation after 1878, the embossed *Tivoli Bottling Co.* bottle listed can be dated to the Clasen and Grube's first year in business.

John A. Van Buskirk and David Henry

Bitters ~ Schnapps:

(1) **VANBUSKIRK & HENRY / CHICAGO ILL'S** // *(clock-face image, hands set at 11:00)* //   [amber]
UNIVERSAL BEVERAGE / OR TONIC SCHNAPPS.

»SB w/ keyed hinge mold; beveled-edge square; short neck w/ tapered collar; 9 1/2" tall. *[Figure 619]*

Ale:

(1) **V & H. / CHICAGO.** //   [amber]
(heel): **A & D. H. C.**

»SB w/ keyed hinge mold; quart ale; ring-neck tapered collar; 9 1/4" tall. *[Figure 620]*

Soda ~ Mineral Water:

(1) **V & H / CHICAGO** //   [aqua]
(heel): **A & D. H. C.**

»SB w/ keyed hinge mold; blob-top soda. *[Figure 621]*

According to the Chicago CDs, John Van Buskirk and David Henry began their partnership in the "wholesale wines and liquors" business at 20 State Street in 1864. They placed a large introductory ad for their new business on the back cover of that years' directory.

Figure 619

Figure 620

Figure 621

Van Buskirk was senior partner in the company for just three years: 1864–66, and the listed bottles were likely produced during this time.

During 1867 and 1868, Van Buskirk became junior partner, and the company name was changed from *Van Buskirk & Henry* to *David Henry & Co.* In 1869, David Henry was listed alone in the CDs, with an office at the same address.

In 1870, David Henry moved to 79 Wabash, where he was listed as a distiller. This was his final listing in the liquor business. In the early 1870s, Van Buskirk was listed in partnership with Dugal Steward as storage and commission merchants at 131 Kinzie.

Wallace's Tonic Stomach Bitters

See the George Powell & Co. listing.

T. Irving Walker and Newton S. Taylor

Medicine:

(1) **WALKER & TAYLOR / CHICAGO / ILL** *[aqua, clear]*

»SB; nearly rectangular; base wider than shoulders; 3 sunken panels & one flat side; flared lip; 4 1/2" tall. *[Figure 622]*

(2) **BROWN'S / ORIENTAL / HAIR RENEWER /** *[aqua]*
WALKER & TAYLOR / CHICAGO. ILLS

»SB w/ keyed hinge mold; beveled-edge rectangle; 4 flat panels; square flared-ring collar; 5" tall. *[Figure 623]*

444   Chicago — T. Irving Walker and Newton S. Taylor

(3) **PRATT & BUTCHER / MAGIC OIL // WALKER & TAYLOR / PROPRIETORS / CHICAGO ILL**   [aqua]

»SB w/ keyed hinge mold; oval; square flared-ring lip; 3 3/4" tall. [Figure 624]

(4) **SLOAN'S / OINTMENT // WALKER & TAYLOR / PROPRIETORS // CHICAGO. ILL // //**   [aqua]

»SB w/ keyed hinge mold; beveled-edge square (1/2 x 3/4"); squat, wide-mouth salve jar; flared-ring collar; 1 3/4" tall. [see Sloan's listing above, Figure 578, top and bottom left]

(5) **SLOAN'S / OINTMENT // WALKER & TAYLOR / PROPRIETORS // CHICAGO. ILL // //**   [aqua]

» SB w/ keyed hinge mold; beveled-edge square (1 x 1"); squat, wide-mouth salve jar; flared-ring collar; 2 3/4" tall. [see Sloan's listing above, Figure 578, center and right]

(6) **WALKER & TAYLOR / CHICAGO ILL //**   [blue teal]
(side): **DR. WEAVER'S // COMPOUND EXTRACT / OF FIREWEED OR //**
(side): **HUMOR & LIVER SYRUP**

»SB w/ keyed hinge mold; beveled-edge rectangle; 3 cathedral sunken panels: 1 large panel on front, 2 side-by-side smaller panels on reverse; 8 1/2" tall. [Figure 625]

In the 1865 Chicago CD, Thomas Irving Walker and Newton S. Taylor were both listed as employees at the H. Scovill wholesale patent medicine company at 76 Randolph (see N. S. Taylor listing, this chapter). Interestingly, they are also listed in the patent-medicine summary in the 1865 directory as having started their own company at the same address. From 1866 to 1876, Walker and Taylor

Figure 622

Figure 623

Figure 624

Figure 625

were partners in the patent medicine and veterinary drug business from new headquarters at 134 & 136 Wabash (1866–71), then 65 W. Lake (1872–73), and finally at 79 Market Street (1874–76).

In 1866, they bought the rights to Walter B. Sloan's Chicago patent medicine line (*see Sloan listing, this chapter*) and in 1869 republished Walter B. Sloan's book, *Sloan's Complete Farrier and Cattle Doctor*. The volume was the most recent edition of W. B. Sloan's original 1849 publication, *The Complete Farrier or Horse Doctor* (which had gone through several earlier editions).

Walker & Taylor's 1866 Chicago CD ad (on the patent-medicine summary page), announced that they had become sole proprietors of Sloan's *Condition Powders, Horse Ointment, Family Ointment, and Instant Relief*. It also indicated that they had acquired and were bottling and selling ***all*** of the listed *Brown's, Pratt & Butcher, Dr. Weaver's*, and *Sloan's* embossed patent medicine products by the end of the Civil War. To date, the only Sloan's product they are known to have continued selling in embossed bottles was *Sloan's Ointment* (*see bottles #4 and #5*). However, bottle #1 is generically embossed and may have contained any of their advertised medicines.

In fact, except for *Sloan's Ointment* and *Pratt & Butcher Magic Oil*, none of the Walker & Taylor stable of patent medicine products are known in any earlier embossed version. Other pre–Civil War Professor Brown's hair products (restoratives and dyes) are known from Rochester, NY, and other early Dr. Weaver's embossed medicines are known from New London, CT (Odell 2007:52, 365). In the 1989 Greer-collection auction sale (Harmer Rook # 1417), there was a listing for an open-pontiled *PRATT & BUTCHER / MAGIC OIL / BROOKLYN NY*. From 1857 to 1859, Pratt

& Butcher advertised their *Magic Oil* in Brooklyn as a cure for "for pains, headache, toothache, sprains, sore throat" (Odell 2007:285). The advertised function of Sloan's medicines can be found in the Sloan listing. The use of *Browns Hair Renewer* and Dr. Weaver's *Humor & Liver Syrup* is fairly clear from the names.

Walker & Taylor's 1866 Chicago ad also mentioned an additional product in their patent-medicine line that has not yet been found individually advertised or in embossed-bottle form: *Irving's Giant Pain Curer.*

Walker & Taylor are listed as patent-medicine manufacturers and wholesalers in the 1870 USC. Walker died in September, 1880.

Dr. C. D. Warner

Medicine:

(1) **WARNER'S // WINE OF LIFE**   [amber, aqua]
 (*base*): **L&W**
 Note: *Some of these bottles do not have the* **L&W** *base mark.*

 »SB w/ post mold; 3-piece-mold whiskey-style bottle; high-shoulder cylinder w/ long-neck; ring-neck tapered collar; shoulder-embossed only; 11 1/4" tall. *[Figure 626]*

(2a) (*side*): **WARNER'S // WHITE WINE / AND / TAR SYRUP //**   [aqua]
 (*side*): **CHICAGO // //**
 (*base*): **A & D H C.**

 »SB w/ keyed hinge mold; rect.; 3 embossed sunken panels; flat panel for paper label; doubling-ring collar; 8 1/2" tall. *[Figure 627]*

Figure 626

Figure 627

(2b) (side): **WARNER'S // WHITE WINE / AND / TAR SYRUP //** [aqua]
  (side): **CHICAGO**
  (base): **N** *(embossed 5-pointed star)*

  »SB w/ post mold; rect.; 3 embossed sunken panels; 4th sunken panel for paper label; tapered tooled collar; 8 1/2" tall. *[Figure 628]*

Liquor~Remedy:

  (1) **WARNER'S / AROMATIC / BRANDY** [amber]

  »SB w/ post mold; 3-piece mold; high-shoulder whiskey cylinder; long-neck; ring-neck tapered collar; shoulder-embossing only; 11" tall. *[No photo — similar to Warner's Wine of Life bottle]*

  (2) (shoulder): **WARNER'S /** [amber]
  (slanted across body): **AROMATIC / BRANDY**

  »SB w/ post mold; 3-piece mold; high-shoulder whiskey cylinder; long-neck; ring-neck tapered collar; 11 1/2" tall. *[Figure 629]*

  (3) **WARNER'S / IMPORTED // ENGLISH GIN** [aqua]

  »SB; square beveled-edge square case bottle; tapered collar; 8 3/4" tall. *[Figure 630]*

Although some Dr. Warner's remedies are embossed "Chicago" (*see #2a and #2b*), and Warner's bottles are commonly found in the Chicago area, these products were only directly marketed in the city from 1870 to 1873. During this time, they were distributed by a local branch of

Figure 628

Figure 629

Figure 630

Warner's Proprietary Medicine Co., established at 619 State Street and run by Alfred and William Strickland (*see discussion under Strickland listing, this chapter*). The Warner medicine company in Chicago was *unrelated* to the large William R. Warner medicine company of Philadelphia and the world-famous Hubert H. Warner "Safe Cure" company of Rochester, NY.

Dr. C. D. Warner was a Michigan physician who received his medical training at the Detroit Commercial College during the Civil War, and in the late 1860s opened a medical practice in the Reading, MI, area, where he developed and sold his blood and liver pills and *White Wine and Tar Syrup* (cough remedy: *see Figure 628*) until he relocated to Coldwater, MI, in 1889.

When he began the manufacture of Warner's White Wine of Tar he walked from house to house, selling his medicines, which he carried in a grip sack. Later he was enabled to purchase a horse and buggy, and so satisfactory did his remedies prove that his business grew rapidly and has now reached extensive proportions, making Dr. Warner one of the substantial citizens of Coldwater. [Collin 1906:616]

In Coldwater, Warner "devoted his attention largely to the White Wine of Tar—a remedy which has become known all over the world" (Collin 1906:616). In 1916, the AMA determined that the primary active ingredients of this widely sold medicine were opium and alcohol (Crump 1921:106).

When the Strickland brothers decided to focus their patent-medicine marketing activities on Warner's medicines in 1870 (*see their listing, this chapter*), their advertising efforts included a six-month newspaper ad campaign in the *Chicago Tribune* from June through November of that year. The following medicines were listed for sale in the ads (interestingly, the *White Wine & Tar Syrup* was *not* included in the ads, and the glass-maker marks on the *WW&T* embossed bottles listed above suggest they were not marketed in Chicago until the later 1870s):

Warner's Wine of Life—"a splendid appetizer and tonic, and the finest thing in the world for purifying the blood…. Those who wish to enjoy good health, and a free flow of lively spirits, will do well to take the Wine of Life."
Price $1 in quart bottles.

Warner's Cough Balsam—"The extraordinary power it possesses in immediately relieving, and eventually curing, the most obstinate cases of Coughs, Colds, Sore Throat, Bronchitis, Influenza, Catarrh, Hoarseness, Asthma, and Consumption is almost incredible."
Price $1 in large bottles.

Warner's Emmenagogue—"the only article known to cure the Whites…. It is also a cure for female irregularities, and may be depended upon in every case where monthly flow has been obstructed through cold or disease."
Price One Dollar.

Warner's Dyspepsia Tonic—"It strengthens the stomach and restores the digestive organs to their healthy state. Weak, nervous and dyspeptic persons should use [it]."
Price One Dollar.

Warner's Pile Remedy—"has never failed (not even in one case) to cure the very worst cases of Blind, Itching or Bleeding Piles.
Price One Dollar.

In addition to the Warner's medicines listed in the 1870 ads (only the first of which is known from embossed bottles), Warner's-brand alcoholic tonic beverages were sold in the Chicago area at about the same time.

These are listed separately as "*Liquor~Remedy*" products because they were not advertised in the 1870 Warner's ads (part of the Warner's Wine of Life ad is the claim that it was "far superior to brandy, whiskey, wine, bitters, or any other article") and because they were basically distillery products rather than medicines. However, they were specifically marketed as "remedies" (see Wilson and Wilson 1968:145 re: *Warner's Aromatic Brandy*), and an 1873 court-case decision in nearby Terre Haute, IN, hinged on the fact that *Warner's Imported English Gin*, sold to two minors, was "prescribed" by a physician as a remedy for illness.

No other business selling these distillery products could be found in the 1870s Chicago CDs, with one possible exception. In the 1872 through 1874 CDs, an individual named Major L. Warner is listed as an independent distiller. He had previously (1870–71) been employed as a distiller at the Thomas Lynch Distillery in Chicago, and he may conceivably have bottled the light-alcoholic brandy and gin products listed and attributed (by us) to Dr. C. D. Warner.

Dr. Weaver's Extract of Fireweed
See the Walker & Taylor listing.

Timothy L. Welch and William L. Goggin
Beer:

(1) **WELCH & GOGGIN / 150 / DEARBORN / ST. / CHICAGO // THIS BOTTLE / NOT TO / BE SOLD**   [aqua]
(base): **C & I**

»SB w/ post mold; slope-shouldered blob-top beer; manufacturer's name and address embossed within oval slug plate outline; ca. 9" tall. *[No photos. See Robert Kott rubbings of front-side, rear-side, and base embossing: Figure 631. Bottle form similar to Davies and Seipp tall beers from Chicago illustrated in Figures 234 and 568.]*

———————— ❖ ————————

This bottle slightly postdates the ca. 1880 end of this volume's coverage, but it is an important one to include because of the base-embossed glassmaker's mark

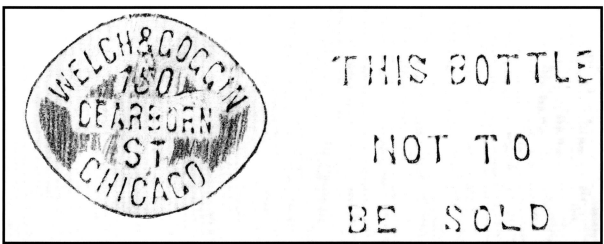

Figure 631

(Cunningham & Ihmsen) and because a round "slug plate" insert was used for the embossed lettering.

From 1878 to 1880, William Welch and a partner named P. L. Hansen ran the *Hansen & Welch* saloon at 150 Dearborn. During this time William Goggin worked as a clerk at 9 State Street (1878), and then as a "beer agent" (brand-name distributor) and "contractor" from his residence (1879–80). The 1880 USC lists Goggin as a 30-year-old beer bottler.

Beginning in 1881, Goggin bought out Mr. Hansen and became junior partner with Welch in the saloon at 150 Dearborn—where they also attempted to develop their own beer-bottling partnership (Goggin may have brought his earlier beer-bottling contracts with him to the partnership). In 1882, they are listed separately at the same address: Welch as a salon operator, and Goggin as "Contractor 35" bottling Schlitz Beer there. He also placed an ad in the 1882 CD indicating he was "sole agent for Schlitz Milwaukee Export and Bottled Beer and Besley's Waukegan Ale."

By 1883, Welch and Goggin have gone their separate ways, and Goggin is once again listed as a contractor (probably as a brand-name beer distributor) from his residence.

Given this business history, the Welch & Goggin embossed bottles must date from 1881 (and perhaps 1882). However the Pittsburgh CDs indicate that the *C&I* glass company split up into *Cunninghams & Co.* and the *Ihmsen Glass Co.* in 1878. At present, this contradiction cannot be resolved.

Edwin M. and Alvin T. Wells

Medicine:

(1) **WELLS / GERMAN / LINIMENT // CHICAGO / ILLS**   [aqua]
»OP; shouldered cylinder; rolled lip; 3 3/4" tall. *[Figure 632]*

(2) **WELLS / GERMAN / LINIMENT // CHICAGO / ILLS**   [aqua]
»IP; shouldered cylinder; tapered collar; 5 3/4" tall. *[Figure 633]*

(3) // // **PECTORAL SYRUP / OF / WILD CHERRY** // //   [aqua]
(upper panel): **WELLS** / (lower panel): **CHICAGO / ILLS**
»oval IP; beveled-edge rectangle; large front sunken panel; small upper and lower rear sunken panels; unembossed sunken side panels; double-ring collar; 7 1/4" tall.*[Figure 634]*

❖

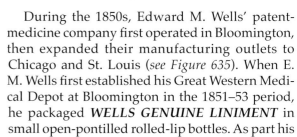

During the 1850s, Edward M. Wells' patent-medicine company first operated in Bloomington, then expanded their manufacturing outlets to Chicago and St. Louis (*see Figure 635*). When E. M. Wells first established his Great Western Medical Depot at Bloomington in the 1851–53 period, he packaged **WELLS GENUINE LINIMENT** in small open-pontilled rolled-lip bottles. As part his Chicago/St. Louis expansion in the late 1850s, he changed the product's name to **WELLS GERMAN LINIMENT** and packaged it in larger-sized, tapered-collar containers (*see Wells' Bloomington listing*).

Figure 632

In the late 1850s, he moved his family and his E. M. Wells patent-medicine base of operation to Chicago. Wells' first Chicago listing was in the 1858 CD. That year, the E. M. Wells patent-medicine company was listed in the patent-medicine summary at 93 S. Dearborn. That first year, his brother Alvin Wells was not separately listed. But E. M. Wells died in April 1859 (*Bloomington Daily Pantagraph*, April 27, 1859, p.3), and although his Chicago company was listed in the 1859 CD, at 134 S. Water, he himself was not listed in the alphabetical index. Instead, *A. T. Wells & Co.*, "patent medicines," was listed, with his brother and widow named as principals, operating at 108 Randolph. *A. T. Wells & Co.* was also listed at the same address in the Chicago CDs for 1860, 1861, and 1862, but Mrs. Wells is no longer named and Alvin Wells is listed as "boarding" at 233 State and then 227 W. Lake, so Mrs. Wells may have left Chicago after her husband's death. From 1863 onward, there were no listings for the Wells medicine company in Chicago. Since all known Wells' bottles are pontilled, the company may have been selling off existing stock after 1859. The history of his brother Alvin's association with the E. M. Wells company has been researched and summarized by Mike Burggraaf, who discovered a number of Wells pontiled medicine bottles in the yard of an early residence in Fairfield, IA, that had been the property of Alvin Thayer Wells:

Figure 633

> Some detailed research revealed that during 1853, A. T. Wells was employed as a traveling salesman for his brother's drugstore in Bloomington, Illinois. The following year he came to Fairfield, Iowa, where he operated his own drugstore. He remained in the druggist business at Fairfield through 1857. During that same year he moved back to Illinois to join his brother in operating the patent medicine business, which was [*then relocating to*] Chicago. Later in [*1859*] his brother died and A. T. Wells remained in Chicago to continue managing the medicine business. [*Later*] he moved back to Fairfield and operated the patent medicine company from his old drugstore. [Burggraaf 2003:36]

From the last almanac that Edward Wells produced while he was in Chicago (Wells 1860; *see also* Wells' Bloomington listing), his *German Liniment*

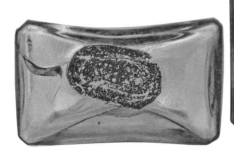

Figure 634

Figure 635

(embossed *"Genuine" Liniment* on the earlier Bloomington bottles) was touted as a cure for " the Worst Old Sores, Wounds, Poll-Evil, Sweneys, Snake-Bite, Spavin, Rheumatism, Frozen Limbs, Burns, Scalds, Colic, &c." His *Pectoral Syrup of Wild Cherry* was sold as a cure for consumption. According to a discussion in the *Chicago Daily Times* that Wells reproduced in the 1860 *Almanac*:

> Some six years since, E. M. Wells, of Chicago, thought a remedy could be prepared that would not only cure diseases of the lungs, but restore the general health, and thus remove the cause. He kept this object steadily in view, studied and experimented day and night, visited the Eastern cities several times, and consulted the best physicians, chemists, and pharmaceutists, and when he was satisfied the object was attained, if it could be, **commenced a course of treatment on himself for what had been considered hereditary Consumption**. Every symptom was removed, the general health restored, and full confidence established in the remedy. It was then dispensed as a prescription for about a year, when it became exceedingly popular, and was brought out under the name of **Wells' Pectoral Syrup of Wild Cherry**, and is the most popular cure for lung diseases ever introduced.

Perhaps, considering Wells' death as the 1860 *Almanac* was being prepared, the cure was not as effective as he had hoped.

He also produced and sold *Wells' Great Western Vegetable Pills*, *Wells' German Condition Powder*, and (as wholesale agent) *Denie's Electro-Magnetic Plaster* (in a "tin box").

Henry N. Wheeler and Henry M. Bayless

Fruit Jar:

(1) **MADE . FOR / WHEELER / & / BAYLESS / CHICAGO / I^{LL} // MASON'S / PATENT / NOV 30TH / 1858** [aqua]

»SB; cylindrical quart preserve jar; threaded ground lip; 7 1/2" tall. *[Figure 636]*

The firm of *Wheeler & Bayless*, 8 Dearborn Street, is listed in the 1862 IRS assessment for Chicago and first appeared in the 1862 Chicago CDs (with *"Bayliss"* misspelled). They began as oil-lamp dealers only, and their initial 1862 ad indicated they were "manufacturers and jobbers of oils and lamps, and every description of coal oil goods." In the 1863 CD, they were again listed as "oil and lamp" dealers.

But beginning in 1864 they expanded their business ads to include glass. In 1865 and 1866, they identified themselves as "wholesale dealers in lamps, oils, and glassware." Also 1866 was the last year they were listed as business partners. Henry Bayless had already moved away and was listed as residing in St. Louis, but they must have still had a business relationship for that season.

From 1867 into the early 1870s, Henry Wheeler remained in the "lamps, oils, and glassware business on his own. Then in the later 1870s he acquired a new partner (Wheeler, Chapman & Co.) and focused on lamp and lantern sales.

Interestingly, the *only* time fruit jars were mentioned in Wheeler's CD ads was 1867—the year after the *Wheeler & Bayless* partnership ended. His ad that year read, "oils, lamps, glassware, fruit jars, &c." This probably indicates that the *Wheeler & Bayless* embossed jars were produced late in the partnership (ca. 1865–66).

Joseph Willard

Medicine ~ Pharmacy:

(1) (side): **J. WILLARD** // // [aqua]
   (side): **CHICAGO**

»IP; beveled-edge rectangular medicine or pharmacy bottle; embossed on sunken side panels only; ca. 8" tall. New find reported by Dan Puzzo in Antique Bottle Club of Northern Illinois Newsletter; bottle not yet seen or photographed by us, but likely has double-ring collar or tapered collar; ca. 8" tall.

James Willard emigrated to Chicago from Massachusetts in 1853. For a five-year period from 1854 to 1858 he was listed in Chicago CDs as a druggist operating a retail pharmacy located at 64 N. Clark in the city. He no longer appeared in Chicago business directories after 1858.

The embossed bottle listed was likely used by Willard as a generic prescription container for the apothecary products he mixed and sold at his pharmacy.

Wolf's Stomach Bitters

See the Samuel Hogeboom & Robert Wolf listing, this chapter.

Jacob A. Wolford

Whiskey:

(1) (shoulder): **JACOB A. WOLFORD** // **CHICAGO** [amber]
   (body): **WOLFORD** / **Z** *(in a decoratively shield shaped sunken panel)* //
   WHISKEY *(arched above large flattened raised-glass "dimple," but not in a sunken panel)*
   Encircling Base: (outer ring): **A & D H CHAMBERS / PITTSBURGH, PA**
   (inner ring): **PAT_ AUG. 6ᵀᴴ 72**

»SB; ridged-barrel shape w/ 4 horizontal raised rings on both the upper body and lower body; inside screw-thread lip; ring-neck lip finish; 8 3/4" tall. [*Figure 637*]

Figure 636

Figure 637

Jacob Wolford first appeared in the 1870 Chicago CD, listed as a wholesale and retail liquor and cigar dealer with a wholesale warehouse located at 123 Clark and a retail store at 122 State Street. From 1871 to 1876, the CDs listed him as being in the "wine and liquor" business, moving from 122 State to 223 Washington after 1871.

From 1877 to 1881, Wolford ran a saloon at the 123 Clark Street location. From 1882 onward, he was listed only at his residence, with his occupation variously given as "capitalist," "broker," and "commission merchant."

No ads for "**WOLFORD'S Z**" could be located to learn when Jake Wolford first produced it, but interestingly, articles were located indicating that he raced in regattas in a sailboat named "Zephyr." Also, as a chronological hint for dating the liquor's appearance, from 1874 to 1878 a local race horse (owned by John F. Smith of Freeport) was named "Wolford's Z"—perhaps because of its "kick"? Also, a humorous description of a local 1875 masquerade ball in the *Chicago Tribune* (Feb. 7, 1875, p.5, col. 5) mentions the product, indicating its popularity at the time:

> The masquerade given by the Independents, at Martine's South Side Hall, proved a very pleasant affair.... Old Tom Gin, Sour Mash Whisky, and Wolford Z. entered together. They were warmly welcomed by the gentlemen, and sometimes repulsed, sometimes accepted, by the ladies. A "religieuse" was seen holding on to the handle of one of these overlarge bottles, and afterwards whirling around with it in a rather dizzy manner.

Notable, the distinctively styled "Wolford's Z" bottle is the same design as *Chapin & Gore's* "Sour Mash 1867" bottle (*see their listing*). Both whiskey bottles were manufactured and used during the early 1870s, and the two products may have been direct competitors.

Dr. George W. Wolgamott

Medicine:

(1) **D^R. WOLGAMOTT'S / APERIENT AND / BLOOD PURIFIER /**   *[amber]*
(*front heel*): **CHICAGO ILLS.**
(*reverse*): Embossed outline image of Maltese Cross.
»SB; tall, oval, straight-sided "flask" shape; double-ring collar; 8 1/4" tall. *[Figure 638]*

Dr. Wolgamott was first listed in the Chicago CDs in 1878, as a physician at 126 State Street. During 1879–80 he took on a partner, Dr. George Hewitt, and the two of them had offices at 103 State. From 1881 until his death in 1900, Dr. Wolgamott was listed alone as a physician at various State Street addresses until 1891, and then at office quarters in the Masonic Temple.

Dr. Wolgamott's Aperient and Blood Purifier was advertised in the *Chicago Tribune* for only four years after his arrival in Chicago. It was offered by Dr. Wolgamott alone in 1878, by *Wolgamott & Hewitt* in 1879 and 1880, and then finally by Dr. Wolgamott alone again in 1881. During the final three years of the ads, Dr. Wolgamott had renamed his practice the *Garden City Rheumatic Institute*. His ads featured testimonials from successfully cured patients:

THE PLAGUE STAYED.
And Disease and Suffering Through the Northwest
Wholly Avoided.
Dr. Wolgamott's Wonderful Aperient and Blood Purifier,
and the Remarkable Cures It Is Effecting.
Rheumatism, Neuralgia, and Liver Disease Entirely Cured,
and Health and Happiness Restored.
[*Chicago Tribune*, Oct. 5, 1878, p. 7, col. 5]

A Radical Cure for Rheumatism
In Its Various Forms Has Been Found,
And Messrs. Wolgamott & Hewitt, the Discoverers,
Have Located in Chicago,
Establishing the Garden City Institute for
The Treatment of Rheumatism and all Blood Diseases.
[*Chicago Tribune,* Feb. 16, 1879, p. 5, col. 7]

GOOD HEALTH.
How the Greatest Blessing of God
May Be Restored to Suffering Humanity.
Rheumatism, Neuralgia, Gout, Liver Complaint,
Dyspepsia and General Blood Troubles
May Be Easily Overcome by the Use of
Dr. Wolgamott's Blood Purifier and Aperient.
**The Most Remarkable Remedial Agents
of the Nineteenth Century.**
[*Chicago Tribune,* Nov. 29, 1879, p. 5, col. 7]

Figure 638

Dr. Wolgamott's bottle obviously directly copied the style of the popular Connell's Brahminical Moonplant remedy (*see Connell's listing, this chapter*), and no doubt was intended to capture some of the Connell's Aperient market in Chicago.

Gottlieb Wurster and Co.

Ale~Porter:

    (1) **G. WUSTER & Co / CHICAGO.**   [amber]
         (base): **GW & Co**
         Note: **R** omitted in "**WUSTER**"

         »SB w/ post mold; 2-piece mold quart ale; ring-neck tapered collar; 9" tall. *[Figure 639]*

    (2) **G. WURSTER / CHICAGO.**   [dark amber]
         (base): **G. W**

         »SB w/ post mold; 2-piece mold quart; ring-neck tapered collar; 9 1/4" tall. *[Figure 640]*

Soda ~ Mineral Water:

    (1) **GOTTLIEB WURSTER / & CO.**   [aqua]
         (base): **W**

         »SB w/ keyed hinge mold; blob-top soda. *[Figure 641]*

    (2) **WURSTER & Co / CHICAGO //**   [aqua]
         (heel): **A & D. H. C.**
         (base): **W & Co–**

         »SB; oval-blob-top soda. *[Figure 642]*

Figure 639

Figure 640

Figure 641

458   Chicago — Gottlieb Wurster and Co.

(3) **G. WURSTER / CHICAGO //** [aqua]
(heel): **A & D. H. C.**
(base): **G.W**

»SB w/ keyed hinge mold; oval-blob-top soda. [Figure 643]

(4) **GOTTLEIB WORSTER** [aqua]
(base): **W**
Note: "**WORSTER**" instead of **WURSTER**

»SB w/keyed hinge mold; oval-blob-top soda. [Figure 644]

(5) **G. WURSTER** [aqua]
(base): **G.W**

»SB w/ keyed hinge mold; oval-blob-top soda. [Figure 645]

(6a) **G. WURSTER / CHICAGO / ILLINOIS** [aqua]
(embossed vertically down)

»SB w/ post mold; tall cylindrical soda w/ semi-round base; shouldered blob top; 9" tall. [Figure 646, left and left-center]

(6b) **G. WURSTER / CHICAGO / ILLINOIS** [aqua]
(embossed vertically down)
(heel): **A & D. H. C.** (very small embossing)

»SB w/ post mold; tall cylindrical soda w/ flat base, very short neck, and lozenge-shaped blob top; 7 3/4" tall. [Figure 646, right and right-center]

(7) **G. WURSTER / CHICAGO / ILL^S //** [aqua]
(heel): **A & D. H. C.**
(encircling base): **KELLY PAT FEB 5 1878**
(center base): **G.W.**

Figure 642

»SB w/ post mold; Kelly-patent Codd-style gravitating-stopper soda bottle; size unknown (but see similar Wm. A. Hausburg Kelly-patent bottle illustrated in Figure 334). [Bottle recorded and sketched by Robert Kott: Figure 647]

Quart Soda ~ Cider:

(8) **G. WURSTER / CHICAGO / ILLS** [aqua]

»SB; squat, high shouldered "Chicago style" quart soda; shouldered-blob top; 9" tall. [Figure 648]

(9) **G. WURSTER / CHICAGO ILL<u>S</u> //** [amber, green]
(heel): **A & D. H. C.**

»SB; slope-shouldered cider-finish quart; ring-neck collar (*green*) and shouldered-blob-top (*amber*) varieties; 12" tall. [Figure 649; Figure 650]

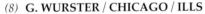

Gottleib Wurster was first listed in the Chicago CDs in 1868, as a grocer operating from his residence at 550 N. Clark. In the 1869 and 1870 CDs, he was listed in the alphabetical index as a bottler of "mineral water" (from his 1869 residence at 730 N. Wells) and as a

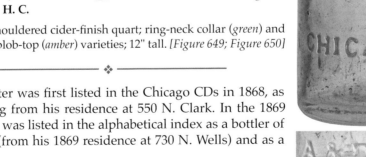

Figure 643

Figure 644

Figure 645

Figure 646

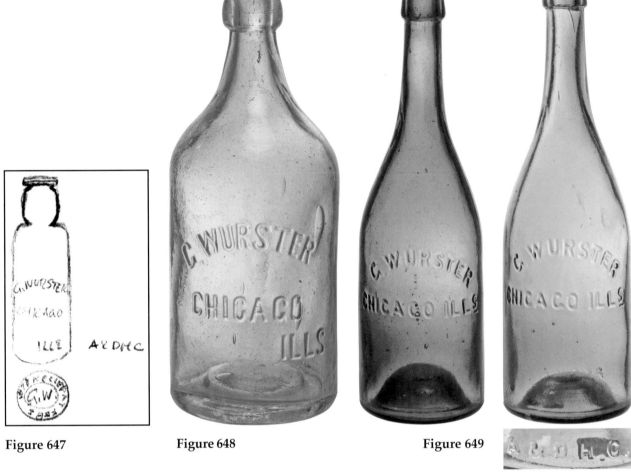

Figure 647 Figure 648 Figure 649 Figure 650

"mineral water dealer" (from his 1870 residence at 55 Sigel). He was also listed in the 1870 USC as a 34-year-old "mineral water manufacturer."

In 1871, Wurster was first listed as a "soda water bottler" in the Chicago CDs, but he was not listed in the bottler summary that year. However, in 1872 he was listed under *Wurster & Huttenlocher*, with John Huttenlocher, as "mineral water manufacturers." The following year (1873), Wurster was senior partner in *Gottleib Wurster & Co.* at his 55 Sigel Street residence. His junior partner in the new bottling works that second year was still John Huttenlocher, who in 1871 had bottled in Chicago on his own (*see Huttenlocher listing, this chapter*).

From 1874 onward, Wurster bottled alone at the same address, so all of the "& Co." embossed bottles date to his early years in business in partnership with Huttenlocher. Wurster was listed in the CDs as a "soda water manufacturer" at 55 Sigel from 1874 through 1882, after which the business was closed. For some reason, in the 1877 CD only, George and Frederick Wurster were listed as operating the bottling works, and Gottleib was not mentioned. In 1878, he was back (with his name misspelled "Wuster") as a bottler of "mineral waters."

The Wurster Company is listed in the 1880 IL Industrial Census as having operated full-time 12 months a year, producing an annual value of bottled products estimated at $12,000. According to the 1880 USC, Wurster was 45 years of age at the time. Toward the end of the 1870s, he switched over to using the new-style Hutchinson internal stopper bottles, four embossed styles of which are known (Harms 1978; Wright 1992). Since these bottles postdate the end of the bottling era studied in this volume, they are not listed.

It should be noted that although two embossed ale-style bottles and a cider-style bottle are known for the Wurster Company, the company was never listed in the CDs as having brewed or bottled ale or porter or cider. One of these bottles has the "& Co." embossing, so Wurster's foray into ale bottling was probably confined to the early 1870s, during his first few years in business. Also, the cider-style quart listed as #9 above probably did contain cider. Wurster's handwritten 1870s Illinois trademark registration form notes that "All mineral water, cider, porter, and soda water bottles are marked 'G. Wurster, Chicago Ills.' All siphon bottles are marked 'G. Wurster' on the heads and are also marked 'G. Wurster Chicago Ills' on the site of each bottle."

Siphon bottles were generally not manufactured until 1880 and later, along with Hutchinson-style soda bottles (both of which were used by Wurster toward the end of his bottling career), so neither of these two bottle categories is included in the present volume.

Ypsilanti Spring Co.

Mineral Spring Water:

(1) **YPSILANTI / SPRING / COMPANY / CHICAGO / ILLS.**   [yellow amber]
   (base): **AGWL**
   Note: Side embossing within sunken "tombstone-shaped" slug-plate panel.

   »SB w/ post mold; Saratoga-style quart mineral-spring bottle; ring-neck tapered collar; 9¾" tall. *[Figure 651]*

The *Ypsilanti Spring* bottling company postdates 1880, so by our 1840—1880 guideline for inclusion it would normally be excluded from the present study. However, our study borders are somewhat fluid *(see Introduction, this volume)*, depending on whether particular bottle styles and production technologies used in the late 1870s and early 1880s represent the end of old traditions or the beginning of new ones.

The *Ypsilanti Spring* bottle listed represents the last use of a Saratoga-style mineral spring water bottle in Illinois. Such bottles were a midwestern adaptation of a very popular eastern bottle style and bottling tradition (see, e.g., Tucker 1986), similar to early black-glass ales. Such Saratoga mineral water bottles were most popular in the Illinois and Mid-Mississippi Valley area just before, during, and after the Civil War (see listings for *Buck & Rayner* and *Stephen Israel* in Chicago, and *Perry Springs*).

Two popular mineral water sanitariums were established about 1880 at Ypsilanti, in Washtenaw County Michigan—using water from local wells—and they continued in operation up until the World War I era (see Ypsilanti Historical Society website). By the mid-1880s, these waters had become popular enough to be profitably bottled and sold, and in 1885 the *Ypsilanti Mineral Water Company* was established at 88 Randolph Street (later, at the company's "drug store" at 92 East Van Buren) in Chicago. The use of Saratoga-style mineral water bottles was probably designed to give the impression that Ypsilanti Water was an "old fashioned" remedy.

Figure 651

And the bottles' contents certainly would have tasted medicinal. Most bottled Saratoga waters were from iron springs or magnesium springs (*see Perry Springs listing*). But *Ypsilanti Spring* water (from the "Owen" well) was drawn from an artesian sulphur spring. This may well have inhibited the popularity of *Ypsilanti Spring* water, which was only advertised in Chicago for about 12 months (May 1885–April 1886), and the company was only listed in the Chicago CDs during those two years. It was sold in $15 barrels, in five-gallon and 10-gallon kegs ($3 and $5), and in cases of 24 bottles each: pints $4/case, quarts $6/case. So far, we have been able to document only the quart-size embossed bottles.

> Your life may depend on a correct regulation of the bowels, stomach, liver, or kidneys. There is no remedy on earth, nor has there ever been one, that could regulate and cleanse these organs as can and does Ypsilanti Mineral Spring Water.... [This] is not a table water, but is **a most powerful mineral water, that no one will drink for the pleasure of drinking, as it tastes bad and smells bad**, but if you wish to be healed of a catarrh that smells worse, or a debilitated state of an unhealthy body, caused by blood poison, uric acid, mercurial or quinine effects, **in the name of all that's good get rid of your Miss Nancyisms**, and regain ambition, strength, and health by a liberal use of this natural water, unaltered by the hand of man. [*Chicago Tribune*, May 30, 1885, p. 15, col. 4]

COLEHOUR
(Cook County)

[1870–80 population unknown: *see discussion that follows*]

Henry C. Kassens

Soda ~ Mineral Water:

(1) **HENRY KASSENS / COLEHOUR / ILL. //**   [aqua]
    (*heel*): **De S. G. Co.**

»SB w/ post-mold; oval-blob-top soda. [*Figure 652*]

William S. Smith

Soda ~ Cider:

(1) **W^M S. SMITH / COLEHOUR / ILL. //**   [amber]
    (*encircling base*): **WIS. G. Co. / MILW.**
    Note: Side embossing in circular slug plate. Although the bottle likely contained soda or cider, it may be an early beer bottle.

»SB w/ post-mold; quart shouldered-blob-top soda/cider. [*Figure 653*]

From the stylistic and technological attributes of the two bottles listed above, the **KASSENS** soda likely dates to the late 1870s and the **SMITH** quart soda/cider likely dates to the early 1880s. Thus, the **SMITH** bottle—considered alone—would have been produced too late for inclusion in the present study. However, since both bottles were probably produced sequentially over a short time

Figure 652

for use at the same bottling works, and since they are the only two known late-1870s-style embossed Colehour bottles, both are listed for this summary. A later-style (Hutchinson internal stopper) embossed **KASSENS** soda bottle has also been documented that is not listed or illustrated above.

The "Mythical Hamlet"

Colehour was an ill-fated town development effort in southeastern Cook County (near the Indiana border) by real-estate developers William H. Colehour, Charles W. Colehour, and Edward Roby. Although a "Colehour" post office existed from 1875 to 1894 (Adams 1989:326), the town development effort was a going concern only during the late 1870s and early 1880s. It was referred to in the *History of Cook County* as a "somewhat mythical hamlet" whose boundaries "were never clearly defined" (Goodspeed and Healy 1909:316). The developers paid only $10,000 cash for the land, and took out a loan for the remaining $86,000 due, backed by a deed to the property. In the mid-1870s, "a part of the land was subdivided and improved by grading streets, making ditches, etc., and a part sold, freed from the lien created by the deed of trust" to help pay development costs. But the effort failed to turn a quick profit and, in mid-1878, William Colehour filed for bankruptcy with debts exceeding $800,000. The remaining history of Colehour was largely a record of court cases—concerned with questions of "who-owed-who-what." By May 1892, the case had made it all the way to the U.S. Supreme Court (146 US 153 Roby v. Colehour). The area eventually became part of Chicago, and was called the "Ironworkers' Addition" to the city.

Henry Kassens

No direct evidence has been found for Kassens' soda bottling works in Colehour, but the 1870 USC listed him as a 24-year-old carpenter living in Baltimore, MD, and the 1880 USC listed him as a 34-year-old saloonkeeper living in Colehour. An 1881 *Chicago Tribune* article (Oct. 13, 1881, p. 578, col. 2) noted that Henry C. Kassens was one of the original property deed holders from the 1875–78 land development offering. The bottle listed is an early DeSteiger Glass Works product, manufactured not long after the company was first established in LaSalle in 1877 (*see DeSteiger LaSalle listing*). So, Kassens' Colehour bottling operation was likely established ca. 1877-1880.

Kassens left Colehour and moved to Hyde Park in the early 1880s—he is noted as a resident of Hyde Park in a March 27, 1883, *Chicago Tribune* article (p. 2, col. 3), and in a March 1, 1884, *Hyde Park Herald* article (p. 3). Soon after his arrival there (in 1884), Kassens made a one-year effort to establish a brewery in Hyde Park, but his effort was apparently unsuccessful (Van Wieren 1995:77). He was still a Hyde Park resident as late as 1888 (*Hyde Park Herald*, Aug. 3, p. 2, col. 3), but during the early 1890s he had moved to Chicago, where he operated a saloon at 10324 3rd Ave. (1891–95 Chicago CD listings). He must have died in 1895, since Mrs. Maria Kassens ran the tavern in 1896.

William Smith

The embossed William S. Smith Colehour bottle listed was manufactured by the *Wisconsin Glass Co.* This company did not begin producing glass bottles in Milwaukee until 1881, and was only in operation for four years, 1882–85 (Toulouse 1971:541). William Smith was not listed in the 1880 USC for Colehour, so he must have arrived in the area shortly afterwards. Barring the unlikely possibility that there were two contemporary bottling operations in Colehour, Smith probably took over Kassens' bottling works ca. 1881–82 after Henry Kassens moved to Hyde Park. Examples of Smith's embossed bottles are rarely found, and a William S. Smith was listed as a Chicago bartender in the 1883 CD (and as a city "resident" in 1884). If this is the same man, perhaps he only bottled in Colehour during the 1881 or 1882 season.

Figure 653

COLLINSVILLE
(Madison County)

[1860–70 = Twp. census only]
1880 census: 2,900

Henry Baierlein

Soda ~ Mineral Water:

(1) H. BAIERLEIN / COLLINSVILLE / ILL. [aqua]

»SB w/ post-mold; oval-blob-top soda. *[Figure 654]*

Figure 654

———————————❖———————————

The 1880 USC lists Henry G. Baierlein as a 56-year old soda water manufacturer in Collinsville. However, Baierlein does not appear in an 1882 Collinsville commercial directory (Brink & Co. 1882:460).

From the post mold construction of Baierlein's blob-top embossed soda bottles, they appear to have been manufactured during the 1870s. Since only a single embossed-bottle style is known, his bottling business was probably active in Collinsville ca. 1875–1880.

Bernard Bischoff

Soda ~ Mineral Water:

(1) B. BISCHOFF / COLLINSVILLE / ILL [aqua]
(base): **B B**

»SB w/ post-mold; shouldered-blob-top soda. *[Figure 655]*

———————————❖———————————

Bernard Bischoff immigrated to the St. Louis area from Germany during the early Civil War years, and first settled in Belleville, where he learned the bottling business. He later went to O'Fallon, where he was engaged in the soda and mineral water bottling business on his own by the mid-to-late 1860s (Miller

Figure 655

1989, 2007a; *see O'Fallon entry*). He was listed as a 42-year-old O'Fallon soda water manufacturer in the 1870 USC, and the *St. Clair County Atlas* (Warner and Beers 1874) contained an ad indicating that Bischoff was a bottler of ale, soda, cider, and mineral water in O'Fallon. So he had continued bottling in O'Fallon until at least the mid-1870s (long enough to have four different styles of embossed soda bottles manufactured for his use (*see Bischoff O'Fallon entry*).

After that, sometime in the late 1870s, he moved his bottling operation to Collinsville. From the style and manufacturing technology of his single embossed Collinsville bottle style, it dates to the late 1870s and likely marks the initiation of his bottling works there. Stylistically, the Collinsville bottle is identical to a late Bischoff O'Fallon embossed bottle (*see Figure 655b*), with only the town name recut to **COLLINSVILLE**. According to his Collinsville obituary in *The Advertiser* [*October 16, 1915*]: "When a young man he came to America and settled in the vicinity of Belleville, where he became familiar with the soda water business, and later on established that business in O'Fallon and about 35 years ago came to Collinsville and engaged in the making and selling of soft drinks."

The 1900 USC for Collinsville still listed 70-year-old Bernard Bischoff as owner of a soda factory there, with his son Henry as an employee of the bottling plant. Miller (2007a:71) believes that Bernard effectively retired in the early 1900s after which his son took over operation of the factory—which he ran (*according to his 3/16/1928 obituary*) until the mid-1920s.

COLUMBIA
(Monroe County)

1870 census: 1,200
1880 census: 1,300

John Gundlach

Soda ~ Mineral Water:

(1) **JOHN GUNDLACH / COLUMBIA / ILL //** [aqua]
    (*heel*): **L. G Co**
    (*base*): **J.G**

»SB w/ post-mold; oval-blob-top soda. [*Figure 656*]

---❖---

The 1880 USC for Columbia lists John Gundlach as a 51-year old brewer born in Germany. He was also listed as a brewer there in the 1870 and 1880 USC and in the 1875 ISD. He was born in Cronberg, Germany, in the 1920s, where his father was also a brewer. The family immigrated to the United States in 1844, came to Belleville, and took up farming (Columbia Centennial Committee 1959). He left the farm to establish his first brewery operation in Belleville with Fidel Stoelze 1855. But in 1856 he removed to Columbia with his brother Philip, built a family residence and brewery compound there during 1856 and 1857 (in the vicinity of what is now 625 N. Main Street—*see National Register of Historic Places nomination form for the residence*). The brick brewery building in Columbia was known as the *Monroe Brewery*. It was built at a cost of about $30,000 and in 1883 had an annual capacity of about 6,000 barrels of beer. Beneath the structure were beer caves capable of storing about 3,500 barrels (McDonough & Co. 1883b:452).

From the style and technological attributes of the embossed bottle listed above, Gundlach apparently added soda water bottling to his business for a short time during the 1870s. This was confirmed by published statements from 1875 and 1883:

The year 1855, and part of 1856, were spent in Belleville. At this latter date he removed to the young and growing town of Columbia, in Monroe County, where he still continued his pursuit of a brewer. This he has followed ever since, **uniting it with the manufacture of soda**, and having built up a large and profitable trade. [Brink & Co. 1875:30]

The Monroe Brewery...There is also a malt house and mineral water establishment connected with the business. [McDonough & Co. 1883b:452]

The Gundlach mineral water bottling operation is further documented by a listing in the 1880 USC. One resident in the Gundlach family household was a nonfamily member boarding there: 18-year-old Charles "Koemsker." He was listed by the census-taker as a "soda water bottler."

Figure 656

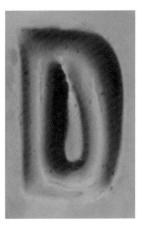

DANVILLE
(Vermilion County)

1850 census: 700
1860 census: 1,600
1870 census: 4,800
1880 census: 7,700

Patrick Carey

Ale:

(1) **P. CAREY / DANVILLE / ILL** [amber]

» SB w/ post mold; 2-piece mold quart ale; ring-neck tapered collar; 9 1/4" tall. [Figure 657]

Patrick Carey was listed as a liquor dealer in Danville in the 1870 USC and a saloonkeeper in the 1880 USC. From the bottle's style and manufacturing technology, it was made during the 1870s. According to the 1879 *History of Vermilion County* (Beckwith 1879:447), a man named Alexander Moore was a bookkeeper for Carey's liquor business for "five or six years" during the late 1860s and early 1870s. The likely age of Carey's embossed ale would be sometime during this period.

Figure 657

William Galtermann

Soda ~ Mineral Water:

(1) **WILLIAM GALTERMANN / DANVILLE / ILL'S** [aqua]

»SB w/ keyed hinge mold; high sloped shoulders; short, cylindrical neck; flat, lozenge-blob-top soda. *[Figure 658]*
Note: Matthews Gravitating Stopper–style soda closure, but lacking the embossed MGS patent data on the base (see, e.g., Buff & Kuhl Alton listing above).

Figure 658

William Galtermann could not be found in any Vermilion County historical studies or USC data, but from the fact that this bottle has a keyed hinge mold and that the Matthews stopper 1864 patent data is no longer being cited on the base, it likely dates from the very late 1860s (ca 1868–69). Galtermann may have bottled mineral water for a short-lived tavern operation, or he may have established Danville's first bottling works—i.e., the "steam bottling works" being operated by William Grabs in the early 1870s prior to the arrival of Charles Shean in the mid-1870s (*see his listing, this chapter*).

Charles Shean

Soda ~ Mineral Water:

(1) **C. SHEAN / DANVILLE / ILL.** [light blue]

»SB w/ post mold; short-neck, oval-blob-top soda. *[Figure 659]*

(2) **C. SHEAN / DANVILLE / ILL.** [aqua]

»SB w/ post mold; squat, Chicago-style quart soda; oval-blob-top; 9" tall. *[Figure 660]*

Figure 659

Figure 660

According to the *History of Vermilion County* (Williams 1930, v. 2:241), Charles Shean first ran a mineral water bottling business in Mattoon and then moved to Danville in 1875, where he purchased an existing bottling works. The only known "existing" soda Danville bottling operation during the 1870s would have been the steam bottling works operated by William Grabs on West Main Street prior to 1879 (see Beckwith 1879:339). As far as is known, Mr. Grabs did not use embossed bottles at his steam bottling works. But the use of unembossed soda bottles was rare in Illinois before 1880, so embossed Grabs sodas may yet turn up.

After arriving in Danville in 1875, Mr. Shean may have worked at Grabs bottling shop for a few years before buying him out. This would fit with the style and technological attributes of the Shean-embossed blob-top sodas listed, which appear to be very late examples, dating to the end of the 1870s. The bulk of Shean's embossed bottles were made in the later Hutchinson (internal stopper) style of the early and mid-1880s, and are too late for listing in the present volume.

John Stein

Beer:

(1) **JOHN STE[IN] / CITY BREWERY / DANVILLE. / I[LL.]**   [yellow amber]   **Figure 661**
Note: Bottle fragment only—bracketed letters broken away.

»SB w/ probable post mold; embossing in circular slug plate; probably slope-shouldered quart w/ oval blob top (estimated from fragment); probably ca. 11 1/2" tall. *[Figure 661]*

Ernst Blankenburg established an early brewery in Danville, which he sold to John Stein in 1875 when he decided to go into the liquor, wine, and cigar business (Beckwith 1879:236). Stein may have come to Danville from Amboy, IL, in Lee County, where there was a *Stein & Helm Brewery* in 1874–75. John Stein's brewery at 441 East Van Buren Street in Danville became a successful operation, which he ran from the mid-1870s until 1899 (Van Weiren 1995:70, 82).

Considering Stein's quarter-century career in the Decatur brewing business, the fact that only a single fragment early embossed Stein bottle has been found is somewhat surprising. From the bottle's early slug-plate embossed style and technological attributes, it looks like a mid-1870s product like those manufactured by Cunningham & Ihmsen (in Pittsburgh) during the later 1870s. Stein may have ordered the bottles when he first entered the brewing business in Decatur ca. 1876–77 to advertise his new product to the local market (*see, e.g., Bradbury Jacksonville listing*).

DARMSTADT
(St. Clair County)
[1870–80 census by Twp. only]

Christian Gross

Soda ~ Mineral Water:

(1) **CHRIST GROSS / DARMSTADT / ILL //**   [aqua]
(heel): **L.G. Co.**

»SB w/ post mold; oval-blob-top soda; variable neck lengths. *[Figure 662]*

Christian Gross was a soda water bottler in Darmstadt during the 1870s and 1880s. Gross arrived in St. Clair County in 1867 (Miller 2007a:204) and was listed as a brickmaker in the 1870 USC. He entered the soda-bottling business in Darmstadt during the late 1860s with a partner named John Fleischman (*see their listing*) by buying out the J. H. Sanders bottling works there (*see Sanders listing, this chapter*). "The Gross & Fleischman partnership did not last very long. By the 1870 Census Mr. Fleischman had left the county.... Also, Gross was the only [*Darmstadt*] soda water manufacturer listed in that census" (Miller 2007a:204).

Gross was listed alone as a soda bottler in the 1873 and 1874 CAR and in the 1880 USC. By 1881, he produced 4,000 cases of bottled soda water annually (Brink, McDonough & Co. 1881b:369).

The embossed bottle listed, known in both short and long-neck slope-shouldered oval and flat-topped oval-blob-top varieties (Miller 1989), was manufactured by the Lindell Glass Company across the river in St. Louis during the mid-to-late 1870s.

Christian Gross and John Fleischman

Soda ~ Mineral Water:

(1) **GROSS & FLEISCHMAN / DARMSTADT / ILL //** *[aqua]*
 (heel): **A & D. H. C.**

 »SB w/ keyed hinge mold; shouldered-blob-top soda. *[Figure 663]*

Figure 662

———————— ❖ ————————

This bottle appears older than the one produced when Gross was in business by himself. Gross and Fleischman were likely only partners for a short time in the late 1860s (*see Christian Gross listing, this chapter*). They bought out the Sanders & Co bottling works in Darmstadt ca. 1867–68, and Fleischman was no longer living in the area by the time of the 1870 USC (Miller 2007a:204).

Figure 663

J. H. Sanders and Co.

Soda ~ Mineral Water:

(1) **J. H. SANDERS & Co. / DARMSTADT / ILL //**   [aqua]
    (heel): **A & D H. C**

  »SB w/ keyed hinge mold; shouldered-blob-top soda. *[Figure 664]*

This is the earliest of the Darmstadt soda bottles. But little is known about Mr. Sanders or his "& Co." partner(s). Miller (2007a) dates the *Sanders & Co.* bottling business to a brief period in the 1860s. Sanders was not listed in either the 1860 or the 1870 USC for Darmstadt.

From the style and technological attributes of his embossed bottle, J. H. Sanders likely established his Darmstadt bottling works (on Race Street between Marissa and Jackson) after the end of the Civil War and was bought out by Gross & Fleischman ca. 1867–68 (see Miller 2007a:204, 447).

Figure 664

DECATUR
(Macon County)

1860 census: 3,800
1870 census: 7,100
1880 census: 9,500

Fuller's Banner Ink

Ink:

(1) **FULLER'S / BANNER INK / DECATUR, ILL**   [aqua]
Note: An identically embossed bottle with a different embossed town name is also known (see Springfield listing). The Springfield also has an **I.G.Cº** *base mark.*

  »SB w/ post mold; cylinder master ink; sloped shoulders; raised (ridged) upper shoulder below cylindrical neck; flared square-ring collar; 7 1/2" tall. *[Figure 665]*

The identity of the company or individual that manufactured *Fuller's Banner Ink* and marketed it in both Springfield and Decatur is something of a mystery. Thus far, we have not been able to document this business in the commercial or census records for either town. From the style and technological attributes of both the Springfield and Decatur bottles, they were likely produced and marketed for a short period sometime during the later 1870s.

Caspar's 1889 Dictionary of the American Book, News, and Stationery Trade includes what appears to be an orphan listing for *The Fuller Mfg. Co.* under its heading for manufacturers of colored ink and copying fluids (Caspar 1889:603). But according to an online manufacturing history, this Fuller ink company was first started in Chicago in

Figure 665

1886 by Harvey B. Fuller as a sales outlet for *Fuller's Liquid Fish Glue*. He was 21 at the time. Mr. Fuller moved to Minneapolis the following year and expanded his manufacturing line to laundry bluing. He also manufactured and supplied ink to the local schools. Harvey Fuller is never mentioned in the Springfield or Decatur histories, and anyway he would have been too young to have manufactured Fuller's Banner Ink in central Illinois in the mid-1870s.

A possible Springfield/Decatur candidate as manufacturer of *Fuller's Banner Ink* is Charles H. Fuller, who lived and worked in both towns during the 1850s–1880s. We have found no mention of him as an inkmaker, but according to his obituary (Decatur *Daily Republican*, Dec. 26, 1896) he worked in a variety of jobs in local government and in the private sector after being injured in the Civil War. He was city clerk in Decatur from 1867 until 1872, and in 1884 he was appointed pension examiner. We have not determined his occupation from 1872 to 1883, during which time *Fuller's Banner Ink* would have been produced.

Fredrick D. Kuny

Soda ~ Mineral Water:

(1) **F. KUNY / DECATUR / ILLS //** [aqua]
(*heel*): **A. & D. H. C**

»SB w/ post mold; shouldered-blob-top soda. [Figure 666]

(2) **F. KUNY & C° / DECATOR / ILL** [aqua, green]
Note: **DECATOR** is misspelled.

»SB w/ keyed hinge mold; shouldered-blob-top soda. [Figure 667]

Fred Kuny immigrated to the United States from Germany and began his Decatur bottling operation in 1866. Kuny began his bottling works with a partner named Klohr (*see their listing, this chapter*), probably for just the first season of operation (1866). The bottling works was located at the northeast corner of East Main and Jackson streets.

According to the 1880 McDonough County history, Kuny was a manufacturer of soda and mineral waters and a bottler and distributor of export beer and cider (Brink, McDonough & Co. 1880a:122). He was listed in the 1870 USC, the 1872 TW & W Railway directory, the 1875 ISD, and the 1875 Decatur CD as a soda water manufacturer. The Fred Kuny bottling works was also listed in the 1880 IIC. At that time Mr. Kuny had a maximum of six employees a year (five adult males and one child) making from $1.00 to $1.50 per day. They worked 10 hour days from May to November and eight hour days the remainder of the year. Annual payroll costs were $800, materials costs were $3,000, and the annual value of bottled product was estimated at $9,000.

Kuny continued in the bottling business in Decatur after 1880, and later Hutchinson-style (internal rubber stopper) soda bottles embossed with his company name are also known. These were used too late for inclusion in the list. Kuny sold his bottling business to his son-in-law Gus Aherns in 1887 (Nelson 1910:36).

Figure 666

Figure 667

Fredrick Kuny and Klohr

Soda ~ Mineral Water:

(1) **KUNY & KLOHR / DECATUR / ILL //**   [aqua]
(heel): **A & D. H. C.**

»SB w/ keyed hinge mold; shouldered-blob-top soda. *[Figure 668]*

Based on the hinge-mold construction of the embossed bottle listed, the partnership of Kuny & Klohr dates to the Civil War era and predates Kuny's sole ownership of the business (*see his listing, this chapter*). The firm of *Kuny & Klohr* is listed in the June and July 1866, IRS tax assessment. The partnership appears to have been dissolved later that year, since the 1880 McDonough County history indicated that Kuny assumed sole ownership of the bottling works "in 1866."

Figure 668

DIXON
(Lee County)

1860 census: 2,200
1870 census: 4,000
1880 census: 3,650

Frank J. Finkler and Co.

Soda ~ Mineral Water:

(1) **F. J. FINKLER & Co / DIXON / ILLS //**   [aqua, blue]
(heel): **C&I**

»SB; oval-blob-top soda. *[Figure 669]*

Frank J. Finkler was the brother of Alex and John Finkler, who bottled soda water in LaSalle, Ottawa, and Streator during the 1860s and 1870s (*see their listings, this chapter*). Frank worked for his brother Alex in the La Salle soda water factory in 1860 (USC). After serving in the Civil War, Frank Finkler moved to Dixon and opened his own soda water business. He is listed in the 1870 and 1880 USC as a soda water manufacturer in Dixon.

The embossed soda bottle listed was produced and used during the 1870s. The last year the *Cunningham & Ihmsen* glassworks of Pittsburgh (*C&I*) produced embossed bottles was 1878.

George L. Herrick

Medicine:

(1) **FARMERS & FARRIERS / LINIMENT //**   [aqua]
    G. L. HERRICK / DIXON ILLS
    Note: two embossed dots under raised **S** *in* **ILLS**.

»SB w/ keyed hinge mold; rectangular; 4 sunken panels; applied double-ring collar; 5 3/4" tall. *[Figure 670]*

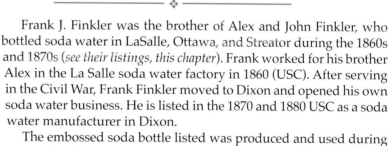

Figure 669

Figure 670

Bitters:

(1) G. L. HERRICK / DIXON, ILLS // // KEYSTONE / TONIC BITTERS // // [olive amber]
(base): **W. McC & Co**
Note: two embossed dots under **S** *in* **ILLS**

»SB w/ keyed hinge mold; beveled-edge square case bottle; flat sides; tapered collar; 9¼" tall. *[Figure 671]*

❖

George Herrick was born in Vermont in 1815 and came to Illinois in 1837. In 1851, when he was 36 years old, he moved from Grand Detour to Dixon (on the Rock River in northern Illinois). From the keyed hinge mold on his embossed Dixon *Farmers & Farriers Liniment* bottle, it was likely produced during the Civil War era.

Herrick began bottling and marketing bitters as early as 1855 (according to an ad in the *Ballston* [NY] *Democrat*, March 1855). His embossed *Keystone Tonic Bitters* is an early smooth-based bottle that was probably used during the 1860s, after the Civil War.

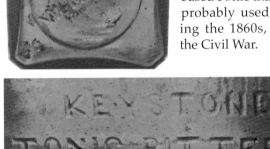

Figure 671

Later, during the 1870s, *"Key-stone" Tonic Bitters* was bottled and sold by the McLain Brothers of Forreston, IL (in adjoining Ogle County about 20 miles north of Dixon). Their embossed bottle was made from a similar mold as the earlier Herrick variety, so the McLain brothers may have bought the brand and the recipe from Mr. Herrick (*see their Foreston listing*). No ads for Keystone Bitters have been located for either proprietor to indicate the purported health-renewing attributes of the product.

From temporally significant features of the bottle bases, Herrick clearly produced this bitters brand before the McLain brothers did. The Herrick embossed bottle has a keyed hinge mold typical of the 1860s, while the McLain brothers' bottle had an 1870s post-mold base.

DUNTON

(Cook County)

[Renamed Arlington Heights, 1874]
[no 1870 town census]

Louis Sass and Frederick W. Müller

Soda ~ Mineral Water:

(1) **L. SASS / DUNTON / ILL**   [aqua]
(*heel*): **C & I**
(*base*): **L S**

»SB w/ hinge mold; slope-shouldered, short-neck, oval-blob-top soda. [*Figure 672*]

(2) **SASS & BRO / DUNTON / ILL //**   [aqua]
(*heel*): **C & I**
(*base*): embossed outline of soda bottle

»SB w/ post mold; slope-shouldered, short-neck, oval-blob-top soda. [*Figure 673*]

❖

Dunton was renamed Arlington Heights in 1874. But in 1871–72, less than two years before Dunton was renamed, Chicago bottler Louis Sass (a partner in the Chicago bottling firm of *Sass & Haffner* (*see their listing*), and his half-brother Frederick W. Müller, established the *Sass & Bro.* bottling works there at 116 S. Dunton Avenue. Previously Müller had been an employee with *Sass & Haffner* in Chicago.

During their first Dunton bottling season, Müller was still an employee (bottle #1). By the second season, he had become a partner in the Dunton bottling operation (bottle #2).

Louis Sass was still listed as a bottler in Dunton in the 1873 CAR, but he likely remained associated with the Dunton bottling operation only during its start-up year or two (ca. 1871–72), after which Müller bought

Figure 672

out his half-brother's interest and Sass returned his full attention to the Chicago partnership (*see Arlington Heights Memorial Library information posted on Digital Past*). Beginning in 1873–74, when the town name changed to Arlington Heights, the bottling operation there was renamed the *F. W. Müller* bottling works (*see Müller's Arlington Heights listing*).

DuQUOIN
(Perry County)

[Changed to Du Quoin, 1931]
1870 census: 2,200
1880 census: 2,800

Edwin Hayes

Soda ~ Mineral Water:

(1) EDWIN HAYES / DUQUOIN / ILLS   [aqua]

»SB w/ keyed hinge mold; slope-shouldered soda w/ very long neck and oval blob top; 7 3/4" tall. *[Figure 674]*

Edwin Hayes emigrated with his family from England to America as a 10-year-old child. He was 17 when he settled in DuQuoin at the end of the Civil War:

Ed Hayes was born in Tinsley, Lancashire, England, December 2, 1848, and when a lad of ten summers crossed the Atlantic. He soon began work in the mines, where he was employed until 1865, which year witnessed his removal westward. On reaching Illinois he cast his lot with the citizens of DuQuoin, and engaged in clerking in the grocery store of his uncle for about five years. On the expiration of that period he went with his uncle to Mobile, Ala., where he continued to work as a salesman for one year. He then returned to DuQuoin, where in connection with his brother he engaged in the bottling business, manufacturing soda waters.

In 1875 Mr. Hayes established the bottling works in Murphysboro, and continued business along that line for seventeen years. [Biographical Pub. Co. 1894b:702–3]

Since the bottle listed is technologically and stylistically older than the embossed bottle known from Edwin's bottling partnership with his brother Thomas (*see their listing, this chapter*), it was likely produced for sale during the late 1860s when he worked in his uncle's grocery store in DuQuoin.

Although Ed Hayes moved his bottling operation about 20 miles south to Murphysboro in the mid-1870s, no pre-1880 blob-top sodas are known from there embossed with his name. He is listed in the 1880 IIC as a soda water bottler in Murphysboro, and continued in the bottling business there (at the Murphysboro Steam Bottling Works) until selling out to an employee (Joseph Steinle: *see Biographical Pub. Co. 1894b:236–7*) just prior to the 1892 bottling season. The embossed

Figure 673

Figure 674

bottles Hayes may have used from 1880 to 1891 were produced too late for inclusion in the present study.

Thomas and Edwin Hayes

Soda ~ Mineral Water:

(1) **T H & BRO / DuQUOIN ILL_ //**   *[amber]*
   (heel): **W. M^cC & Co**

   »SB w/ post mold; slope-shouldered, shouldered-blob-top soda. *[Figure 675]*

Figure 675

Thomas and Edwin Hayes began a soda water manufacturing company in DuQuoin ca. 1872, and continued in partnership through 1875, when Edwin left DuQuoin to establish a bottling works in Murphysboro (*see his Murphysboro listing*). Edwin Hayes was probably the senior partner early on (their soda bottling business is listed as *"Edwin Hayes & Bro."* in the 1873 CAR), but when he left for Murphysboro ca. 1875, Thomas likely became the senior partner for the DuQuoin bottling works, and the sole operator there

Their partnership was dissolved ca. 1875, and Edwin continued in the soda water business on his own in DuQuoin during the later 1870s (*see his listing, this chapter*).

The embossed bottle listed above is of post-mold construction and thus dates to the 1870s. Its squat, no-neck style suggests it was produced and used in the late 1870s. Thomas also took on other jobs to make ends meet, and he was killed in the summer of 1883 while timbering a mine shaft about 15 miles north of town.

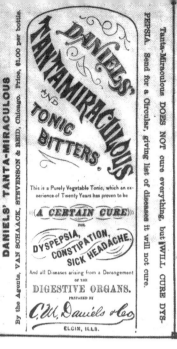

EAST ST. LOUIS
(St. Clair County)

1860 census: no data
1870 census: 5,600
1880 census: 15,000

Christian Lutt and Johanna Spannagel

Ale:

(1) **C. LUTT & Co / EAST S^T LOUIS / ILL^S** [aqua]

»SB w/ post mold; 2-piece mold quart ale; ring-neck tapered collar; 9" tall. *[Figure 676]*

Cider:

(1) **C. LUTT & Co / EAST S^T LOUIS / ILL** [amber, olive green]

»SB w/ post mold; slope-shouldered, ring-neck, cider-style quart; 12" tall. *[Figure 677]*

Soda ~ Mineral Water:

(1a) **CHRIST LUTT & Co / EAST S^T LOUIS** [aqua]
(base): **E·S·L.** *(arched embossing)*

»SB w/ post mold; slope-shouldered oval-blob-top soda. *[Figure 678]*

(1b) **CHRIST LUTT & Co / EAST S^T LOUIS** [aqua]
(base): **E.S.L.** *(straight-line embossing)*

»SB w/ post mold; slope-shouldered oval-blob-top soda. *[No photo, but see Miller 2007a:294. Same side embossing as in Figure 678]*

(2) **C. LUTT & Co / EAST S^T LOUIS** [aqua]
(base): **E.S.L.** *(straight-line embossing)*

»SB w/ post mold; slope-shouldered, oval-blob top soda. *[Figure 679]*

(3) **C. LUTT & CO / EAST ST. LOUIS / ILL. //** [aqua]
(heel): **I. G. Co**

»SB w/ post mold; short-neck, slope-shouldered, oval-blob-top soda. *[Figure 680]*

Figure 676

479

Figure 677

Figure 678

Figure 679

Figure 680

Lutt & Co. began in the soda and cider bottling business in East St. Louis in the spring of 1871, when Christian Lutt was 31. The company's first ads in the *East St. Louis Gazette* and the *East St. Louis Herald* that spring characterized them as "Soda and Cider Manufacturers" located at the corner of Broadway and Main Street; the *Gazette* ad indicated the "& Co." coinvestor in Lutt's bottling business was Mrs. Johanna Spannagel, who was 27 years old when they began in business together (Miller 2007a:291).

In the *History of Saint Clair County, Illinois* (Brink, McDonough & Co. 1881b), the *Lutt & Co.* factory was described as being 40 x 80 feet in size and two stories high. The factory was said to have cost $4,000 to build. At the time of the publication, the company employed four people, produced up to 400 cases of soda/cider per day, and had annual retail sales totaling about $25,000. The Thomas Schoenbein soda and mineral water bottling company in East St. Louis was dissolved at the end of the 1860s (*see Schoenbein listing below*), so Lutt & Co. may have taken over (and expanded?) this earlier bottling works.

The Lutt and Spannagel association during the 1870s is interesting and somewhat enigmatic. Christian Lutt did not arrive in the area until after 1870, and was apparently not a resident of East St. Louis. The 1880 USC listed him as a resident of Jerseyville, some 30 miles to the north. However, his partner Johanna Spannagel and her husband, George, did reside in East St. Louis during the 1870s. The curious aspects of the Lutt & Spannagel association are as follows:

1. During the 1870s, there was a soda bottling operation in Jerseyville with bottles embossed **J. N. SPANNAGEL** *(see Johanna Spannagel Jerseyville listing)*. Managed by Christian Lutt in Jerseyville (?).
2. During the 1870s, *Christian Lutt & Co.* bottled in East St. Louis (using the above listed embossed bottles). Managed by Johanna Spannagel in East St. Louis (?).
3. During the early 1870s, Johanna's husband George N. Spannagel sold his brewery operation in East St. Louis and "purchased $2,000.00 worth of machinery for the manufacturing of soda and mineral waters" (Miller 2007a:507), and was assumed to have begun his own bottling works. However, the address given for the operation is essentially the same as that given for the Lutt & Co. factory, and all known **G. N. SPANNAGEL** bottles postdate 1880 (thus he and his bottling works are not listed in the present volume). So he purchased machinery for the Lutt & Co. operation and joined the "& Co." Lutt partnership. He is not known to have begun bottling on his own until the Lutt operation closed down, on the death of his wife Johanna (see the later discussion), in the early 1880s.

Lutt was listed as a soda water manufacturer in the 1873 CAR and in the 1875 ISD. The 1880 IIC indicated that *Lutt & Co.* operated full-time 12 months a year with five workers making from $1 to $2/day. The company had a yearly payroll of $4,000 and an annual product value of $19,162.50.

C. Lutt and Co. was bankrupted and went into court receivership in September 1883, just a few months after the death of Johanna Spannagel (Miller 2007a:291). At about the same time, embossed bottles for the George (G. N.) Spannagel bottling works started to appear in East St. Louis.

During the early 1880s, just before the Lutt & Co. bankruptcy, the company used one collared-blob–style internal stopper embossed soda manufactured by the Illinois Glass Co. in Alton. This bottle is not listed here as it was produced too late for inclusion in the present study.

Thomas Schoenbein

Soda ~ Mineral Water:

(1) **TH^{os} SCHOENBEIN & Co / EAST S^T LOUIS / ILL //**   [aqua]
    (heel): **A. & D H. C**
    Note: *Single embossed dot below raised* **T** *in* **S^T LOUIS**.

      »SB w/ keyed hinge mold; slope-shouldered oval-blob-top soda. *[Figure 681]*

Thomas Schoenbein was a saloonkeeper in East St. Louis during the 1860s and 1870s. At the end of the Civil War, Schoenbein formed a soda water bottling company with his brother-in-law, Henry Croissant. Their company "was established soon after Illinoistown was renamed East St. Louis, probably in 1865 or 1866" (Miller 2007a:466). The *East St. Louis Gazette* for May 6, 1869 reported that the soda water business partnership of T. Schoenbein & Company had been dissolved. So their bottling works was in operation for no more than three seasons in the late 1860s.

Henry Croissant became a tollgate keeper on the St. Clair County Turnpike, while Thomas continued as the owner of a saloon.

Edward Schroeder

Ale:

(1a) **E. SCHROEDER / EAST. S^T LOUIS //**   [*many colors**]
(heel): **A & D. H. C.**
Note: As shown in Figure 682, the A&DHC variety of this bottle is known to occur in at least 6 color shades: red-amber, amber "black" glass, yellow, olive-amber, green, and olive-green.

»SB; 2-piece mold quart ale; ring-neck tapered collar; 9" tall. [*see Figure 682, left three bottles*]

(1b) **E. SCHROEDER / EAST. S^T LOUIS //**   [*aqua*]
Note: This bottle has the same side embossing shown in the Figure 682 bottles, but it has an elongated narrower neck and lip and lacks the **A & D. H. C.** *heel embossing.*

Figure 681

»SB; 2-piece mold quart ale; ring-neck tapered collar; 9 1/4" tall. [*Figure 682, third from right*]

Cider ~ Soda:

(1) **E. SCHROEDER / EAST S^T LOUIS / ILL //**   [*amber*]
(heel): **B. G. Co**
(base): **E S**
Note: Side embossing low on body near heel. Rear-heel maker's mark indicates bottle blown by the Belleville Glass Co.

»SB w/ post mold; slope-shouldered quart cider or soda; shouldered blob top; 11 3/4" tall. [*Figure 683*]

Soda ~ Mineral Water:

(1) **E. D. SCHROEDER / EAST S^T LOUIS / ILL**   [*aqua*]
(heel): **A & D. H. C**

»SB w/ keyed hinge mold; slope-shouldered, shouldered-blob-top soda. [*Figure 684*]

(2) **E. SCHROEDER / EAST S^T LOUIS / ILL //**   [*aqua*]
(heel): **A & D. H. C**

»SB w/ hinge mold; slope-shouldered, oval-blob-top soda. Both off-centered and centered name varieties indicate sequential redesign of mold embossing after removal of the "D." used in variety #1b above. [*Figure 685*]

(3) **E. SCHROEDER / EAST S^T LOUIS / ILLS**   [*aqua*]
(base): **M.G. Co**

»SB; slope-shouldered, shouldered-blob-top soda. [*No photo; illustrated by Miller 2007a:482*]

Figure 682

Figure 683

Figure 684

Figure 685

(4) **E. SCHROEDER / EAST S**ᵀ **LOUIS / ILL** *[aqua]*
 (base): **C & I**

» SB w/ keyed hinge mold; slope-shouldered, oval-blob-top soda. *[Figure 686]*

(5) **E. SCHROEDER / EAST S**ᵀ **LOUIS / ILL** *[aqua]*

» SB w/ post mold; slope-shouldered, oval-blob-top soda.
 Note: Variant with smaller embossed lettering. [Figure 687]

(6) **E. SCHROEDER / EAST ST LOUIS / ILLS //** *[aqua, amber, blue]*
 (heel): **I.G. Co.**
 (base): **E S**

» SB w/ post mold; slope-shouldered, short neck, oval-blob-top soda. *[Figure 688]*

(7) **M**ᴿˢ **E. SCHROEDER / EAST ST LOUIS / ILL //** *[aqua]*
 (heel): **I.G. Co.**
 (base): **E S**

» SB w/ post mold; slope-shouldered, short neck, oval-blob-top soda. *[Figure 689]*

(8) **E. SCHROEDER** *(flat)* **/ EAST S**ᵀ **LOUIS / ILL.** *[aqua]*
 Note: Two raised embossed dots beneath the superscript **T** *in* **S**ᵀ **LOUIS.**

» SB w/ keyed hinge mold; narrow cylindrical "Selters Water" form; short neck; oval blob top; 7" tall. *[Figure 690]*

Figure 686

Figure 687

Figure 689

Figure 688

Figure 690

Edward Schroeder bottled soda water in Cairo for a few seasons in the late 1850s in Cairo (*see his Cairo listing*), then relocated to St. Louis during the early Civil War years, where he became associated with pioneer St. Louis bottler John Cairns (as part of *John Cairns & Co.*) in a grocery and soda-bottling operation with outlets in both St. Louis, MO, and East St. Louis, IL. Cairns was a long-time soda water bottler in St. Louis (1857–69). Schroeder's son George indicated in his East St. Louis biographical sketch (Chapman Brothers 1892a) that his father was first affiliated with Cairns' St. Louis bottling works, (becoming involved in Cairns East St. Louis operation near the end of the Civil War (Miller 2007a:469).

Edward Schroeder moved to East St. Louis permanently toward the end of the Civil War, where he first remained involved with the Cairns business: "Newspaper advertisements in the *East St. Louis Herald* from 1865–1868 listed **JOHN CAIRNS & COMPANY, GROCERY STORE AND SODA FACTORY** at the 10th Street & Illinois Ave. address," where his son's biographical sketch says he first bottled soda on the Illinois side of the river (Miller 2007a:469). There is an unconfirmed statement that an E. SCHROEDER / ILLINOISTOWN / ILLS soda bottle exists. "Illinoistown" was renamed East St. Louis on April 22, 1864 (Adams 1989:399).

In 1869, when he was in his mid-30s, Edward Schroeder apparently bought out Cairns East St. Louis bottling operation. Soon after he started out on his own, he rebuilt the bottling works at its earlier site at Illinois and 10th (later 919 Illinois Avenue). The new building was a two-story brick edifice whose grand opening was announced in the *East St. Louis Gazette,* August 19, 1871.

At first, he produced only nonalcoholic products:

> EDWARD SCHROEDER,
> Manufacturer and Dealer in
> SODA, SARSAPARILLA,
> **Pure Apple Cider**
> AND
> **SELTZER WATER**
> The proprietor guarantees his liquids to be pure
> and free from all noxious adulterations.
> [*East St. Louis Gazette,* Oct. 1, 1870]

We direct the attention of our readers to the advertisement of Mr. Schroeder, who devotes his time to the manufacture of drinks of a temperate nature, to be found among the new ones this week. He has been in the business for four years past. The increase of his business—necessitating the extension of his manufactory and new and improved machinery—is proof positive of the estimation in which he is held by this community and surrounding country. [*East St. Louis Gazette,* Sept. 24, 1870]

But he soon expanded his line to include ale, tonic beer, iron phosphate, and "imperial nerve food" (Miller 2007a:470).

He is listed in the 1872 TW & W Railway directory and in the 1875 ISD as a soda water manufacturer. According to the 1879–80 IIC, Schroeder employed eight full-time workers year-round, who worked 10-hour days and were paid from $2.00 to $2.50 per day each. He estimated the annual value of his bottled products that year at $20,000. In 1881, Schroeder's factory could produce and bottle 400 cases of soda water daily.

Edward Schroeder died in mid-1887, after which his wife and son George continued the business until 1901. Schroeder's post-1880 internal-stopper Hutchinson and collared-blob style bottles were used too late for inclusion in the present volume, as were those produced by his widow and son. All of the listed Schroeder embossed bottles would have been manufactured and used from ca. 1869 to 1880. He also opened a branch bottling works in Memphis in the early 1870s, where embossed Schroeder bottles are known.

EDWARDSVILLE
(Madison County)

1850 census: 700
1860 census: 2,000
1870 census: 2,200
1880 census: 2,900

Frank Harles

Soda ~ Mineral Water:

  (1a) **FRANK HARLES. / EDWARDSVILLE / ILLS_ //**   [aqua]
  (heel): **A & D. H. C.**

   »SB w/ post mold; slope-shouldered, oval-blob-top soda. *[Figure 691]*

  (1b) **FRANK HARLES. / EDWARDSVILLE / ILLS_**   [aqua]

   »SB w/ post mold; slope-shouldered, oval-blob-top soda. *This bottle is a later variant of #1a, with the A&DHC heel mark peened-out but still visible. [Figure 691]*

According to the 1882 *History of Madison County* (Brink & Co. 1882:344), wagonmaker Frank Harles opened his soda water business in spring 1871 with a production capacity of three to four thousand cases of filled bottles per annum. Harles was listed in the 1872 TW & W Railway directory and in the 1875 ISD as a bottler of soda water in Edwardsville. The style and technological attributes of the listed bottles indicates they were manufactured and used during this 1870s period. The 1879–80 IIC indicated that Harles had two adult male employees at his horse-powered soda manufactory operated full time (10 hours per day) year-round. Harles paid $300 a year in wages and $175 a year in production materials. He estimated his annual income at $1,000.

Harles was still listed in the 1880 USC as a soda and ice dealer, but he later turned the bottling business over to his son Henry and returned to his wagonmaker trade.

Figure 691

ELGIN
(Cook and Kane Counties)

1850 census: 2,400
1860 census: 2,800
1870 census: 5,400
1880 census: 8,800

Casper Althen

Beer:

  (1) **CASPER ALTHEN / EAGLE BREWERY / ELGIN, ILLS. // THIS BOTTLE /**   [amber]
  IS NEVER SOLD /
  (heel): **A. & D. H. C**

   »SB w/ post mold; slope-shouldered quart; narrowed shouldered-blob top; 9 3/4" tall. Front embossing in circular slug plate. *[Figure 692]*

Charles Tazewell opened a brewery in Elgin in the west side of Kimball Street in 1849. Casper Althen purchased a controlling interest in the brewery in 1868, bought out his partner in 1870, and then moved the bottling works to a site just north of the old West Side Distillery. In a series of renovations and additions, earlier wooden buildings were replaced with brick structures, eventually raising production capacity to between 5,000 and 6,000 barrels per year. "Casper's Superior" was one of the most popular brands (Waite 1908:71–72). This brand has also been seen advertised on a postcard photo ad for the brewery. According to a *Chicago Tribune* notice (January 24, 1881), Althen also operated an icehouse in Elgin.

From the style and technological attributes of the embossed bottle listed above, it was likely manufactured and used during the late 1870s or early 1880s.

Althen (and later his three sons, who joined the business in the 1890s) operated the brewery well into the twentieth century. According to the *History of County* (Waite 1908:71), the name of the brewery during the 1870s and 1880s (*Casper Althen & Co.*) was changed to the *Elgin Eagle Brewery* in 1894.

Charles M. Daniels and Co.

Medicine:

(1) **C. M. DANIEL'S & Co / ELGIN. ILL.** // //   [amber]
    DANIELS' / TANTA MIRACULOUS // //
    (base): **W. McC**

»SB w/ keyed hinge mold; square; 3 sunken panels; tapered collar; 9" tall. [Figure 693]

Dr. Charles M. Daniels came to the rural Elgin area in 1839, when he was 26 years old. In the mid-1840s, he moved to southern Illinois and studied medicine. He later returned to Elgin to practice medicine and was listed in the 1864–65 ISD as a physician there (Hawley 1893). He continued to reside and work as a physician in Elgin into the 1880s (1880 USC).

According to Hawley (1893) and E. C. Alft (1980s *Elgin Daily Courier* newspaper column, "Days Gone By"), Daniels' Tanta Miraculous was

Figure 692

Figure 693

sold widely "all over the United States" for many years as a "vegetable tonic for dyspepsia. During the 1870s, the Chicago distribution agent (*Van Schaak, Stevenson & Reid*) listed the medicine as "Daniels' Tantamiraculous and Tonic Bitters." One of the bottle's paper labels (Figure 694) from this Chicago distributor described the contents as a "Purely Vegetable Tonic, which an experience of Twenty Years has proven to be *A CERTAIN CURE FOR DYSPEPSIA, CONSTIPATION, SICK HEADACHE,* And all Diseases arising from a Derangement of the DIGESTIVE ORGANS" (Ring 1980:158). Tantamiraculous was also listed for sale in the 1878 *Colburn, Burk & Co.* drug catalog.

From the style and technological attributes of the embossed bottle listed, it was likely manufactured and used during the 1860s.

William Dettmer

Soda ~ Mineral Water:

(1) **WM. DETTMER / ELGIN, ILL.** [amber]
(base): **C & I / X**
Note: embossed in depressed circular slug plate. The large "**X**" symbol on the base is embossed in outline only, and the **C & I** embossing is arched above it.
»SB w/ post mold; slope-shouldered quart; shouldered-blob-top soda; 12" tall. *[Figure 695]*

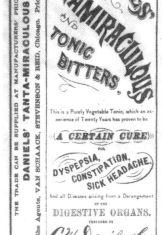

Figure 694

No direct documentation of Dettmer's brief Elgin bottling operation has yet been located. Dettmer had earlier been a junior partner with Gustave Hausburg in a soda-bottling works at Blue Island, IL (*see their Blue Island listing*), and according to the 1880 USC he returned to soda bottling when he set up shop in Elgin. However, none of the pint bottles more commonly associated with soda bottling have yet been documented, so aqua pint Elgin Dettmer soda bottles may yet show up.

Slug-plate embossed bottle logos were first used in Illinois ca. 1876 (*see Jacksonville Bradbury listing*), and the *Cunningham & Ihmsen (C&I)* glass works manufactured bottles until ca. 1878, so the Dettmer quart bottle listed above likely was manufactured during this two-year period. William Dettmer was still listed as a soda bottler in Elgin in the 1880 USC.

Elgin Ink

Ink:

(1) **ELGIN INK** [aqua]
»SB w/ post mold; cone-shaped ink; squared flared-ring collar; 2 1/2" base diameter; 2 1/2" tall *[Figure 696]*

We have been unable to locate any historic documentation of the *Elgin Ink Company* or its proprietors in county histories of the area. So, likely it was only a short-lived and/or unsuccessful enterprise.

Figure 695

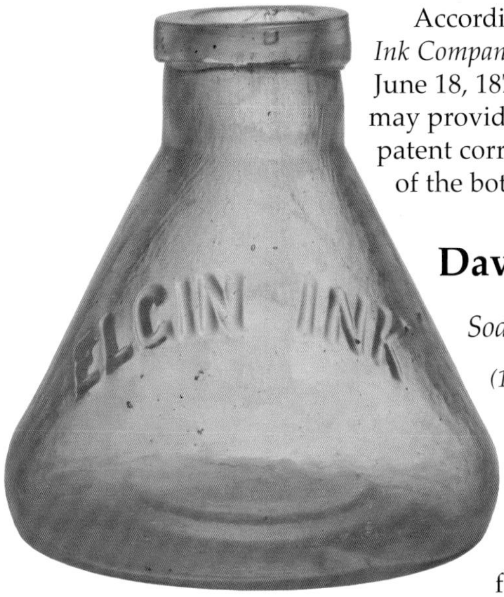

According to the U.S. Patent Office *Official Gazette* (Vol. 13, p. 1048), the *Elgin Ink Company* of Elgin, Illinois patented a label design for *"Elgin Black Ink"* on June 18, 1878. The design patent application itself has not been located, but it may provide information on who operated the company. The date of the label patent corresponds well with the late 1870s style and technological attributes of the bottle.

David Henry and John Van Buskirk

Soda ~ Mineral Water:

(1) **H & V** / **ELGIN ILL** [aqua]

»SB w/ keyed hinge mold; slope-shouldered, oval-blob-top soda. *[Figure 697]*

❖

Figure 696

Figure 697

With only embossed initials for the manufacturers, we have thus far been unable to locate any direct Elgin historical information on this bottling operation. However, a possible connection to a nearby contemporary Chicago bottling works can be postulated (as a "best guess" with the information at hand).

The Chicago soda bottle was embossed **V & H / CHICAGO**. The proprietors there were John Van Buskirk and David Henry (*see their Chicago listing*).

According to the Chicago CDs, *Van Buskirk & Henry* began their partnership in the "wholesale wines and liquors" business in 1864. However, Van Buskirk was senior partner in the company for just three years: 1864–1866, when the Chicago soda bottles were produced.

During 1867 and 1868, Van Buskirk became junior partner, and the company name was changed to *David Henry & Co.* In 1869, David Henry was listed alone in the CDs, with an office at the same address. Possibly during 1867–68, Henry & Van Buskirk made an effort to expand their bottled soda market into the Elgin area. The style and technological attributes of the above Elgin bottle are closely consistent with the two-year period during which Henry was the senior member of the partnership.

Hausburg Brothers and Co.

(William, Gustave, and Charles)

Soda ~ Mineral Water:

(1) **HAUSBURG BROS & CO / ELGIN ILLS //** [aqua]
(heel): **A & D. H. C.**

»SB w/ post mold; slope-shouldered shouldered-blob-top soda. *[Figure 698]*

Cider ~ Soda:

(1) **HAUSBURG BROS & CO / ELGIN ILLS. //** [amber]
(heel): **A & D. H. C.**

»SB w/ post mold; slope-shouldered quart with ring-neck cider finish; 11" tall. *[Figure 699]*

❖

In Chicago, William, Gustave, and Charles Hausburg were listed as bottlers of mineral water from their residence at 112 Claybourn Avenue in the CDs for 1872 and 1873 (*see their Chicago listings*). Prior to this time, one of the brothers, likely Gustave, had been senior partner in two soda or mineral water bottling operations in the south Chicago suburb of Blue Island (*see Chicago Hausburg & Innes and Hausburg & Dettmer listings*).

By the time the 1874 Chicago CD was published, Gustave had left the Chicago partnership, probably to help set up a family bottling works at Elgin. By the late 1870s, he had moved back to Blue Island and established the *G. H. Hausburg* bottling works in Blue Island (*see his listing, this chapter*).

William and Charles Hausburg bottled together in Chicago until 1878, after which William took over sole management of the Chicago bottling operation (*see his Chicago listing*), while Charles presumably moved to Elgin to join Gustave (or more likely, take over for him when he moved on to Blue Island) at the Hausburg brothers' bottling operation there.

But Charles was soon bottling in Blue Island as well. We have recorded both Charles Hausburg–embossed Chicago bottles (an early 1880s aqua quart with a *C&Co* maker's mark) and Gustave Hausburg–embossed Blue Island bottles (an early 1880s blob-top amber quart with a *C&Co* maker's mark) that date to the 1880s (both of which are too late for inclusion in the current study). Later-1880s and 1890s Charles Hausburg-embossed Blue Island bottles have also been seen in collections (including a narrow, cylindrical, amber Baltimore-Loop-closure pint and Hutchinson-closure sodas), as have early twentieth-century Charles Hausburg "South Chicago" crown-top sodas.

Figure 698

Figure 699

So at some time during the late 1870s and the early 1880s, both Charles and Gustave Hausburg moved to Blue Island in order to pursue the bottling trade in the south Chicago area.

No direct records of the Elgin Hausburg bottling works have yet been located, but the *Hausburg Bros.* regional bottling history indicates that Gustave Hausburg (later probably with Charles' help) managed the *Hausburg Bros.* bottling business in Elgin during the mid- and late-1870s. The "*& Co.*" member(s) of the Elgin business are as yet unknown.

Joseph and Henry Shure

Soda ~ Mineral Water:

(1) **SHURE BRO'S / ELGIN. ILL** [aqua]

»SB w/ keyed hinge mold; slope-shouldered oval-blob-top soda. *[Figure 700]*

❖

The Shure Bothers, listed in Elgin in the 1875 ISD, also bottled soda water in Chicago, where they were listed in the 1870 USC as Chicago soda and mineral water bottlers (*see their Chicago listing*).

But Joseph and Henry Shure were only listed as bottlers in the Chicago CDs in 1869, at which time they operated the *Shure & Bro.* "cider manufactory" from their residence at 835 S. Halsted.

Figure 700

They were in business in Elgin by 1875 (ISD), where they were also listed as soda water manufacturers. Perhaps after their 1869–70 attempt to establish a bottling business in Chicago, they moved to Elgin to open a similar business. Their Elgin bottling effort probably dated to ca. 1871–75.

Joseph reappeared in the Chicago CDs in the late 1870s, where he was listed first as a feed store operator, then as a salesman, then as a saloon operator. The Shure brothers were no longer in the soda-bottling business by the time of the 1880 USC.

EVANSTON
(Cook County)

1860 census: 800
1870 census: 3,000
1880 census: 4,400

John G. Westerfield

Preserves:

(1) **J. G. / WESTERFIELD // // EVANSTON / ILL. // //**   [aqua]
(encircling central depression in base): **C & I PITTS PA.**

»SB w/ keyed hinge mold; beveled-edge square pickle/preserve bottle; 4 flat panels with embossed wreath at arched top of each panel; one panel has lower wreath (arched up) as well, probably to surround paper label; lettering embossed at the heels of two opposing panels only; ring-top rounded thickened collar; 10" tall. [*Figure 701*]

John Gedney Westerfield (the surname was anglicized by JGW from "Westervelt", according to a family genealogy website) immigrated to the Wilmette area of Illinois near Evanston in the 1850s. He is widely (and apparently anecdotally) believed by local historical sources to have "built pickle and vinegar factories in the area in 1857." He was involved in platting the original village of Wilmette in the late 1860s and was a village trustee when it was incorporated in 1872 (Andreas 1884). Westerfield's early association with the Evanston area and

Figure 701

the pickle business is also supported by an August 31, 1988 *Chicago Tribune* article titled "First Settlers Found Pastoral Wilmette Ideal for Family Life":

> John Westerfield, who first saw Wilmette in 1847 while on a business trip from New York, bought the section with the Ouilmette cabin in 1865, when he razed it due to erosion damage....
>
> Westerfield platted the city and became its first village president in 1872, when it incorporated. He also opened a pickle factory near his home where he grew cucumbers and developed a special strain of seed purchased by other growers.

Westerfield is listed in the 1870 USC as a pickle dealer in the Evanston area (New Trier Township). *The Cunningham & Ihmsen Co.,* manufacturers of the Westerfield embossed pickle bottle, began their glassmaking business in Pittsburgh in 1865. From the style and technological attributes of his embossed pickle bottle, as well as the *C&I* maker's mark, his pickle factory likely had an Evanston address prior to the incorporation of Wilmette and was probably in operation during the late 1860s.

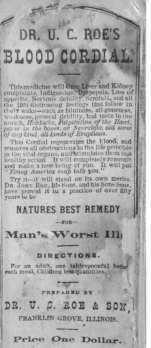

FARMINGTON
(Fulton County)

1850 census: 1,400
1860 census: 2,200
1870 census: 2,100
1880 census: 1,100

Riley Bristol, Charles Tyler, and Co.

Medicine:

(1a) (side): **BRISTOL TYLER & Co** // **DR STREETER'S** / **MAGNETIC** / **LINIMENT** //   [aqua]
(side): **FARMINGTON ILL** // //

»SB w/ keyed hinge mold; 3 sunken panels; slope shoulder; squared flare-ring collar; 6" tall. *[Figure 702, left]*

(1b) **DR STREETER'S** / **MAGNETIC** / **LINIMENT** // // //   [aqua]

»SB w/ keyed hinge mold; 3 sunken panels; slope shoulder; squared flare-ring collar; 5¾" tall. *[Figure 702, center]*

(1c) **DR STREETER'S** / **MAGNETIC** / **LINIMENT**   [aqua]

»SB; 4 beveled-edge flat panels; slope shoulder; squared flare-ring collar; 6" tall. *[Figure 702, right]*

Riley Bristol was listed as a druggist in Farmington in the 1870 USC, while Charles Tyler was listed in the same census as a patent medicine dealer there. The "*& Co.*" (bottle #1a) member of the business early on may have been W. E. Streeter, a successful Fulton County farmer who "followed peddling in 1864–5" (Chapman & Co. 1879:767).

Bristol began his pharmacy business in 1856 (Andreas, Lyter & Co. 1871:47), and sold the business in 1883. Based on technological characteristics of the bottles listed,

Figure 702

Dr. Streeter's Magnetic Liniment was likely bottled and sold by Bristol's drugstore in Farmington during the late 1860s and the 1870s.

Dr. Streeter's liniment was marketed widely as a cure for rheumatism in people and lameness in horses. An 1870s trade card for the product listed "unsolicited testimonials" extolling its virtues for relieving sprains, stiff joints, and cramps.

FORRESTON
(Ogle County)

[previously: Foreston]
1860 census: 1,100
1870 census: 2,200

Samuel McLain and Brother

Bitters:

(1) **M^CLAIN BRO^S / FORRESTON ILL^S** // // **KEY_STONE / TONIC BITTERS** // //   [amber]
(*base*): **A & D. H. C.**

»SB w/ post-mold; square case bottle; flat sides; beveled corners; tapered collar; 9" tall.
[Figure 703]

Keystone Tonic Bitters was first bottled and marketed by George Herrick of Dixon, IL, during the 1860s (*see his Dixon listing*). Herrick's bottling operation was located about 20 miles south of Forreston.

Later, during the 1870s, *Key-stone Tonic Bitters* was bottled and sold by the McLain Brothers of Forreston, IL. Their embossed bottle was made from nearly the same mold as the earlier Herrick variety, so the McLain brothers likely have bought the brand and the recipe from Mr. Herrick. No ads for Key Stone Bitters have been located for either proprietor to indicate the health-renewing powers claimed for the product.

From temporally significant features of the bottle bases, Herrick clearly produced this bitters brand before the McLain brothers did. The Herrick embossed bottle has a keyed hinge mold typical of the 1860s, while the McLain brothers' bottle had an 1870s-style post-mold base.

Two lines of evidence suggest that the McLain brothers may not have begun bottling *Key-stone Tonic Bitters* in Forreston until the late 1870s. First, the spelling change that put two R's in "Forreston" was not officially instituted until 1877–78 (Adams 1989:364); second, Samuel McLain was still characterized as "a wholesale manufacturer of specialties in medicine" as late as 1899 (Clark and Co. 1899b:205).

Samuel McLain's brother was not listed as a partner in the medicine business during these later years. Samuel had at least three brothers: John, Alexander, and Andrew. Jesse H. was the only other McLain of appropriate age listed as residing in Forreston in 1878 (Kett 1878a:725). The 1880 USC listed John as a retired farmer in Forreston.

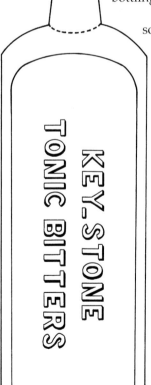

Figure 703

FRANKLIN GROVE
(Lee County)

1870 census: 760
1880 census: 730

Dr. Uriah C. Roe and Sons

Medicine:

(1) (side): **DR U. C. ROE & SONS** // // [aqua]
    (side): **PREPARATIONS** // //

»SB w/ hinge mold around central circular depression; beveled-edge rectangle w/ 3 sunken panels; squared flared-ring collar; 7 3/4" tall. [*Figure 704*]

Uriah Chittenden Roe (*Figure 705*) came from a family of physicians. His father was Dr. John Roe, a pioneer Illinois physician who moved his family to Springfield, IL, from Kentucky in 1827 "and began the study of medicine" (Hill & Co. 1881:598). Dr. John Roe settled in Ogle County in 1835. His practice "extended from Roscoe on the north, to Dixon on the south, and from Sycamore on the east, westward to the hamlet of Buffalo. He was one of the first settlers in the county, and was under the necessity of going to Princeton, Bureau County, a distance of fifty miles, to do his milling. Having little competition in his profession, he was a very busy man, and on occasions, visited as many as forty-two patients in a single day" (Kauffman and Kauffman, eds., 1909:1004). One of U. C. Roe's brothers, physician Dr. Malcolm C. Roe, was a life-long resident of Chana in Ogle County.

When Uriah Roe was 20 years old, he decided to follow in his father's footsteps and pursue a career in medicine:

> In 1845 he entered the Ohio Botanico-Medical College, and attended one term of lectures. After sixteen years practice of medicine he received a diploma from this institution. Upon his father's removal to Chicago, about 1845, [Uriah] entered into a large and lucrative practice.... In the spring of 1854 Mr. Roe came to Franklin Grove and lived here until 1860, and returned again in 1870. In 1860, he began the manufacture and sale of medicine, in which business with his sons he is now engaged. He has a large number of teams and wagons out through the country in the sale of his medicines. The doctor also treats

Figure 704

chronic diseases, and lectures on phrenology and the laws of health. [Hill & Co. 1881:598–599]

This biographical sketch indicates that U. C. Roe first began selling his patent medicines ca. 1860—about the time he left Franklin Grove after his first residence there. He may have been caught up in events at the outbreak of the Civil War, but we do not know his whereabouts until he came to Fairfax, IA, in 1864: "Dr. U. C. Roe came to Fairfax in 1864 for the practice of medicine. He also sold drugs. The business finally drifted into a grocery store, as it seems that the settlers preferred sugar and prunes to sugar and quinine" (Brewer and Wick 1911:87–88).

As indicated in the Lee County, IL, biographical sketch above, U. C. Roe returned to Franklin Grove ca. 1870, when he was 45 years old, and renewed his patent-medicine business—eventually involving two of his sons (first Frederick, and later Nathaniel as well) in the new endeavor. The new business was called the *Roe Family Medicine Company* (see Lee County Historical Society website). According to a small booklet about the town printed in 1870 by the *Telegraph and Herald Book and Job Print*, of Dixon, IL, Franklin Grove had three doctors: an Allopathic practitioner, an Eclectic School physician and newly-arrived Dr. Roe. "What school he follows we are not advised, though he differs from either of the others."

Figure 705

From the style and technological attributes of the listed embossed bottle, it was produced to contain the various products of Dr. Roe's medicine line at about the time of his return to Franklin Grove: ca. 1870. One of the embossed bottles we have recorded still retains its paper label:

DR. U. C. ROE'S
BLOOD CORDIAL

This medicine will Cure Liver and Kidney complaints, Indigestion, Dyspepsia, Loss of appetite, Nervous debility, Scrofula, and all the 1001 distressing feelings that follow in their wake—such as faintness, all-goneness, weakness, general debility, bad taste in the mouth, *Headache, Palpitation of the Heart, pains in the bones, or Neuralgia, old sores of any kind, all kinds of Eruptions.*

This cordial regenerates the blood, and removes all obstructions to the life principle in the vital organs, and stimulates them to a healthy action. It will completely renovate and make a new being of you. It will put *"Young America snap into you."*

Try it—it will stand on its own merits. Dr. John Roe, his Sons, and his Sons Sons, have proved it in a practice of over fifty years to be

NATURE'S BEST REMEDY
— FOR —
Man's Worst Ills

This *Blood Cordial* was sold for $1.00 per bottle in 1870. Its label indicates that the family remedies had been handed down from father (John Roe), to sons (U. C. Roe, et al.), to grandchildren (Frederick and Nathaniel). In this light, a 1905 pharmacist's-catalog list of medicines for sale—from "Frederick U. Roe & Sons" of Franklin Grove—suggests that the tradition continued, and their product list likely reflects the range of Dr. U. C. Roe medicines of the early 1870s. The F. U. Roe medicines included: *Roe's Ague Pills, Roe's Balm of Gilead Ointment, Roe's Blood Cordial, Roe's Physic, Roe's Cough Drops,* and *Roe's Hair Balsam.*

FREEBURG
(St. Clair County)

1860 census: 550
1870 census: 900

Frederick Darmstatter

Soda ~ Mineral Water:

(1) **DARMSTATTER / FREEBURG / ILL'S** [aqua]

»SB; slope-shouldered oval-blob-top soda. *[Figure 706]*

———————— ❖ ————————

Figure 706

Miller (2007a:120) indicates that Frederick Darmstatter ("Fridrich" in the 1870 USC) moved to Freeburg from rural St. Clair County soon after his marriage in 1863. Darmstatter was listed in the 1870 USC as a miller there.

By 1874 (*St. Clair County Atlas: Freeburg References*), Fred Darmstatter was listed as "Proprietor of the Freeburg Hotel. West side Square." A period photo of the hotel is included in the atlas. "Hotels in those days had bars, and bars needed soda water for the ladies, etc. Perhaps it was during the early 1870s when Frederick briefly tried to manufacture his own soda water. The Darmstatter bottle's style suggests an 1860s–early 1870s vintage" (Miller 2007a:120).

Miller's observation is supported by the fact that Frederick Darmstatter was listed as a soda water manufacturer in both the 1880 USC and the 1879–80 IIC. Technologically and stylistically, as Miller also notes, the bottle listed seems to have been manufactured in the early 1870s, likely when Darmstatter first began his bottling efforts. The 1879–80 industrial census indicates that Frederick himself was his only employee, that the bottling works operated six months per year (May to November), and that he invested $600 annually in materials and earned $936 in sales of his bottled product.

FREEPORT
(Stephenson County)

1860 census: 5,400
1870 census: 7,900
1880 census: 8,500

Thomas and Lawrence Crotty

Soda ~ Mineral Water:

(1) (*heel*): **A & D. H. C.** // [aqua]
(*base*): **CROTTY BROS __ FREEPORT. ILL**[S]
Note: There are two raised embossed dots beneath the superscript **S** *in* **ILL**[S].

»SB w/ heel mold; shouldered-blob-top soda. *[Figure 707]*

———————— ❖ ————————

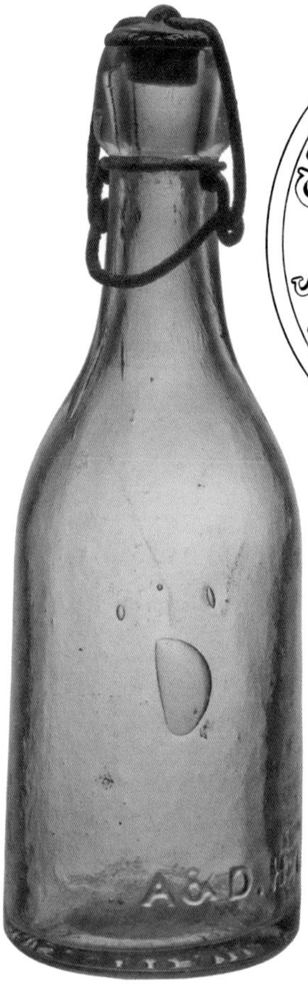

The Crotty Brothers also had a bottling operation in Ottawa, IL, where they used very similar embossed bottles (*see Ottawa listing*). They were listed in 1869 and 1870 Ottawa CDs and in the 1873 and 1874 Illinois CAR. *Coutant & Co.* (*see Coutant listing*) and the Crotty brothers were sequential Ottawa bottlers—likely at the same bottling plant—during the later 1860s and early 1870s. In fact the Crotty Brothers were probably the "*& Co.*" partners in the *Coutant & Co.* bottling works during the late 1860s, buying out B. W. Coutant and operating it themselves from 1869 onward.

These two Ottawa companies are the only Illinois soda water manufacturers known to have used bottles embossed on the base only. Since early soda bottles were placed upside-down in wooden cases (to keep the corks moist) base embossing would certainly have been easily readable.

By the time of the 1870 USC, Lawrence Crotty was listed as a soda manufacturer in Freeport. So perhaps during the early 1970s, the Crotty brothers expanded their bottling operations to Freeport, with Lawrence moving there to oversee Freeport operations while Thomas stayed in Ottawa to manage their bottling operation there. According to information in the 1880 Stephenson County history (Western Historical Co. 1880:480–481), it appears that Lawrence Crotty only operated the brothers' Freeport bottling outlet during 1870 and 1871:

> *Soda Water Factory*—Maintained by Galloway & Snooks, occupies a building near the corner of Jackson and Walnut streets, erected forty years ago.... The present business was established in 1872, as a successor of Crotty Brothers, and includes a patronage extending throughout the city and adjoining country.
>
> The line of manufacture is soda water, champaign cider, root beet, etc., of which an aggregate of 2,000 gross are put upon the market annually, furnishing employment to four men, at a monthly cost of $125, and doing a business of $3,000 per year.

Figure 707

The Crotty Brothers Ottawa bottling operation was likely bought out in the mid-1870s by Alex Finkler, who operated the Ottawa bottling works after that (*see his listing*). We have thus far been unable to locate embossed bottles for the successor *Galloway & Snooks* bottling operation at Freeport, but it would be surprising if embossed bottles were not used by the company.

Francis A. Read and Frederick R. Bartlett

Soda ~ Mineral Water:

(1) **REED & BARTLETT / FREEPORT / ILLS**   [aqua]

»SB w/ keyed hinge mold; slope-shoulder shouldered-blob-top soda. [*Figure 708*]

The firm of Read & Bartlett was listed in the May 1866 IRS assessment for Freeport under "retail dealers and manufacturers." Both men were involved in a variety of businesses in Freeport during the 1860s and 1870s. The 1860 ISD lists Churchill and Bartlett in the hardware business there, and an early twentieth century *Read Genealogy* volume (Dodd 1912:274) noted that Read was "an extensive and prosperous

Figure 708

merchant" in Freeport, involved in the dry goods, millinery and carpet business. The volume also noted that the two were brothers-in-law: Frederick Bartlett was married to F. A. Read's sister. From the style and technological attributes of the embossed bottle listed, the two men were also involved in the soda-bottling business in Freeport for a short time during the 1860s. This may have been only a single-season effort, since Read's name is misspelled on the bottle and no corrected embossed varieties of the bottle are known.

Surprisingly, another unrelated **REED & BARTLETT** embossed soda of approximately the same age is known. It is embossed **SODA WATER** beneath the company name. But has no embossed town of origin. It is also base-embossed **DB WILLIAMS / & CO.** This bottle was produced by an Indianapolis bottling partnership, and appears to be unrelated to the Freeport bottle listed.

FROGTOWN
(Clinton County)
[1870–80 population unknown]

Henry Burger

Soda ~ Mineral Water:

(1) **H. BURGER / FROGTOWN / ILL** *[aqua]*

»SB w/ keyed hinge mold; slope-shoulder shouldered-blob-top soda. *[Figure 709]*

❖

Henry Burger came to Clinton County in 1859 and was listed in both the 1860 and 1870 USC as a general store keeper with a post office box in Carlyle, IL (about 10 miles E of the Frogtown locale). In the 1870 IIC, he was listed as a soda water manufacturer with a PO box in Breese (about two miles SW of Frogtown). He was also listed in the 1873 CAR as the proprietor of a general store in Frogtown (*Note*: the Frogtown post office was established in 1877, and Henry Burger was postmaster there, likely at his general store, during the late 1870s). In the 1880 USC, he was listed as a 51-year-old grocer with a Breese PO box. From the census data (and from early plat maps), Burger apparently moved to farmland property in the Frogtown Pond area just north of Breese in around 1870, relocating his Carlyle general store operation there at the time as well. He seems to have begun bottling his own soda water for the store at about this same time.

His soda water bottling works is listed in the 1870 IIC for Breese Township, Clinton County. Burger is listed in the census as having $500 capital invested in his business, with four employees paid a total of $450 per year. The business operated six months a year. He filled 1,000 boxes (24,000 bottles!) of soda water each season with a product retail value of $800 (or about 3¢ per bottle wholesale).

It is interesting to note that on the reverse side of the bottle from the **BURGER** embossing are ghost embossed letters of a peened-out Baltimore soda water bottler's name: **P. BABB / BALT°**. Chris Rowell, on his Baltimore antique bottle website, indicates that Babb was listed as a bottler in Baltimore CDs for only a short period, from 1853 to 1857. This appears to be an extreme illustration of the fact some glass houses occasionally recycled old iron bottle molds for extended periods of time by removing earlier merchant embossing from the molds.

Figure 709

GALENA
(Jo Daviess County)

1850 census: 6,000
1860 census: 8,200
1870 census: 7,000
1880 census: 6,500

Joseph Evans

Ale~Porter:

(1) **J. EVANS // GALENA. ILL** [olive "black" glass]
Note: There are two raised embossed dots beneath the second **L** in **ILL**
»IP/SB; front and rear shoulder-embossed; 3-piece mold; quart ale; 9 3/4" tall. *[Figure 710, left]*

(2) **J EVANS / GALENA ILL** [olive "black" glass]
»IP/SB; shoulder embossed on one side; 3-piece mold; quart ale; 9 3/4" tall. *[Figure 710, center]*

(3) **J. EVANS // GALENA ILL** [olive "black" glass]
»IP/SB; front and rear shoulder-embossed with larger, cruder lettering than bottle #1 above; 3-piece mold; quart ale; 9 3/4" tall. *[Figure 710, right]*

Soda ~ Mineral Water:

(1a) **J. EVANS / GALENA / ILL** [cobalt]
»IP; oval-blob-top soda. *[Figure 711]*

(1b) **J. EVANS / GALENA / ILL** [aqua, cobalt]
»SB w/ keyed hinge mold; shouldered-blob-top soda. *[Figure 712]*

Joseph Evans was a partner with his two brothers, Evan Evans and David H. Evans, in the wholesale ale and porter business in St. Louis during the mid-to-late 1840s. From the mid-1840s to the Civil War years, the three brothers used several different embossed three-piece-mold black glass quart bottles (purchased from Pittsburgh glass houses) for their respective businesses (*see Figure 713*).

Figure 710

Figure 711

Figure 712

Figure 713. Examples of Evans Brothers embossed black-glass ales used in St. Louis during the late 1840s and early 1850s.

Joseph moved his part of the bottling business upriver to Galena ca. 1848. Advertisements in the *Weekly Western Gazette* in 1851 and 1854 list J. Evans as the proprietor of the *Buckeye Ale and Porter Depot* at No. 29 Levee Street in Galena. Evans is last listed in the Galena city directory in 1859.

His 1854 advertisement states that he was a:

Wholesale Dealer In

Cincinnati, Dayton, Pittsburgh, Philadelphia and Kennett's third proof
ALE AND PORTER
ALSO
A large assortment of Imported and Domestic Wines and Liquors.

The ale bottles used by Joseph Evans in Galena are listed in their approximate order of use. Interestingly, the town-name rear shoulder embossing that appears on his earliest bottle is identical to the rear half of the bottle mold used in Galena by *Volz & Glueck* (*see their listing, this chapter*). Perhaps Evans bought out the ale-bottling operation of this short-lived Galena company when he moved north from St. Louis.

Black-glass ale bottles of the 1850s often have refined (refired?) bases that make it difficult to tell if they were manufactured with pontil-rod technology, but stylistically all of the Evans ales listed look like they were made during the 1850s decade. Evans only smooth-based snap-case bottles are the sodas listed as #1b, and these were made using the same bottle mold as the #1a pontilled example, so it appears his final years in the bottling business in Galena were the early Civil War years (ca. 1860–61).

Figure 714

Anton Oetter

Soda ~ Mineral Water:

(1) **A OETTER / GALENA / ILL^S //**   [aqua]
(*heel*): **W^m McCully & Co**
(*base*): **X**

»SB w/ keyed hinge mold; slope-shouldered, shouldered-blob-top soda. [*Figure 714*]

(2) **A OETTER / GALENA / ILL^S //**   [aqua]
(*heel*): **A & D. H. C.**
(*base*): **X** (*same as Figure 714*)

»SB w/ keyed hinge mold; slope-shouldered, oval-blob-top soda. [*Figure 715*]

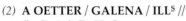

Figure 715

The earliest known appearance of Oetter's name on embossed soda bottles is on Kentucky pontiled sodas dating to the late 1850s: **A. OETTER / PADUCAH / KY**. Shortly thereafter, Oetter moved his bottling operation to St. Louis. Embossed **A. OETTER** St. Louis bottles have an **FA&Co** heel mark, indicating they were made in Pittsburgh by *Fahnstock, Albree & Co*. Such bottles were produced by FA&Co only ca. 1860–63. No Oetter bottles are known from the mid-1860s, but perhaps he became caught up in the Civil War and left the bottling business for a while.

From the technology and style details of the bottles listed, Oetter next appeared as a soda bottler during the late 1860s and/or early 1870s, in Galena, IL. He was listed as a soda water bottler there by both the 1870 and 1880 USC. The 1875 Illinois state directory also lists "A. Etter" as a soda water manufacturer in Galena, and an 1878 Jo Davies County history (Kett 1878b:647) also mentions him as a soda water manufacturer.

George Volz & John F. Glueck

Ale~Porter:

(1) **VOLZ & GLUECK // GALENA. ILL** [olive "black" glass]
   Note: *There are two raised embossed dots beneath the second* **L** *in* **ILL**.

   »IP/SB; front and rear shoulder-embossed; 3-piece mold; quart ale; 9 3/4" tall. *[Figure 716]*

--- ❖ ---

George Volz (also transcribed as Foltz and Voltz in the 1855 and 1865 Illinois state census records) and John Glueck were brothers-in-law who immigrated to the United States in the 1840s from Wurttemberg, Germany, and Stockholm, Sweden, respectively. Voltz established a brewery in Galena toward the end of the 1840s, which he continued to operate during the 1850s and 1860s there "on the east side" (according to his *Galena Weekly Gazette* obituary of June 29, 1883). Early on, he operated the brewery in association with his brother-in-law as brewmaster. Their short-lived bottled-ale operation appears to date to the very beginning of their business association (ca. 1849–50: *see discussion under Joseph Evans Galena listing, this chapter*).

According to the *Master Brewers Association of America Communicator* (Vol. 41#2:209, 2004), Glueck left the Galena brewery business during the late 1850s and traveled to California, where he established the National Brewing Company at San Francisco. He remained in business from 1861 to 1877.

Figure 716

GALESBURG
(Knox County)

1850 census: 900
1860 census: 5,000
1870 census: 10,200
1880 census: 11,400

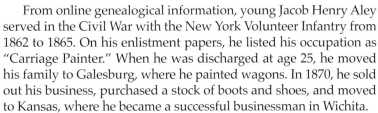

Jacob H. Aley

Soda ~ Mineral Water:

(1) **J. H. ALEY / GALESBURG / ILL //**   [aqua]
    (heel): **A & D H. C**

»SB w/ keyed hinge mold; slope-shouldered, shouldered-blob-top soda. *[Figure 717]*

❖

From online genealogical information, young Jacob Henry Aley served in the Civil War with the New York Volunteer Infantry from 1862 to 1865. On his enlistment papers, he listed his occupation as "Carriage Painter." When he was discharged at age 25, he moved his family to Galesburg, where he painted wagons. In 1870, he sold out his business, purchased a stock of boots and shoes, and moved to Kansas, where he became a successful businessman in Wichita.

No evidence of Jacob Haley's Galesburg soda bottling operation was found in Knox County historical documents, and only a single example of a *J. H. ALEY* embossed bottle is known. From the evidence of the bottle itself, however, he did briefly try his hand at the bottling business.

Figure 717

From the technological and style attributes of the bottle, it appears to date from the mid-to-late 1860s. Given its rarity, Haley may have only bottled soda water for a brief time (perhaps as a single-season effort) soon after his Civil War service. He may have attempted to restart the soda manufactory in Galesburg that had previously been operated during the Civil War years by James Eaton (*see Eaton's listing, this chapter*).

Charles Brechwald and Co.

Whiskey:

(1) **C. BRECHWALD & Co / GALESBURGH. ILL**   [amber]
    Note: Bottle decorated with a ring of 14 vertical shoulder panels interrupted by the embossing on the front shoulder.

»SB; whiskey cylinder; fluted shoulder; long neck; tapered collar; 13" tall. *[Figure 718]*

❖

Charles Brechwald was listed in the 1880 USC as a 50-year-old wholesale merchant in Galesburg. The 1878 *History of Knox County* adds that he was a Galesburg liquor dealer who immigrated to the United States from Germany in 1853 and settled in Galesburg in 1856. "In 1863 he engaged in the retail liquor trade, and in 1872 in

Figure 718

the wholesale trade" (Chapman & Co. 1878:656). During the 1870s and early 1880s, *Brechwald & Co.* operated their wholesale liquor business from the ground floor of the Galesburg opera house building, which the company owned (*New York Times*, Dec. 30, 1886; Perry 1912:68). From the style and technological attributes of the embossed Brechwald bottle, it was manufactured for the company's use during the 1870s, perhaps to introduce Brechwald's wholesale liquor business.

James E. Eaton

Soda ~ Mineral Water (late 1850s):

(1) **J. E. EATON / GALESBURGH / ILL** [aqua]

»IP; shouldered-blob-top soda. *[Figure 719]*

(2) **JAMES E. EATON / GALESBURG / ILL** [aqua]

»IP; (shouldered?)-blob-top soda. *[Figure 720]*

Soda ~ Mineral Water (early 1860s):

(3) **JAS. E. EATON / GALESBURG / ILLS //** [aqua]
(*side heel*): **FA & Co**

»SB w/ keyed hinge mold; shouldered-blob-top soda. *[Figure 721]*

(4) **JAMES. E. EATON /** [aqua]
GALESBURG / ILL //
(*heel*): **A & D. H. C.**

»SB w/ keyed hinge mold; shouldered-blob-top soda. *[Figure 722]*

Figure 719

Figure 720

Figure 721

Figure 722

(5) Paul and Parmalee (1973:90) list an otherwise unknown **JAS. E. EATON / GALESBURG / ILL** aqua quart soda with an **FA & Co heel mark**. *We have not been able to confirm the existence of this bottle, which would be the only known example of an Illinois embossed FA&Co quart soda. The P&P record is likely an inaccurate listing of bottle #3.*

James Eaton had a long and varied career bottling soda and mineral water in several towns in northern Illinois and southern Wisconsin. He began as a soda water manufacturer and bottler in the Galesburg/Knoxville area from 1857 to 1861. He was listed in the 1857 Galesburg CD as a soda water manufacturer at the corner of West and Simmons, and in the 1861 Galesburg CD as a mineral water manufacturer at the corner of Brooks and First Street. His hand-powered soda water manufactory was listed in the 1860 IIC for Knox County, which indicated that he had four adult male workers (his average monthly labor costs were $60) and that he filled 5,000 dozen bottles of soda per year with a retail value of $3,000.

Eaton established a similar bottling business in Chicago in the early 1860s *(see his Chicago listing)*, but soon left the bottling operation there in charge of a relative (George Eaton) and returned to Galesburg, where he is listed as just a resident in the 1865 city CD.

He later bottled for several years in southern Wisconsin before returning to establish another bottling operation in Peoria beginning in 1879 *(see his Peoria listing)*, where he remained active in the bottling business until his death in 1888.

Early on, James Eaton's family had married into the Hickey soda bottling family (Hiram, Sephreness [*a.k.a. Stephen*], John, and Ransom), who were based in Peoria. Two of James Eaton's sons were named Hiram and Sephreness. The Hickeys also sold bottled products elsewhere in northern Illinois *(see Hickey listings in Bloomington, Galesburg, Joliet, La Salle, and Peoria, this volume)* and in Wisconsin, where they and their extended family marketed bottled soda products in several cities and towns during the 1860s and 1870s). Hickey family members had Wisconsin bottling operations in Fond du Lac, Green Bay, Janesville, Milwaukee, Monroe, Oshkosh, and Whitewater (Peters 1996).

Perhaps as a result of his in-law associations, James Eaton moved to Wisconsin with at least one son (Sephreness M.) and established bottling operations there after the Civil War. He bottled soda water at Beloit, Burlington, and Waukesha from the mid-1860s to the late 1970s (Peters 1996:24, 28, 183), before moving back to Illinois in the late 1870s and setting up shop in the old Hickey Bottling Works factory in Peoria from the late 1870s into the 1880s.

It is also possible that it was Eaton who first attracted the Hickey family to the potential of the southern Wisconsin bottling market. A pontiled James Eaton soda bottle is known from Whitewater, WI (Peters 1996:200), and perhaps he began bottling in Whitewater before moving south to Galesburg, IL.

George Edwards

Soda ~ Mineral Water:

(1) **GEORGE EDWARDS / GALESBURG / ILL //**   [aqua]
(heel): **L&W**

»SB; slope-shouldered shouldered-blob-top soda. [Figure 723]

George Edwards was listed in the 1870 USC for Galesburg as a manufacturer of soda pop. A native of Illinois, Edwards was 28 years old at the time. He seems to have followed James Haley *(see later discussion)*—who followed James Eaton *(see his listing)*—in the soda bottling business in Galesburg. According to the *Portrait and Biographical*

Figure 723

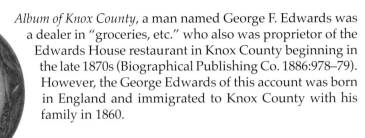

Album of Knox County, a man named George F. Edwards was a dealer in "groceries, etc." who also was proprietor of the Edwards House restaurant in Knox County beginning in the late 1870s (Biographical Publishing Co. 1886:978–79). However, the George Edwards of this account was born in England and immigrated to Knox County with his family in 1860.

[Isaac?] R. Edwards

Soda ~ Mineral Water:

(1a) **I. R. EDWARDS / GALESBURG / ILL //** [aqua]
(heel): **I. G. Co**
(base): **E**

»SB w/ post mold; high-shouldered, long-neck "Selters water"-style, oval-blob-top soda. *[Figure 724]*

(1b) **I. R. EDWARDS / & CO. / GALESBURG / ILL /** [aqua]
(heel): **I. G. Co**
(base): **E**

»SB w/ post mold; high-shouldered, long-neck "Selters water"-style, oval-blob-top soda. *[No photo: same mold and bottle shape as #1a, with "& CO." inserted below the bottler's name.]*

❖

The I.R. Edwards bottles are something of a mystery. They are shaped like "Selters Water"-style mineral water bottles, and from their technological attributes they appear to have been made and used during the mid-1870s. However, no I. R. Edwards company could be located in the historical records or census documents, and bottler George Edwards (*see his Galesburg listing*) does not appear to have had any family members with the initials "I. R." The only possible connection we could locate was an 1870 USC listing for a doctor named Isaac Edwards, who was living and practicing in Galesburg at the time. Perhaps he bottled local mineral spring water as a health drink.

Figure 724

GALVA
(Henry County)

1855 census: 75
1860 census: 1,000
1870 census: 2,200
1880 census: 2,100

Clarence M. Ayres

Soda ~ Mineral Water:

(1) **C. M. AYRES & Co / GALVA ILLS. // GALVA / MINERAL WATER / WORKS //** [cobalt]

(heel): **A & D. H. C.**
(base): **C. M. A & Co** *[embossed in half circle near edge]*
   »SB; slope-shouldered oval-blob-top soda. *[Figure 725]*

———————— ❖ ————————

According to the *Atlas of Henry County* (Warner and Beers Co. 1875b), Clarence M. Ayres was proprietor of a Galva restaurant in 1875, at which time he was in his mid-20s. He appears to have bottled soda and mineral water as a sideline (perhaps for use in his restaurant) for a season or two during the later 1870s.

GREENVILLE
(Bond County)

1850 census: 400
1860 census: 1,000
1870 census: 2,000
1880 census: 1,900

Edward A. Floyd

Medicine:

(1) **FLOYD.S / AMERICAN / PENETRATING / LINIMENT**   [aqua]
   »OP; shouldered cylinder; short neck; rolled lip; 4" tall (small size). *[Figure 726]*

(2) **FLOYD S / AMERICAN / PENETRATING / LINAMENT**   [aqua]
   Note: "Liniment" is misspelled.
   »OP; shouldered cylinder; short neck; rolled lip; 4 3/4" tall (large size). *[Figure 726]*

Figure 725

———————— ❖ ————————

Edward Floyd was listed as a merchant in Bond County in the 1850 USC. He placed an ad in the April 26, 1850, *Greenville Journal* announcing that he was a sales agent for *H. G. Farrell's Arabian Liniment* (see H. G. Farrell Peoria listing), so perhaps Floyd was a druggist. An 1852 ad in the same paper announced that he was selling *McLean's Volcanic Oil Liniment* (a St. Louis company).

The first ad found for his own liniment is from the *Greenville Journal* for October 8, 1852:

FLOYD'S AMERICAN PENETRATING LINIMENT:
   For the cure of the following diseases: rheumatism, swellings, limp, sprains, bruises, old sores, pain in the back, headache, ringworms, &c, &c, upon mankind, where an external means is required. ALSO fistula, poll evil, wind galls, sweeny, founder and almost all diseases of horses....
   It has been manufactured at a great expense, and by one who has thoroughly tested its virtues. Only 25 cents.
   Edward A., Floyd, Proprietor Greenville, Illinois.

The product was available at E. A. Floyd's Liniment Depot in Greenville. Ads for Floyd's penetrating liniment occurred with some frequency in the regional newspapers during 1853

Figure 726

and 1854. After that, the ads tapered off, and by 1857 Floyd had become a "dental surgeon" in Greenville: "The subscriber...would respectfully inform the citizens of this and adjoining counties that he is prepared to perform all the different branches of Dental Surgery. Teeth extracted, filled, cleaned, inserted on pivot or plate in the most approved style, all diseases of the gums as may arise from decayed teeth, successfully treated (*Greenville Journal*, July 30, 1857).

By the time of the 1660 USC, Floyd had moved to Macomb, presumably to continue his dental practice.

All of Floyd's *Penetrating Liniment* bottles are pontilled, and judging from his newspaper ads, the bottles were filled and the liniment was sold from 1852 to 1855–56.

P. Herman Grafe

Soda ~ Mineral Water:

 (1) **P. H. GRAFE / GREENVILLE / ILLS //**   [aqua]
   (heel): **I.G. Co.**

   »SB w/ post mold; slope-shouldered, oval-blob-top soda. *[Figure 727]*

No direct documentation of Mr. Grafe's bottling works in Greenville has been located. P. H. Grafe was listed as a butcher in Greenville in both the 1870 and 1880 USC records. But he appears to have made an unsuccessful effort to bottle soda water for a short time during the mid-to-late 1870s. The first year of bottle-manufacture for the *Illinois Glass Co.* (**IG Co**) in Alton was 1874, and from the style and technological attributes of the embossed Grafe bottle listed, it was likely produced during one of the first few years the new Alton glassworks was in operation.

By 1880, Grafe was listed as a 33-year-old butcher in the USC for Greenville, and during the early 1880s he became involved in launching the Greenville Building and Savings Association (Carson 1905:116).

Figure 727

 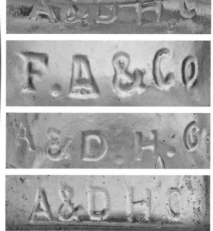 

HIGHLAND

(Madison County)

1850 census: 700
1860 census: no data
1870 census: 1,800
1880 census: 2,000

Highland Soda ~ Mineral Water Bottling Chronology

The history of soda and mineral water bottling in Highland provides a wonderful single-line bottling chronology for the region. Beginning in the mid-1850s, soda and mineral water products were put-up in embossed glass bottles from a single bottling-works location for over 100 years. Only the names of the proprietors changed as the business changed hands over the years. A town street map produced for the large Madison County plat map published in 1861 shows the location of the "M&Co Soda Mfg" bottling works facing Mulberry Street (across the street from C. Schott's Brewery) between Washington and Pestalozzi (on Washington Street Lot #1), near the northwest edge of town (*Figure 728*). Street names changed over the years, but the location of the bottling works remained the same.

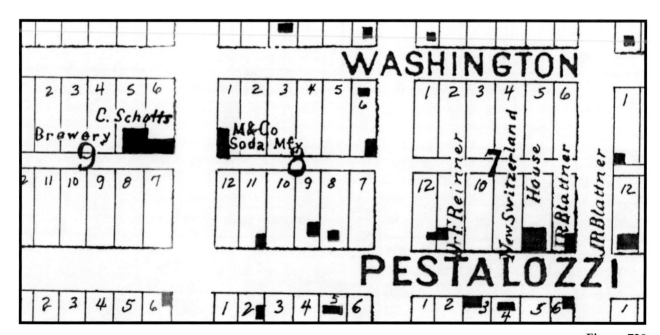

Figure 728

Limiting the list to the 1840–80 focus of this study, the early bottling works proprietors, whose names were embossed on their bottles, were as follows:

1. Weber & Mueller: ca. 1855–56
2. Mueller & Beck: ca. 1857–60
3. Beck & Brother: ca. 1860/61–66
4. Amsler Brothers: ca. 1867–72
5. Christian Schott: ca. 1873–80

These bottling companies and the descriptions of their embossed soda and mineral water bottles are listed alphabetically below.

John and Samuel Amsler

Soda ~ Mineral Water:

(1) **T. AMSLER & C°** / **HIGHLAND** / **ILLS** //   [aqua]
(heel): **A. & D H. C**
*Note: The "**T**" initial appears to be a mold flaw or embossing error, corrected in bottle #2. Also, there is a single raised embossed dot beneath the superscript* **O** *in* **C°**.

»SB w/ keyed hinge mold; slope-shouldered, oval-blob-top soda. *[Figure 729]*

Figure 729

Figure 730

Figure 731

(2) **S. AMSLER & C⁰** / **HIGHLAND** / **ILLS** //   [aqua]
(heel): **A. & D H. C**
Note: *As with bottle #1, there is a single raised embossed dot beneath the superscript* **O** *in* **C⁰**.

»SB w/ keyed hinge mold; slope-shouldered, oval-blob-top soda. [Figure 730]

(3) **J. AMSLER.** / **HIGHLAND** / **ILLS** //   [aqua]
(heel): **A. & D H. C**
Note: *The* **J. AMSLER.** *embossing is off center to the left (***&C⁰** *was likely peened out), so this bottle was probably produced after #2.*

»SB w/ keyed hinge mold; slope-shouldered, oval-blob-top soda. [Figure 731]

The 1870 USC for Highland listed a "Jack" Amsler as a 47-year-old Swiss-born soda factory operator in Highland. The same census listed Samuel Amsler ("Amster") as a 42-year-old Swiss-born cooper there. There was no evidence from the census records of a third brother ("**T. AMSLER**") or other relative in Highland who may have been involved in the business, so probably the **T. AMSLER** variant was an early **J. AMSLER & C⁰** mold error.

Alfred and Alexander Beck

Soda ~ Mineral Water:

(1) **BECK & BRO** / **HIGHLAND** / **ILLS** //   [aqua]
(arched around edge of base): **F A & Co**

»SB w/ keyed hinge mold; shouldered-blob-top soda. [Figure 732]

Figure 732

Alfred Beck was listed as a soda water manufacturer in the 1860 USC for Highland. His brother Alexander was listed in the 1866 *Gazetteer of Madison County* as operating a soda water factory in Highland. The Beck & Brother partnership also appeared in the IRS tax assessments for 1864 and 1866.

By the time of the 1870 USC, Alfred had become a cooper and Alexander was listed as a business agent.

Anton Mueller and Alfred Beck

Soda ~ Mineral Water:

(1) **MUELLER** / **&** / **BECK** / **HIGHLAND** / **JLL**   [aqua]
Note: *"***JLL***" misspelling of* **ILL***; also a large rectangular area with 4 rivet impressions is apparent on rear body where (earlier?) embossing has been removed.*

»IP; shouldered-blob-top soda. [Figure 733]

(2) **MUELLER & BECK** // **HIGHLAND** / **ILL**ˢ   [aqua]

»IP; shouldered-blob-top soda. [Figure 734]

(3) **MUELLER & BECK** / **HIGHLAND** / **ILLS** //   [aqua]
(heel): **F.A & Co**

» SB w/ keyed hinge mold; shouldered-blob-top soda. [Figure 735]

Paul and Parmalee (1973:92) also list (but do not illustrate) a fourth *Mueller & Beck* bottle variety, which they characterize as a Hutchinson-style bottle with an A&DHC makers mark. Hutchinson style bottles were not patented until 1878 and were not widely produced until ca. 1880 and later, and no A&DHC maker's mark was found on any of the Mueller & Beck bottles documented by us, so this is likely a misattribution.

Anton Mueller & Alfred Beck were listed as partners in the soda water manufacturing business in Highland in the 1860 USC and the 1860 ISD. (This latter source mislisted the company as "*Miller & Peck*.") Mueller & Beck were also listed on the summary business directory that accompanied the large 1861 Madison County plat map with the inset Highland street map showing the Mulberry Street location of their business—which was labeled "M&Co Soda Mfy" on the map (*see Figure 728*).

The *Mueller & Beck* hand-powered soda factory was included in the 1859–60 IIC, which indicated they had $5,000 invested in the business and had used 85 barrels of sugar, 40 bottles of oil vitriol, 25 barrels of marble dust, 30 barrels of lemon oil, 125 barrels of tartaric acid, 1,050 pounds of corks, and 400 pounds of twine during the period from June 1859 to June 1860. They had six male employees, with an average monthly labor cost of $100, and filled 160,000 bottles of soda during the year, with a retail value of $6,700 (or about 4¢ per bottle).

Figure 733

Figure 734

Figure 735

The *Mueller & Beck* partnership succeeded a soda water business begun in Highland during the mid-1850s by Mueller and Jacob Weber (*see Weber & Mueller listing, this chapter*).

Christian Schott

Soda ~ Mineral Water:

 (1) **C. SCHOTT / HIGHLAND / ILLS /**   [aqua]
   (*front heel*): **A & D H. C**

   » SB w/ post mold; oval-blob-top soda. *[Figure 736]*

 (2) **C. SCHOTT. / HIGHLAND / ILL _ //**   [aqua]
   (*rear heel*): **A & D. H. C.**

   » SB w/ post mold; oval-blob-top soda. *[Figure 737]*

 (3) **C. SCHOTT / HIGHLAND / ILLS**   [aqua]

   » SB w/ post mold; oval-blob-top soda. *[Figure 738]*

Christian Schott is listed as a soda water manufacturer in both the 1873 and 1874 CAR volumes for Illinois, and in the 1875 ISD. In the 1860 and 1870 USC, he had been listed as a brewer. In the late 1850s, Gerhardt Schott and his two sons, Martin and Christian, had taken over the earlier Bernay's Highland Brewery in town, located immediately across the street from the soda factory (*see introductory discussion, this chapter*).

Figure 736   **Figure 737**   **Figure 738**

After the Civil War, the father returned to Germany and his two sons ran the brewery operation. In 1870, Martin bought out his brother and ran the brewery on his own (Kious and Roussin 1998). In the early 1870s, Christian acquired the soda bottling works from the Amslers and relocated across the street to bottle soda. He appears in the 1880 USC as a 43-year-old soda factory operator, and according to the 1879–80 IIC he was running the factory full-time 6 months a year with two adult and two child laborers, who were paid $2 and $1 per day, respectively. He had $600 in annual labor costs and $1,600 in material expenditures, and he produced bottled soda valued at $3,000 during the 12-month period.

Jacob Weber and Anton Mueller

Soda ~ Mineral Water:

(1) **WEBER & MILLER / HIGHLAND / ILLS** [lime green]

»IP; paneled (10-sided); vertically embossed on 3 adjacent panels; slope-shouldered, shouldered-blob-top soda. *[Figure 739]*

Anton Mueller and Jacob Weber are said to have built a soda water factory in Highland in 1856 (Coats 1983:60). They were the first operators of what became a century-long business in Highland (*see introductory discussion*). Note that Mueller's name is misspelled (or anglicized?) "MILLER" on the embossed bottle listed.

The Weber and Mueller partnership was a brief one. By 1860, Mueller had formed a soda-bottling partnership with Alfred Beck. A brief Highland business directory printed on a large Madison County plat wall-map in 1861 lists Jacob Weber as proprietor of the "Highland House and Highland Beer Saloon."

Figure 739

JACKSONVILLE
(Morgan County)

1850 census: 2,700
1860 census: 5,500
1870 census: 9,200
1880 census: 11,000

John E. Bradbury and Co.

Ale~Beer:

(1) **J. E. BRADBURY & Co / JACKSONVILLE. ILL. // THIS BOTTLE / NOT TO / BE SOLD //**   [amber]
(*base*): **C&I**

»SB w/ post mold; quart beer; obverse embossing in circular slug plate; oval-blob top; 11 3/4" tall. *[Figure 740]*

John Bradbury was listed in the Jacksonville CD for 1874/75 as a bartender. In the 1876 and 1877 directories, Bradbury was listed as both a saloonkeeper and the owner of a wholesale and retail liquor business. He placed a full-page advertisement on page 8 of the 1876 CD listing his wholesale and retail stock as "foreign and domestic wines, liquors, and cigars." The ad highlighted the fact that the proprietor also offered "foreign and domestic *ales and porter*." His business was located on the north side of the city square, under the Park House.

There is no mention of his wholesale and retail liquor trade after 1877, but he continued as a saloonkeeper at different locations from 1878 through 1886, first on the west side of the square and later on west Court Street.

> Mr. Bradbury, though not o'er young in years, is extremely old in business qualifications. Ever ready to surround difficulties,...and his genial manners and friendly disposition, bespeaks well of his future success. It is such enterprising and go-ahead sort of men as Mr. Bradbury which go far in making a business community. Give him a call and sample his stock, with which his shelves are constantly loaded. [1876CD:49]

Since the Pittsburgh bottle manufacturer Cunningham and Ihmsen closed in 1878, the bottle described above likely dates to 1876–77 when

Figure 740

Bradbury owned the liquor business. Perhaps his personal brand of bottled beer or ale was a grand-opening advertising device. Notably, because the business operated for such a short time, the bottle closely documents an early date for circular slug-plate embossing technology (*see introduction, this volume*).

Philip Braun and Co.

Soda ~ Mineral Water:

(1) **P. BRAUN & Co / JACKSONVILLE / ILLS_ //**   [aqua]
(*heel*): **W. McC & Co.**

»SB w/ keyed hinge mold; slope-shouldered, oval-blob-top soda. [*Figure 741*]

There were two competing soda-bottling operations in Jacksonville during the late 1860s and 1870s: one run by Buell & Schermerhorn (*see their listings*); the other operated sequentially by Braun & Killian (ca. 1870–71), Braun & Co. (w/ Kershaw 1872–74), Kershaw (1875), Ricks (1876–77), and Kershaw again (1878-80) (*see their listings*).

Braun & Co. is listed as a soda water manufacturer in the 1872 *TW & W* Railway Directory, the 1873 CAR, and the 1874 Jacksonville CD. According to the 1874 directory (p. 63), Albert Kershaw was the "*& Co.*" partner in this business. The company also provided soda-fountain service:

> Phil Braun's horse ran away with his delivery wagon on North Sandy St. yesterday, and spilt out a soda fountain. Little damage was done. [*Jacksonville Daily Journal,* June 20, 1874, p. 4, col. 1]

The company was first mentioned in the *Jacksonville Daily Journal* during May 9 and 10, 1872:

> **BOTTLING WORKS**
> BOTTLED BEER,
> BOTTLED SODA,
> BOTTLED CIDER.
> We are making a specialty of Bottled Beer, the very best quality delivered in any quantity to any part of the city. P. Braun & Co. [*JDJ,* May 9, 1872, p. 2]

> BOTTLED UP—In addition to cider and soda-water, P. Braun & Co., under Hatfield's, are now making a specialty of bottled beer. This delightful tonic beverage they also deliver to all parts of the city for family use. They have new copper fountains, and all the late improvements for the business. [*JDJ,* May 9, 1872, p.2, col. 4]

> Not bad to take—We were yesterday indebted to P. Braun & Co. for samples of their soda water, bottled cider and bottled beer, and find these articles worthy of all commendation. Persons who enjoy these refreshing beverages should test the productions of this enterprising firm. [*JDJ,* May 10, 1872]

As of this writing, no examples of Braun & Co. embossed cider or beer bottles are known—only the listed soda-bottle style. According to the *Jacksonville Journal* (Oct. 15, 1874, p. 4, col. 3), the Braun & Kershaw partnership was dissolved after the 1874 bottling season, with Kershaw continuing to bottle soda, cider, and beer on his own (*see Kershaw listing*):

> DISSOLUTION NOTICE—The partnership heretofore existing Philip Braun and Albert Kershaw is this day dissolved by mutual consent. All persons knowing themselves indebted to the firm in any way, will please call at once and settle. Parties having bills against the firm will please present them at office. BRAUN & KERSHAW

> TO CONTINUE IN BUSINESS—Having purchased the business interest of Philip Braun in the above firm, I will continue in the basement rooms in Hatfield's building, northwest

Figure 741

corner square. Bottled beer, cider &c., constantly put up and furnished to suit customers. None but the best quality of goods sent out. Leave orders. ALBERT KERSHAW

Soon after leaving the bottling business, Philip Braun became Morgan County coroner. His recorded cases date from December 1876 to September 1878 (Robert W. Dalton, personal communication 1996).

Philip Braun and John Killian

Soda ~ Mineral Water:

(1) **BRAUN & KILLIAN / JACKSONVILLE / ILLS_ //** [aqua]
   (heel): **W. McC & Co.**

   »SB w/ keyed hinge mold; slope-shouldered, oval-blob-top soda. *[Figure 742]*

Figure 742

There were two competing soda-bottling operations in Jacksonville during the late 1860s and 1870s: one run by Buell & Schermerhorn (*see their listings, this chapter*) and the other operated sequentially by Braun & Killian (ca. 1870–71), Braun & Co. (w/ Kershaw 1872–74), Kershaw (1875), Ricks (1876–77), and Kershaw again (1878–80) (*see their listings, this chapter*).

Braun & Killian were short-term partners in the soda water business, likely in late 1870 and 1871, since they are not listed in the annual city directories. Braun & Co. had taken on a new bottling partner by 1872 (Albert Kershaw). Braun & Killian were among the earliest companies to use embossed soda bottles in Jacksonville (*see also Buell & Schermerhorn, this chapter*). Killian was an established grocer and confectioner and likely provided the capital and a sales outlet for their business. After the 1871 bottling season, Killian left the partnership to focus his attention on the grocery business, and *Braun & Co.* took on a new bottling partner: Albert Kershaw (*see Braun & Co. listing, this chapter*).

Killian's grocery business did not fare well, and he declared bankruptcy in early 1878 (Morgan County, Misc. Book A, p. 551). He died that fall (*Jacksonville Journal*, Oct. 4, 1878, p. 4, col. 3).

Charles H. Buell and Charles Schermerhorn

Soda ~ Mineral Water:

(1) **BUELL / & / SCHERMERHORN //** [light green]
   (heel): **A. & DH. C**

   »SB w/ keyed hinge mold; tall, cylindrical, high-shouldered soda, oval blob top; 8" tall. *[Figure 743]*

(2) **BUELL / & / SCHERMERHORN / JACKSONVILLE / ILL. //** [light blue]
   (heel): **A. & DH. C**

   »SB w/ keyed hinge mold; tall, cylindrical, high-shouldered soda, oval blob top; 8 1/4" tall. *[Figure 744]*

There were two competing soda-bottling operations in Jacksonville during the late 1860s and 1870s: one run by Buell & Schermerhorn (*see Schermerhorn listing, this chapter*) and the other operated sequentially by Braun & Killian (ca. 1870–71), Braun & Co. (w/ Kershaw 1872–74), Kershaw (1875), Ricks (1876–77), and Kershaw again (1878–80) (*see their listings*).

Buell & Schermerhorn, proprietors of the Jacksonville Bottling Works, first opened their soda bottling manufactory in May 1869:

A NEW MANUFACTORY—On West Morgan street, Messrs. Buell & Schermerhorn have just received the machinery and other essentials for manufacturing soda water, sarsaparilla, ginger ale, &c. These gentlemen expect to furnish all the soda &c. that is necessary for this *[section]* of the state. And we *[wish]* for them in their new enterprise liberal patronage from all dealers in, and drinkers of, these refreshing yet harmless beverages. [*Jacksonville Daily Journal*, May 1, 1869, p. 3, col. 3]

THE JACKSONVILLE BOTTLING WORKS
WEST MORGAN STREET, NEAR THE SQUARE

The subscriber will be prepared in a few days to supply the city and surrounding villages with
DELICIOUS SUMER BEVERAGES
INCLUDING
Ginger Ale,
Sarsaparilla,
Soda, Etc.
OUR GINGER ALE
IS SOMETHING
Never Introduced in This Section Before,
But is of great reputation in the east. Try it. It is
HEALTHY AND TEMPERATE!
Our arrangements are such that we will be able to fill all orders with dispatch.
BUELL & SCHERMERHORN [*Jacksonville Daily Journal*, May 3, 1869, p. 2, col. 6]

IN RUNNING ORDER—Messrs. Buell & Schermerhorn, the gentlemanly proprietors of the Jacksonville Bottling Works, have received all their apparatus and are now in complete running order, turning out a superior article of sarsaparilla soda and ginger ale. They have also a handsome new wagon, which seems to be just the thing exactly. [*Jacksonville Daily Journal*, May 13, 1869, p. 3, col. 4]

Figure 743

Ask for Buell & Schermerhorn's sarsaparilla and soda, put up in such extraordinary large bottles and selling for the same price as the ordinary sized bottles. [*Jacksonville Daily Journal*, May 18, 1869, p. 3, col. 3]

On May 12, 1870, the *Jacksonville Journal Courier* (p. 3, col. 2) noted that Buell & Schermerhorn's bottled beverages were "flavored with lemon, sarsaparilla, and strawberry." On April 25, 1872 (p. 3, col. 3), the same paper noted that Buell & Schermerhorn were "putting up a hundred or more dozen bottles of lemon and sarsaparilla daily." On April 18, 1873, the paper noted that the company had added to its line the "best cider vinegar, our own manufacture, at 25 cents a gallon."

More details on the nature and scale of Buell & Schermerhorn's bottling business were found in the R. G. Dun credit rating files curated by the Harvard Business School's Baker Library (Vol. 166, p. 258; RGD abbreviations written out in brackets):

#20665 (July 26, 1869) ..... Own no [real estate]. Have a [capital] as they represent to me of [$1300], of which sum "S" has furnished nearly all. They are [industrious] & as far as I can judge making a start as though they intend to make money. Have horse & wagon to deliver.

#20665 (Jan. 20, 1870) ..... No [real estate]. Have about $2000 between them. Now selling flour and corn until soda water season.

Figure 744

A.W.M. (May 9, 1871) ..... No means outside of business, which is small. Seem to hold out pretty well and attend to their business (would think them fair risk for small [amounts]).

#20665 (Feb. 3, 1872) ..... No [real estate]. [Industrious] & they pay as they go, so far as we know here.

#20665 (Jan. 11, 1873) ..... Have now opened a grocery store in connection with above, borrowed of a Mr. Wells [$1500] and gave mortgage on their stock. No [real estate]. They are [industrious] & of [good] habits but have not a great margin over their indebtedness. Wells is the father-in-law of one of them.

#20665 (Jan. 12, 1874) ..... No [real estate]. Mortgage on stock renewed to Wm. Wells [father-in-law] of one of them. Mortgage not against [creditors], as it covers stock they may buy [&] allows them to sell.

By the 1874 bottling season, Charles Schermerhorn is listed alone as a grocer and soda water manufacturer (*see his listing in this chapter*).

Dr. William Hamilton

Medicine:

 (1a) **HAMILTON**ˢ **/ VEGITABLE // COUGH. BALSAM**   [aqua]
  Note: "VEGETABLE" is misspelled.

  »OP; beveled-edge rectangle (1 3/8" x 1 3/4"); short neck; tapered collar; 7" tall. *[Figure 745]*

 (1b) **HAMILTON**ˢ **/ VEGITABLE // COUGH. BALSAM**   [aqua]
  Note: Nielson (1978: #275) lists as square-cross-section variety of bottle 1a, with the "VEGETABLE" misspelling corrected.

  »OP; beveled-edge square (2 1/4" square in cross-section); short neck; tapered collar; 7" tall. *[no photo]*

 (2a) **Dᴿ HAMILTON'S // COUGH BALSAM**   [aqua]
  Note: Two embossed dots (..) under raised "R" in **Dᴿ**.

  »OP; oval; short neck; rolled lip; large size: 6 3/4" tall. *[Figure 746a–d]*

 (2b) **Dᴿ HAMILTON'S // COUGH BALSAM**   [aqua]

  »OP; oval; short neck; rolled lip; small size: 5 1/2" tall. *[Figure 746e–g]*

 (2c) **Dᴿ HAMILTON'S // COUGH BALSAM**   [aqua]

  »OP; oval; short neck; paper-thin flared lip; 5 1/2" tall. *[Figure 746h–k]*

 (2d) **Dᴿ HAMILTON'S // COUGH BALSAM**   [aqua]

  »SB w/ keyed hinge mold; oval; short neck; thickened flared-ring collar; 5 1/2" tall. *[Figure 747]*

 (3a) **Dᴿ HAMILTONS // SYRUP OF BLACKBERRY / & SASSAFRAS**   [aqua]

  »OP; oval; short neck; rolled lip; 4 3/4" tall. *[Figure 748a–d]*

 (3b) **Dᴿ HAMILTONS // SYRUP OF BLACKBERRY / & SASSAFRAS**   [aqua]

  »SB w/ straight-line hinge mold; oval; short neck; rolled lip; 4 3/4" tall. *[Figure 748e–g]*

Figure 745

Figure 746

Figure 747

Figure 748

(3c) **DR HAMILTONS // SYRUP OF BLACKBERRY / & SASSAFRAS // //** [aqua]
(base): **C&I**

»SB w/ keyed hinge mold; beveled-edge rectangle; 4 sunken panels; short neck; double-ring collar; 4 3/4" tall. *[Figure 749]*

(4a) **DR. HAMILTON'S // INDIAN LINIMENT** [aqua]

»OP; oval; short neck; rolled lip; 5 3/8" tall (listed by Nielsen 1978: #274). *[no photo]*

(4b) **DR. HAMILTON'S // INDIAN LINIMENT** [aqua]

»SB w/ keyed hinge mold; oval; long neck; thick flared lip; 5" tall. *[Figure 750]*

(4c) **DR. HAMILTON'S // INDIAN LINIMENT** [clear]

»SB w/ keyed hinge mold; oval; sunken oval area in base; short neck; thick (wider) flared lip; 4 3/4" tall. *[Figure 750]*

Dr. Hamilton's first patent-medicine products were put up in some of the earliest embossed bottles known from Illinois. The first advertisement we have found for Hamilton's Vegetable Cough Balsam (see #1a–c) is from the *Sangamo Journal* (Springfield) for September 2, 1842:

Important to the Afflicted!
HAMILTON'S VEGETABLE COMPOUND

COUGH BALSAM

A CERTAIN CURE FOR
Coughs, Colds, Consumption, Bronchitis, Asthma, Dyspepsia,
Hooping Cough, and all Diseases of the Lungs and Windpipe.

This Medicine is a certain cure for all of the above-named diseases, as has been satisfactorily tested by the proprietor, he having administered it to hundreds within the past four years with complete success. It has never failed in cases of the most obstinate character of BRONCHITIS, (a disease that baffles the skill of the most experienced physicians,)—this disease is hurrying thousands to a premature grave, under the mistaken name of *Consumption*—the usual symptoms are cough, soreness of the lungs and throat, hoarseness, Asthma, hectic fever, and sometimes spitting of blood. Let all diseased TRY IT.

Morgan County, Ill., May 1842

Persons afflicted with any of the above mentioned diseases, and have given up all hopes of ever being cured, if they will place themselves under my charge in Jacksonville, I pledge myself to perfect a cure, or no charge will be made for either medicine or attendance.

WM. HAMILTON, Proprietor.
For sale on the north side of the public square, by
J. M. HUCKHARDT, Agent.

This medicine—its name shortened by 1847 to "*Hamilton's Cough Balsam*" (*Morgan Journal*, Jan. 29, 1847, p. 3, col. 6: see #2a–c)—was widely advertised in regional newspapers

Figure 749

Figure 750

during the 1840s and 1850s, and Hamilton had several regional sales agents in central Illinois and elsewhere. From the evidence of bottle #2c, his *Cough Balsam* was still being marketed in the area during the Civil War era.

Hamilton later added two other products to his patent-medicine line: his *Syrup of Blackberry & Sassafras* (#3a–b) and his *Indian Liniment* (#4a–b). The earliest ads we have found for Hamilton's "bowel syrup" of blackberry and sassafras (advertised as a specific for cholera and diarrhea) date to 1847–48. From the smooth-based bottle (#3b), this product was sold into Civil War times as well. The product was still being marketed (by others?) during the late 1860s (bottle #3c). Hamilton's final entry into the patent-medicine arena was his *Indian Liniment*. From the rarity of pontiled examples, his liniment appears to have been introduced during the later 1850s. The only newspaper ad we have located was printed in the *Weekly Madison Press* (Edwardsville) for Aug. 11, 1858. The liniment appears to have been produced during the ensuing decade, but was not heavily advertised.

Hamilton's company also produced medicinal pills beginning in the late 1840s, but these were apparently not packaged in embossed bottles. They included *Anti-bilious Pills*, *Vegetable Cathartic Sugar-Coated Pills*, and *Vegetable Ague Pills*. There are also occasional mentions during the later 1850s (usually from other regional retail drug outlets) of a product called *"Hamilton's [Never Failing] Ague Tonic."* This may well have been another William Hamilton product, but little is known about it—the only newspaper ad found was in the *Daily Illinois State Register* (Springfield) for July 7, 1857—and no embossed bottles have been documented.

During the 1840s, Hamilton's medical-practice offices were located "under the Morgan House" in Jacksonville. The 1860 Jacksonville CD listed the patent-medicine business of Dr. William and George W. Hamilton at the northwest corner of Fayette and Jordan. By 1868, William Hamilton was just listed as a physician at his residence.

Robert and John Hockenhull

Medicine:

(1) **R & J. HOCKENHULL // DRUGGIST // JACKSONVILLE // ILLS.** [aqua]

Note: "**DRUGGIST**" *is singular in the embossing. It is possible that a pre-1847 variety of this embossed bottle may be found dating from the 3-year period before John joined his brother Robert the business.*

»OP; beveled-edge rectangle; 4 sunken panels, all embossed downward; tapered collar; 5 3/8" tall. *[Figure 751]*

(2) **HOCKENHULL** [aqua]

»OP; beveled-edge rectangle (nearly square); double-ring collar; 2 sunken panels, one embossed; 5 3/4" tall. *[Figure 752]*

Robert Hockenhull opened a drug store in Jacksonville in 1844. His brother John joined him as a partner in 1847. The partnership was dissolved in 1855. Robert Hockenhull continued alone in the retail and wholesale drug business from 1855 to 1860.

According to his biographical sketch (Chapman Bros. 1889:191–192), Robert Hockenhull was born in England and first traveled in America in 1838, when he was 22 years old. He liked the Illinois country well enough to settle here, and returned to England and convinced his parents to advance him enough money to start a business here. He and his brother John settled in Jacksonville in 1839, hoping to establish themselves as apothecaries.

In 1840, Robert was a clerk for the Jacksonville druggist firm of *Reed & King*. Four years later, at age 28, he opened his own drugstore on the north side of the square. By 1847, he advertised himself as a wholesale and retail druggist and dealer in foreign and domestic hardware and cutlery at the same location (*Morgan Journal,* July 24, 1847, p. 3, col. 4). At first, John was a clerk in his brother's drugstore. But from contemporary newspaper ads, he became a junior partner in the business by mid-1847 (*Jacksonville Sentinel,* June 12, 1847). John left the partnership in 1855, after just eight years in the druggist trade with his brother.

As they do today, drugstores of the pre–Civil War era carried a wide assortment of goods, includ-

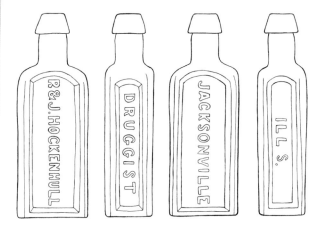

Figure 751

ing drugs, patent medicines, hardware, paints, oil, varnishes, perfumery, window glass, glassware, and "fancy articles" (*Morgan Journal* Jan. 29, 1847, p. 3, col. 5). But in an unusually extensive and detailed 1859 advertisement in the *Jacksonville Sentinel* (May 13, 1859, p. 2, col. 6), Robert Hockenhull announced that he had greatly enlarged the druggist portion of his business, and intended to close up his hardware department. The ad was accompanied by an extensive list of the generic medicines and national and regional brand-name medicinal products he stocked, including *Hockenhull's Extract of Ginger*, listed with other brands in a section for medicines that cured diarrhea. If Robert and John started preparing and bottling their own extract of ginger while they were still partners (prior to 1855), it is possible that the two bottles listed contained this extract brand. However, they may have also served as general prescription containers for the Hockenhull pharmacy.

Robert Hockenhull's business did not long outlast his 1859 ad. It is likely that Hockenhull gave up active control of the retail trade by 1860, since that is the last time he was listed as the proprietor of his store on the east side of the town square between State and Court. He had become part of the banking partnership of *Hockenhull, King & Elliott* in Jacksonville by 1865 (Chapman Bros. 1889:191), but still kept his hand in the pharmacy business for awhile as a member of the firm of *Hockenhull & Young* (Donnelley, Loyd & Co. 1878a:498).

Albert Kershaw

Soda ~ Mineral Water:

(1) **A. KERSHAW / JACKSONVILLE / ILLS_ //** [aqua]
(*base*): **C & I**

»SB w/ keyed hinge mold; slope-shouldered, oval-blob-top soda. *[Figure 753]*

(2) **A. KERSHAW / JACKSONVILLE / ILLS_ //** [aqua]
(*heel*): **D. O. C.**

»SB w/ keyed hinge mold; slope-shouldered, oval-blob-top soda.

Figure 752

There were two competing soda-bottling operations in Jacksonville during the late 1860s and 1870s: one run by Buell & Schermerhorn (*see their listings*) and the other operated sequentially by Braun & Killian (ca. 1870–71), Braun & Co. (w/ Kershaw 1872–74), Kershaw (1875), Ricks (1876–77), and Kershaw again (1878–80) (*see their listings*).

In October 1874, Albert Kershaw bought out his partner, Philip Braun, and became the sole proprietor of what was formerly their soda water business (*see Braun & Co. listing, this chapter*). Kershaw ran the business for just a single season, using bottle #1, and then sold the bottling works to John Ricks on September 9, 1875 (*see Jacksonville Grantor/Grantee Index, Book 10, p. 389, and Ricks listing, this chapter*).

After Ricks died on February 12, 1878, Kershaw re-acquired the bottling works. According to the 1879–80 IIC, Kershaw had four male and two female full-time employees, who he paid $1.00 to $1.50 per day. His materials costs during that 12-month period were $400, and his wholesale/retail sales were $1,500. He continued bottling soda into the early 1880s (using bottle #2 above), but had closed the business down by 1884. In the later 1880s, Kershaw ran a saloon in Jacksonville, and then a "wholesale malt liquor" establishment (Robert W. Dalton, personal communication 1996).

The Kershaw bottle manufactured by *D.O.C.* appears to have been made from the old *C&I* mold.

Figure 753

Figure 754

John W. Ricks

Soda ~ Mineral Water:

(1) **J. W. RICKS / JACKSONVILLE / ILLS_ //**   *[aqua]*
(*heel*): **D. O. C.**

»SB w/ keyed hinge mold; slope-shouldered, oval-blob-top soda. *[Figure 754]*

There were two competing soda-bottling operations in Jacksonville during the late 1860s and 1870s: one run by Buell & Schermerhorn (*see their listings*) and the other operated sequentially by Braun & Killian (ca. 1870–71), Braun & Co. (w/ Kershaw 1872–74), Kershaw (1875), Ricks (1876–77), and Kershaw again (1878–80) (*see their listings*).

Henry and John Ricks moved to Jacksonville from Springfield in 1866 and established the Morgan Brewery there, about a mile north of the city (Andreas, Lyter & Co. 1872a). According to the *Jacksonville Journal-Courier*, they dissolved their brewery partnership on April 14, 1869, after which Henry ran the brewery alone. John ran a saloon in Jacksonville (Robert W. Dalton, personal communication 1996) until he bought Kershaw's soda bottling business in September 1875 (*see Kershaw listing, this chapter*). When John Ricks died in February 1878, Kershaw reassumed ownership and operation of the soda manufactory.

John Ruf

Soda ~ Mineral Water:

According to the *Jacksonville City Directory* for 1860–61, an individual by the name of John Ruf manufactured soda water on College Street between West and Church in 1860. No embossed bottles have yet been documented for this operation. According to the 1860 USC, John "Ruff" (50), his wife Sofia (37), and their teenaged daughter were Jacksonville residents at the time. Notably, in the later 1860s, bottles embossed **JOHN RUF** and **J & S RUF** are known from Waterloo, IL (*see John Ruf's Waterloo listing*).

Charles Schermerhorn

Soda ~ Mineral Water:

(1) **SCHERMERHORN / JACKSONVILLE / ILL.** //   [aqua]
(heel): **.D.O.C**
Note: Leading period in front of **.D.O.C** heel mark left over from peening out and recutting a portion of an earlier **A &. D.H.C** maker's mark (see Buell & Schermerhorn listings, this chapter).

»SB w/ keyed hinge mold; tall, cylindrical, high-shouldered soda, oval-blob top; 8 1/8" tall.
[Figure 755]

The obverse-side embossing on the bottle listed is unchanged from the embossed bottles of the *Buell & Schermerhorn* partnership (*see their listing*), except that the "**BUELL &**" portion has been peened out.

There were two competing soda-bottling operations in Jacksonville during the late 1860s and 1870s: one run by Buell & Schermerhorn (*see their listing, this chapter*) and the other operated sequentially by Braun & Killian (ca. 1870–71), Braun & Co. (w/ Kershaw 1872–74), Kershaw (1875), Ricks (1876–77), and Kershaw again (1878–80) (*see their listings*).

Buell & Schermerhorn, proprietors of the *Jacksonville Bottling Works*, first opened their soda bottling manufactory in May 1869. Schermerhorn appears alone as proprietor of a grocery store and a cider and soda-water bottling business in the 1874 Jacksonville CD, so the last bottling season for the two partners together was 1873. Charles Schermerhorn remained in the grocery and soda-bottling business in Jacksonville for at least two decades. The *Jacksonville Journal-Courier* for March 14, 1893 noted that "Charles Schermerhorn is having gas put into his bottling works on West Morgan street preparatory to using a gas engine. It will be quite an improvement on the old method of creating pressure for bottling" (Robert W. Dalton, personal communication 1996). After ca. 1880, Schermerhorn used embossed Hutchinson-style internal-stopper bottles for his soda products, but these later bottles are too recent for inclusion in the current study.

The bottle listed is something of an enigma. It is a pre-1880 corked blob-top style that seems likely to have been used in the mid-to-late 1870s, when Schermerhorn first began bottling on his own. However, it was made from a modified version of the earlier Buell & Schermerhorn mold by the Dominick O. Cunningham plant in Pittsburgh. Business records indicate that this company did not begin manufacturing bottles until 1882. But they must have been in business by at least 1876–77 (*see John Ricks entry, this chapter*), and probably earlier. If DOC did not start in business during the 1870s, there are no known Schermerhorn embossed bottles dating to the period from 1874 to 1878–79. In this case, perhaps Charles Schermerhorn

Figure 755

continued using Buell & Schermerhorn bottles (or unembossed bottles) when he was first in business on his own in the 1870s. But he may not have done much bottling during the 1870s when he was developing his grocery business. The 1879–80 IIC for Jacksonville indicated that Schermerhorn had no bottling employees and only bottled five months a year, using only $300 of materials and producing only $800 worth of bottled product.

JERSEYVILLE
(Jersey County)

1850 census: 800
1860 census: 2,600
1870 census: 2,600
1880 census: 2,900

Charles M. Boyle

Soda ~ Mineral Water:

(1) **C. M. BOYLE / JERSEYVILLE / ILL^S //**   [aqua]
(heel): **A & D. H. C.**

»SB w/ heel mold; slope-shouldered, oval-blob-top soda. *[Figure 756]*

(2) **C. M. BOYLE / JERSEYVILLE / ILL^S //**   [aqua]
(base): **I.G. C^o**

»SB w/ post mold; slope-shouldered, oval-blob-top soda. *[Figure 757]*

Figure 756

According to the *History of Greene and Jersey Counties* (Continental Historical Co. 1885:499), Charles M. Boyle carried on three lines of trade at the time the *History* was published: a grocery store, an ice-house, and a bottling business. The same source goes on to state: "He commenced the bottling business in 1868, and his grocery store was established in 1875." This statement is confirmed by a notice published in the *Jersey County Democrat*, July 30, 1868 (p. 4, col. 3):

JERSEYVILLE ALE AND PORTER DEPOT
Mr. Charles Boyle, late of the firm of Boyle and Connifry, has established an Ale and Porter Depot in the building lately occupied by Fitzgeralds, opposite Swartz's Livery Stable. Mr. Boyle will deliver Ale and Porter to any part of the city or in the country, by leaving orders at his Ale Depot. He will furnish dealers in the city or country at wholesale prices. Give him a call.

There are no known embossed Boyle ale or porter bottles, and he must have expanded into bottling

Figure 757

soda and mineral water sometime after he first opened his *Jerseyville Ale & Porter Depot*. He was listed in the 1870 USC for Jerseyville as a 24-year-old wholesale liquor dealer.

From the style and technological attributes of the embossed soda bottles listed, the bottles were made and used during Boyle's early bottling years (during the 1870s), perhaps soon after he opened his grocery store in 1875. The 1885 county history indicated he was still bottling that year, and he was listed as a bottler in the 1887 ISD. But if he continued bottling soda beverages after the late 1870s (for sale in his grocery business), he no longer used embossed bottles. The 1891 Jerseyville CD listed him as a hotelkeeper.

Dr. John D. Freeman

Medicine:

(1) **J. D. FREEMAN.S // SUDORIFIC.OR / SWEATING.DROPS** [aqua]

»OP; cylinder; rolled lip; embossed vertically downward; 5 1/8" tall. *[Figure 758]*

(2) **DR J. D. FREEMAN'S / VERMIFUGE // // // //** [aqua]

»OP; rectangle w/ distinctive concave beveled corners and flat panels; rolled lip; 4 5/8" tall. *[Figure 759]*

(3a) **DR J. D. FREEMAN'S // COUGH BALSAM // // //** [aqua]

»IP; beveled-edge square case bottle; embossed vertically downward on two adjacent flat panels; tapered collar; 8 1/4" tall. *[Figure 760]*

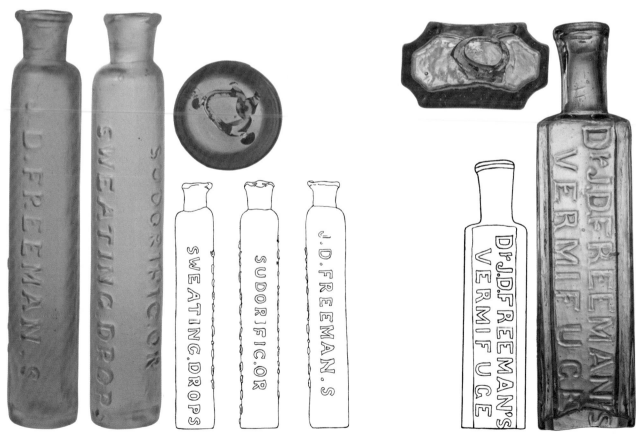

Figure 758

Figure 759

(3b) **Dr. J. D. FREEMAN'S // // COUGH BALSAM // //** [aqua]

»OP; beveled-edge rectangular bottle w/ 4 sunken panels; embossed downward on 2 opposing wide panels; tapered collar; 8 1/4" tall. [Reported by Odell (2007:137) w/ only **COUGH BALSAM** panel illustrated: *see Figure 761*]

Note: The two cylindrical bottles listed reflect the fact that, in the mid-1850s, Dr. Freeman expanded his regional sales market to St. Louis, MO—across the Mississippi River from his Alton, IL, sales area—although his manufacturing operation was still based at Jerseyville (see discussion that follows). These particular medicines seem to have been specifically targeted for the St. Louis market. Both are shouldered cylinders with vertical embossing.

(5) **DR. J. D. FREEMAN'S / DIARHEA, SYRUP, OR / CHOLORA MIXTURE / ST. LOUIS** [aqua]

»OP; shouldered cylinder; embossed downward; rolled lip; 5 1/8" tall. *[Only known example documented in 1988 Greer collection auction sale catalog (see Odell 2007:137).]*

(6) **DR J. D. FREEMAN'S / ECLECTIC / LINIMENT / S^T LOUIS.** [aqua]

Note: Both the embossed superscript **R** in **D^R** and the embossed superscript **T** in **S^T** have two embossed dots beneath them.

»OP; shouldered cylinder; embossed downward; rolled lip; 5 1/8" tall. *[Figure 762]*

───────── ◆ ─────────

J. D. Freeman was the proprietor of the Eclectic Medical Depot in Jerseyville from about 1846 through the mid-1850s. He brought his family to west-central Illinois from Michigan when he was in his mid-30s. He was a strong advocate of the Eclectic Medicine movement in Illinois at the time:

At a meeting in Springfield a state Medical Society was organized in June, 1850. There was a fine spirit of public service in the profession which usually worked incessantly largely on a credit basis. Chicago had a large city hospital under joint allopathic and homeopathic management. Dr. J. D. Freeman of Jerseyville had undertaken to propagate an eclectic movement through the state with an organ, the Eclectic Advertiser, "devoted to medical reform and foreign and domestic news" in which he vigorously attacked the "shocking barbarity" of old school medical practices. [Cole 1919:218]

Figure 760

The earliest advertisement found for Dr. Freeman and his medicines first appeared in west-central Illinois newspapers in early June 1849. The ad indicated he had been in practice in Jerseyville for "three past years." The following is the version of the ad found in the *Greenville Journal*, Aug. 31, 1849 (p. 4, col. 4):

MEDICAL NOTICE.

DR. J. D. FREEMAN takes this method of offering his Professional service, to the inhabitants of Greenville and vicinity, in all its various branches. As his system of practice is not properly understood here, we would merely remark, that **he will be governed by the principles of the Eclectic or Reform Botanic School of New York**. As he claims the name of Eclectic, the meaning of which is not to be confined to any narrow set of principles, theories or rotation of remedies, but to search every direction for information, and to gather such remedies as we are satisfied are requisite to enable

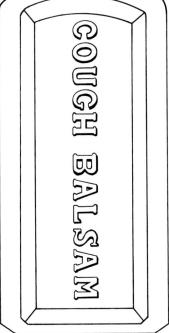

Figure 761

us to cure the various diseases which we have to encounter. Consequently, far different from the Thomsonian or steam system, and the limited idea that Calomel and Lancet are adequate to cure all diseases.

He has constantly on hand a preparation to cure the Ague, which he warrants never to fail, provided the directions are strictly followed.

In regard to his charges, as he has heretofore been governed by the principle of equality—of making such charges as will compare with other operations in the same Country—consequently he will be governed by the same principle here, and will aim to make a general settlement once every year, and intends taking, as far as possible, such kinds of produce as the country produces for his pay.

As **he has a large assortment of medicines on hand, suitably selected for the various diseases of the Country**, he hopes to be able to supply all calls in this line, and by strict attention to his business he hopes to merit and to receive a liberal share of the public patronage.

Office—East side Public Square

Jerseyville, April 28, 1849

We, the undersigned, take pleasure in stating to the Public, that **Dr. J. D. FREEMAN has sustained a large and laborious practice in Jersey County, together with considerable extent of practice in Calhoun and Macoupin Counties, Ill., for three past years**, with the utmost success. He has admirably succeeded in curing diseases that have baffled the skill of other physicians.
[Signatures follow; highlighted portions ours]

Figure 762

Several of the medicines mentioned here were discussed in more detail in a newspaper ad Dr. Freeman placed in the *Alton Telegraph* for January 9, 1854 (p. 6, col. 2). The ad also noted that the medicines were discussed in more detail in a 100-page booklet he had prepared, titled *Guide to Health, and Family Physician*. The booklet was described as a family guide to "the principles of preserving health," "the nature, cause, and symptoms of the diseases which mostly prevail in the country," and the way his medicines should be applied to cure disease. Several of the medicines he listed for sale are not known from embossed bottles. These include *Aperient Anti-Bilious Pills* (said to "eradicate all foul and bilious matters from the system"), *Anti-Ague and Fever Pills* ("for the purpose of strengthening the system"), *Liver Pills* ("for liver complaints, jaundice, dyspepsia, indigestion, and headache"), *Compound Emetic Powders* ("for relieving the stomach from the corruptions accumulating previous to an attack of fever...[and] all obstructions of the glands of the neck, which invariably relieves the lungs"), *Sweet Balsam of Life* (for "curing headache, colics, pain in the breast and side, coughs, colds, spitting of blood, shortness of breath, small-pox, measles, scarlet fever, flux, purging, vomiting, &c."), and the newly developed *Dr. J. D. Freeman's Alterative Balsam and Mother's Relief* (for "regulating the system and purifying the blood, and to keep the female system in a natural and healthy state, while bearing her offspring").

Several of the embossed medicines listed were also described in the 1854 ad, along with an embossed Freeman medicine that may have been marketed only in St. Louis (see discussion below):

Sudorific or Sweating Drops. "These drops have become much celebrated for the purpose of removing obstructions of the capillary system or skin, and consequently producing perspiration, making them a sovereign remedy in fever. Also for relieving pain in any part of the system that may be caused by colds, &c. In congestive diseases they are almost a sure remedy, on account of the wonderful power they possess to drive fluids from the centre to the surface."

Diarrhea or Cholera Mixture [see below]. "This medicine never fails to cure all of the following diseases: cholera Infantum, diarrhea, dysentary or flux, looseness of the bowels, &c., and for Asiatic cholera I have used it with excellent success.... During the season of 1851, in the vicinity of Jerseyville, I had a very large practice among the dysentary or flux, which for success I defy any physician within our country to produce or parallel."

Tonic Vermifuge. "This medicine I have prepared for the express purpose of expelling worms from children and adults without experiencing any weakness whatever."

Regional sales agents listed in Dr. Freeman's 1854 Alton ad included his own Eclectic Medical Depot in Jerseyville, Bacon & Hyde (St. Louis), Schultze & Eggers (St. Louis), A. S. Barry & Co. (Alton), Hewit & Co. (Upper Alton), C. Freeman (Springfield), Julius L. Barnstack (Edwardsville), C. J. Ellis (Belleville), H. W. Ashley, Lehman, Upham & Co. (Chesterfield), P. C. Higgins (Bunker Hill), Griggs & Loveland (Brighton), and Docker & Ferring (Shawneetown).

Dr. Freeman expanded his regional sales area to St. Louis for a short time during the late 1850s, by means of his local sales agents. Two of his preparations, sold in shouldered open-pontiled cylindrical bottles, were vertically embossed "**DR. J. D. FREEMAN'S / DIARHEA, SYRUP, OR / CHOLORA MIXTURE / ST. LOUIS**" and "**DR J. D. FREEMAN'S / ECLECTIC / LINIMENT / S^T LOUIS**."

J. D. Freeman was listed in an 1855 ISD as a manufacturer of drugs and patent medicines in Jerseyville. But sometime in the 1855–56 period, Dr. Freeman moved to Springfield, IL, where he became involved in the building of the St. Nicholas Hotel. He was listed in the 1857 Springfield CD as a dealer in drugs and medicines "beneath" the St. Nicholas Hotel. In the later 1850s, he shifted his focus to the hotel business. He left Springfield in about 1864 for southern Illinois, and by the 1870 USC had become a banker in Carbondale.

No smooth-based (snap-case) bottles embossed with Dr. Freeman's name are known, indicating that he had ceased his patent-medicine business by the start of the Civil War. A biographical sketch of Freeman company salesman Benjamin Pepper (National Historical Co. 1883:1203) indicates he was an Alton-based salesman for Freeman's drug company from 1858–60, after which he took a job at the Alton prison. This provides further indication that Freeman had ceased patent-medicine production and sales by the outset of the Civil War.

Johanna Spannagel

Soda ~ Mineral Water:

(1) **I. SPANNAGEL / JERSEYVILLE / ILLS**   [aqua]
Note: "**I.**" **SPANNAGEL** *appears to be either a spelling error or a mold flaw in one of the* **J. SPANNAGEL** *bottle orders—otherwise identical to the corrected variety below.*
»SB w/ keyed hinge mold; slope-shouldered, oval-blob-top soda.

(2) **J. SPANNAGEL / JERSEYVILLE / ILLS**   [aqua]
»SB w/ keyed hinge mold; slope-shouldered, oval-blob-top soda. *[Figure 763]*

Johanna Spannagel was a partner in *Christian Lutt & Co.* of East St. Louis during the 1870s (*see Lutt's East St. Louis listing*). *Lutt & Co.* began in the soda and cider bottling business in East St. Louis in the spring of 1871, when Christian Lutt was 31 and Johanna Spannagel was 27. The company's first ads in the *East St. Louis Gazette* and the *East St. Louis Herald* that spring characterized them as "Soda and Cider Manufacturers" located at the corner of Broadway and Main Street; the *Gazette* ad indicated the "& Co." co-investor in Lutt's bottling business was Mrs. Johanna Spannagel (Miller 2007a:291).

The Lutt and Spannagel association during the 1870s is interesting and somewhat enigmatic. Christian Lutt was apparently not a resident of East St. Louis. The 1880 USC listed him as a resident of Jerseyville, about 30 miles to the north. However, his partner Johanna Spannagel and her husband, George, resided in East St. Louis during the 1870s. The curious aspects of the Lutt & Spannagel association are as follows:

(1) During the 1870s, there was a soda bottling operation in Jerseyville with bottles embossed **J. N. SPANNAGEL**. Was this bottling operation managed by Christian Lutt in Jerseyville?

Figure 763

(2) During the 1870s, *Christian Lutt & Co.* bottled in East St. Louis. Was this bottling operation managed by Johanna Spannagel in East St. Louis?

Lutt was listed as a soda water manufacturer in the 1873 CAR and in the 1875 ISD. *Lutt and Co.* was bankrupted and went into court receivership in September 1883, just a few months after the death of Johanna Spannagel (Miller 2007a:291).

JOLIET
(Will County)

1850 census: 2,700
1860 census: 7,100
1870 census: 7,300
1880 census: 11,700

John E. and Hiram M. Hickey

Soda ~ Mineral Water:

(1a) **J & H HICKEY / JOLIET / ILL** [aqua]
»IP; shouldered-blob-top soda. *[Figure 764]*

(1b) **J & H HICKEY / JOLIET / ILL** [aqua]
»SB w/ keyed hinge mold; otherwise same mold as bottle #1a; shouldered-blob-top soda.

John Hickey and his brother Hiram bottled soda and mineral water in Joliet during the late 1850s and early 1860s "on Merchants Row along Bluff Street." J. D. Paige (*see his Joliet listing, this chapter*) worked at the Hickey Joliet bottling works from 1857 to ca. 1861–62 before buying it out (Le Baron & Co. 1878b:703; Whiteside 1990). Paige's tenure as a worker at the Hickey bottling works likely represents the entire time the Hickey bottling plant was in business in Joliet. The two Hickey-brothers bottles in this listing would have been made and used during the late 1850s (#1a) and the early 1960s (#1b).

John and Hiram Hickey were members of the Hickey~Eaton extended-family "clan" who established numerous soda and mineral water bottling operations in several northern Illinois and southern Wisconsin cities and towns from the 1850s through the late nineteenth century (*see Hickey and Eaton listings under Bloomington, Galesburg, La Salle, and Peoria*). In southern Wisconsin, Eaton family members bottled soda water at Beloit, Burlington, and Waukesha from the mid-1860s to the late 1970s. Hickey family members had Wisconsin bottling operations in Fond du Lac, Green Bay, Janesville, Milwaukee, Monroe, Oshkosh, and Whitewater (Peters 1996).

After leaving Joliet, John and Hiram Hickey later bottled at Fond du Lac, Wisconsin; Hiram also went on to run soda bottling operations at Milwaukee, Janesville, and Oshkosh (Peters 1996)

Figure 764

John D. Paige

Soda ~ Mineral Water:

(1) **J. D. PAIGE / JOLIET ILL //**   [aqua]
    (heel): **A&DHC**

    »SB w/ keyed hinge mold; shouldered-blob-top soda. *[No photo: According to Ron Ridder rubbing, the bottle is side-embossed identically to Bottle #2.]*

(2) **J. D. PAIGE / JOLIET ILL //**   [aqua]
    (heel): **TW & Co**

    »SB w/ post mold; shouldered-blob-top soda. *[Figure 765]*

(3) **PAIGE'S BOTTLING HOUSE / JOLIET, ILLS.**   [aqua]
    (heel): **De S. G. Co.**

    »SB; large embossed lettering; oval-blob-top soda. *[Figure 766]*

Ale ~Beer~Cider:

(1) **ALE** *(on shoulder)* / **J. D. PAGE / JOLIET ILL //**   [amber]
    (heel): **T. W. & CO.**

    »SB w/ post mold; high-shouldered squat quart; shouldered-blob top; 10$1/2$" tall. *[Figure 767]*

(2) **SPRUCE BEER** *(on shoulder)* / **J. D. PAGE / JOLIET ILL //**   [olive amber]
    (heel): **T W & CO**

    »SB w/ post mold; high-shouldered squat quart; shouldered-blob top; 10$1/2$" tall. *[Figure 768]*

(3) **CIDER** *(on shoulder)* / **J. D. PAGE. / JOLIET. ILL //**   [aqua]

    »SB w/ post mold; high-shouldered squat quart; shouldered-blob top; 8$1/2$" tall. *[Figure 769]*

(4) *(unembossed shoulder)* / **J. D. PAIGE / JOLIET ILL**   [olive-amber]
    (heel): **T W & CO**

    »SB w/ post mold; high-shouldered squat quart; shouldered-blob top; 10$1/2$" tall. *[Figure 770]*

(5) *(unembossed shoulder)* / **PAIGES. BOTTLING**   [amber, lime green]
    HOUSE / JOLIET. ILL. //
    (heel): **DES G Co**

    »SB w/ post mold; high-shouldered squat quart; shouldered-blob top; 9$1/4$" tall. *[Figure 771]*

Figure 765

Figure 766

John Dean Paige first came to Joliet in 1857 when he was 20 years old. He reportedly (Le Baron & Co. 1878b; Whiteside 1990) "walked all the way...from Jefferson County, Wis." When he arrived, he immediately "went to work at the bottling works run by John Hickey or Merchants Row along Bluff Street. The industrious worker eventually bought out his employer." He purchased the Hickey brothers bottling business (*see their listing, this chapter*) shortly after returning to Wisconsin to be married in 1861 (Whiteside 1990). During the 1860s and early 1870s, Page expanded his bottling interests—first to include the nearby town of Wilmington (about 15 miles to the southeast of Joliet) and later to include Marshalltown, IA, and Grand Rapids, MI (Le Baron & Co. 1878b:703). Paige sold his Joliet bottling business to

Figure 767

Figure 768

Figure 769

John Ranft in 1883–84, and went on to become a major player in Joliet city politics.

Early on, J. D. Paige had two "& Co." partners who went on to become prominent bottlers on their own. Horace W. Blood first came to Illinois from New York in 1863 (at the age of 20) and farmed for two years before taking employment in 1865 at J. D. Paige's bottling house in Joliet. He remained in Joliet for two seasons learning the bottling business, then moved south to Wilmington to oversee operations at J. D. Paige's bottling depot there, having purchased a half-interest in the operation. By 1870, Blood had acquired a controlling interest in the Wilmington bottling works (*see his Braidwood and Wilmington listings*).

John D. Vail, Paige's brother-in-law, was also an early J. D. Paige employee. He was an "& Co." worker at Paige's bottling works during the 1863–65 seasons (Burggraaf and Southard 1998:458), before moving to St. Louis to start an cider and small beer business there from his residence at 1712 Broadway. With his brother Delos W. Vail, he was *J. D. Vail & Co.* in 1868—bottling spruce beer in 10-sided Merrill-patent stoneware bottles stamped with the company name. From 1869 through 1872, he again ran the business alone, from a variety of addresses, before closing down in 1873 to return to Joliet and again bottle with his brother-in-law J. D. Paige.

Their short renewed Joliet partnership lasted for just a season or two (*see Paige & Vail listing, this chapter*). After that Vail moved to Marshalltown, IA, and established a soda-bottling operation there (see Burggraaf and Southard 1998:458–459). There is no record that Paige ever bottled at Marshalltown, so he likely participated in Vail's business as a silent financial partner. Perhaps Paige had a similar (financial) relationship at his claimed bottling operation in Grand Rapids, MI.

During their short mid-1870s Joliet partnership, Paige & Vail obviously shared ideas on their different bottling experiences (Paige with sodas, and Vail with ale and beer). While they were together, they expanded to spruce beer, using a 10-sided Merrill-patent stoneware bottle shoulder-stamped "**PAIGE & VAIL**" (drawing in Kott 2005b: *see Figure 772*), and after Vail left

Figure 770 Figure 771 Figure 772

for Marshalltown (to focus his efforts on soda and mineral water bottling), Paige continued to bottle ale and small beer, using an 8-sided Merrill-patent stoneware bottle shoulder-stamped "**J. D. PAIGE**" (*see Figure 773*) and the embossed amber bottles listed under *Ale~Beer~Cider*.

John D. Paige and John D. Vail

Soda ~ Mineral Water:

   (1) **PAIGE & VAIL / JOILET, ILL //**   [aqua]
       Note: Bottle embossing has "JOLIET" misspelled.
      »SB w/ keyed hinge mold; shouldered-blob-top soda. *[Figure 774]*

   John D. Vail was J. D. Paige's brother-in-law (*see Paige listing, this chapter*), and was first a J. D. Paige employee early in the company's history. He was an "& Co." worker at Paige's bottling works during the 1863–65 seasons (Burggraaf and Southard 1998:458), before moving to St. Louis to start an cider and small beer business there from his residence at 1712 Broadway (*see Figure 775*). He closed the St. Louis bottling operation down in 1873 and returned to Joliet to again bottle for a time with his brother-in-law J. D. Paige.

   Their short renewed Joliet partnership lasted for just a season or two, after which, Vail took over Paige's branch bottling operation at Marshalltown, IA, and moved there to run it (Burggraaf and Southard 1998:458–459).

   During their short mid-1870s Joliet partnership, Paige & Vail obviously shared ideas on their different bottling experiences (Paige with sodas, and Vail with ale and beer). While they were together, they expanded to spruce beer, using a 10-sided Merrill-patent stoneware bottle

Figure 773   **Figure 774**   **Figure 775**

shoulder-stamped "**PAIGE & VAIL**" (*see Figure 772*), and after Vail left for Marshalltown (to focus his efforts on soda and mineral water bottling), Paige continued to bottle ale and small beer, using an 8-sided Merrill-patent stoneware bottle shoulder-stamped "**J. D. PAIGE**" (*see Figure 773*), and the embossed amber bottles listed under *Ale~Beer~Cider*.

JONESBORO
(Union County)

1860 census: 800
1870 census: 1,100
1880 census: 900

Dr. Luther K. Parks

Medicine:

(1) (side): **EGYPTIAN** // DR **L. K. PARKS** //   [aqua]
     (side): **ANODYNE** // //
     Note: There are 2 embossed dots below the superscript **R** in **D**^{R}. *[Figure 776]*

  »OP; beveled-edge rectangle; 4 sunken panels; rolled lip; 5 1/8" tall. Bottle embossed vertically downward.

Figure 776

Dr. Luther Kitchell Parks was a 30-year-old physician in Jonesboro in 1850 (USC). According to a biographical sketch (Lam et al. 1954:165), he decided on a medical career as a young man: "After seeing the effects of the great cholera epidemic of New Orleans in 1832–1833, he determined to study medicine and read with Dr. N. H. Torbet for three years. He attended St. Louis Medical College, and graduated from that institution. He did a post graduate course in Philadelphia, and later took a special course in St. Louis to learn tooth extraction." He arrived in the Jonesboro area ca. 1846, and practiced medicine there during much of the 1850s. Dr. Parks had moved to nearby Anna by the 1860 USC. He remained a physician in Anna until his death in February 1872, at 53 years of age (*Jonesboro Gazette*, Feb. 24, 1872). "In his diary he records many trials of a pioneer doctor, the greatest of which was a lack of ready cash. In July, 1868, he wrote 'Cash receipts for practice amount to just 2 bits today' " (Lam et al. 1954:165).

The June 6, 1854, the *Cairo Times* contained an advertisement for Dr. Parks "Egyptian Medicines," including his *Egyptian Anodyne* and an *Egyptian Collyrium*: "Dr L. K. Parks Far-famed Egyptian Medicines. Egyptian Anodyne, .50¢ bottle, Egyptian Collyrium, .25¢ vial **for sore eyes.** The Egyptian Medicines are sold by Dr. Parks in Jonesboro and by his agents in the following towns: Dr. W. Wood—Cairo; Spence & Bradford—Caledonia; John Hodges—Thebes; L. B. McCreary—Metropolis; James Hammond—Vienna; James Morgan—Murphysboro."

The likely source of the "Egyptian" Medicine name, was the fact that the Jonesboro area in deep southwestern Illinois along the Mississippi River was known colloquially as "Egypt" or "Little Egypt"—with local town names like Cairo and Karnak. Embossed Parks' *Egyptian Collyrium* bottles have not been documented, but may yet be found. "Collyrium" was a general term for an eye wash or eye cleanser. An "anodyne" was anything that relieved stress or pain, and in this era usually contained opium.

An example of the listed bottle could not be located for photography, but it has been recorded at auction sales and documented by Fike (1987:176).

KANKAKEE
(Kankakee County)

1860 census: 3,000
1870 census: 5,200
1880 census: 5,700

[Dietrich] H. Kammann

Soda ~ Mineral Water:

(1) **D. H. KAMMANN. / KANKAKEE. / ILLS** ·   *[aqua]*
(base): (outer ring) **GRAVITATING STOPPER / MADE BY /**
(inner ring) **JOHN MATTHEWS N Y /**
(center) **PAT^D / OCT 11 / 1864**

»SB w/ post mold; Matthews-style flat-topped blob soda with short neck and Matthews gravitating internal-glass-stopper closure. *[Figure 777]*

(2) **D. H. KAMMANN / KANKAKEE, ILL.**   *[amber]*
Note: embossed in circular slug plate.
(base): **8**

»SB w/ post mold; slope-shouldered quart soda; oval-blob top; 11" tall. *[Figure 778]*

According to an 1883 ad in the local newspaper, D. H. Kammann arrived in Kankakee in 1869 and established a bottling works on Indiana Avenue near the river. He was listed as a soda bottler in the 1875 ISD, and as a manufacturer of "pop beer" (at the southeast corner of East Avenue and Chestnut Streets) in the 1876 Kankakee CD. Kammann was listed in the 1879–80 IIC as a mineral water manufacturer in Kankakee, in operation 12 months a year with two full-time male employees and two full-time female employees. He paid his skilled workers $2.50/day and his laborers $0.75/day. During the 12-month census period, he paid $450 in wages, had $300 in material expense, and produced an amount of bottled product valued at $2,000.

He advertised bottled mineral water for sale in Kankakee until after the turn of the century, and bottled soda in both Hutchinson-style bottles and crown-top bottles after using those listed at the beginning of this entry during the 1870s. His later bottle styles were produced too recently for listing in the current study.

Figure 777

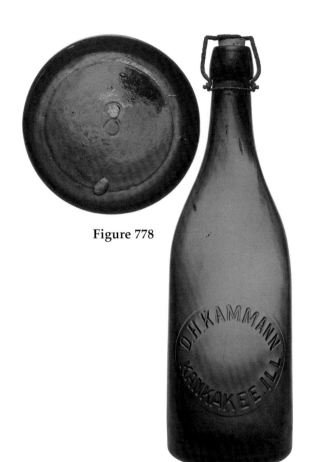

Figure 778

KEWANEE
(Henry County)

1860 census: 1,500
1880 census: 2,700

Daniel H. Spencer

Soda ~ Mineral Water:

(1) **D. H. SPENCER / KEWANEE ILL. //** [aqua]
      (heel): **D.O.C.**
      (base): **S**

»SB w/ probable post mold; slope-shouldered oval-blob-top soda. *[Figure 779: Photo courtesy of John Paul and Illinois State Museum photographer Doug Carr. Originally published as part of Figure 74 in Paul and Parmalee (1973:94).]*

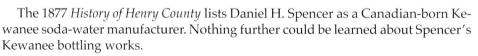

The 1877 *History of Henry County* lists Daniel H. Spencer as a Canadian-born Kewanee soda-water manufacturer. Nothing further could be learned about Spencer's Kewanee bottling works.

Figure 779

LACON
(Marshall County)

1850 census: 1,000
1860 census: 1,900
1870 census: 2,100
1880 census: 1,800

[Henry] E. Rupert

Soda ~ Mineral Water:

(1) **H. E. RUPERT / LACON, / ILL'S /** [aqua]
(left front heel): **FA & CO**
(base): **R**

»SB w/keyed hinge mold; shouldered-blob-top soda. *[Figure 780]*

H. E. Rupert's first soda-bottling operation was at Lacon, IL, a small town on the Illinois River about 30 miles downstream from Pekin. From the *Fahnstock, Albree & Co.* glass-house mark on the heel of his smooth-based Lacon bottles, they were produced and used at Lacon ca. 1861–62.

Rupert moved further downriver to Alton soon afterward (*see his Alton listing*). His Civil War–era Alton bottles were made from the same glass mold and embossed exactly the same as his earlier Lacon bottles, except for the fact that the Alton bottles were produced by a different Pittsburgh glass house (*Lorenz, Wightman & Co.*).

H. E. Rupert was not listed in the only Civil War–era Alton CD (1866) or the 1868 Alton CD, so it appears he did not come to Alton until the later 1860s. An Alton digger active in the 1970s and 1980s has indicated he found (but did not save) broken Rupert bottles embossed **SPRINGFIELD, ILL'S** (*not seen by us*), so perhaps Rupert made an interim stop there after leaving Lacon and before setting up shop in Alton in the late 1860s.

Rupert was in his late 20s when he was listed in the 1870 USC as an Alton manufacturer of soda water.

Figure 780

LASALLE
(La Salle County)

1860 census: 4,000
1870 census: 5,200
1880 census: 7,800

Dr. A. L. Babcock

Medicine:

(1) **PUT. UP. BY / D**^{R} **A. L. BABCOCK / LASALLE / ILLINOIS**   [aqua]
   Note: single embossed dot below superscript **R** *in* **D**^{R}.

   »OP; beveled-edge rectangle; 4 flat panels with embossing on one panel only; tapered collar; 5 3/4" tall. *[Figure 781]*

———— ❖ ————

From its stylistic and technological attributes, this bottle appears to have been manufactured and used during the late 1850s only. It was likely a generic bottle for a number of different medicine recipes prescribed and formulated by Dr. Babcock.

DeSteiger Glass Company

Whiskey Flask:

Figure 781

(1a) *[embossed "Sheaf of Wheat" design]* //   [amber]
   (base): **D S G Co**

   »SB; half-pint shoofly flask; 2 raised rings at base of neck; ring-neck tapered collar; about 6" tall.

(1b) *[embossed "Sheaf of Wheat" design]* //   [amber, aqua]
   (base): **D S G Co**

   »SB; pint shoofly flask; 2 raised rings at base of neck; ring-neck tapered collar; 7 1/4" tall. *[Figure 782]*

(1c) *[embossed "Sheaf of Wheat" design]* //   [aqua]
   (base): **D S G Co**

   »SB; quart shoofly flask; 2 raised rings at base of neck; ring-neck tapered collar; ca. 8" tall.

———— ❖ ————

This series of shoofly-like flasks are hexagonal in cross-section and have an overall debossed quilted surface, except for a panel on the front face, which is embossed with a "Sheaf of Wheat" design. These appear to have been early presentation flasks, made for advertising purposes by the *DeSteiger Glass Co.* They were similar in function to

Figure 782

the advertising flask made by the Alton Glass Company on its opening in 1873–74 (*see Alton Glass Co. listing*).

The DeSteiger flasks were likely made about three years later, and date to the later 1870s (ca. 1877–78), when the DeSteiger brothers first began manufacturing glass bottles in LaSalle (*see introduction, this volume and Lockhart et al. 2007:31*).

Ben Doll

Soda ~ Mineral Water:

(1) **B. DOLL / LASALLE / ILL //**   [aqua]
(heel): **De S. G. Co.**
(base): **D**
Note: Embossed in bottle mold, not removable slug plate.

»SB w/ post mold; slope-shouldered, oval-collared-blob-top soda. *[No photos: images recorded from on-line auction sale]*

❖

The listed bottle is a late-1870s collared-blob style soda (*see Bolin & Co. Lincoln listing*). According to online genealogical information, Ben Doll was born in Germany in 1845. He emigrated to the United States and settled at LaSalle, where he died in 1895. From the evidence of his embossed bottle styles (including 1880s Hutchinson-style sodas manufactured too late for inclusion in the present study), he operated his LaSalle soda factory during the later 1870s and the 1880s. He was listed in an 1877 LaSalle County history as a "manufacturer of sodawater" (Kett 1877a:603). From genealogical information, his soda factory was located at Ninth & Tonti Street in LaSalle.

Louis Eliel and Co.

Lager Beer:

(1a) **L. ELIEL & Co. / LA SALLE / ILL. /**   [amber]
(front heel): **De S. G. Co.** //
THIS / BOTTLE / NOT TO BE / SOLD

»SB; tall pint; oval blob top; 9" tall. *[Figure 783]*

(1b) **L. ELIEL & Co. / LA SALLE ILL //**   [amber]
THIS / BOTTLE / NOT TO BE / SOLD
(base): **DeS & CO.**

»SB; tall pint; oval blob top; 9 1/2" tall.

(2) **L. ELIEL & Co / LA SALLE / ILL //**   [aqua]
THIS / BOTTLE / NOT TO BE / SOLD

»SB; long-neck, slope-shoulder quart; shouldered blob top; 11 1/2" tall. *[Figure 784]*

❖

The *Eliel Brewery* was first established by Jeremiah Eliel in 1854, when he was 26 years old, just two years after LaSalle was incorporated as a city. Shortly thereafter his brother Louis joined the business in a managerial capacity, and it was listed as the *Louis Eliel Brewery* in the 1860 IIC—which indicated that the company used 20,000 bushels of barley and 60 bales of hops to produce 6,500 barrels of "Lager Beer" that year, with a sales value of $40,000.

Figure 783

The Eliel brothers operated the brewery continuously throughout the remaining period of our study. By the late 1870s, Eleil's brewery was "making over one hundred barrels per day, and employing thirty hands" (Kett 1877a:302).

By the Civil War years, Jeremiah and his two brothers (Louis and Gustavus) were also involved in a tannery and leather business as *Eliel & Co.* By 1864, Louis and Gustavus had relocated the tannery business to Chicago (at Hawthorne and North Branch), with offices at 115 Kinzie. Chicago CDs indicate that they operated the tanning company in Chicago from 1864 through 1871. Then in 1872, Louis and Jeremiah incorporated the brewery in Chicago as *L. Eliel & Co.* Gustavus continued to focus his attention on the family tannery business in Chicago, with a variety of partners, during the mid- and late-1870s.

After their first year of incorporation in Chicago as *L. Eliel & Co.*, Louis and Jeremiah primarily managed the company in LaSalle, at the site of the brewery itself. They were listed in the 1877 LaSalle business directory as "Brewers of Lager Beer." At the time, they still maintained a Chicago (sales?) office at 247 Lake Street (Kett 1877a:620).

Stylistically and technologically, the *L. Eliel & Co.* embossed bottles listed date from the period of the mid- and late-1870s.

Figure 784

Alexander Finkler and Co.

Soda ~ Mineral Water:

(1) **A. FINKLER & C⁰ / LA SALLE / ILL //**   [cobalt, aqua]
    (heel): **A. &. DH. C**

»SB w/ keyed hinge mold; shouldered-blob-top soda. [Figure 785]

❖

Alex and John A. Finkler were owners of a LaSalle soda water bottling company in the 1860 and 1870 USC and in the 1867 ISD. Early on, John had also been a silent "& Co." partner with J. B. St. Clair in 1864 (according to the IRS assessment that year) in a Peoria bottling business there. A third brother, Frank J. Finkler, also worked at the LaSalle bottling plant according to the 1860 USC, before leaving to serve in the Civil War.

After serving in the Civil War, Frank had expanded the brothers' bottling partnership to a new bottling operation in Dixon (*see his Dixon listing*), where he bottled during the late 1860s and early 1870s. At about this same time, John set up a soda bottling works in Streator (*see his Streator listing*). Alex had moved the LaSalle bottling operation to Ottawa (*see his Ottawa listing*) ca. 1872, "at Riverside; P.O. Streator," according to his 1877 LaSalle County CD listing (Kett 1877a:419, 604). The 1880 USC

Figure 785

listed Alex and John as soda water manufacturers in Ottawa and Streator, respectively. Alex Finkler died in Ottawa in 1883, after which his son (Alexander C.), who was just 14 when his father died, carried on the Finkler soda water bottling business for 11 years, before finally selling out in 1894 (Lewis Publishing Co. 1900:437–438).

Ransom E. Hickey and Co.

Soda ~ Mineral Water:

   (1) **R. HICKEY & CO'S / LASALLE ILL**   [aqua]
       »IP; shouldered-blob-top soda. *[Figure 786]*

   (2) **R. HICKEY & CO / LA SALLE / ILLS**   [aqua]
       »IP; shouldered-blob-top soda. *[Figure 787]*

Although Ransom Hickey is usually associated with early soda and mineral water bottling in Peoria (*see his Peoria listing*), the style, manufacturing technology, and embossing of the two pontiled bottle varieties listed confirm that he bottled in LaSalle during the late 1850s. Since the LaSalle bottling operation became the *S. Hickey* bottling works from 1860 to 1862, and since there are two different embossed varieties of R. Hickey LaSalle bottles, *R. Hickey & Co.* likely bottled there from ca. 1856/57 to 1859.

The large Hickey clan and their in-laws, the James Eaton family (*see Eaton's Galesburg, Chicago, and Peoria entries elsewhere in this volume*), bottled mineral and soda water in several localities in Illinois and Wisconsin from the early 1850s into the 1880s. During the 1850s, four members of

Figure 786   **Figure 787**

the Hickey family were in the regional bottling business: Hiram in Peoria and Joliet, Ransom in Peoria and LaSalle, John in Joliet and Galesburg, and Stephen in Milwaukee (*see their listings elsewhere in this volume, and also Stephen's 1860–62 LaSalle listing below*). Stephen may also have been the "*& Co.*" partner of Ransom Hickey in LaSalle during the late 1850s (*see Stephen's listing below*).

Stephen M. Hickey

Soda ~ Mineral Water:

(1a) **S. M. HICKEY / SUPERIOR / SODA WATER / LA SALLE ILLS** [teal]

»IP; shouldered-blob-top soda. *[Figure 788]*

(1b) **S. M. HICKEY / SUPERIOR / SODA WATER / LA SALLE ILLS** [teal, aqua]

»SB w/ keyed hinge mold; central raised dimple on base; shouldered-blob-top soda. *[Figure 788]*

Stephen M. Hickey (*aka:* "Sephremus" or "Sephreness," named after his father) was a member of the large Hickey family who bottled soda and mineral water in Illinois (*see the Joliet, LaSalle, and Peoria entries*) and several towns in Wisconsin (Peters 1996). S. M. Hickey is listed in the 1860 ISD and the 1860 IIC as a soda water bottler in LaSalle. The 1860 industrial census indicated he had invested $1,000 in real estate and $2,000 in raw materials that year, and that he paid his six adult male workers an average of $96/month. The value of his bottled product that year (described as "soda waters and other beverages") was listed at $10,000.

Figure 788

S. M. Hickey was listed as a bottler in Bloomington, 40 miles south of LaSalle, in the IRS tax assessments for 1863 and 1864. The IRS tax assessments indicate that Stephen Hickey had moved to Freeport, IL, by 1865. These same documents also list Ransom Hickey in Bloomington (*see the Bloomington listing*) in 1864 as an "ale dealer," so he may have been associated in business with Stephen there, as well as in LaSalle (*see the discussion of Hickey family relationships under the Peoria listing*). Thus the LaSalle "**R. HICKEY & Co**" bottle in the previous listing above may have been a partnership between Ransom and Stephen (as a silent financial partner from Milwaukee).

During the late 1850s, Stephen Hickey was a soda and mineral water bottler in Milwaukee (Peters 1996). The embossing on his Milwaukee bottle is identical to that on LaSalle bottles #1a and #1b (same mold), except that the "**S. HICKEY**" line on the Milwaukee bottle has been closed up to accommodate his middle initial and the bottom-line "**MILWAUKEE WISC**" has been replaced by "**LA SALLE ILLS**." Since both smooth-base and pontiled varieties of the LaSalle bottle are known, and since he was in Bloomington by 1863, Hickey's LaSalle bottling operation likely dates to ca. 1860–1862.

LEBANON
(Madison County)

1850 census: 500
1860 census: 1,700
1870 census: 2,100
1880 census: 1,900

Charles J. Reuter

Soda ~ Mineral Water:

(1) **CHARLES REUTER / LEBANON.ILL //**   [aqua]
(heel): **A & D. H. C**

»SB w/ keyed hinge mold; shouldered-blob-top soda. *[Figure 789]*

❖

Charles Reuter immigrated to the St. Louis area with his family ca. 1849 when he was about 10 years old. They moved to Lebanon, IL, just before the Civil War years where his father operated a grocery. Charles left the area for several years during the Civil War era "to seek his fortune," but he returned in 1868, when he was about 30 years old and "built a framed, steam powered soda/mineral water factory, probably the first one in the old town. Business was good, and Charles built his family a large, Greek Revival style home at 516 South Herman St. After twenty-seven years in business, Charles retired" (Miller 2007a:426; Bateman 1907:1102).

Reuter was listed in the 1873 CAR and in the 1875 ISD as a bottler of soda water in Lebanon. The embossed bottle listed is the only style that Reuter is known to have used in his business prior to 1880. Three styles of Reuter-embossed collared-blob and Hutchinson-internal-stopper soda bottles have also been documented (Miller 2007a:429–431), but these likely date to the 1880s and early 1890s—too late for inclusion in the present volume.

Reuter's soda water manufactory was recorded in the 1879–80 IIC. At the time, he used a seasonal maximum of five laborers who he paid a total of $800 annually. His outlay for production material that year was $1,200, and he estimated the annual

Figure 789

value of his bottled products at $2,800. His steam-powered factory was run by a single-engine, six-horse-power boiler.

Henry H. Smith

Medicine:

(1) **H. H. SMITH'S // INDIAN / BALM //** [aqua]
LEBANON. ILLINOIS

»OP; beveled-edge rectangle; 4 sunken panels; short neck; rolled lip; 4" tall. *[Figure 790]*

This is a very early west-central IL patent medicine. Very little is known about Henry Smith: he was listed in the 1840 USC in Lebanon but was not listed in the 1850 census. No ads have yet been discovered to record the specific healing powers claimed for his *Indian Balm*. However, "Indian Balm" is an herbal folk-medicine name (*a.k.a.* beethroot, cough root, stinking Benjamin, *Trillium erectum*). According to herbal medicine guides, the plant's root has a faint turpentine smell, and a faint sweet astringent initial taste. But it becomes bitter and acid when chewed, causing salivation. It was used as an antiseptic, astringent, and expectorant, and as an active ingredient in alterative and pectoral syrups.

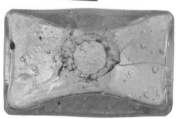

Figure 790

LEMONT
(Cook County)

1860 census: 1,400
1870 census: 3,600
1880 census: 2,100

John G. Bolton

Soda ~ Mineral Water:

(1a) **J. G. BOLTON / LEMONT / ILL^S //** [dark cobalt]
(heel): **A & D. H. C.**
Note: 2 embossed dots beneath superscript **S** *in* **ILL^S**

»SB w/ keyed hinge mold; high shoulder soda w/ shouldered blob top. *[Figure 791]*

(1b) **J. G. BOLTON / LEMONT / ILL^S //** [light and dark cobalt]
(heel): **W. M^cC & Co .**
(base): **J.G.B**

»SB w/ post mold; slope shouldered, shouldered-blob-top soda. *[Figure 792]*

(2) **J. G. BOLTON / LEMONT / ILLS. /** [aqua]
(heel): **ARTHUR CHRISTIAN //**
(rear heel): **A & D H C / PAT. APR. 13TH, 1875**

Figure 791

Figure 792

    *Note: Two embossed dots beneath superscript "***TH***" in ***13***TH*.

    »SB w/ post mold; cylindrical, Arthur Christin-style gravitating-rubber-stopper soda; oval blob top; 7 3/4" tall. *[Figure 793]*

Cider:

    (1a) **J. G. BOLTON. / LEMONT. / ILL.**   *[amber, green]*
        (*unembossed concave base w/ central glass dimple*)

        »SB w/ deep basal kick-up; ring-neck cider-finish quart; 12" tall. *[Figure 794]*

    (1b) **J. G. BOLTON. / LEMONT. / ILL. //**   *[amber, green]*
        (*base, around margin*): **W. McCULLY & CO. PITTSBURGH, PA.**

        »SB w/ post mold; ring neck cider-finish finish quart; 12" tall. *[No photo. Two sources indicate that some of the quart ciders embossed like #1a have a McCully & Co. maker's mark embossed around the flattened outer edge of the base: see Kott 2005b; Sineni 1978:10.]*

Ale ~ Porter:

    (1) **J. G. BOLTON / LEMONT / ILL**   *[amber]*
        (*central base*): 2 embossed rings crossed by large embossed X /
        (*base, around margin*): **W. McCULLY & Co. / PITTSBURGH.**
        Note: Side-embossed, not shoulder-embossed.

        »SB; 2-piece-mold quart ale; long neck; ring-neck tapered collar; 10 3/4" tall. *[Figure 795]*

Figure 793

Figure 795

Figure 794

Figure 796

John G. Bolton knew (and very worked with) John A. Lomax in the Chicago soda and mineral water bottling business prior to the Great Chicago Fire of October 1871 (see Sineni 1978:8). Both men immigrated to America from the same city (Bury) in England (*see the John Lomax Chicago listing*). Bury is located only a few miles east of the town of Bolton, and the main road between the two is Bolton Road. In Chicago, the Bolton family lived on Charles Place (near the Lomax bottling works), and John Lomax's son George married J. G. Bolton's sister Nellie (*Midwest Bottled News* 1987:5–6; *Chicago Tribune* Oct. 18, 1872, p. 8, col. 6).

Unfortunately the Chicago CDs do not reflect this association. In the CDs, John and James Bolton show up in Chicago just after the Civil War listed as stonecutters, boarding at 23/24 Charles Place, and elsewhere, with John not listed at all between 1870 and 1872. Their last Chicago listing was 1874.

Apparently, sometime after the 1871 Chicago fire, J. G. Bolton and his father, James B. Bolton, moved to Lemont to establish their own bottling operation. Early on, the company was known as *J. B. Bolton & Co.*, and an early 1870s stoneware "small-beer" bottle with this embossing is known (*see Figure 796a–b*). By the mid-1870s, his father had passed away (Kott 2005b), and John Bolton carried on the Lemont bottling business on his own. He is listed alone as a Lemont manufacturer of soda water in the 1875 ISD.

John Bolton died in June of 1883. Afterward, Bolton's bottling business was managed by his wife, and by Edward Wold, until 1888. Post-1880 Bolton-embossed Hutchinson-style internal stopper bottles are known (see Sineni 1978), but they were produced too late for inclusion in the present study.

Twelve-sided, 11" tall, stoneware "small-beer" bottles, marked **J.G.B.** (sometimes in a debossed rectangle) were also used by Bolton during the later 1870s (*Figure 796c*). Such bottles usually contained lightly alcoholic "small-beers," such as spruce beer (particularly popular in the Chicago area) and lemon beer.

LEWISTOWN
(Fulton County)

1860 census: 2,900
1870 census: 1,200
1880 census: 1,800

The Medicine Company

Medicine:

(1) *(side):* **THE MEDICINE COMPANY** // [aqua]
    (side): **LEWISTON ILL** // // //

»SB w/ post mold; beveled-edge rectangle; 4 sunken panels, embossed on opposite narrow sides; tapered collar; 6 3/4" tall. [*Figure 797*]

Figure 797

From its style and technological attributes, this bottle was likely produced and used during the late 1870s. A post-1880 paper-labeled medicine bottle sold by the same firm indicated the bottle contained *"Sweet Elixir for Diarrhoea, etc. etc."* The paper label listed the company's address in Lewistown as 88 Main Street. Note that "Lewistown" is misspelled on their late-1870s bottle listed.

LINCOLN
(Logan County)
1860 census: 5,700
1880 census: 5,600

William G. Gesner and John N. Lipp

Soda ~ Mineral Water:

(1) **GESNER & LIPP / MINERAL WATER / LINCOLN. ILLS** [aqua]

»SB w/ keyed hinge mold; shouldered-blob-top soda. *[Figure 798]*

Figure 798

❖

The partnership of William G. Gesner and John N. Lipp was listed in the IRS tax assessments between 1862 and 1864. Mr. Lipp was listed alone as a retail liquor dealer in Lincoln the following year (1865). They are listed in the 1864–65 ISD as bakers and confectioners in Lincoln.

The Lincoln *Herald* for January 11, 1866, carried an advertisement for the City Bakery owned by *John N. Lipp & Brother*. The advertisement went on to state that the Lipp brothers also bottled and distributed mineral and soda water. Based on this notice, it appears that the partnership of Gesner & Lipp, which from the listed bottle clearly included a soda bottling operation, dated to the later Civil War years: ca. 1862–1865.

Jacob, John, and Herman Bollin

Soda ~ Mineral Water:

(1) **J. BOLLIN & Co / LINCOLN / ILLS. //** [aqua]
    (heel): **A & D. H. C.**

»SB w/ post mold; late 1870s oval collared-blob-top soda; lettering cut into mold, not removable slug plate. *[Figure 799]*

❖

Bollin brothers Jacob and John, under the direction of eldest brother Jacob, ran a saloon and wine and liquor sales company in Lincoln during the later 1860s and 1870s. According to a short biographical essay in an 1878 Logan County history (Donnelley, Loyd & Co. 1878b:345), Jacob first opened his wine and liquor business in Lincoln in 1866 at 412 Pulaski Street. By 1878, *J. Bollin & Co.* included his brother John, and from the evidence of the listed bottle, their wine and liquor business was also marketing bottled mineral waters.

Figure 799

According to a later county history (Inter-State Publishing Co. 1886:456–458), John Bollin died in 1881 and Jacob Bolin turned his full attention to running his saloon in Lincoln during the early 1880s. At about that time, youngest brother Herman Bollin took over the earlier soda-manufacturing business. No Herman Bollin embossed bottles have been located, but the earlier J. Bollin & Co. soda bottles likely date to the ca. 1878–80 period.

LITCHFIELD
(Montgomery County)

1860 census: 1,600
1870 census: 3,900
1880 census: 4,300

William Gerhardt

Soda ~ Mineral Water:

(1a) **Wm GERHARDT / LITCHFIELD / ILL //** [aqua]
(heel): **A & D. H. C.**

»SB w/ post mold; slope shouldered, oval-blob-top soda. *[Figure 800]*

(1b) **Wm GERHARDT / LITCHFIELD / ILL //** [aqua]
(base): **L & W**
Note: Same mold as bottle #1a, but with a small L&W maker's mark on the base instead of the A&DHC heel mark.

»SB (w/ heel mold?); slope shouldered, oval-blob-top soda. *[no photo: same side mold as Figure 800]*

(2) **W^M GERHARDT / LITCHFIELD / ILL^S //** [aqua]
(heel): **M. G. C^O**
Note: There are two embossed dots under the raised "**M**" in **W^M**.

»SB w/ post mold; slope shouldered, oval-blob-top soda. *[Figure 801]*

William Gerhardt came to Litchfield in 1868 and was listed in the 1870 USC as a vinegar maker there. Soon afterward he turned his attention to bottling soda and mineral water, as he was listed in the 1872 *TW & W* Railway directory, the 1874 Montgomery County Atlas (Brink, McCormack & Co. 1874b:32), and the 1875 ISD as a Litchfield manufacturer of soda water. He died on October 12, 1877, from a mysterious blow to the head that was officially attributed to his being kicked by a horse. Suspicions were raised about the accident after an account in the nearby Hillsboro newspaper suggested that they "had it on good authority" that Frederick Weber, a competitor of Gerhardt in the Litchfield bottling business (*see Weber's listing, this chapter*), had "hired a negro" to do Gerhardt in.

Figure 800

However, after detailed investigation by a coroner's inquiry, nothing could be proven and the case was dropped:

> Friday afternoon, Wm. Gerhardt, of this city, deceased near Wright's Ford. He left home with his team in the morning to get a load of wood belonging to him.... About a third of a load had been put on the wagon, when he received a blow near the top of his head, and extending downward on the left side. Blood was found on the wheel and on a whiffletree. Mr. Gerhardt started homewards on foot, pursuing for a few hundred yards the wagon road, and then striking into a cornfield. Here his tracks could be seen, and it is evident that he staggered as he went. Coming to a gap in the fence, he seated himself on the sloping rails for a time, and at last fell over dead. His hat was found near, partially filled with blood, and cut through by the blow which killed him. There were other blows received, but the one on his skull was the fatal one. The deed was committed not long after he reached the wood, and his dead body was not discovered for several hours thereafter.
>
> He may have been killed by a kick from his horse, or he may have been murdered. His gloves, and seat, and cushion were found on the wagon, but no weapon could be discovered, and no sign of a struggle.
>
> He was about fifty years of age, and leaves a wife and four small children in indigent circumstances. [*The Litchfield Monitor*, Oct. 17, 1877, p. 3, col. 3]

Joseph F. Neuber

Soda ~ Mineral Water:

(1) **J. NEUBER / LITCHFIELD / ILL**ˢ // [aqua]
 (heel): **M. G. Cº**
  »SB w/ post mold; slope shouldered, oval-blob-top soda. *[Figure 802]*

Figure 801

According to Joseph Neuber's obituary notice in the *Litchfield News Herald* for January 1, 1915, he immigrated from "Bohemia" in 1866 and first settled in St. Louis, after which he went to San Francisco for a while "until 1881 when he came to his present home in Litchfield and engaged in the grocery business, retiring from same some 16 years ago."

This death notice was likely in error regarding Neuber's Litchfield business history. He made only one small order of soda bottles, which are very rarely encountered today, and from their style and technological attributes, the bottles appear to have been made and used sometime in the late 1870s. In fact, they look like a mold modification of bottle #2 in the William Gerhardt listing. Apparently, Neuber first came to Litchfield at about the time of Gerhardt's death (1877–78)—perhaps because he heard that a vacant soda manufactory was being offered for sale—and took it over. From the rarity of his bottles, he made a very short, and unsuccessful, attempt to start up his own bottling business in ca. 1878–79. That failing, he went into the local grocery business.

Frederick Weber

Soda ~ Mineral Water:

(1) **F. WEBER & Co. / LITCHFIELD / ILL.** [aqua]
 Note: This first Weber embossed bottle has the peened-out remnants of an earlier mold embossing on the reverse side. The remnant lettering can faintly be read as **J. SCHWEINHART / PITTSBURGH**. *Pontiled embossed Scheinhart Embossed sodas are known from Pittsburgh.*

 »SB w/ keyed hinge mold; long-neck, high-shouldered, shouldered-blob-top soda. *[Figure 803]*

Figure 802

(2) **F. WEBER / LITCHFIELD / ILLS //** [teal]
(base): **F.W**
Note: Mold recut with different embossed lettering style.

»SB w/ keyed hinge mold; oval-blob-top soda. *[Figure 804]*

(3) **F. WEBER / LITCHFIELD / ILLS //** [light blue]
(heel): **A & D. H. C.**
Note: Mold again recut with new embossed lettering style.

»SB w/ post mold; shouldered-blob-top soda. *[Figure 805]*

(4a) **F. WEBER / LITCHFIELD / ILLS //** [aqua]
(heel): **A & D. H. C.**
Note: Mold again recut from #3 above with new embossed lettering style.

»SB w/ post mold; sloped shoulders w/ collared-blob-style neck; shouldered-blob-top soda. *[Figure 806]*

(4b) **F. WEBER / LITCHFIELD / ILLS //** [aqua, green]
(heel): **A & D. H. C.**

»SB w/ post mold; sloped shoulders w/ collared-blob-style neck; short "lozenge" style blob-top soda. *[Figure 807]*

———————— ❖ ————————

Fred Weber was listed as a soda water manufacturer in Litchfield in the 1873 CAR and in the 1875 ISD. According to an 1882 biographical sketch published in a history of Bond and Montgomery counties (Perrin 1882b:183–184), he came to St. Louis from Bavaria in 1856 as a young man of 18, and first settled in St. Louis. He served as a

Figure 803

Figure 804

Figure 805

Figure 807

Figure 806   military baker during the Civil War, afterward worked as a baker in St. Louis for a year, and then moved to Litchfield:

> In June, 1867, he started a factory for the manufacture of soda and seltzer waters, to which he still devotes his attention. In 1881, he manufactured during the season an average of one hundred dozen of bottles per day, and these were shipped to various points in the State. These waters have become very popular as a healthful beverage, and, in consequence, demand has steadily increased from year to year. For the past two years, Mr. Weber has run a steam cider press with good success. During the busy season, his business requires the services of eight active men. [Perrin 1882b:184]

According to Weber's obituary notice, published in the *Litchfield Daily Herald* for Sept. 18, 1909, he continued to run his factory at the corner of Monroe and Edwards streets in Litchfield until 1890, when he sold his bottling operation to Charles A. Tolle and retired to Hillsboro.

Fred Weber used at least three varieties of true-blob soda bottles during the 1860s and 1870s. Based on attributes of style and technology, they are listed in chronological order at the beginning of the entry. Bottle #1 was used during the late 1860s. The identity of the "& Co." member(s) of the business during Weber's first few years in operation are not yet known. Bottle styles #2 and #3 were used during the early and later 1870s, respectively. During the 1880s, Weber is known to have used two collared-blob soda varieties and two Hutchinson internal-stopper soda styles. These were used too late for inclusion in the present study.

Fred Weber's Litchfield "Soda Water Manufactory" was included in the 1879–1880 IIC, which indicated that he employed three adult males, two adult females, and a child at the bottling works, paying them a total of $800 in wages annually. The works operated full time three months a year, two-thirds time four months a year, half time two months a year, and were idle for three months. During the census year, he had invested $1,800 in materials and labor, and his retail sales amounted to $3,500.

LOUDEN CITY
(Fayette County)

1860 census: 1,200
1870 census: 1,200

Dr. John C. Wills

Medicine:

(1) (side): **DR J. C. WILLS // WORLD //** [aqua]
    (side): **WORM CANDY // //**

»SB w/ keyed hinge mold; beveled-edge rectangle; 4 flat panels; thickened-ring lip; 3 1/2" tall. *[Figure 808]*

Dr. Wills was listed in the USC for 1860 and 1870 as a resident of Louden City in northeastern Fayette County. He was listed as a 34-year-old physician in the 1860 USC. He and his family came to the area during the late 1850s from Coshocton, OH. By the time of the 1880 USC, he was listed as a physician in adjacent Effingham County.

The technological attributes of the bottle listed indicate that it was produced and used during the 1860s. Some of Dr. Wills embossed vermifuge bottles have been found in Decatur, so his children's worm medicine must have been popular enough during the Civil War era to find a market as far away as 40 miles to the north.

An 1874 business directory for Marion County, IN, noted that by that time the local firm of *Shackelford & McCoy* were "sole agents and proprietors in the West for Dr. J. C. Wills' World Worm Candy; manufactured at Indianapolis." This suggests that Dr. Wills only marketed the medicine himself in Illinois during the 1860s.

Figure 808

MARYLAND
(Ogle County)
1860 census: 1,200
1870 census: 1,200

Samuel Haldeman

Medicine:

(1) *(monogram):* **SH** / *(vertically):* **S. HALDEMAN / MARYLAND / ILLS** [aqua]

»SB; rectangular w/ 4 sunken panels; short neck; square, flared-ring collar; 6 1/4" tall. Embossed on one side only. *[Figure 809]*

Samuel Haldeman moved his family from Pennsylvania to Whiteside County, IL, near the end of the Civil War years. In the mid-1870s, they moved to adjacent Ogle County and settled at Maryland. Haldeman is listed there in the 1880 USC for Ogle County as a 59-year old manufacturer of medicine.

According to a biography of Samuel's son Benjamin (who moved to Nebraska and then Morrill, KS, in the mid-1870s), his father was the sole Illinois general agent for the sale and distribution of *Younce's Indian Cure Oil and Pain Destroyer*, and when his son went west, Samuel appointed him "agent for this popular remedy for Morrill and vicinity" (online transcription of biographical sketch from Cutler 1883). This *Indian Cure Oil* was clearly Haldeman's principal (and perhaps only) patent-medicine product, and was likely the product contained in the bottle listed.

According to the W. H. Beers & Co. history of Miami County, OH (Beers & Co. 1880:756), Dr. D. A. Younce was a 43-year-old physician in Covington, OH, where he had earlier been a farmer. His cure oil is said to have originated when "while lifting a heavy box, he broke a ligament in his hip, by which he lost the entire use of his leg; a kind friend prescribed a remedy, and, by adding several oils, he effected a cure; during his lameness, he began the study of medicine, by the addition of several more oils, he effected many great cures; he is the patentee of D. & A. Younce's Indian Cure Oil, which is sold in seventeen different states besides his own; he treats all chronic diseases."

The Haldeman bottle listed above likely contained this product during the mid- to-late 1870s.

Figure 809

MASCOUTAH
(St. Clair County)

1850 census: 400
1860 census: 2,100
1870 census: 2,800
1880 census: 2,600

Nikolaus J. Bassler

Soda ~ Mineral Water:

(1a) **N. BASSLER & C⁰ / MASCOUTAH / ILLS //** [aqua]
(heel): **A & D. H. C**
Note: The mold for this bottle was made from an earlier Mascoutah bottle mold.
F. HAROLD *was been peened-out (see Herold listing, this chapter) and over-cut with*
N. BASSLER & Co. *The* **MASCOUTAH / ILLS** *embossing was not changed.*

»SB w/ keyed hinge mold; shouldered-blob-top soda. *[Figure 810, left]*

(1b) **N. BASSLER & C⁰ / MASCOUTAH / ILLS** [aqua]
Note: Later mold of bottle #1 with larger lettering and lacking the **A&D.H.C**
heel embossing.

»SB w/ keyed hinge mold; shouldered-blob-top soda. *[Figure 810, right]*

(3) **N. BASSLER / MASCOUTAH / ILLS //** [aqua]
(heel): **A. & D. H. C**

»SB w/ post mold; shouldered-blob-top soda. *[Figure 811]*

The *Mascoutah Herald* for November 20, 1885 contained the obituary for Nikolaus Johann Bassler following his death at age 58 (see Miller 2007a:54). Bassler had first owned a tailor shop after his arrival in Mascoutah ca. 1855. Then in 1866, Bassler purchased Ferdinand Herold's bottling business with junior partners August Koob and Philip Pfeifer. Koob had married into the Bassler family in 1863 when he took Helen Bassler as his wife. Bottles #1 and #2, were made and used during the late 1860s (#1) and early 1870s (#2).

By the mid-1870s, Koob had bought out his partners and was running the bottling works on his own for a time (*see his listing, this chapter*). Then, in 1878 (as men-

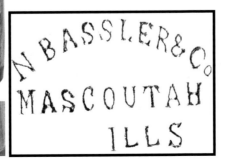

Figure 810

Figure 811

tioned in Bassler's 1885 obituary notice), Koob decided to sell out and move to Belleville to establish a new bottling works there. Bassler then reacquired the Mascoutah bottling works from Koob, and again operated it for a time in the late 1870s (probably with his son-in-law Philip Pfeifer, Jr., who had married his daughter Anna), using bottle #3. When Bassler died, the business was taken over by his son-in-law.

The somewhat entangled history of Mascoutah bottling works is summarized below:

1859/60. Herold establishes the Mascoutah bottling works (*see his listing, this chapter*).

1866. Herold sells out to the *Bassler & Co.* partnership, composed of Bassler, August Koob, & Philip Pfeifer, Sr. The Herold bottle mold is re-cut by A. & D. H. Chambers to reflect the change in ownership.

Mid-1870s. Koob gains primary ownership of the company and a bottle embossed with his name is made (*see his listing, this chapter*).

1878. Bassler buys out Koob's interest in the company. A bottle is made by A. & D. H. Chambers embossed with Bassler's name alone. Koob moves to Belleville.

1885. Bassler dies and the bottling works is taken over by his son-in-law, Philip Pfeifer, Jr., son of Philip Pfeifer, Sr., his former partner.

Ferdinand Herold

Soda ~ Mineral Water:

(1) **F. HEROLD / MASCOUTAH / ILLS //**   [aqua]
»SB w/ keyed hinge mold; shouldered-blob-top soda. *[Figure 812]*

Figure 812

Ferdinand Herold was listed in the 1860 ISD as a soda water manufacturer in Mascoutah. According to Miller (2007a:215), Harold was also listed in the 1862 U.S. Military Census for Mascoutah (he was 32 years old at the time) and in a local business census made in 1866. Harold sold his bottling business that year to *N. Bassler & Co.* (*see their listing, this chapter*).

Herold was married in Mascoutah in 1857, but he was not included in the 1860 USC (Miller 2007a). However, the 1859–1860 IIC did include Herold's "Soda Water Manufactory," indicating that he began his bottling business there shortly before the Civil War, so a pontiled version of the listed smooth-base bottle may yet be discovered.

The 1860 IIC recorded Herold as "Fred'k" Herold, with the handwritten annotation of "6 Months," perhaps indicating that he had not started his bottling business until the middle of the June (1859)–June (1860) recording period. The census indicated that Herold had invested $2,500 in the business, and had on hand 3,500 pounds of sugar, 15 carboys of vitriol, and "10,000 doz. bottles" (the latter valued at $500). That year he had five adult male employees, paid $125/month. His product ("10,000 doz. Soda") was valued at $3,000, about 2 1/2 ¢ per bottle.

August G. Koob

Soda ~ Mineral Water:

(1) **A. KOOB / MASCOUTAH / ILLS. //**   [aqua]
    (heel): **A & D. H. C.**
»SB w/ keyed hinge mold; shouldered-blob-top soda. *[Figure 813]*

August Gustave Koob is listed in the 1870 USC of Mascoutah as a 33-year-old soda water manufacturer. He had first become involved in soda water bottling in 1866, when Nikolaus Bassler purchased Ferdinand Herold's bottling business with August Koob as a junior partner (*see Bassler and Herold listings, this chapter*).

By the mid-1870s, Koob had bought out his *Bassler & Co.* partners and ran the bottling works on his own for a few seasons. Then, in 1878 (as mentioned in Bassler's 1885 obituary notice), Koob decided to sell out and move to Belleville to establish a new bottling operation there. At that time, Bassler reacquired the Mascoutah bottling works from Koob, and again operated it in the late 1870s (*see his listing, this chapter*).

In Belleville, Koob purchased the bottling works Louis Ab Egg had started there (*see his Belleville listing*) from Ab Egg's widow, Josephine, prior to the 1879 bottling season (Miller 2007a:266). All of Koob's late-nineteenth-century embossed Belleville soda bottles are post-1880 collared-blob styles and Hutchinson internal-stopper sodas (three varieties: see Miller 2007a:272–273).

Koob was recorded as living next door to the Abegg's in the 1880 USC, and Abegg's son Charles worked in his bottling business. Koob's successful bottling business lasted into the early twentieth century. After his death in 1888, his widow and children (including August G. Koob, Jr.) continued the Koob bottling business.

Figure 813

MATTOON
(Coles County)

1860 census: 2,000
1870 census: 5,000
1880 census: 5,700

John Fife and Co.

Soda ~ Mineral Water:

(1) **J. FIFE & Co. / MATTOON / ILL**[S] //   [aqua]
    (heel): **A & D. H. C**

»SB w/ keyed hinge mold; short neck; shouldered-blob-top soda. [Figure 814]

❖

From their stylistic and technological attributes, these embossed Fife bottles were likely manufactured and used during the late 1860s or early 1870s. However, John Fife has not been found listed in the 1870 USC for Mattoon, nor was he listed in an 1879 taxpayer list for Mattoon Township (Perrin 1879).

Apparently his residence in the area, and his short-lived bottling business there, were confined to the middle or later 1870s (although his embossed soda does have a hinge mold, it also has a very late-looking short neck). There is a Coles County marriage record for John Fife and Margaret Berviller dated June 4, 1879. But the Fifes were recorded in the 1880 USC as residing in Washington, IN, where he was listed as a 33-year-old "pop maker" born in Scotland.

Figure 814

MENDOTA
(La Salle County)

1860 census: 1,900
1870 census: 3,500
1880 census: 4,100

Frederick Molln and Co.

Soda ~ Mineral Water:

(1) **F. MOLLN** *(arched)* / **& Co** / **MENDOTA. ILL**ˢ /
(front heel): **C&I**   [aqua, blue]

»SB w/ keyed hinge mold; shouldered-blob-top soda.
[Figure 815]

Frederick Molln was listed in the 1880 USC for Mendota as a 56-year-old retired saloonkeeper. From its stylistic and technological attributes, the embossed Molln bottle appears to have been manufactured and used during the 1860s. Molln likely bottled soda water on a part-time basis during the mid- to late 1860s, and perhaps sold the soda water in his saloon.

During the post–Civil War era, saloons often provided bottled soda and mineral waters for mixing drinks. The "& Co." partner in Molln's sideline bottling business has not been determined *(see Wohlers listing, this chapter).*

Figure 815

Manuel J. Viera

Shampoo:

(1) *(side):* **VIEIRA'S** / **TOILET SHAMPO** // **M. J. VIEIRA** / **MENDOTA ILL'S** // // //   [aqua]
Note: *The only available study bottle is broken. The neck, lip, and one side panel are missing.*

»SB w/ keyed hinge mold; beveled-edge rectangle w/ 4 sunken panels; lip form unknown; total height unknown. *[Figure 816]*

Manuel J. Viera was listed in the 1870 USC for the city of Mendota as a 28-year-old barber. Mr. Viera apparently bottled his own proprietary shampoo formula for his customers. From the rarity of the bottles, his bottled-shampoo business effort was likely short-lived.

Edward Wohlers

Soda ~ Mineral Water:

(1) **E. WOHLERS** *(arched)* / **MENDOTA. ILL**ˢ /
(front heel): **C&I**   [aqua]

»SB w/ keyed hinge mold; shouldered-blob-top soda. *[Figure 817]*

Edward Wohlers came to the United States from Germany in 1860 and was listed in the 1880 USC of Mendota as a soda water manufacturer.

Figure 816

Figure 817

From the stylistic and technological attributes of his single embossed bottle variety, the Wohlers bottles were probably made and used during the late 1860s. Perhaps Wohlers was the junior partner in the F. Molln & Co. soda bottling business listed above, and later took over the operation from Molln. The Molln and Wohlers bottles look like they were made using the same mold, with Molln's name peened out and replaced by the **E. WOHLERS** embossing (with "**& Co**" obviously peened out below).

Wohlers' two sons, Ed (Jr.), and Fred, were running the family bottling business in Mendota by the time of the 1900 USC.

METAMORA
(Woodford County)

1860 census: 1,000
1870 census: 700

Bottling House

Soda ~ Mineral Water:

(1) **BOTTLING. HOUSE / METAMORA / ILLs** [aqua]
(base): **J E**
*Note: A variant of this bottle has a backwards "**J**" in the **J E** base mark (see Figure 818).*

»SB w/ keyed hinge mold; shouldered-blob-top soda.

There is no known "**J E**" glassmaker's mark, so the base embossing on this bottle likely represents the bottler's initials. One other "BOTTLING HOUSE" bottle, also with "J E" base embossing is known from Bloomington (*see the Bloomington Bottling House listing*).

No direct documentary evidence has surfaced for a bottler with the initials "**J E**" in historical records from Bloomington. But a soda factory operator with these initials *has* been documented for Metamora. A soda-factory operator there named Joseph Erl was recorded in the 1870 USC ("Soda Water Maker"), the 1880 USC ("keeps Soda Factory"), and the 1879–80 Illinois Industrial Census. In the 1879–80 IIC, Erl was listed as operating the factory, with four adult male employees, full-time eight months a year (the factory was idle for the remaining four months). His workers were paid from $1.25 to $1.75 per day. He paid $550 in wages and $2,000 for bottling materials and supplies during the census year, and estimated the annual retail value of his bottled beverages at $3,100.

There was only one known soda-bottling factory in Bloomington during the 1840–80 period of our study, at the corner of Madison and Mulberry streets. J. E.'s *Bottling House* in Bloomington probably operated from that manufactory during the only time the building was not being used by other soda water manufacturers (*see the Bloomington Bottling House entry*).

The Bloomington soda factory would only have been available for use by Joseph Erl during the later 1860s (ca 1865–68), before he left (ca. 1869) to establish his new bottling works in Metamora.

Also, during ca. 1865–66 (before relocating to Bloomington), Erl may have been one of the partners in the *E & Z* bottling works at Pekin (with Gottleib Zerwekh: *see E & Z Pekin listing*). Pekin is situated just 20 miles southwest of Metamora.

Figure 818

MINOT

Clarke's/Compound Mandrake/Bitters

There is not now, and never has been, an Illinois town named "Minot" (Adams 1989). Nevertheless, publications on embossed bitters bottles (e.g., Ring 1980:135; Ring and Ham 1998:158) persist in mistakenly attributing an oval, aqua, late-1870s bottle embossed *Clarke's Compound Mandrake Bitters* to this fictional town. The closest (similar-sounding) Illinois town is Minonk, in Woodford County. But there are several towns actually named "Minot" in the eastern United States—including one in Maine, one in Massachusetts, one in Mississippi, and one in North Dakota.

The Reverend Walter Clarke was a Baptist minister in Minot, ME, who advertised his patent-medicine concoctions in Baptist publications. His first ads in the 1879 and 1880 editions of *The Freewill Baptist Register* touted the benefits of Clarke's European Cough Remedy, Clarke's Rheumatic Elixir, Clarke's Excelsior Panacea, Clarke's Family Vegetable Pills,…and *Clarke's Compound Mandrake Bitters*.

His first full-page ad, in *The Freewill Baptist Register* issue XLVII for 1879, refers to his bitters as "a new remedy for bilious and liver complaints...prepared after long study and careful experiment."

Clarke's Compound Mandrake Bitters,

... are composed of some of the most effective remedies modern science has yet been able to produce from the vegetable world, two of the most powerful being extracted from MANDRAKE and LEPTANDRIN, which...form one of the most powerful remedies for Bilious and Liver disorders, as it certainly is one of the greatest BLOOD PURIFIERS ever compounded. Many diseases...proceed from a morbid and disordered condition of the Liver. The whole class of what are called Skin Diseases, such as Eruptions, Boils, Blotches, Pimples, &c., are simply indications of impurity of the Blood; and Foul-Stomach, Sick Headache, Dizziness, Dysentery, Diarrhoea and Colic, and many cases of Dyspepsia, have the same origin, and will speedily yield to the judicious use of the
COMPOUND MANDRAKE BITTERS.

MONEE
(Will County)

1860 census: 900
1870 census: 600
1880 census: 500

Jacob and Adam Vatter

Soda ~ Mineral Water:

(1a) **J. & A. VETTER / MONEE //** [aqua]
(*heel*): **A & D. H. C.**
Note: On this first embossed Vatter bottle, their last name is misspelled.

»SB w/ keyed hinge mold; shouldered-blob-top soda. *[Figure 819]*

(1b) **J. & A. VATTER / MONEE //** [aqua]
(*heel*): **A & D. H. C.**
Note: Corrected spelling of embossed name.

»SB w/ keyed hinge mold; shouldered-blob-top soda. *[Figure 820]*

Adam Vatter Jr. and Charles Merz

Soda ~ Mineral Water:

(1) **VATTER & MERZ / MONEE / ILLS //** [aqua]
(*heel*): **A & D. H. C.**

»SB w/ keyed hinge mold; shouldered-blob-top soda. *[Figure 821]*

Figure 819

Figure 820

Figure 821

Adam Vatter Jr.

Soda ~ Mineral Water:

(1) **A. VATTER JR / MONEE ILLS //**   *[amber]*
   (base): **A.V.**

   »SB w/ post mold; quart blob-top soda; embossed in circular slug plate; 11½" tall. *[Figure 822]*

Adam Jr.'s uncle and father, Jacob and Adam Vetter, Sr., began the Vatter bottling business in Monee near the end of the 1860s, and they were listed there as bottlers in the 1870 USC. After they first came to Monee in 1855, the two brothers were carpenters (1860 USC). During his early years there, Adam Vatter, Sr., "gave most of his attention to the erection of churches; nearly all of the German churches in this, Greengarden, Peotone, and Crete Townships are works of his" (Le Baron 1878b:573).

The 1878 *History of Will County* (Le Baron 1878b:902) described Adam Vatter, Jr., as an "insurance and lightning rod agent, and partner in the firm of Vatter & Merz, pop manufacturers, Monee; one of the early settlers; was born in Germany March 11, 1852; came to the United States in 1855, and to this state and settled in Monee with his father's family in the same year." Fraternal-order records indicate that Adam Vatter and Charles Merz were two of the five charter members of Monee Lodge #660 of the I.O.O.F. on April 8, 1878.

In an 1878 listing of real estate owners in Will County (Le Baron 1878b:945), Adam, Jr. was listed as a "capitalist," and his father was listed as "Vatters (sic), Adam, factory." So his father may have still owned the bottling works that Jr. was then operating with Charles Merz.

Adam Vatter, Jr., carried on the bottling business after the break-up of the Vatter and Merz partnership. The short partnership appears to have lasted for only a season or two: the **A. VATTER JR** quart bottle listed above dates to the late 1870s or early 1880s.

Figure 822

MORRIS
(Grundy County)

1860 census: 2,100
1870 census: 3,100
1880 census: 3,500

Dr. David LeRoy

Medicine:

(1) (side): **D^R LE ROY'S // ANTIDOTE / TO MALARIA //**   [aqua]
(side): **MORRIS ILL // //**
Note: There are 2 embossed dots beneath the superscript **R** *in* **D^R**.

»OP; beveled-edge rectangle; 3 indented panels; double-ring collar; 6 3/16" tall. *[Figure 823]*

(2) (side): **DR LEROY'S // COUGH / SYRUP //**   [aqua]
(side): **MORRIS. ILL. //**
Note: There are 2 embossed dots beneath the superscript **R** *in* **D^R**.

»OP; beveled-edge rectangle; 3 indented panels; double-ring collar; 6 7/8" tall. *[Figure 824]*

Dr. David LeRoy was an early Grundy County physician who first came to Morris to establish a medical practice about 1855 (Palmer, *ILGenWeb* project, http://grundycountyil.org/history.

Figure 823

Figure 824

medical.php, 2009). Technologically and stylistically, the listed medicine bottles were produced and used during the late 1850s (since both are open pontiled). Their function is obvious from the embossed contents-description. Dr. LeRoy's practice in Morris was essentially limited to the late 1850s. He was listed as a Morris physician in the 1864–65 ISD, but he enlisted as a surgeon in the 91st Illinois Infantry Regiment on August 15, 1862, and was not mustered out until July 12, 1865 (Vance 1886:314).

When Dr. LeRoy returned to Morris after the Civil War, he apparently became a merchant (Palmer, *ILGenWeb* project, http://grundycountyil.org/history.medical.php, 2009)—perhaps he sold patent medicines during his merchant years, but no postwar medicines are known that were embossed with his name.

MT. VERNON
(Jefferson County)

1850 census: 400
1860 census: 700
1870 census: 1,200
1880 census: 2,300

Dr. Charles M. Lee

Medicine:

(1) (side): **DR. LEE'S // RHEUMATIC / LINEMENT //**   [aqua]
   (side): **MT. VERNON // // //**
   Note: "**LINEMENT**" *is misspelled in the embossing.*

»OP; beveled-edge rectangle; 3 sunken panels; 6" tall.
[Figure 825]

---❖---

An advertisement published in the Mt. Vernon, IL newspaper, *The Jeffersonian* (May 1852) stated that Dr. "H. M." Lee, oculist "would say to those afflicted with blindness, sore, weak, or inflamed eyes...[that they] would do well to call on Dr. Lee at his residence at Mrs. Martha L. Barker's, 15 miles west of Shawneetown on the road leading to McLeansboro. He will also attend to the curing of *rheumatism*, swelled joints...etc." From the bottle's style and technological attributes (open pontiled), it was produced and used during the 1850s when Dr. Lee lived in the Mt. Vernon area and sold medicines there.

Figure 825

However, the ad was in error as to the doctor's initials. Dr. Charles M. Lee, "Oculist and Aurist," moved his practice to Quincy, IL, in 1859. He is listed there in the 1860 USC as a 60-year-old physician born in New Jersey. Dr. Lee's numerous Quincy newspaper ads indicate that he moved there in 1859, and make it clear that "The Doctor also treats Rheumatism, Cancers, Scrofula, and all Chronic Diseases" (*Quincy Daily Herald*, Aug. 16, 1862, p. 1). The connection is further enriched by the fact that the only known example of Dr. Lee's embossed Mt. Vernon bottle was excavated from a privy in Quincy.

To date, there are no known examples of Dr. Lee medicines embossed from Quincy. His newspaper ads indicate that his Quincy practice lasted from 1859 to ca. 1868.

MURPHYSBORO
(Jackson County)

1870 census: (Twp = 3,460)
1880 census: 2,100

Frederick A. Kuehle

Soda ~ Mineral Water:

(1) **A. F. KUEHLE / MURPHYSBORO / ILLS**   [aqua]
(*encircling base*): **K H & G Z O**

»SB w/ post mold; shouldered-blob-top soda. *[Figure 826]*

A short 1894 biographical essay on Kuehle's son, Frederick A. C. Kuehle, (Biographical Publishing Co. 1894b:623) indicates his father was born in Germany, where he was a shoemaker. "At the age of seventeen he crossed the Atlantic to New Orleans, and after a short time went to Cape Girardeau, Mo., and thence to Cairo, Ill., where he engaged in merchandising. In 1870 he came to Murphysboro, and as a member of the firm of *Kuehle & Son* still carries on mercantile pursuits." In 1893, "the Kuehle family lived next door to their store *[also the post office, as Kuehle was postmaster at the time]* at 1209 walnut Street" (Fishback 1982:31).

Kuehle's "mercantile pursuits" were probably undertaken in the form of a general store, which sometime during the 1870s included bottling his own brand of soda or mineral water. From the bottle's rarity (one example known) and style, it likely dates to 1875 or later. Sometime in the late 1870s the bottling works were taken over by Ed Hayes, earlier of DuQuoin *(1880 IIC—see the Hayes DuQuoin listing)*.

An interesting aspect of his embossed bottle is the glassmaker's mark embossed on the base. **KH&G ZO** was the maker's mark for Kearns, Herdman & Gorsuch (**KH&G**) of Zanesville, Ohio (**ZO**). This particular partnership of the Kearns glass companies was in business from 1868 to

Figure 826

1885/86 (Lockhart et al. 2008:3, 5–6). In the 1840 to 1880 scope of the present study, the listed Murphysboro bottle is one of only two known examples of an embossed **KH&G ZO** maker's mark (*see the John M. Berry Prentice listing*).

There is some genealogical confusion as to Kuehle Sr.'s first and middle names. The biographical summary refers to him as "Fred A." and some genealogical websites identify him as Frederich August, indicating he was born in Hanover, Prussia, in 1838, was married in Cairo in 1862, and died in Murphysboro in 1921. However, a Kuehle who arrived in Cape Girardeau at about the right time is also identified as "August Ferdinand." This is probably the same man, but the confusion is noted because the listed bottle is embossed "A. F." Kuehle. Perhaps he was known as "Fred" even though it was his middle name. If Fred was his first name, then the embossing on the bottle is a glassmaker's error.

NAPERVILLE

(Du Page County)

1860 census: 2,600
1870 census: 1,700
1880 census: 2,100

Aylmer Keith

Medicine:

(1) **A. KIETH'S / PERSIAN** // *(side):* **LINIMENT // NAPERVILLE / ILL^S** // // [aqua]

»OP and IP; beveled-edge rectangle; 4 sunken panels; tapered-collar and rolled-lip varieties; 4 3/4" tall. *[Figure 827]*

Aylmer Keith was born and raised in Oneida County, NY, where in his adult years he operated a grocery store and was involved in village politics. He moved to Illinois with his family and settled in Naperville ca. 1840, when he was about 38 years old. He was appointed Du Page County clerk in 1841, and was involved in production of the local newspaper ca. 1852–54 (Richmond and Vallette 1857:51). "While a resident of Naperville, he was a merchant, druggist, and banker. He was also one of the original investors in the plank road that was built from Naperville to Chicago.... He was appointed as Director of the Merchants and Mechanics Bank of Chicago in 1851.... He died November 15, 1855, in Naperville" (*Euclid Lodge No. 65 Trestleboard*, Vol. 1, No. 1, 2008:2—*see Figure 828*).

Aylmer Keith was listed as a druggist in Naperville in an advertisement in the February 1847 *Illinois State Journal* (Springfield). He was also listed as a 48-year-old Naperville druggist in the 1850 USC.

In the early 1850s, Keith advertised not only his *Persian Liniment* but also a *German Itch Ointment*. In the same early Naperville privy pits containing the *Persian Liniment* bottles, Naperville resident and historical bottle digger Tom Majewski (see Richmond and Valette 2006) has found examples of a small (2 1/8" tall x 1 1/2" dia.), open-pontiled but unembossed wide-mouthed, high-shoulder, eight-paneled salve-style jar with a rolled lip which may well have contained Keith's *German Itch Ointment* (T. Majewski, personal communication 2009).

AYLMER KEITH
EUCLID'S FIRST WORSHIPFUL MASTER

Figure 828

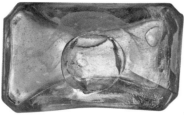

Figure 827

Keith's Persian Liniment was apparently advertised in three sizes—3, 6, and 12 ounces (S. Garrow, personal communication, 2008)—but thus far only the 4 3/4" size has been documented. The larger size(s) were likely sold as veterinary medicine, while the smaller size(s) were marketed for people. Keith sold his medicines as part of two known early-1850s partnerships. *A. & C. W. Keith* are listed in Naperville in the 1854 ISD as sellers of "Drugs and Medicines," and a later 1854 *Woodstock Sentinel* newspaper ad indicated that the "general agents" for his liniment were *Keith & Kimball* of Naperville:

LOOK HERE!
KEITH'S PERSIAN
LINIMENT
Is the Best Remedy known, for
BOTH MAN AND BEAST.

For particulars, call at any Druggist's and get the "American Horse Doctor" gratis, and don't forget to buy a bottle of the liniment.
General agents for the U.S.,

KEITH & KIMBALL,
Naperville, Ill.

Keith's Persian Liniment was also advertised in the August 5, 1854, *Weekly Chicago Democrat*, sold "in large bottles" by *C. W. Keith & Co.* (Baldwin 1973). By this time, Aylmer Keith may have retired from the company due to failing health. A few years after Aylmer Keith's death (ca. 1858–59), two of his nephews (Charles W. and William F.), one of them a former partner in Keith's earlier medicine business, formed a "drug business" partnership for a year at Joliet (Wm. F. Keith, Will County IL GenWeb biography). They may have continued to sell Keith's medicines there just prior to the Civil War.

Morris B. Powell and Charles W. Stutenroth

Bitters:

- (1) *(upper shoulder):* **PAT. APPLIED FOR // FAVORITE BITTERS /**   [amber]
  POWELL & STUTENROTH

    »SB w/ keyed hinge mold; tenpin-shape; ribbed, "barber-pole" spiral embossing, upper left to lower right; tapered collar; 9 1/4" tall. *[Figure 829, left]*

- (2) *(upper shoulder):* **PATD JULY 28 1868 // PEOPLES FAVORITE BITTERS /**   [amber]
  POWELL & STUTENROTH

    »SB w/ keyed hinge mold; taller, narrower tenpin-shape than #1; ribbed "barber-pole" spiral embossing, upper left to lower right; tapered collar; 10 5/8" tall. *[Figure 829, right]*

This highly ornate bottle was designed by Morris B. Powell and Charles W. Stutenroth of Naperville, IL, who described the design as "adapting the line of beauty to any glass or any other kind of bottle, combining elegance of appearance with great durability" (Design Patent #3,141, awarded July 28, 1868). Bottle #1 was used by *Powell & Stutenroth* before the patent was granted, and was embossed as shown on the patent application *(see Figure 830)*. Bottle #2 was used after the patent was granted, with the word "PEOPLES" added to the embossing.

According to a printer's block for the paper label or an ad, documented by Ring and Ham (2004:90), described the manufacturer's claims regarding the purported health benefits of *People's Favorite "Tonic" Bitters*: "Invariably Cures Loss of Appetite, Flatulency, Dizziness, Jaundice, Nausea, and Dyspepsia. Also Biliousness and Bilious Colic. It is a specific for Nervous and Bilious Headache."

Figure 829

An extensive Naperville newspaper ad for *Powell & Stuttenroth's* bitters from April 21, 1869, has been reproduced online at the *Antique Bottle and Glass Collector* website:

PEOPLE'S FAVORITE BITTERS!

A medicine so potent, palatable and pure that wherever known it becomes at once the Favorite of the people. In all cases of Dyspepsia, Liver Complaint, Headache, Loss of Appetite, Humors in the Blood, Blotches and Pimples on the Face, Weakness of the Kidneys, General Debility, Nervous Complaints, Impurities of the Blood, Chills, Fevers, and Diarrhea, The People's Favorite Bitters should be used. The unimpeachable testimony of ladies, farmers, mechanics, commercial and professional men who have practically tested the mixture of The People's Favorite Bitters, prove it to be beyond question, the best medicinal compound for the above diseases now offered to the public.

If a reasonable attention to diet be given, and The People's Favorite Bitters be taken strictly according to the directions, and no relief obtained in any or all the ills mentioned above, **THE PRICE OF THE BITTERS WILL BE REFUNDED** by the proprietors, or any of their authorized agents.

The reason The People's Favorite Bitters have met with much unprecedented success, and are now so generally and highly esteemed, is because they **ASSIST** nature, instead of exhausting her; they give tone to the stomach and the digestive organs, enrich the blood, and tend to build up rather than tear down; in short they import vigor to the entire frame, and help the system to throw off disease.

Figure 830

The following are the ingredients of **The People's Favorite Bitters**: *Sarsaparilla, Dandelion, Wild Cherry, Buchu, Cubeba, Orange Peel, Anise, Gentian, Columbo and Sassafras [emphasis ours]*. Any one, at all acquainted with medicines, will know at once that no better or more wholesome Bitters could be made.

One trial will convince the most skeptical that The People's Favorite Bitters are all that they are recommended. For **LADIES** they are excellent; for **TRAVELERS** invaluable, for mechanics, farmers, persons of sedentary habits; for professional men, overtaxing the brain at the expense of the body, The People's Favorite Bitters are a necessity. During sudden changes in the weather, in time of general sickness, these Bitters can be taken with advantage. At any time they are useful, and should be constantly on hand. If you want to save money, preserve your health, do the most good, and get the greatest happiness, use The People's Favorite Bitters.

Price, $1.00 per Bottle
POWELL & STUTENROTH,
Proprietors and Manufacturers,
Naperville, **ILL.**

People's Favorite Bitters seems to have been marketed only during the late 1860s and early 1870s. By 1875 (Chicago CD), Charles W. Stutenroth is listed alone in Chicago as a saloon proprietor at 44th and Halsted.

John Stenger

Ale ~ Beer:

(1) **JOHN STENGER // NAPERVILLE IN_** [amber]
(base, around large central "dimple"): **W McCully & Co Pittsburgh PA**
*Note: The "***IN_***" embossing is an apparent mold flaw or mold-maker's error.
It should read "***NAPERVILLE ILL***".*

»SB; 3 piece mold quart ale; shoulder and base embossed; ring-neck tapered collar; 10" tall. Bottle known only from restored fragments. *[Figure 831]*

(2) **JOHN STENGER'S / LAGER BIER / NAPERVILLE /** *[aqua]*
ILL // THIS (arched) **/ BOTTLE / NOT TO BE / SOLD**

»SB w/ post mold; slope-shouldered blob-top beer; front embossing in circular slug plate; 9½" tall. Bottle known only from restored fragments. *[Figure 832]*

Peter Stenger started the Stenger Naperville Brewery ca. 1848–1849, soon after immigrating from Germany, and settled there with his family. Two of his sons, John and Nicholas, took over the business in the early 1850s and ran it together until Nicholas died in 1867 (see 1865 letterhead: *Figure 833*). John then managed the brewery alone until he sold it to the Chicago firm of *Henn & Gabler* in 1893, and *H & G* closed it down soon thereafter. The brewery was located on the north side of Franklin Avenue, between Webster and Main streets (*Du Page County Historical Society Research by Beverly Stenger and Lester Schrader, Naperville*). A detailed description of the brewery itself, and its operations, is available at *NaperAle.com*.

During the heyday of its operation, the Stenger Brewery was known for its distinctive and famous German-style "superior lager beer." John and Nicholas Stenger built up an $80,000/year business during the late 1850s and early 1860s making true lager beer in the "Old Country" way. The beer was primarily brewed and shipped in barrels. They hauled their barreled beer to saloons as far away as Elgin, Ottawa, and Chicago, and according to Naperville historian Les Schrader, the deliveries were made overnight for fear that the barrels would burst if they were exposed to the sun for too long a time during shipment.

Figure 831

Figure 832

From the stylistic details, embossed Pittsburgh glassmaker's mark, and manufacturing technology of the embossed Stenger ale-style bottles (*see bottle #1*), these appear to have been manufactured and used by John Stenger during the late 1860s, soon after he took over sole management of brewery operations. They may have been produced to advertise the new firm name and management in 1867–68, after Nicholas Stenger passed away.

From the manufacturing technology and style details of the embossed aqua beer bottle (*see bottle #2*), it appears to have been manufactured and used during the mid-1870s. At this time, the advent of pasteurization made bottled-beer sales more feasible. Stylistically, removable embossed slug plates for use with bottle molds also first appeared at this time (*see introduction, this volume*). A Naperville newspaper archives search by Naperville Heritage Society Research Associate Brian Ogg has demonstrated that a local "middleman" bottler named Jacob Keller advertised in 1872–73 that he was selling John Stenger's popular Bavarian lager beer in glass bottles (see 2008 Naperville Community Television production "Two Brothers, One Beer and the American Dream"). Perhaps Keller or another merchant was still doing so using slug-plate embossed bottles ca. 1875–76 when bottle #2 was produced and used.

Figure 833

NAPLES
(Scott County)

1850 census: xxx
1850 census: xxx

John Handy

Stone-Bottled Medicine?:

(1a) **HANDY'S / COMPOUND** *[yellow, brown, and black glazes]*

»Hand-turned quart stoneware bottle; embossed above sharp shoulder; blob top; ca. 9 1/4–9 1/2" tall. *[Figure 834, left and center]*

(1b) **HANDY'S / COMPOUND / J.N.EBEY // WINCHESTER** *[brown glaze]*

»Hand-turned quart stoneware bottle; embossed above and below sharp shoulder; blob top; 9 1/2" tall. *[Figure 834, right]*

The Thomas and John Handy were grocers in west-central Illinois from the 1850s to the 1870s. John Handy is listed in the 1870 USC as a 30-year-old grocer in Naples, IL; the same census lists Thomas as a 40-year-old grocer living in Springfield, IL. The two men may have been brothers, but that assumption may be faulty since the 1870 USC lists Thomas' place of origin as England and John's as Ireland.

Figure 834

In any case, all known examples of the **HANDY'S** stamped stoneware bottles have been found in the Naples area, so they were probably sold from the grocery there. This is supported by the fact that—from the one stamped example—the bottles were made at the John Neff Ebey pottery works in nearby Winchester (about 10 miles SE of Naples). J. N. Ebey manufactured stoneware in the Manchester and then Winchester areas during the 1840s and 1850s (Mounce 1989), before relocating his pottery works to White Hall in Greene County in 1864. From the technological and style attributes of the listed bottles, they appear to have been produced in the 1850s or early 1860s.

The use of the word "Compound" in the product name would seem to suggest that the bottles contained some kind of medicinal mixture, but no advertising (or paper-labeled bottles) have been located to confirm this possibility.

NELSON
(Lee County)
1870 census: 600

James J. Allison

Soda ~ Pop Beer:

(1) **J. J. ALLISON / NELSON. ILL.** // *(heel)*: **DeS. G. Co.**   [aqua]

»SB w/ post mold; unusually tall, high-shouldered, cylindrical soda w/ flattened blob top; 7 1/4" tall. Bottle style similar to that of Matthews Gravitating Stopper bottles (*see the Bower Olney listing*). [Figure 835]

———❖———

James Allison was born in Scotland in 1830 and immigrated to the US in 1851 (1870 USC). He lived and farmed in Wisconsin for about 10 years, but had been trained as a mason, carpenter, and joiner (from his daughters historical recollections), and he later moved to the Sterling area in northern Illinois in 1862 to work on a bridge being constructed across the Rock River. He was listed in the 1870 USC as a resident of Rock Falls in Whiteside County, Illinois.

During the 1870s he moved eight miles upriver to the town of Nelson, in western Lee County. He moved back to Rock Falls after his wife's death in Nelson in 1879, where he was listed as a "pop beer" manufacturer in Rock Falls by the 1880 USC. He continued as a retail pop bottler until after the turn of the century (1900 and 1910 USC); he died in Rock Falls in 1912.

Since the *DeSteiger Glass Co.* did not begin to manufacture glass bottles until 1877 (*see their LaSalle listing*), and Alison left Nelson in 1879, it looks like he began his bottling operation there in 1877 or 1878, initially using embossed bottles for his pop-beer manufactory. California Pop or "pop beer" was patented by Charles C. Haley of Troy, New York, on October 29, 1872 (U.S. Patent #132,574). It was similar to spruce beer and other popular "small beers" of the nineteenth century, but combined oil of spruce, sassafras, and wintergreen with ginger root, hops, malt, and "pure spirits."

No later embossed bottles are known from his Rock Falls bottling works, so he likely used unembossed, paper-labeled bottles in later years

Figure 835

NEW ATHENS
(St. Clair County)
1880 census: 700

Peter Baumann

Medicine:

> (1) **BAUMANN'S, / ARCANUM // LINIMENT / PATENTED AUGUST 13 / 1867 // // //**   [aqua]
> »SB w/ keyed hinge mold; beveled-edge rectangle; 4 sunken panels; thick flared-ring collar; 4 1/2" tall. There is also a larger size (5 1/4" tall) with the same embossing. *[Figure 836]*

Peter Baumann, Jr., was a long-time merchant in New Athens, according to data in the 1860, 1870, and 1880 USC. In 1867, he patented his "Improved Liniment" in the name of *P. Baumann & Brothers* (Patent # 67,706). According to the patent description: "The liniment I shall describe has been found very efficacious in curing strains, galls, scratches, and other diseases in horses, and it may be used with success in many inflammatory diseases of the human body.... The ingredients composing the mixture consist of spirits camphor, petroleum, oil, spike, lavender, oil of turpentine, ammonia water, oil of linseed, liquid pitch."

From the hinge-mold technology used in manufacturing the bottles, *Baumann's Liniment* appears to have been put-up and sold during the late 1860s and early 1870s.

Figure 836

NOKOMIS
(Montgomery County)

1870 census: 900
1880 census: 1,100

Dennis P. Brophy

Bitters:

(1) **BROPHY'S BITTERS** *(within crescent moon above large embossed 5-pointed star)* /   *[aqua]*
    TRADE MARK /
    (heel): **NOKOMIS / ILLINOIS**

»SB; beveled-edge square case bottle; short neck; squared, flared-ring collar; 7 1/2" tall. *[Figure 837]*

Dennis P. Brophy, the manufacturer of *Brophy's Bitters,* is listed as a Nokomis druggist in the 1870 USC. Brophy's biographical sketch in the 1874 Montgomery County Atlas states that:

> He also has done considerable business in the drug and medicine line, which he has but recently discontinued. During his experience in compounding medicines he succeeded in making a new preparation of bitters, having a sedative character. It is very highly spoken of as an anti-febrile and an anti-dyspeptic. This preparation he puts up himself, from the best and purest drugs, and keeps a supply on hand for the benefit of the afflicted. [Brink, McCormick & Co. 1874b:55]

A woodcut image of Brophy's store appears on page 105 of the Atlas (*Figure 838*). A sign painted on the false-front second story façade of the building (above the post office) reads: "Depot of Brophy's Celebrated Domestic Tonic Bitters." A trademark drawing of a down-turned crescent over a star, with the words "Brophy's Bitters" in the crescent, is drawn below. Below the trademark symbol is "Dennis P. Brophy, Prop.r"

From the manufacturing technology and style of the embossed bottle listed, the bottles appear to have been manufactured and used ca. 1870–75.

Figure 837

Figure 838

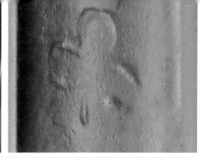

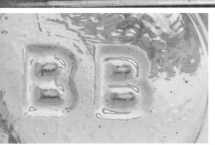

O'FALLON
(St. Clair County)
1880 census: 900

John Bayet and Williams

Soda ~ Mineral Water:

(1) **BAYET & WILLIAMS / O.FALLON / ILL.** [amber]
»SB w/ post mold; oval-blob-top soda. *[Figure 839]*

Figure 839

Thomas Miller (2007a:57), who has conducted an extensive search of St. Clair County period-documents, has been unable to find any mention of this firm. The style and manufacturing technology of the embossed bottles indicate that this apparently short-lived bottling business was in operation during a season or two in the mid- or late 1870s.

A John Bayet from the O'Fallon area served in the Civil War (Miller 2007a:57). In 1887, the U.S. Patent Office granted a patent #375,527 to John Bayet of O'Fallon, IL, for a ratchet and lever mechanism (specific use not stated). The 1891–92 *St. Clair County Gazetteer* and the *1900* USC for O'Fallon listed John Bayet as a coal miner.

Bernard Bischoff

Soda ~ Mineral Water:

(1) **B. BISCHOFF / O'FALLON / ILL ^{s} //** [aqua]
(*rev. heel*): **A & D. H. C**
Note: There are two embossed dots beneath the raised S in ILLS.

»SB w/ keyed hinge mold; oval-blob-top soda. [Figure 840]

(2) **B. BISCHOFF / O'FALLON / ILL.** [aqua]
(*base*): **W-B**

»SB w/ heel mold; oval-blob-top soda. [No photo: side embossing and bottle shape closely similar to bottle #3 below (see Miller 2007a:74)]

(3) **B. BISCHOFF / O.FALLON / ILL.** [aqua]
(*base*): **B B**

»SB w/ heel mold; oval-blob-top soda. [Figure 841]

❖

Bernard Bischoff married Anastasia Chenot on April 15, 1866 in St. Louis. Her 1933 obituary (*O'Fallon Progress* October 19; see also Miller 2007a:72) stated that Bernard Bischoff "was a soda water manufacturer in Collinsville until his death which occurred Sept. 24, 1915." Bischoff's own obituary (*The Collinsville Advertiser*, Oct. 16, 1915) indicated that he was born in Germany in 1829. "When a young man he came to America and settled in the vicinity of Belleville, where he became familiar with the soda water business, and later on established that business in O'Fallon and about 35 years ago came to Collinsville and engaged in the making and selling of soft drinks. He met with great success and became wealthy." The business chronology in the obituary indicates that Bischoff operated his O'Fallon bottling business prior to ca. 1880, when he relocated the business to Collinsville.

The style and manufacturing technology of the soda bottles listed indicate that Bischoff's O'Fallon soda bottling business likely began in the late 1860s. He was listed as a soda water manufacturer there in the 1870 UISC.

The 1874 *Illustrated Historical Atlas of St. Clair County* (Warner

Figure 840

Figure 841

and Beers 1874) contained an advertisement stating that Bischoff was a bottler of ale, soda, cider, and mineral water.

Bischoff was still listed as bottling mineral water in O'Fallon by the 1879–80 IIC, which recorded that he operated five months a year full-time and seven months a year half-time. It listed the value of material used to produce his product during that 12-month period at $500 and the sales value of his bottled mineral water at $2,500. No wages were listed, so apparently Bischoff worked alone at his bottling business at the time. The only indication of a possible temporary partner in Bischoff's bottling operation is the **W-B** base embossing on bottle #2. The **W** could possibly be the surname initial of a partner—perhaps Mr. Williams from the short-lived *Bayett & Williams* partnership in O'Fallon (*see their listing, this chapter*).

The earliest (blob-top) Bischoff bottle embossed **COLLINSVILLE,** a town located in adjoining Madison County, likely dates to ca. 1880–81 (*see the Bischoff Collinsville listing*). In later years (ca. mid- to-late 1880s), his son Henry took over management of the Bischoff Collinsville bottling business.

OLNEY
(Richland County)

1860 census: 1,500
1870 census: 2,700
1880 census: 3,500

William Bower

Soda ~ Mineral Water:

(1) **Wm BOWER, / OLNEY, ILLS. //**   [aqua]
(vertical down): **THIS BOTTLE / IS NEVER SOLD**
(base): (outer ring) **GRAVITATING STOPPER / MADE BY /**
(inner ring): **JOHN MATTHEWS / NEW YORK /**
(center): **PAT. / OCT 11 / 1864**

»SB soda w/ heel mold; Matthews-style flat-topped blob soda with short neck and Matthews gravitating internal-glass-stopper closure. *[Figure 842]*

William Bower was born in Olney on May 21, 1842. He began to make his own way in business when he was only 14, trying his hand at teaching, marble-cutting, watchmaking, and tinner's trades before enlisting in the Eighth Illinois Volunteer Infantry early in 1861 to fight in the Civil War (Battey & Co. 1884:730–731; Bowen & Co. 1909:219–221).

He returned to Olney in fall 1863, and went into the druggist business with his

Figure 842

brother-in-law that December. He bought out his partner's interest 15 months later "and, by industry and energy, has built up a large business. During the last seven years he has been extensively engaged in the manufacture of cigars.... He carries a stock of from $12,000 to $15,000 of goods, and his annual sales, in all departments, amount to from $50,000 to $60,000 (Battey & Co. 1884:731). Bower continued to operate his successful pharmacy business ("a substantial two-story structure, twenty-five by one hundred and sixty-five feet, running from Main to Market streets") until well after the turn of the twentieth century, finally turning the operation over to his son, Ernst, in 1912.

Bower's embossed Matthews gravitating stopper-style bottle was likely manufactured and used soon after he bought out his partner and took over the local drug store business on his own (during the later 1860s or early 1870s) to sell mineral water or flavored soda water in his pharmacy. Even later in his career, Bower was noted for his early introduction of new flavored sodas and confectionary drinks at his drug store: "Fruit lemonade originated at Bower's, as did the Coca Cola drinks '76' and '61'" (King 1945).

OQUAWKA

(Henderson County)

1850 census: 800
1860 census: 1,600
1870 census: 1,400
1880 census: 800

Dr. Andrew C. McDill

Medicine:

(1) **D**^R **A. C. M**^C**DILL / VERMIFUGE**   *[aqua]*
Note: there are 2 embossed dots beneath superscript **R** *in* **D**^R.

»OP; sharp-shouldered cylinder; rolled lip; 4 3/4" tall, 7/8" diameter. *[Figure 843]*

Dr. Andrew C. McDill was an Ohio-born physician, living and practicing medicine in Ohio in 1840 (USC). He was still listed as an Ohio physician in the 1850 USC, at which time he was 41 years old. But there is some evidence he may have practiced in Illinois for a time during the 1840s: an 1883 biographical sketch of physician John H. McDill indicated he "commenced reading medicine in the office of his brother, A. C. McDill, of Monmouth, Ill. before graduating from medical school in 1845; in the same year he came to Henderson County, Illinois, where he commenced the practice of medicine" (Union Publishing Co. 1883).

During the 1850s, Andrew McDill moved permanently from Ohio to Henderson County along the Mississippi River in northwestern Illinois, where he was listed in the 1860 USC as living and practicing medicine in the rural Oquawka area. He may have come to Henderson County to join his brother in medical practice there. In a March 1855 issue of the *Golden Age*, T. W. McDill (another brother?) was listed as a patent-medicine distributor in Oquawka. By 1870 (USC), Andrew McDill had moved to nearby Biggsville, where he was retired from medicine and had taken up farming.

From the stylistic and technological attributes of the embossed bottle listed, it was likely manufactured and used during the 1850s. Thus the McDill brothers may well have distributed Andrew's vermifuge both in Butler, OH, and in Henderson County, IL.

Figure 843

OTTAWA
(La Salle County)

1850 census: 3,200
1860 census: 6,500
1870 census: 7,700
1880 census: 7,800

Dr. William W. Cavarly

Bitters:

(1) **DR CAVARLY'S / AGUE. BITTERS / OTTAWA. ILL.** [aqua]
  Note: there are 2 embossed dots beneath superscript **R** in **D^R** and the embossed **S** in **BITTERS** extends partly off the heel base.

»OP; round; shouldered broad cylinder with thickened, flared collar; 2 1/2" in diameter, 5" tall. [Figure 844]

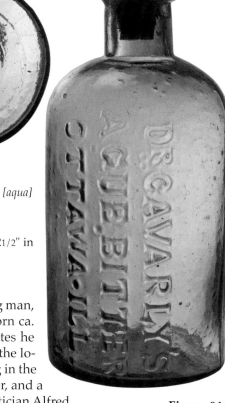

Figure 844

William Cavarly was a local physician who died while still a young man, and thus had a very short professional career in Ottawa. He was born ca. 1822 according to the 1850 USC, and according to genealogy websites he died in August 1855, when he was in his early 30s. He was raised in the local area but likely left for a time to receive medical training, returning in the early 1850s with a young wife (Juliana Cavarly—nine years his junior, and a distant relation). Her father was prominent regional lawyer and politician Alfred W. Cavarly, who also moved up to Ottawa from Carrollton in 1853, perhaps to remain near his daughter.

Dr. Cavarly was inducted into the local Masonic Lodge (in 1854) soon after his marriage (Milligan 1907:139) and died about a year later.

Bitters such as Dr. Cavarly's were sold widely as cures for fever and ague during the late 1850s and early 1860s (*see Van Deusen's Ague Bitters Springfield listing*). From the style and technological attributes of Dr. Cavarly's bitters bottle, it was likely produced and sold for just a short time between his marriage and death, in the early-to-mid 1850s.

B. W. Coutant and Co.

Soda ~ Mineral Water:

(1) (base): **COUTANT & CO / OTTAWA ILL'S** [aqua]

»SB w/ heel mold; shouldered-blob-top soda. [Figure 845]

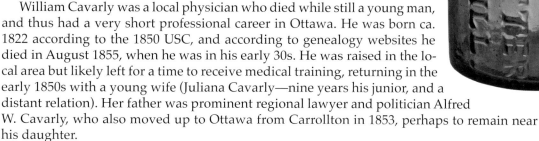

The April 1866 IRS tax assessment for LaSalle listed *B. W. Coutant & Co.* as a local soda water manufacturing business. The firm was also listed in the summary annual IRS assessment for 1866 in both Ottawa and LaSalle (these Illinois River towns are only about 15 miles apart). The "*& Co.*" partners in the *Coutant & Co.* bottling works during the late 1860s may well have been the Crotty Brothers (*see their*

Figure 845

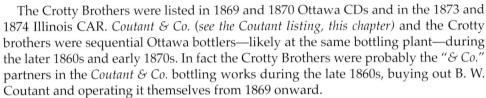

listing), who bought out Coutant in 1869 and operated the bottling works themselves for the next several years before selling their Ottawa bottling operation to Alex Finkler in the mid-1870s (*see his listing*).

Thomas and Lawrence Crotty

Soda ~ Mineral Water:

(1) (heel): **A & D. H. C.** //   [aqua]
    (base): **CROTTY BROS. OTTAWA. ILL**

»SB w/ heel mold; shouldered-blob-top soda. *[Figure 846]*

The Crotty Brothers were listed in 1869 and 1870 Ottawa CDs and in the 1873 and 1874 Illinois CAR. *Coutant & Co.* (*see the Coutant listing, this chapter*) and the Crotty brothers were sequential Ottawa bottlers—likely at the same bottling plant—during the later 1860s and early 1870s. In fact the Crotty Brothers were probably the "*& Co.*" partners in the *Coutant & Co.* bottling works during the late 1860s, buying out B. W. Coutant and operating it themselves from 1869 onward.

These two Ottawa companies are the only Illinois soda water manufacturers known to have used bottles embossed on the base only. Since early soda bottles were placed upside-down in wooden cases (to keep the corks moist), base embossing would certainly have been easily readable.

The Crotty brothers expanded their bottling operations to Freeport for a short time during the early 1870s (*see their Freeport listing*). Their bottling operation was likely bought out in the mid-1870s by Alex Finkler, who operated the Ottawa bottling works after that (*see his listing, this chapter*).

Figure 846

Alexander Finkler

Soda ~ Mineral Water:

(1) **A. FINKLER / OTTAWA. / ILL.**   [aqua]

»SB w/ post mold; high-shouldered, stubby-neck, blob-top soda; 8" tall. *[Figure 847]*

Alex and John Finkler were listed as owners of a LaSalle soda water bottling company in the 1860 and 1870 USC. The 1870 census listed them both as soda water makers in La Salle County. At the time, they were 42 and 33 years old, respectively. A third brother, Frank J. Finkler, also worked at the LaSalle bottling plant according to the 1860 USC, before leaving to serve in the Civil War.

Frank later expanded the brothers' bottling partnership to a new bottling operation in Dixon (*see his Dixon listing*), where he bottled during the late 1860s and early 1870s. At about this same time, John set up a soda bottling works in Streator (*see his Streator listing*). Alex had moved the LaSalle bottling operation to

Figure 847

Ottawa "at Riverside; P.O. Streator," according to his 1877 LaSalle County CD listing (Kett 1877a:419, 604).

The 1880 USC listed Alex and John as soda water manufacturers in Ottawa and Streator, respectively. Alex Finkler died in Ottawa in 1883, after which his son (Alexander C.), who was just 14 when his father died, carried on the Finkler soda water bottling business for 11 years, before finally selling out in 1894 (Lewis Publishing Co. 1900:437–438).

Post-1880 Hutchinson-style soda bottles embossed "**A. FINKLER**" are also known, but were manufactured too late for listing in the present study. The listed bottle has the same embossing as the later Hutchinson-style sodas (Ron Ridder rubbing), but according to Ridder's notes, it has a short-neck oval blob top and is nearly 2" taller than Hutchinson-style bottles.

Edward Y. Griggs

Medicine:

(1) **DR. JONES** / **RED** / *(embossed cloverleaf design)* / **CLOVER** / **TONIC** // //   [amber]
    E. Y. GRIGGS / **OTTAWA, ILLS.** // //

   »SB w/ post mold; beveled-edge rectangle; 4 sunken panels; horizontal front embossing & vertical rear embossing; 9" tall. *[Figure 848]*

(2a) **DR JONES** / **RED** / *(embossed cloverleaf design)* / **CLOVER TONIC** // //   [amber]
    GRIGGS & CO / **OTTAWA. ILLS.** // //

   »SB w/ post mold; beveled-edge rectangle; 4 sunken panels; horizontal front embossing & vertical rear embossing; 8 1/2" tall. *[Figure 849, top left]*

(2b) **DR JONES** / **RED** / *(embossed cloverleaf design)* / **CLOVER TONIC** // //   [amber]
    GRIGGS & CO. / **OTTAWA ILLS.** // //

   »SB w/ post mold; beveled-edge rectangle; 4 sunken panels; horizontal front embossing & vertical rear embossing; 8 3/8" tall. *[Figure 849, top right, bottom]*

Edward Young Griggs was born in Baltimore in 1818. As a young man, he and his wife lived in Springfield, OH. They moved to Ottawa, IL, in April 1849. He worked as a clerk for J. G. Nattinger until 1850, "and then opened a drug and book store in the three-story brick building where the National City Bank now stands. In 1853 he opened his drug store where he is now doing business, and has since been one of the leading merchants of the place" (Lewis Publishing Co. 1900:22–23). Ads for *Griggs Sarsaparilla* appeared in 1855 Ottawa newspaper ads and a city directory the same year, but no embossed bottles have been found, and this early Griggs product may have been bottled with paper labels only.

E. Y. Griggs was listed in the 1871 ISD as a druggist at 123 Main in Ottawa. E. Y. Griggs' son, Oakley, graduated from Michigan State University as a pharmaceutical chemist in 1875, and then worked in his father's business. In 1878 he joined the firm, and *Griggs & Co.* expanded their operations to Streator (Hoffman 1906).

From the style and technological attributes of the embossed bottles listed, *Dr. Jones Red Clover Tonic* was first bottled and sold by Griggs during the mid-to-late 1870s: bottle #1 was used in the 1875–78 period before the company became *Griggs & Co.*, and bottles #2a and #2b were used during the later 1870s (*Figure 849, top left*) and early 1880s (*Figure 849, top right, bottom*). (See Devner

Figure 848

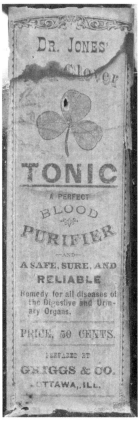

Figure 849

1968 and Baldwin 1973 for a discussions of early ads for the product.) We have been unable to locate a "Dr. Jones" in the LaSalle County area during this time, so the name may have just been used as a marketing technique. *Griggs' Red Clover* tonic was marketed primarily as an appetizer, blood purifier, and dyspepsia cure:

THE TRUE TEST.

If a man is hungry within an hour more or less after a meal he is dyspeptic, it shows his stomach is not able to dispose of what he has eaten, but to eat again, and thus impose more work, is absurdity. Take **Dr. Jones Red Clover Tonic,** which cures dyspepsia, and all stomach, liver, kidney, and bladder troubles. It is a perfect tonic, appetizer, blood purifier, a sure cure for ague and malarial diseases. Price 50 cents. [*The Bryan (Ohio) Democrat,* Aug. 30, 1882, p. 2, col. 4]

John M. Kagy

Bitters:

(1) **KAGY'S / SUPERIOR / STOMACH BITTERS** // // // //   [amber]

»SB w/ post mold around central cup-like depression; beveled-edge square "case" bottle; 4 flat panels; embossed on one side only; short neck; tapered collar; 9 1/2" tall. *[Figure 850]*

Figure 850

Two or three of these scarce bottles have been found in the Ottawa area (see e.g., Ring 1980:275), and in the wider west-central Illinois region. From their style and manufacturing technology, the bottles appear to have been made as a single order in the late 1860s and would have been used for a few years during the 1865–70 period.

John M. Kagy was born in 1834 in Fairfield County, OH, and came to Ottawa, IL, during the Civil War years. He was married in Ottawa on September 19, 1865, when he was 31 years old. His son was born there in October 1867.

The 1866–67 Ottawa CD listed John Kagy as operating a wholesale liquor business in Ottawa with his brother-in-law Jacob M. Walters (*Walters & Co.*). 1866 was the first year Kagy was involved with the firm, since the 1865 state census for Ottawa listed the firm as *Walters & Church*, and Kagy was not yet recorded as an Ottawa resident. The 1869–70 Ottawa CD listed J. M. Kagy as a member of the grocery firm of *Kagy & Trumbo*. So the listed embossed bottle would likely have been manufactured for and used by *Walters & Co.* ca. 1866-68, using John Kagy's bitters recipe.

According to LaSalle County histories, Aaron A. Kagy was a local farmer and livestock dealer in the local Ottawa area during this same period (he was John Kagy's brother). Both men were included in the 1866 IRS income-tax document for Ottawa, but only Andrew had an occupation listed: that of *retail* liquor dealer. Possibly Aaron Kagy was involved in retail sales of the *Kagy's Stomach Bitters* being manufactured and sold wholesale by *Walters & Co.*

By the time of the 1870 USC, J. M. Kagy was listed as a farmer in the Kankakee area. According to USC records, John Kagy later moved West. And he died in Montana.

Dr. J. Stickels

Medicine:

(1) **DR. J. STICKELS** // **COMPOUND / SYRUP OF CEPHALANTHUS** //   [aqua]
OTTAWA ILL // //

»SB w/ keyed hinge mold; square; 3 sunken panels; flared square-ring collar; 6 1/2" tall. *[Figure 851]*

Figure 851

From the style and technological attributes of the listed bottle, it would seem to date to the late 1860s or early 1870s. However, we have been unable to locate any LaSalle county resident (or physician) with a name similar to "Stickels" from our searches of historical records, genealogical records, or census records of this period. Our attempts to locate newspaper ads for the product have been unsuccessful as well. So, for the time being, "Dr. Stickels" (or "Dr. Stickel") is an Ottawa phantom.

However, the likely function of the Stickels medicine can be addressed. *Cephalanthus* syrup was a fairly popular folk herbal remedy of the time, and herbal medicine guides agree that *C. occidentalis* (common buttonbush) was primarily used as a purgative or vomit-inducing medicine, febrifuge, cathartic, and diuretic. *Cephalanthus* syrup, made from the flowers and leaves, was often administered as a mild tonic and laxative.

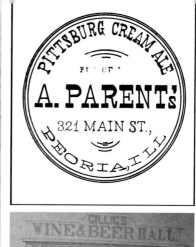

PANA

(Christian County)

1860 census: 700
1870 census: 2,200
1880 census: 3,000

Frederick and Jacob Weber

Soda ~ Mineral Water:

(1) **F. WEBER & BRO / PANA. ILL^S //**
    (heel): **A & D. H. C.**
    Note: *There are 2 raised embossed dots beneath the superscript* **S** *in* **ILL^S**.

    »SB w/ keyed hinge mold; short-neck, shouldered-blob-top soda. *[Figure 852]*

Frederick Weber and his brother, Jacob, both immigrants from Bavaria, were listed as soda water manufacturers in Pana by the 1875 ISD and the 1880 USC. An 1880 history of Christian County (Brink, McDonough & Co. 1880b:200) also listed the firm of *F. "Webber" & Brother* as Pana "pop" manufacturers.

Frederick Weber also operated a soda and mineral water business in Litchfield (*see his Litchfield listing*), where he was listed as a soda water manufacturer in the 1873 CAR and in the 1875 ISD. According to an 1882 biographical sketch published in a history of Bond and Montgomery counties, he came to St. Louis from Bavaria in 1856 as a young man of 18 and first settled in St. Louis. He served as a military baker during the Civil War and afterward worked as a baker in St. Louis for a year, after which he moved to Litchfield: "In June, 1867, he started a factory for the manufacture of soda and seltzer waters, to which he still devotes his attention" (Perrin 1882b:183–184).

Very likely, Fred Weber's success in the Litchfield soda bottling business led to the idea of expanding his business into the Pana market, where

Figure 852

his brother and father both lived, about 30 miles northeast of Litchfield. Pana was connected by railway to a wider Illinois market for his soda products, and the railroad line ran through Hillsboro, just 10 miles east of Litchfield.

So, in the mid-1870s, Fred Weber's brother, Jacob Weber, Jr., was enlisted as the junior partner in the expanded business, and Jacob became the on-site manager of the Pana soda manufactory during the late 1870s.

PEKIN
(Tazewell County)

1850 census: 1,700
1860 census: 3,500
1870 census: 5,700
1880 census: 6,000

Dr. Alfred Brown

Medicine:

(1) (side): **D^R. A. BROWN'S** // // (side): **BLOOD PURIFIER** // // [aqua]
(base): **C & I**

»SB w/ keyed hinge mold; beveled-edge rectangle; 3 sunken panels; tooled tapered collar; 8 3/4" tall. [*Figure 853*]

❖

Dr. Alfred Brown was listed in the 1860 USC as a 52-year-old physician in Pekin. March 1854 issues of the *Illinois Journal* (Springfield) carried ads for Dr. A. Brown's "Female Alterative or Woman's Friend," a medicine said to be a sure remedy for maladies described in the book *Married Woman's Private Medical Assistant*. It was prepared by Dr. A. Brown of Pekin, IL, and was available for sale at *Corneau & Diller's* drug store in Springfield for $5.00 per bottle (This medicine has not been found in embossed bottles).

The listed bottle was dug from a privy in Peoria, IL, about 10 miles north of Pekin, and thus may well be one of Dr. Alfred Brown's Pekin medicine products. From the stylistic and technological attributes of the embossed bottle, its "blood-purifying" contents were packaged and sold by Dr. Brown during the 1860s. Since the *Cunningham & Ihmsen* **(C&I)** did not begin to manufacture glass bottles in Pittsburgh until 1865, Dr. Brown's *Blood Purifier* would postdate the Civil War.

Gottlieb J. Zerwekh

Soda ~ Mineral Water:

(1) **E & Z / PEKIN ILL.** [aqua]
Note: *Zerwekh first entered the soda-bottling business with an unknown partner (see discussion that follows).*

»SB w/ keyed hinge mold; long neck, shouldered-blob-top soda. [*Figure 854*]

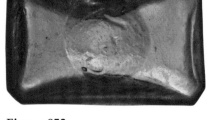

Figure 853

(2a) **G. J. ZERWEKH** *(arched)* / **PEKIN ILL.** [aqua]
   Note: not seen; listed in Oertle (2008). Likely closely similar embossing to #2b below.
   »SB w/ keyed hinge mold; long neck, shouldered-blob-top soda.

(2b) **G. J. ZERWEKH** *(arched)* / **PEKIN ILL.** [aqua]
   (base): **G.J.Z**
   »SB w/ keyed hinge mold; long neck, shouldered-blob-top soda. *[Figure 855]*

(3) **GJZ** *(within an embossed triangle)* // [aqua]
   (heel): **C & I**
   (base): **P** *(within an embossed triangle)*
   »SB w/ post mold; oval-blob-top soda. *[Figure 856]*

(4) *[No side embossing]* [aqua]
   (heel): **C & I**
   (base): **G J Z**
   »SB w/ post mold; shouldered-blob-top soda. *[Figure 857]*

(5) *[No side or heel embossing]* [aqua]
   (base): **Z**
   Note: Attribution of this minimally embossed soda to Zerwekh is based on the fact that these base-marked plain bottles have been excavated together with embossed **JGZ** bottles.
   »SB w/ post mold; shouldered-blob-top soda. *[Figure 858]*

Figure 854

Figure 855

Figure 856

Gottlieb Zerwekh was born in Wurtemburg, Germany, in 1833. He was trained as a locksmith apprentice in Germany prior to immigrating to America in 1853. He came to Peoria in 1854, and moved across the Illinois River to Pekin in 1855, working as a locksmith. Then in 1866, the year after he moved to Pekin,

> he began in business for himself, manufacturing soda and pop on a small scale. Being the only one in the county thus engaged he was soon compelled to increase his business and is now occupying three floors of the large brick building located at No. 230 Court Street. [Biographical Publishing Co. 1894a:260–261]

Despite the best efforts of regional researchers who have examined historical documents and newspaper microfilm (Ben Oertle and Jim Searle, personal communications 2008; see also Oertle 1990, 1995, 2008), no information has yet been located to identify Zerwekh's initial business partner (the "**E**" of **E & Z**) when he first launched his soda bottling business in 1866. Searle and Oertle have suggested that Mr. "**E**" may have been a financial backer or an experienced soda maker who helped Zerwekh get his bottling business "up and running" in its first year of operation. William T. Edds was a prominent Pekin businessman during this time. In 1853, he "engaged in the manufacture of pork, lard and whiskey barrels. He was very successful in this venture, and five years later found him the proprietor of four shops in different parts of the city, and at the same time he was engaged as a wholesale liquor dealer and retail grocer. These enterprises he abandoned in 1873" (Biographical Publishing Co. 1894a:686–687).

Two other strong candidates for Mr. "**E**" also come to mind. James Eaton ran bottling works in nearby Galesburg during the late 1850s and in Chicago during the early 1860s (*see his Galesburg and Chicago listings*), before returning to Galesburg (about 40 miles NW of Pekin) by 1865, where he is listed as just a resident. He was bottling in Wisconsin by the end of the 1860s, but we have no record of his activities ca. 1866–67. He returned to Peoria as a bottler in the late 1870s. Mr. "**E**" may also have been Joseph Erl, who ran a bottling house in Bloomington in the mid-1860s before coming to Metamora to establish his bottling house there, ca. 1969 (*see his Bloomington and Metamora listings*). Again, we have no record of his activities ca. 1866–67. But it is interesting to note that Metamora is located just 20 miles northeast of Pekin.

Figure 857

The embossed bottles used by Zerwekh from 1866 to ca. 1880 are listed in approximately the order in which they were manufactured. From style and technological attributes, styles #1 and #2 date to the later 1860s, and styles #3–#5 date to the decade of the 1870s. The company's post-1880 bottles were produced and used too late for inclusion in the present volume.

Zerwekh was listed as a Pekin soda water manufacturer in the 1870 USC, the 1872 Toledo, Wabash & Western Railway Directory, and the 1875 ISD. According to the *History of Tazewell County* (Bateman 1905:1103), Zerwekh operated the bottling works with his sons, Adam and William, until his death in 1889. After that, Adam became sole owner of the business. A turn-of-the-century ad for *G. J. Zerwekh & Sons Co.* described their carbonated beverages as "the drink that cheers but not inebriates...they are absolutely pure and healthful in every way. A great deal of the drinking water is not pure, but not a particle of contamination can ever be found in our make of soft drinks."

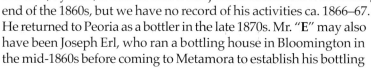

Figure 858

PEORIA
(Peoria County)

1844 census: 1,600
1850 census: 5,900
1860 census: 14,000
1870 census: 23,000
1880 census: 29,000

James D. Bastow

Medicine:

(1) *(side):* **J. D. BASTOW'S // COUGH CURE //**   [aqua]
(side): **PEORIA ILL**

»SB w/ keyed hinge mold; beveled-edge rectangle; 4 sunken panels; double-ring collar; 6 1/4" tall. *[Figure 859]*

Figure 859

James D. Bastow was a 35-year old druggist in Peoria in 1870 (USC). He was first listed in Peoria CDs during the late 1860s (he was not listed in the 1865 CD, and there was no 1866 CD). His first (1867) Peoria CD entry listed him as a "dispensing chemist" and dealer in paints, oils, and glass, on Adams Street above Pecan. His ad in the 1870 CD (p. 71) added that he was an agent for the *American and China Tea Company* and that he was a "Dealer in Drugs, Perfumery and Toilet Articles, and Approved Medicines for Family Use."

In the Peoria CDs of the early 1870s, he was generally listed as a druggist, and by 1875 he was operating his pharmacy from his residence at 413 Main Street. He continued as a druggist there from 1875 to 1880. By 1880 (USC) he had left Peoria and moved west to Colorado Springs, CO.

From its stylistic details and its keyed-hinge-mold technology, the bottle listed appears to have been manufactured and used at the beginning of Bastow's Peoria druggist business during the late 1860s.

Orrin P. Bissell

Bitters:

(1) **GREAT / WESTERN.TONIC.BITTERS / PATENTED.JANY.21.1868 // //**   [amber]
O. P. BISSELL & Co / PEORIA ILL // //
(base): **L&W**
Note: This is the earliest variety of embossed Bissell's Tonic Bitters. The Lorenz & Wightman (L&W) glass house of Pittsburgh closed in 1871.

»SB w/ keyed hinge mold; beveled-edge square "case" bitters; flat panels; tapered collar; 9" tall. *[Figure 860]*

(2) **BISSELL'S / TONIC.BITTERS / PATENTED.JANY.21.1868 // //**   [amber]
O. P. BISSELL / PEORIA, ILL. // //
(base): **L&W**
Note: This is the second variety of embossed Bissell's Tonic Bitters. **GREAT WESTERN** *has been peened out and replaced with* **BISSELL'S** *on one embossed panel and* **& CO.** *has been peened out on the second embossed panel. The Lorenz & Wightman (L&W) glass house of Pittsburgh closed in 1871.*

»SB w/ keyed hinge mold; beveled-edge square "case" bitters; flat panels; tapered collar; 9" tall. *[Figure 861]*

Figure 860

Figure 861

(3) **BISSELL'S / TONIC BITTERS / PATENTED JANY 21.1868 // //
O. P. BISSELL / PEORIA ILL. // //**   [amber]

Note: *This is the third variety of embossed Bissell's Tonic Bitters. From its stylistic attributes and post-mold construction, this bottle was manufactured later (ca. mid-to-late 1870s) than bottles #1 or #2. [Figure 862]*

»SB w/ post mold; beveled-edge square "case" bitters; flat panels; tapered collar; 9 1/4" tall.

Figure 862

Orrin P. Bissell & Co. was established in Peoria by 1864 as a wholesale and retail dealer in dry goods and "Yankee notions." During his company's early years, Bissell was not involved in bitters bottling—and he did not patent the *GREAT WESTERN TONIC BITTERS* he later marketed and sold. As Bissell noted on his bottle's opposite-panel paper labels (one in English, one in German: *see Figure 863*), his bitters recipe "was originally used in an extensive practice by one of the oldest and most celebrated physicians of the Northwest. The recipe for preparing it was only procured from the old doctor after much solicitation for the purpose of supplying the public with an innocent, useful and elegant TONIC. A wineglassful three times a day before meals is a sure cure for Dyspepsia, Debility of the Digestive Organs, Female Diseases and all other chronic complaints. It is the best appetiser in the world, and a certain preventative of Ague."

The January 21, 1868, patent date embossed on the bottle was for U.S. Patent #73,552, awarded to Dr. James T. Stewart of Peoria for an "Improved Medical Compound." It was one of the few late-nineteenth-century bitters whose detailed recipe was not a closely guarded secret:

> My invention has for its object to furnish an improved tonic stomach-bitter, and, as a secondary effect, blood-purifier, which is applicable in all cases of debility, especially those resulting from and following ague and other malarial fevers, and which may be taken freely, and for a great length of time, without producing headache or other unpleasant symptoms. It also counteracts the effect of malaria upon the system, thereby preventing ague and other fevers. It also acts as an appetizer, and benefits those suffering from chronic diseases which have produced a debilitated condition of the system; and is especially beneficial in consumption.
>
> The ingredients, and the proportions of each used in this compound, are as follows: Whiskey, one gallon; water, one half gallon; white sugar, one half pound; tincture of orange peel, four ounces; tincture of gentian, (compound,) two ounces; tincture of cardamom, (compound,) two ounces; tincture of curcuma, two ounces; and oil of lemon, sixty-four drops.
>
> In preparing the compound, I first cut the oil of lemon with a little alcohol, and add it to the whiskey; I

Figure 863

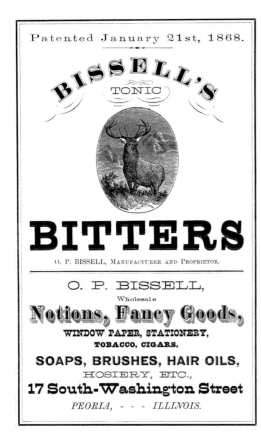

Figure 864

then dissolve the white sugar in the water, and add the solution, with the other ingredients, to the oil of lemon and whiskey. The compound is then filtered, which completes the process. [Stewart 1868: U.S. Patent text]

Dr. Stewart was a Peoria physician during the 1850s, 1860s, and 1870s who served as a Union Army surgeon during the Civil War. As noted in the 1876 Peoria CD (p. 79): "Dr. Stewart is a thorough student, and has, during his professional life, obtained by degrees an exceedingly rare and valuable collection of plants. The Doctor is a member of the Scientific Association of Peoria. He has his fine office at 102 South Adams street." He did not market his own patent medicine compounds, but "assigned" them (sold the rights to their manufacture) to Samuel R. Whitlow, a doctor and patent rights dealer in the rural Peoria area (1870 USC), who also designed his own medicine compounds (e.g., Whitlow's *Golden Age Cough Mixture*, 1868 Patent #81,446, and Whitlow's *Democratic Bitters*, 1871 Patent #113,713). Apparently, Whitlow sold the recipe and bottling rights for Dr. Stewart's patented bitters to Bissell sometime in 1869. Bissell, in turn, packaged and sold the bitters at his dry goods and notions store beginning in 1870—the first year he was listed in the Peoria CDs as a bitters manufacturer.

The 1871 Peoria CD (p. 59) listed Bissell as a "wholesaler dealer in notions, and **sole proprietor and manufacturer of Great Western Tonic Bitters**, 17 South Washington and Commercial Place." Bissell's company was listed in the 1871 ISD and in the 1880 USC as a Peoria patent medicine and bitters manufacturer. He had changed the name to *Bissell's Tonic Bitters* by the time of the 1872 Peoria CD, in which he placed a full-page ad opposite page 104, as its "Manufacturer and Proprietor" (Figure 864).

He continued to list himself in the Peoria CDs throughout the 1870s and into the early 1880s as a bitters manufacturer. A post-1880 clear-glass variety of the bottle (too recent for listing in this volume) is also known. It was embossed **O. P. BISSELL'S / / / / TONIC BITTERS.**

Blue Jacket Stomach Bitters

See the Francis & Spier listing in this chapter.

John Cairns and Ransom E. Hickey

Soda ~ Mineral Water:

(1) **CAIRN & HICKEY / PEORIA. ILLS** [aqua, teal, green]
Note: The base of a teal example of this bottle (Figure 865) shows the only evidence we have seen for the use of both an iron-pontil rod and an open glass-tube pontil rod in the production of the same bottle.

»IP; slope-shouldered soda with "petal paneled" front shoulder; shouldered-blob-top soda. [Figure 865]

The senior member of this early Peoria soda-bottling partnership was very likely John Cairns, a well-known and highly successful pre–Civil War St. Louis bottler. As his business grew during the 1850s, Cairns is known to have expanded his bottling interests to two other regional towns as well: East St. Louis, IL (*see the Schroeder East St. Louis listing*) and Keokuk, IA (see Burggraaf and Southard 1998). Cairns may well have provided the financial backing for the Hickey family to launch their first Peoria bottling operation in the early 1850s. The "**CAIRN**" embossing on the listed bottle was likely a glassmaker's error.

According to Peoria Historical Society information (e.g., Donath 1991; Kenyon 1976; Stalling 1979), John Cairns and Ransom Hickey became partners in their new Peoria mineral water bottling venture at Seventh and Spencer Streets shortly after Ransom purchased the Peoria Mineral Spring in 1853. During the early and mid-1850s, Ransom Eaton Hickey was joined in the Peoria bottling operation, off and on, by at least two of his brothers, Hiram M. Hickey and Sephreness ("Stephen") M. Hickey. All three were listed in an 1855 Peoria County Census as bottlers in a soda manufactory, and the 1857 USC listed both Ransom and Hiram as soda water manufacturers. According to Donath (1991:2), "Ransom Hickey...diversified the uses of the Spring waters by making ginger ale and other *[flavored]* drinks. First the drinks were bottled *[at his residence]*, with the aid of a water line directly from the Springs to the lower level of the house. Later on, he bottled in the barns across the street, and finally a brick factory was built there in 1855."

Figure 865

During the later 1850s and early 1860s, Ransom's three brothers (including elder brother John E. Hickey) and their uncle James E. Eaton moved away from the area to establish bottling works at a number of other towns in northern Illinois and southern Wisconsin (see Peters 1996; *see Hickey brothers listings for Bloomington, Joliet and LaSalle, and Eaton listings for Chicago, Galesburg, and Peoria*). S. M. Hickey was also found listed as a bottler in Freeport, IL in 1865, but no embossed Hickey bottles have yet been documented from Freeport.

According to Peoria Historical Society information, Ransom Hickey bought out Cairns share of the Peoria bottling business by 1857, after which stayed on as a bottler in Peoria during the early and mid-1860s. Ransom Hickey died in 1868 (*see his separate listing, this chapter*).

William A. Callender and Son

Bitters:

(1) D^R. CALLENDER & SON^S / CELEBRATED // // LIVER BITTERS /   [amber]
    PEORIA ILLINOIS / PATENTED APRIL 14 1874 // //

»SB w/ post mold; beveled-edge square "case" bitters; tapered collar; 9 1/2" tall *[Figure 866]*

Figure 866

The company of William A. Callender & Son (John W. Callender) was first listed in the 1875 Peoria CD as manufacturers of bitters at 1365 S. Adams (their residence was 1371 S. Adams). The listed bottle indicates that its contents were patented in 1874, and the U. S. Patent record for an "improvement in medical compounds or bitters" (Patent #149,573) has been located:

> Be it known that I, William A. Callender, of the city of Peoria,...have invented a medicine or mixture for the cure of liver complaint and other malarious diseases;...the following is a true description of the ingredients, quantities thereof, and mode of preparing or compounding, and administering the same.
>
> Take of senna leaves, one ounce,; aloes, half an ounce; rhubarb root, one ounce; Peruvian bark, one ounce; cardamom seeds, half an ounce; cubebs, half an ounce; blessed thistle, two ounces; wormwood, two ounces; licorice root, half announce; sarsaparilla root, one ounce; juniper berries, two ounces; agaricus, one ounce. These ingredients are steeped in ten gallons of pure spirits, of 100° gravity, or proof, at a heat of about 140° Fahrenheit, for twelve hours in an air-tight vessel. After this time, draw off the liquid and let it settle until clear. Then add white sugar until the gravity of the spirits is reduced to 84°. Then add one ounce of burnt sugar coloring, and for every gallon add one-fourth of an ounce of allspice, and it is then fit for use.
>
> The dose for an adult is two table-spoonfuls, three times a day, before meals, for four or five days, after which time a dose once a day will be sufficient until the cure is effected. For adult females, or girls over twelve years of age, one table-spoonful, three times a day, before eating. For children under twelve years, one tea-spoonful three times a day, reduced by sweetening to the child's taste.
>
> I call the medicine "Liver Bitters."

William Callender was noted in the 1876 Peoria CD (pp. 85–86) as "manufacturer of the celebrated *Callender Liver Bitters*, and he is therefore proprietor of an immense business. These bitters are prepared to act upon the liver, to purify the blood, act upon the kidneys, and are a great preservative of the health generally. These bitters were patented in April, 1874, and in June of that year he commenced their manufacture and sale, since which time a remarkably large number of bottles of the medicine have been sold. They have most rapidly grown into popularity, their merit and powerful curative qualities carrying conviction to the minds of the people of the great worth of the article, and their inventor, Mr. C., may be rated as a great public benefactor."

The Callender bitters company was listed in each subsequent Peoria CD into at least the early 1880s, although John Callender does not appear as a partner with his father after 1877, when William Callender is listed alone as a "druggist" in one of the two 1877 CDs. By 1878, the retail distributor for his bitters was *Colburn, Birks & Co.* in Peoria, whose drug catalogs and product labels referred to him as "Dr." Callender (*Figure 867*).

Figure 867

Figure 868

Callender's Liver Bitters, produced just before the *Colburn, Birks & Co.* distributorship, have been found bottled in embossed **REED'S 1878 GILT EDGE TONIC** bottles with Callender's paper labels pasted over the embossing (*Figure 868*). The *Gilt-Edge Tonic* was a very popular late-1870s product bottled in New Haven, CT, by the George Reed bitters company there (see Baldwin 1973:408; Wilson and Wilson 1971:134).

A clear, cylindrical Callender's bottle with a tooled lip (and with decorative indented flutes around the heel and shoulder) is also known. It was embossed **DR CALLENDERS / LIVER BITTERS** above the fluting on the upper shoulder (Ring 1980:117). It appears to have been produced during the early 1880s—too late for formal inclusion in the present volume.

Frederick Cleveland

Baking Powder:

(1) (side): **F. CLEVELAND** // // (side): **PEORIA ILLS.** // //   [clear]

»SB; rectangular; 4 sunken panels; mid-neck ring w/ squared, flared-ring collar; 8 1/2" tall. [*Figure 869*]

Frederick Cleveland was never listed in business in any of the pre-1880 Peoria CDs, but he did appear in the 1870 USC for Peoria as a 32-year old manufacturer of baking powder. He later

Figure 869

moved to New York, where he died in 1897, and where a biographical sketch was found summarizing his Illinois business history:

> At the age of twenty-four, in 1862, he entered the firm, B. S. Cory & Company, druggists of Waukegan, Illinois *[see Cory's Waukegan listing]*, and there continued in business until 1868, having different partners, one of whom was his brother George. In 1868, he closed out his drug business in Waukegan and re-opened in Chicago, under the firm name, Cleveland & Rice, druggists *[no embossed bottles known]*. There his health utterly failed and by his physician's advice he moved to Peoria, Illinois, in 1869. **In Peoria he became interested in manufacturing and placed upon the marked Cleveland's Baking Powder, a brand which won the Western market and eventually became one of the three leading brands of baking powder in the country**. In 1874 his brother George again became his partner, and as Cleveland Brothers they built up an enormous business. The firm prospered abundantly for many years but in time became a member of the consolidation which included the Royal Cleveland and Price Companies. Frederick Cleveland sold his interest in Cleveland Brothers in 1889 and retired from business. [Fitch 1923:160–161]

From this business summary, the embossed bottle listed was likely manufactured and used ca. 1870–73.

William H. Davis

Medicine:

(1) (side): **DAVIS'S INIMITABLE // HAIR RENEWER //**   *[aqua, amber]*
(side): **PEORIA. ILL // //**
(base): **L & W**

»SB w/ keyed hinge mold; rectangular; 3 sunken panels; double-ring collar; 7 1/2" tall. *[Figure 870]*

William H. Davis apparently bought out E. J. Humphrey's drugstore business in the late 1850s. Humphreys was listed at 55 Main Street in the 1856/57 Peoria CD. He disappears from local records after that, and W. H. Davis was listed in the Peoria CD for 1859 at the same address. Davis' younger brother, Robert S. Davis, was trained in the pharmacy field by working at his brother's drugstore after the Civil War, and in 1872 he entered the Peoria druggist business on his own at 125 Main Street. According to Frank Donath (personal communication), records

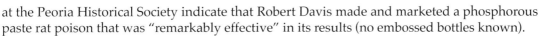

Figure 870

at the Peoria Historical Society indicate that Robert Davis made and marketed a phosphorous paste rat poison that was "remarkably effective" in its results (no embossed bottles known).

From the style, technological attributes, and L&W maker's mark of the embossed bottle listed above, it appears to date to the Civil War era.

James E. Eaton

Soda ~ Mineral Water:

   (1) **J. E. EATON / PEORIA / ILL.**   *[blue teal, shades of green]*

      »SB; blob-top quart soda; embossing in circular slug plate; 12" tall. *[Figure 871]*

James E. Eaton was an uncle of the Hickey brothers who bottled soda and mineral water in Peoria and elsewhere in northern Illinois and southern Wisconsin (*see, e.g., Cairn & Hickey listing and Ransom Hickey listing, this chapter*). Eaton had been a soda water bottler in Galesburg during the late 1850s (*see Galesburg listing*) and in Chicago during the Civil War years (*see Chicago listing*), before moving to southern Wisconsin during the later 1860s and 1870s to establish bottling works with various family members there (Peters 1996).

James Eaton moved back to the Peoria/Knox County area in 1879 to reopen the old Hickey bottling factory in Peoria at 704 Seventh Street (*Figures 872 and 873*). He was first listed bottling soda water there in the 1880 Peoria CD, and the earliest (1879–80) embossed bottles he used are listed at the beginning of the entry. After Ransom Hickey (who died in

Figure 871

Figure 872. Ca. 1880 historical photo of James E. Eaton (center with buttoned jacket) and soda factory workers in front of R. E. Hickey's Soda Water Factory. At the time, Eaton had recently taken over the Hickey bottling plant in Peoria to begin his own soda-bottling operation there. (Photo courtesy of Frank Donath.)

Figure 873. Ca. 1880 historical photo of James E. Eaton (third from right, behind blacksmith Shack Dye) and two of his soda factory employees (including Alex Hickey) on one of his soda wagons, at Dye's Practical Horse Shoer shop in Peoria. (Photo courtesy of Frank Donath.)

1868) relinquished control of bottling-plant operations in the early 1860s, the plant was operated for a year by *Lorentz & Co.*—a.k.a. Bohl, Singer & Lorens (see Bateman, ed., 1902: no embossed bottles known)—and then by Karl Gillig and Charles Singer (*see their listings, this chapter*) until Gillig retired in 1875 (Donath 1991:3).

James Eaton was first listed bottling soda water at the Hickey plant in the 1880 Peoria CD, and the earliest (1879–80) embossed bottles he used are listed at the beginning of the entry. Eaton was in his late 60s at the time. He was recorded in the 1879–80 IIC as having $1,500 invested in the business and was operating five months/year with four employees paid from 80¢ to $1 per day for their work. That year, he paid $250 in wages, used $800 of materials and supplies, and produced $1,500 worth of bottled soda water.

Eaton's daughter Julia had married a young Boone County, IL, attorney named Preston Clark. "Her parents became so lonesome for their daughter, they persuaded Clark and his wife to come to Peoria about 1885. Mr. Eaton then took his son-in-law into the partnership which became known J. E. Eaton & Co." (Donath 1991:3). After Eaton died in 1888, Clark assumed sole operation of the Peoria mineral spring bottling plant. Eaton's obituary in the *Peoria Daily Transcript* for May 28, 1888 (p. 6, col. 4) painted a vivid picture of his personality:

> James E. Eaton of Eaton & Co., the mineral water firm, died yesterday morning of pleurisy at his residence. He belonged to that class of men who have the faculty of accumulating money and then losing it again in some unfortunate speculation. Several times in his life Mr. Eaton was a wealthy man and as often did he lose his all and begin at the bottom again.... *[Early in his adult life]* he was a banker in New York State and afterwards Ohio. He went up to Columbia County, Wisconsin, and was County Clerk and Postmaster.... In 1840 he went to California and amassed another fortune. He then returned to Illinois and located in Galesburg.... When he came to Peoria it was just after losing a fortune in a Mexican speculation. In his nine years residence here he had accumulated quite a good deal of world's goods and was on the high road to prosperity when the pruning knife of time cut him down.

Varieties of post-1880 Hutchinson-style (internal rubber stopper) embossed bottles are known, but these were produced and used too late for inclusion in the present volume.

Hiram G. Farrell

Medicine:

(1) **FARRELL'S / FEVER & AGUE / DROPS.** [aqua]
Note: Based on manufacturing technology (**square** keyed hinge mold), heavy embossing with seriffed forward-slanted (lines 1 & 3) and back-slanted (line 2) lettering, and the fact that HGF is not specifically named (see listings below), this Peoria-excavated bottle may have contained the first Farrell medicine marketed in embossed bottles.

»OP; shouldered cylinder w/ vertical embossing; flared lip; 4 1/2" tall and 1 1/2" diameter. [Figure 874]

(2a) **H G FARRELLS / ARABIAN / LINIMENT / PEORIA** [aqua, olive-green]

»OP w/ straight-line hinge mold; shouldered cylinder w/ vertical embossing; rolled lip; 3 3/4" tall (1-ounce size). [Figure 875a–e]

(2b) **H G FARRELLS / ARABIAN / LINIMENT / PEORIA** [aqua]

»OP w/ straight-line hinge mold; shouldered cylinder w/ vertical embossing rolled lip; 3 3/4" tall (1-ounce size). Like #2a, but with mold lettering recut. [Figure 875f–g]

(3a) **H. G. FARRELL.S / ARABIAN / LINIMENT / PEORIA** [aqua]

»OP w/ half-circle hinge mold; shouldered cylinder w/ vertical embossing; rolled lip; 4 1/2" tall (2-ounce size). [Figure 876a–c]

(3b) **H G FARRELL S / ARABIAN / LINIMENT / PEORIA** [aqua]
Note: This variety has recut mold lettering, with the period after **FARRELL** removed; vertical embossing begins just below the bottle shoulder.

»OP w/ half-circle hinge mold; shouldered cylinder w/ vertical embossing; rolled lip; 5 1/4" tall (2-ounce size). [Figure 876d–g]

(4) **H. G. FARRELL.S / ARABIAN / LINIMENT / PEORIA** [green]

»OP w/ half-circle hinge mold; shouldered cylinder w/ vertical embossing; rolled lip 5 1/4" tall (4-ounce size). [Figure 877]

(5) (heel) **H. G. FARRELL** // (heel) **PEORIA ILL** [aqua]

»OP; shouldered oval cylinder; horizontal heel embossing; flared lip; 4 3/4" tall. [Figure 878]

Figure 874

(6) **H. G. FARRELL'S / ARABIAN / LINIMENT / PEORIA** [aqua]

»SB w/ keyed hinge mold; shouldered cylinder w/ vertical embossing; flared square-ring collar; 4" tall (2-ounce size). *[Figure 879a–b]*

(7) **H. G. FARRELLS / ARABIAN /** [aqua, clear]
LINIMENT / PEORIA

Note: This variety has recut mold lettering, with the apostrophe in **FARRELL'S** *removed.*

»SB w/ post mold; shouldered cylinder w/ vertical embossing; flared square-ring collar; 4" to 4 1/4" tall (2-ounce size). *[Figure 879c]*

Stylistically and technologically, the first embossed *Farrell's Arabian Liniment* bottles from Peoria are among the earliest embossed bottles produced for any Illinois bottler, dating to the mid-1840s. *Farrell's Fever & Ague Drops* were likely bottled and sold even before their *Arabian Liniment*. If their fever medicine was produced and sold in the late 1830s or early 1840s, when Hiram and William Farrell were still working together in their pioneer Peoria druggist business, it may well have been packaged in the first embossed bottles ever used by any Illinois merchant. The Farrells' linseed oil factory and pharmacy business are platted (*see location "T" on Figure 880, left*) and advertised (*Figure 880, right*) in the first Peoria city directory, published in 1844.

Hiram G. Farrell was listed as a Peoria druggist in the USC of 1850, 1860 and 1870 (as well as later USC records: see later discussion). His embossed bottles date ca. 1840–1900. The examples listed, in approximate chronological order, are only those manufactured and used during the 1840–1880 period of the present study.

H. G. Farrell and his older brother William B. Farrell were very early partners in an extremely successful Peoria patent-medicine and pharmacy business. Their business eventually turned almost its entire attention to the promotion of *Farrell's Arabian Liniment*, a widely popular liniment

Figure 875

Figure 876

Figure 877

Figure 878

in the mid-to-late 1840s and 1850s. They eventually had an acrimonious business falling-out, and William moved to Chicago in 1849 (see *his Chicago listing*)—where he continued to sell his version of the Arabian Liniment in competition with his brother's Peoria version (with each calling the other a "base counterfeit").

The Farrell brothers migrated to Peoria from Mt. Pleasant, PA, in 1836. Soon after their arrival, William (who was 21) opened a drugstore on Main Street. Hiram (who was 14) worked as a clerk in their store. There were no local banks, and early on the Farrell brothers weathered uncertain financial times in their rustic pioneer outpost, in part with help from relatives back home. By 1837, they had diversified their business and were dealers in "drugs, medicines, dyestuffs, chemicals, varnishes, patent medicines, window glass, paints and oils, and surgical instruments." They also sold port and Madeira wines for "medicinal purposes" (Bogard 1982:99–100). In 1840, after serving a four-year apprenticeship, Hiram joined his brother's firm, and at the same time they expanded their Main Street operation to include an oil factory for the production of castor oil (a lubricant and cathartic), linseed oil (a drying agent for paints and varnishes), and lard oil (lamp fuel). But their income could not overcome their debt, and by 1844

> the Farrells began urgently requesting persons indebted to them to pay with such available articles as beeswax, white beans, dry hides or pork. In May they advertised that they...would receive every kind of country produce in exchange for articles in their drugstore.
> Nevertheless, in June William mortgaged his real estate on Main Street and the oil factory property to the administrators of *[his principal creditor]*. The last date on which "W. B. and H. G. Farrell"

Figure 879

advertised in the *Peoria Democratic Press* was July 17, 1844. Not long afterwards William withdrew from the drugstore.... By then the oil factory must have been shut down. [Bogard 1982:99–100]

Less than two months later, the first ads appeared for "Farrell's Arabian Nerve and Bone Liniment" in the *Peoria Democratic Press*, "manufactured only by W. B. Farrell" and good for man or beast. In 1846, Hiram started selling the liniment in his drug store, using ads and paper labels showing a wounded Arab and his horse being treated with the medicine. The following year, he tried and failed to copyright the liniment, but did copyright a booklet about the "celebrated" liniment, proclaiming it a "certain cure for all kinds of rheumatism,...bruises, cramps,...goiter,... paralysis,...inflammations and tumors...and most of the diseases of horses" and noting, "Price 37 1/2 cents. None genuine without the signature of H. G. Farrell" (*Figure 881*).

The earliest advertisement (and testimonials) we have been able to locate for *Farrell's Arabian Liniment* was found in the *Illinois State Journal* (Springfield, IL) for August 20, 1846 (p. 2, col. 6):

**The Great Remedy for
Man or Beast
FARRELL'S ARABIAN LINIMENT**

FARRELL'S Arabian Liniment is a compilation of balsams from plants peculiar to that country, and it is by the use of the articles composing this remedy, that the Arabian Physicians so celebrated in the healing art, perform cures that are considered almost miraculous by those who have not studied so deeply all the curatives nature has given us.

The Arabian Liniment acts with force and energy, penetrating the flesh to the bone, releasing cramped cords and sinews, restoring action to limbs which have long lost their use. **Scarcely one year has passed since this article was introduced, and more than fifty thousand bottles have been sold.**

IN Drugs, Patent Medicines, Paints, Oils, Varnishes, Surgical Instruments, Dye-stuffs, Glassware, Perfumery, &c.; manufacturers of Castor, Linseed and Lard Oils; will always pay cash for Castor Beans, Flaxseed and Lard; and every kind of country produce received in exchange for articles in their line. They have reduced their prices to suit the times, and are determined to sell as low as any house in the state.   Peoria, May 1, 1844.

Figure 880. Detail from 1844 Peoria city plat map showing the location of W. B. & H. G. Farrell's Linseed near the Illinois River (Location "T" on North Fayette Street). The map was produced by town surveyor S. DeWitt Drown for inclusion in the first city directory.

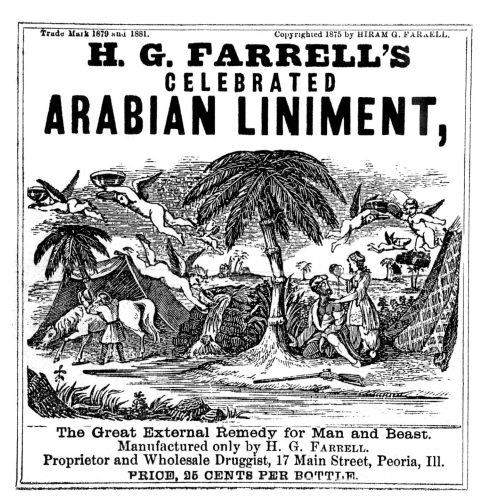

Figure 881

Thousands are ready to attest to its excellent qualities. It has gained a growing reputation where it has been used in most cases which require an external remedy—such as Rheumatism, soreness and swelling of the breast, pain and weakness in the back and joints, sprains, bruises, cramps,...frozen limbs, bites of insects, old sores, fever, ague, goiter or swelled neck, paralysis, wounds, burns, all swellings, salt rheum, sore eyes, run-rounds, etc. We daily hear of the most astonishing cures on animals by this excellent remedy; and we hazard nothing in saying that there has never been its equal as a curative for horses.

In the late 1840s, William moved to Chicago and the battle of the brothers began over who was the proprietor of the "authentic" *Farrell's Arabian Liniment* and who was selling the "base counterfeit" (*Figure 882*).

In 1852 Hiram warned the public that the most dangerous of all the counterfeits was W. B. Farrell's Arabian Liniment. He denounced his brother, "This W. B. Farrell not only has the audacity to call his stuff ARABIAN Liniment, but with unparalleled impudence asserts that it is the original and only genuine, and that he introduced it to the public in 1839, when the truth is no one ever saw a bottle of it before the spring or summer of 1840." Years later Hiram gave both 1843 and 1842 as the year that the liniment originated. [Bogard 1982:103–104]

William sold his version of the medicine in Chicago only until 1858 (*see his Chicago listing*). However, Hiram's version of the liniment was marketed and sold for more than 50 years, eventually becoming the consuming focus of his business.

The wonderful curative powers of this article have given it a fame that has reached Europe. The demand for them is such that they ship unlimited quantities to European ports, and much more extensive are the sales throughout the United States, as the patrons of this well known firm are increasing beyond computation. Mr H. G. Farrell is about to discontinue the drug business, and henceforth proposes to give his undivided attention to this specialty. The firm will thus be enabled to introduce it in all civilized countries in the world. [Edwards Peoria City Directory 1876:106]

H. G. Farrell continued to give "his entire personal attention" to the *Arabian Liniment* business until he retired in 1903 at the age of 80. The embossed bottles listed are only those that would have been manufactured and used prior to 1880. They are listed in approximate chronological order of manufacture and use.

In addition to the *Arabian Liniment*, and the very early *Fever & Ague Drops*, embossed bottle #5 above may have contained a third Farrell's patent medicine product. Paper labels found associated with unembossed pontiled bottles indicate they contained Farrell's *Cough Syrup* "For Asthma, Coughs, Colds, Hoarseness, Pain and Soreness of the Breast, difficulty of Breathing, &c., &c." "In bad Coughs and soreness of the breast, a full dose taken on going to bed, and a flannel cloth saturated with H. G. FARRELL'S Arabian Liniment, and laid on the breast over night, will have a very beneficial effect" (*Figure 883*). Additional Farrell's paper-labeled products include toothache drops (*Figure 884*) and ink (*Figure 885*).

Figure 882

Figure 883

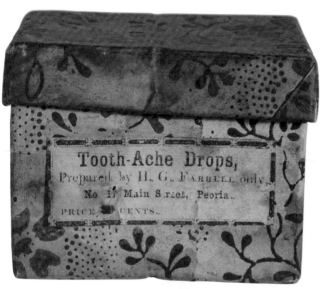

Figure 884

Figure 885

DeWitt C. Farrell and Richard S. Cox

Medicine:

(1) (heel): **FARRELL & COX** // (heel): **PEORIA** [aqua]
Note: *This bottle is nearly identical to H. G. Farrell bottle #4 in the Hiram G. Farrell listing, with the heel embossing modified slightly.*

»OP; shouldered oval cylinder; horizontal heel embossing; flared lip; 5" tall. *[Figure 886]*

The partnership of Richard S. Cox and DeWitt Clinton Farrell was first listed in the Peoria CD for 1857, which contained a full-page ad on p. 17 announcing their new business venture ("successors to W. A. Herron") as "Wholesale & Retail Druggists, and dealers in Paints, Oils, Glass, Varnishes, Dye-Stuffs, Glass-Ware, Etc., No. 27 Main, corner of Washington Street." The new company's ad also noted they were "Agents For All Popular Patent Medicines."

Farrell & Cox were also listed in the 1859 CD as druggists located at 27 Main Street, just 10 doors up from the business of Hiram G. Farrell, DeWitt's successful older bother (DeWitt was six years younger than Hiram), whose wholesale and retail drug business was headquartered at 17 Main. In the 1860 Peoria CD, DeWitt Farrell is listed as sole proprietor of his druggist business (pg. 47), Cox having joined the local wholesale druggist firm of Johnson & Cox (p. 50). Thus, the listed bottle dates between 1857 and 1860.

From 1863 onwards, "Clint" Farrell, as he was known locally, turned his attention to the distillery and milling business, establishing the Diamond Mills Distillery in Peoria. According to Peoria Historical Society files, "he engaged a Japanese yeast maker who completely doubled the whisky yield from a bushel of grain by making yeast out of rice. This discovery built a large fortune for Mr. Farrell." He built a famous residence during the Civil War years at the corner of Madison and Hancock, with steam heat and silver-plated hinges and hardware. "The home for years was a center of Peoria's gay social whirl."

The similarity of the *Farrell & Cox* bottle listed, in terms of size, shape, and the unusual placement of the embossing, with one of H. G. Farrell's bottles (*#5 in his listing, this chapter*) is striking, and suggests they were made by the same glass house, perhaps part of the same order. D. C. Farrell is listed as sole proprietor of the business in the 1860 Peoria CD (pg. 50), indicating that this bottle dates between 1857 and 1860.

Figure 886

Hubert Fellrath

Soda ~ Mineral Water:

(1a) **H• FELLRATH / PEORIA / ILLS•** // [blue teal]
(base): **F** *(large seriffed font)*

»SB w/ keyed hinge mold; long-neck, shouldered-blob-top soda. *[Figure 887]*

(1b) **H• FELLRATH / PEORIA / ILLS•** // [aqua, blue teal]

»SB w/ keyed hinge mold; long-neck, shouldered-blob-top soda. *[Figure 888]*

(2) **H. FELLRATH / PEORIA ILL _** // [aqua]
(heel): **A. & D. H. C**
(base): **F** *(large seriffed font)*

»SB w/ keyed hinge mold; long neck, slope-shouldered soda; shouldered-blob top. *[Figure 889]*

Figure 887

Figure 888

Figure 889

(3a) **H. FELLRATH / PEORIA ILLS. //** [aqua]
(heel): **A & D. H. C.** (small embossing)

»SB w/ keyed hinge mold; short-neck soda w/ Matthews gravitating stopper-style collar; 7 1/8" tall. The maker's mark suggests the bottle postdates Matthews patent restrictions, and thus dates to the late 1860s or early 1870s. [Figure 890, left]

(3b) **H. FELLRATH / PEORIA ILLS. //** [aqua]
(heel): **I. G. Co.**

»SB w/ post mold; short-neck soda w/ Matthews gravitating stopper-style collar; 7 1/8" tall. Mold lettering recut from #3a. The maker's mark suggests that Fellrath's "Selters Water"-style bottle continued in use into at least the late 1870s. [Figure 890, right]

Ginger Ale ~ Seltzer:

(1) **PEORIA / SELTZER WATER** [cobalt]
(base): **F**
Note: Bottles attributed to Fellrath because they are only found in Peoria, because he advertised seltzer water, and because the bottles have an **F** embossed on the base (see sodas #1a and #2).

»SB w/ keyed hinge mold; shouldered-blob-top soda. [Figure 891]

(2) **H. FELLRATH / PEORIA / ILLINOIS** [aqua]
(heel): **A & D. H. C.**
Note: Vertical embossing spaced around cylindrical bottle. Bottle has semi-round base but will stand on its own.

»SB; high-shouldered bottle with semi-round base; shouldered-blob top; 9 1/4" tall. [Figure 892]

Figure 890

Figure 891

According to the Peoria CDs, Hubert Fellrath began the manufacture of soda water in Peoria in 1867 from his residence at Gay and Water streets, having been a saloonkeeper before that. According to a German-language biographical sketch (Bek 1906:49), he had established a separate soda-bottling factory opposite his residence in 1870. He is listed as a Peoria manufacturer of soda water in the 1875 ISD. By 1880, Fellrath's bottling works produced 600 to 700 cases of soda water and ginger ale weekly (in addition to seltzer water, etc.) during the summer months (Johnson and Co. 1880:647).

According to the 1979–80 IIC, he had $12,000 invested in the business and a maximum of five employees (4 adult males and a child) during the year (six months full-time; six months half-time). Workers were paid from $1.50 to $2.00/day. His annual expenses were $1,500 in wages and $5,000 in materials and supplies. He estimated the sales value of his bottled soda water product that year at $10,000.

He continued as a Peoria soda, mineral water, cider, and beer bottler until he retired from business in 1899. Although he later used Hutchinson-style internal-wire-shank blob top rubber-stopper bottles, the embossed bottles listed are only those that were likely manufactured and used during the 1867–80 period that is the scope of the present study.

Figure 892

John H. Francis and James G. Spier

Bitters:

(1) **BLUE JACKET / STOMACH BITTERS // //** [amber, olive green]
    FRANCIS & SPIER / PEORIA ILL'S // //
    (base): W^M M^{C}C & Co

»SB w/ keyed hinge mold; beveled-edge square case bitters; short neck; tapered collar; 9 1/4" tall. *[Figure 893]*

According to the Peoria CDs from 1868 to 1870 and the 1870 USC, John H. Francis and his two brothers, Lewis N. and Willis, worked as distillers in Peoria, and by 1870 John Francis was co-owner of the Phoenix Distillery there (with Philip Zell). The 1870 USC listed James G. Spier's occupation as whiskey rectifier. Spier first appeared as a distiller in the 1867 Peoria CD and the 1870 CD listed *Spier & Co.* as wine and liquor dealers and "Bitters Manufacturers." *Spier & Co.* was no longer listed in the CDs after the 1870–71 edition. The *Francis & Spier* partnership was never listed in the CDs and must have been short-lived.

Because he was co-owner of the Phoenix Distillery, we have guessed that John Francis might have financed the *Blue Jacket Bitters* initiative (although it is possible that Lewis and/or Willis were involved as partners). The product was likely marketed by Spier for a short time in the early 1870s. The rarely found embossed bottle likely dates to this time.

No ads have yet been located to indicate the purported medicinal curative powers of *Blue Jacket Bitters*.

German Wine Bitters

See the Robert Strehlow listing in this chapter.

Charles E. Gillig and Charles J. Singer

Soda ~ Mineral Water:

(1) **GILLIG & SINGER / PEORIA / ILLS /** [aqua]
    (left-front heel): F.A & Co
    (base): **G & S** *(large-lettered variety)*

»SB w/ keyed hinge mold; oval-blob-top soda. *[Figure 894]*

(2) **GILLIG & SINGER / PEORIA / ILLS /** [aqua]
    (left-front heel): F.A & Co
    (base): **G & S** *(small-lettered variety)*

»SB w/ keyed hinge mold; shouldered-blob-top soda. *[Figure 895]*

Figure 893

As discussed in this volume's introduction, *Fahnstock, Albree & Co.* manufactured glass bottles for just a few years in Pittsburgh: beginning in 1860 (when they took over operation of the bottling works from Frederick R. Lorenz) and ending after 1863 (when *Lorenz & Wightman* took over operation of the glass factory). During the early 1860s in Peoria, Charles E. Gillig was proprietor of the St. Louis Exchange Hotel and Charles J. Singer was proprietor of the nearby Railroad Exchange Hotel (CDs; Bateman 1902). The two men were never listed as soda and mineral water

Figure 894   **Figure 895**

bottlers in the Peoria CDs of this time, but they clearly got together to manufacture these beverages for use in their respective hotels. In 1864, Singer left the hotel business to bottle soda and mineral water full-time as a member of the firm of *Bohl, Singer & Lorens*, using water from the Peoria Mineral Springs (a.k.a. Lorentz & Co.—see Donath 1991:3) and occupying the bottling works of Ransom Hickey (*see his listing, this chapter*). As far as is known, this one or two-season company did not use embossed bottles.

After the end of the Civil War, Singer continued bottling at the same location with a new partner: Karl G. Gillig (*see the folllowing listing*), son of his earlier partner Charles E. Gillig, who had also left the hotel business during the Civil War years, and was running a saloon, wine, and beer hall in Peoria (CDs) during the mid-to-late 1860s and the early 1870s.

Karl G. Gillig and Charles J. Singer

Soda ~ Mineral Water:

(1) **GILLIG & SINGER / PEORIA / ILLS //**   [amber, aqua]
    (heel): **L & W**
    (base): **G & S** *(medium-lettered variety)*

   »SB w/ post mold; short-neck, oval-blob-top soda. *[Figure 896]*

(2) **GILLIG & SINGER / PEORIA / ILLS //**   [aqua]
    (heel): **C & I**
    (base): **G & S** *(small-lettered variety)*

   »SB w/ post mold; oval-blob-top soda. *[No photo. Closely similar mold to #1.]*

Charles E. Gillig, who had been Singer's bottling partner during the early 1860s when the two were involved in similar hotel business in Peoria (*see their listing, this chapter*), was listed as running a saloon, wine, and beer hall in Peoria (CDs) during the mid-to-late 1860s and the early 1870s (*see Figure 897*). In the 1870 USC, he and his family were listed as living next door to the Charles Singer family.

Karl G. Gillig, Charles Gillig's son, became Singer's partner in the soda bottling business in Peoria soon after the Civil War, and they remained in the bottling business together at the old Hickey bottling works from ca. 1867/68 to 1874 (CDs). The embossed bottles listed would have

Figure 896

Figure 897

been manufactured and used during this time. In 1875, Karl Gillig retired from the soda bottling operation (see Bateman 1902), after which he was no longer listed in the CDs.

From 1875 onward, Charles Singer became sole proprietor of the bottling business at the old Hickey plant (*see his listing, this chapter*).

John M. Gipps, Thomas Cody, and Co.

Ale ~ Beer:

(1) **GIPPS, CODY & C^O / PEORIA ILL • ^{S}** [amber]
   (base): **A. G. W. Co.** *(around edge of inner central kick-up)*

   »SB w/ post mold; ring-neck cider finish, slope-shouldered quart; 12 1/2" tall. *[Figure 898]*

(2) **RETURN THIS BOTTLE / TO / GIPPS CODY & Co / PEORIA / ILL** [amber]
   (base): **L&W**

   »SB w/ post mold; slope-shouldered quart; lozenge-style oval blob top; 12 1/2" tall. *[Figure 899]*

❖

Peoria is notable for the early efforts of local brewers to bottle their beers and ales in glass containers (*in addition to the Gipps companies, see listings for Metz & Pletscher, Parent, Weber, and Wichman, this chapter*). A summary history of the *Gipps & Co.* partnerships up to the mid-1870s was published in the 1876 Peoria CD (pp. 93–94):

> The product of the breweries of Peoria is of a superior quality of both beer and ale. This, together with the situation that the city occupies respecting the Western States, offers a vast inducement to the

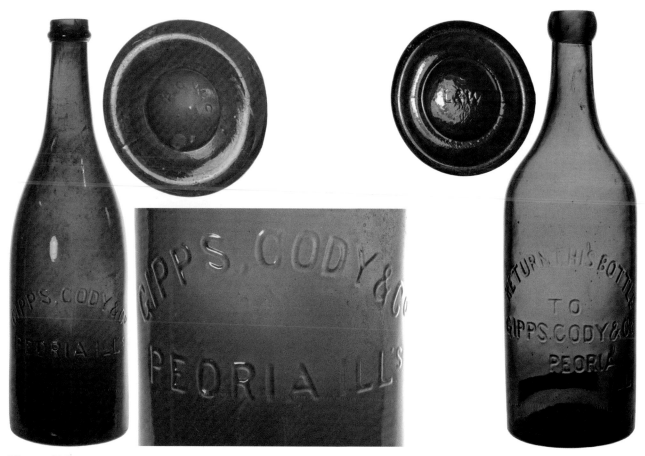

Figure 898 **Figure 899**

trade throughout Central Illinois and adjoining States to deal permanently here. The firm of Gipps & Co...was established in 1864 by W. H. Hine and John M. Gipps, under the firm name of Hine & Co. In 1867 Mr. Hine sold out to H. Howe. The brewery was then conducted under the firm name of Gipps, Howe & Co. In 1869 Mr. Howe sold his interest to Mr. N. Shurtleff; this firm was known as Gipps & Shurtleff. **In 1871 Mr. Shurtleff dissolved his connection with the firm, and Messrs. Thomas Cody and Willis Balance became associated with the firm as Gipps, Cody & Co. In 1874 Mr. Cody retired, and Leslie Robinson became a partner** [see Gipps & Co. listing, this chapter].

The business is splendidly located, being situated between the railroad and the river...adjacent to Union Depot. The brewery building...is 60 by 100 feet, and two stories. The malt house and kiln is 25 by 75 feet. Two fermenting cellars, each 20 by 75 feet, and two large beer cellars, each 40 by 75 feet, with an ice house 16 feet high covering all the cellars.... The valuation of the brewery and contents is placed at about $75,000. There are nine experienced men employed about the buildings.... They brew four times a week in winter and six times or more in summer. Fifty barrels of beer or ale are brewed at a time, and they can malt three hundred bushels of grain a week. Their sales amount to about five hundred barrels a month during the winter and double this number during the summer. The firm have extended their trade outside of Peoria, and within a radius of one hundred miles on all the railroads leading out of the city.

Beginning as early as the late 1860s (CDs), the firm was known as the Eagle Brewery. The embossed bottle described above dates to ca. 1871–74, the three-year period during which Thomas Cody was associated with the brewery.

John M. Gipps and Co.

Ale ~ Beer:

(1) **GIPPS & Co. / PEORIA / ILLINOIS** [yellow and olive amber]
(base): **D. S. G. Co.** (encircling large embossed X)

»SB w/ post mold; slope-shouldered quart; embossed in circular slug plate; narrow blob top; 12" tall. *[Figure 900a–b]*

(2) **GIPPS & Co. / PEORIA / ILLINOIS** [amber]
(base): **L&W**

»SB w/ post mold; slope-shouldered quart; embossed in circular slug plate; lettering style indicates different mold; narrow blob top; 11 1/2" tall. *[Figure 900c–d]*

(3) **GIPPS & Co. / PEORIA / ILLINOIS** [yellow amber]
(base broken away on studied example, but probably):
D. S. G. Co. (encircling large embossed X)

»SB w/ post mold; slope-shouldered tall pint; embossed in circular slug plate (same style as #1); narrow blob top; 9 1/2" tall. *[Figure 901]*

As discussed in the *Gipps, Cody & Co.* entry, John M. Gipps began a brewing business during the late Civil War years. He continued in the brewing business until his death in 1882 (see Biographical Publishing Co. 1890; Oertle 2008). From the technological and stylistic attributes of the embossed Gipps bottle listed, and from the DeSteiger Glass Company maker's mark on the bottle, it was manufactured and used in the late 1870s, soon after Thomas Cody had left the company. During the 1870s (1870–71 CD ad), *Gipps & Co.* were "manufacturers of beer, stock and cream ales, porter, vinegar, and all kinds of malt." Beginning as early as the late 1860s (CDs), the firm was known as the Eagle Brewery. A woodcut image of the brewery complex published in the 1880s (1887 CD: opposite p. 120) shows the scale of Gipps late-nineteenth-century brewing operations (*Figure 902*).

With a hiatus in operation during the Prohibition Era, the Gipps brewery continued in operation long after John Gipps passed away—well into the mid-twentieth century (Oertle 2008).

Figure 900

Figure 901

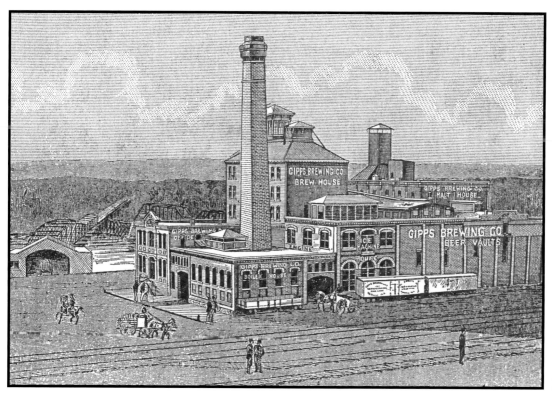

Figure 902

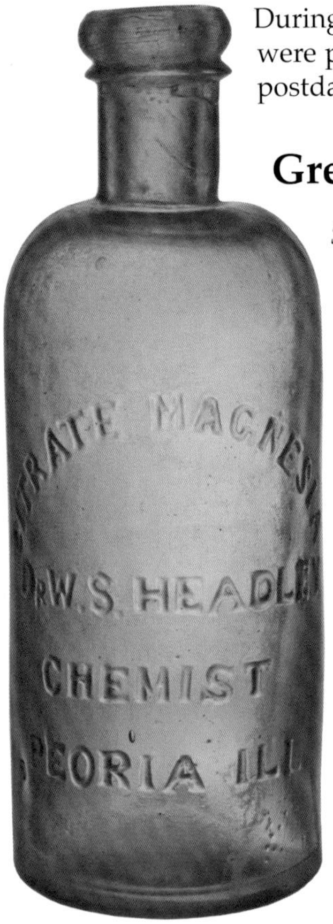

During the late 1800s and early 1900s, several additional varieties of embossed Gipps bottles were produced and used. But these are not listed as they were made after 1880 and thus postdate the end of our present study.

Great Western Tonic Bitters

See the O. P. Bissell listing in this chapter.

Dr. William S. Headley

Medicine:

(1) **CITRATE MAGNESIA / DR W. S. HEADLEY / CHEMIST / PEORIA ILL**   [aqua]
»SB w/ keyed hinge mold; high-shouldered cylinder; short neck; double ring collar; 7 1/4" tall. *[Figure 903]*

William S. Headley placed his first CD ad as a "Druggist & Chemist" in the 1861 Peoria City Directory, as "successor to Dr. B. F. Miles" at 54 Main (*Figure 904*). But the ad must have been placed late, because he was not yet listed as a druggist in the body of the CD. He appeared in subsequent CDs through the Civil War years and was last listed in the CD for 1865.

Headley was listed as a physician in the 1864 ISD, and was listed as "Dr." Headley in the 1863 and 1865 CDs, but his primary occupation appears to have been that of a pharmacist. His embossed citrate of magnesia was likely produced and sold ca. 1862–64.

Figure 903

Ransom E. Hickey

Soda ~ Mineral Water:

(1) **R. HICKEY / PEORIA / ILLS**   [aqua]
»IP; long-neck, shouldered-blob-top soda. *[Figure 905]*

Figure 904

Figure 905

(2) **R• E• HICKEY // PEORIA / ILL•** [aqua, blue teal]

» "Pebbled" smooth base, sometimes referred to as "sand pontiled"; long neck; oval-blob-top soda. *[Figure 906]*

(3a) **R. HICKEY // PEORIA / ILLS /** [aqua]
(heel): F.A & Co

» IP; long-neck, shouldered-blob-top soda. *[Figure 907]*

(3b) **R. HICKEY // PEORIA / ILLS /** [aqua, green]
(heel): F.A & Co

» SB w/ keyed hinge mold; long-neck, shouldered-blob-top soda. *[Figure 907]*

(4) **R. HICKEY // PEORIA / ILLS /** [aqua]
(heel): F.A & Co
(base): H

» SB w/ keyed hinge mold; shouldered-blob-top soda. *[Figure 908]*

(5) **R. HICKEY // PEORIA / ILLS** [aqua]

» SB w/ keyed hinge mold; long-neck, shouldered blob-top soda. *[Figure 909]*

(6) **R. E. HICKEY / PEORIA / ILL** [aqua]

» SB w/ keyed hinge mold; oval-blob-top soda. *[Figure 910]*

John Cairns and Ransom Hickey became partners in Hickey's first Peoria mineral-water-bottling venture at Seventh and Spencer Streets shortly after he purchased the Peoria Mineral Spring in 1853 (*see Figure 911*:1844 plat map detail). Cairns, a prominent St. Louis, MO, soda and mineral water bottler bottler, was probably an absentee financial backer of the operation (*see Cairn & Hickey listing, this chapter*). During the early and mid-1850s, Ransom Eaton Hickey was joined in the Peoria bottling operation, off and on, by at least two of his brothers, Hiram M. Hickey and Sephreness ("Stephen") M. Hickey. All three were listed in an 1855 Peoria County Census as bottlers in a soda manufactory, and the 1857 USC listed both Ransom and Hiram as soda water manufacturers. From Peoria Historical Society information, Hickey bought out Cairns' share of the Peoria bottling business by 1857, after which stayed on as a bottler in Peoria during the early and mid-1860s. His embossed sodas are listed in approximate chronological order of use: the pre-1860 pontiled sodas are listed first, followed by the smooth-base Civil War–era sodas.

Hickey's Peoria soda water manufactory was listed in the 1860 IIC, which indicated that between June 1859 and June 1860 he used 11,500 pounds of sugar, 2,000 gross corks, 2,000 pounds of twine, and "other materials" with a total value of $2,300 in his hand-powered factory to produce 2,000 gross bottles of soda water valued at $8,400. He employed 10 male laborers and paid them an annual average of $180 each for their work.

According to Donath (1991:2), "Ransom Hickey...diversified the uses of the Spring waters by making ginger ale and other *[flavored]* drinks. First the drinks were bottled *[at his residence]*, with the aid of a water line directly from the Springs to the lower level of the house. Later on, he

Figure 906

Figure 907

Figure 908

Figure 909

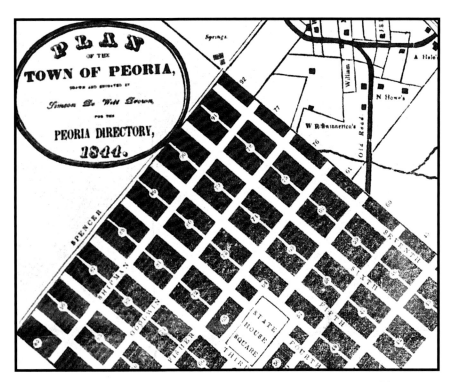

Figure 911

bottled in the barns across the street, and finally a brick factory was built there in 1855."

During the later 1850s and early 1860s, Ransom's brothers, and his uncle James E. Eaton (*see Eaton's Galesburg listing*), moved away from the Peoria area to establish bottling works at a number of other towns in both northern Illinois (*see Hickey brothers listings for Bloomington, Joliet and LaSalle, and Eaton listings for Chicago*) and southern Wisconsin (see Peters 1996).

Figure 910

Ransom Hickey died in 1868. He apparently relinquished management of bottling-plant operations after 1863. Afterward, the plant was operated for a year or two by *Lorentz & Co.*—a.k.a. Bohl, Singer & Lorens (see Bateman, ed., 1902, no embossed bottles known)—and then by Karl Gillig and Charles Singer (*see their listing, this chapter*) until Gillig retired in 1875 (Donath 1991:3). Hickey seems to have remained somewhat active in bottling-plant operations during the mid-1960s, however, since he was listed as a soda manufacturer in the 1865 and 1867 CDs.

After Gillig's retirement, Singer (*see his separate listing, this chapter*) continued to bottle at the Hickey plant until 1879 when Hickey's uncle James Eaton returned to the Peoria/Knox County area to set up his own bottling operation at the old Hickey Peoria Springs bottling factory at 704 Seventh Street (*see Eaton's listing and Figure 872*).

Edward J. Humphreys

Soda ~ Mineral Water:

(1) **E. J. HUMPHREYS / N⁰ 55 / MAIN Sᵀ / PEORIA**   [aqua]
   Note: There are two embossed dots under the superscript **O** *in* **N⁰** *and the superscript* **T** *in* **Sᵀ**.

   »IP; shouldered-blob-top soda. [*Figure 912*]

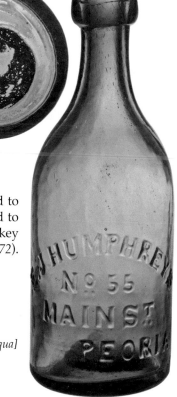

Figure 912

According to the 1850 USC, Edward Humphreys was a 25-year-old Peoria druggist. Humphreys was also listed in the 1850 Peoria CD, as a druggist from his residence on Main between Washington and Adams. The 1850 CD also contained an advertisement (p. 182) stating that he sold "the best of Mineral Water in its season, with syrups that cannot be surpassed."

The next available Peoria CD was published in 1856. The 1856 CD also listed Humphreys as a druggist whose business was located at 55 Main Street. Humphries was no longer listed in the 1857 CD, or at any time after that.

The embossed bottle listed was thus manufactured and used by Humphreys during the early 1850s.

Dr. Hero Kruse

Medicine:

(1) (side): **D^R. H. KRUSE^S** // // (side): **MALARIA ANTIDOTE** // //   [aqua]
(base): **L&W**
Note: Beneath the underlined raised **S** *in* **KRUSE^S** *are two diagonal embossed lines (/ \).*

»SB w/ keyed hinge mold; beveled-edge rectangle w/ sunken embossed side panels; long neck; tapered collar; 6 1/4" tall. *[Figure 913]*

Dr. Hero Kruse was first listed as a Peoria physician in the 1864–65 ISD. He was listed as a physician, surgeon, and "occulist" in the 1867 Peoria CD. The Peoria CDs continued to list him as a doctor (eventually focusing on his occulist practice) up to 1873, first at No. 9 Bridge and later at South Adams Street "next to the City Hotel." After that, he became a "druggist," with a store at 511 South Adams. According to the CDs, Kruse then ran his pharmacy in Peoria throughout the 1870s and into at least the early 1880s:

The Doctor is a graduate of the Medical University of Heidelberg, Germany, and practiced as a physician for many years in Peoria, and is thus fully qualified to put up prescriptions of all kinds. About four years ago the Doctor went into the drug business and has been doing finely, his store is one of the most elegant in the city, and his stock is large and well selected, embracing everything known to the drug trade. During his long and constant attention to this intricate branch of business, **he has made some very valuable discoveries in the mysteries of pharmacy, thus enabling him to bring relief and make whole the many human creatures whose lives are made wretched by disease.** Among those wonderful rem-

Figure 913

edies is Dr. Kruse's Malaria Antidote for fever and ague, which of late has been gaining a wide and lasting reputation, and the Doctor's sales are increasing rapidly; he also keeps a large stock of imported roots and herbs constantly on hand. [Edwards Peoria City Directory 1876:99]

Given this product history, and the embossed bottle's keyed-hinge-mold construction, it likely dates to ca. 1867–70, while Kruse was still in medical practice in Peoria. The medicine's success may have been the impetus for him to shift his attention to the drug trade.

Adolph L. Matthies

Soda ~ Mineral Water:

(1) **A. L. MATTHIES / CHEMIST / & / APOTHECARY / PEORIA. ILLS** [cobalt]
Note: Several examples of a similar-shaped unembossed cobalt-blue pontiled soda bottle have also been dug in Peoria. Since the Matthies smooth-base soda is the only other early cobalt Peoria soda, perhaps the pontiled cobalt bottle was made for his use when he first launched his pharmacy business in 1859.

»SB w/ keyed hinge mold; shouldered-blob-top soda. *[Figure 914]*

Figure 914

According to city directories, Adolph L. Matthies opened an apothecary shop in Peoria in 1859. He is listed as a druggist in the 1860 USC and in Peoria CDs through 1875. His drug store was located at 317 S. Washington Street.

From the style and technological attributes of the Matthies embossed soda bottle listed, it dates to the Civil War era.

Peoria Seltzer Water

See the Hubert Fellrath listing in this chapter.

George Metz and John C. Pletscher

Ale ~ Beer:

(1a–b) **METZ & PLETSCHER / PEORIA ILL'S / THIS BOTTLE / NEVER TO BE SOLD** [amber]
(base): **M G Co**
Note: Bottles from this mold were finished with two different closure styles: shouldered-blob top and ring-neck "cider" finish.

»SB w/ post mold; slope shouldered quart; 10 3/4" tall. *[Figure 915]*

From a Peoria CD search, John C. Pletscher (*see his separate listing, this chapter*) began to bottle "malt liquors and wines" from his residence at 107 Fulton in 1874 or 1875 (1875 CD—there was no CD for 1874, and Pletscher sold "pumps" in 1873). He was also listed alone in the CDs as a bottler from 1876 to 1878, a few doors away from his residence, at 414–416 Fulton. He was no longer listed after 1878, and no one named Metz was ever listed in the late 1870s CDs as a brewer, bottler, or alcoholic-beverage dealer.

From the shapes and applied-lip styles of the embossed bottles, the two *Metz & Pletscher* beer/ale bottles listed are likely to have been manufactured earlier (mid-1870s) than the late-1870s ale bottle in the *J. C. Pletscher* listing. But there were only three listings for anyone named Metz in the 1875 CD: John Metz (a policeman), Rudolph Metz (a harness-maker), and Johann Metz, ("resident").

Perhaps Mr. Metz was an early nonlocal financier of Pletscher's bottling operation. The most likely senior-partner candidate is George Metz, who had previously been the senior partner

Figure 915

Figure 916

in the *Metz & Stege Brewery* of Chicago (*see A. Stege's Aurora listing*). According to Chicago city directories, Metz left the *Metz & Stege* Chicago brewery partnership after the 1874 season—just the time to join the short-lived *Metz & Pletscher* partnership when it was formed in Peoria in 1875. There is little doubt that he became the senior partner/financier of the *Metz & Pletscher* Peoria brewery, although he was not a city resident.

Anthony Parent

Ale ~ Porter:

    (1) **A. PARENT / PEORIA. ILL**   [dark amber;
        (base): **L&W** amber "black glass"]

        »SB w/ keyed hinge mold; two-piece mold quart ale; ring-neck tapered collar; 10" tall. *[Figure 916]*

    (2) **A. PARRENT. / PEORIA / ILL**   [amber]
        Note: Light-strike embossing; "**PARRENT**" misspelled.

        »SB w/ keyed hinge mold; two-piece mold pint ale or porter; ring-neck tapered collar; 8 1/2" tall. *[Figure 917]*

Several varieties of unembossed two-piece ale bottles have also been found in 1870s Peoria contexts (*see Figure 918, left*), one of which still retains its paper label. These bottles were also likely used in Anthony Parent's bottling business there. They include:

    (3a) (Applied decorative paper label):   [amber "black glass, amber, olive amber]
        PITTSBURG CREAM ALE / A. PARENT'S / 321 MAIN ST., / PEORIA, ILL

        »SB w/ crude keyed hinge mold; two-piece mold quart ale; ring-neck tapered collar; 10" tall. *[Figure 918, right]*

(3b) (base): **L&W** [amber]

»SB w/ keyed hinge mold; two-piece mold quart ale; ring-neck tapered collar; 10" tall.

(3c) (unembossed) [lime green]

»SB w/ refined keyed hinge mold; two-piece mold quart ale; ring-neck tapered collar; 9 1/2" tall.

(3d) (unembossed) [amber, green]

»SB w/ keyed hinge mold; two-piece mold quart ale; ring-neck tapered collar; 10" tall.

(3e) (unembossed) [olive green]

»SB w/o mold lines, flat heel, concave base, and central glass bump; two-piece mold quart ale; ring-neck tapered collar; 10" tall.

The 1870 USC listed Anthony Parent as a bookkeeper in Peoria. In 1871 or 1872, Parent began bottling "Peoria lager beer, Pittsburgh stock and cream ale and porter, cider and native wines, for family use" (1873 CD, p. 160) at 91 Main [*"new 319"*] in Peoria. He was listed in the Peoria CD for 1875 as having opened a restaurant at 319 Main. He had moved his bottling operation to 109 Main (*Figure 919*), where he was listed as a bottler of "beer, ale, porter, cider and wine, and wholesale dealer in cider and manufacturer of cider vinegar" (1875 CD, p. 170). Parent was no longer listed in the Peoria CDs after 1875. He had moved to Detroit, MI, by 1880, where he was a real estate dealer (1880 USC).

The two embossed keyed-hinge-mold bottles described both date to ca. 1871–72. The unembossed ales listed likely date to the 1873–75 period.

Figure 917

Figure 918

Figure 919. Hand-colored print of the interior of A. Parent's beer and ale depot at 321 Main Street in Peoria. (Published by A. T. Andreas as part of his 1873 Atlas Map of Peoria County, Illinois.)

John C. Pletscher

Ale ~ Porter:

(1) **J. C. PLETSCHER / PEORIA ILL** [dark amber]
 (base): **C&I**

 »SB w/ post mold; 2-piece mold quart ale; ring-neck tapered collar; 9½" tall.
 [Figure 920]

John Pletscher began to bottle "malt liquors and wines" from his residence at 107 Fulton in 1874 or 1875 (1875 CD—there was no CD for 1874, and Pletscher sold "pumps" in Peoria in 1873). He was also listed as a bottler in the CDs from 1876 to 1878, a few doors away from his residence, at 414–416 Fulton. He was no longer listed in the CDs as in business in Peoria after 1878.

From its stylistic and technological attributes, the embossed bottle listed likely dates to the later years of Pletscher's Peoria bottling effort. During the

Figure 920

first season or two, his embossed bottles indicate he bottled with a (silent?) partner named Metz (*see Metz & Pletscher listing, this chapter*).

Augustus W. H. Reen

Soda ~ Mineral Water:

(1) **MINERAL WATER / BY / A. W. H. REEN**   [aqua]

»SB w/ raised ring near heel; shouldered-blob-top soda. *[Figure 921]*

❖

According to historical-society research by Frank Donath, Augustus Reen was trained in the pharmacy business in Prussia, where he was born in 1825. He came to the United States in 1858 and came to Peoria in 1860, where he was listed as a clerk in the 1861 CD.

He was first listed as a druggist in the CDs in 1863, and was in partnership with Fritz Reuter in the pharmacy business until 1868, when Reen was first listed alone as an apothecary and druggist at 75 Main.

From 1870 to 1872, his ads listed him as a prescription druggist who also sold "wines, liquors, oils, etc." In the later 1870s, he moved his pharmacy to 303 Main and was listed only as a druggist/chemist. The embossed Reen bottles are rarely found, and from their style and technological attributes they likely were manufactured and used ca. 1870–72, when Reen advertised that he was selling other bottled products as well.

Figure 921

James B. St. Clair [and John A. Finkler]

Soda ~ Mineral Water:

(1) **J. B. S**^{T} **CLAIR & C**^{O} **/ PEORIA / ILLS /**   [aqua]
(*left front heel*): **F.A & C**^{O}
*Note: The raised **T** in **S**^{T} has two embossed dots beneath it.*

»SB w/ keyed hinge mold; ling neck; shouldered-blob-top soda. *[Figure 922]*

❖

James St. Clair was a bottler in Peoria during the Civil War era. He was first listed as a partner in a Peoria bottling works with John A. Finkler in the IRS tax assessment of May 1864. Finkler had also earlier been involved with bottling businesses in LaSalle (*see his LaSalle listing*) and later in Streator (*see his Streator listing*).

Fahnstock, Albree & Co. was a glass manufacturing business that produced embossed soda bottles in Pittsburgh for only four years: 1860–63 (*see introduction, this volume*). Thus the St. Clair & Co. bottle was likely manufactured in 1863, when they began their bottling business in Peoria (they likely set up shop too late for inclusion in that year's Peoria CD). There was no 1864 CD, but St. Clair was listed alone as a bottler in the 1865/66 Peoria CD as the operator of a soda factory at the corner of Bluff and Knox roads. When the next Peoria CD was published in 1867, he was no longer in business there.

His location and activities during the later 1860s are unknown, but from at least 1871 into the early twentieth century he operated a soda-bottling works at Muscatine, IA (Burggraaf and Southard 1998:479–480).

Figure 922

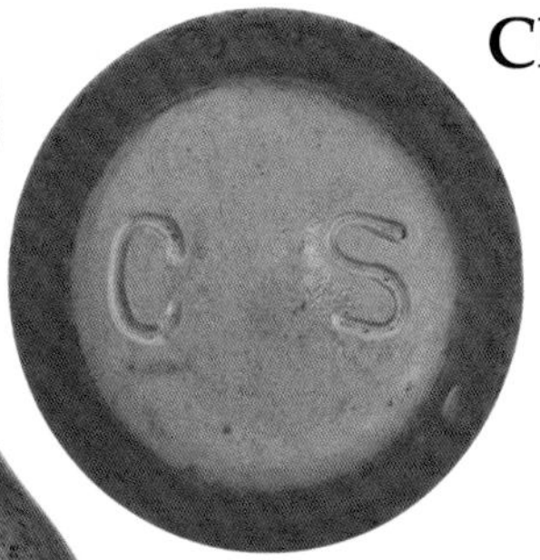

Charles Singer

Soda ~ Mineral Water:

(1) **CHARLES SINGER / PEORIA / ILLS** [aqua]
  (base): **C S**

  »SB; oval-blob-top soda. *[Figure 923]*

(2a) **CHA^S SINGER / PEORIA / ILLS //** [aqua]
  (heel): **A. & D. H. C**
  (base): **C.S**
  Note: There are two embossed dots under the raised **S** *in* **CHA^S**.

  »SB w/ post mold; short neck, shouldered-blob-top soda. *[Figure 924, left]*

(2b) **CHA^S SINGER / PEORIA / ILLS //** [aqua]
  (heel): **C & I**
  (base): **C.S**
  Note: There are two embossed dots under the raised **S** *in* **CHA^S**.

  »SB w/ post mold; oval-blob-top soda. *[Figure 924, right]*

(2c) **CHA^S SINGER / PEORIA ILLS //** [aqua]
  (heel): **C & C^O**
  (base): **C.S**
  Note: There are two embossed dots under the raised **S** *in* **CHA^S**.

  »SB w/ post mold; short neck, oval-blob-top soda. *[Figure 925]*

Figure 923

Figure 924

Singer came to Peoria in 1855 and operated a saloon and a hotel for a number of years during the late 1850s and early 1860s (Biographical Publishing Co. 1890:766). He was a partner in a "sideline" soda water bottling business with Charles Gillig when both men were Peoria hotel operators during the Civil War years (*see their listing, this chapter*). Then near the end of the war years, he turned his attention to bottling full time at the old Hickey bottling factory near the Peoria Mineral Spring. For a season or two he was part of a three-man partnership that apparently did not use embossed bottles (*see Ransom Hickey listing, this chapter*).

Karl Gillig, Charles Gillig's son, became Singer's partner at the Hickey plant ca. 1867 or 1868 (*see their separate listing, this chapter*). Karl Gillig retired in 1875, after which Singer operated the business alone until his death in 1887. He vacated the Hickey factory in the late 1870s when Hickey-relative James Eaton returned to bottle there (*see Gillig & Singer and Eaton listings, this chapter*), afterwards running his bottling operation from new headquarters at 105–107 Fulton Street.

He was listed there in the 1879 CD, at the time of his move, as a "manufacturer of Selters Water, Soda Water and Ginger Ale." He was also listed in the 1879–80 IIC as a Peoria soda water manufacturer who employed five adult males and one child year-round and paid out $1,296 in wages during the year (at the rate of $1.25–$1.50 per day). Singer estimated the cost of production materials that year at $600, and the retail value of his bottled products at $3,800.

After his death in 1887, his sons continued the business. Hutchinson-style embossed bottled bottles with internal rubber stoppers were used by the business during the 1880s, but these were used too late for inclusion in the present volume.

Figure 925

Robert Strehlow

Bitters:

(1) **ROBERT STREHLOW // GERMAN / WINE BITTERS // PEORIA ILL_ // //**   *[amber]*

   »SB w/ keyed hinge mold; beveled-edge square bitters; flat panels; short neck; tapered collar; 9" tall. *[Figure 926]*

---❖---

Robert Strehlow is listed in the 1860 ISD as a spirits rectifier, distiller, and dealer in liquor. He appears in the 1856–1861 Peoria CDs as a liquor merchant on Fulton between Water and Washington. In the 1863 and 1865 CDs, he was listed at 7 & 9 Fulton as a "wholesale and retail liquor merchant, and with a partner named Sieber, as a "rectifier" and "distiller." There was no 1866 CD produced, but Strehlow was no longer listed in 1867 or later as a Peoria businessman. From the bottle style and manufacturing technology, and from the CD listings, his bitters were likely produced during the final few years of the Civil War. No newspaper ads or paper labels have yet been seen to document the claimed health benefits of *Strehlow's German Wine Bitters*.

George Weber

Beer:

(1) **GEO. WEBER / PEORIA. / ILL. //**   *[amber, olive-amber]*
   (base): **D S. G. Co.** (arched around embossed "**7**")

   »SB w/ post mold; slope shouldered quart; ring-neck tapered collar; 10¾" tall. *[Figure 927]*

644 PEORIA — GEORGE WEBER

Figure 926

Figure 927

(2) **GEO. WEBER / PEORIA. / ILL. //** [amber]
   (heel): **DeS. G. Co.**

   »SB w/ post mold; slope shouldered quart; oval blob top; 10 3/4" tall. *[Figure 927]*

In the 1871–75 Peoria CDs, *Weber & Co.* (probably Frederick Weber and W. Cockle) were listed as operating the Union Brewery (see Ogg 1996:53). The 1875–76 CD still listed Frederick Weber "at brewery" at 1381 Washington Street, and George Weber was listed for the first time as a "beer bottler" from his residence at 166 N. Fayette. In the CDs of the later 1870s, the brewery was no longer listed, but George Weber was still listed as a beer bottler from 1876 to 1878 and as a "wholesale and retail dealer and bottler of Joseph Schlitz Milwaukee beer" in 1879.

In the 1880 USC, George Weber's occupation is given as beer bottler and saloonkeeper. The 1880–81 Peoria CD listed him only as a saloonkeeper.

From the glassmaker's mark on the embossed bottles listed, and from their stylistic and technological details, they were likely manufactured and used ca. 1877–79. Based on the changing maker's mark style, the bottles are listed in probable order of manufacture.

John Wichmann

Beer ~ Ale:

(1a) **JOHN WICHMANN / PEORIA / ILL //** [amber]
   (heel): **A. & D. H. C**

   »SB w/ post mold; slope shouldered quart; oval blob top; 10 3/4" tall. *[Figure 928]*

(1b) **JOHN WICHMANN / PEORIA / ILL //** [olive green]
   (heel): **A. & D. H. C**

   »SB w/ post mold; slope shouldered quart; ring-neck cider finish; 10 3/4" tall. *[No photo; embossed as #1a]*

(2) **JOHN WICHMANN / PEORIA / IL** [amber]
   (heel): **C & I**
   Note: An olive-green variety of this bottle is known as well, but it is broken at the neck so lip finish is unknown.

   »SB w/ post mold; slope shouldered quart; ring-neck cider finish; 10 3/4" tall. *[Figure 929]*

According to the early Peoria CDs, John Wichmann was involved in several business ventures during the later 1850s and early Civil War years. He was variously listed as a boot-and-shoe

Figure 928

Figure 929

maker (1856), a grocer (1857, 1861), a saloonkeeper (1859–60), and the proprietor of a combined grocery and saloon at the corner of New and Plank (in 1863).

By the time of the next Peoria CD (1865), when he was in his early 40s, he had gone into the brewery business (with Franklin Gabler) as proprietor of the *Central Brewery* on Madison near Linden. By 1867, he had a new partner at the *Central Brewery* (H. Lammers) and the two men were listed as "beer and lager beer" brewers. Their brewery partnership continued until 1875.

During the final few years of *Central Brewery* operation, 1876 through 1877, Lammers had left the partnership, and Wichmann ran the brewery on his own. After that, it was no longer listed in the CDs, and seems to have closed its doors. From 1878 through 1880 (CDs; USC), Wichmann was listed as a saloon operator and beer bottler.

From the style and manufacturing technology of his embossed bottles, they were likely made and used ca. 1875–1880, after the end of his *Central Brewery* partnership with Lammers.

PERRY

(Pike County)

1850 census: 400
1870 census: 800
1880 census: 770

Perry Springs

Mineral Water:

(1) **CHALBEATE WATER / I / PERRY SPRINGS ILL'S** [amber]

»SB w/ keyed hinge mold; pint; Saratoga-style mineral-water bottle; 7 1/2" tall. *[Figure 930a]*

(2) **CHALYBEATE WATER / M / PERRY SPRINGS ILL'S** [aqua]

»SB w/ keyed hinge mold; pint; Saratoga-style mineral-water bottle; 7 1/2" tall. *[Figure 930b]*

As part of their unpublished report on an archaeological excavation study conducted in 1993 at the site of the famed Perry Springs resort hotel and health spa in rural Pike County, Farnsworth et al. (1995) summarized the history of the resort, and its three mineral springs, as follows:

> In the late 19th century, Perry Springs was an immensely popular mineral-water spa and 150-room hotel complex (Figure 930c), advertised by its owners as the "Saratoga of the West" (Hollenberg 1979; Landrum 1968, 1980, 1986a; Norris 1978). It was first developed between 1856 and 1861 by Zachariah Wade, who built a two-story, 52-room stucco hotel on the site. After the Civil War, the complex was purchased and expanded by Springfield confectioner Benjamin Watson and business partner Noah Divelbiss. By 1867, Watson had added a four-story guest residence and entertainment facility that was connected to the original hotel via an elevated covered walkway. Watson's new structure tripled the guest capacity of the spa: according to a note in the January 19, 1867 *Jacksonville Daily Journal* "the two hotels at Perry Springs were jammed to the gills last week, with 300 guests from nine states registered. The place now has 15 bath houses. [Watson 1870, 1878]

The Perry Springs resort was located about two miles east of Perry in northeastern Pike County on a small tributary to the Middle Fork of McKee Creek. It was developed amid three floodplain springs along a 500-foot stretch of the stream. As shown in Figure 930d (detail from Pike County township map published by Andreas, Lyter & Co. 1872b), they were identified respectively as the Iron Spring, Magnesia Spring, and Sulphur Spring:

> Each not only tastes differently, but operates differently, and what a wise provision of Providence is here illustrated—three springs, but a few rods apart, all strongly medicated and having

Figure 930

each different medicinal properties; and of all the diseases that affect the human family [there are] but very few of them but what one of these springs would not greatly relieve, if not wholly cure. [Andreas, Lyter & Co. 1872b:8]

Some 'over educated idiots'...computed that if you drank enough [Iron and Magnesia] water to float a battleship [you might get] a theraputic dose of the element. One of them said that you could get more iron by sucking an old-fashioned hand cut nail than...by drinking a cistern full of the water. But there was absolutely no gain saying anything about the sulphur spring. You could smell the sulphur dioxide a quarter of a mile away—and more than that if the wind was in the right direction" (Tendick 1973:26).

Each of the three springs was credited with different curative powers. But the Sulphur Spring, located across the road from the hotel near the stables, was not developed as elaborately as the other two. According to a detailed 1866 description of the newly completed resort complex:

The water from the Iron Spring is efficacious for all chronic diseases. That of the Magnesia, for liver complaints; while that of the Sulphur has been found to be remarkably useful to those afflicted with "sore eyes."
Two of the springs, Iron and Magnesia, are covered with elegant pagodas...surrounded by basins made of Joliet marble finely cut, and the water flows up into urns of the same material. The spring houses and basins are more elegant than the famous 'Congress Spring' at Saratoga. ["AMICUS" 1866:3]

Embossed bottles of *Perry Springs* water are so far known from only two of the three springs: an aqua variety with a large embossed "**M**" (for Magnesia Spring water) and an amber bottle with a large embossed "**I**" (for Iron Spring water). If Watson also bottled water from the Sulphur Spring, the bottle would likely be embossed with a large "**S**" and would likely be color-coded a pale yellow amber (*see Chicago's Ypsilanti Spring sulphur water bottles*).

From the style and technological attributes of the embossed *Perry Springs* bottles, Watson likely shipped *Perry Springs* water in such bottles from the late 1860s to the late 1870s. An advertisement he placed for the Perry Springs Hotel in the August 12, 1876, *Quincy Daily Herald (p. 2, col. 7)* also noted: "waters from these springs will cure all cases of dyspepsia, diabetes, Bright's Disease, Kidney, Stomach, and bowel diseases. **Water shipped in glass or wood**."

Benjamin Watson was the proprietor of Perry Springs until 1880, when he sold the resort.

PERU

(La Salle County)

1860 census: 3,100
1870 census: 3,700
1880 census: 4,600

Kasper Haas

Soda ~ Mineral Water:

(1) **K. HAAS / PERU, ILL.** *(flat embossing)* [aqua]
   (base): **C&I** *(large embossing)*

»SB w/ post mold; oval-blob-top soda; w/ high sloped shoulders extending to blob top (no neck). [*Figure 931*]

Figure 931

The 1860 USC listed a 50-year-old brewer and cooper named "Christian" Haas in Peru, IL. Christian Haas was also listed in the 1864 *Illinois State Gazetteer and Business Directory* as a brewery operator in Peru.

Kasper Haas (probably Christian's son) was first listed in the May 1865 IRS tax assessment for Peru as a retail liquor dealer. Kasper was not found in the 1870 USC, but he was listed in the 1880 USC for Peru as a 47-year-old soda water manufacturer.

From the style and technology attributes of his embossed soda bottle, it appears to have been produced and used during the 1870s—likely the later 1870s. Since the *Cunningham & Ihmsen Co.* (**C&I**) ceased manufacturing embossed bottles in Pittsburgh in 1878 (Lockhart et al. 2004:2–8), Kasper Haas likely went into the soda bottling business ca. 1875–78.

PITTSFIELD

(Pike County)

1850 census: 600
1860 census: 2,100
1870 census: 1,600

"Dr. Thomson" [Thomas Grimshaw & Co.]

Medicine:

(1) DR THOMSON^S // GALVANIC // LINIMENT // PITTSFIELD. ILL^S [blue aqua]
Note: *superscript* **R** *in* **D**^{R} *has single embossed dot beneath it.*

»OP; beveled-edge square; 4 sunken embossed panels; thin flared lip; 4 3/4" tall; 1 1/4" square.
[Figure 932]

The earliest ad found in regional newspapers for *"Dr. Thomson's" Galvanic Liniment* appeared in Pittsfield's *Pike County Free Press* on May 22, 1851. It announced that *Thomas C. Grimshaw & Co.* were the sole Pike County agents for *Thomson's Galvanic Liniment.* (An April 17, 1850 ad in the same paper indicated that Grimshaw & Co. were successors to Pittsfield druggist John Grimshaw.) A search of historical, genealogical, and census records gives no hint that there was ever a "Dr." Thomson in Pike County, although there was apparently a blacksmith named John D. Thomson living in the area in 1853. Also, the *Illinois State Gazetteer* for 1858 lists a Thomas Thomson as proprietor of Pittsfield's Atlas Hotel. Thomas Grimshaw may well have created the "Dr. Thomson" name as a marketing aid for his liniment.

By 1855, Grimshaw was actively expanding his sales market for the liniment. His *Pike Free Press* ad on April 26 that year indicated that *TGL* was intended for "the lame, stiff jointed and rheumatic patient" and that "this liniment can be got of any store in the county and other counties. Agents wanted in every town in the western states; all letters must be directed to T. C. Grimshaw, Pittsfield."

During the 1850s, *Thomson's Galvanic Liniment* was marketed as far away as Quincy in Adams County, where the most detailed ads were placed by local wholesale and retail agent H. H. Hoffman:

DR. THOMSON'S GALVANIC LINIMENT
We have just received a full and large supply of this very valuable liniment, and would call particular attention to it as a sovereign and thoroughly tried remedy for Rheumatism,

Figure 932

Swellings, Sprains, Toothache, Sweeney, Big Head, Poll Evil, Galls, &c., &c. [*Quincy Daily Whig,* July 11, 1853]

THOMSON'S GALVANIC LINIMENT
It is now generally admitted that the liniment is a little ahead of any other preparation now before the people, for the cure of Rheumatism, Swellings, Sprains, Lame Backs, Chilblains, Burns, Ulcers, etc. It acts like a charm; the galvanic properties strike to the root to the disease and give immediate relief. [*Quincy Weekly Herald,* March 23, 1857, p. 3, col. 8]

The last advertisement noted for *TGL* was published by the *Pike Free Press* for January 24, 1861:

Dr. Thomson's GALVANIC LINIMENT
[has?] a curative power over man and beast unequaled by any preparation of the kind...has stood the test of ten years use...cures Sprains, Bruises, Cuts, Ring Bone, Poll Evil, Big Shoulder, Big Head, Wind Chills, and will instantly relieve Colic...cures Rheumatism, Neuralgia, Sprains, Bruises, Chilblains, and Frosted Feet.... It never fails when used according to instructions. It is no Humbug, nor is it a worthless Nostrum. It is based upon scientific principles, and cannot fail to give satisfaction. Price 25 cents. This celebrated LINIMENT is manufactured by T. C. Grimshaw, Pittsfield, Ill.

The term "galvanic," of course, implies chemical production of an electric current. Perhaps some ingredient(s) in *Dr. Thomson's Liniment* produced a tingling sensation on the skin.

PONTIAC

(Livingston County)

1850 census: 30
1860 census: 700
1870 census: 1,700
1880 census: 2,200

James Caldwell and Charles McGregor

Ink:

(1) **CALDWELL & McGREGOR'S / INK / PONTIAC, ILL.**   [aqua]

»SB w/ heel mold; cylinder ink w/ 2 raised-glass rings at both heel and shoulder; slightly domed shoulder; short neck; flared rounded lip; 3" tall, 1 7/8" diameter body. [*Figure 932.5*]

Figure 932.5

James Caldwell was a 29-year-old druggist and Charles McGregor was a 25-year-old retail book and stationery dealer according to the 1870 USC for Pontiac. The 1875 ISD listed the firm as booksellers and stationers. An 1880 issue of *The St. Louis Medical Society Journal* (V. 38, No. 12) included the company in the regional Druggists Directory as "booksellers, Pontiac." In 1889, *Caspar's Dictionary of the American Book, News, and Stationery Trade* still listed Caldwell & McGregor as booksellers, stationers, and druggists in Pontiac (Caspar 1889:688).

According to a biographical sketch of Caldwell printed by the S. J. Clarke Publishing Company (1900:90):

> In July, 1865, Caldwell came to Pontiac, Illinois, and embarked in the drug business on West Madison street in partnership with John A. Fellows, under the firm name of Fellows & Caldwell. They continued in business together for some 3 years and were finally succeeded by the firm of Caldwell & McGregor, who for twenty-five years carried on a most successful business, theirs being by far the oldest drug store or business firm in the city. To their stock of drugs and books they later added jewelry and built up a good trade in that line. They built the block on the northeast corner of Mill and Madison streets, then the finest business block in the city.... In 1895 the partnership was dissolved and Mr. Caldwell retired from the drug trade.

From descriptions of the style and technological attributes of their embossed master-ink bottle, it appears to have been produced and used ca. 1875–1880.

PRENTICE

(Morgan County)

[1860–70 population unknown]
Township and countywide population data only.
Very small town (see the following discussion)

John M. Berry

Bourbon Bitters:

(1) (side): **JOHN M. BERRY / PRENTICE ILL'S // //**   [olive amber]

(side): **UNCLE MARB'S / OLD BOURBON BITTER'S // //**
(base, encircling central depression): **K H & G Z O**

»SB w/ keyed hinge mold; **quart** shoo-fly flask (w/ vertical mid-side ridges; vertical side embossing of the two flat panels on each side); unembossed, flat front and back panels; ring-neck cider finish; 9 1/4" tall, 2 1/8" deep. *[Figure 933]*

(2) **UNCLE MARB'S / O. B. / BITTERS** ["amber"]

»SB, likely w/ keyed hinge mold; **pint** shoo-fly flask (w/ vertical mid-side ridges); embossed only on one flat face, w/ unembossed, side and back panels; ring-neck cider finish; 7 1/8" tall, 1 1/2" deep.

The pint variety of this bottle (#2) has not been seen by us, but was documented and illustrated as a line-drawing (*see Figure 934*) by Carlyn Ring (1980:465), who worked from a photograph sent to her by the owner. She was unable to document the mold-line seam on the base (it was probably a keyed hinge mold), and she recorded the bottle color as "amber" (it was likely olive-amber like the quart). Probably because of the photo quality, she also referred to the flask as "oval" (i.e., a "coffin" style), rather than as a shoo-fly flask (with vertical mid-side ridges like the quart), and misread the initials **O. B.** (for "Old Bourbon") on the pint bottle as "**C. B.**"

In the 1860 and 1870 Morgan County USC records, John M. Berry was listed as a merchant in Mauvaise Terre Township. He was also a charter member of Odd Fellows Lodge No. 341, organized at Prentice on October 9, 1867 (Perrin 1882a:142). During these years, the village of Prentice itself was so small as to be almost nonexistent—little more than a cluster of buildings around an early mill. An 1861 Scott and Morgan County plat map, with an inset map of Prentice, showed the place to be comprised of just 52 residential lots and 14 business lots, and listed only

Figure 933

six business subscribers in the town: a hotel operator, a physician, a shoe-maker, a blacksmith, a grocer, and a general merchant named H. W. Boyce. John Berry was not listed (C. D. Moody and F. Hess 1861).

    This town lies twelve miles north of Jacksonville, on the Chicago and Alton Railroad. It was laid out June 27, 1857.... In an early day Mr. Hall's father bought an old horse mill, near this place, and, after remodeling it, ran it for many years. To this mill settlers came from all parts of the country. The old mill, after good service, wore out, and was replaced by one more modern, which was used until February, 1878, when it was destroyed by fire. [Donnelley, Loyd & Co. 1878a:432]

Berry was not mentioned, and he was no longer listed in any of the Morgan County business directories or resident lists when the Scott and Morgan County history was published in 1878, although M. S. Berry (a son?) was listed as a farmer residing at Prentice, and mention was made that both parents (unnamed) of Columbus Berry, who lived "14 miles from Jacksonville," had died in 1872 (Donnelley, Loyd & Co. 1878a).

An interesting aspect of embossed bottle #1 is the glassmaker's mark embossed on its base. **KH&G ZO** was the maker's mark for Kearns, Herdman & Gorsuch (**KH&G**) of Zanesville, Ohio (**ZO**). This particular partnership of the Kearns glass companies operated from 1868 to 1885/86 (Lockhart et al. 2008:3, 5–6). In the 1840 to 1880 scope of the present study, this bottle is one of only two known Illinois examples of an embossed **KH&G ZO** maker's mark (*the other is a Frederick A. Kuehle soda—see his Murphysboro listing*).

From the style, maker's mark, and technological attributes of the embossed *Uncle Marb's* bottles, they were likely produced, filled, and marketed during the late 1860s or early 1870s.

No ads or paper labels for *Uncle Marb's Bitters* have as yet been located, so the claimed medicinal value of the product is not known. But the fact that it was called *"Bourbon" Bitters* and was packaged in a whiskey flask, rather than a traditional bitters-shaped square quart bottle, suggests it was primarily marketed as liquor.

Figure 934

PRINCETON
(Bureau County)

1855 census: 2,000
1860 census: 2,500
1870 census: 3,300
1880 census: 3,400

Friedrich Althoff

Soda ~ Mineral Water:

  (1) **C. ALTHOFF / PRINCETON / ILLS.**   *[blue aqua]*

      »SB w/ keyed hinge mold; large embossed lettering; long neck; shoulder-blob-top soda. *[No photos]*

  (2) **F. ALTHOFF• / & CO. / PRINCETON / ILLS**   *[aqua]*

      »SB; large embossed lettering; long neck; shoulder-blob-top soda. *[Figure 935]*

Figure 935

From the stylistic and technological attributes of the two embossed soda varieties listed, they were manufactured and used during the Civil War era. Fred Althoff was found listed as a Princeton soda-water manufacturer in the 1873 CAR and in the 1880 USC, where he was listed as a 53-year-old "popp" maker born in Prussia. He is known to have lived in the Princeton area as early as 1860, when the USC listed him as a resident of Bureau, IL, located on the Illinois River about eight miles SE of Princeton.

At the time of the 1880 census, Fred Althoff had a 21-year-old son named Charles (who went on to become a famous duck-decoy painter in the Princeton area and whose hand-painted decoys are now highly sought-after by collectors). Since one of the 1860s bottles listed above was embossed **C. ALTHOFF**, perhaps Fred had a father or brother, possibly named Charles (not found in the historic records), who started the business. Alternatively, his middle name may have been Charles (again, not documented historically), or the bottle may just reflect a glassmaker's mold error.

Fred Althoff apparently continued bottling soda water during the 1870s and early 1880s, but we have not seen embossed bottles from this time.

Dr. A. R. Bodley

Medicine:

(1) **DOCTOR BODLEY'S / FEVER.AND.AGUE // BALSAM // // //** [aqua]

»OP; beveled-edge rectangle; 4 sunken panels; embossed on two adjacent panels; short neck; double-ring collar; 7 1/2" tall. *[Figure 936]*

------- ❖ -------

From the bottle's stylistic and technological attributes, this seldom-seen embossed medicine was likely produced and sold for just a year or two during the later 1850s. Its manufacturer and town of origin would likely be a mystery, if not for the coincidental publication of a volume of historical sketches of Princeton, IL, and surrounding Bureau County by Isaac Smith in 1857. Smith's publication also contained an 1857 Princeton business directory, accompanied by the following ad (Smith 1857:61):

PEOPLE'S DRUG STORE.

BODLEY & WILSON,
Druggists and Apothecaries,

Dealers in Paints, Oils, Varnishes, Camphene, Turpentine, Brushes, Cigars and Fancy articles; all of the popular Patent Medicines of the day. Proprietors of **DR. BODLEY'S FEVER AND AGUE**

Figure 936

BALSAM, which is unequaled as a cure for all Fevers of an Intermittent character. Garden, Field and Flower SEEDS, constantly on hand. Agents for **CLARK'S FEMALE PILLS**.
Princeton, Illinois.

A. R. BODLEY, M.D. H. WILSON

Dr. Bodley will attend to calls in his profession. Office at the Drug Store.

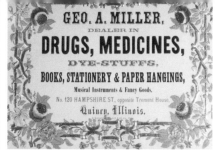
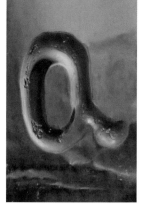

QUINCY
(Adams County)

1850 census: 6,900
1860 census: 13,700
1870 census: 24,000
1880 census: 27,300

Herman and William Boschulte

Soda ~ Mineral Water:

(1) **BOSCHULT & BRO_ / QUINCY / ILL //**   *[aqua]*
(heel): **A & D. H. C**
Note: Name spelled "Boschulte" in business listings and census data, but "Boschult" on the bottles.

»SB w/ keyed hinge mold; shouldered-blob-top soda. *[Figure 937]*

―――――――― ♦ ――――――――

Figure 937

Two of the five Boschulte brothers, Herman and William, had a long-time involvement in the Quincy soda and mineral water bottling business. Their father, German immigrant Peter Henry Boschulte, had come to Quincy ca. 1852 and established a soda-bottling business with (Lewis?) Lampe, circa 1855. Soon afterward, Peter passed away, and his 20-year-old son Herman took his place in the business (Landrum 1967; Wilcox 1919:402). The two men were listed together as soda makers in the 1857–58 Quincy CD (*see Lampe & Boschulte entry, this chapter*). By 1861 (CD), *Fischer & Boschulte* (including Boschulte brothers Herman and William) were listed as Quincy soda manufacturers at a new address (*see their entry, this chapter*).

Sometime during the early Civil War years (no CDs available), Henry Fischer dropped out of the business and it became *Boschulte & Bro*. Their listing in the 1864 CD indicates their bottling operation was located at 146 State (rear). In the 1866 CD, Herman was listed as manager of the company.

By 1868, William and younger brother Charles were listed as operators of the family bottling works, along with new partner Adam Knauf, as *Boschulte & Co.* (*see their entry, this chapter*). The CDs indicate that *Boschulte & Co.* continued in business through 1872, after which the company closed its doors. From 1873 onward, the two senior Boschulte brothers (Herman and William) and Adam Knauf were listed as employees of *Durholt & Co.* (*see Henry Durholt listing, this chapter*).

William & Charles Boschulte and Co.

Soda ~ Mineral Water:

(1a) **BOSCHULT & Co. / QUINCY / ILL //** [aqua]
(*heel*): **A & D. H. C**
Note: *Mold modification of* **BOSCHULT & BRO_** *bottle, with* **BRO_** *peened out and replaced by* **Co.**

»SB w/ keyed hinge mold; shouldered-blob-top soda. [*Figure 938*]

(1b) **BOSCHULT & Co / QUINCY / ILL //** [aqua]
(*heel*): **A & D. H. C.**
Note: *Newly cut mold for bottle #1a (rear heel mark also smaller and differently punctuated* (**A & D. H. C.**).

»SB w/ keyed hinge mold; shouldered-blob-top soda. [*Figure 939*]

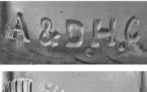

Figure 938

Figure 939

Various combinations of the five Boschulte brothers, occasionally with other partners, were involved in the Boschulte family bottling business from the late 1850s to the early 1870s *(see Boschulte & Bro, Fischer & Boschulte, and Lampe & Boschulte listings, this chapter).*

Beginning in 1868 (CD), William and younger brother Charles were listed as operators of the family bottling works, along with new partner Adam Knauf, as *Boschulte & Co*. The 1870 IIC of Quincy included the "Boschulte & Knauf" soda water works, which was in operation six months a year for two years, using from one to four adult male employees who were paid a total of $700 between June of 1869 and June of 1870. Bottling supplies—including 2,500 bottles ($200), 100,000 corks ($600), 1,500 lbs vitriol ($50), marble dust ($160), and other materials ($160)—cost $1,170 during the census year, and the value of their bottled soda water product (4,500 cases) was estimated at about $3,400.

The CDs indicate that Boschulte & Co. continued in business through 1872, after which the company closed its doors. From 1873 onward, the two eldest Boschulte brothers (Herman and William) and Adam Knauf were listed as employees of *Durholt & Co.* (see Henry Durholt listing, this chapter).

Dr. D[avid] Bunnel

Medicine:

(1) **BUNNEL'S / GERMAN // // LINIMENT // //** [aqua]

»Strange "puddled" unmolded SB; shouldered, beveled-edge rectangle; 4 flat panels; embossed on opposite-side wide panels; short neck w/ very narrow opening (ca. 0.3"); early 1850s-style thin flared lip; about 4" tall. *[Identified from fragments: See Figure 939.5.]*

───────── ❖ ─────────

Several damaged examples of the embossed bottles listed were discovered in disturbed construction context at a Quincy building site. Dr. Bunnel's Quincy newspaper ads for his *German Liniment* all date to 1853 and 1854.

AS MOSES lifted up the serpent in the wilderness, that the children of Israel might be healed, even so man shall be healed by using Bunnel's German Liniment, then why will ye suffer the

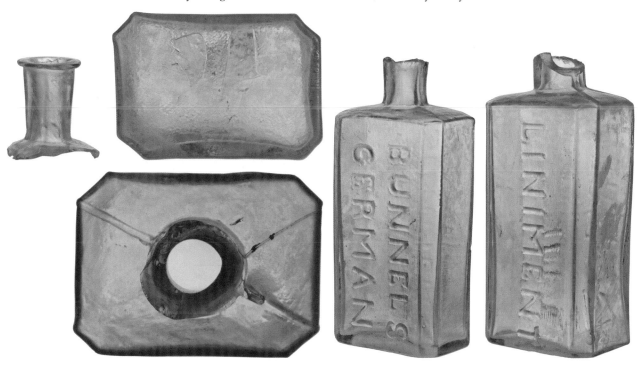

Figure 939.5

excruciating pains that the human family is subject to, when 25 cents will relieve you. Therefore be ye prepared, for no man knoweth the day or the hour that affliction cometh. The German Liniment is an entire eradication of all pains, no family should be without it.—This liniment has established a reputation where it has been used, which cannot be surpassed by any in use. The use of one bottle will do more to convince you of its efficacy than all the advertisements in the world. Warranted to give satisfaction when used according to directions, or money refunded. Prepared only by the proprietor.

D. BUNNEL,
Quincy, Illinois

For sale by nearly all the respectable dealers throughout the state of Illinois. Agents wanted in every town and village where there is not one already established. All orders addressed to the proprietor will be promptly responded to. [*The Quincy Whig*, April 4, 1853, p. 4]

The newspaper ad was printed in the local papers from early 1853 to early 1854. By July 1853, Dr. Bunnel formed a partnership with Dr. J. E. Kirkpatrick to open the Bunnel & Kirkpatrick eye infirmary in Quincy on Hampshire Street at the northwest corner of the town square, advertising themselves as "Occulists & Surgeons." By January 1854, Dr. Bunnel was operating the *Eye Infirmary* on his own. But by mid-year he was out of business, and a local attorney advertised in the *Quincy Daily Whig* that he had been empowered to collect debts owed to "the late firm of Dr. Bunnel and Co."

The Bunnel's bottle fragments shown above are hard evidence for the ***only*** example we have documented of a smooth-based (rather than glass-rod pontiled or iron-pontiled) early-1850s Illinois product bottle. The technology used for this bottle's manufacture, and the glass house that made it, are unknown. From the bottle's size, weight, and narrow neck, it was likely made by a flint-glass tableware manufacturer who normally produced refined clear-glass hollow wares such as tumblers, perfumes, and chemist bottles.

Alonzo, David, and John Burleigh

Medicine:

(1) *(side):* **BURLEIGHS** // **25C** // *(side):* **ELECTRIC BALM** // **GRATIS** [aqua]

»OP; beveled-edge rectangle; 3 full sunken panels (sides, face); sunken half-panel (op. face); short neck; rolled lip; 4 5/8" tall. [*Figure 940*]

(2a) **BURLEIGH & BRO'S** // // **ELECTRIC BALM** // **QUINCY / ILL** *(top of panel)* [aqua]

»SB w/ keyed hinge mold; 1 1/2" x 3/4" beveled-edge rectangle; 4 sunken panels; 3 1/4" tall to shoulder (lip and neck broken away). [*Figure 941a(left), d–e*]

(2b) **BURLEIGH & BROS** // **QUINCY / ILL** *(top of panel)* // **ELECTRIC BALM** // // [aqua]

»SB w/ keyed hinge mold; 1 1/2" x 1 1/8" beveled-edge rectangle; 4 sunken panels; 3 3/4" tall to shoulder (lip and neck broken away). [*Figure 941a(right), b–c*]

During the later 1850s, when bottle #1 would have been produced and used, none of the Burleigh brothers were based in Quincy. According to the 1860 USC, Alonzo S. Burleigh was living and selling the brothers' *Electric Balm* in New Albany, IN (just across the Ohio River from Louisville) and was listed as a patent-medicine vendor. [Odell (2007:59) notes that the Burleighs' *Electric Balm* was advertised in *Indianapolis Banner* on May 11, 1853.] David B. Burleigh and his brother John Q. Burleigh resided together in Palmyra, MO (just across the Mississippi River, about 20 miles southwest of Quincy), where they were both listed as patent-medicine vendors. They obviously marketed their *Electric Balm* across the River to the Quincy and west-central area, since examples of the bottle have been recovered in both Quincy and Petersburg. Illinois must

Figure 940

Figure 941

have been a lucrative market for their liniment-style pain medicine, because the two brothers in Palmyra relocated to the Quincy area during the Civil War years.

The earliest Illinois listing for *Burleigh & Bros.* patent-medicine business was found in the November 11, 1865, edition of the *Quincy Whig Republican*:

> **BURLEIGH & BRO., who are manufacturing** medicine at Mendon, Ill., hereby warn their patrons in Adams and the adjoining counties against a man by the name of NATHAN E. MORRIS passing through the country on foot, reporting himself as our agent, which is false, also collecting money from those who have our medicine. He has not been in our employ since June, and if he has any of our accounts to collect they are stolen.

A December 1866 tax assessment listing for Quincy, IL, also lists the Burleigh brothers' business in operation there. Alonzo Burleigh is listed as a 47-year-old physician in the 1870 USC of Quincy. The 1870 USC listed David B. Burleigh as living in Mendon, IL, a town located about 10 miles northeast of Quincy in Adams County.

According to a Quincy newspaper notice (*The Weekly Whig and Republican* May 29, 1869, p. 3, col. 6), the firm ceased to sell its *Electric Balm* and was dissolved in 1869:

> **Dissolution.**
> The firm heretofore known as Burleigh & Brother is this day dissolved by mutual consent. All claims against the said firm will be settled by
> D. B. BURLEIGH
>
> May 15th, 1869 Mendon, Ills.

However, Dr. A. S. Burleigh continued to practice medicine in Quincy after the dissolution of the partnership. He first renounced patent medicines altogether and advertised himself as an "Odyllic Physician" for chronic diseases, with offices on Spring between 6th and 7th, healing the sick without medicine:

> Dr. Burleigh gives no medicine but cures by a new and scientific method which is far superior to any yet discovered.... So powerful and so natural is this system, that by it many...who have for years suffered untold miseries in the back and head have been entirely cured in one week. Dr. Burleigh's practice is very effective in all spinal diseases; also cures Cancers of all kinds in from 10 to 15 days, and that too without leaving the patient poisoned, but his blood will be cleansed so that it will never again return. [*Daily Quincy Herald,* August 31, 1867, p. 4, col. 2]

However, Dr. Burleigh and his practice continued to "evolve." Less than a decade later, a half-page ad in the 1876 Quincy CD (p. 24), headed "**QUINCY NOT BEHIND, BUT IN ADVANCE**," announced that "Dr. A. S. Burleigh is engaged in manufacturing a general assortment of FAMILY MEDICINES at 61 N. 5th Street" (*Figure 942*). These included the earlier *Electric Balm* and Indian Life Pills, Worm Candy, Condition Powders, Egyptian Salve, Blood Purifier, Ready Relief, English Horse Liniment, Cough Balsam, Hair Cream, and a "general assortment" of Flavoring Extracts.

As of this writing, no embossed bottles of mid-to-late 1870s age have been documented for any of these products (including the *Electric Balm*).

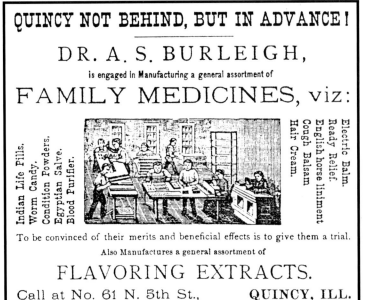

Figure 942

Henry Durholt

Soda ~ Mineral Water, Early Varieties (ca. 1865–70):

 (1a) **HENRY DURHOLT // QUINCY ILL.**   [aqua]

  »SB w/ keyed hinge (*heel*) mold; shouldered-blob-top soda. *[Figure 943]*

 (1b) **HENRY DURHOLT // QUINCY ILL.**   [green]

  »SB w/ keyed hinge (*base*) mold; blob-top-era soda (w/ neck and lip broken away). *[No photo. Same mold embossing as #1a.]*

 (2) **HENRY DURHOLT // QUINCY ILL. /**   [aqua]
  (*heel*): **A & D. H. C.**

  »SB w/ keyed hinge mold; shouldered-blob-top soda w/ shorter neck than #1a. *[Figure 944]*

Soda ~ Mineral Water, Late Varieties (ca. 1878–80):

 (3) **HENRY DURHOLT / QUINCY ILL //**   [aqua]
  (*heel*): **A. &. D H. C**
  Note: Reused old mold with **"& Co"** peened out.

  »SB w/ keyed hinge mold; shouldered-blob-top soda. *[Figure 945]*

 (4) **HENRY DURHOLT / QUINCY ILL //**   [aqua]

  »SB w/ post mold; squat slope-shouldered oval-blob-top soda (as much as 1/2" difference in neck lengths of individual bottles—see *Figure 946*).

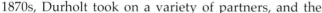

  Henry Durholt first entered the Quincy soda bottling business in 1855 at the age of 29. He likely had substantial financial backing from his father-in-law—prominent St. Louis soda and mineral water bottler Henry Grone—with whom Durholt had apprenticed when he lived in St. Louis from 1847–55 after immigrating to America from Germany, and from whom he had learned the bottling trade (*see Grone & Durholt listing, this chapter*). Durholt married Henry Grone's daughter Catherine in 1849 (Chapman Bros. 1892b:469). Grone, of course, stayed in St. Louis when Henry Durholt moved to Quincy to start the new bottling business, but two of Grone's sons came along with Durholt as partners in the new *Grone & Durholt* business.

  Henry Durholt, who was a soda bottler in Quincy for some 30 years (*see also Henry Durholt & Co. listing, this chapter*), is first listed in city directories as sole owner of the bottling works in 1863, when he bought out his in-laws' interest in the earlier Grone & Durholt partnership. His earliest embossed bottles, used during the later 1860s, are listed in approximate chronological order as #1–#2. During the early and middle 1870s, Durholt took on a variety of partners, and the

Figure 943

Figure 944

Figure 945

Figure 946

bottling works were known as *Henry Durholt & Co.* (*see listing, this chapter*). During the final few years of the 1870s, the company was again managed solely by Durholt. The embossed bottles used then are listed as #3–#4.

Throughout his long bottling career, Henry Durholt's factory was located along the 100 block of 7th Street (in the area bordered by 6th and 7th streets and State and Kentucky streets) (*Figure 947*). His bottling company continued in business during the final two decades of the nineteenth century as well (Landrum 1967). The embossed bottles used during these final decades of the factory's use were manufactured too late for inclusion in the present volume.

Durholt's business was eventually taken over by his son-in-law Frederick C. Hoffman, who continued to operate the business out of Henry Durholt's early factory building behind his residence (*see Figure 947c*: historic photos provided courtesy of Carl Landrum).

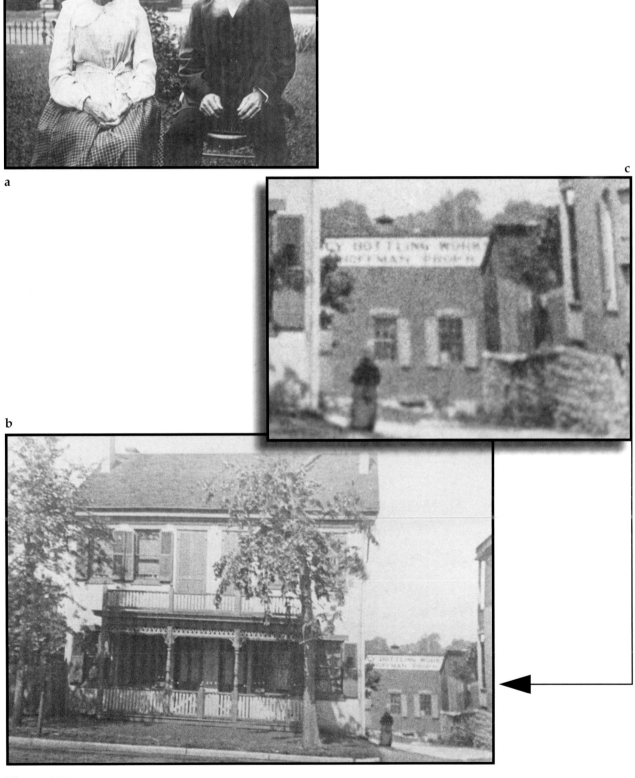

Figure 947

Henry Durholt and Co.

Soda ~ Mineral Water:

    (1) **HENRY DURHOLT & C⁰ / QUINCY ILL //**   [aqua]
         (heel): **A. &. D H. C**

        »SB w/ keyed hinge mold; shouldered-blob-top soda. *[Figure 948]*

    (2) **HENRY DURHOLT & CO / QUINCY ILL //**   [aqua]
         (heel): **C & I**

        »SB w/ post mold; shouldered-blob-top soda. *[Figure 949]*

    (3) **HD & Co** (*very large embossing*) **//**   [aqua]
         (heel): **C & I**

        »SB w/ post mold; shouldered-blob-top soda. *[Figure 950]*

Henry Durholt & Co. was listed as manufacturing soda water in Quincy beginning with the 1869 CD and the 1870 IIC. The Industrial Census that year listed 44-year-old Henry Durholt as a manufacturer of "Soda Water & Cider" using two bottling machines, a generator, and a hydrolic pump. They operated 12 months/year with 2–6 adult male employees. Durholt incurred $4,945 in costs that year for about 18,000 lbs. sugar ($2,500), 12,000 lbs. vitriol ($400), 35 barrels marble dust ($105), three barrels of tartaric acid ($140), and $1,800 for "corks, bottles, etc." The factory produced 17,143 cases of soda water valued at $12,000—and 900 dozen bottles of cider valued at $800.

Figure 948   **Figure 949**   **Figure 950**

According to the Quincy CDs, the "& Co." designation was used by Durholt from 1869 to ca. 1877–78. The 1869 and 1871 CDs indicated his business partners were Edward Hutmacher (*see Hutmacher listing, this chapter*) and Caspar Grone; the 1873 through 1876 CDs indicated his business partners were two of the Boschulte brothers (*see their listing, this chapter*) and Adam Knauf.

In the period from ca. 1878 to 1880, the CDs indicate that although Knauf was still associated with the company, Durholt had again taken on its sole management/ownership. During these few years, the Durholt embossed bottles did not include the "& Co." designation (*see Henry Durholt listing, this chapter, bottles #4 and #5*).

The 1880 IIC again listed the company name as *H. Durholt & Co.* and indicated that the factory produced soda water year-round (five months full-time, seven months half-time). That year, Durholt had seven adult male and four adult female employees who were paid a total of $1,700/yr for their work, at the daily rate of $1.25 to $1.50 each. He incurred $2,800 in annual materials costs and annually produced $800 worth of bottled soda water.

In the 1882 CD and throughout the 1880s, the Boschulte brothers were again partners in the business, and the "& Co." designation returned. These later bottles postdate the focus of the present volume and are not listed.

Henry Fischer and Herman & William Boschulte

Soda ~ Mineral Water:

(1) **FISHER & BOSCHULT / QUINCY / ILL** [aqua]
(heel): **A. &. D H. C**
Note: Senior partner spelled **FISCHER** in the CDs and **FISHER** on the bottles. Also, the Boschulte name is spelled with a final "e" in the CDs, but not on the bottles.
»IP; shouldered-blob-top soda. *[Figure 951]*

The Fischer & Boschulte soda manufactory was only listed in the 1861 Quincy CD, and was likely only a one-season or two-season business endeavor. But it served as Herman and William Boschulte's introduction to the soda-bottling business—a business they continued to pursue throughout much of the remainder of the late nineteenth century after Henry Fischer's departure from this early partnership (see *Boschulte & Bro., Boschulte & Co., Durholt & Co.,* and *Lampe & Boschulte* listings elsewhere in this chapter).

Figure 951

John J. Flynn and Co.

Soda ~ Mineral Water:

(1) **J. J. FLYNN / QUINCY / ILL'S.** [aqua, cobalt]

»SB w/ post mold; oval-blob-top soda. *[Figure 952]*

(2) **J. J. FLYNN / & Co / QUINCY / ILL'S.** [aqua]
(heel): **E H E**

»SB w/ post mold; oval-blob-top soda. *[Figure 953]*

(3) **J. J. FLYNN / & Co. / QUINCY / ILL** [aqua]
(base): **F**

»SB w/ post mold; oval-blob-top soda. *[Figure 954]*

(4) **J. J. FLYNN & Co / QUINCY ILL. //** [aqua]
(heel): **I. G. Co**
(base): **F**

»SB w/ post mold; oval-blob-top soda. *[Figure 955]*

❖

Although he did not have business listings in the late-1870s Quincy CDs, John J. Flynn (whose family were Irish immigrants, and who had earlier worked in a cotton mill in Massachusetts) first came to Quincy and began bottling "white soda" there (probably spruce beer and root beer) in May 1875 when he was 21 years old (*Quincy Daily Journal* 1889:120; Collins and Perry 1905:666). When he first arrived, he lived with his uncle, James Flynn, "at the foot of Washington Street" (Landrum 1986b). "Mr. Flynn, being a chemist, made his own formula for the white soda. He had the oil of lemons shipped in from Italy" (*Quincy Herald-Whig* 1970: historic business biography).

The next spring he turned his attention to the manufacture of root beer, starting at Front and Delaware. He operated as his limited capital would permit. As the business grew and he needed larger quarters, he moved to 708 Jersey where he bottled white soda from his own formula.... By 1881 he had moved to 200 Maine, where he erected a modern building and operated under the name J. J. Flynn. The firm made soda water and root beer.

There is some confusion on the early location of the company. Flynn's obituary stated that at one time he was on Ninth between Maine and Jersey and associated with Adam Knauf. It also stated that he started his first bottling works with spruce and root beer. [Landrum 1986b:4E]

According to the Quincy CDs, *J. J. Flynn & Co.* was established at 2nd and Maine by 1882, and Adam Knauf was his "& Co." partner from then until 1886. Flynn (*Figure 956*) and his delivery company (*Figure 957*) became

Figure 952

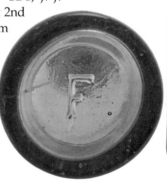

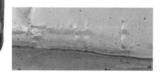

Figure 953

Figure 954

Figure 955

Figure 956

very successful in the soda water line, and by 1889 he had built an impressive bottling works factory complex along the 200 block of Maine, which continued to grow and prosper over the years (compare Figure 958 top *[late 1880s]*, Figure 958 middle *[early 1900s]*, and Figure 958 bottom *[1930s]*).

One of the finest factory buildings erected in Quincy this year is that occupied by J. J. Flynn & Co., carbonators and bottlers. This building is one hundred by forty feet, three stories high, and is supplied with all the new and machinery required in the business of making and bottling

Figure 957

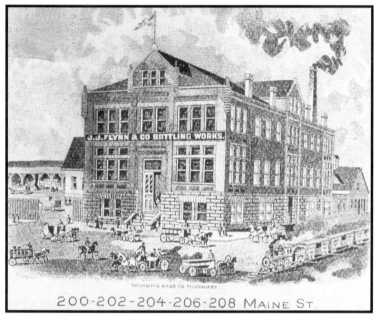

200-202-204-206-208 Maine St.

carbonated waters, extracts, sodas, ales, syrups, etc., etc. Between thirty-five and fifty men are employed during the year in the Quincy works. Branch houses are established in Illinois, Iowa and Missouri cities, and the goods put up by this firm are shipped into almost every town and hamlet in the above named states. J. J. Flynn & Co.'s goods are shipped as far south as New Orleans. [*Quincy Daily Journal*, 1889:120]

The business had operated at a much smaller scale only a decade earlier. The 1879–80 IIC recorded that Flynn's soda-water factory operated with 10 adult male and six adult female workers, but operated full-time for only four months a year and 1/3 time for the remaining eight months. Flynn paid annual wages of $2,600 ($1.25 to $1.50/day per worker), purchased $3,000 worth of bottling supplies annually, and produced an annual bottled product with a sales value estimated at $6,500. John J. Flynn continued to expand his business and his factory complex at 2nd and Maine throughout the late nineteenth century, and his sons and family carried on the business after he died in 1907. Given the 1840–80 scope of the present volume, the embossed soda bottles listed, in their approximate chronological order of use, are only those manufactured and used ca. 1875–80.

During the later nineteenth century and early twentieth century, Flynn and his company were great believers in advertising the company and its products (*Figure 959*), and their letterhead included a whimsically flamboyant company logo (*Figure 960*). The miniature etched seltzer-style bottle shown in Figure 959 (right) is only about 3" tall.

Figure 958

Figure 959

Figure 960

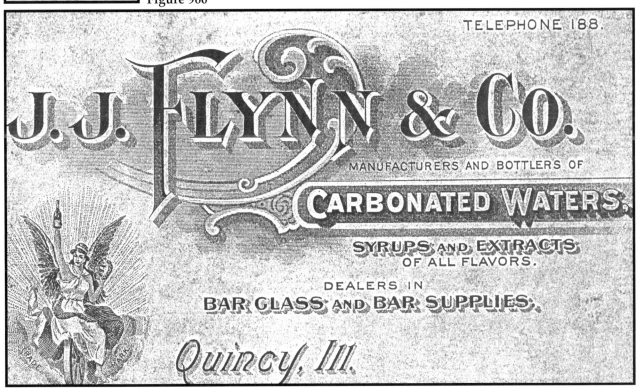

Caspar H. & Henry Grone and Henry Durholt

Soda ~ Mineral Water:

(1) **GRONE & DURHOLD // QUINCY ILL.**　　　　　[aqua]

»IP; shouldered-blob-top soda. *[Figure 961a–b]*

(2) **GRONE & DURHOLD // QUINCY ILL.**　　　　　[blue teal]

»SB w/ keyed hinge mold; shouldered-blob-top soda; longer neck than #1 above. *[Figure 961c–f]*

In 1855, when Henry (Johann Heinrich) Durholt was in his late 20s, he immigrated to the United States from Germany, first settling at St. Louis (Landrum 1967), where his father-in-law was in business. Durholt (misspelled "**DURHOLD**" on the early bottles) first came to Quincy and entered the Quincy soda bottling business in 1856, probably with substantial financial backing from his father-in-law—prominent St. Louis soda and mineral water bottler Henry Grone, Sr.—with whom Durholt had apprenticed and learned the trade. During the *Grone & Durholt* years, two of Henry Grone's sons, Caspar and Henry Jr. (see Smith 2005:20–22), were also partners in the Quincy business and worked at the Quincy bottling works. However, Caspar left the partnership after the 1859 bottling season (he may have died—see Chapman Bros. 1892b:469):

DISSOLUTION OF COPARTNERSHIP

THE COPARTNERSHIP HERETOFORE existing between C. H. Grone, Henry Grone and Henry Durholt, is this day dissolved by mutual consent. The business of the late firm will be continued by Henry Grone and Henry Durholt, (C. H. Grone having retired) under the name and style of Grone and Durholt. All accounts against and all accounts due the late firm will be settled by the new firm.

Quincy, Nov. 28th, 1859 **Henry Grone, Henry Durholt.**
[*The Quincy Daily Herald*, Dec. 6, 1859, p. 2, col. 7]

The *Grone & Durholt* company was listed in the 1860 IIC for Quincy as manufacturers of soda water, with two adult male employees who were paid a total of $50/month. Supply costs included $600 for 6,000 lbs. sugar, $50 for 100 lbs. tartaric acid, and $150 for "corks, twine, etc." That year the factory produced 12,000 dozen bottles of soda water valued at $6,000.

The *Grone & Durholt* partnership in Quincy ended during the early 1860s, when Durholt bought out his brother-in-law and assumed sole management of the company. The final listing for the *Grone & Durholt* partnership was in the 1861 Quincy CD. There was no CD issued in 1862, but by the time the 1863 CD was published, Henry Durholt was operating the Quincy bottling business on his own (see listings for *Henry Durholt* and *Durholt & Co.*, this chapter).

Figure 961

Edward Hutmacher

Soda ~ Cider:

(1) **HUTMACHER & RICHTER / QUINCY. ILL //**   [aqua]
(heel): **A. & D. H. C**
Note: For a short time when he first started bottling, Hutmacher was in partnership with Richter. Mr. Richter has not been identified in the CDs.

»SB w/ keyed hinge mold; oval-blob-top soda. *[Figure 962]*

(2) *H* (large letter in embossed circle) //   [aqua]
(heel): **C & I**
Note: Examples of this embossed bottle are only recovered in Quincy, in contexts dating to the time Hutmacher was in business (cf. **HD & Co** bottle style under Durholt listing).

»SB w/ post mold; oval-blob-top soda. *[Figure 963]*

Edward Hutmacher is all but invisible in searches of Quincy commercial records, CDs, and newspaper archives. The CDs indicate that in 1866 he ran a "soap and grease" factory. His next CD entries are in the directories of 1869 and 1871, and in the 1870 USC, when Hutmacher was an "& Co." partner at the *H. Durholt & Co.* soda-bottling plant. He was next listed as a grocer and cider manufacturer in the 1872 CD. Following 1872, he no longer appeared in the Quincy city directories.

Hutmacher's unaccounted for business years are 1867 and 1868. So his short partnership with Mr. Richter (identity unknown: see bottle #1) likely occurred during one of these two seasons. But the partnership must have been very short-lived, since the two men were never listed together in city directories.

Embossed bottle #2 was likely produced and used as a cider bottle in 1872, when Edward Hutmacher made a one-year effort to establish a grocery and cider manufactory in Quincy.

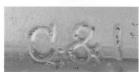

Figure 963   **Figure 962**

Edward Hutmacher has sometimes been confused with a major Quincy dealer in Mississippi River ice during the same period, since they had the same last name, and since bottling and ice-production businesses sometimes overlapped. But the *Hutmacher & Kreitz* river-ice business was run by Rudolph ("Rud") Hutmacher (Landrum 1969:119–121). The two men may have been related, but they were in independent businesses in the late 1860s and early 1870s.

Daniel Kaiser

Soda ~ Mineral Water:

(1) **DAN^L KAISER / QUINCY // THIS BOTTLE IS / NEVER SOLD** [green, blue, aqua]

»IP; early slope-shouldered tapered-collar soda. *[Figure 964]*

(2) **DAN^L KAISER // QUINCY ILL** [green?, aqua]

»IP; ca. early-to-mid 1850s tapered collar or shouldered blob top; 9" tall (quart-size?) pontiled soda (No photos. Rubbing of front and rear side-embossing only). *[Figure 965]*

Daniel Kaiser was listed in the earliest Quincy city directories (CDs for 1855 and 1857) as a manufacturer and bottler of soda water. In 1855, his bottling works was located on south State between 9th and 10th streets; in 1857 the address was given as 146 State. The first Quincy CD was published in 1855, but from the style of bottle #1, Kaiser had clearly established his bottling works there well before 1855. The tapered-collar closure style of the bottle dates its manufacture to between ca. 1845 and 1852–53 (*see introduction, this volume*).

Since Daniel Kaiser immigrated from Germany to the United States in 1851, and came to Quincy soon afterward, the evidence of the bottle indicates he first opened his soda bottling works there in 1852 or 1853. An 1855 Quincy census undertaken partly to enumerate local militia members listed Daniel Kaiser as a 30–40 year old head of a nine-person household that also included three males and a female aged 20–30, a 10–20 year

Figure 964

Figure 965

old female, and three small children. All four adult males were militia members, and Daniel Kaiser was listed as a soda water manufacturer. The annual value of his bottled product was estimated at only $300. This suggests that bottling was just a small-scale, part-time business for him in the mid-1850s, and explains why he was also said to have "made hubs for wagon wheels (Wilcox 1919:316).

His bottling works was no longer listed in the 1859 Quincy CD. He himself was listed for a final time in the 1860 CD as a Quincy "resident" only. Civil War–era smooth-based Daniel Kaiser bottles are known that are embossed "Keokuk, Iowa" (Burggraaf and Southard 1998:402), so Kaiser clearly moved his bottling operation some 30 miles upriver to Keokuk in 1860 or 1861. The glassmaker's mark on the Keokuk bottles (**F. A & Co**) indicates that the bottles were produced in Pittsburgh by *Fahnstock, Albree & Co* (soda bottle manufacturers there ca. 1860–63: *see introduction, this volume*).

Absalom J. Kalb

Medicine:

(1) *(side):* **KALBS** // *(front):* **EYE WATER** // // //   [aqua]

»OP; beveled-edge rectangle; 3 sunken panels; rolled lip; 2 3/4" tall. *[Figure 966]*

(2a) **KALB°S** // **VEGETABLE** // // **QUICK RELIEF**   [aqua]

»OP; beveled-edge rectangle; 2 sunken embossed side panels; sunken **KALB°S** half panel on upper front face; double-ring collar; 4 7/8" tall. *[Figure 967]*

(2b) **KALB°S** // **VEGETABLE** // // **QUICK RELIEF**   [aqua]

»SB w/ keyed hinge mold; beveled-edge rectangle; 2 sunken embossed side panels; sunken **KALB°S** half panel on upper front face; double-ring collar; 4 7/8" tall. *[Figure 968]*

Absalom J. Kalb first appears in the Quincy CDs in 1857 as a supplier of groceries and provisions at 118 Maine. He was likely making patent medicines as part of his business. These became popular enough that he listed himself as a patent medicine manufacturer in the 1859 CD, at 80 1/2 Hampshire. He continued to produce his flagship *Quick Relief* medicine during much of the late nineteenth century. His ads indicated that the *Vegetable Quick Relief* was intended "For Internal and External Uses, for **MAN** and **BEAST**. It will Kill Pain in from 3 to 15 Minutes."

Figure 966

Figure 967

During much of the 1870s and 1880s, his manufactory was located at 5th and Hampshire. From his 1861 CD ad, he was producing and selling several "vegetable remedies"—including *Pile Liniment, Blood Purifier, Vermifuge, Cough Syrup, Whooping Cough Alleviator, Corn & Wart Salve,* and *Sanative Vegetable Pills*—in addition to his *Eye Water* and *Quick Relief*, but the latter two (listed earlier) are the only ones known to have been sold in embossed bottles. A full-page ad he placed in the 1876 Quincy CD highlighted several additional vegetable remedies—including *Ague Tonic, All-Healing Salve,* and *Egyptian Rheumatic and Strengthening Salve*—but no embossed bottles have been located for these later medicines either. His business continued to be listed in Quincy, at different address, through 1891.

Kalb advertised himself in the Quincy papers as "Dr." A. J. Kalb, a physician with success curing cancers, as early as 1857 (*Quincy Whig*, February and March issues). The earliest ad found for Kalb's *Quick Relief* was published in the *Quincy Daily Whig & Republican* on Oct. 2, 1858 (p. 3, col. 1):

KALB'S QUICK RELIEF
For the cure of the Tooth-ache, Head-ache, Rheumatism, Neuralgia pains, Cuts, Sprains, Bruises, Dysentery, Diarrhoea, weak and inflamed eyes. This medicine is what its name implies—it cures quickly. It should be kept in every family.... Manufactured and sold wholesale or retail by
A. J. KALB
No. 80, north-west corner Square, Quincy, Ill.

The bottle style is an obvious knock-off of the widely popular *Davis Vegetable Pain Killer* medicine marketed over much of the United States during the later 1840s and 1850s from the company's headquarters in Rhode Island.

Figure 968

The Kalb's *Eye Water* bottle, with its rolled lip and open-pontiled base, dates to pre–Civil War times as well, but the earliest Quincy newspaper mention we have located for Kalb's *Eye Water* dates to May 14, 1860.

[Lewis] Lampe and Herman Boschulte

Soda ~ Mineral Water:

(1) **LAMPE & BOSCHULT / QUINCEY / ILL** [aqua]
   Note: "**QUINCEY**" *is misspelled on the embossing.*
   »IP; shouldered-blob-top soda. *[Figure 969]*

The soda-water bottling firm of Lampe & Boschulte is listed in the 1857 Quincy CD and again in the 1860 ISD, but their partnership was dissolved that year.

Their business venture was recorded in the June 1859–June 1860 IIC for Quincy, which indicated that their soda water factory had two adult male employees and labor costs of $50/month. They listed the value of supplies on hand as 6,000 lbs. sugar ($600), 100 lbs. tartaric acid ($50), and "corks, twine, etc." ($150). They filled 12,000 dozen bottles of soda water during the census year, valued at $6,000 (about 4¢ per bottle).

Herman Boschulte then entered into a new Quincy bottling partnership with his brother William and with Henry Fischer (ca. 1861–62; *see their listing, this chapter*). After Henry Fischer left this second short-lived partnership, the Boschulte brothers went on to establish a successful Quincy soda-water manufactory of their own from the later Civil War years through 1872 (*see Boschulte & Bro. and Boschulte & Co. listings, this chapter*). During the later 1870s and 1880s, they continued in the Quincy bottling business as part of Henry Durholt & Co. (*see Durholt listings, this chapter*).

Mathias R. and Henry W. Lundblad

Soda ~ Mineral Water:

(1) **M R & H W / LUNDBLAD // QUINCY** [aqua, green]
   »IP; double-ring embossing at shoulder; squat cylindrical body with decorative spiraled glass ribbing below raised shoulder rings (separated by spaces for embossing); long neck; tapered collar. *[Figure 970]*

This highly ornate Lundblad bottle is very similar to, and perhaps was adapted from, the design for a late-1840s iron-pontiled soda bottle used by George Eagle in New York City (see McKearin and Wilson 1978:241). The two very ornate bottles differ only in the number of raised shoulder rings and the direction of the spiraled body ribbing.

The Lundblad bottle was the earliest embossed mineral water bottle manufactured for use in Quincy, and the company far predated the earliest city directories. In fact, the bottle was one of the first few embossed soda or mineral water bottles used by Illinois

Figure 969

Figure 970

bottlers anywhere in the state (*see listings for Dinet in Chicago, Clark in Belleville, and DePuyt in Waterloo*).

The bottling works that Mathias and Henry Lundblad established in Quincy from 1849 to 1851 had a very short and ill-fated business history, but thanks to probate records, the company is unusually well documented historically. We are indebted to Kermit Wisterburg of the *Swenson Swedish Immigration Research Center* at Augustana College in Rock Island, IL, for his help tracking down the Lundblad brothers' arrival and early history in Illinois, and for translating the following from the original Swedish for us:

> Next to Gustaf Flack and Mr. Ostrom, who have been previously mentioned, [Henry] Lundblad (from Gotland) would have been the Swede who earliest began his own business. Lundblad arrived [*in Chicago*] in 1847, and the following year began a soda water factory, which he kept for several months. He then traveled to the city of Quincy in Illinois, where he died, but his wife returned here [*to Chicago*] and died at the hospital in Jefferson. The couple had two sons, who are likely employed at the Northwestern Railway Company here in Chicago.
>
> Later, the soda water factory came into the hands of Mr. Anders Larson. Mr. C. J. Sundell, later the consul, was also interested in it at the same time. [Johnson and Peterson 1880:242]

The Lundblad brothers both originally came to America by ship: MRL (24-year-old farmer) arriving in New York August 16, 1843; HWL (29-year-old merchant) arrived in New York with his wife and two children December 9, 1846." Mathias Lundblad first settled in New Orleans, but Henry Lundblad went to Chicago, where in 1847 he established himself in a business of bottling carbonated beverages. By 1850, he was living and bottling in Quincy, IL, where he and his brother Mathias had established a bottling works together.

According to the Chicago CDs, the Chicago bottling business that Henry Lundblad started in 1847 (at 137 Illinois Street) was not taken over by Sundell & "Lawson" (*a.k.a. Andrew Lawson—see*

Sundell & Lawson Chicago listing, this chapter) until 1854. But Henry Lundblad had relocated to Quincy with his brother after the 1849 Chicago bottling season (*see later discussion*). The occupants of his former Chicago factory, if any, from 1850 to 1853 are unknown.

Two Chicago CDs were published during the time Lundblad was bottling there. He was not listed in the 1847 CD; the only 1849 CD available to us was missing the mid-alphabet pages, and he did not place a business ad that year. His bottling operation may have been very small and home-based. There are no known embossed Lundblad Chicago bottles, which is perhaps not surprising since he and his brother used unembossed bottles during their first season in Quincy as well (*see later discussion*).

After the 1849 bottling season in Chicago, Henry and Mathias Lundblad relocated to Quincy to establish their bottling works there.

> **SODA APPARATUS**—The subscribers at this place, on Maine street, between Third and Fourth, have completed an apparatus for making Mineral and Bottled Soda Water, by which they will be able to furnish Soda of very superior quality to Refectorers **and in bottles** to our citizens, at cheaper rates than it can be obtained elsewhere. **Their forge pumps are very powerful, which will enable them to charge Soda more highly with carbolic acid gas than can be done by any machinery**. They also have an apparatus for filling and corking bottles, that is very ingenious. Persons who have any curiosity about the matter will do well to call and see the practical operation of their apparatus. M. R. & H. W. LUNDSBLAD. [*Quincy Whig*, March 19, 1850, p. 3, col. 3]
>
> **Mineral and Bottled Soda Water.**
> The subscribers are prepared to furnish a very superior article of **Mineral and Bottled Soda Water** at Wholesale, through the warm season. We would also inform our friends and citizens that our **Bottle Soda and Mead Fountains** are in successful operation, and will send their pure and sparkling streams to all our customers.
> N.B. All orders from the country for Bottle Soda Water, promptly attended to and carefully packed. M. R. & H. W. LUNDBLAD [*Quincy Whig*, March 26, 1850, p. 3, col. 4]
>
> **SODA WATER MANUFACTORY**—An establishment for the manufacture of the above article on an extensive scale has just been erected in this city. From the large quantity brought to the city from St. Louis last summer, we have no doubt the demand here will be sufficient to afford a liberal encouragement to the liberal and enterprising and ingenious proprietors—Messrs. Lundblad. They are located on the south side of Maine Street, between Third and Fourth. [*Quincy Herald & Argus*, April 19, 1850, p. 3, col.2]

The Lundblad brothers' first season in business seems to have been a successful one. Unembossed, pontiled, taper-top aqua soda bottles begin to appear in Quincy excavation contexts of this age for the first time. The brothers were listed in the 1850 USC for Quincy, residing (as renters) in the house where their bottling operation had been established. Henry and his wife Caroline were 33 at the time, with two young boys aged 6 and 2. Mathias, age 31, lived in the household, as did their soda-factory laborer George Gray (age 11).

However, the following year disaster struck. On June 11, just as the 1851 bottling season was getting under way, Henry Lundblad died. There was a cholera epidemic in Quincy at the time, and death was an everyday occurrence. Each week, the *Quincy Weekly Whig* listed those who had died the previous week, and their cause of death. But the issue for the week Henry Lundblad died was lost when the newspapers were microfilmed. He may have been killed by cholera, or he may have been the victim of some sort of misadventure—at the soda factory or elsewhere (*see newspaper excerpt about the Lundblads' "powerful" forge pumps, that enabled them to "charge Soda more highly with carbolic acid gas than can be done by any machinery"*).

Whatever the case, he had made no will at age 34, and the probate records occasioned by his passing shed an unusually detailed light on the details of his bottling business.

Bottling Partnership and Hired Hands

An April 28 payment-schedule letter to the bottling works from which Henry Lundblad had just received his 1851 order of elaborately embossed soda bottles (*see later discussion*) included the following comment: "I have taken the whole Soda Water Business here alone, and bought mine brother out."

Two laborers at the bottling works submitted bills to the estate for their work in the spring of 1851 (both continued to work for one week after Henry Lundblad died). Young George Gray was paid $9.00 for work from May 21 to June 18, and Simpson Nelson was paid $9.23 for 16 days work @ $15/month.

Bottling Works Equipment and Supplies

The bottling works was located in the house the Lundblad family also rented from Hiram Rogers as a residence (at $7.90/month, from probate receipts), and its "personal estate" contents were inventoried and valued on June 14 as part of the probate process:

One Horse Power at	$ 40.00
4 Soda Fountains at	100.00
Soda pumps & machinery for Bottling	135.00
1 Copper Still worm & head	25.00
25 Large Boxes and 50 Small Boxes for Sodas	40.00
44 Gross Soda Bottles at 4.00 [per gr.]	176.00
[small] lot of Bottle Corks	1.50
[small] lot of Syrup	10.00
[small] lot of Twine	3.00
1 one-horse Waggon	25.00
2 Barrels Marble dust	5.00

Embossed and Unembossed Bottles

The 44 gross soda bottles (6,336 bottles) valued at 2 3/4¢ apiece as part of the Lundblad estate are substantially more bottles than the 1851 bottle order. According to the 1851 bottle invoice and Lundblad's response letter, the 1850 bottle order was from a Pittsburgh glass house (not named), while the bottles ordered 1851 were manufactured in St. Louis by *Sell, Hale & Co.*—i.e., the recently established Missouri Glass Works (*Figure 971*). The company's April 21, 1851, invoice to M. R. & H. W. Lundblad indicated they had shipped two lots of "Minerals" (mineral water bottles): one consisting of 23 barrels of bottles, each containing 144 bottles (1 gross/bbl); the second consisting of five barrels of bottles containing 4 3/4 gross bottles. The bill for these 3,996 bottles was $180.38 (plus $2.00 in drayage fees), or 4 1/2¢ each. In November 1850, the Lundblad brothers had shipped the Missouri Glass Co. several casks of "broken glass" and several containers of "chipped and damaged" bottles from their first season in operation as cullet, so the company deducted $5.09 form their bill (less freight costs) for this recycled material. The bill asked for payment within 60 days.

Henry Lundblad's response indicated he was pleased with their product, but tentative about the payment schedule:

Quincy Ills 28 April 1851
Messrs. Sell, Hall & Co.
St. Louis
Gentlemen
   I have rec'd the 28 Barrels of Soda Bottles and your letter of 22th inst: with account of the same, amounting to $179.59. Some I will have ready to pay the 22 of next June. The Bottles seem, and I hope, to be much better [than those] from Pittsburgh.
   I have taken the whole Soda Water Business here alone, and bought mine brother out.
Yours Respectfully,
H. W. Lundblad

Considering the invoice, estate valuation, and correspondence together, it seems likely that the bottles enumerated in the estate, just six weeks after Lundblad's death, fall into three groups: 23 gross (3,312) of the new St. Louis embossed, spiral-swirled pint bottles (aqua and green); 4 3/4 gross (684) of new St. Louis taper-topped quart sodas (probably unembossed: the broken quart pontiled soda illustrated under the Daniel Kaiser discussion above could conceivably be an example of one of these Lundblad quarts—*see Figure 965*); and 16 1/4 gross (2,340) of left-over taper-topped, aqua, pint unembossed sodas from the 1850 Pittsburgh order.

Figure 971. Detail from ca.1856 half-plate daguerreotype photo by Thomas M. Easterly of the recently established St. Louis Glass Works at the N.W. corner of Broadway and Monroe streets (see Kilgo 1994:Fig. 4-55). (Missouri Historical Museum, St. Louis)

There is no indication in the Lundblad probate records that the St. Louis bottle order was ever paid for. The rarity of the embossed Lundblad bottles today may reflect the fact that they were repossessed by the manufacturer and returned to St. Louis, where they were added to the vats as "cullet" for reuse. Perhaps Lundblad's remaining Pittsburgh bottles were sold to the Missouri Glass Co. in bulk as well, for the same purpose.

Soda Bottle Labels and Advertising
One invoice in the Lundblad probate file is particularly interesting in light of the fact that we have never seen a midwestern soda bottle (of the 1840–80 era) with any part of a preserved paper label attached. Considering the decorative detail of Lundblad's embossed bottles, they seem even less likely than most to have been paper-labeled. But an April 15, 1851, bill to the Lundblad estate from *S. M. Bartlett & Co.* is for "two packs of *[advertising?]* Cards **and Soda Labels** $5.00."

Lundblad Sales Accounts
From invoices in the Lundblad probate files, his 1851 bottled-soda orders were not only from Quincy firms, although there were 15 of those, but also to merchants in seven other towns—all located adjacent to the Mississippi River itself from Keokuk (35 miles upriver from Quincy) to Louisiana (40 miles downriver from Quincy). These included one account in Keokuk, IA; four in Warsaw, IL; two in Canton, MO; two in LaGrange, MO; two in Marion City, MO; one in Hannibal, MO; and an isolated account 20 miles further downstream in Louisiana, MO.

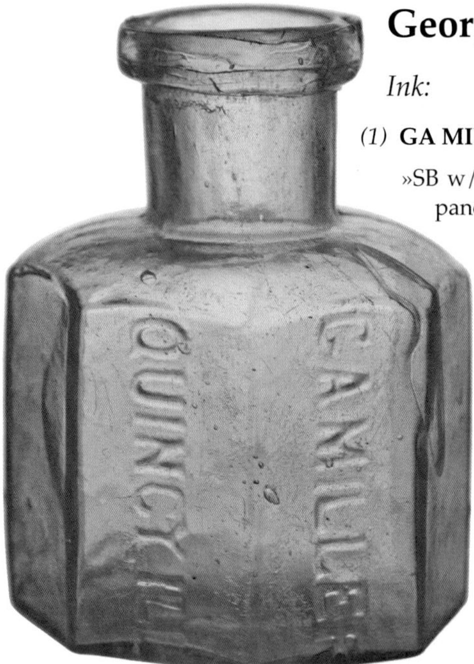

George A. Miller

Ink:

(1) GA MILLER / QUINCY. ILL [aqua]

»SB w/ keyed hinge mold; octagonal; vertically embossed on two adjacent panels; flared ring collar; 2 1/2" tall. *[Figure 972]*

(2) G. / A. / M / I / L / L / E / R // [aqua, green]
Q / U / I / N / C / Y // ILL

»SB w/ keyed hinge mold; octagonal; flared ring collar; 2 1/2" tall. G. A.. MILLER and QUINCY are embossed one above the other, one letter per panel. "ILL" is embossed on one panel following a blank panel. *[Figure 973]*

❖

George A. Miller was junior partner with his father in a pioneer Quincy druggist business as early as summer 1836, when *O. F. & G. A. Miller* "opened the first regular drug store in the place" and built a castor and linseed oil factory on the west side of Front Street (Collins and Perry 1905:91). They placed several newspaper ads for their merchandise in the earliest preserved issues of the Quincy Whig in 1838 and 1839.

Figure 972

The first post office was not established in Quincy until 1826, and by the mid-1830s, when Quincy was first incorporated, the town was still just a collection of modest buildings (the 1858 *Illinois State Gazetteer* indicated that Quincy's population in 1835 was just 700 souls). Benjamin Willis, a young Quincy lawyer, wrote the following observation about the town's development in a Christmas letter dated December 26, 1834 (*personal-correspondence curated at the Illinois State Historical Library*):

> During the past season our town (Quincy) has been one of the most healthy places in the west. And I know of no reason why it should not continue to be so. It is evident the sickness in years past has been owing to the practice of unskillful physicians & exposure in uncomfortable dwelling houses. We have now two or three Doctors who understand their business & the people are erecting better houses, & nothing except the lack of building materials prevents an unexampled *[growth?]* of the town.

Figure 973

In the early days, the Millers' wholesale and retail business efforts focused on importing national-brand patent medicines and bitters (*Butler's, Chapman's, Green's, Lee's, Miles', Moffatt's, Sappington's, Dr. Wistar's*) and specialized glassware:

> Just Received by O. F. & G. A. Miller...Window glass [*many sizes listed*], pint flasks, quart bottles, 1/2 gall. do., 1 do. do., varnish do., assorted vials, prescription do., porter bottles, castor oil bottles, wine do., jars all sizes, Tincture bottles, Salt Mother do., Show Globes, Graduated measures, breast pipes, nipple shells, cupping glasses, nurse bottles, funnels, all kinds of tumblers & decanters. [*Quincy Whig,* April 6, 1839]

George Miller's father left the business in the late 1840s, and from at least 1850 onward G. A. Miller ran the pharmacy alone—still maintaining his company's focus on being the Western distributor of established brand name medicines bottled in the East.

By the early 1860s, Miller's retail drug store was located on Hampshire Street (first at 65 Hampshire and later at 120 Hampshire). During the later 1850s and early 1860s, he expanded the store's focus to include books, stationery, wallpaper, and—from the embossed bottles listed—bottled ink. His great-granddaughter has provided us with an example of one of George Miller's colored pressed-paper advertising signs, manufactured at the time for placement in store windows (*Figure 974*). From the manufacturing technology and style of Miller's embossed ink bottles, they were likely made and used during the Civil War years.

George Miller continued to advertise and sell brand-name patent medicines throughout the 1860s in his retail and wholesale drug business. But by 1870, his interests were turning to real estate and politics, and he formed a new company to pursue his various business interests: Miller, Montgomery & Co. George A. Miler died of a stroke while at work in his office in the building he owned at Sixth & Hampshire on January 17, 1888. He was 76 years old, and his Quincy druggist and mercantile efforts had spanned a half-century.

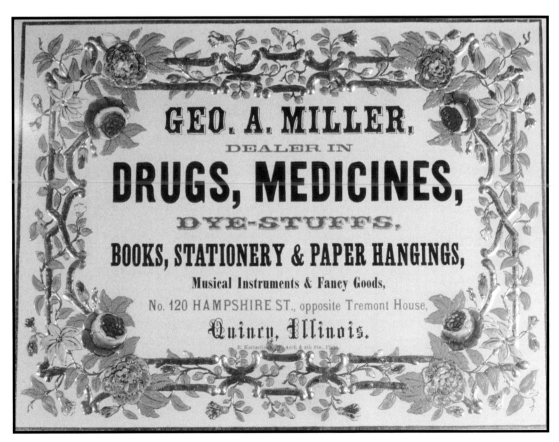

Figure 974

August Schrader

Whiskey:

(1) **SCHRADER'S / KENTUCKY / WHISKY.** [clear]

»SB w/ central circular depression; strap-side "Union" style flask; flared ring collar; 7 1/4" tall. *[Figure 975]*

Figure 975

Little is known about August Schrader, and only two whole or fragmentary examples of his heavy, clear-glass (slight orange tint) embossed whiskey flask have been found in Quincy. He was listed as a day laborer in the 1860 USC for Quincy, but the 1870 / USC listed him as an engineer in a distillery there. In the 1880 USC, he was listed in Quincy simply as an engineer. An 1866 IRS assessment list for Quincy businesses listed the firm of *Griffith, Bush & Schrader* as "Rectifiers," and the separate firm of *Griffith & Schrader* as "Wholesale Liquor Dealers."

It seems likely that August Schrader's *Kentucky Whisky* recipe was manufactured, marketed, and sold in Quincy for one or two years during the late 1860s.

A few fragmentary examples of a square amber bitters embossed **SCHRADER'S BITTERS** have also been found in Quincy excavation contexts of about the same age as the **SCHRADER'S KENTUCKY WHISKY**. It is possible that these were another product made from an August Schrader recipe, but the evidence at hand tends to suggest a different manufacturer for the bitters (*see Frank Schrader listing, this chapter*).

Frank Schrader

Bitters:

(1) **SCHRADER'S // // BITTERS // //** [amber]

»SB w/ keyed hinge mold; beveled-edge square case bottle; flat panels; fragmentary (sherds only), but probably ca. 9" tall; applied tapered collar. *[No photos: sherds only]*

There is some possibility that the bottle listed was the late-1960s product of a Quincy rectifying and distilling company involving August Schrader (*see his listing, this chapter*). But the evidence available to us thus far suggests otherwise.

In December 1867, a wholesale liquor company involving Frank Schrader began placing ads in Quincy newspapers as "Proprietors of Schrader's Highland Bitters":

IRA N. MALIN   JAS. D. MALIN
FRANK SCHRADER

MALIN & SCHRADER

Wholesale Dealers in Foreign and Domestic
Wines, Liquors, Cigars and Bar Fixtures
No. 91 Hampshire Street, Quincy, Illinois.

We have now on hand a large stock of Pure Imported Wines, Brandies, Gins, Cordials, Rum, Irish and Scotch Whiskies; also Old Bourbon, Rye and Domestic Whiskies, Catawba Wine, Case Goods, Cigars, Bar Fixtures, &c. All brands Speer's California Wines and Brandy. **Proprietor's of Schrader's Highland Bitters.** [*Quincy Daily Herald*, March 22, 1868, p. 3]

The company issued a "dissolution notice" on page 1 of the July 29, 1868 *Quincy Daily Herald*, announcing that from that time on the firm would be known as *Ira N. Malin & Son*, but that "Mr.

Schrader will remain with the firm as a salesman." The new company's ads continued otherwise unchanged for the remainder of 1868, still advertising that they were "Proprietors of Schrader's Highland Bitters." The ads were no longer run in 1869 or later years.

No advertisings or paper labels for *Schrader's Highland Bitters* have yet been located to learn the specific therapeutic or curative powers claimed for the product by its manufacturers.

Herman Schroeder

Medicine:

(1) **SCHROEDER'S / IMPERIAL / BALM** [white "milk" glass]

»SB; beveled-edge rectangle; flat panels; rounded shoulders; short neck; flared tapered pharmacy-style lip; 4 1/2" tall. *[Figure 976]*

The earliest newspaper ad found in the *Quincy Herald* for Schroeder's Drug Store was from the July 6, 1869 issue, noting that the new drug store was located "two doors east of the Post Office." From the early newspaper ads placed by Schroeder, he focused on attracting a "refined," largely female clientele by stocking a line of elegantly packaged toiletries, perfumes, and drugs, and imported soda and mineral waters (e.g., Saratoga Springs, Kissinger, Vichy, Arctic Soda Water). In the early 1870s, he started packaging and advertising his own brand name recipes for toiletries. His 1871 *Quincy Herald* newspaper ads touted *Schroeder's Freckletine* (a skin treatment for pimples, tan and skin roughness), *Schroeder's Improved Tooth Powder* (QDH, Aug. 23, 1871), and *Schroeder's Imperial Cologne* ("equal" to celebrated German types: QDH, Nov. 9, 1871). He even sold his own (embossed) brands of bone-handled toothbrushes. In 1872, he advertised *Schroeder's Cold Cream* (QDH, Jan. 12, 1872), *Schroeder's Imperial Cologne Water* (QDH, Jan. 12, 1872), and *Schroeder's Sparkling Soda Water* (QDH, May 9, 1872). In the later 1870s, *Schroeder's Camphorated Cream* was added to the stable of Schroeder's-brand products (QDH, Dec. 13, 1878).

Herman Schroeder's *Quincy Daily Herald* ads for Schroeder's *Imperial Balm* were only run from January 19 to June 6, 1871. The June 6 QDH ad (p. 4) listed its uses: "For complexion use Imperial Balm, it removes pimples, tetter, tan and roughness." The bottle's paper label (*see Figure 976*) indicated that *Imperial Balm* should be used "for beautifying the complexion, eradicating freckles, eruptions, sunburn and tan," and declared that the product was "Patronized by Actresses and Opera Singers."

To date, this is the only Schroeder's-brand toiletry known to have been packaged in an embossed bottle. The paper label indicated the product was "Entered According to Act of Congress in the Year 1871 by H. Schroeder, Quincy, Ill." This was advertiser-jargon of the time for the fact that Mr. Schroeder secured a trademark-design patent on his paper label (No. 317: "Bust of woman, & words *Schroeder's Imperial Balm*).

Schroeder continued in the Quincy pharmacy business until the mid-1880s, and was prominent enough in the trade that the first annual (1882) Illinois Board of Pharmacy listing (of State-qualified pharmacists) listed Schroeder as one of the members of the Pharmacy Board.

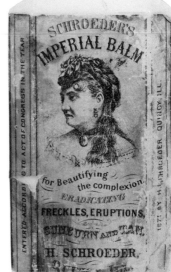

Figure 976

RED BUD

(Randolph County)

1870 census: 900
1880 census: 1,300

Excelsior Brewing Co.

Soda ~ Mineral Water:

(1a) **EXCELSIOR / BRG CO / RED BUD / ILL /**   [aqua]
    (heel): **L & W**

    »SB w/ post mold; oval-blob-top soda. *[No photos. Same mold as L.G. Co bottle listed below.]*

(1b) **EXCELSIOR / BRG CO / RED BUD / ILL. //**   [aqua]
    (heel): **L.G. Co**

    »SB w/ post mold; oval-blob-top soda. *[Figure 977]*

Figure 977

Emil Berger purchased a brewery in Red Bud from John Meyer in 1866 (Miller 2009). Berger leased the brewery to the Excelsior Brewing Company of St. Louis from for a short time from 1875 to 1878. No embossed ale or beer bottles are known that can be attributed to the Red Bud brewery.

Henry Buettner, who had managed the Red Bud business for Excelsior, bought the brewery from Berger in 1878 when the Excelsior lease was terminated. As suggested by the style of the embossed bottles listed, Mr. Buettner changed the focus of the Red Bud Brewery operation to soda-water manufacturing during the late 1870s. Buettner's interest in soda-water manufacturing is confirmed by an entry in an 1883 history of Randolph, Monroe, and Perry Counties, Illinois:

> *Soda Water Factory.*—This business was commenced in the spring of 1878, by Henry Buettner, and is situated in the middle of the block, east of the post-office, on the north side of East Market street. It has two fountains and a generator, with a capacity for manufacturing fifteen hundred boxes of soda water every month. The building is a two story brick, and was originally built for a brewery. [McDonough & Co. 1883b:401]

Since the bottles listed were made during the period when the Excelsior Brewing Company leased the Red Bud business, they must date between 1875 and 1878. These glass containers are clearly soda water bottles. It seems likely that Buettner began bottling soda water in Red Bud during his tenure as the manager of the brewery while it was under lease to Excelsior.

ROCKFORD
(Winnebago County)

1850 census: 2,100
1860 census: 7,000
1870 census: 11,000
1880 census: 13,100

John Chisholm

Soda ~ Mineral Water:

(1) **J. CHISHOLM / ROCKFORD / ILL //**   [aqua]
(heel): **W. McC & Co**

»SB w/ no visible mild lines on base; shouldered-blob-top soda. *[Figure 978]*

The 1870 USC for Rockford listed Chisholm is listed as a 46-year old saloonkeeper born in Canada. Chisholm was also listed in a Winnebago County CD listing published in the 1877 county history (H. F. Kett 1877b:472). At that time, he resided at 416 S. Main in Rockford and his business was listed as "cigars, etc." He probably died or left town soon afterward, as he was not listed in the 1880 Rockford CD.

Chisholm's embossed soda bottle is seldom seen, although he may have made more than one bottle order since some Chisholm bottles reportedly have a *C&I* glass-works heel mark. His bottles were likely produced as part of a one- or two-year bottling effort and used for a short time during the late 1860s or early 1870s, perhaps in the context of Chisholm's Rockford saloon operation.

Figure 978

Freeman and David Graham

Whiskey:

> (1) **GRAHAM'S DISTILLERY / ROCKFORD. ILL'S**   [amber]
> (base): **M^cC**
>
> »SB w/ post mold; 2-piece mold quart; long-neck squat cylindrical liquor bottle; shoulder-embossed only; ring-neck tapered collar; 8 1/2" tall. *[Figure 979]*

Freeman Graham & Co. was first established in Rockford in 1865. Initially the company operated a water-powered cotton mill, but after the mid-1870s (ca. 1876), Graham and his son David established and began operating a distillery in East Rockford, at 1310 South Main Street (Kett & Co. 1877b:423, 489).

According to Johnson (2004:116), the Graham's Rockford distillery operated from 1876 to 1915. *Graham's Distillery* became Graham Bros. Distillery in 1892, under the supervision of sons David and Byron, and soon expanded to two distillery operations (Biographical Publishing Co. 1892:1071–1072). An 1890 summary of the operating scale of Graham's distillery indicated that Julius, Freeman, and Byron Graham managed the firm, "with a capital stock of $150,000. They manufacture sour mash whiskeys. The annual product is $300,000. They employ 45 hands with an annual payroll of $30,000" (Brown and Rowe 1891:51).

The style attributes and technological details of the embossed bottle listed suggest it was manufactured and used during the late 1870s, perhaps to herald the appearance of Graham's retail liquors in Rockford.

Figure 979

ROCK ISLAND
(Rock Island County)

1850 census: 1,700
1860 census: 5,100
1870 census: 7,900
1880 census: 11,700

Henry Carse and John Elder

Soda ~ Mineral Water:

(1) **CARSE & ELDER / ROCK ISLAND / ILL'S**   [aqua]
    (base): **C & E**

»SB w/ keyed hinge mold; shouldered-blob-top soda. *[Figure 980]*

This was the earliest of Henry Carse's Rock Island soda and mineral water bottling partnerships. He first went into the bottling business with a junior partner named John Elder in 1868 (Rock Island Preservation Society 1992:8). The partnership lasted for just a season or two. *Carse & Elder* were listed together in an 1868–69 CD as "soda water" manufacturers at Ontario and Exchange streets in Rock Island (Schmeiser 1970:127). A casual newspaper mention of the company was found in the November 30, 1868, edition of the *Rock Island Union*, which reported a hunting accident involving "a young man employed in Carse & Elder's soda factory." This news note indicates the company had at least one daily-wage employee, and that they were still in operation during the cold months of the year.

Henry Carse was a native of Belfast, Ireland, who came to America with his parents in 1848, while he was still a young child. They first settled in Pittsburgh, where Henry learned the bottling business as a young man. After his father's death, Henry and his mother relocated to Rock Island in 1868 (Clark 1897:66). When they arrived, and *Carse & Elder* first opened their soda bottling works, Henry was 26.

Elder's junior partnership interest was taken over by George Lamont sometime during 1869 (*see Carse & Lemont listing, this chapter*).

Figure 980

Henry Carse and George Lamont

Soda ~ Mineral Water:

(1) **CARSE & LAMONT / ROCK ISLAND / ILL'S**   [aqua]
    (base): **C & L**

»SB w/ keyed hinge mold; oval-blob-top soda. *[Figure 981]*

The *Carse & Lamont* soda bottling partnership began sometime in 1869, when George Lemont took over John Elder's junior partnership interest in the firm (*see Cares & Elder listing, this chapter*). George Lamont was listed in the 1870 USC for Rock Island as a 36-year-old manufacturer of soda water.

Figure 981

Henry Carse appears in the same USC listing as a 28-year-old soda water manufacturer living in the Lamont household.

It was in 1870 that Carse purchased property on 11th Street between 4th and 5th Avenues where a new factory was to be located. At the time of purchase, the area was described as more or less a swamp. But the land was filled, the factory built, and the company went on to bottle at least 11 beverages, including a regional favorite, Black Hawk ginger ale. [Rock Island Preservation Society 1992:8]

Lamont sold his share the business to John Ohlweiler in 1872 (*see Carse & Ohlweiler listing, this chapter*). The announcement of the partnership change appeared in Rock Island newspapers in early May 1872.

Henry Carse and John Ohlweiler

Soda ~ Mineral Water:

(1) **CARSE & OHLWEILER / ROCK ISLAND / ILL'S** [aqua]
(base): **C&O**

»SB w/ post mold; shouldered-blob-top soda. *[Figure 982]*

(2) **TRADE / C & O / MARK / ROCK ISLAND / ILL. //** [aqua]
CODD'S PATENT / 4 /MAKERS / RYLANDS & CODD / BARNSLEY

»SB; pinched-neck, Codd Marble–style internal gravitating-stopper soda; tapered-collar blob top; 8¼" tall. This bottle was made in England. *[Figure 983]*

(3) **CARSE & OHWEILER / ROCK ISLAND / ILL. /** [aqua]
(heel): **H. CODD PAT. / JULY23 & APR29 //**
(reverse heel): **1872 & 1873.**
(base): **C & O**

»SB; pinched-neck, Codd marble–style internal gravitating-stopper soda; sharp-shouldered blob top; 8¼" tall. This bottle is probably American-made (see discussion below). *[Figure 984]*

Soda ~ Cider:

(1) **CARSE & OHLWEILER / ROCK ISLAND. ILL** [amber]
(base): **C & I**

»SB w/ post mold; quart-size soda or cider; shouldered-blob-top; 11½" tall. *[Figure 985]*

Figure 982

Figure 983

John Ohlweiler joined Henry Carse in 1872 as the third in a series of junior partners in Carse's soda bottling business (*see also Carse & Elder and Carse & Lemont listings, this chapter*), and the firm name was changed to *Carse & Ohlweiler*—"the name under which the company would continue for over fifty years" (Rock Island Preservation Society 1992:8).

Carse & Ohlweiler were the proprietors of the Rock Island Bottling Works at 425 Eleventh. Their ad in the 1876 CD gave their sales

Figure 984

office address as the corner of Swan and Canal Street, and noted they were "Bottlers of Ale, Lager Beer, Porter, Cider, &c." In April 1879, a local newspaper mentioned that "Carse and Ohlweiler have made some improvements at their bottling establishment. The reputation of their bottled cider and champaign is fast extending." The caption accompanying a nineteenth-century photo of Jacob Ohlweiler's saloon mentions that "featured brands of liquid refreshment came from Jacob's brother, John Ohlweiler of Carse and Ohlweiler Bottling Works. Their line included soda water, seltzer, ginger ale, birch beer, and other flavored soft drinks **all made from water drawn from the Blackhawk Springs**" (*Rock Island County Historical Society* collections).

No embossed *Carse & Ohlweiler* ale, porter, or beer bottles have been documented for the period of the 1870s. During the later 1870s, their champagne cider may have been bottled in the embossed quart amber bottles listed, and their signature Black Hawk Ginger Ale may have been put up in the embossed Codd Marble–style bottles listed as #2 and #3 (probably to highlight the British/Irish roots of ginger ale). The company actually ordered their earliest such bottles from England (#2), later finding a regional midwestern glass house to make a facsimile variety of the bottle style (#3). By then the product was popular enough in the Quad Cities area that bottling houses on both sides of the river began ordering and using bottles of the same style for their ginger-ale products (*Figure 986*).

Figure 985

Figure 986. Broken examples of the distinctive sharp-shouldered blob-top Codd Marble gravitating-stopper sodas produced by an unknown U.S. glassmaker for three Davenport, IA, companies during the late 1870s (and perhaps early 1800s: see Burggraath and Southard 1998:195, 214–216; Burggraath 2010:132–133). These bottles mirror the later-style Codd sodas used by the Carse & Ohlweiler bottling works in Rock Island during the 1870s. No other Codd gravitating-stopper sodas are known to have been used by any other bottling works in Illinois or surrounding states, except for a British-style example used for a short time by the Sheboygan Mineral Water Company in Wisconsin (Peters 1996:171–171). Examples of these unusual bottles are usually found broken, perhaps because the bottle design made it susceptible to breakage at the neck constriction. It has also been pointed out that "young boys often times broke the Codd bottles to obtain the marble inside, preventing the reuse of the bottle" (Burggraaf 2010:133).

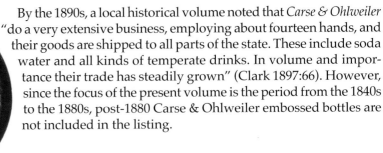

By the 1890s, a local historical volume noted that *Carse & Ohlweiler* "do a very extensive business, employing about fourteen hands, and their goods are shipped to all parts of the state. These include soda water and all kinds of temperate drinks. In volume and importance their trade has steadily grown" (Clark 1897:66). However, since the focus of the present volume is the period from the 1840s to the 1880s, post-1880 Carse & Ohlweiler embossed bottles are not included in the listing.

Julius Junge and Victor Beck

Ale ~ Beer:

(1) **JUNGE & BECK /**   [aqua]
    ROCK ISLAND / ILLS
    (base): Interlaced **JB** *monogram.*

   »SB w/ post mold; slope-shouldered tall pint beer; blob top; ca. 9" tall. *[No photos]*

(2) **JUNGE & BECK /**   [amber]
    ROCK ISLAND / ILL
    Note: There is also abstract decorative embossing above and below **ROCK ISLAND**.

   »SB w/ post mold; slope-shouldered quart beer or ale; embossed in circular slug plate; blob top; 11 1/2" tall. *[No photos]*

(3) **JUNGE & BECK. /**   [amber]
    ROCK ISLAND / ILLS.

Figure 987

(base): **D S G Co** *(embossed around central embossed* X*).*

»SB w/ post mold; slope-shouldered quart beer or ale; embossed in circular slug plate; blob top; 12 1/2" tall. *[Figure 987]*

Julius Junge and Victor Beck established a brewery in Rock Island during the middle 1870s. They are both listed as brewers in the 1877 *History of Rock Island County* (Kett 1877c). Junge had previously worked at a Rock Island bottling works for a year, and applied his experience to bottling ale and beer in partnership with Beck during the late 1870s (Bateman 1914:1223).

William B. Sargent

Cider ~ Soda:

(1) **W. E. SARGENT / ROCK ISLAND / ILLS_ //**   [olive-amber]
    (heel): **A & D. H. C.**

   »SB w/ post mold; slope-shouldered quart; ring-neck "cider" finish; 12 1/2" tall. *[Figure 988]*

William Sargent was listed as a Rock Island grocery store clerk in the 1860 USC. By 1865, he was listed in an Illinois statewide CD as a partner in a grocery store there with Harry K. Williams. In 1873 (CAR), 1874–75 (Rock Island CD), and 1875 (statewide CD), Sargent was listed alone as a grocer in Rock Island. The Rock Island CDs did not list him in 1873 or 1876, and the 1875 CD listed him only as a resident. A 1908 county history (Kramer & Co. 1908:120) summarizes the early history of Sargent's Rock Island grocery businesses:

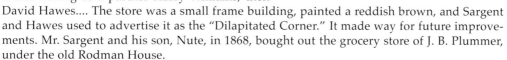

> W. B. Sargent started a small grocery store in 1860 on the corner of Illinois and Washington Streets, where the Peoples National Bank stands, afterwards having as a partner Harry Williams, then David Hawes.... The store was a small frame building, painted a reddish brown, and Sargent and Hawes used to advertise it as the "Dilapitated Corner." It made way for future improvements. Mr. Sargent and his son, Nute, in 1868, bought out the grocery store of J. B. Plummer, under the old Rodman House.

Throughout Sargent's business history, his middle initial was sometimes listed as "E" and sometimes as "B"—but he was William "B" in the census records. From the style and technological details of Sargent's embossed quart cider (or soda?) bottle, it was manufactured and used toward the end of his career in the grocery business: ca. 1871–75.

Figure 988

SANDWICH

(De Kalb County)

1870 census: 1,800

Abel, Humiston and Co.

Bitters:

(1) **"MOUNTAIN / ROOT BITTERS"** // //   [amber, olive green]
 ABEL HUMISTON & Co. / SANDWICH ILL• // //

»SB w/ keyed hinge mold; vertically embossed bitters-style square case bottle; tapered collar; 9 1/2" tall. *[Figure 989]*

❖

Abel and Humiston were listed together as retail liquor dealers in the April, 1866, IRS tax assessment for Sandwich (the actual 1866 IRS listing was "Humiston & Able"). The partnership was comprised of John Abel and Lenson (*aka:* "Lanson") S. Humiston.

At the same time, Abel and Humiston's *Mountain Root Bitters* was being marketed by *S. J. Smith & Co.* in Chicago (*see the Sylvester J. Smith & Co. Chicago listing*). Both Smith and Abel/Humiston used the same embossed bottle mold for the product, changing only the embossed name of the distributor on its opposite face.

Sylvester J. Smith and his father Orrin were listed in the Chicago CDs from 1865 through 1867 as commission forwarding merchants (i.e., product marketing and distribution middlemen). The Smith Company's wholesale sales effort for their *Mountain Root Bitters* seems to have been limited to 1866 and 1867, after which they were no

Figure 989

longer listed in the Chicago CDs. Apparently, they bought the Chicago distribution rights for *Mountain Root Bitters* from Abel and Humiston, who were marketing it on their own in Sandwich at the same time. Of course, the opposite could be true as well: Abel & Humiston may have purchased the Sandwich distribution rights for *Mountain Root Bitters* from Smith and Co.

In 1867, another group of Chicago commission merchants (*Plows, Harris & Upham*) first appeared in the Chicago CDs. They also focused their wholesale distribution business on "alcohols, wines, liquors, &c." By 1868, W. J. Plows had moved the business to 154 Dearborn and had taken on two new partners: John Abel and Lenson (or "Lanson") S. Humiston. In 1869, Abel and Humiston became associated with a large Chicago distiller and rectifier (*J. A. Montgomery & Co.*), and may have continued a short-lived bottling and marketing effort for their *Mountain Root Bitters* at that time (perhaps still labeled *Abel, Humiston & Co.*, Sandwich, IL, since Sandwich was located just 40 miles southwest of Chicago).

No newspaper ads or paper labels for *Mountain Root Bitters* have yet been located to indicate the specific supposed health benefits claimed for the bitters.

SHELDON

(Iroquois County)

1870 census: 200

Jacob S. Miller

Soda ~ Mineral Water:

(1) **J. S. MILLER / SHELDON / ILLS //** [aqua]
(*heel*): **A & D. H. C.**
(*base*): *large, faceted, heavily embossed* **5-pointed star**.
»SB w/ keyed hinge mold; oval-blob-top soda. *[Figure 990]*

Jacob S. Miller was the proprietor of a soda water bottling works in Sheldon during the 1870s and early 1880s. Miller was listed as a 65-year-old soda manufacturer in the 1880 USC. From the style and technological attributes of the listed embossed bottle, it was likely produced and filled during the early 1870s.

Figure 990

SPRINGFIELD
(Sangamon County)

1845 population: 3,000
1850 census: 4,500
1860 census: 9,300
1870 census: 17,400
1880 census: 19,700

Joel B. Brown and Co.

Drugs/Medicine:

(1) **J. B. BROWN & Co / WHOLESALE DRUGGISTS / SPRINGFIELD. ILLS.** // // // //   [aqua]

»SB w/ post mold; 2 1/8" beveled-edge square; long tapered neck; double-ring collar; 4 sunken panels; 8 1/2" tall. *[Figure 991]*

According to Springfield CDs and histories (Power 1871, 1876), the Brown family drugstore was first established in Springfield during the early Civil War years by brothers John H. and William B. Brown, as *John H. Brown & Bro*. The 1864 CD also listed the company as a patent-medicine manufacturer. Their business was located on the town square at the corner of 5th and Adams. They were last listed as partners in the druggist business there in the 1869–70 CD.

Several of John and William's sons also were involved as apprentices in the family pharmacy, and they eventually took over management of the business. The CDs of the early 1870s listed the Brown drugstore partners as Daniel C. and Joel B. Brown (sons of William), at 131 S. 5th. By 1876 (CD), Dwight Allen Brown (son of William) and Joel B. Brown (son of John) had established a second drugstore on the square nearby (115 S. 5th). The Springfield CD for 1867 indicated that Daniel Brown had two employees, 31 years experience in the pharmacy business (he had earlier run a second family drugstore in nearby Decatur with his cousin Joel: see Power 1876:149), and that he had operated the Brown drugstore in Springfield for eight years. The same 1877 CD indicated that Joel Brown sold books and drugs, and that he had been in the drug or book line ("or both") for the past 14 years in Springfield, and that his current store was at the "sign of the Golden Lion" on the west side of the town square.

By 1877–78, Joel Brown had become senior partner with D. Allen Brown as *J. B. Brown & Co.* at the sign of the Golden Lion. By the 1880 CD, J. B. Brown and D. Allen Brown were listed as operating separate pharmacy business in Springfield, so their listed embossed bottle appears to date from the three-year period they were in business together as *J. B. Brown & Co.*, ca. 1877–79. The bottle's contents are thus far unknown, but it appears to have been manufactured for a specific medicinal product, rather than as a generic pharmacy bottle.

Figure 991

Roland W. Diller

Drugs/Medicine:

(1) **R. W. DILLER** *(side)* // **SPRINGFIELD. IL** //   [aqua]
TURNER'S // **LOTION** //

»OP; beveled-edge rectangle; 4 sunken panels; short neck; rolled lip; 4 1/4" tall. *[Figure 992]*

(2) **R. W. DILLER** *(side)* // **SPRINGFIELD. ILL** //   [aqua]
TURNER'S // **LOTION** //

»IP; beveled-edge rectangle; 4 sunken panels; longer neck; double-ring collar; 6 1/2" tall. *[Figure 993]*

Roland Weaver Diller and his cousin, Jonathan Roland Diller, were both involved in the pharmacy business in Springfield early in the town's history. J. R. Diller was a very early pioneer Springfield druggist in the firm of *Wallace & Diller* (with William S. Wallace). Detailed banner ads for the firm of *Wallace & Diller* have been found in the Springfield *Sangamo Journal* beginning as early as September 23, 1837 (*Figure 994*). They were listed as "wholesale and retail dealers in drugs, medicines, paints,

Figure 992

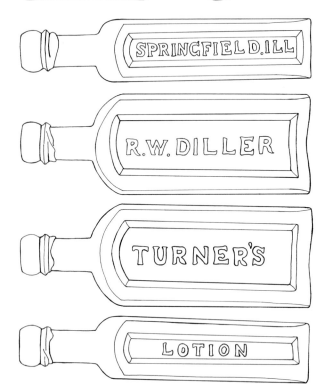

Figure 993

Figure 994

oils, dye-stuffs, perfumery, &c., &c." and their detailed ads from 1837 to 1841 listed over 100 products for sale—including 53 patent medicines, 28 of which included the proprietary manufacturer's name. Interestingly, during this early period, all of the medicines were produced by eastern manufacturers. By 1842, they began advertising their own brand of *Medicated Horehound Candy* for hoarseness and sore throat (*Illinois Register,* Feb. 4, 1842, p. 1, col. 4), and by 1845 they had added a febrifuge called *Wallace & Diller's Western Tonic or Fever and Ague Bitters*—a powdered product "neatly put up in small packages which will make from a quart to a half gallon of strong bitters" (*Illinois Register,* June 20, 1845, p. 1, col. 5).

J. R. Diller died in July 1849, and Wallace turned the business over to Diller's cousin Roland W. Diller (who had come to Springfield in 1844) and Charles S. Corneau, both of whom had been clerks in the earlier *Wallace & Diller* business. Corneau was business agent during the 1849 transition, and continued to sell Wallace & Diller's *Western Tonic* mixture along with their more recent *Cholera Mixture*. By 1850, Corneau & Diller's ads included not only the products of the earlier firm, but also *Corneau & Diller's Bottled Soda* (*Illinois Journal,* July 9, 1850, p. 3, col. 2: no embossed bottles known) and *Corneau & Diller's Diarrhoea Mixture* (*Illinois Journal,* July 7, 1851, p. 3, col. 1: no embossed bottles known). The earliest identifiable Corneau & Diller bottle is an unimbossed (but paper-labeled) *Caster Oil* (*Figure 995*). By 1855, their patent-medicine sales lists included a number of products concocted and bottled by Illinois medicine men (e.g., Hamilton's/Jacksonville, Farrell's/Peoria, Sloan's/Chicago).

When Corneau died in 1860 Diller became sole proprietor of the firm. Diller (*Figure 996*) continued to run his drugstore in Springfield at the same location (on the east side of the Square at 122 S. 6th: *Figure 997*) until he retired in 1899, at which time his son Isaac took over the business. (Inter-State Publishing Co. 1881; Wallace 1904:1497–1498) His early partner, Corneau, suffered from a lengthy illness, and Diller alone ran the business through the mid-to-late 1850s. Both of Diller's early embossed bottles (see listing) date to this pre–Civil War period. Although both of the embossed bottles are embossed *Turner's Lotion*, no ads for the product (likely a skin lotion) have been found, and the identity of Mr. Turner is unknown. Several other embossed "Turner's" medicinal products are known from the later 1850s (including *Balsam, Bitters, Fever & Ague Cure, Gonyza & Styllingia,* and *Sarsaparilla*), but they appear to be unrelated. The surname was a common one, and these other embossed medicines were manufactured and sold in New York, North Carolina, and Cincinnati (Odell 2007:352).

Figure 995

Figure 996

Figure 997

Andrew J. Dunning

Bitters:

> (1) **A. J. DUNNING'S / PURE BITTERS** // // // // [green, olive amber]
> »SB w/ keyed hinge mold; beveled-edge square case bottle; short neck; tapered collar; 9 1/2" tall. *[Figure 998]*

From USC data, A. J. Dunning came to Springfield from New York during the 1850s. In 1860, at age 27, he was listed as a railway conductor living in Springfield. In 1870, he was listed as a Springfield merchant. By the time of the 1880 census, he was once again a railway conductor.

In 1866, Dunning was named Eminent Commander of the Masonic Elmwood Commandery (Knights Templar) in Springfield (Inter-State Publishing Co. 1881:622). Two years later, he placed his only known bitters ad in the 1868 Springfield CD (p. 210):

> **A. J. DUNNING,**
> **Dealer in Wines, Brandy, Whisky**
> And Proprietor of the Celebrated Orange Bitters,
> **Nos. 1, 2, 3, and 4 Nicholas Block,**
> Springfield, Illinois.

Considering the rarity of his bitters bottles (about 3 complete and nearly complete examples known, all from the same mold), Dunning's *Celebrated Orange Bitters* manufacturing venture was likely a limited success, with the product only being sold through his liquor business while the initial order of bottles lasted (see *Figure 999*: envelope with advertising logo, cancelled in 1865). From the style and manufacturing technology of his embossed bottles, they were likely manufactured and used during the late 1860s. No ads have yet been located in the Springfield papers to document the purported health-promoting attributes of Dunning's *Orange Bitters*.

Figure 998

Figure 999

John P. Fixmer

Bitters:

(1) **JOHN P. FIXMER / SPRINGFIELD ILLS. // // MFR. & • PROP. OF / BONEKAMP / STOMACH BITTERS // //** [light aqua, clear]

»SB w/ heel mold; beveled-edge square case bottle; 4 flat panels; short neck; tooled ring-neck tapered collar; 8 3/4" tall. *[Figure 1000]*

John P. Fixmer was first listed in the Springfield CDs in 1876 as a wholesale liquor dealer at 128–130 E. Jefferson. That same CD issue (p. 288) contained a full-page ad and testimonials introducing Fixmer's *Bonekamp Stomach Bitters* to the people of Springfield:

TO THE PEOPLE!
HEALTH and VIGOR
Secured by the use of
BONEKAMP'S STOMACH BITTERS.
The Great German Health Restorer of the Age!

The manufacturer of this excellent medicine has been engaged in the manufacture of Bonekamp's Bitters in Germany for 40 years, and the reputation of the same is celebrated both in the United States and Germany, among his countrymen.

This medicine is one of the most speedy remedies for the restoration of health, the loss of which has been caused by bilious and other diseases, attended by poverty of blood, loss of appetite, dyspepsia, headache, and general derangement of the blood and stomach.

All of the accompanying testimonials were dated from November 1875 to January 1876. One of them (by J. H. Beam, confectioner) was particularly colorful: "As an appetizer and stimulant, I find nothing better than the above. It acts upon the kidneys and liver, purifies the blood, tones and enlivens the whole system. Having had the dropsy, and been tapped some fourteen times, and relieved of sixty gallons of water, I am now able to attend to my every-day work, on an average of fifteen hours per day; sometimes I feel as though I could work twenty-four hours."

The ad ended by noting that *Bonekamp's Bitters* were manufactured by John P. Fixmer, northwest corner of Jefferson & 2nd Street, Springfield, and were for sale at Fosselman's drug store "and all the principal drug stores of the city." The price was $1 per bottle, $8 per dozen wholesale.

Mr. Fixmer and Cornelius A. Jansson also patented their bitters compound. The patent application was filed February 9, 1876, and the patent (No. 193,326)—"Improvement in Medical Compounds—was granted July 24, 1877:

The object we have in view is the production of a medical compound to be used as a tonic, and for the cure of fever and ague, liver-compliant, neuralgia, and bilious diseases generally.

The ingredients used and their preparations are as follows: Alcohol, fifteen (15) gallons; water thirty-three (33) gallons; bitter-apple (colocynth) eighteen (18) pounds;

Figure 1000

sugar twenty (20) pounds; gentian-root, twelve (12) pounds; orange-peel, six (6) pounds; cascarilla bark, three (3) pounds; curcuma, three (3) pounds; anise-seed, three (3) pounds; cassia, two (2) pounds; rhubarb-root, one and one-half (1 1/2) pound; cloves, one (1) pound.

In the preparation of this medicine we take the bitter-apple, gentian-root, orange-peel, cascarilla-bark, curcuma, and rhubarb, all well pulverized, and allow them to stand in the alcohol fourteen (14) days. The liquor is then strained off, and to this liquor is added the sugar, water, anise-seed, cassia, and cloves.

The preparation of ingredients, as well as the manner of preparing them, may be changed to some extent without departing from the nature and scope of our invention, it being understood that the preparations named we prefer to use in the manufacture of our medicine.

The 1877 CD listed Fixmer as a "rectifier of spirits" and wholesale liquor dealer at 2nd and Jefferson, and summarized some of his Springfield business history: He first arrived in 1853; for some time was a railway baggage master; entered the grocery business in 1861; and "is now manufacturer of the celebrated Bonekamp stomach bitters, the great German fever and ague cure." (Fixmer was listed as a grocer in the 1873 CAR.)

John Fixmer's wholesale liquor business was listed Springfield CDs throughout the 1880s and 1890, until 1898. He continued to manufacture his Bonekamp Bitters during this time (at the same address), as evidenced by the fact that he applied for and received a trademark patent (No. 25,540 dated Nov. 20, 1894) on his **JPF** monogram logo.

The listed embossed bottle is found only in late 1870s excavation contexts, and thus was used when Fixmer's bitters was first distributed. Two later Fixmer's *Bonekamp Stomach Bitters* bottles are known, but they were produced and used after the 1880 end-date of our current study. An amber quart with a tall, "lady's leg"–style neck and tooled cider-style finish was used during the 1880s. It was paper-labeled, but had an embossed **JPF** logo on the base. During the 1890s, a clear pint medicine-style bottle was used that was paper-labeled only. An unembossed aqua 7" tall square bottle is also known that was produced with a colored paper label in a sunken panel, under glass.

Charles M. Foster

Medicine:

(1) (side): **FOSTERS // INDIAN / HEALTH RENEWER //** [aqua]
    (side): **CHICAGO ILL'S // //**

»SB w/ keyed hinge mold; beveled-edge rectangle; 3 embossed sunken panels; tapered collar; 9" tall. *[see Figure 282 in Foster's Chicago listing]*

"Dr." Charles Foster's name appears in the 1874 *Atlas of Sangamon County* (pg. 103), which illustrates a drawing of his home in Springfield. The picture caption states that Dr. C. M. Foster is the manufacturer of *Foster's Indian Health Renewer* and another drug named *Foster's Child's Relief* (no embossed bottles known). Since no Springfield-embossed bottles for these medicines have been found, they may have been put up in unembossed bottles with paper labels only. More likely, the Chicago-embossed bottles were used in the Springfield area (*see later discussion*). Prior to 1871–72, when he is first listed as a physician in Springfield, Foster worked there as a butcher and in the hide and leather trade. In the 1873 Springfield CD, Foster was listed as "proprietor and manufacturer, *Indian Health Renewer* and *Childs Relief*" from his residence at 1st and Madison. In the 1874 CD, he was listed as manufacturing bitters at the same address. In the 1875 CD, he was listed only as a Springfield resident.

From Chicago CD evidence, Foster's first Chicago connection was during 1869 and 1870. In 1869, he moved up to Chicago (boarding at the Keystone House) and joined the "general commission [forwarding] merchant" firm of J. C. Barnes & Co. By 1870, he was a partner in the firm (Barnes, Foster and Stone), but his address was listed as Springfield, IL. It is important to note that the third new member of the firm was R. R. Stone: the manufacturer of *Manzanita Bitters* (*see Stone listing, this chapter*). This shows that the Barnes firm was involved in patent-medicine wholesaling, and suggests that the reason Foster joined the firm was to market his newly concocted *Indian Health Renewer* and perhaps his *Child's Relief*. This effort seems not to have been particularly successful, since the firm is no longer listed in the CDs after 1870, at which time Foster seems to have returned to Springfield as a self-styled "doctor" and focused his medicine-sales efforts downstate (many of his Chicago-embossed bottles have turned up in the Springfield area). At the same time, a new commission-merchant company continued to try to market Foster's *Health Renewer* in Chicago:

WE CAN RECOMMEND FOSTER'S INDIAN

Health Renewer for strengthening the system against the blasts of winter. It tones the stomach, cleanses the liver, and regulates the kidney and bowels. Sold by REILAND & RANNEY, 143 West Madison-st. (*Chicago Daily Tribune*, 11/1/1874, p. 15, col. 7)

However, Foster did not prosper in Springfield either. He lost his Springfield house (illustrated in the 1874 *Atlas: see Figure 283 in his Chicago listing*) to a sheriff's sale in 1876 and moved to back Chicago, where he was then listed as a physician in the 1876 through 1880 Chicago CDs and in the 1880 U.S. census. In the 1877 Chicago CD, he is listed as "mnfr. Foster's Indian health renewer" at 528 Michigan Ave. This is his only post-Springfield listing offering the *Health Renewer* for sale in Chicago, but apparently the embossed bottles were still available for sale as late as 1877:

THE CHILDREN __

A Torpid State of the Liver and Kidneys is the great cause of Nearsightedness so general among our children. FOSTER'S Indian Health Renewer is known to be a safe and sure cure for this evil. (*Chicago Daily Tribune*, 5/6/1877, p. 15, col. 7)

The 1874 Sangamon Co. Atlas indicated that Foster's *Indian Health Renewer* would cure "all diseases of the Liver, Lungs, Stomach, and Kidneys. It has never failed to cure Heart-Disease, Asthma, Dyspepsia, Costiveness, and Nervousness." He called his *Child's Relief* a "vegetable preparation" and declared that "Where it has been used, a child has never been known to die with Cholera Infantum or Summer-Complaint. Will cure Chronic Diarrhoea."

Fuller's Banner Ink

Ink:

(1) **FULLER'S / BANNER INK / SPRINGFIELD, ILL** [aqua]
   (base): **I.G.C°**
   *Note: An identically embossed bottle with a different embossed town name is also known (see Decatur listing above). The Decatur bottle lacks the **I.G.C°** base mark.*

   »SB w/ post mold; cylinder master ink; sloped shoulders; raised (ridged) upper shoulder below cylindrical neck; square ring collar; 7½" tall. [Figure 1001]

The identity of the company or individual that manufactured *Fuller's Banner Ink* and marketed it in both Springfield and Decatur is something of a mystery. Thus far, we have not been able to document this business in the commercial or census records for either town. From the style and technological attributes of both the Springfield and Decatur bottles, they were likely produced and marketed for a short period sometime during the later 1870s.

Caspar's 1889 Dictionary of the American Book, News, and Stationery Trade includes what appears to be an orphan listing for *The Fuller Mfg. Co.* under its heading for manufacturers of colored ink and copying fluids (Caspar 1889:603). But according to an online manufacturing history, this Fuller ink company

Figure 1001

was first started in Chicago in 1886 by Harvey B. Fuller as a sales outlet for *Fuller's Liquid Fish Glue*. He was 21 at the time. Mr. Fuller moved to Minneapolis the following year and expanded his manufacturing line to laundry bluing. He also manufactured and supplied ink to the local schools. Harvey Fuller is never mentioned in Springfield or Decatur histories, and anyway he would have been too young to have manufactured *Fuller's Banner Ink* in central Illinois in the mid-1870s.

A possible Springfield/Decatur candidate as manufacturer of *Fuller's Banner Ink* is Charles H. Fuller, who lived and worked in both towns during the 1850s–80s. We have found no mention of him as an inkmaker, but according to his obituary (Decatur *Daily Republican*, Dec. 26, 1896) he worked in a variety of jobs in local government and in the private sector after being injured in the Civil War. He was city clerk in Decatur from 1867 until 1872, and in 1884 he was appointed pension examiner. We have not determined his occupation from 1872 to 1883, during which time *Fuller's Banner Ink* would have been produced.

John Johnson and Co.

Soda ~ Mineral Water:

   (1a) **J. JOHNSON & Co / SPRINGFIELD / ILL //** [aqua]
   (left side heel): **F.A & Co**

   »IP; long neck; shouldered-blob-top soda (7 3/4" tall). *[Figure 1002a–c]*

   (1b) **J. JOHNSON & Co / SPRINGFIELD / ILL //** [aqua]
   (left side heel): **F.A & Co**

   »SB w/ hinge mold; shorter neck; shouldered-blob-top soda (7 1/4" tall). *[Figure 1002a–b, d]*

   (2) **J. JOHNSON & Co / SPRINGFIELD / ILL //** [aqua]
   (heel): **A & D. H. C**

   »SB w/ hinge mold; shouldered-blob-top soda. *[Figure 1003]*

Soda water bottling works involving John Johnson and Charles Peterson had a long and varied history in Springfield during much of the late nineteenth century (*see their joint listing, this chapter*). In the summer of 1860, just at the beginning of the Civil War years, the *Henry Korf & Co.* bottling works was established on Fifth Street between Jefferson and Madison. Charles Peterson was listed in the 1860 Springfield CD as a soda water manufacturer at the same address, and he was no doubt the "*& Co.*" partner at the Korf Bottling works. Peterson had learned the soda-bottling trade in St. Louis between 1853 and 1860 (Wallace 1904:953). In fact, there was a John "Petterson" soda-bottling works in St. Louis at the time (perhaps a relative?), so Charles Peterson's training may have been there (see Smith 2005:31–32). Korf had previously bottled near St. Louis as

Figure 1002

well (see later discussion), so the two men likely came together from the St. Louis area to Springfield to start their new Illinois soda-bottling venture:

> BOTTLED SODA WATER—We would invite the attention of the citizens of Springfield and vicinity to our Soda Manufactory, on North Fifth street, one door north of J. C. Planck's grocery store, where we are prepared to furnish a superior article of bottled soda water on the most reasonable terms. Saloons and Restaurants can be supplied daily by leaving orders at the *Factory*. Orders promptly attended to. **HENRY KORF & CO.** *[Illinois Journal, June 7, 1860, p. 2, col. 6 — ad placed May 12]*

Figure 1003

Henry (Heinrich Dietrich) Korf, had previously operated a bottling works in St. Charles, MO, for a short time (ca. 1859–60) with Jacob Zeisler, but had sold his interest in the operation after only two months (Conard 1901:552) and relocated to Springfield. His short-lived St. Charles business used pontiled soda bottles embossed **KORF & ZEISLER / S^T CHARLES / M^O**. Korf was involved in the Springfield operation for a similarly brief period. Sometime during the 1860 season, he sold his interest in the Springfield soda works to John Johnson, who immediately ordered bottles embossed **J. JOHNSON & CO.** from the newly established Fahnstock, Albree & Co. glassworks in Pittsburgh. No embossed bottles have been found for the single-season or partial-season Korf bottling operation, but since embossed bottles were used during his short stay in St. Charles, **HENRY KORF & CO.** Springfield bottles may yet be found.

John Johnson, a native of Sweden, came to America in 1851 when he was 26 years old. He settled in St. Louis for a few years, and then relocated to Springfield in the mid-1850s. He became involved in the ice business in Springfield, and entered the soda water bottling business in 1860 when he bought out Henry Korf's interest in the Korf & Peterson bottling works. Beginning with the 1861 bottling season, the company became **J. JOHNSON & CO.** (see embossed bottle list). During the *Johnson & Co.* years, the company operated from the 5th and Jefferson location first established by *Korf & Co.* During this time, Johnson and Peterson constructed a factory building a few blocks away (at 4th and Carpenter) that was specifically designed as a factory to produce bottled soda water. They moved into their new quarters at 531 N. 4th Street in 1864, at about the time the company name was changed to *Johnson & Peterson* (see their listing, this chapter).

Johnson and Peterson's partnership in their Springfield bottling business for 31 years. But after their first few years in operation, Johnson became afflicted with a paralytic disease and did not take a particularly active role in the day-to-day business (Obituary, *Daily Illinois State Journal*, May 13, 1892, p. 4, col. 4). In fact, Charles Peterson was viewed (and acted) as principal manager of the bottling operation from the very beginning of the partnership:

> PETERSON'S BOTTLED SODA WATER
> Having enlisted for the season in the Soda Factory on North 5th street, I am prepared to furnish Hotels, Saloons, and Families with a first rate article of Soda Water in bottles, or to charge Soda Fountains for customers in the city or country, at any hour of the day. Charges moderate—terms cash.
> May 11th, 1861 CHAS. PETERSON
> *[Illinois State Journal, May 11, 1861, p. 2, col. 7]*

> SODA WATER.—We take pleasure in calling attention to the advertisement of CHARLES PETERSON, who has established Soda Works on North Fifth street.... We have to thank him for two dozen bottles. Those wanting Soda Water can do no better than to leave their orders with him.
> *[Illinois State Journal, May 11, 1861, p. 3, col. 2]*

After a few years (ca. 1864), the company became **JOHNSON & PETERSON** (*see their listing, this chapter*). Peterson became sole operator of the bottling works after Johnson's death in 1892 (*see later discussion*). Peterson continued bottling at the 4th Street address until he retired from business in 1903 at the age of 68.

A fascinating addition to our historical knowledge about Johnson's Springfield soda-bottling business has recently been reported by Mike Burggraaf (2010) in his new update to *The Antique*

Bottles of Iowa (Burggraaf and Southard 1998). Burggraaf's research reveals that Johnson's physical incapacity, while removing him from the day-to-day operation of the Johnson & Peterson bottling works in Springfield, also freed him to develop and financially back similar businesses elsewhere:

> John E. Johnson established a grocery business and a bottling works in Keokuk in 1865. Mr. Johnson, who was actually a resident of Springfield, Illinois, had taken in John Burk as his business partner in operating the bottling works. Mr. Burk had previously been in the dry goods business in Keokuk since 1850. The Keokuk Soda Water Manufactory was under the John Johnson & Company firm name until 1870, when the Firm name was changed to Johnson & Burk. [Burggraaf 2010:341]
>
> During this partnership, the residency of Mr. Johnson was always listed at Springfield, Illinois. In November of 1876, Mr. Johnson sold out his share in the bottling works to Henry Bemoll. [Burggraaf 2010:327]

John Johnson and Charles J. Peterson

Soda ~ Mineral Water:

(1) **JOHNSON & PETERSON / SPRINGFIELD / ILL //**   [pale blue]
   (*heel*): **A. & D H. C**

   »SB w/ hinge mold; shouldered-blob-top soda. *[Figure 1004]*

Figure 1004

(2) **JOHNSON & PETERSON / SPRINGFIELD / ILL //**   [pale blue]
   (*heel*): **L. &. W**

   Note: Same mold as #1, so J&P, not the glassmaker, must have owned the mold.

   »SB w/ hinge mold; shouldered-blob-top soda. *[Figure 1005]*

(3) **JOHNSON & PETTERSON / SPRINGFIELD / ILL //**   [aqua]
   (*base*): **J & P.**
   Note: PETERSON is misspelled, and the base-embossed "J" is embossed backwards.

   »SB w/ post mold; oval-blob-top soda. *[Figure 1006]*

(4) **JOHNSON & PETERSON / SPRINGFIELD / ILL //**   [aqua]
   (*heel*): **B. F. G. Co.**
   (*base*): **J&P**

   »SB w/ keyed hinge mold; oval-blob-top soda. *[Figure 1007]*

(5) **JOHNSON & PETERSON / SPRINGFIELD / ILL //**   [aqua]
   (*heel*): **A & D. H. C.**
   (*base*): **J&P**
   Note: This bottle is an example of a late-1870s or early-1880s "collared blob" style blob-top soda.

   »SB w/ keyed hinge mold; very short neck; shouldered-blob-top soda. *[Figure 1008]*

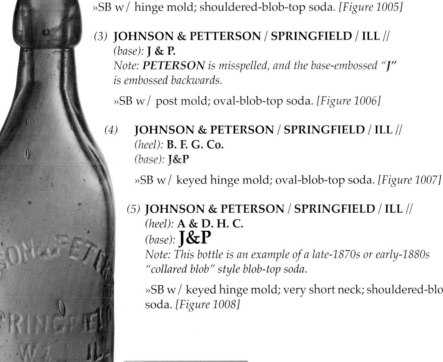

Figure 1005

Figure 1006

Figure 1007

Figure 1008

(6) **JOHNSON & PETERSON / SPRINGFIELD / ILLS //** [aqua]
(heel): **D. O. C.**
Note: *The Dominick Cunningham glassworks produced this bottle in the early 1880s (see volume introduction), but it is listed here because of its unusual "old fashioned" true-blob design.*

»SB w/ heel mold; short neck; oval-blob-top soda. *[Figure 1009]*

❖

As discussed in the *J. Johnson & Co.* section, John Johnson and Charles Peterson began bottling soda water together in Springfield in 1861. Although the partnership lasted for more than three decades, Johnson's ill health limited his active participation in the business, and Peterson ran the day-to-day operations of the company.

Johnson and Peterson began their business bottling from a temporary factory located near the intersection of north 5th Street and Jefferson, but three years later they moved into a building at the SW corner of 4th Street and Carpenter that they had constructed as permanent quarters for their bottling works. The embossed bottles listed above were manufactured for and used by the *Johnson & Peterson* company from 1864 through the late 1870s. Bottles #1–#3 were used from the mid-1860s to the early 1870s; they are listed in their approximate chronological order of manufacture. Bottle #4 was made and used during the later 1870s. Bottle #5 is included in the list because it is an unusual "traditional" true-blob style, although it was manufactured in the early 1880s at the Dominick Chambers Pittsburgh glassworks.

During the 1880s and early 1890s, at least six varieties of embossed collared-blob and Hutchinson-style (internal rubber stopper) bottles were made for the company's use, but these were used too late for inclusion in the present study (*see introduction, this volume*).

Charles Peterson and his brother Alfred also established a branch bottling works in Bloomington that operated under Alfred's supervision during the early 1870s (*see their Bloomington listing*). During this period, the two brothers returned to the traditional "Petterson" spelling of their last name. They were listed that way in the CDs, and the Bloomington bottles are embossed with this spelling. The Springfield bottle produced during this time (#4 above) was also made with the "double-T" spelling. After the Bloomington business closed, the double-T spelling was no longer used on Charles Peterson's later Springfield embossed bottles.

According to the 1879–80 IIC for Springfield, the Johnson & Peterson soda-water factory had four employees that year, two of whom were adult males. Employees worked year-round an average of 10 hours/day, and were paid $1.50/day each. The company

Figure 1009

Figure 1010

paid a total of $350 in wages for the year, spent $2,500 on production materials and supplies, and produced bottled soda valued at $5,000.

After Johnson's death in 1892, Peterson took over sole management of the bottling works and the corporate name was changed to *Charles J. Peterson Company*. At least five additional varieties of Hutchinson-style bottles (embossed **C. J. PETERSON**) were made and used by Peterson from 1892 until he retired from business in 1903.

The bottling-works building that Johnson & Peterson built during their first few years in operation and moved into in 1864 must have been particularly well designed and constructed. The factory was in continuous use by Peterson for nearly 40 years until he retired in 1903. Between 1903 and the World War I era, it was used as a soda factory by *Lauterbach & Reisch*, then by *F. X. Reisch*. In 1912, Reisch sold the building to Henry Dresch, who used it as a soda-bottling works in Springfield until his beverage company finally closed its doors in 1988 (*Figure 1010*). This record of 124 years of continuous use is impressive by any standard.

Jeremiah A. Kelleher and Co.

Ale, Porter & Spruce Beer:

(1) **KELLEHER & CO** [tan stoneware]
Note: salt-glazed stoneware bottle, not glass.

»12-sided Merrill-patent bottle; salt-glaze quart; shouldered-blob top; 10 3/4" tall. *[Figure 1011]*

There are several listings elsewhere in this volume for bottlers who used embossed glass ale bottles and put up spruce beer or other Weiss beers in stoneware bottles as well. The Kelleher listing is unusual in that no embossed glass bottles are known for the company, only the stamped stoneware bottle listed here. Kelleher first appeared in an 1876 CD ad under bottlers (but not in the alphabetical text listings):

J. A. KELLEHER,
City Bottling Works,
Dealer in
BOTTLED LAGER and ALE,
217 MONROE ST.

His second (and final) listing was in the 1877 CD, which listed Jeremiah A. Kelleher as a spruce beer manufacturer, noting that he came to Springfield from Massachusetts in 1875: "Is now manufacturing spruce beer; bottler of ale and porter. Sells at wholesale exclusively, and is the only factory of the kind in the state outside of Chicago. Factory 318 N. 5th."

Figure 1011

The 1877 Springfield CD also listed a James Kelleher as a member of the bottling business that year. James was likely the "& Co" member of the short-lived bottling operation. The *Illinois Journal* of April 17, 1877, also included a Kelleher ad (page 4, col. 5):

**CITY
BOTTLING WORKS,**
J. A. Kelleher, Proprietor
Bottled Ale, Porter, Lager & Spruce Beer

The undersigned announces to the citizens of Springfield and vicinity that he is prepared to furnish Ale and Beer of the best brands, brewed expressly for my trade and bottled to order.

All Goods Warranted

to give satisfaction and delivered to any part of the city free of cartage. Orders from adjacent towns promptly filled.

Works, 217 Monroe Street

Apr17dtc A. J. Kelleher

Henry Korf and Co.

See the John Johnson & Co. listing, this chapter.

Thomas Lewis

Medicine:

(1a) **LEWIS** (quarter panel) // **EVERYBODY'S** (side) // // **LINIMENT** (side)   [aqua]

»OP; small beveled-edge rectangle (7/8" x 15/8"); 4 sunken panels; short neck; rolled lip; 4 3/4" tall. *[Figure 1012a]*

(1b) All ditto #1a, but larger size: 1 1/4" x 1 3/4", 5 1/4" tall. *[Figure 1012a, f)]*

(1c) **LEWIS'** (quarter panel) // **EVERYBODY'S** (side) // // **LINIMENT** (side)   [aqua]

»IP; large-size of #1a and #1b; beveled-edge rectangle (1 1/2" x 2 3/4"); 4 sunken panels; longer neck; double-ring collar; 7 3/4" tall. *[Figure 1012a–e]*

Thomas Lewis was born and raised in New Jersey but in 1837, when he was 29, he and his two brothers and sister and their families (27 people in all) moved west in search of a better life and settled in Springfield. Lewis was "an active, energetic business man (Power 1876:455) who became involved in a number of entrepreneurial Springfield-area business ventures. During the 1840s, for instance, Lewis and a machinist named John W. Johnson built and operated the Everybody's Mill and the Fair Play Foundry and Machine Shop (Wallace 1904:473). Beginning in the later 1840s, Lewis advertised the establishment of his Everybody's Store selling general merchandise "East of the Capitol," and an auction house he ran next door (see *Sangamon Journal*, June 10, 1846).

Local newspaper ads in 1860 and 1861 promoted Lewis' "*Everybody's Liniment* — for man and beast" — with testimonials indicating the liniment had been offered for sale for the past eight years. It was likely not sold through Everybody's Store, since the ads indicated Thomas Lewis manufactured his liniment for wholesale distribution only, and that it was sold retail in Springfield by Miss Hannah G. Griffith and by pharmacists E. Owen, S. H. Melvin (*see his listing, this chapter*), M. Alphonse, and J. K. Lewis.

Figure 1012

J. K. Lewis was likely Thomas Lewis' brother John, who in the early 1840s had been a member of the pharmacy partnership of *Grubb & Lewis* in Springfield. Notably, the *Sangamo Journal* of July 29, 1842 (p. 4, col. 6) contained extensive ads for two patent-medicine products prepared and distributed solely by *Grubb & Lewis*: their eight-ounce *North American Specific* (for "a multitude of diseases" including King's Evil, White Swellings, Indolent Ulcers, Illinois Mange "&c., &c."); and their one-ounce *Smith's Cholera Mixture*. Embossed bottles are not known for either of these preparations, but they are noted here because the *Grubb & Lewis* ads indicate that these products were put up in bottles "capped with wax, and stamped with the initials *G. & L.*"

Since Lewis' liniment was still being sold during the early Civil War era, it is possible that a nonpontiled embossed bottle with a keyed-hinge-mold base may be found as well.

Alexander and Morris Lindsay

Medicine:

 (1) **A. &. M. LINDSAY'^S // LINIMENT**   [aqua]

  »OP; cylinder; short flared neck w/ lip; 1 1/8" diameter; 4 1/4" tall. *[Figure 1013]*

 (2) **A & M. LINDSAY'S / LINIMENT.**   [aqua]

  »OP; cylinder; short flared neck w/ lip; 1 3/8" diameter; 6" tall. *[Figure 1014]*

Figure 1013

Figure 1014

In 1840, Alexander and Morris Lindsay first marketed one of the earliest Illinois medicine products ever packaged in embossed bottles (if not *the* earliest—cf. *Farrell's Arabian Liniment*, Peoria). The Lindsay brothers were born and raised in Pennsylvania and came to Springfield separately in the early and mid-1830s.

From early newspaper advertisements, Alexander was the first of the brothers to arrive in Springfield. According to Lang (1981:5), "the first mention of Alexander Lindsay in Springfield was on April 20, 1833, when he bought the livery and blacksmith shop of J. J. Allender and also went into the hauling business with the same." Alexander would have been just over 30 years old at the time. By June 12, 1835, he had entered a blacksmithing partnership with William Dillard. By February 5, 1836, he announced that he had become sole operator of the blacksmith shop and had "purchased the shop of Mr. James B. Redman nearly opposite the stage stable of R. Allen, Esq." (*Sangamo Journal,* Feb. 10, 1836).

At about this time, younger-brother Morris Lindsay arrived in town (Wallace 1904:1396). Given Alexander's other Springfield business interests, Morris (who was 23 years old in 1836) likely took on the task of running their new general store. Alexander Lindsay assumed the added duties of Justice of the Peace in Springfield that summer, but his business pressures were growing and he soon resigned the office. By the following spring (May 13, 1837), he announced in the *Sangamo Journal* that he needed his creditors to pay up, and that "I have started three fires in my [*blacksmith*] shop, where I expect to be every day.... I shall hereafter keep my office at the point of the anvil where I may be found when health will permit."

By the following spring, the brothers' commercial attention was focused primarily on growing the family grocery:

Farmers & Mechanics
FAMILY GROCERY.
A. LINDSAY & BROTHER,

...are now receiving a large assortment of Family Groceries at the stand formerly occupied by J. J. Allender, one door east of Esq. Clement's office. They intend keeping all the requisites of a family Grocery, and hope to receive a share of public patronage. [*Sangamo Journal,* May 5, 1838, p. 3, col. 5]

An ad the Lindsay brothers placed in the *Sangamo Journal* May 1, 1839, listed a wide variety of grocery products for sale, and also noted: "P.S. Just received a supply of the Sanative for the cure of the consumption; Morrison's Hygean Pills; Brandeth's Pills; Nerve & Bone Liniment; Godfrey's Cordial; Bateman's drops for children; and remedies for Fever and ague, with a number of other celebrated patent medicines." Their ads in fall 1839 listed a similar variety of "name-brand" eastern patent medicines for sale. These same ads were run by A. &. M. Lindsay until October 1840, at which time their own liniment was first announced in a large, half-column ad, accompanied by testimonials from recent users:

LINDSAY'S LINIMENT.
IMPORTANT TO THE AFFLICTED.
A cure for all diseases that require an external application,
EITHER ON MAN OR BEAST.

Its efficacious effect upon all bone complaints, weak joints, and those that have been dislocated, broken or otherwise injured have been well tested. It is positively the best remedy, if thoroughly applied, for sprains, burns, bruises, and for many cases of Rheumatism; for horses chafed by the saddle, sprained shoulders, scratches, swelled limbs, cuts, wounds of any kind, weakness in the back, &c....

It will greatly relieve a pain in the breast or side, and perhaps there is no better cure for the tetter or ring worm if thoroughly applied several times a day...

We are making arrangements to supply the increasing demand...Price fifty cents per single bottle.
A. LINDSAY & BRO.,
North side public square.
[*Sangamo Journal*, Oct. 30, 1840, p. 3, col. 4]

Similar ads in early 1841 listed six sales agents in nearby towns, and by March of that year *"LINDSAY'S NERVE AND BONE PILLS"* were also being advertised. The Lindsay's 1841 ads also included much more detailed references to their liniment as horse medicine—to cure sweeney, fistula, big head, big jaw, spavin, splints, sprain, trembling or weakness, etc. Beginning in June 1941, the *A. & M. Lindsay's Liniment* ads were much larger and included elaborate decorative banners (*Figure 1015*), and by 1842 the Lindsay brothers listed sales agents in six states, including agents in 48 towns in Illinois. Their patent-medicine venture had become so large by the early 1840s that the Lindsays decided to abandon their grocery business, placing the following ad in the *Sangamo Journal* on April 20, 1843: *A. Lindsay & Bro.*, having "purchased the entire stock of drugs and medicines of the late firm of Brookie & Bush, beg leave to state to the public that they have changed their business and are now dealing in drugs and medicines...at their old stand, NW cor. square."

Figure 1015

But this business shift apparently also involved a break-up of the brothers' partnership. By 1844, Alexander Lindsay—alone—was advertising four new medicinal and pharmaceutical products, "for sale at my office in Joseph Lewis's shoe store, North side of the square" (*Illinois State Register*, May 3, 1844, and Sept. 6, 1844). These included *A. Lindsay's "Anti-Dispeptic" Tonic and Pills; A. Lindsay's "Anti-Billious" Pills; A. Lindsay's "Aramatic" Tooth Powder; and A. Lindsay's Western Nerve Tonic* (for coughs, hoarseness, colds, and "general weakness"—in 50¢ and 75¢ bottles). No embossed bottles have yet been found for either of Alexander Lindsay's tonics.

After 1844, ads for Lindsay's medicines no longer appeared in

Springfield newspapers. We know that Morris Lindsay remained in Springfield for about 10 years—the 1855 CD listed him as a clerk for *Freeman & Co.*, and the 1857 CD listed him as a member of the *Broadwell & Lindsay* grocery firm. He also served as city councilman and postmaster in the late 1850s and early 1860s before moving to Carbondale in 1863 (Wallace 1904:1396).

But Alexander Lindsay disappeared from the Springfield business scene altogether. The 1854 ISD listed *A. Lindsay & Co.* as distributors of wholesale drugs and medicines at Naples, IL (a major shipping point on the Illinois River 50 miles west of Springfield). From additional fragmentary evidence—including the fact that Alexander Lindsay's daughter died of cholera at Naples in 1850 (*Illinois State Journal*, July 25, 1850), and the fact that he was postmaster at Naples in 1853 and died there in 1857 (Lang 1981:6)—it seems likely that he moved his patent-medicine wholesaling business to Naples in 1845.

An 1859 advertisement in the *Jacksonville Sentinel* by druggist Robert Hockenhull is the last mention of *Lindsay's Liniment* being offered for sale in the region. But from the evidence at hand, it appears that the Lindsay brothers only worked together to bottle and sell their liniment during the period from 1840 to 1843. So the embossed bottles listed were likely manufactured and used during this four-year period. Only a single ad has been found that mentions embossed bottles (*Sangamo Journal,* July 23, 1841, p. 3, col. 6): "None are genuine but those that have "**A. & M. Lindsay's Liniment**" blown in the bottle." This is by far the earliest printed reference to an embossed Illinois product bottle. The ads from this period mention only a "50¢" size, but the larger bottle variety may have been sold specifically as horse medicine.

James J. Lord and Ebenezer H. Topping

Medicine ~ Bitters (Civil War era):

(1) **DR TOPPING'S / ALTERNATIVE & / CATHARTIC SYRUP // //
J.J. LORD & Co / SPRINGFIELD. ILL // //**
Note: two raised dots under superscript **R** *in* **D^R**.

[aqua, pale blue]

»SB w/ keyed hinge mold and concave central depression; beveled-edge square case bottle; 4 flat panels; short neck; sharp-edged double-ring collar; 8" tall. *[Figure 1016]*

Medicine ~ Bitters (Post-1880 era):

(2) **COLUMBIAN /**   [yellow amber]
CATHARTIC BITTERS // // LORD & Co. / SPRINGFIELD, ILLS. // //
Note: From the style, manufacturing technology, and excavation contexts of this bottle, the manufacture and use of this bottle substantially postdates 1880, and Lord's active involvement in the company. It is only included here for comparative purposes and for use in the business-history discussion below.

»SB w/ post mold; beveled-edge rectangle; 4 sunken cathedral panels; rounded shoulders; heavily tooled, slightly tapered ring-neck collar; 7 3/4" tall. *[No photos: postdates study period]*

Curtis Mann (personal communication 2007) has referred to James Judson Lord as "The Patent Medicine Poet." According to Lord's 1905 obituary, he was born in 1829 in Burwick, ME, and came to Waverly, IL, in 1851 (about 20 miles

Figure 1016

southwest of Springfield), where he met and married his wife and engaged in the mercantile business. He came to Springfield about 1860 (when he was first listed in the Springfield CD as a patent-medicine manufacturer and proprietor of *Dr. Topping's Alterative Syrup* from his residence on Edwards between 2nd and Spring). Both Lord and Ebenezer H. Topping were listed in the 1860 USC as living in the same Springfield household. E. H. Topping and Mrs. Lord were both from Ohio, via Waverly, so perhaps the two were related.

A daily ad for Dr. Topping's syrup was first placed in the *Illinois State Journal* (Springfield) on July 15, 1858:

READ THIS! READ THIS!!
A New Remedy for Bilious Diseases
Dr. Topping's Alterative and Cathartic Syrup.

This medicine is purely vegetable, and is an infallible remedy for the Flux or Dysentary, Diarroehea, Cholera Morbus, Cholera Infantum and Summer Complaints with children....

Just received *[in Springfield]* and for sale by J. B. Fosselman, Canedy & Johnston, and Corneau & Diller. Price $1 per bottle.

Lord was not mentioned in the ads, and an accompanying Waverly testimonial indicated the medicine had been in use there since 1855, so perhaps Topping marketed it there through Lord's business establishment. It apparently did well enough regionally for them to bring it to Springfield in 1860—as business partners—and to have embossed bottles (#1 above) made for the new Springfield partnership.

Ebenezer H. Topping was a partner in the Springfield medicine business only until late 1862, when he enlisted as a captain in the 110th Illinois Infantry (He listed his occupation as "lawyer" when he entered the army, but the 1862 military census for Sangamon County lists him as a "Patent Medicine Man"). After the war, he moved to Kansas and took up farming.

J. J. Lord continued alone in the Springfield patent-medicine business—and continued to manufacture *Dr. Topping's Syrup*—until 1877, at which time he took on a partner named Albert Hayden, and expanded his mercantile efforts into the china, glass, and queensware trade (1877 CD). From the style and technological attributes of the embossed *Dr. Topping's* bottle listed, Lord's 1860 glass order was the only one involving embossed bottles. But as late as the late 1870s and early 1880s, the medicine was still manufactured by Lord (design patent #2810 was awarded Nov. 14, 1882, for a *Dr. Topping's* paper-label) and offered for sale (by proprietors Richards & Hudson, according to a J. J. Lord company letterhead used on March 18, 1879: *see Figure 1017*).

Surprisingly, a much-later (post-1880) embossed *Lord & Co.* bottle is also known. No ads or information regarding Lord's *Columbian Cathartic Bitters* (see listing) have yet surfaced, but from the style and technological attributes of the embossed bottle it was likely manufactured and used during the mid-to-late 1880s, too late for formal inclusion in the present volume. This bottle was produced more than 20 years after Lord's *Alterative & Cathartic Syrup*, after Lord himself was no longer actively involved in company business. The fact that the *Columbian Cathartic Bit-*

Figure 1017

ters bottle was embossed suggests that late embossed varieties of Dr. Topping's bottles *may* have been made and used as well, but none have yet been recorded. More likely, the *Lord & Co.* cathartic medicines produced during this period were paper-labeled, without embossing. The late embossed bottle was likely a one-shot advertising initiative tried in the 1880s after Richards & Hudson took over the company.

Lord's "patent-medicine poet" handle is based a single high-profile event. He submitted a poem to a competition for the Lincoln Monument dedication, won, and read it at the September 10, 1869, dedication ceremony: *"Not to the Dust, but to the Deeds alone / A grateful people raise th' historic stone"* (see Olroyd 1915:183–184).

William W. McNeil

Medicine:

(1) **DR. W. W. M^{C}NEIL'S // JUSTLY CELEBRATED // PAIN EXTERMINATOR // SPRINGFIELD // ILLS** [aqua]

» SB (heel mold?); 12-sided, 11/4" diameter paneled shouldered cylinder; embossed vertically on 5 adjacent panels; short neck; thickened flared lip; 41/2" tall. *[Figure 1018]*

Figure 1018

William McNeil and his older brother Francis were born and raised in Maryland and both graduated from medical colleges in the East before coming to Illinois in the late-1830s (Francis at the University of Maryland and William at Jefferson Medical College in Philadelphia: see Power 1876:509–510). Francis began medical practice in Springfield in 1835 (when he was 26), and William first settled in Petersburg in 1839 (when he was 28). Early on, they bottled and sold a rheumatic liniment together that was advertised in the June 20, 1845 issue of Springfield's *Illinois Register* (p. 4, col. 5):

RHEUMATIC LINIMENT,
PREPARED BY
DOCTORS F. A. & W. M'NEILL.

The above medicine is offered to the public as a certain cure for Rheumatism in all its various forms. Individuals laboring under that most painful, and hitherto almost incurable disease, would do well to give it a trial. Price $1.25 pr. bottle.

No embossed bottles are known for this early McNeill medicine. William McNeill's relocated his medical practice to several central Illinois towns during the course of the mid-nineteenth century, including Petersburg, Mechanicsburg, Springfield, and Taylorville. He was only listed in Springfield in the 1875 and 1876 CDs. In 1875, he was listed as a patent-medicine maker. He appeared in the 1876 CD as a physician. From the style and technological attributes of the listed embossed *Pain Exterminator* bottle, it was likely the patent medicine he was making and selling in Springfield in 1875.

Samuel H. Melvin

Medicine:

(1a) **S. H. MELVIN / SPRINGFIELD / ILL_** [aqua]

» SB w/ keyed hinge mold; round-shouldered cylinder; vertical embossing; short neck; thickened flared lip; 3" tall. *[Figure 1019, center]*

(1b) **S. H. MELVIN / SPRINGFIELD / ILL** [aqua, clear]

» SB w/ keyed hinge mold; round-shouldered cylinder; vertical embossing; short neck; thickened flared lip; 4" tall. *[Figure 1019, left]*

Figure 1019

(1c) **S. H. MELVIN / SPRINGFIELD / ILL_** [clear]

»SB w/ keyed hinge mold; round-shouldered cylinder; vertical embossing; short neck; thickened flared lip; 5 1/8" tall. *[Figure 1019, right]*

Dr. Sam Houston Melvin (see 1874 portrait: *Figure 1020*) was born in Pennsylvania in 1829 and educated in Ohio, where he studied medicine and graduated from the Medical Hall Institute of Steubenville. He entered the wholesale drug business there in the early 1850s and came to Springfield in 1858 to follow the same business (Lewis Pub. Co. 1892:293–295). In 1859, Melvin purchased a substantial 3-story Springfield drugstore that had recently been constructed by local pharmacist J. B. Fosselman at the corner of 5th and Washington (Russo et al. 1995:100—*see Figure 1021*).

Dr. Melvin was first listed in the 1860 CD as operating the drugstore, and as a "wholesale and retail dealer in drugs, medicines, paints, oils, perfumery, surgical & dental instruments, &c." He was a neighbor and friend of Abraham Lincoln, who offered him a position in his administration when he was elected president, but Melvin chose to remain in Springfield. When Lincoln was assassinated, Melvin was one of the 11 delegates sent to Washington to escort the remains back home. He was listed in the Springfield CDs as a druggist at the 5th & Washington location from 1860 through 1868. By 1869, he had joined the pharmacy partnership of *Melvin & Glidden*, which soon became *Glidden & Co.*

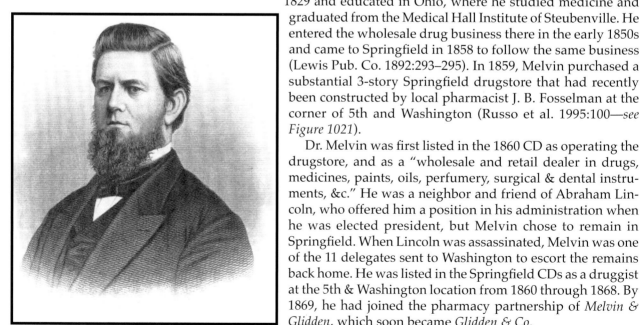

Figure 1020

From 1870 onward, Melvin was no longer involved in the druggist business, having turned to banking and railroad development. In 1875, Dr. Melvin relocated to the Napa Valley in California where he became half owner in a stock ranch. He eventually became president of the College of Pharmacy at the University of California.

The embossed *S. H. Melvin* pharmacy bottles listed were obviously manufactured and used between 1860 and 1868.

Henry C. Myers

Medicine:

Figure 1021

(1) **MYERS // EMBROCATION // SPRINGFIELD, ILL. // //**   [aqua]
   Note: one example known; no photos available.

   »SB w/keyed hinge mold; beveled-edge rectangle (13/8" x 21/2") w/ 3 sunken panels; long neck with central ring; thickened square-ring collar; 61/2" tall.

--- ❖ ---

Henry Myers was born in Pennsylvania in 1817 and first came to Springfield in 1838. He temporarily relocated to Boone County, MO, during the early 1840s, but returned ca. 1949–50. He entered the grocery business in Springfield in 1850 when he bought out Fitzpatrick's grocery store (see *Illinois Daily Journal*, May 20, 1850; Inter-State Pub. Co. 1881:697–698; Power 1876:537).

He first entered the patent-medicine business with his family-recipe *Embrocation* in the late 1840s, while still a resident of Fayette, MO. From the newspaper ad reproduced here, he marketed the *Embrocation* in both Springfield and St. Louis early on, and apparently embossed pontiled examples are known (Beeler, personal communication 1995):

MYERS' EMBROCATION
A Grandfather's Legacy.
Myers' Embrocation—A safe and certain cure for rheumatism. This sovereign and effectual cure for that painful disease was discovered upwards of a half century ago, by the grandfather of the undersigned, and has never been made known out of the family.... Prepared only by H. C. Myers, Fayette, Mo. For sale by Wallace & Diller, Springfield, and Dr. Easterly, St. Louis. [*Illinois Journal*, Feb. 17, 1848]

By the late 1860s, Myer's son Frank had joined him in the grocery business (which became *H. C. Myers & Son*); Frank took over sole operation of their "Wonder Store" on the northeast corner of the square when Henry Myers died on January 24, 1871.

From the stylistic and technological attributed of the listed embossed bottle, it was likely used by Myers during the later Civil War years.

Alfred A. North

Medicine:

(1) **NORTH'S / PECTORAL // // BALSAM OF / HOARHOUND. // //**   [aqua]

   »OP; beveled-edge 11/4" x 2" rectangle; 4 flat panels; rounded shoulders; short neck; tapered collar; 51/2" tall. *[Figure 1022]*

(2) **NORTH // // // //**   [pale aqua]

   »OP; beveled-edge 11/2" x 2" rectangle; 4 sunken panels; rounded shoulders; short neck; tapered collar; 53/4" tall. *[Figure 1023]*

--- ❖ ---

Figure 1023

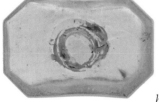

Figure 1022

Alfred North was born in Pennsylvania in 1823. In 1840, at the age of 17, he went to Mobile, Alabama, "for the purpose of learning the drug business; remained five years; then went to Cincinnati, Ohio, where he embarked in the same business.... In *[1850]* he came to *[Sangamon]* county, Illinois, where he represented his district in the legislature for three terms; his health failed, and he had to give up politics, *[where he]* was known as one of the 'Long Nine,' being six feet and four inches in height" (*Illinois State Journal* obituary, Oct. 11, 1892, p. 5, col. 6; Inter-State Pub. Co. 1881:698–699).

North was involved in the patent-medicine trade in Springfield only during the 1850s, enlisting as an officer in the 10th Illinois Cavalry Regiment during the Civil War. Following the war, Major North was elected twice to the office of Springfield Assessor and Collector, afterward becoming involved in "the grain trade" (Inter-State Pub. Co. 1881:699).

Alfred North's first Springfield medicine ad ran in *The Daily Journal* in the summer of 1851, indicating he was the agent for a "Life & Health Insurance Agency, At No. 4, North side of the Square" and that he also carried "a complete assortment of Drugs, Medicines, Dye stuffs, Paints, Oils, Patent medicines, &c., &c." His next ad, that fall (for the "Western Agency," at the same address), enumerated the line of his own medicines available:

No. 4, North side Square.
SHOULDER BRACES, of the most approved patterns.
Dr. BANNING'S Female Pills, for diseases of Females.
Dr. NORTH'S Fever and Ague Killer, a sure and certain cure.
Dr. BARTRAM'S nerve and Bone Liniment for sprains, rheumatism, &c.
Dr. NORTH'S Balsam of Horehound and Naptha for coughs and colds.
Dr. NORTH'S Sarsaparilla and Blood Pills, for impurities of blood, &c.
Dr. BARTRAM'S Dysentary and Diarrhoea Cordial, never known to fail.
Dr. NORTH'S Expectorant Cough Candy, for coughs and colds.

The above valuable remedies, which every family should be supplied with, can be had at the Western Agency, No. 4, North side square, Springfield, Ill.
ALFRED A. NORTH, Ag't.
[*Illinois State Journal*, Sept. 7, 1851, p. 3, col. 2]

Although North continued to sell his medicines during the early and mid-1850s, newspaper ads indicate he also expanded into (and focused his advertising upon) other businesses—becoming a partner in the dry goods business with S. B. Ayres and taking over Vredenburgh's Old Stand to sell tobacco and cigars later in 1851 and during 1852. By 1853–55, he had dissolved his partnership with Ayres and again focused his attention on selling his patent-medicine products (*Figure 1024*).

In 1855, Alfred North formed the partnership of *North & Bristol* in Springfield ("Family Medicine Depot, Opposite the Post Office") to further diversify his business in patent medicines, pharmaceuticals, perfumery, and soaps. An extensive (nearly full-column) *North & Bristol* product ad in 1855 indicated that "Dr. North's" own product line that year included his *Cough Candy, Cherokee Liniment* (a.k.a. *Cherokee Balsam of Horehound*, a.k.a. *Pectoral Balsam of Horehound*), *Sarsaparilla Blood Pills, Cholera Drops, Ague Pills, Mother's Balm,* and *Pain Killer*. The 1855 ad further noted that his *Cherokee Liniment* was "the best embrocation now in use for Bruised, Burns, Pain in the Back, Rheumatism, Tooth Ache, Sore Throat, &c., &c.... Also one of the best compounds for Horse flesh extant, for the cure of Galled Shoulder, Strains, Scratches, Sweeney, Big Head, and all Sores, Bruises, or cuts" (*Daily Illinois State Register*, July 3, 1855, p. 3, col. 5).

North's *Cherokee Balsam of Hoarhound* was found advertised regionally as late as 1856. That year, the *Decatur Daily Chronicle* for September 25 indicated it could be purchased for 25¢ per bottle.

It should be noted that the embossed bottles for North's medicines were designed and manufactured early in his patent-medicine career, well before he arrived in Springfield. Since his bottles fortuitously did not name the hometown of the medicine's manufacture, the mold did not have to be changed as Alfred North moved from one city to another. Thus, embossed examples of North's patent-medicine bottles identical to those found in Springfield are also known from Mobile, AL (where North operated during the early 1840s, and where a second/larger size of the generic medicine bottle simply embossed "*NORTH*" has also been found). Since North moved to Cincinnati during the later 1840s, it is probable that similar bottles will be found there as well.

Figure 1024. Daguerreotype photo of Alfred North's elaborate 1850s patent-medicine sales and delivery wagon. (Family-owned photo published in a 1930s Springfield newspaper. Location of original unknown.)

Thomas J. V. Owen

Medicine:

(1) (side): **OWEN'S BUCHU** // // [pale aqua]
    (side): **SPRINGFIELD ILL'S** // //

   »SB w/ keyed hinge mold; beveled-edge rectangle; 4 sunken panels; short neck w/ raised mid-neck ring; thickened flared lip; 63/4" tall. *[Figure 1025]*

(2) **T. J. V. OWEN / DRUGGIST / SPRINGFIELD. ILL** [aqua]
    Note: *semicircular embossing above and below "**DRUGGIST**"; also oval embossed paper-label border on reverse side.*

   »SB w/ keyed hinge mold; strap-sided oval bottle; no slug plate; short neck; thick flared collar; 31/2" tall. *[Figure 1026]*

Thomas Jefferson Owen was born in 1824 in Kaskaskia and came to Springfield in the summer of 1840 "where he commenced the study of medicine" (Power 1876:552). He served as a military hospital steward and assistant surgeon during the early 1840s, and returned to Springfield to enter the physician's apothecary trade in 1843 at the tender age of 19:

NEW APOTHECARY HALL
The undersigned would inform his old friends and the public generally, that he has purchased the entire stock belonging to the late firms of Drs. Betts & Frazier, and that, by the splendid assortment of medicines he has added to the above, he is prepared to fill all orders in his line at the shortest notice. He flatters himself that having served through a regular course of apprenticeship he can be relied upon in filling prescriptions for Physicians with neatness and accuracy. His medicines are pure and unadulterated. He invites physicians to call and examine his stock, before purchasing elsewhere, as he is confident he can satisfy the most accurate judges.
T. J. V. OWEN, 3 doors east of Lamb's store. [*Illinois State Register ad,* first placed June, 16, 1843]

Figure 1025

Look here everybody.
The very best purely vegetable compound cough expectorant is kept at Tom Owen's, apothecary three doors southeast public square. Just one trial is all it wants, and if it don't cure the disorder it pretends to, why we are mistaken. [*Illinois State Register*, Jan. 26, 1844, p. 4, col. 7]

By 1849, Thomas Owen had become Caleb Birchall's junior partner in the druggist firm of *Birchall & Owen* (No. 10, south side square"). Their earliest ad was published in the local paper in spring 1849, announcing their new "depository of Medical articles...destined to produce a great revolution in the matter of proprietary medicines, in that the *best remedies* which can be produced, may be obtained for the *lowest prices*." The medicines they advertised for sale initially were "Prepared solely by W. A. Holton, Alton, Ills." and included *Nerve and Bone Liniment, Western Cough Syrup, Diarrhoea Syrup, Indian Salve, Holton's Hair Restorer*, and *Wild Cherry Pulmonic* (*Illinois State Register*, March 9, 1849, p. 3, col. 3). None of these products are known to have been put up in embossed bottles.

By 1856 (from *Illinois State Register* ads), the *Birchall & Owen* partnership had become *T. J. V. Owen & Brother* (with William F. Owen) at the same address. Their business had diversified (e.g., paints, dyes, window glass, glassware, spices, etc.), but continued to focus on providing regional and national-brand patent medicines to the Springfield populace. By the time of the 1959 CD, Owen's brother had left the business and Thomas Owen was in the druggist business on his own. Owen enlisted as a soldier in the Civil War, but had returned to the Springfield druggist by 1864. He ran a pharmacy at No. 16, on the south side of the public square, through the late 1860s.

The Springfield CDs then listed him as a druggist from his residence in 1871 and 1872, at which time he was first listed as the proprietor of *Owen's Extract of Buchu*. He sold his Buchu extract in Springfield until 1874, when he moved to Decatur. The 1875 Decatur CD listed him as a patent-medicine manufacturer; Thomas Owen died there in 1876.

Thus the embossed bottles listed were manufactured and used in Springfield from 1870 to 1874. No ads have yet been found to indicate the medicine's advertised function, but Buchu is a South African plant extract (*Agathosma betulina*) used historically as an herbal medicine to treat urinary and reproductive-system disorders.

Figure 1026

Matthew Rinehardt and Co.

Ale ~Beer:

(1) **R & C^{O}** [dark amber]

»SB w/ keyed hinge mold; 2-piece mold quart ale; ring-neck tapered collar; 9" tall. [Figure 1027]

--- ❖ ---

Two broken examples of this minimally embossed ale have been found in Springfield (and not elsewhere) over the years, but the bottler has remained something of a mystery. From the excavation contexts where the broken bottles were found, and the style and manufacturing technology of the ale bottles themselves, they would seem to date to the middle or late 1860s. From their rarity, they were likely made as a single order for an unsuccessful bottling enterprise.

Figure 1027

As luck would have it, a Springfield CD search produced a single candidate manufacturer. There were two different 1866 directories published for Springfield. In one of them, there was a listing for *Reinhardt, Mayer & Co.*, rectifiers and wholesale dealers in wines and liquors at Fifth and Jefferson Street. The proprietors were listed as M. Reinhardt, Adolph Mayer, and Ferd. Bauer. In the competing 1866 CD, Matthew Reinhardt was listed as a saloon operator. In the 1863 and 1864 CDs, Reinhardt was listed as a clerk; he no longer appeared in the CDs after 1866. By 1868, Adolph Mayer was part of the firm of Ensel & Mayer, rectifiers. It is likely that in 1866 Mayer (the brewer/rectifier) teamed up with Reinhardt (the retail sales outlet) to bottle and sell their own brand of ale (or schnapps or brandy) and that they were singularly unsuccessful.

Samuel A. Slemmons and James C. Conkling

Spices:

(1) **SPICE MILLS // SLEMMONS & / CONKLING // SPRINGFIELD. ILL. // //**   [aqua]

»SB w/ keyed hinge mold; beveled-edge rectangle (1 1/2" x 3/4"); 4 sunken panels; ring-neck lozenge top; 5 1/4" tall. *[Figure 1028]*

Samuel Slemmons was born in Ohio in 1842. He came to Springfield in 1859 and served in the military during the early years of the Civil War before returning to Springfield to engage in business. James Conkling was born in New York City in 1816, relocating to Springfield in 1838 after having apprenticed in a law office in the East. He developed a law practice in Illinois, was elected mayor of Springfield in 1845, and later served in the state legislature. During the 1870s and 1880s, he "became actively engaged in business pursuits. He built several blocks in the city, including a spice and hominy mill, and erected the finest private residence in Springfield" (Wallace 1904:54). According to Power (1876:216), Conkling's role in business was primarily that of a developer and financier: "James C. Conkling, more than any other capitalist of Springfield, uses his wealth in extensive building enterprises, and for the encouragement of manufactures."

Figure 1028

Figure 1029. 1876 Springfield city directory image (p. 37) of the *Illinois Hominy and Spice Mills* at 301-303-305-307-309-311 South 4th Street, operated by J. C. and J. J. Conkling and S. A. Slemmons.

In their joint business venture, which began in 1870, young (28-year-old) Samuel Slemmons was "was one of the original projectors of the Springfield [a.k.a. Illinois] Spice and Hominy Mills, and aided in building and running the same" (Power 1876:461), while 54-year-old politically connected James Conkling was the developer who funded the business start-up and eventually owned a controlling interest in the factory.

Slemmons retired from principal management of the original business on the 300 block of south 4th Street (*Figure 1029*) in 1876, and then started a new venture known as the Globe Spice Mills between 4th and 5th Street: "The concern roasts and prepares coffee, prepares spices, manufactures baking powder and roasts peanuts" (Inter-State Pub. Co. 1881:576).

Figure 1030

The Illinois Hominy & Spice Mills operation was first known as *Slemmons & Conkling* (ca. 1870–73), then as *Slemmons, Conkling & Co.* (ca. 1873–75), then, from 1876 on, as *Conkling (J. C. & J. J.) and Slemmons* (see *Figure 1030*). From the style and technological attributes of the embossed bottles listed, they were probably manufactured and used when the company first began in business, ca. 1870–71.

Dr. Topping's Alterative & Cathartic Syrup

See James J. Lord & Co. listing, this chapter.

Turner's Lotion

See Roland W. Diller listing, this chapter.

Martin M. Van Deusen

Bitters:

(1) D^R VAN DEUSEN'S / AGUE BITTERS / SPRINGFIELD ILL // // // //   [aqua]
   Note: two embossed dots beneath superscript **R** in D^R

   »IP; beveled-edge 2½" square "case" bitters bottle; 4 flat panels w/ slight depression and slug-plate outline on embossed face; short neck; double-ring collar; 8½" tall. [*Figure 1031*]

(2) *"Urban Legend" bottle*
   (*Not confirmed and likely perpetuated by repeated hearsay misinformation*)

An embossed cylindrical *Van Deusen's Ague Bitters* bottle, with the same embossing as bottle #1 (except that it is embossed "**ILL.**" with a period), has been illustrated and described in earlier publications (Ring 1980:468; Ring and Ham 1998:549). The cylindrical bottle is described as being the same height and aqua glass color as the square bottle.

Figure 1031

Figure 1032

The bottle illustration used in the earlier publications is a line drawing from an unknown source (*see Figure 1032*), not a photograph, and extensive inquiry has failed to locate any actual examples or fragments of this bottle variety. All of the several broken examples of embossed *Van Deusen's Ague Bitters* bottles known to have been found in Springfield excavated contexts are rectangular (like #1).

Martin Van Deusen was born and raised in New York and first came to Springfield and opened his drug store on the town square in November 1851, when he was 36 years old:

M. M. VAN DEUSEN,
Druggist and Apothecary,
West side the Public Square.
SPRINGFIELD, ILLS.

WOULD respectfully invite the attention of Physicians, Country Merchants, and the Public in general to his LARGE, NEW and well selected assortment of East India, Mediterranean and European **DRUGS AND MEDICINES,** French, English and American **CHEMICALS** of all kinds, **PAINTS, OILS, DYE-STUFFS AND GLASSWARE.**
A full assortment of the most popular
FAMILY MEDICINES
of the day.
Also, **WINES** and **BRANDIES** for medicinal purposes, which are warranted pure and of the most approved brands.
Perfumery, Toilet Soap, Combs, Brushes, Fancy Articles, &c., &c.
Prescriptions dispensed accurately by competent assistants at all hours. Please call and examine my stock and priors before purchasing elsewhere, as I will sell on more favorable terms than ever before offered in the city. Every article warranted pure and genuine.
M. M. VAN DEUSEN [*Illinois Journal*, Dec. 12, 1851, p. 2, col. 5]

By summer 1852, Van Deusen began advertising his own patent-medicine brand, *Van Deusen's Cholera Specific*, along with directions for its use—"This medicine has been prepared with great care, and will be found to give instant relief in all cases of Diarrhoea, Dysentery, Colic, Cholera Infantum, and that human scourge the Asiatic Cholera, and should be in every family.... Prepared only by M. M. Van Deusen, Druggist and Apothecary, west side public square, Springfield. Price 50 cents" [*The Daily Journal*, Aug. 3, 1852, p. 3, col. 3]. No embossed bottles are known for Van Deusen's *Cholera Specific*.

Van Deusen's newspaper ads during the mid-1850s do not mention his own brand medicines, so perhaps his *Cholera Specific* did not sell well. His ads during this time focus on promoting the popular and widely known brand-name eastern patent medicines he offered for sale. In 1856, he was also elected as an alderman for the City of Springfield (Inter-State Pub. Co. 1881:566).

Van Deusen's Ague Bitters was first advertised (with testimonials) in the *Illinois State Journal* during the fall 1859 [e.g., Oct. 27, p. 1, col. 5–6] and spring and summer 1860:

THE BEST
PREVENTATIVE AND POSITIVE CURE
EVER OFFERED TO THE PUBLIC

DR. M. M. VAN DEUSEN'S
CELEBRATED
AGUE BITTERS!
A Certain Cure for Fever and Ague, Dumb Ague, &c.

...During many years residence in the South and West, I have seen Chills, Fever and Ague, Dumb Ague, and all Billious Diseases in their various forms. Of the numerous preparations styled "Fever and Ague Specifics," etc., I found that many would temporarily relieve the patient; but on the slightest predisposing cause, the affection would return as bad or worse than before. I therefore devoted myself to the object of discovering something that would...thoroughly eradicate the disease from the system, and act as an effective preventative.

For the last four years I have been using this medicine with unvarying success; and during the year 1859, I retailed from my own counter, *over one hundred dozen bottles*, and in no instance has it failed to effect a complete cure, when taken according to directions.

These bitters are also the best medicine you could possibly use for Indigestion, loss of Appetite, Weakness, etc., in fact in all cases where a tonic is required. The leading physicians of Springfield speak in the highest terms of the value of this medicine, and prescribe it for their patients.

It is put up in bottles containing a pint, and sold at One Dollar.

Be careful to observe that my name is blown on the side of the bottle, as follows: "*Dr. Van Deusen's Ague Bitters, Springfield, Ill.*" The above is prepared only by me, under my own personal supervision, at my Laboratory, West side Square, Springfield, Ill.

M. M. VAN DEUSEN [*Illinois State Journal,* July 6, 1860, p. 1, col. 5–6]

This July 6, 1860, ad was followed by testimonials from retailers in Auburn, Virden, Richland, and Bird's Point, indicating that each of them had sold several dozen bottles over the past year. The ad also listed regional sales agents in Decatur, Mechanicsburg, Auburn, Virden, Girard, Carlinville, Lincoln, Pleasant Plains, Richland, New Berlin, Bement, Clear Creek, Naples, Danville, West Urbana, Homer, Buffalo Station, Chatham, Williamsville, and Meredosia.

The 1859 Springfield CD listed Martin Van Deusen as a druggist on 5th Street between Washington and Adams. The same CD also listed "A. Van Deusen" (Abraham Van Deusen) as a patent medicine dealer on 8th Street between Gemini and Cancer. According to the 1860 USC records, both men were from New York and both were a similar age (Martin was 45 and Abraham, who had moved to Chicago by 1860, was 40), so perhaps they were brothers. Abraham's only Springfield CD listing was in 1859, and the 1859–60 period may have been the only time that the *Van Deusen's Ague Bitters* was actually bottled (since all of the known bottles and fragments are pontiled). M. M. Van Deusen was still listed as a Springfield druggist in the 1860 and 1863 CDs, but after that he was gone as well.

From the ads and the CD data, Martin Van Deusen sold *Ague Bitters* in his drugstore only in 1859 and the early 1860s. An isolated 1864 newspaper ad by D. L. Gold & Co. [*Daily Illinois State Register,* Sept. 20, 1864, p. 1, col. 4] indicated that **both** *Van Deusen's Ague Bitters* and *Van Deusen's Cholera Specific* would be "on Sale by Sept. 17, 1864." This looks like a sale notice for secondhand or remaindered stock purchased at auction from Van Deusen's recently closed drugstore.

William W. Watson and Son

Soda ~ Mineral Water

Although no embossed bottles are known for this company, W. W. Watson and his son Benjamin A. Watson did advertise that they bottled soda water in Springfield during the 1850s, and excavated unembossed soda and mineral water bottles are all but unknown from Springfield excavations. The possibility that embossed pontiled Watson soda bottles may exist is further suggested by the fact that Watson's son Benjamin bottled mineral springs water in embossed bottles after he moved to Pike County just after the Civil War years and developed the Perry Springs Mineral Water Spa and resort there (*see his Perry Springs listing, this chapter*).

William Watson was born in New Jersey in 1794, then moved to Nashville, TN, after the War of 1812, where his son Benjamin was born in 1834 just before the family moved to St. Louis. The Watsons came to Springfield in 1836, and by the early 1840s William and Benjamin had opened a bakery and confection shop in town, as W. W. Watson & Son (see *Sangamo Journal,* Sept. 2, 1842, p. 3, col. 6). They placed Springfield newspaper ads throughout the 1840s and early 1850s for their bakery, "ice cream saloon," confectionary products, and ice delivery service. The suggestion that embossed Watson & Son soda water bottles may be found is based on the following ad, first placed in *The Daily Journal* on May 27, 1851:

Bottled Soda Water.
We are now manufacturing Bottled Soda Water,—by a new and complete set of machinery,—that we will warrant equal to any in the State. We will perfect arrangements to have it delivered to families at 50 cts, per doz., in a few days. Orders respectfully solicited.
W. W. WATSON & SON.

STAUNTON
(Macoupin County)
1880 census: 1,400

Christian Hoffmann

Soda ~ Mineral Water:

(1) **CHR. HOFFMANN / STANTON / ILLS //**   [aqua]
(base): **C.H**
Note: "Staunton" is misspelled on the embossed bottle.

»SB w/ post mold; very short neck, collared-blob style soda; embossed in bottle mold, not slug plate; oval-blob-top. [Figure 1033]

No record of either Mr. Hoffman or his bottling works could be found in Macoupin County historical volumes or census records, but from his bottle's technological and stylistic attributes, it was likely manufactured (and used?) for a short time in the later 1870s. Perhaps Messrs. Knemoeller and Stille took over operation of the Hoffmann bottling works very shortly after it was first established.

Rudolph Knemoeller and [Ernst R.] Stille

Soda ~ Mineral Water:

(1) **KNEMOELLER / & / STILLE /**   [aqua]
    STAUNTON / ILLS //
(heel): **L G C°**
(base): **K & S**

»SB w/ post mold; oval-blob-top soda. [Figure 1034]

Little is known of the *Knemoeller & Stille* bottling works, but from the style and technological attributes of the embossed bottle, their partnership likely dated to the

Figure 1033

mid- or late 1870s. Rudolph Stille was listed as a 56-year-old soda water bottler in the 1900 USC for Staunton, and both collared-blob-style and Hutchinson-style Staunton sodas embossed **R. STILLE** are known (dating ca. 1880–1900, too late for inclusion in the present volume).

Herman W. Stille and his brother Ernst Rudolph ("Rudolph"?) Stille are known from genealogical records to have resided in the Staunton area at about this time. Rudolph Knemoeller could not be directly connected with the soda bottling business, but he was about the right age to have been Stille's early partner: from genealogical records, his son Henry was born in Staunton in 1874, and was a life-long Staunton resident.

STREATOR
(La Salle County)

1870 census: 1,500

John A. Finkler

Soda ~ Mineral Water:

(1) **J. A. FINKLER / STREATOR / ILL**ˢ // [blue, aqua]
(heel): **A & D. H. C.**
(base): **J. A. F.**
Note: There are 2 embossed dots under the raised **S** in **ILL**ˢ

»SB w/ keyed hinge mold; oval-blob-top soda. *[Figure 1035]*

Figure 1034

John Finkler was in the soda water bottling business with his brother Alex in nearby LaSalle during the 1860s (*see his LaSalle listing*). They are listed as residing there together in the 1863 IRS tax assessment records and in the 1860 and 1870 USC.

The 1870 census listed them both as soda water makers in La Salle County. At the time, they were 42 and 33 years old, respectively.

John Finkler first became involved in the soda bottling business during the later Civil War years. An 1864 IRS tax assessment listed him as the silent "*& Co.*" partner of J. B. St. Clair in a Peoria bottling business.

John relocated to establish a bottling works on Main Street in Streator "sometime before 1872," according to the *Streatorland*

Figure 1035

Historical Society (Richard O'Hara, *Streator Times Press*, July 22, 2002), and he was still a "mineral water manufacturer" there in 1888 (at 17 N. Second Street).

His brother Alex moved his bottling operation from LaSalle to Ottawa (*see his Ottawa listing*) ca. 1872, "at Riverside; P.O. Streator," according to his 1877 LaSalle County CD listing (Kett 1877a:419, 604). He was also listed there as a soda bottler in the 1880 USC. A third brother, Frank J. Finkler, also established a bottling works for a time in Dixon, IL (*see his Dixon listing*).

From its technological and style attributes, the embossed bottle listed should date to ca. 1869–71, when John Finkler first established his Streator bottling operation.

  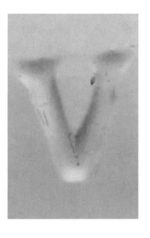

VIRDEN
(Macoupin County)

[1860 Twp. census: 750]
[1870 Twp. census: 3,100]
1880 Virden census: 1,600

Ernest Achilles and Hille

Soda ~ Mineral Water:

(1) **ACHILLES & HILLE / VIRDEN. ILLS**   [aqua]
Note: 2 embossed dots under S in **ILLs**

»SB w/ keyed hinge mold; oval-blob-top soda. Large central Raised-glass "bump" on base. *[Figure 1036]*

(2) **ACHILLES & HILLE / VIRDEN ILLS**   [aqua]

»SB w/ post mold; narrow, cylindrical, oval-blob-top soda. *[Figure 1037]*

Figure 1036

Achilles & Hille were proprietors of a general store in Virden according to the 1873 (CAR). The following year, the 1874 CAR continued to list them in the general store business, but added that they also manufactured soda water.

Louis Achilles was listed in the 1860 USC as a 52-year-old Virden merchant. The 1860 Statewide CD listed him as a dry goods merchant and baker. In the 1870 USC, Ernest Achilles (son of Louis?) was listed as a Virden dry goods merchant. From bottle style and manufacturing technology, soda #1 was produced and used during the late 1860s; soda #2 was made and used during the early to mid-1870s.

No record or identity could be uncovered for Ernest's partner, Mr. Hille, in the Virden grocery and soda business they operated during the late 1860s and early 1870s.

Figure 1037

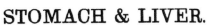

WATERLOO

(Monroe County)

1850 census: 800
1860 census: 1,400
1870 census: 1,500
1880 census: 1,800

Henry Boeke

Soda ~ Mineral Water:

(1) **H. BOEKE / WATERLOO, ILL^S_ //**
(heel): **I. G. CO**

»SB w/ post mold; oval-blob-top soda. *[Figure 1038]*

Henry Boeke was listed in the 1880 USC for Waterloo as a 40-year-old operator of a soda bottling works. An 1883 Monroe County history also listed him as the operator of a soda factory in Waterloo. From the style and technological attributes of the embossed bottle listed, and from its *Illinois Glass Co.* maker's mark, it was likely manufactured and used during the late 1870s (ca. 1875–80). The fact that he was still bottling during the early 1880s (beyond the scope of the present study), suggests he may have later used one or more varieties of embossed Hutchinson-style (internal rubber stopper) soda bottles as well.

[aqua]

Figure 1038

George DePuyt

Soda ~ Mineral Water:

(1) **GEO. DE.PUYT / WATERLOO / ILL.** [aqua, olive green]
»IP; tapered-collar soda. *[Figure 1039]*

From the style and technological attributes of the embossed mineral water bottle listed, it was likely manufactured and used ca. 1845–52. The 1850 USC for Waterloo listed George DePuyt as a 35-year-old "barkeeper" born in Germany. In the 1870 USC for Waterloo, George DePuyt (pronounced "DePyte") was listed as a 54-year-old saloonkeeper from Hesse Darmstadt, Germany. According to the McDonough & Co. 1883 county history (p. 318), DePuyt was also active in Waterloo politics. He was a member of the town's first elected Board of Trustees in 1859, and was reelected to serve on the town board in 1874.

DePuyt was involved in the Waterloo saloon trade for at least 20 years, but the embossed mineral water bottle listed was produced at the very beginning of this time—and perhaps earlier. When he first set up in business in Waterloo during the late 1840s, he may have begun as a mineral water bottler before moving to the saloon business, or he may have produced mineral water in the early days as a drink mixer to sell in his saloon.

According to a modern DePuyt family descendent (Launius personal communication, 2005), the family has preserved a recipe George DePuyt used for preparing (and perhaps bottling?) bitters, which he may also have marketed as part of his Waterloo saloon trade. No embossed DePuyt bitters bottles are known, however.

Figure 1039

[Sophia] and John Ruf

Soda ~ Mineral Water:

(1) **S & J. RUF / WATERLOO / ILL** [aqua]

»SB w/ keyed hinge mold; oval-blob-top soda. *[Figure 1040]*

The *S. & J. Ruf* partnership has been found listed as an ongoing Waterloo business one year only: 1864 (IRS tax assessment for May of that year). After that, John Ruf is listed alone as a Waterloo soda water manufacturer (*see his listing, this chapter*).

The possible identity of "**S**" **RUF** is a guess based on a thin thread of evidence. According to the Jacksonville (IL) City Directory for 1860–61, an individual named John Ruf manufactured soda water in 1860 in Jacksonville, on College Street between West and Church. (No embossed bottles have yet been documented for this operation, but unembossed pontiled sodas are occasionally found in Jacksonville privies.) According to the 1860 USC, John "Ruff" (50), his wife Sofia (37), and their teenaged daughter were Jacksonville residents at the time. If the Jacksonville and Waterloo bottlers were the same John Ruf, perhaps the family moved to Waterloo during the early 1860s, and the *S. & J. Ruf* company in 1864 were Sophia and John.

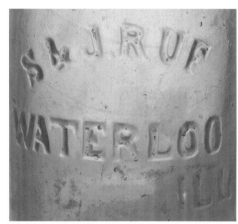

Figure 1040

Coincidentally, a biographical sketch of one of Ruf's grandsons (Smith 1912:1093) indicated that his grandfather came to the Midwest from Germany in 1852, first settling in St. Louis. Jacksonville was not mentioned, but the sketch indicated that Ruf lived in Waterloo from 1863 to 1878 before returning to Germany to live out the final two years of his life.

John Ruf

Soda ~ Mineral Water:

(1) **J. RUF / WATERLOO / ILLS :** [aqua]
Note: No maker's mark on bottle, but DeSteiger-style embossed lettering.

»SB w/ keyed hinge mold; shouldered-blob-top soda. *[Figure 1041]*

(2) **JOHN. RUF / WATERLOO / ILL** [aqua, blue]
Note: Bottle shape and embossing identical to S & J. RUF bottle listed, except S & J has been peened out and replaced with JOHN.

»SB w/ keyed hinge mold; shouldered-blob-top soda. *[Figure 1042]*

Figure 1041

Figure 1042

The History of Southern Illinois (Smith 1912:1093) stated that John Ruf settled in Waterloo in 1863 and lived there until he returned to his native Germany in 1878. Ruf is listed as a bottler in the 1864–66 IRS tax assessments for Waterloo, the first year in a partnership: *S. & J. RUF*, perhaps with his wife, Sophia (*see their listing, this chapter*). Ruf also appears in the 1870 USC for Waterloo as a 66-year-old soda water manufacturer. A John Ruf was a soda water manufacturer in Jacksonville in the 1850s; however no embossed **JOHN RUF** Jacksonville bottles are known (*see* **[Sopia] & J RUF** *listing, this chapter*).

The strongest contrasting bit documentary evidence regarding a John Ruf Jacksonville–Waterloo connection is an 1865 Illinois commercial census (*p. 58*) for Jacksonville indicating that a person named "John Ruff" was in business there that year producing products (the products are not named) with an annual estimated value of $2,300.

WAUKEGAN

(Lake County)

1850 census: 300
1860 census: 3,400
1870 census: 4,500
1880 census: 4,000

Dr. Benjamin S. Cory and Son

Bitters:

(1) **DR. CORY'S** (*in small arched sunken panel*) // **STOMACHIC** // [aqua]
 (*unimbossed flat panel*) // **BITTERS**
 Note: Dr. Cory's embossed bottle is only known from a few broken examples. The bottle's shape and style of embossing emulate that of the popular, contemporary Ayer's Cherry Pectoral medicine bottle. Dr. Ayer's Medicine products were regionally marketed in the Midwest from the company's base of operations at Lowell, MA (see Baldwin 1973:46).

»IP; beveled-edge rectangle; 3 sunken panels; double-ring collar; ca. 6" to 7" tall. *[No photo; described from fragmentary excavated example]*

Benjamin Sayre Cory was born in 1805 and was educated as a physician and surgeon in New York State. After finishing his education, he established a medical practice in Canada. During the later 1830s, "he was a regularly commissioned surgeon for the Prince Edward Co. Regiment, raised for active service against the Rebels" (Dickinson 1914:31). In 1854, B. S. Cory moved his family from Wellington, Ontario, to Waukegan, IL, where a number of members of the "Cory clan" already resided.

B. S. Cory & Son (likely his eldest son, James) list themselves as "druggists and apothecaries" in their ads, and appear to have carried on a two-pronged attack to market their *Stomachic Bitters* in northeastern Illinois in 1859. According to their 1859 newspaper and CD ads, they simultaneously opened wholesale and retail distribution outlets in both their Lake County hometown area of Waukegan (at 47 Washington Street) and in downtown Chicago (at 174 South Clark Street: *see Cory Chicago listing*). Except for the business addresses, the large ads were fundamentally the same. They included a very ornate coat of arms (*Figure 1043*) with the inscription "**BIBE ET VIVE**" ("Drink and Live"), and advertised:

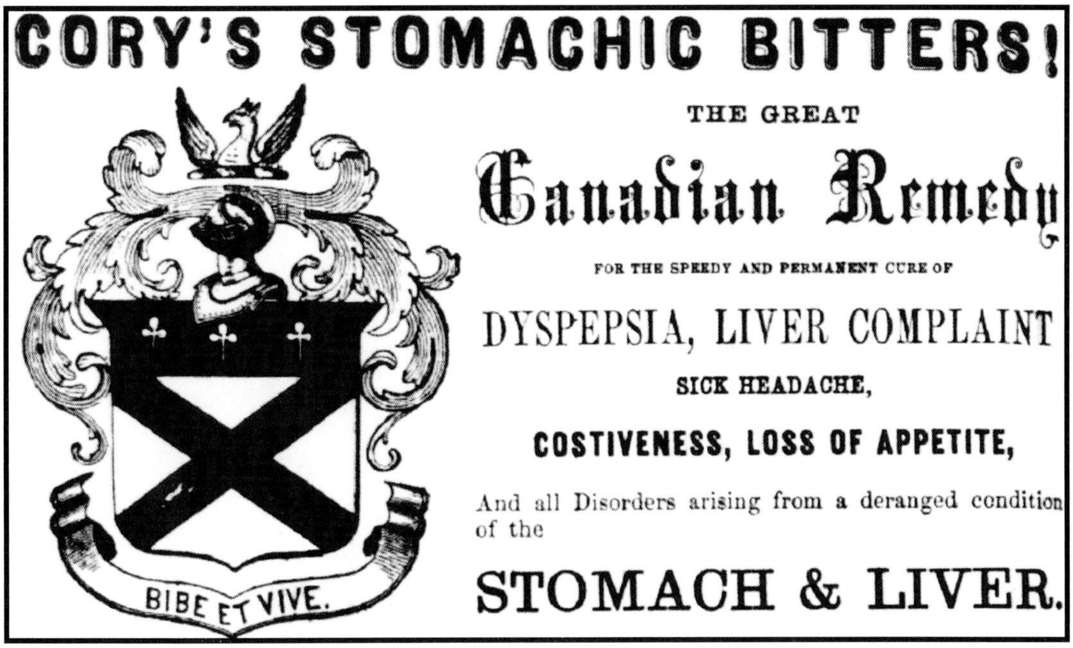

Figure 1043

CORY'S STOMACHIC BITTERS
The Great *Canadian Remedy*
for the speedy and permanent cure of
DISPEPSIA, LIVER COMPLAINT,
SICK HEADACHE, COSTIVENESS, LOSS OF APPETITE
And all disorders arising from a deranged condition of the
STOMACH & LIVER.

PRICE ONE DOLLAR PER BOTTLE — SOLD BY ALL RESPECTABLE DRUGGISTS.

An advertisement placed by the company in the *Woodstock Sentinel* beginning in mid-January, 1859, provided more detail on the claimed curative powers of *Dr. Cory's Stomachic Bitters*:

> These bitters are compounded from Gums long known to the Profession as possessing peculiar curative powers in cases of Dyspepsia, Liver Complaint, &c., which operate powerfully upon the Stomach, Bowels, &c., and the Digestive Organs, restoring them to a healthy and vigorous action, and thus by the simple process of nature, enable the system triumph over Disease.
> Its adaptation to the cure of all afflictions of the Billiary or Digestive Organs is peculiar to itself.... **For this disease, every physician will recommend Bitters of some kind, then why not use an article known to be infallible? Every country has its Bitters as a preventative of disease and for strengthening of the system in general, and among them all there is not to be found a more healthy people than the Canadians**, from whom this preparation emanated, based upon scientific experiments which have tended to advance the destiny of this great preparation in the grand medical scale of science.... And as it neither creates nausea nor offends the palate, and rendering unnecessary any change of diet, or interruption to the normal pursuits of life, but promotes sound sleep and healthy digestion, the complaint is thus removed as speedily as is consistent with the production of a thorough and permanent cure.

The Cory & Sons firm was also listed in Civil War–era IRS assessments (in 1862 as apothecaries & physicians and in 1863 as an apothecaries and retail liquor dealers). However, the ads appear to have been a one-shot effort, and in Chicago CDs they are only listed (with their ad) in 1859. They are not listed at all, even as Chicago druggists, in 1858 or 1860.

In the IRS assessment for June, 1866, B. S. Cory is listed alone as a physician in Waukegan.

WEST URBANA
(Champaign County)

1855 census: 420
1860 census: 1,700
(Incorporated 1855, renamed *Champaign* in 1860)

William Doane

Soda ~ Mineral Water:

(1) **WM. DOANE / WEST URBANA / ILLS** [aqua]
»IP; shouldered-blob-top soda.

We have not yet been able to locate the only known example of this embossed pontiled soda bottle to photograph and measure it, but details of the bottle's embossing, style, and technological attributes, provided by two independent observers, suggest it was manufactured between 1853 and 1860. The fact that Champaign was first incorporated as West Urbana during only a five-year period (1855–59) further restricts the period of the bottle's manufacture and use.

We have not yet found any specific documentation of William Doane as a West Urbana soda and mineral water bottler during the late 1850s. But a detailed lot-by-lot street map of the town has been located, showing all of its buildings and naming many of its public structures (Alexander Bowman 1858). One of the labeled buildings on the map is the Doane House hotel (*Figure 1044*), a combination "depot, hotel, telegraph office, and social center for the community" located downtown along the Illinois Central Railroad tracks at the Main Street crossing between Walnut and Oak. According to the Champaign-Urbana Mass Transit District website, the Doane hotel was built in 1856.

The Doane who was the hotel's namesake owner when it was first built has not yet been confirmed to be William, but for the present it is a best-guess possibility. Likely, Mr. Doane did not run the facility for long. An April 16, 1857, issue of the *Urbana Union* newspaper indicated that John Campbell had sold his interest in the hotel to a "Mr. Hubbard." The hotel, which was shown a three-story structure in a historic painting by Lisle Wiseman titled "West Urbana Illinois—1858," burned to the ground 40 years later.

If William Doane was the first manager of the hotel, he likely bottled flavored soda for a short time prior to the Civil War (ca. 1856–57) to serve to his customers in the hotel restaurant and bar.

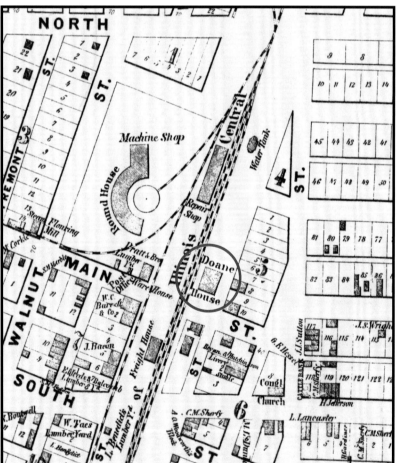

Figure 1044

WESTERN SARATOGA
(Union County)

1850–70 census data unknown:
See discussion below.

Dr. Hiram Penoyer

Medicine:

(1) **D^R H** / **PENOYER'S** // *(side):* **CELEBRATED** // **EXTRACT FROM** / **THE** / **HYGEAN. SPRINGS** // *(side):* **OF. ILLS**_   [aqua]

Note: There are two embossed dots beneath the raised **R** in **D^R**

»SB w/ keyed hinge mold; beveled-edge rectangle; 4 embossed sunken panels; double ring collar; 6 3/4" tall. *[Figure 1045]*

The village of Western Saratoga (*aka* Saratoga Village or Saratoga) was located about six miles northeast of Anna, IL. It was essentially a "paper town" laid out around a mineral spring development on a 160-acre tract purchased in 1838 by Dr. Hiram Penoyer. The original 1841 town plat was comprised of numerous residential lots centered on a park adjacent to another public square surrounding the principal spring. But the "town" was occupied largely by seasonal visitors to the springs and "never amounted to much except on paper" (Perrin 1883:427). It was described in the 1858 *Illinois State Gazetteer* as a "post village...owes its rise to a medicinal spring, which attracts numerous visitors in the warm season."

The town's name was chosen to reflect the grandeur of the famous Eastern Saratoga Springs Mineral Water Spa in New York (Zeuch 1927:115). A boarding house built near the spring was in operation for several years to entertain summer guests (Leonard 1941), and very early on (March 1, 1845) an act was approved by the Illinois State Legislature to incorporate the *Hygean Spring Seminary*

Figure 1045

at Western Saratoga, the object of which would be "the promotion of the general interest of education" (Laws of the State of Illinois 1845:113). This institution appears to have been largely a "blessed hope"; however during the 1840s and 1850s the boarding house and a bathhouse were built there, along with a store and a grist mill:

> Penoyer made the mistake of platting his town and dedicating...a square to the public [*containing the spring*]...precluding anyone from taking hold of it and developing it as it deserved. Another error...was placing upon the lots so high a price that no one felt they could afford to invest. However, about 1840, a man named Bradley purchased a small tract, and erected a boarding house.... Dr. Penoyer and a man named Harkness...built a bathing-house, about forty rods from the spring, and connected with it by a series of pipes. This bathing-house was about one hundred feet long and nine feet wide. This was used for some time, but gradually falling into disuse it rotted down. As long as people could get accommodation, they flocked here in great numbers.... [*When*] the Doctor discovered his mistake, [*he*] had what was called a deed from the public to himself made, conveying the spring back to himself. This curious document was signed by the visitors who, from time to time, were attracted to the place. [*Perrin 1883:234–235*]

From the stylistic and technological attributes of the embossed *Hygean Springs* bottle listed, it was likely manufactured and filled the early 1860s. It contained a mineral extract from the springs, first produced by Dr. Penoyer during the 1850s by boiling down the spring waters:

> It is a tolerably strong sulphur water, and contains sulpurated hydrogen, a small quantity of Sulphate of lime, carbonate of soda, chloride of sodium, and, perhaps, a little alumina and magnesia. The water is said to be a specific for dyspepsia and chronic diseases of the skin. It is also said to be beneficial in cases of scrofula.
> Dr. Penoyer seems to have been a poor manager, and yet the waters were shipped and sold by him, in quantities, to many parts of the country. **For some years he made a practice of boiling it down and bottling and peddling it about the country, and shipping it to those wanting it at a distance.**
> **The Doctor's method of boiling the water was to take 100 gallons, and boil it until only one remained. This one gallon was quite thick, and tasted like soft soap suds, or very strong soda water.** It was about the time that the Doctor was engaged in making this **medicine**...that there was an epidemic of flux. It was very fatal, and the physicians gave up many cases which Dr. Penoyer was able to cure with his medicine, in every instance in which it was given a fair trial. [Perrin 1883:235; see also Zeuch 1927:116]

The *Perry Springs* mineral water springs, spa, and hotel near Perry in west-central Illinois (*see its listing*) were a similar Civil War–era tourist attraction.

WILMINGTON

(Will County)

1870 census: 1,800
1880 census: 1,100

Horace W. Blood and Co.

Soda ~ Mineral Water:

(1) **H. W. BLOOD & Co. / WILMINGTON / ILL**^{S} //   [aqua]
(*heel*): **A & D. H. C.**
Note: There are two embossed glass dots beneath the superscript **S** in **ILL**^{S}.

»SB w/ keyed hinge mold; oval-blob-top soda. [*Figure 1046*]

Figure 1046

Horace W. Blood first came to Illinois from New York in 1863 when he was 20 years old. He farmed for two years before taking employment in 1865 at J. D. Paige's bottling house in Joliet (*see Paige's Joliet listing*). Blood remained in Joliet for two seasons learning the bottling business, then moved 15 miles south to Wilmington to oversee operations at J. D. Paige's bottling depot there (*see Paige's Joliet listing*), having purchased a half-interest in the operation. He also engaged in the ice business in Wilmington.

In 1870, he purchased a controlling interest in Paige's Wilmington bottling works (Kott 1978:5; Le Baron 1878b:751) and commenced soda and mineral water bottling there as *H. W. Blood & Co.*

In 1873, he moved his operation five miles further southwest to Braidwood and set up a new bottling works there (*see his Braidwood listing*). The listed embossed bottle was thus manufactured and filled only during two or three bottling seasons, ca. 1870–72.

WOODSTOCK
(McHenry County)

1860 census: 1,300
1870 census: 1,600

Dr. Holland W. Richardson

Bitters:

(2) DR H. W. RICHARDSON / WOODSTOCK, ILL. // // [amber]
 WAHOO & WILD CHERRY / BITTERS // //

 »SB w/ keyed hinge mold; beveled-edge square bitters-style case bottle; flat panels; tapered collar; 9 1/2" tall. *[Figure 1047b, d]*

(2) D^R .RICHARDSON / WOODSTOCK, ILL. // // [light olive]
 WAHOO & WILD CHERRY / BITTERS // //

 »SB w/ keyed hinge mold; beveled-edge square bitters-style case bottle; flat panels; tapered collar; 9 1/2" tall. *[Figure 1047a, c]*

Advertisements placed in 1867 and 1868 editions of the *Woodstock Sentinel* newspaper stated that *Wahoo Bitters* was manufactured and sold by Dr. H. W. Richardson, who operated the Corner Drug Store, located at "No.1 Excelsior Block, Woodstock, Ill. Established 1854" (*see Figure 1048*).

Dr. Richardson was trained as a physician and surgeon, and he came to Woodstock ca. 1854 to practice his trade. His *Wahoo & Wild Cherry Bitters* was enough of a commercial success that he entered the druggist trade at about the time of the Civil War, buying an established drugstore in the 100 block of Cass Street. He later took on a partner named Anderson Murphy, and eventually "sold his interest in the business to Luman (L. T.) Hoy and the firm became Murphy and Hoy, around 1869" (Woodstock Historic Preservation Commission brochure, "Walking Tour of the Historic Woodstock Square"—#19). During ca. 1868–70, he refocused his attention on his medical practice, relocating to Chicago in 1871 to establish a practice as a physician and surgeon there (Chicago CDs).

From the style and technological attributes of the embossed bottle listed, it was manufactured and used during the mid-to-late 1860s.

Figure 1047

Dr. Richardson's spring and early summer 1867, *Woodstock Sentinel* ads noted that he also made and sold *Dr. Richardson's Peruvian Balsam* (no embossed bottles known), and detailed the curative powers he attributed to his *Wahoo Bitters*:

DR. RICHARDSON'S
WAHOO
BITTERS
For the permanent cure of Dyspepsia, Liver Complaint,
Nervous Debility, Constipation, Diseases of the Kidneys
and Pulmonary Affections. Liberal discount to the trade.
H. W. R.

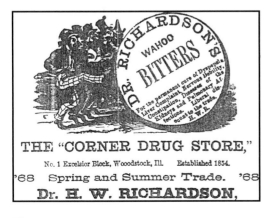

Figure 1048

Bibliography

Adams, James N.
1989 *Illinois Place Names.* Illinois State Historical Society, Occasional Publications No. 55. Springfield, IL.

Adams, Samuel Hopkins
1907 *The Great American Fraud: Articles on the Nostrum Evil and Quackery Reprinted from Collier's.* American Medical Association Press, Chicago, IL.

"AMICUS"
1866 Perry Springs. *Jacksonville Daily Journal,* June 29, 1866, page 3.

Andreas, Alfred T.
1884 *History of Chicago: From the Earliest Period to the Present Time.* Vol. I.—Ending with the Year 1857. A.T. Andreas Co., Chicago, IL.
1886 *History of Chicago: From the Earliest Period to the Present Time.* Vol. Ill.—From the Fire of 1871 until 1885. A.T. Andreas Co., Chicago, IL.

Andreas, Lyter, and Co.
1871 *Atlas Map of Fulton County, Illinois.* Andreas, Lyter & Co., Davenport, IA.
1872a *Atlas Map of Morgan County, Illinois.* Andreas, Lyter & Co., Davenport, IA.
1872b *Atlas Map of Pike County, Illinois.* Andreas, Lyter & Co., Davenport, IA.
1873 *Atlas Map of Peoria County, Illinois.* Andreas, Lyter & Co., Davenport, IA.

Baldwin, Joseph K.
1973 *Patent and Proprietary Medicine Bottles of the Nineteenth Century.* Thomas Nelson, Inc., Nashville, KY.

Balfour, Edward
1885 *The Cyclopaedia of India and of Eastern and Southern Asia, Commercial, Industrial, and Scientific; Products of the Mineral, Vegetable, and Animal Kingdoms, Useful Arts and Manufactures.* Volume 3. Bernard Qaritch, London.

Baskin, O. L., and Co.
1883 *History of Alexander, Union, and Pulaski Counties, Illinois.* O. L. Baskin & Co. Chicago, IL.

Bateman, Newton (ed.)
1902 *Historical Encyclopedia of Illinois and History of Peoria County.* 2 Volumes. Munsell, Chicago, IL.
1905 *Historical Encyclopedia of Illinois and History of Tazewell County, Illinois.* 2 Volumes. Munsell, Chicago, IL.
1907 *Historical Encyclopedia of Illinois and History of Saint Clair County, Illinois.* 2 Volumes. Munsell, Chicago, IL.
1914 *Historical Encyclopedia of Illinois and History of Rock Island County.* 2 Volumes. Munsell, Chicago, IL.

Battey & Co.
1884 *Counties of Cumberland, Jasper, and Richland, Illinois. Historical and Biographical.* F. A. Battey & Co., Chicago, IL.

Beckwith, H. W.
1879 *History of Vermilion County, Together With Historic Notes on the Northwest.* H. H. Hill & Co., Chicago, IL.

Beers and Co.
1880 *History of Miami County, Ohio.* W. H. Beers & Co., Chicago, IL.

Bek, F. B.
1906 *Eine Populare Gefchichte der Stodt Peoria.* Peoria, IL.

Bielski, Ursula
1998 *Chicago Haunts: Ghostlore of the Windy City.* 2nd revised ed. Lake Claremont Press, Chicago, IL.

Biographical Publishing Co.
- 1886 *Portrait and Biographical Album of Knox County, Illinois.* Biographical Publishing Co., Chicago, IL.
- 1890 *Portrait and Biographical Album of Peoria County Illinois.* Biographical Publishing Co., Chicago, IL.
- 1892 *Portrait and Biographical Record of Winnebago and Boone Counties, Illinois.* Biographical Publishing Co., Chicago, IL.
- 1894a *Portrait and Biographical Album of Tazewell and Mason Counties, Illinois.* Biographical Publishing Co., Chicago, IL.
- 1894b *Portrait and Biographical Record of Randolph, Jackson, Perry and Monroe Counties, Illinois.* Biographical Publishing Co., Chicago, IL.

Biographical Review Publishing Co.
- 1892 *Biographical Review of Cass, Schuyler and Brown Counties, Illinois.* Biographical Review Publishing Co., Chicago, IL.

Blackmar, Frank W. (ed.)
- 1912 *Kansas, a Cyclopedia of State History, Embracing Events, Institutions, Industries, Counties, Cities, Towns, Prominent Persons, Etc.* Vol. 3, Pt. 2. Standard Publishing Co., Chicago, IL.

Blakelee, George E.
- 1884 *Blakelee's Industrial Cyclopedia. A Simple, Practical Guide for the Mechanic, Farmer, Housewife and Children of every Thrifty Household in Town or Country.* The Baker & Taylor Co., New York.

Blasi, Betty
- 1974 *A Bit about Balsams. A Chapter in the History of 19th Century Medicine.* Farley Goepper Printing Co., Louisville, KY.

Bogard, Mary O.
- 1982 Peoria's Pioneer Druggists, the Farrells, and Farrell's Arabian Liniment. *Pharmacy History* 24(3):99–105.

Bowen & Co.
- 1909 *Biographical and Reminiscent History of Richland, Clay and Marion Counties, Illinois.* B. F. Bowen & Co., Indianapolis, IN.

Bowman, Alexander
- 1858 Map of Urbana and West Urbana, Illinois. Large-scale wall map showing lot locations and annotated property-improvements.

Bradford, Thomas L.
- 1905 Homoeopathy in Illinois. In *History of Homoeopathy and Its Institutions in America*, Volume 1, by William H. King. pp. 343–362. The Lewis Publishing Co., Chicago, IL.

Brewer, Luther A., and Barthinius L. Wick
- 1911 *History of Linn County Iowa from Its Earliest Settlement to the Present Time.* Pioneer Publishing Co., Chicago, IL.

Brink & Co.
- 1874 *Illustrated Atlas Map of Cass County, Illinois.* W. R. Brink and Co., Edwardsville, IL.
- 1875 *The Illustrated Historical Atlas Map of Monroe County, Illinois.* W. R. Brink and Co., Edwardsville, IL.
- 1882 *History of Madison County Illinois [1682–1882]. Illustrated. With Biographical Sketches of many Prominent Men and Pioneers.* W. R. Brink and Co., Edwardsville, IL.

Brink, McCormick and Co.
- 1874a *Illustrated Atlas Map of Sangamon County, Ill. Carefully Compiled from Personal Examinations and Surveys.* Brink, McCormack & Co., Edwardsville, IL.
- 1874b *Illustrated Atlas Map of Montgomery County, Ill. Carefully Compiled from Personal Examinations and Surveys.* Brink, McCormack & Co., Edwardsville, IL.

Brink, McDonough and Co.
- 1880a *History of Macon County, Illinois. With Illustrations Descriptive of Its Scenery and Biographical Sketches of Many Prominent Men and Pioneers.* Brink, McDonough & Co., Philadelphia, PA.
- 1880b *History of Christian County, Illinois. With Illustrations Descriptive of Its Scenery and Biographical Sketches of Many Prominent Men and Pioneers.* Brink, McDonough & Co., Philadelphia, PA.
- 1881a *History of Marion and Clinton Counties, Illinois. With Illustrations Descriptive of Its Scenery and Biographical Sketches of Many Prominent Men and Pioneers.* Brink, McDonough & Co., Philadelphia, PA.
- 1881b *History of Saint Clair County, Illinois. With Illustrations Descriptive of Its Scenery and Biographical Sketches of Many Prominent Men and Pioneers.* Brink, McDonough & Co., Philadelphia, PA.

Burggraaf, Mike, and Tom Southard
- 1998 *The Antique Bottles of Iowa: 1846–1915.* Volume 1. Privately published by the authors. Ohio Wholesale Copy Service, Northfield.

Burggraaf, Mike
- 2003 Digging for Wells. *Antique Bottle & Glass Collector* 20(8):34-36.
- 2010 2010 Update to The *Antique Bottles of Iowa: 1846–1915.* Volumes Three & Four. Privately

published by the author. Frontline Printing and Design, Fairfield, IA.

Burnham, John H.
1879　*History of Bloomington and Normal in McLean County, Illinois.* J. H. Burnham, Bloomington, IL.

Calumet Press
1895　*Album of Genealogy and Biography, Cook County, Illinois.* Calumet Book & Engraving Co., Chicago, IL.

Cannon, Richard
1993　Boyd and Beard. *Antique Bottle & Glass Collector* 10(1):140–15.
1995　Medicinal Pairs. *Antique Bottle & Glass Collector* 12(6):46.
2000　Two in One. *Antique Bottle & Glass Collector* 17:40–41.

Carson, Gerald
1961　*One For a Man, Two For a Horse. A Pictorial History, Grave and Comic, of Patent Medicines.* Bramhall House, New York.

Carson, Will C. (ed.)
1905　*Historical Souvenir of Greenville, Illinois. Being a Brief Review of the City from the Time of its Founding to Date.* The LeCrone Press, Effingham, IL.

Caspar, Carl Nicolaus
1889　*Caspar's Dictionary of the American Book, News and Stationery Trade. Wholesale and Retail.* C. N. Caspar's Book Emporium, Milwaukee, WI.

Chambers, Robert, Jr.
1884　*Chambers Journal of Popular Literature, Science and Arts.* July 12th issue.
1887　Occasional Notes. *Chambers Journal of Popular Literature, Science and Arts.* June 25th issue.

Chapman & Co.
1878　*History of Knox County Illinois.* C. C. Chapman & Co. Peoria, IL.
1879　*History of Fulton County Illinois.* C. C. Chapman & Co. Peoria, IL.

Chapman Brothers
1887a　*Portrait and Biographical Album of McLean County, Illinois.* Chapman Brothers, Chicago, IL.
1887b　*Portrait and Biographical Album of Coles County, Illinois.* Chapman Brothers, Chicago, IL.
1889　*Portrait and Biographical Album of Morgan and Scott Counties.* Chapman Brothers, Chicago, IL.
1892a　*Portrait and Biographical Record of Saint Clair County, Illinois.* Chapman Brothers, Chicago, IL.
1892b　*Portrait and Biographical Record of Adams County, Illinois.* Chapman Brothers, Chicago, IL.

Chicago Historical Society
2005　*The Electronic Encyclopedia of Chicago.* Online resource maintained by the CHS, and including a *Dictionary of Leading Chicago Businesses (1820–2000)* prepared by Mark R. Wilson, Stephen R. Porter, and Janice L. Reiff.

Clark and Co.
1897　The *Biographical Record of Rock Island County Illinois.* S. J. Clarke Publishing Co., Chicago, IL.
1899a　The *Biographical Record of McLean County, Illinois.* S. J. Clarke Publishing Co., Chicago, IL.
1899b　The *Biographical Record of Ogle County, Illinois.* S. J. Clarke Publishing Company, Chicago, IL.
1900　The *Biographical Record of Livingston and Woodford Counties, Illinois.* S. J. Clarke Publishing Co., Chicago, IL.
1912　*Chicago: It's History and Builders.* Volume 4: *A Century of Marvelous Growth.* S. J. Clarke Publishing Company, Chicago, IL.

Coats, Betty Spindler
1983　*The Swiss on Looking Glass Prairie, A Century and a Half, 1831–1981.* Friends of Lovejoy Library, Edwardsville, IL.

Cole, Arthur C.
1919　*The Centennial History of Illinois.* Volume 3: *The Era of the Civil War 1848–1870.* Illinois Centennial Commission. A. C. McClurg & Co., Chicago, IL.

Collin, Henry P.
1906　*A Twentieth Century History and Biographical Record of Branch County, Michigan.* Lewis Publishing Co., Chicago, IL.

Collins, William H., and Cicero F. Perry
1905　*Past and Present of the City of Quincy and Adams County, Illinois.* S. J. Clarke Publishing Co., Chicago, IL.

Columbia Centennial Committee
1959　*History of Columbia and Columbia Precinct, Monroe County, Illinois 1859–1959 and Centennial Celebration, Columbia, Illinois, July 3-4-5, 1959.* Document prepared for the event by the Centennial Committee.

Conard, Howard L.
1901　*Encyclopedia of the History of Missouri. A Compendium of History and Biography for Ready Reference.* The Southern History Co., Proprietors, St. Louis, MO.

Continental Historical Co.
1885　*History of Greene and Jersey Counties, Illinois.* Continental Historical Co., Springfield, IL.

Cronkite, Larry
  1998　*Some Branches of the Cronk, Cronkite, Cronkhite Family Tree.* Privately published and distributed by the author.

Crump, Arthur J.
  1921　*Nostrums and Quackery,* Volume 2. *Articles on the Nostrum Evil, Quackery, and Allied Matters Affecting the Public Health...from the Journal of the American Medical Association.* AMA Press, Chicago, IL.

Cundall, Lloyd
  1923　*From the Little Ash Tray the Mighty Illinois Glass Company Grew.* Unpublished twelve-page study reproduced and distributed by the Illinois Glass Company in 1973, on the occasion of their manufacture of a centennial replica of the 1873 ash tray. [Accompanied by an historical introduction by Professor Marijean Seelbach, of the State University College at Brockport, NY, titled *Mr. Smith's "Glass Ash Tray"*].

Cutler, William G.
  1883　*History of the State of Kansas.* A. T. Andreas Co., Chicago, IL.

Daniels, Daisy Paddock
  1971　*Prairieville U.S.A.* Historical Society and Museum of Arlington Heights, IL.

Davis, David J.
  1955　*History of Medical Practice in Illinois. Volume II: 1850–1900.* Illinois State Medical Society. Lakeside Press, R. R. Donnelley & Sons, Chicago, IL.

Deiss, Ronald W.
  1981　The Development and Application of a Chronology for American Glass. Unpublished M.A. Thesis, Department of Anthropology, Illinois State University, Normal, IL. [Printed version bound and distributed in 1982 by ISU's Midwestern Archaeological Research Center.]

Denzin, Don
  1993　Bottles, Bung Holes, and Booze! American Figural Spirits Barrels. *American Bottle & Glass Collector* 10(7):46–51.

Devner, Kay
  1968　*Patent Medicine Picture.* Tombstone Epitaph printers, Tucson, AZ.

Dickinson, Harriet Cory
  1914　*Some Chronicles of the Cory Family.* Tobias A. Wright, NYC.

Dodd, Edwin D.
  1973　*The Centennial Bottle.* Commemorative brochure by the president of Owens-Illinois, describing the circumstances of manufacture of the centennial replica of the embossed *Yoerger & Bro.* soda bottle.

Dodd, Henry M.
  1912　*Read Genealogies.* Henry M. Dodd, Clinton, NY.

Donath, Hollie
  1991　*History of Peoria Mineral Springs.* Unpublished academic paper made available by the author.

Donnelley, Gassette, and Loyd
  1879　*History of Greene County, Illinois: Its Past and Present.* Donnelley, Gassette, and Loyd Publishers, Chicago, IL.

Donnelley, Lloyd & Co.
  1878a　*History of Morgan County Illinois: Its Past and Present.* Donnelly, Lloyd &Co., Chicago IL.
  1878b　*History of Logan County, Illinois: Its Past and Present.* Donnelly, Lloyd & Co., Chicago, IL.

Douglas, J. Webb
  1886　Wizard Oil. *The Medical World. A Practical Medical Monthly.* Volume 4. C. F. Taylor, Philadelphia, PA.

Draperies & Window Coverings magazine
  2001　Paper with a Past and Future: S. A. Maxwell Co. Celebrates Its 150th Anniversary. *D&WC Magazine* 20(11): November 2001 editorial column.

Duis, E.
  1874　*The Good Old Times in McLean County, Illinois.* The Leader Publishing and Printing House, Bloomington, IL.

Ebert, Albert E.
  1905　*Early History of the Drug Trade of Chicago, Complied from the Records of the Chicago Veteran Druggists Association.* Illinois State Historical Society, Publication No. 10. Illinois State Historical Society Transactions for 1905.

Farnsworth, Kenneth B., Karen Atwell, and Raymond W. Perkins
  1995　*Sulphur Spring Excavations, Perry Springs, Pike County, Illinois.* Report of Investigations. Center for American Archeology, Contact Archeology Program, Kampsville, IL.

Faulkner, Ed, and Lucy Faulkner
  2006　A New Idea for Ink Bottles—Paper. *Bottles and Extras* 17(4):48–49.

Federal Publishing Co.
1905   *Memoirs of the Lower Ohio Valley, Personal and Genealogical, with Portraits.* Volume 2. Federal Publishing Co., Madison, WI.

Fike, Richard E.
1987   *The Bottle Book. A Comprehensive Guide to Historic, Embossed Medicine Bottles.* Peregrine Smith Books, Salt Lake City, UT.

Fishback, Woodson W.
1982   *A History of Murphysboro, Illinois 1843–1982.* Jackson County Historical Society. Quail Ridge Press, Brandon, Miss.

Fitch, Charles Elliott
1923   *Encyclopedia of Biography of New York. A Life Record of Men and Women Whose Sterling Character and Energy and Industry have made them Preeminent in their own and many other States.* The American Historical Society, New York.

Fowler, Ron
1984   Speaking Soda Confidentially: The Survivors. *Old Bottle Magazine* 17(6):3–7.

Goodspeed, Weston A., and Daniel D. Healy
1909   *The History of Cook County Illinois—Being a General Survey of Cook County History, including a Condensed History of Chicago and Special Account of Districts outside the City Limits; From the Earliest Settlement to the Present Time.* The Goodspeed Historical Association, Chicago, IL.

Graci, David
1994   More on Stoneware Bottles. *American Bottle & Glass Collector* 11(6):2–3.
1995   *American Stoneware Bottles.* Calem Publishing Co., South Hadley, MA.
2001   *(More) American Stoneware Bottles.* Privately published by the author, South Haldey, MA.

Graci, David, and David Rotilie
2003   Smith vs. Merrill. *American Bottle & Glass Collector* 20(4):36–39.

Guest, Gary
2004   New York City Bottle Legacies, Part III: Knickerbocker, William & George Eagle, the Crystal Palace and More. Article from ca. 2004 issue of *Antique Bottle & Glass Collector* Magazine, reproduced on the *AB&GC* web site.

Gums, Bonnie L., Eva Dodge Mounce, and Floyd R. Mansberger
1997   *The Kirkpatrick's Potteries in Illinois: A Family Tradition.* Transportation Archaeological Research Reports 3. Illinois Transportation Archaeological Research Program, Department of Anthropology, University of Illinois, Champaign-Urbana, IL.

Hall, Jim
1976   Bottle of the Month. *Midwest Bottled News* 7(1):5.

Harms, Robert
1978   Harms' Chicago Hutchies. *Midwest Bottled News* 9(10):8–18.

Hawley, Gertrude
1893   *History of Hampshire.* Series of articles originally published in the Hampshire Register newspaper.

Hill & Co.
1881   *History of Lee County: Together with Biographical Matter, Statistics, etc.* H. H. Hill Co., Chicago, IL.

Hill, Luther B.
1910   *History of Benton County, Iowa.* The Lewis Publishing Co., Chicago, IL.

Hinchcliffe, John
1870   *Historical Review of Belleville, Illinois, From Early Times to the Present, With a Glance at its Business, Present and Prospective.* Kimball & Taylor, Belleville, IL.

Hoffman, Urias J.
1906   *History of La Salle County, Illinois.* S. J. Clarke, Chicago, IL.

Hollenberg, Judith
1979   Perry Springs, the "Saratoga of the West." *Illinois Times,* June 8–14, 1979, pp. 4–7.

Holst, Jim
1994   *Pontiled Medicine Price Guide.* Privately produced and distributed by the author.

Hoste, Ray
1978   The W. H. Hutchinson Company, Their Closures and Their Bottles. *Old Bottle Magazine* 11(4):6–9.

Hughes, Byron
1970   The Lomax Story. *Midwest Bottled News* 1(9):2.

Inter-State Publishing Co.
1881   *History of Sangamon County, Illinois.* Inter-State Publishing Co., Chicago, IL.
1886   *History of Logan County, Illinois.* Inter-State Publishing Co., Chicago, IL.

Johnson and Co.
1880   *History of Peoria County Illinois.* Johnson & Co., Chicago, IL.

Johnson, Eric A.
2004   *Rockford: 1920 and Beyond.* Postcard History Series. Arcadia Publishing Co., Chicago, IL.

Johnson, Eric, and C. F. Peterson
1880   *Svenskarne i Illinois. Historiska Anteckningar.* Printed by W. Williamson, Clark Street, Chicago, IL.

Jordan, Richard E.
1981   Samuel Smith – Auburn Soda Bottler. *Old Bottle Magazine* 14(3):5–8.

Kauffman, Horace G., and Rebecca H. Kauffman (eds.)
1909   *Historical Encyclopedia of Illinois and History of Ogle County.* Volume 2. Munsell Publishing Co., Chicago, IL.

Kemp, Bill
2009   Patent Medicine Makers Claimed to Cure All. *Bloomington Pantagraph,* October 31, 2009.

Kenyon, Theo Jean
1976   An Invitation. Mineral Springs. *Peoria Journal Star,* June 23, 1976, D3.

Ketcham, Steve
2009   Recent Finds. *Bottles & Extras* 20(2):4.

Kett, H. F.
1877a   *Past and Present of La Salle County, Illinois.* H. F. Kett, Chicago, IL.
1877b   *History of Winnebago County, Illinois.* H. F. Kett, Chicago, IL.
1877c   *Past and Present of Rock Island County, Illinois.* H. F. Kett, Chicago, IL.
1878a   *History of Ogle County, Illinois.* H. F. Kett, Chicago, IL.
1878b   *History of Jo Daviess County, Illinois.* H. F. Kett, Chicago, IL.

Kilgo, Dolores A.
1994   *Likeness and Landscape. Thomas M. Easterly and the Art of the Daguerreotype.* Missouri Historical Society Press, St. Louis, MO.

King, Ann Weesner
1945   Memories of early Olney businesses, transcribed and made available online at http://olneymemories.blogspot.com

Kious, Kevin, and Donald Roussin
1997   Breweries of Belleville, Illinois. *American Breweriana Journal* #89:4–13.
1998   Schott Brewing Company Highland, Illinois. *American Breweriana Journal* article posted on line at http://breweriana.com/history/historyschott.html

Knudson, W.
2006   *Nostrums and Quackery. Articles on the Nostrum Evil and Quackery Reprinted from the Journal of the American Medical Association.* AMA Press, Chicago, IL.

Komorowski, Ray
1977   Lomax's Corner. *Midwest Bottled News* 8(2):10. [This is one of a series of short Komorowksi data reports on the Lomax bottling works printed in several issues of *MBN* during 1977 and 1978.]

Kott, Robert J.
1978   Blood Bottles. *Midwest Bottled News* 9(9):1
1981   John A. Lomax and his Chicago Locations. *Midwest Bottled News* 12(3):3–5.
2005a   *Chicago Sodas and Mineral Waters.* Ring-bound notebook with documentation and recordation of Chicago-area sodas during a study period spanning ca. 1965–2005.
2005b   *Ex-Urbus Soda and Mineral Waters.* Ring-bound notebook with documentation and recordation of Chicago-area sodas during a study period spanning ca. 1965–2005.

Kramer and Co.
1908   *Historic Rock Island County: History of the Settlement of Rock Island County from the Earliest Known Period to the Present Time.* Kramer & Co., Rock Island, IL.

Lam, Howard, George E. Parks, and Robert L. Rich
1954   *100 Years of Progress. The Centennial History of Anna, Illinois.* Anna Centennial Committee publication. Missourian Printing & Stationery Co., Cape Girardeau, MO.

Landrum, Carl
1967   A Century Ago in Quincy. Soda Water Made Long Ago. *The Quincy Herald-Whig,* January 22, 1967 (page unknown). Printed in Landrum's newspaper column, "From Quincy's Past."
1968   When Mineral Springs Prospered. *The Quincy Herald-Whig,* March 10, 1968. Printed in Landrum's newspaper column, "From Quincy's Past."
1969   *Historical Sketched of Quincy—The First 100 Years.* Royal Printing Co., Quincy, IL.
1980   Heyday of the Pike County Spa. *The Quincy Herald-Whig,* February 3, 1980. Printed in Landrum's newspaper column, "From Quincy's Past."
1986a   Perry Springs Once Thrived as a Health and Social Resort. *The Quincy Herald-Whig,*

August 24, 1986. Printed in Landrum's newspaper column, "From Quincy's Past."
1986b Flynn Family and Coke have been Associated since 1905. *The Quincy Herald-Whig*, May 4, 1986. Printed in Landrum's newspaper column, "From Quincy's Past."

Lang, Jim
1981 Springfield Pontil Medicines. *Midwest Bottled News* 12(5):5–8.

Le Baron, William, Jr.
1878a *The Past and Present of Kane County, Illinois.* William Le Baron, Jr., and Co., Chicago, IL.
1878b *The History of Will County, Illinois.* Wm. Le Baron, Jr., and Co., Chicago, IL.
1879 *The History of McLean County, Illinois.* William Le Baron, Jr., and Co., Chicago, IL.

Leonard, Lulu
1941 *History of Union County.* Published by the author, Anna, IL.

Lewis Publishing Co.
1889 *An Illustrated History of Los Angeles County, California.* Lewis Publishing Co., Chicago IL.
1892 *The Bay of San Francisco.* Volume 2. Lewis Publishing Co., Chicago, IL.
1900 *Biographical & Genealogical Record of La Salle County Illinois.* Volume 2. Lewis Publishing Co., Chicago, IL.

Leybourne, Douglas M.
1993 *The Collector's Guide to Old Fruit Jars.* Produced and distributed by the author, North Muskegon, MI.

Lindsay, Bill
2009 *Historic Glass Bottle Identification and Information Website. Part III: Types of Bottle Closures.* Bill Lindsay's massive, deeply layered, and highly illustrated labor-of-love website on historic-bottle style and technology. Hosted by the Society for Historical Archaeology. [We used it primarily in 2009, but it is a constantly expanding research resource.]

Lockhart, Bill, Pete Schulz, Carol Serr, Jay Hawkins, and Bill Lindsey
2008 The Dating Game: William Frank & Sons, Pittsburgh, Pennsylvania (1866–1875). *Bottles and Extras* 19(2):32–36.

Lockhart, Bill, Carol Serr, and Bill Lindsey
2007 The Dating Game: De Steiger Glass Co. *Bottles and Extras* 18(5):31–37.

Lockhart, Bill, Pete Schulz, Carol Serr, Bill Lindsey, and David Whitten
2008 The Dating Game: The Kearns Glass Companies. *Bottles and Extras* 19(24): 50–58.

Lockhart, Bill, David Whitten, Bill Lindsey, Jay Hawkins and Carol Serr
2004 The Dating Game: Cunningham Family Glass Holdings. *Bottles and Extras* 16(3): 2–8.

Mack, Edwin F.
1914 *Old Monroe Street: Notes on the Monroe Street of Early Chicago Days.* Central Trust Co. of Illinois, Chicago, IL.

Malik, Carl
1980 The Tremont House. *The Midwest Bottle News* 11(8):5–6

Marshall, W. G.
1882 *Through America; or, Nine Months in the United States.* Sampson Low, Marston, Searle, & Rivington, London.

McDonough & Co.
1883a *Combined History of Edwards, Lawrence and Wabash Counties, Illinois [1682–1883]. With Illustrations Descriptive of Their Scenery and Biographical Sketches of Some of Their Prominent Men and Pioneers.* J. L. McDonough & Co., Philadelphia, PA.
1883b *Combined History of Randolph, Monroe and Perry Counties, Illinois. With Illustrations Descriptive of Their Scenery and Biographical Sketches of Some of Their Prominent Men and Pioneers.* J. L. McDonough & Co., Philadelphia, PA.

McKearin, Helen, and Kenneth M. Wilson
1978 *American Bottles & Flasks and their Ancestry.* Crown Publishers, Inc. NY.

Midwest Bottle News, The
1977 Sanford Ink Co: Chicago, Illinois. *The Midwest Bottle News* 8(10):19–20. Article reprinted from the Middle Tennessee Bottle Collectors' Club of Nashville, Tennessee (June 1977 issue).

Miller, Thomas R.
1989 *A Survey of Early Soda/Mineral Water Manufacturing in St. Clair Co. A Glimpse of Illinois History Through Glass.* Thomas R. Miller, Smithton, IL.
2007a *Book I (A–Z). St. Clair County, Illinois, Soda and Related Beverage Bottles (Includes Biographical Sketches and other Information).* Thomas R. Miller, Smithton, IL.

2007b   *Book II (A–Z). St. Clair County, Illinois, Prescription and Medicine Bottles (Includes Biographical Sketches and other Information)*. Thomas R. Miller, Smithton, IL.

2008   *Book IV. St. Clair Co., ILL. Beer and Distilled Products Bottles & Jugs (A–Z). (Includes Histories, etc.)*. Thomas R. Miller, Smithton IL.

2009   Excelsior Brewing Company, Red Bud, Illinois. Unpublished 4-page manuscript submitted by Thomas Miller.

Milligan, William L.
1907   *The White Apron. A Compilation of the History of Occidental Lodge No. 40, A. F. & A. M., Ottawa, Illinois*. Republican-Times Printers, Ottawa, IL.

Miner, Edward
1905   *Past and Present of Greene County Illinois*. S.J. Clarke, Chicago, IL.

Moody, C. D., and F. Hess
1861   *Map of the Counties of Morgan and Scott*. Wall-sized (36" x 60") county plat map printed by Edward Mendel, Chicago, IL.

Mounce, Eva Dodge
1989   *Checklist of Illinois Potters and Potteries*. Foundation for Historical Research of Illinois Potteries, Circular 1(3).

National Historical Co.
1883   *History of St. Clair County, Missouri*. National Historical Co., St. Joseph, MO.

Nelson, William E. (ed.)
1910   *City of Decatur and Macon County, Illinois. A Record of Settlement, Organization, Progress and Achievement*. 2 Volumes. Pioneer Publishing Co., Chicago, IL.

Nielsen, R. Frederick
1978   *Great American Pontiled Medicines*. Published by the author, Medford, NJ.

Norris, Jennifer
1978   Deserted Health Spa in Pike County Once Top Western Resort. *Jacksonville Journal-Courier*, August 20, 1978, page 18.

Odell, John
2007   *Pontil Medicine Encyclopedia*. 2nd Edition. Digger Odell Publications, Mason, OH.

Oertle, Ben (ed.)
1990   *Central Illinois Bottles and Glasses*. Pekin Illinois Bottle Collectors Association.

1995   *Supplement to Central Illinois Bottles and Glasses: July 15, 1995*. Pekin Illinois Bottle Collectors Association.

2008   *Central Illinois Bottles and Glasses with Special Listing of Illinois Mini-Jugs. Supplement #2*. Pekin Illinois Bottle Collectors Association.

Ogg, Bryan J.
1996   *Peoria Spirits. The Story of Peoria's Brewing and Distilling Industries*. Peoria Historical Society.

Olroyd, Osborn H. (ed.)
1915   *The Poets' Lincoln. Tributes in Verse to the Martyred President*. Published by the editor at "The House Where Lincoln Died," Washington, D.C.

Panek, John
1977   Chapin & Gore. *Midwest Bottled News* 8(12):21–22.

Paul, John R., and Paul W. Parmalee
1973   *Soft Drink Bottling. A History with Special Reference to Illinois*. Illinois State Museum Society, Springfield, IL.

Perrin, William H. (ed.)
1879   *The History of Coles County, Illinois*. Wm. LeBaron, Jr., and Co., Chicago, IL.

1882a   *History of Cass County, Illinois*. O. L. Baskin and Co., Chicago, IL.

1882b   *History of Bond and Montgomery Counties, Illinois*. O.L. Baskin, & Co., Chicago, IL.

1883   *History of Alexander, Union and Pulaski Counties, Illinois*. O.L. Baskin & Co., Chicago, IL.

Peters, Roger M.
1996   *Wisconsin Soda Water Bottles 1845–1910*. 3rd Edition. Wild Goose Press, Madison, WI.

Power, John Carroll
1871   *History of Springfield, Illinois, Its Attractions as a Home and Advantages for Business, Manufacturing, etc.* Springfield Board of Trade. Illinois State Journal Printers, Springfield.

1876   *History of the Early Settlers of Sangamon County, Illinois. Centennial Record*. Edwin A. Wilson & Co. Publishers, Springfield, IL.

The Quincy Daily Journal
1889   J. J. Flynn & Co.'s Bottling Works. In *Quincy Illustrated: a Sketch of Early Quincy and a Description of the Quincy of To-Day*, p. 20. Published by the *Quincy Daily Journal*.

Religious Philosophy Publishing Association
1866   *Chicago as It Is. A Stranger's and Tourist's Guide to the City of Chicago. Containing Reminiscences of Chicago in the Early Days. An Account of the Rise and Progress of the City*. Religious Philosophy Publishing Association, Chicago, IL.

Richmond, C. W., and H. F. Vallette
1857 *A History of the County of Du Page, Illinois. A History Containing an Account of Its Early Settlement and Present Advantages, a Separate History of the Several Towns, Notices of Religious Organizations, Education, Agriculture, and Manufactures, with the Names and Some Account of the First Settlers in Each Township.* Steam Presses of Scripps, Bross and Spears, Chicago, IL.

Richmond and Arnold
1904 *Biographical Record: This Volume Contains Biographical Sketches of Leading Citizens of Macoupin County Illinois.* Richmond and Arnold, Chicago, IL.

Riley, John J.
1957 *A History of the American Soft Drink Industry. Bottled and Carbonated Beverages 1807–1957.* American Bottlers of Carbonated Beverages, Washington, D.C.

Ring, Carlyn
1980 *For Bitters Only.* The Nimrod Press, Inc., Boston, MA.
1984 *For Bitters Only. Update.* Privately printed with author copyright.

Ring, Carlyn, and William C. Ham
1998 *Bitters Bottles.* Privately printed with author copyright by Boyertown Publishing Co., Boyertown, PA.
2004 *Bitters Bottles Supplement.* Privately printed with author copyright by Boyertown Publishing Co., Boyertown, PA.

Rock Island Preservation Society
1992 *Chicago Addition Walking Tour, Rock Island Illinois.* Rock Island Preservation Commission.

Russo, Edward J., Melinda Garvert, and Curtis Mann
1995 *Springfield Business. A Pictorial History.* G. Bradley Publishing Co., St. Louis, MO.

Schmeiser, Alan
1970 *More Pop.* Michalan Press, Dixon, CA.

Simon, George
1971 History of the F. W. Muller Company. *Midwest Bottled News* 2(7):1.

Sineni, Russ
1978 Old J. G. Is All We Had. *Midwest Bottled News* 9(4):7–10. [Reprinted in 1979 as "Meet Old J. G., Lemont's Late and Great King of Sodas. OK?" in *Antique Bottle World* 6(10):26–28].

Smith, Brandon
2005 *Blob Top Sodas: Guidebook to St. Louis Missouri.* Privately printed and bound by the author.

Smith, George Washington
1912 *A History of Southern Illinois. A Narrative Account of Its Historical Progress, Its People, and Its Principal Interests.* The Lewis Publishing Co., Chicago, IL.

Smith, Isaac B.
1857 *Sketches of the Early Settlement and Present Advantages of Princeton Illinois, including Valuable Statistics, etc. Also, a Brief Sketch of Bureau County and a Business Directory.* I. B. Smith, Princeton, IL.

Sperry, F. M.
1904 *A Group of Distinguished Physicians and Surgeons of Chicago.* J. H. Beers & Co., Chicago, IL.

Spoelstra, Fred
1994 Dr. Cronk...The Complete Story! *Canadian Bottle & Stoneware Collector* No. 7:20–26.

Stalling, Judy
1979 Quenching an Historic Thirst. *West Bluff Word,* June 1979, p. 5.

Tendick, Cecil
1973 Our Western Saratoga. *Jacksonville Journal-Courier,* July 8, 1973, p. 26.

Toulouse, Julian Harrison
1971 *Bottle Makers and Their Marks.* Thomas Nelson Inc., New York.

Tucker, Donald
1986 *Collector's Guide to the Saratoga Type Mineral Water Bottles.* Donald & Louise Tucker, Inc., North Berwick, ME.

Union Publishing Co.
1883 *History of Hardin County, Iowa.* Union Publishing Co., Springfield, IL.

Vance, J. W.
1886 *Report of the Adjutant General of the State of Illinois.* Volume 5: *Reports for the Years 1861–1866.* H. W. Rokker, State Printer, Springfield, IL.

Van Wieren, Dale P.
1995 *American Breweries II.* Eastern Coast Breweriana Association, West Point, PA.

Von Mechow, Tod
2008 *Soda and Beer Bottles of North America.* Detailed, database website maintained and frequently

updated and expanded by von Mechow at www.sodasandbeers.com/index.html

Waddy, George
1996 Who Is Hanbury Smith? In "Mineral Waters from Yankee Country," Waddy's ongoing column in *American Bottle & Glass Collector* July:42–43.
1997 New Finds. In "Mineral Waters from Yankee Country," Waddy's ongoing column in *American Bottle & Glass Collector* March:38.

Wagner, Douglas
1980 Pontiled Sodas from Chicago. *Antique Bottle World* 7(10):21–24.

Waite, Joslyn R.
1908 *History of Kane County, Illinois.* Pioneer Publishing Co., Chicago, IL.

Wallace, Joseph
1904 *Past and Present of the City of Springfield and Sangamon County, Illinois.* 2 Volumes. S. J. Clarke Publishing Co., Chicago, IL.

Walthall, John A.
1993 Chicago Stoneware Bottles. *Antique Bottle & Glass Collector* 10(2):12–13.

Warner and Beers Co.
1874 *An Illustrated Historical Atlas of St. Clair Co., Illinois. Compiled, Drawn and Published from Personal Examinations and Surveys by Warner & Beers.* Union Atlas Co., Chicago, IL.
1875a *Atlas of Macoupin County and the State of Illinois.* Union Atlas Co., Chicago, IL.
1875b *Atlas of Henry County Illinois.* Union Atlas Co., Chicago, IL.

Waterman, A. B.
1908 *Historical Review of Chicago and Cook County and Selected Bibliography.* Volume 2. Lewis Publishing Co., Chicago, IL.

Watkins, Lee H.
1969 John S. Harbeson: Pioneer San Diego Bee Keeper. *Journal of San Diego History* 15(4).

Watson, Benjamin A.
1870 *PERRY Mineral Springs, Pike County, Illinois. Magnesia and Chalybeate Waters.* Sixteen-page advertising brochure printed by Britt and Cassatt, Springfield, IL.
1878 *PERRY Mineral Springs! Pike County, Ill. MAGNESIA and CHALYBEATE Waters.* Fourteen-page advertising brochure printed by Cadsgon and Gardner, Quincy, IL.

Watson, Richard
1965 *Bitters Bottles.* Thomas Nelson & Sons, NY. [see also *Supplement to Bitters Bottles*, 1968, by the same author and publisher]

Weber, Sharon, and Marcia Mackenbrock
2006 *Dig It! Privy Artifacts A–Z.* Enthusia Small publishers, Naperville, IL.

Welko, Paul
1973a J. A. Lomax—Revisited. *Midwest Bottled News* 4(2):1–2.
1973b Bottle of the Month. *Midwest Bottled News* 4(5):9.
1974 Chicago Soda Scene—A. J. Miller. *Midwest Bottled News* 5(1):5.
1977a Did You Know about Hutchinsons? *Antique Bottle World* 4(3):24–26.
1977b Bottle of the Month. *Midwest Bottled News* 8(4):9–10.
1978 Did You Know About...Arthur Christin's Patent? *Antique Bottle World* 5(3):12–14.
1979 Bottle of the Month. *Midwest Bottled News* 9(5):9–10.
1982 J. A. Lomax—A Giant in the Early Chicago Bottling Industry. *Antique Bottle World* 9(1):6–7.

West Publishing Co.
1894 *The Northeastern Reporter (Vol. 36), Containing all the Current Decisions of the Supreme Courts of Massachusetts, Ohio, Illinois, Indiana, Appellate Court of Indiana, and the Court of Appeals of New York: February 16–May 11, 1894.* West Publishing Co., St. Paul, MN.

Western Historical Co.
1880 *The History of Stephenson County, Illinois.* Western Historical Co., Chicago, IL.

Whiteside, John
1990 A Leader of His Time also Dug a Hole in the Hillside. *Joliet Herald-News* column, January 2, 1990.

Whitten, David
2009 *Glass Factories That Manufactured Insulators.* Online resource assembled by the author, at www.myinsulators.co./glass-factories/

Wichmann, Jeff
1999 *The Best of the West. Antique Western Bitters Bottles.* Pacific Glass Books, Sacramento, CA.

Wilcox, David F. (ed.)
1919 *Quincy and Adams County History and Representative Men.* 2 volumes. Lewis Publishing Company, Chicago, IL.

Wiley, W. H.
1892　*Food and Food Adulterants.* U. S. Department of Agriculture, Division of Chemistry, Bulletin 18. U.S. Government Printing Office, Washington, D.C.

Williams, Jack Moore
1930　*History of Vermilion County, Illinois.* Volume 2. Historical Publishing, Topeka, KS.

Wilson, Charles E. (ed.)
1906　*Historical Encyclopedia of Illinois and History of Coles County.* Munsell Publishing Co., Chicago, IL.

Wilson, Rex L.
1981　*Bottles on the Western Frontier.* University of Arizona Press, Tucson.

Wilson and St. Clair Co.
1868　*Biographical Sketches of the Leading Men of Chicago.* Wilson and St. Clair, Publishers, Chicago, IL.

Wilson, William L., and Elizabeth Wilson
1968　*Spirits Bottles of the Old West.* Henington Publishing Co., Wolfe City, TX.
1971　*19th Century Medicine in Glass.* 19th Century Hobby & Publishing Co., Eau Gallie, FL.

Wright, Craig
1992　*Chicago Hutchinson Sodas.* Privately reproduced and distributed by the First Chicago Bottle Club.

Wright and Potter Co.
1865　*The Tenth Exhibition of the Massachusetts Charitable Mechanic Association, at Faneuil and Quincy Halls, in the City of Boston, September, 1865.* Wright and Potter, Boston, MA.

Young, James Harvey
1961　*The Toadstool Millionaires. A Social History of Patent Medicines in America before Federal Regulation.* Princeton University Press, Princeton, NJ.

Zeuch, Lucius H.
1927　*History of Medical Practice in Illinois.* Volume 1: *Preceding 1850.* Illinois State Medical Society. The Book Press Inc., Chicago, IL.

Abridged Index

A complete unabridged index for *Bottled in Illinois* is available at http://isas.illinois.edu/bottlebook-index-full.pdf. The index included in this volume is only a fraction of the information available in the unabridged version. Heel stamps, maker's marks, cities, counties, diseases and traumas, collar styles, molds, and many more categories of information not available in this index can be found in the unabridged index. In addition, the unabridged index includes the embossed lettering on every bottle in this volume, so a researcher working from only a bottle fragment can search and find every bottle with that sequence of lettering in order to identify their bottle.

A page number followed by *f* indicates figure and *t* indicates table.

A

A. & C. W. Keith (Naperville), 580
A. & F. X. Joerger (Alton), 102
A. & M. Lindsay (Springfield), 714–717
A. & M. Lindsay's Liniment, 30, 75, 714–717
    as earliest embossed bottle, 29, 715
A. & W. W. Strickland (Chicago), 429
A. Finkler & Co. (LaSalle), 550–551, 550*f*
A. J. Miller (Chicago), 50, 50*f*
A. Leibenstein & Co. (Chicago), 330–331, 331*f*
    externally threaded closures, 29
A. Lindsay & Bro. (Springfield), 715–716
A. Lindsay & Co. (Springfield), 717
A. Lindsay's "Anti-Billious" Pills, 716
A. Lindsay's "Anti-Despeptic" Tonic and Pills, 716
A. Lindsay's "Aramatic" Tooth Powder, 716
A. Lindsay's Western Nerve Tonic, 716
A. M. Beninger & Co. (New York), 431
A. Mette & Bros. (Chicago), 49–50, 49*f*
A. Mette & Brother (Chicago), 369–371, 370*f*
A. R. Bremer & Co. (Chicago), 378
A. S. Barry & Co. (Alton), 538
A. T. Wells & Co. (Chicago), 451
A. W. Sargent & Co. (Chicago), 394–396, 395*f*
Ab Egg, Charles, 112, 568
Ab Egg, Josephine, 568
Ab Egg, Louis, 45, 110–112, 111*f*, 112*f*, 114, 115–116, 115*f*, 568
Abbott & Kingman (Chicago), 360, 361
Abel, Humiston & Co. (Sandwich), 419, 697–698, 697*f*
Abel, John, 419, 697–698, 697*f*
Achilles, Ernest, 733–734, 733*f*, 734*f*
Achilles, Louis, 734
Achilles & Hille (Virden), 733–734, 733*f*, 734*f*
Adams, Dr. A. L., 186–187, 186*f*
Adams, F. H., 188
Adams, Thompson W., 188, 188*f*
Adler, Simon, 324–325
Adler Distilling Co. (St. Louis), 324–325
Adolph Clasen & Co. (Chicago), 441–442
advertisements
    Alton (IL)
        Alton Glass Co. whiskey flask, 58, 86, 86*f*
        Betts & Russell, 87
        Betts' Ginger Brandy, 87
        Buff & Kuhl, 88–89
        Illinois Glass Co. ashtrays, 89–92, 90*f*
        Keeley & Brother, 94–95
        Magnolia Bitters, 92–93
        Yoerger, Augustin, 102
    Beardstown (IL)
        Ehrhardt, Frederick W., 108–109
        Schneider & Krohe, 109
    Bloomington (IL)
        Bradner's Honey Balsam, 131, 131*f*
        Dr. H. S. Woodard, 155–156
        Lackey & Bro., 131, 131*f*
        Lackey's Iron Bitters, 131, 131*f*
        Mueller & Stein, 134
        Woodard and Howe, 154
        Woodard's Instant Relief, 156
    Cairo (IL)
        Pfifferling's Trionadraphia, 170
        Schutter, William H., 171
    Chicago (IL)
        Bailey & Eaton, 196
        Bennett Pieters & Co. on liquor tax stamp, 384, 385*f*
        Bennett Pieters & Co. Red Jacket Bitters, 386
        Buck & Rayner, 207–208
        Cassily & Co., 212
        Christie's Ague Balsam, 217–218
        Connell's East Indian Remedies, 223
        Cory's Stomachic Bitters, 738, 739, 739*f*
        Doty & Waugh, 239
        Dr. A. L. Adams Liver Balsam, 187
        Dr. J. M. Connell's Brahminical Moonplant. East Indian remedies, 223
        Dr. Syke's Sure Cure for Catarrh, 433
        Dr. Wolgamott's Aperient and Blood Purifier, 455–456
        Drs. C. M. Fitch & J. W. Sykes, 435
        Excelsior Liniment, 211
        Excelsior Medicine, 212
        Farrell, William B., 254, 254*f*
        Farrell's Indian Black Ointment, 255
        Fluid Lightning, 228
        Foster's Indian Health Renewer, 259, 706
        Great Western Wholesale and Retail Patent Medicine Depot, 217
        Hamlin's Wizard Oil, 275–278, 276*f*, 277*f*, 278*f*
        J. H. Reed & Co., 390, 390*f*
        John A. Lomax, 359, 359*f*, 364, 365, 365*f*, 366
        Manzanita Bitters, 427
        Miller, Andrew J. (as barber), 372
        Pain Exterminator, 395
        Paul Rouze & Co. soda water apparatus, 392
        Penton, Fisher & Co., 377
        Perrine's Floating Bee Palace, 379–380
        Professor Lennord's Nectar Bitters, 213
        Rouze, Paul wine and liquor, 392
        Sanford Manufacturing Co., 394
        Sawyer, Sidney, 399
        Sawyer's Fluid Extract of Bark, 399, 400
        Sloan, Walter B., 414, 416, 416*f*, 417
        Steel & Co., 422
        Stomachic Bitters, 226
        Swain's Burbon Bitters, 431
    earliest for Illinois product, 717
    Galena (IL)
        Evans, Joseph, 505
    Greenville (IL)
        Floyd, Edward A. (as dental surgeon), 512
        Floyd's American Penetrating Liniment, 511
        Freeman, Dr. John D., 536–537
    Jacksonville (IL)
        Hamilton's Vegetable Cough Balsam, 528
        Jacksonville Bottling Works, 524
        P. Braun & Co., 522
    Jerseyville (IL)
        Jerseyville Ale and Porter Depot, 534
    LaSalle (IL)
        DeSteiger Glass Co. shoofly-style flasks, 60
    Minot (IL)
        Clarke's Compound Mandrake Bitters, 572
    Naperville (IL)
        People's Favorite Bitters, 582
    patent medicines, 42–43, 74
    Peoria (IL)
        Bissell's Tonic Bitters, 606, 606*f*
        Callender's Liver Bitters, 608, 609*f*
        Farrell's Arabian Liniment, 618–620, 619*f*, 620*f*
        Headley, Dr. William S., 632*f*
    Philadelphia (PA)
        Yockel, for private molds, 72*f*
    Pittsburgh (PA)
        Alexander Arbogast, 57, 403
    Quincy (IL)
        Bunnel's German Liniment, 659–660
        Burleigh, Dr. Alonzo S., 662, 662*f*
        Burleigh & Brothers, 662
        J. J. Flynn & Co., 670, 671*f*
        Kalb's Quick Relief, 676
        M. R. & H. W. Lundblad, 679
        Miller, George A., 683, 683*f*
        O. F. & G. A. Miller, 683
        Schrader, Frank, 684
    Schroeder, Edward, 486
    Springfield (IL)
        A. & M. Lindsay's Liniment, 716, 716*f*, 717
        A. Lindsay & Bro., 715
        Birchall & Owen, 725
        Bonekamp's Stomach Bitters, 704
        Dr. Topping's Alterative and Cathartic Syrup, 718, 718*f*
        Dunning, Andrew J., 703, 703*f*
        Henry Korf & Co., 708
        Kelleher, Jeremiah A., 712, 713
        Myers' Embrocation, 721
        North, Alfred A., 722–723
        Owen, Thomas J. V., 724, 725
        Peterson, Charles, 708
        Rheumatic Liniment, 719
        Van Deusen, Martin M., 728
        Van Deusen's Ague Bitters, 728–729
        W. W. Watson & Son, 730
        Wallace & Diller, 700, 701*f*
    Waukegan (IL)
        Cory's Stomachic Bitters, 738, 739, 739*f*
aerated medicated water. *See* medicated aerated water
Agnew & Co. (Pittsburgh), 55*t*, 57
    Chicago (IL), 383*f*, 384*f*, 385
Ague Killer, 222
Ague Tonic, 676
Aherns, Gus, 472
Ailiff, Noah, 47, 169*f*

Ainsworth, Robert, 188–189, 188f, 335
Ainsworth & Lomax (Chicago), 188–189, 188f, 335, 361, 362
Albertson, Albert
 gravitating stopper patent, 27
Albertson gravitating stopper
 Chicago (IL)
  John A. Lomax, 345
Albree, Robert C., 62–63
ale
 Alton (IL)
  Keeley & Brother, 93–96, 93f, 317
 Antebellum era (1850s), 79
 assemblage patterns, 74
 Belleville (IL)
  Thomas Heberer and Brothers, 118–119
 black glass
  for preservation, 9
 Bloomington (IL)
  O. K., Jr., 128–129, 128f
  Schausten, William, 135–137, 136f
 and body mold lines, 12–13
 body shape definitions, 15, 15f
 Breihan, Henry, 10f, 69, 165–167, 165f, 166f
 Cairo (IL)
  Lohr, Andrew, 168–170, 168f, 169f
 Champaign (IL)
  Miller, Nicholas, 182–183, 182f
 Chester (IL)
  Wenda, John, 184–185, 185f
 Chicago (IL)
  A. Mette & Bros., 49–50, 49f
  Amberg & Stenson, 191, 191f
  Carpenter, Andrew, 212, 212f
  Cook, Egbert C., 224
  Dempsey, Daniel, 232–234, 232f
  Doty, Harvey C., 239, 239f
  Gottleib Wurster & Co., 456–461, 457f
  H & Bro's, 279–281, 281f
  H. F. & Co., 255–257, 256f
  Hausburg, William A., 282–283, 282f
  Hayes Brothers, 284–287, 286f, 287f
  Hennessey, Thomas, 63, 290–291, 290f
  Hunt & Kenyon, 294–295, 294f, 295f
  Hutchinson & Co., 69, 300, 301f, 302f, 303, 304, 305f
  Hutchinson & Dunn, 303
  Hutchinson & Sons, 310, 311f
  J. Grossenheider & Co., 269
  John A. Lomax, 336, 337f, 345, 346f, 355f, 356, 364, 365
  John Anderton & Co., 191, 192f
  Kane, Thomas, 315, 315f
  Keeley, Michael, 317–320, 318–320f
  Keeley & Brother, 315–317, 315f, 316f
  Keeley Bottling Works, 317
  King, 69
  King, Thomas, 320–322, 321f
  Lang, Frederick C., 328–330, 328f
  Sass & Hafner, 396–399, 397f
  Seibt & Haunschild, 407–408, 408f
  Sproat, John, 420–421, 420f, 421f
  Steel & Co., 421–422, 421f
  Stenson, James, 422–424f, 422–426
  Taylor & Bro., 436–438, 436f, 437f
  Tivoli Bottling Co., 441–442, 441f
  Van Buskirk & Henry, 442–443, 443f
 Collinsville (IL)
  Bischoff, Benard, 465
 Danville (IL)
  Carey, Patrick, 467, 467f
 De Kalb (IL)
  Dee, Richard, 52
 East St. Louis (IL)
  C. Lutt & Co., 479–481, 479f
  Schroeder, Edward, 482–486, 483f
 five-piece molds, 13
 Galena (IL)
  Evans, Joseph, 503–505, 504f
  Volz & Glueck, 506, 506f
 Jacksonville (IL)
  J. E. Bradbury & Co., 521–522, 521f
 Jerseyville (IL)
  Boyle, Charles M., 534
 Joliet (IL)
  Paige, J. D., 540–542, 541f, 542f
 Lemont (IL)
  Bolton, John G., 554–557, 555f, 556f
 lip style definitions, 24, 25f
 Metz & Pletscher (Peoria), 637–638, 638f
 Naperville (IL)
  Stenger, John, 582–584, 583f
 O'Fallon (IL)
  Bischoff, Benard, 591
 Peoria (IL)
  Gipps, Cody & Co., 629–630, 629f
  Gipps & Co., 630–632, 631f
  Parent, Anthony, 638–640, 639f, 640f
  Pletscher, John C., 640–641, 640f
  Wichmann, John, 645–646, 645f
 pint. See pint ale
 Pioneer era (1840s), 75
 quart. See quart ale
 Quincy (IL)
  J. J. Flynn & Co., 670
 Rock Island (IL)
  Carse & Ohlweiler, 693
  Junge & Beck, 694, 694f
  Schutter, William H., 171, 171f
 Springfield (IL)
  Kelleher & Co., 712–713, 712f
  Matthew Rinehardt & Co., 725–726, 725f
 study inclusion rule, 35
 Wm. McCully & Co. (Pittsburgh), 70
Ale and Porter Depot (Chicago), 317, 319
Alexander & Austin (London), 58
Alexander Arbogast (Pittsburgh), 55t, 57
 Chicago (IL), 402–403, 402f
 "Particular attention paid to Private Molds," 57, 403
Alexander Finkler & Co. (LaSalle), 53, 242
Aley, Jacob H., 507, 507f
Alfred & William Strickland Co. (Chicago), 428–429, 428f, 429f
 Chicago (IL)
  John A. Lomax, 350, 351f
Alfred Alexander & Co. (London), 55t, 58, 69, 366
Alfred's Ale and Porter (London), 350
Allen, R., 715
Allender, J. J., 715
All-Healing Salve, 676
Allison, James J., 586, 586f
 Matthews gravitating stoppers, 27
Alphonse, M., 713
alterative patent medicine, 43
Althen, Casper, 487–488, 488f
Althoff, Charles, 654
Althoff, Friedrich, 653–654, 654f
Althrop, Thomas, 189–190, 189f
Althrop's Constitutional Tonic, 189–190, 189f
Alton Glass Co. (Alton), 55t, 58, 86, 86f
 LaSalle (IL), 549
Amberg, Adam L., 190, 190f, 191, 193
Amberg, John, 191, 191f, 425–426
Amberg & Green (Chicago), 190, 190f, 191, 193
Amberg & Stenson (Chicago), 190, 191, 191f, 212
American and China Tea Co., 603
American Glass Works (Pittsburgh), 55t, 58
American House hotel (Chicago), 186
American Industrial Development period, as study focus, 1, 3
American Medical Association (AMA), 448
American Medical Co. (St. Louis), 179
American Pontiled Soda Database, 167
American Songbag, The (Sandburg), 279
American Temperance House (Chicago), 50, 372–373, 414
Amsler, John, 516–517, 516f, 520
Amsler, Samuel, 516–517, 516f, 520
Amsler Brothers (Highland), 516–517, 516f
Anderton, John, 191, 192f
Andreas, Alfred T., 359, 362, 363–364
Angelo Mattei & Co. (Chicago), 392
Anheuser, E.
 St. Louis brewery-bottling industry growth, 35
Antebellum era (1850s), 34
 product use patterns, 75, 77f, 78f, 79
Anti-Bilious Physic, 222
Antique Soda and Beer Bottles: Bottle Bases web site, 219
Appel, Adam (aka Apple), 323
Apple, Joseph, 323
Apple & Co. (Chicago), 323
Arbogast, Andrew, 55t, 403
Arctic Soda Water
 Quincy (IL)
  Schroeder, Herman, 685
Arnold, Caleb, Jr., 193, 193f, 194f
Arnold & Green (Chicago), 190, 193, 193f, 194f
Arnold Theller & Co. (New York), 439
Aromatic Golden Bitters, 367–368, 367f
Arthur Christin & Co. (Chicago), 219–221
Arthur Christin gravitating stoppers definition, 26
ashtrays
 Alton (IL)
  Illinois Glass Co., 89–90, 90f
Asthma Relief, 371–372, 371f
Atkinson, Peter J., 292
Atlas Hotel (Pittsfield), 649
August Burgwedel & Co. (Chicago), 387
August Mette & Bro. (Chicago), 49–50, 49f, 371
Ayer's Cherry Pectoral, 225, 738
Ayres, Clarence M., 510–511, 511f
Ayres, S. B., 723
Azzaieline for the Hair, 262, 312, 312f

B
B. S. Cory & Co. (Waukegan), 610
B. S. Cory & Son (Chicago), 225–226, 225f, 738
B. S. Cory & Son (Waukegan), 226, 738–739, 739f
B. W. Coutant & Co. (Ottawa), 593–594, 593f
Mr. Babb, 501
Babcock, Dr. A. L., 548, 548f
Baby, Alfred, 220
Backus, Frank, 254
Bacon & Hyde (St. Louis), 538
Badeau, William C., 194–195, 195f
Badeau's Pure Blood Maker and Liver Cure, 194–195, 195f
Baierlein, Henry, 464, 464f
Bailey, William, 195–196, 196f
Bailey & Eaton (Chicago), 195–196, 196f
Bailey & Eaton (Providence, RI), 196
Baker Bros. & Co. (Baltimore), 55t, 58–59, 343
Balance, Willis, 630
balms, 43
 Burleighs' Electric Balm, 660–662, 661f, 662f
 Dr. North's Mother's Balm, 723
 Indian Balm, 554, 554f
 Schroeder's Imperial Balm, 685, 685f
balsams, 43
 Bradner's Honey Balsam, 124, 124f, 131, 131f
 Christie's Ague Balsam, 216–219, 217f, 218f
 Cough Balsam, 662, 662f
 Dr. A. L. Adams Liver Balsam, 186–187, 186f

Dr. Bodley's Fever & Ague Balsam, 34, 654–655, 655f
Dr. Faloon's Rosin Weed Balsam, 46–47, 46f, 121–122, 121f, 122f, 123f
Dr. Hamilton's Vegetable Cough Balsam, 75
Dr. Howe's Blackberry Balsam, 127, 127f
Dr. Howe's Honey Balsam, 127, 127f
Dr. North's Cherokee Balsam of Horehound, 723
Dr. North's Pectoral Balsam of Horehound, 723
Dr. Richardson's Peruvian Balsam, 744
Dr. Z. Waters' Honey Balsam, 154–156
Dr. Z. Waters' Raspberry Balsam, 154–156
Freeman's Cough Balsam, 535–538, 536f
Freeman's Sweet Balsam of Life, 537
Hamilton's Vegetable Cough Balsam, 525–527f, 525–529
North's Pectoral Balsam of Hoarhound, 721–723, 722f, 723f
Pioneer era (1840s), 75
Roe's Hair Balsam, 498
Turner's Balsam, 702
Wakefield's Blackberry Balsam, 68, 137, 137f, 138f, 146, 148
Baltimore Loop stopper
  Blue Island (IL)
    Hausburg, Charles, 282, 491
  study exclusion rule, 37
Baltimore pharmacy bottles, 73
Barker, Mrs. Martha L., 575
Barnes, Foster and Stone (Chicago), 259, 428, 705
Barnes, Silas, 183–184, 183f
Barnsley, England
  glass factories, 55t
  Rylands & Codd, 55t, 68–69
Barothy, Victor, 47–48, 196–197, 196f, 197f, 224
Barothy & Cook (Chicago), 47–48, 197, 198f
  Matthews gravitating stoppers, 27, 47–48
barrel whiskey
  body shape definition, 13, 14f
  Chicago (IL)
    Chapin & Gore, 63, 70, 214–216, 214f
    Wolford, Jacob A., 453–455, 454f
  closure styles, 29
barrel-effigy bottle
  Belleville (IL)
    Thomas Heberer and Brothers, 118–119
Barrett, William H., 199, 199f
Barrett, William M., 199, 199f
Barrett & Barrett (Chicago), 199, 199f
Barrett & Barrett (Holley, NY), 199, 199f
Bartlett, Frederick R., 500–501, 500f
bases
  bumps, raised
    definition, 22, 22f
  dimples
    definition, 22, 22f
  kick-ups. See kick-ups
  shape definitions, 22, 22f
  sodas base-embossed only, 500, 594
Bassler, Anna, 567
Bassler, Helen, 566
Bassler, Nikolaus J., 566–567, 566f, 568
Bassler & Co. (Mascoutah), 567, 568
Bastow, James D., 603, 603f
Batchelder, George W., 132
Bateman's Drops, 30, 715
Bates fastener, definition, 26, 26f
Bauer, Ferd., 726
Baumann, Peter, 587, 587f
Baumann's Liniment, 587, 587f
Mr. Bay, 406
Bayet, John, 589–590, 589f
Bayet & Williams (O'Fallon), 589–590, 589f

Bayless, Henry M., 452–453, 453f
Beach, C. H.
  lipping tool patent, 72
Beal cork-closure swing stopper
  Chicago (IL)
    John A. Lomax, 345
Beam, J. H., 704
Beaver Falls Glass Co (Beaver Falls, PA), 55t, 59, 67, 343
Bebbington, Edward K., 189, 200–201, 200f, 201f, 248–249, 249f, 334, 335, 360, 361
Bebbington & Co. (Chicago), 189, 200–201, 200f, 201f, 335, 361, 362
Beck, Alexander, 517, 517f
Beck, Alfred, 174, 517, 517f, 518f, 520
Beck, Victor, 694, 694f
Beck & Brother (Highland), 516, 517, 517f
Becker, George, 165–166
beer
  assemblage patterns, 74
  Aurora (IL)
    Stege, A., 106, 106f
  Beardstown (IL)
    Ehrhardt, Frederick W., 107
  Belleville (IL)
    Thomas Heberer and Brothers, 118–119
  blob-top
    variety of, Kewanee (IL), 34
  Bloomington (IL)
    P. Kreis & Co., 129
    Schausten, William, 135–137, 136f
  Cairo (IL)
    Lohr, Andrew, 168–170, 168f, 169f
  Champaign (IL)
    Miller, Nicholas, 182–183, 182f
  Chicago (IL)
    A. Mette & Bros., 49–50, 49f
    Amberg & Green, 190
    Apple, Joseph, 192, 193f
    Davies, George F., 231–232, 231f
    George Hofmann & Brothers, 292–293, 292f
    Grossenheider, 68
    Hutchinson & Co., 297
    J. Grossenheider & Co., 268–269, 269f
    John Anderton & Co., 191
    M. Brand & Co., 202–203, 204f
    Seipp, Conrad, 409–410, 410f
    Tivoli Bottling Co., 441–442, 441f
  Danville (IL)
    Stein, John, 469, 469f
  De Kalb (IL)
    Dee, Richard, 52
  Decatur (IL)
    Kuny, Fredrick D., 472
  disclaimer embossing, 73
  Elgin (IL)
    Althen, Casper, 487–488, 488f
  Jacksonville (IL)
    J. E. Bradbury & Co., 521–522, 521f
    P. Braun & Co., 522
  Joliet (IL)
    Paige, J. D., 540–542, 541f, 542f
    Metz & Pletscher (Peoria), 637–638, 638f
  Naperville (IL)
    Stenger, John, 582–584, 584f
  pasteurization
    brewery versus local bottling, 74
    and brewery-product expanded use, 34, 584
  quart bottle study exclusion rule, 37
  Peoria (IL)
    Gipps, Cody & Co., 629–630, 629f
    Gipps & Co., 630–632, 631f
    Weber, George, 643–645, 644f
    Wichmann, John, 645–646, 645f

  Rock Island (IL)
    Junge & Beck, 694, 694f
  slug-plate embossed
    study inclusion rule, 37
  slug-plate embossed bottles
    study inclusion rule, 37
  small beer. See small beer
  Springfield (IL)
    Matthew Rinehardt & Co., 725–726, 725f
  St. Louis brewery-bottling industry, 35, 66
  Welch & Goggin (Chicago), 449–450, 449f
Belfast Ginger Ale
  Aurora (IL)
    Green Brothers, 106
  Beardstown (IL)
    Ehrhardt, Frederick W., 108–109
  Carlinville (IL)
    Zaepffel, August, 174
  Chicago (IL)
    Christin, Arthur, 219–221
    Cook, Egbert C., 224
    John A. Lomax, 364, 366
Bell, John, 392
Belleville Glass Co., 55t, 59
bell-shaped cone ink
  Chicago (IL)
    Northwestern Chemical Manufacturing Co., 375, 375f
bell-shaped shoulders
  Chicago (IL)
    Landsberg, Moses G., 323–325
Bemoll, Henry, 709
Bennett, Isaac, Jr., 428
Bennett Pieters & Co. (Chicago), 293, 374, 381–386, 382–386f, 387, 405–406
Benson, F. A., 204f
Berger, Emil, 688
Bernard's Radical, 104–105, 105f
Berney's Highland Brewery (Highland), 519
Berney-Bond Glass Co. (PA), 59, 343
Berry, Columbus, 653
Berry, John M., 651–653, 652f, 653f
Berry, M. S., 653
Besant, Daniel J., 179–180, 180f
Besley's Waukegan Ale, 450
Best, Mathias, 409
Dr. Betts, 724
Betts, George H., 86–87, 87f
Betts, William J. S., 86
Betts & Russell (Alton), 86
Betts' Ginger Brandy, 87
Black Hawk Ginger Ale, 691, 693
Blackberry Brandy, 232, 232f
Blackhawk Springs, 693
Blake pharmacy bottles, 73
Blankenburg, Ernst, 469
Bligh, Andrew, 202
Bligh, George, 202
Bligh, James, 202
Bligh, Thomas, 201–202, 202f
Bligh's Tonic, 201–202, 202f
blob-top beer
  Chicago (IL)
    John A. Lomax, 356, 357f, 358
    Welch & Goggin, 449–450, 449f
  Naperville (IL)
    Stenger, John, 582–584, 584f
  Rock Island (IL)
    Junge & Beck, 694, 694f
blob-top bottles. See also blob-top beer; blob-top sodas; flattened blob top; lozenge-shaped blob top bottles; oval blob-top; tooled blob-top lips
  Chicago (IL)
    Cook & Co., 224–225, 224f
    Davies, George F., 231–232, 231f

(continued)

blob-top bottles (continued)
   Chicago (IL) (continued)
      Doty, Harvey C., 239, 239*f*
      George Hofmann & Brothers, 292–293
      M. Brand & Co., 202–203, 204*f*
      Palmer House Hotel, 376–377, 376*f*
      Victor Barothy & Co., 196–197, 197*f*
   Kewanee (IL) beer bottle variety, 34
   long-neck
      variation in, 10
      and study focus, 2
blob-top quart soda
   Peoria (IL)
      Eaton, James E., 611–614, 611*f*
blob-top sodas
   Belleville (IL)
      Fisher, Joseph (aka Fischer), 45, 45*f*
      Koob, August G., 46, 46*f*
      Stoltz, Francis (Franz), 119, 119*f*
   Carlyle (IL)
      Hess & Co., 175, 175*f*
      Hubert, Anthony, 175–176, 175*f*
   Centralia (IL)
      Heiss & Hutter, 180, 180*f*
   Chester (IL)
      Wenda, John, 184–185, 185*f*
   Chicago (IL)
      Barothy & Cook, 197, 197*f*, 198*f*
      Filer, Henry, 255–257, 256*f*
      Gottlieb Wurster & Co., 456–461, 457*f*
      Hausburg, William A., 282–283, 283*f*
      Hunt & Salpaugh, 295–296
      Hutchinson & Co., 307, 307*f*
      J. Grossenheider & Co., 268–269, 269*f*
      John A. Lomax, 345, 346*f*
      Lang Brothers, 48, 48*f*
      Lomax, John A., 58, 69
      Paul Rouze & Co., 391–393, 392*f*
      Sproat, John, 420
      Van Buskirk & Henry, 442–443, 443*f*
   De Kalb (IL)
      Dee, Richard, 52, 52*f*
   Galesburg (IL)
      Eaton, James E., 508–509, 509*f*
   inclusion in study, 35
   Murphysboro, 64
   Ottawa (IL)
      Finkler, Alexander, 594–595, 594*f*
   paper label–free, 41*f*
   Peoria (IL)
      Eaton, James E., 611–614, 611*f*
   Quincy (IL)
      Durholt, Henry, 663–667
   Rock Island (IL)
      Carse & Ohlweiler, 68, 691–694, 692*f*
   Streator (IL), 69
Mr. Block, 95
Blocki, Wm., 261
Blood (Braidwood), 69
Blood, Horace W., 162, 163*f*, 541, 742–743
Blood Purifier, 222, 662, 662*f*, 676
blue grass cure, 199, 199*f*
Blue Jacket Bitters, 626, 626*f*
bluing
   assemblage patterns, 74
   Chicago (IL)
      "Mrs. Schafer," 400, 401*f*
      Garden City Chemical Works, 262
      George Powell & Co., 388
      Gillet, McCouloch & Co., 267, 268*f*
      Gillett Chemical Works, 265–266, 265*f*
      Jacques Chemical Co., 313
      Matthews, Henry B., 369
      P. W. Gillet & Son, 267, 267*f*
      Thomas, Levi H., 439–440

Minneapolis (MN)
   Fuller, Harvey B., 472, 707
Bode (Chicago), 55*t*, 56, 59. *See also* Cronk, Edward Y.
Bodemann, Wilhelm, 374
Bodley, Dr. A. R., 654–655, 655*f*
Bodley & Wilson (Princeton), 654–655
body mold lines, definition, 12–13
body shape. *See also specific shapes*
   and contents, 13, 41–42
   definition of terms, 13–21
   and storage, 22
Boeke, Henry, 735, 735*f*
Mr. Bohl, 613, 635
Bohl, Singer & Lorens (Peoria), 627
boiling water remedy, 199, 199*f*
Bolles, Smith & Co. (Chicago), 217, 218
Bollin, Herman, 558–559, 558*f*
Bollin, Jacob, 558–559, 558*f*
Bollin, John, 558–559, 558*f*
Bolton, James, 557
Bolton, John G., 554–557, 555*f*, 556*f*
   gravitating stoppers, 29
Bolton, Nellie, 557
Bonekamp Stomach Bitters, 704–705, 704*f*
Boppart, William, 330
Boschulte, Charles, 658–659, 658*f*
Boschulte, Herman, 657–658, 659, 667, 677
Boschulte, Peter Henry, 658
Boschulte, William, 657–659, 667, 667*f*, 677
Boschulte & Brother (Quincy), 657–658, 657*f*
Boschulte & Co. (Quincy), 658–659, 658*f*
Boschulte & Knauf (Quincy), 659
bottle dealers. *See* glass bottle dealers
bottle manufacturing changes, 71–73
   crown-top closures, 311
   Hutchinson stoppers, 311
bottle technology terms, 7–29
   basal mold lines, 22, 23*f*
   base shapes, 22, 22*f*
   body mold lines, 12–13
   body shape and style, 13–21, 13–21*f*
   closures, 26–29
   colors, 9
   lip styles, 24–25, 24–25*f*
   manufacturing technology, 7–8, 11–12
   neck shape and style, 23, 24*f*
   shoulder shape and style, 22–23, 23*f*
bottler's mask, 26, 28*f*
   business details, M. R. & H. W. Lundblad, 679–681
   county distribution of, 4–5*t*, 6*f*
   and descriptive format of study, 5
   and immigrant densities, 6*f*, 29–30
   and study focus, 4
Bottling House (Bloomington), 120–121, 121*f*, 125, 129, 135, 571
Bottling House (Metamora), 570–571, 571*f*
bourbon
   Quincy (IL)
      Schrader, Frank, 684
bourbon bitters
   Prentice (IL)
      Berry, John M., 651–653, 652*f*, 653*f*
Bower, Ebenezer A., 202, 202*f*, 203*f*
Bower, Ernst, 592
Bower, William, 67, 591–592, 591*f*
   Matthews gravitating stoppers, 27
Bower & Bryan (Chicago), 202, 202*f*, 203*f*
Boyce, H. W., 653
Boyle, Charles M., 534–535, 534*f*
Boyle & Connifry (Jerseyville), 534
Brace, James F., 255, 255*f*
Bradbury, John, 521–522, 521*f*
Mr. Bradley, 742

Bradley, R. D., 125
Bradner's Honey Balsam, 124, 131, 131*f*
Brand, Michael, 202–203, 204*f*
Brand Brewery (Chicago), 203
Brandeth's Pills, 715
Brandt, William, 227
brandy
   Alton (IL)
      Betts' Ginger Brandy, 86–87, 87*f*
   Chicago (IL)
      DeMor & Co., 232
   Quincy (IL)
      Schrader, Frank, 684
   Springfield (IL)
      Dunning, Andrew J., 703
      Van Deusen, Martin M., 728
brandy finishes
   Chicago (IL)
      Palmer House Hotel, 376–377, 376*f*
   definition, 24
Braun, Philip, 522–523, 522*f*, 523*f*, 531
Braun & Killian (Jacksonville), 522, 523, 523*f*, 524, 531, 532, 533
Brazelton, F. A., 204
Brazelton, William P., 203–204, 204*f*
Brechwald, Charles, 507–508, 507*f*
Breihan, Henry, 10*f*, 69, 165–167, 421
   variable neck lengths, 10*f*
Bremer, Alfred R., 378
brewery products
   beer. *See* beer
   body shape definitions, 15, 15*f*, 16*f*
Brewing Association (E. Anheuser & Co.), 35
Brighton, England
   Cantrell & Co., 225
   Cantrell & Cochrane, 225, 366
Brinckerhoff, John, 377
Brisbane, Australia
   W. Steele, 223
Bristol, Riley, 495–496, 495*f*
Bristol, Solomon W., 120, 120*f*
Bristol, Tyler & Co. (Farmington), 495–496
Bristol's Nerve & Bone Liniment, 30, 120, 120*f*
British dumpy sodas
   Chicago (IL)
      Cook & Co., 224–225, 224*f*
British Oil, 30
Broadwell & Lindsay (Springfield), 717
Brookie & Bush, 716
Brophy, Dennis P., 588, 588*f*
Brophy's Bitters, 588, 588*f*
Browen's Rat Killer (Bloomington), 144, 145*f*, 148, 149
Brown, Dr. Alfred, 600, 600*f*
Brown, Daniel C., 699
Brown, Dwight Allen, 699
Brown, Joel B., 699, 699*f*
Brown, John H., 699
Brown, Thaddeus, 363
Brown, Weber & Graham (St. Louis), 118
Brown, William B., 699
brown stout, 94
   Chicago (IL)
      Cook, Egbert C., 224
      John A. Lomax, 355*f*, 356, 365
Brown's Oriental Hair Renewer, 443–446, 444*f*
Bryan, Frederick A., 205, 205*f*
Bryan, Wesley, 202, 202*f*, 203*f*
Buck, Charles G., 206, 207
Buck, George B., 205–206, 206–207, 207*f*
Buck & Rayner (Chicago), 206–208, 441
Buckeye Ale and Porter Depot (Galena), 505
Bucklen, Herbert E., 208–210, 209*f*
Bucklin, H. E.
   Dr. King, 187

Buckmaster, Henry, 93
Budlong, William C., 104–105, 105f
Buell, Charles H., 523–525, 524f
Buell & Schermerhorn (Jacksonville), 522, 523–525, 524f, 531, 532, 533–534
Buettner, Henry, 688
Buff, Jacob, 88–89
Buff & Kuhl (Alton), 88–89, 88f, 89f, 100
    Matthews gravitating stoppers, 27, 67, 88–89
Buff & Preisig (St. Louis), 89
Buffalo pharmacy bottles, 73
Buffalo Station (IL)
    Dr. Van Deusen's Ague Bitters, 729
Buffington & Bro. (Jerseyville), 93
bulbous-necked bottles
    Chicago (IL)
        Drake, Parker & Co., 241, 241f
Bull's Head Hotel (Chicago), 298
Bunnel, Dr. D[avid], 659–660, 659f
Bunnel & Kirkpatrick (Quincy), 660
Bunnel's German Liniment, 30, 659–660, 659f
Burger, Henry, 1, 501, 501f
Burk, John, 709
Burleigh, Alonzo S., 660–662, 661f, 662f
Burleigh, David B., 660–662, 661f, 662f
Burleigh, John Q., 660–662, 661f, 662f
Burleigh & Brothers (Mendon), 662
Burleigh & Brothers (Quincy), 660–662
Burleighs' Electric Balm, 660–662, 661f, 662f
Burnham, Edward, 406
Burnham & Smith (Chicago), 406
Burton M. Ford & Co. (Chicago), 410
Busch, Adolphus
    St. Louis brewery-bottling industry growth, 35
Bush and Brand Brewing Co. (Chicago), 203
Butt, Charles, 210–212, 210f, 211f
Butt, Margaret, 211–212, 211f
Byrne, J. H., 427

C

C. B. Castle & Co. (Bloomington), 46–47, 46f, 64, 121–122, 121f, 122f, 123, 150
C. Brechwald & Co. (Galesburg), 507–508, 507f
C. Freeman (Springfield), 538
C. G. Wicker & Co. (Chicago), 439
C. J. Ellis (Belleville), 538
C. Lutt & Co. (East St. Louis), 479–481
C. M. Daniels & Co. (Elgin), 488–489
C. O. Perrine & Co., 378
C. Peterson & Co. (Bloomington), 135
C. R. Smith & Co. (Chicago), 418, 418f
C. Ragnet & Co., 213
C. S. Hutchins & Co. (Chicago), 296
C. Schott's Brewery (Highland), 515, 515f
C. W. Keith & Co. (Naperville), 580
C. Wakefield & Co. (Bloomington), 148
Mr. Cairns, 95
Cairns, John, 170, 486, 606–607, 607f, 633
Cairns & Hickey (Peoria), 606–607, 607f
Cairo (ship), 167
Caldwell, James, 651, 651f
Caldwell & McGregor (Pontiac), 651, 651f
California pop and pop beer
    Chicago (IL)
        Dorman & Co., 238–239, 239f
        Geer, Schubael, 238, 264, 264f
        John A. Lomax, 355f, 356
        Lomax, John A., 239
        Paul Rouze & Co., 239, 391–393, 392f
    embossed-glass labeling, 42
    ingredients, 393, 586
    Kankakee (IL)
        Kammann, [Dietrich] H., 545

Nelson (IL)
    Allison, James J., 586, 586f
    patent date, 393, 586
Callender, William A., 607–609, 608f, 609f
Callender's Liver Bitters, 607–609, 608f, 609f
Calvert, D. H., 183–184
Camp Douglas (Chicago), 196, 303
Campbell, John, 740
Canedy & Johnston (Springfield), 718
canning jars
    assemblage patterns, 75
Cantrall-style "Dumpy" Soda
    body shape definition, 15, 17f
Cantrall-style Ginger Ale
    body shape definition, 15, 17f
Cantrell & Co. (Brighton, England), 225
Cantrell & Cochrane (Brighton, England), 225, 366
Carey, Patrick, 467, 467f
Carlsbad mineral waters, 88
Carman, Dr. W. H., 184
Carpenter, Andrew, 212, 212f
Carse, Henry, 690–693f, 690–694
Carse & Elder (Rock Island), 690, 690f
Carse & Lamont (Rock Island), 690–691, 690f
Carse & Ohlweiler (Rock Island), 691–693f, 691–694
    Codd marble stoppers, 27, 691–694, 692f
Carter, Thomas C., 188
Case Gin
    body shape definition, 13, 14f
case-gin bottle shape
    Chicago (IL)
        Henry H. Shufeldt & Co., 410–411, 411f
        Sykes, Dr. John W., 435
Casper Althen & Co. (Elgin), 488
Casper's Superior beer, 488
Cassily, William B., 212–213, 213f
Cassily & Co. (Chicago), 212–213, 213f
Castle, Calvin B., 46–47, 46f, 64, 121–122, 121f, 122f, 123, 150
Cather, Solon L., 263, 263f
Cavanaugh, P., 383, 405
Cavarly, Alfred W., 593
Cavarly, Juliana, 593
Cavarly, Dr. William W., 593, 593f
central base kick-up
    Chicago (IL)
        John A. Lomax, 338, 339f
Central Brewery (Peoria), 646
champagne
    Chicago (IL)
        Barothy & Cook, 47–48, 197
        Victor Barothy & Co., 197
champagne cider
    Aurora (IL)
        Green Brothers, 106
    Chicago (IL)
        Christin, Arthur, 220
        Cook, Egbert C., 224
    Rock Island (IL)
        Carse & Ohlweiler, 693
Champagne finish. *See* Ring-neck bottles
champagne-finish quart
    Chicago (IL)
        Keeley, Michael, 317–320, 319f
champagne-style bottles
    Chicago (IL)
        Victor Barothy & Co., 196–197, 196f
Dr. Chapin, 386
Chapin, Gardner, 214–216, 214f, 215f, 216f
Chapin & Co. (Chicago), 215, 323, 455
Chapin & Gore (Chicago), 63, 70, 214–216
    threaded glass stoppers, 29
Charles J. Peterson Co. (Springfield), 712

Charles Schriber & Co. (Chicago), 404
Chase Valley Glass Co. (Milwaukee), 55t, 60
    Chicago (IL), 307, 308f
chemical products. *See also* ink bottles
    assemblage patterns, 74
Chenot, Anastasia, 590
chestnut flask
    Chicago (IL)
        T. Foerster & Co., 258, 258f
Chicago Ale Depot (Chicago), 317
Chicago As It Is. A Strangers' and Tourists' Guide to the City of Chicago (Religious Philosophy Publishing Association), 384
Chicago Board of Health, 314
Chicago Brewing Co. (Chicago), 293
Chicago College of Pharmacy, 206
Chicago Consolidated Bottling Co. (Chicago), 366–367, 371
Chicago Fountain Soda Water Co. (Chicago), 221
Chicago Medical College (Chicago), 314
Chicago Publishing and Mailing House (Chicago), 190
Chicago Syrup Factory (Chicago), 296
Chicago-style quart sodas
    Chicago (IL)
        A. Mette & Bros., 369–371, 370f
        Barothy & Cook, 197, 198f
        Gottlieb Wurster & Co., 458–461, 460f
        H & Bro's, 279–281, 280f
        Hutchinson & Co., 307, 308, 308f
        J. Grossenheider & Co., 268–269, 269f
        John A. Lomax, 345, 346f
        Lang, Frederick C., 328–330, 329f
        Lang Brothers, 325–328, 327f
        Sass & Hafner, 396–399, 398f
    Danville (IL)
        Shean, Charles, 468–469, 468f
    Industrial Expansion era (1870s), 79
Chisholm, John, 688, 688f
Cholera Tincture, 222
chologogues, 43, 75, 406, 407f
    Chologogue, 406, 407f
    Dr. Lurton's Cholagogue, 367, 367f
Christian, John T., 313
Christian Lutt & Co. (East St. Louis), 481, 538–539
Christie, Dr. Abel H., 216–219, 217f, 218f
Christie, J. J.
    slug plate patent, 72
Christie's Ague Balsam, 216–219, 217f, 218f
Christin, Arthur, 219–221, 219f, 220f, 221f, 283
    torpedo-style gravitating stoppers, 27, 28f, 29, 49, 219–220, 219f, 220f
Christin, Mrs. C., 221
Christin & Co. (Chicago), 220
Christin gravitating stoppers
    Chicago (IL)
        John A. Lomax, 348
    Lemont (IL)
        Bolton, John G., 554–557, 556f
chronological markers
    basal mold lines, 22, 71, 87
    lip styles, 24–25, 71–72
    manufacturing technology, 71–73
    neck length, 23
Churchill & Bartlett (Freeport), 500
cider
    assemblage patterns, 74
    Aurora (IL)
        Green Brothers, 106
        Stege, A., 106, 106f
    champagne neck
        body shape definition, 18, 18f, 23, 24f

(continued)

cider (continued)
  Chicago (IL)
    A. Mette & Bros., 49–50, 49f
    Apple, Joseph, 192, 193f
    Barrett & Barrett, 199, 199f
    Gottlieb Wurster & Co., 456–461, 460f
    H. F. & Co., 257
    Hutchinson & Co., 297, 307–308, 307–309f
    Hutchinson & Dunn, 303
    John A. Lomax, 336, 337f, 338, 339f, 343, 344f, 349–350, 350–352f, 364, 365
    Keeley, Michael, 317–320, 318–320f
    Lang, Frederick C., 328–330, 329f, 330f
    Lang Brothers, 325–328, 327f
    Shure & Bro., 412
    Stenson, James, 422–426, 425f, 426f
  cider cure
    Chicago (IL), 199
  Colehour (IL)
    Smith, William S., 462–463, 463f
  Collinsville (IL)
    Bischoff, Benard, 465
  Decatur (IL)
    Kuny, Fredrick D., 472
  East St. Louis (IL)
    C. Lutt & Co., 479–481, 480f
    Schroeder, Edward, 482–486, 483f
  Elgin (IL)
    Hausburg Brothers & Co., 490–491, 491f
  Jacksonville (IL)
    P. Braun & Co., 522
    Schermerhorn, Charles, 533
  Joliet (IL)
    Paige, J. D., 540–542, 541f, 542f
  Lemont (IL)
    Bolton, John G., 554–557, 555f, 556f
  O'Fallon (IL)
    Bischoff, Benard, 591
  Peoria (IL)
    Parent, Anthony, 639
  Quincy (IL)
    Henry Durholt & Co., 666
    Hutmacher, Edward, 673–674, 673f
  ring-neck
    body shape definition, 18, 18f, 23, 24f
  Rock Island (IL)
    Carse & Ohlweiler, 691–694, 693f
    Sargent, William B., 694–695, 695f
  St. Louis (MO)
    Vail, John D., 542
  swing stopper definitions, 26, 26f
  Wm. McCully & Co. (Pittsburgh), 70
cider vinegar
  Jacksonville (IL)
    Buell & Schermerhorn, 524
cider-finish quarts. See also cider-finish ring-neck quarts
  Chicago (IL)
    Gottlieb Wurster & Co., 458–461, 460f
    John A. Lomax, 336, 337f
    Sass & Hafner, 396–399, 397f
    Stenson, James, 422–426, 426f
  Elgin (IL)
    Hausburg Brothers & Co., 490–491, 491f
  Lemont (IL)
    Bolton, John G., 554–557, 556f
cider-finish ring-neck quarts. See also ring-neck cider finish
  Chicago (IL)
    Christin, Arthur, 219–221, 221f
    Hutchinson & Co., 308, 308f, 309f
    John A. Lomax, 338, 339f, 343, 344f, 349–350, 350f

Joseph Apple & Co., 192, 193f
Lang Brothers, 325–328, 327f
Industrial Expansion era (1870s), 79
cider-style quarts
  East St. Louis (IL)
    C. Lutt & Co., 479–481, 480f
cigar jars, study exclusion rule, 37, 39f
Cincinnati (OH)
  Clark & Kump, 168
  Eclectic Medical College, 122, 132, 150
  Hamlin, John A., 274–276
  North, Alfred A., 722
  Sykes, Charles R., 432, 433, 434
Cincinnati Porter, 94
Citrate Magnesia, 632, 632f
citrate of magnesia bottles, 42
  Chicago (IL)
    Buck, George, 205–206, 205f
    Headley, Dr. William S., 632, 632f
citrates (bottle contents)
  body shape definition, 18, 20f
  Citrate of Magnesia, 42
  Essence of Jamaican Ginger, 42
City Bakery (Lincoln), 558
City Hotel (Peoria), 636
City Park Brewery (Belleville), 118
Claret Wine, 94
Clark, George N., 112, 112f, 119
Clark, Philo M., 167
Clark, Preston, 613
Clark & Kump (Cincinnati, OH), 168
Clarke, Rev. Walter, 571
Clarke's Compound Mandrake Bitters, 571–572
Clarke's European Cough Remedy, 571
Clarke's Excelsior Panacea, 571
Clarke's Family Vegetable Pills, 571
Clarke's Red Drops, 205, 205f
Clarke's Rheumatic Elixir, 571
Clark's Female Pills, 655
Clasen, Adolph D., 441–442
Cleveland, Frederick, 609–610, 609f
Cleveland, George, 610
Cleveland & Rice (Chicago), 610
Cleveland Brothers (Peoria), 610
Cleveland House (Chicago), 362
Cleveland Wooden Ware & Match Co. (Chicago), 189
Cleveland's Baking Powder, 610
Clinton House (Chicago), 362
clock face 11:00, embossed. See Van Buskirk & Henry (Chicago)
closures
  Baltimore Loop stopper
    study exclusion rule, 37
  Christin-patent stopper, 219, 219f
  cork. See Cork stoppers
  crown-top, 311
  definition of terms, 26–28f, 26–29
  glass gravitating. See Glass gravitating stoppers
  Himan Franks threaded internal stopper, 29, 70
  Hutchinson-style internal stoppers. See Hutchinson-style internal stoppers
  Lightning stopper
    study exclusion rule, 37
  Matthews gravitating stoppers. See Matthews gravitating stoppers
  patent dates, 26–29
  swing stoppers. See also specific stoppers
    definition of terms, 26–28f, 26–29
cloverleaf, embossed. See Griggs, Edward Y.
Clowry, Thomas, 221, 221f
Clowry & Fitzgerald (Chicago), 221, 221f

Cockle, W., 645
Codd, Hiram, 68–69
  marble gravitating stopper patent, 27, 29
Codd-marble gravitating stoppers
  Chicago (IL)
    Lang, Frederick C., 48–49, 49f
  definition, 26, 27
  Rock Island (IL)
    Carse & Ohlweiler, 68, 691–694, 692f, 693f
Cody, Thomas, 629–630, 629f
Mr. Coe, 56, 66
Colburn, Birks & Co. (Peoria), 369, 608–609
Colburn, Burk & Co., 489
Colehour, Charles W., 463
Colehour, William H., 463
Coleman, Adams & Bro. (Chicago), 188
Coleman, G. C., 188
Collins, Dr. Silas F., 221–222, 222f
collyrium medicines. See eye water
cologne
  Beardstown (IL)
    Ehrhardt, Frederick W., 108
  Chicago (IL)
    George Powell & Co., 388
    Mars Cologne, 208
cologne bottles
  assemblage patterns, 75
colors of bottles, terms for, 9
Columbian Cathartic Bitters, 717–719, 717f
Columbian Gin, 230–231, 231f
Commercial Agency Register, 8
Complete Farrier or Horse Doctor, The (Sloan, W. B.), 445
Comstock, H., 431
Condition Powders, 662, 662f
cone inks
  body shape definition, 18, 19f
  Chicago (IL)
    Northwestern Chemical Manufacturing Co., 375, 375f
  Elgin (IL)
    Elgin Ink Co., 489–490, 490f
cone-shaped body
  Chicago (IL)
    Paul Rouze & Co., 391–393, 393f
Congo Chemical Co. (Chicago), 254
Congo Liniment, 254
Congress mineral water, 208
Conkling, James C., 726–727, 726f, 727f
Conkling (J. C. & J. J.) and Slemmons (Springfield), 727
Connell, J. M., 222–223, 223f, 456
Connell, Jeremiah M., 222
Connell's East Indian Brahminical Moonplant Remedies, 222–223, 456
Connell's East Indian Remedies, 222–223, 223f
Connell's East Indian Remedies, 223
Connor, Joseph
  swing stopper patent, 26, 26f
contents of bottles, and body shape, 13, 41–42
Cook, Egbert C., 47–48, 197, 197f, 198f, 224–225, 224f
Cook, John S., 374, 385
Cook & Co. (Chicago), 224–225, 224f
Cook & Peabody (Chicago), 225
Cook County Hospital (Chicago), 314
Co-Operative Medicine Co. (Chicago), 388
cordials, 43
  Belleville (IL)
    Norman's Pine Knot Cordial, 117–118
  Bloomington (IL)
    Dr. Faloon's Infant Cordial, 123
    Lackey's Justly Celebrated Blackberry Cordial, 123–124, 124f
    Wells' Blackberry Cordial, 152

Diarrhea and Dysentery Cordial, 222
Dr. Bartram's Dysentary and Diarrhoea Cordial, 722
Dr. U. C. Roe's Blood Cordial, 497–498, 497f
Godfrey's Cordial, 715
Quincy (IL)
　Schrader, Frank, 684
cork stoppers
　blob-top cork-closure inclusion in study, 35
　and Hutchinson-style internal stoppers, 24, 35
　keeping moist, 22, 51
　popularity of, 26
Corn & Wart Salve, 676
Corneau, Charles S., 702
Corneau & Diller (Springfield), 600, 718
Corneau & Diller's Bottled Soda, 702
Corneau & Diller's Castor Oil, 702, 702f
Corneau & Diller's Diarrhoea Mixture, 702
Corner Drug Store (Woodstock), 743, 744f
Cory, Dr. Benjamin Sayre, 225–226, 738–739
Cory, James, 226, 738
Couch, Ira, 240
Couch, James, 240
Cough Balsam, 662, 662f
Cough Killer, 257–258
Cough Syrup, 676
Coutant, B. W., 500, 594
Coutant & Co. (Ottawa), 500, 594
Cox, Lewis, 226–227, 226f
Cox, Lewis H., 227
Cox, Richard S., 622, 622f
Cox & Co. (Chicago), 227
Cox and Brandt (Chicago), 227
Cox Chemical Co. (Chicago), 226–227, 226f
Cram, Dr. John S., 227–228, 227f
Cram & Melcher (Chicago), 228
cream ale
　Peoria (IL)
　　Parent, Anthony, 639
Cree swing stoppers
　definition, 26, 26f
　patent date, 26
crescent moon and star, embossed. *See* Brophy, Dennis P.
Croissant, Henry, 482
Cronk, Albert, 228, 229
Cronk, Edward Y., 59, 228–230, 229f, 230f
Cronk, "Dr." Warren, 228–229
Cronkhite, Warren, 228
Croskey, Abraham F., 230–231, 231f, 410
Crotty, Lawrence, 499, 500f, 594, 594f
Crotty, Thomas, 499, 500f, 594, 594f
Crotty Brothers (Freeport), 499, 500f, 594
Crotty Brothers (Ottawa), 500, 594
crown, embossed. *See* Henry H. Shufeldt & Co. (Chicago)
crown top bottles
　Blue Island (IL)
　　G. H. Hausburg, 158, 160
　　Hausburg, Charles, 282, 491
　Carlinville (IL)
　　Zaepffel, August, 174
　Kankakee (IL)
　　Kammann, [Dietrich] H., 545
　study exclusion rule, 37
Crystal Palace Drug Store (Charleston), 183
cullet, 43–44
Cunningham, Henry C., 183–184, 183f
Cunningham & Barnes (Charleston), 183–184
Cunningham & Ihmsen (Pittsburgh), 55, 55t, 60, 61f, 70, 72, 521, 521f
　Chicago (IL), 231f, 232, 375, 375f, 410, 410f, 441f, 442, 449–450, 449f
　Danville (IL), 469

Dixon (IL), 52f, 53, 473, 473f
Elgin (IL), 489, 489f
Evanston (IL), 492f, 493
last bottles produced, 53, 232, 473, 489, 521, 649
Pekin (IL), 600, 600f
Peru (IL), 648f, 649
slug-plate embossed bottles, 60
study inclusion rule, 37
Cunninghams & Co. (Pittsburgh), 55t, 60, 450
Curtis & Perkins (Bangor, ME), 247, 420
cylinder master inks
　Bloomington (IL)
　　J. W. Maxwell & Co., 132–133, 133f
　Decatur (IL)
　　Fuller's Banner Ink, 471–472, 471f
　Pontiac (IL)
　　Caldwell & McGregor, 651, 651f
　Springfield (IL)
　　Fuller's Banner Ink, 706–707, 706f
cylindrical blob top bottles
　Chicago (IL)
　　Chapin & Gore, 214–216, 214f, 215f
　　Christin, Arthur, 219–221, 219f
cylindrical blob top sodas
　Belleville (IL)
　　Fisher, Joseph (aka Fischer), 113–115, 114f
　Chicago (IL)
　　Barothy & Cook, 197, 197f, 198f
　　Christin, Arthur, 219–221, 219f
　　Cronk, Edward Y., 228–230, 229f
　　Victor Barothy & Co., 196–197, 197f
cylindrical bottles
　Alexander Arbogast (Pittsburgh), 57
　Beardstown (IL)
　　Ehrhardt, Frederick W., 107–109, 108f
　Belvedere (IL)
　　Bristol, Solomon W., 120, 120f
　Bloomington (IL)
　　Wakefield & Co., 137–150, 140f, 142f, 145f
　Carlyle (IL)
　　Gunn, John S., 174–175, 175f
　Chicago (IL)
　　A. Mette & Bros., 369–371, 370f
　　Badeau, William C., 194–195, 195f
　　Bennett Pieters & Co., 381–386, 382f
　　Bryan, Frederick A., 205, 205f
　　Buck, George, 205–206, 205f, 206f
　　Buck & Rayner, 206–208, 207f
　　Chapin & Gore, 63, 214–216, 216f
　　Christin, Arthur, 219–221, 219f
　　Cronk, Edward Y., 228–230, 229f
　　Davies, George F., 231–232, 231f
　　DeMor & Co., 232, 232f
　　Dempsey & Hennessey, 234–236, 235f
　　Emmert Proprietary Co., 244–247
　　Farrell, William B., 251–255, 254f
　　Farwell and Brace, 255, 255f
　　Gillett Chemical Works, 265–266
　　Hayes Brothers, 284–287, 284f
　　Hutchings, William A., 296–297, 296f
　　J. Grossenheider & Co., 268–269, 269f
　　John A. Lomax, 352, 353f, 354f, 358, 359f
　　Kane, Thomas, 315, 315f
　　Moench & Reinhold, 373–374, 374f
　　Paul Rouze & Co., 391–393, 393f
　　Perrine, Charles O., 377–381, 377f
　　Snyder & Eilert, 419–420, 420f
　　Steel & Co., 421–422, 421f
　　Stenson, James, 422–426, 425f
　　Sykes, Charles R., 432–434, 433f
　　W. H. Hutchinson & Son, 309, 309f
　　Wheeler & Bayless, 452–453, 453f
　East St. Louis (IL)
　　Schroeder, Edward, 482–486, 486f

Jerseyville (IL)
　Freeman, Dr. John D., 535–538, 535f
Lemont (IL)
　Bolton, John G., 554–557, 556f
multisided
　Albion (IL), 85, 85f
　body shape definition, 13, 13f
Nelson (IL)
　Allison, James J., 586, 586f
Oquawka (IL)
　McDill, Dr. Andrew C., 592, 592f
Peoria (IL)
　William A. Callender & Son, 609
Quincy (IL)
　M. R. & H. W. Lundblad, 677–681
Rockford (IL)
　Graham's Distillery, 689, 689f
Springfield (IL)
　A. & M. Lindsay, 714–717, 715f, 716f
　McNeil, William W., 719, 719f
　Melvin, Samuel H., 719–721, 720f
　Van Deusen, Martin M., 727–729, 728f
Virden (IL)
　Achilles & Hille, 733–734, 734f
cylindrical medicine vials
　definition of body shape, 18, 19f
cylindrical quart preserve jars
　Chicago (IL)
　　Wheeler & Bayless, 452–453, 453f
cylindrical wax-seal jars
　Chicago (IL)
　　Perrine, Charles O., 377–381, 377f
cylindrical whiskey-style quarts
　Chicago (IL)
　　Chapin & Gore, 214–216, 216f

D

D. L. Gold & Co. (Springfield), 729
Damron, Frank, 363
Daniels, Dr. Charles M., 488–489, 488f, 489f
Daniels' Tanta Miraculous, 488–489, 488f, 489f
Darmstatter, Frederick, 499, 499f
David Henry & Co. (Chicago), 443, 490
Davies, George F., 231–232, 231f
Davis, Robert S., 610
Davis, William H., 610–611, 611f
Davis Vegetable Pain Killer, 676
Davis's Inimitable Hair Renewer, 610–611, 611f
Day-Light Liver Pills, 243, 243f
Dazet, Philip, 392
Dee, Richard, 52, 52f
deep basal kick-up
　Chicago (IL)
　　Barothy, Victor, 196–197, 196f
　Lemont (IL)
　　Bolton, John G., 554–557, 556f
Deep Rock Mineral Water
　Chicago (IL)
　　John A. Lomax, 365
DeMor, Dr. Amos B., 232, 232f
DeMor, Edgar, 232
DeMor & Co. (Chicago), 232, 232f
Dempsey, Daniel, 232–236, 233f, 234f, 235f, 291
Dempsey & Hennessey (Chicago), 233, 234–236, 235f, 291
Denie's Electro-Magnetic Plaster, 154, 452
Depot of Brophy's Celebrated Domestic Tonic Bitters (Nokomis), 588, 588f
DePuyt, George, 736, 736f
DeSteiger Glass Co. (LaSalle), 55t, 60, 62, 67, 350, 463, 548–549, 548f, 630, 631f
　Nelson (IL), 586, 586f
　Waterloo (IL), 737
DeSteiger Glass Co. (Streator), 190
Dettmer, William, 160, 160f, 161, 489, 489f

Develbiss, Noah, 646
Dewar, William, 422
Dewey fastener
 definition, 26, 26f
 patent date, 26
Diamond Mills Distillery (Peoria), 622
diamonds (5) in star shape embossing. *See*
 Paul Rouze & Co.
Diarrhea and Dysentery Cordial, 222
Diarrhoea Syrup, 725
Dickerman, W. A., 441
Dickinson, George, 236, 236f, 269, 270, 270f
Dillard, William, 715
Diller, Isaac, 702
Diller, Jonathan Roland, 700, 702
Diller, Roland W., 700–702, 700–702f
Diller's Drug Store (Springfield), 702, 702f
Dinet, Elizabeth, 238
Dinet, Joseph, 237–238, 237f, 238f, 247, 250
Mr. Dintelman, 112
directories, as information sources, 3, 8
Doane, William, 740, 740f
Doane House hotel (West Urbana), 740, 740f
Docker & Ferring (Shawneetown), 538
Doll, Ben, 549
domed shoulder. *See* slightly domed shoulder
Dominick O. Cunningham (Pittsburgh), 62,
 533, 533f
 Springfield (IL), 711, 711f
Dorman, H. C., 392
Dorman, James A., 238–239, 239f, 264
Dorman & Co. (Chicago), 238–239, 239f
Dorries, Charles, 163, 163f
Dorries, Henry, 163, 163f
Dorries & Bro. (Breese), 163, 163f
Doty, Erastus, 239
Doty, Harvey C., 239, 239f, 317
Doty & Waugh (Chicago), 239
Doty's Hotel (Chicago), 239
Double Philadelphia bottles
 study exclusion rule, 37, 38f
Dr. A. Brown's Blood Purifier, 600, 600f
Dr. A. Brown's Female Alterative or
 Women's Friend, 600
Dr. A. H. Smith's Celebrated Ague Cure, 177
Dr. A. H. Smith's Golden Liniment, 176, 177f
Dr. A. L. Adams Liver Balsam, 186–187, 186f
Dr. Banning's Female Pills, 722
Dr. Bartram's Dysentary and Diarrhoea
 Cordial, 722
Dr. Bartram's Nerve and Bone Liniment, 722
Dr. Beekman's Pulmonic Syrup and Pills, 414
Dr. Bodley's Fever & Ague Balsam, 34,
 654–655, 655f
Dr. Bragg's Indian Queen Sugar Coated
 Pills, 414
Dr. Butt's Excelsior Liniment, 30
Dr. Cavarly's Ague Bitters, 34, 593, 593f
Dr. Clark Johnson's Indian Blood Syrup, 314
Dr. Cory's Stomachic Bitters, 738–739, 739f
Dr. Cronk's beverages, 228–229
Dr. F. A. Sabine's Carbolic Ointment, 393
Dr. F. A. Sabine's Harvest Bitters, 393
Dr. F. A. Sabine's World's Remedy, 393
Dr. Fahrney's Blood Vitalizer, 250–251, 251f
Dr. Fahrney's Panacea, 250–251, 250f
Dr. Fahrney's Uterine, 250–251, 250f
Dr. Faloon's Blackberry Diarrhoea Specific,
 122, 123f
Dr. Faloon's Infant Cordial, 123
Dr. Faloon's Instant Relief, 122–123, 123f
Dr. Faloon's Rosin Weed Balsam, 46–47, 46f,
 121–123, 121f, 122f, 123f
Dr. Faloon's Tonic, 122
Dr. Freeman's Eclectic Liniment, 30

Dr. H. S. Woodard (Bloomington), 154–156
Dr. Hamilton's Indian Liniment, 30
Dr. Hamilton's Vegetable Cough Balsam
 Pioneer era (1840s), 75
Dr. Hoffmann's Red Drops, 393
Dr. Howe's Blackberry Balsam, 127, 127f
Dr. Howe's Egyptian Salve, 127, 127f
Dr. Howe's Honey Balsam, 127, 127f
Dr. Howe's Instant Pain Relief, 127, 127f
Dr. Howe's Liniment, 127, 127f
Dr. Howe's Vegetable Liver Pills, 127, 127f
Dr. J. C. Wills' World Worm Candy, 563, 563f
Dr. J. M. Connell's Brahminical Moonplant.
 East Indian remedies, 223
Dr. J. S. Gunn's Golden Vermifuge, 174–175
Dr. J. Stickels Compound Syrup of
 Cephalanthus, 597–598, 598f
Dr. Jacques German Worm Cakes, 246
Dr. Johnson's Balsam for the Lungs, 314
Dr. Johnson's Liniment Mixture, 314
Dr. King, 187
Dr. King's New Discovery for
 Consumption, 208–210, 209f
Dr. Kruse's Malaria Antidote, 636–637, 636f
Dr. Lee's Rheumatic Liniment, 575–576, 575f
Dr. LeRoy's Antidote to Malaria, 574–575, 574f
Dr. LeRoy's Cough Syrup, 574–575, 574f
Dr. Lurton's Cholagogue, 367, 367f
Dr. Major Quinine Substitute, 129, 130f
Dr. North's Ague Pills, 723
Dr. North's Balsam of Horehound and
 Naptha, 722
Dr. North's Cherokee Balsam of
 Horehound, 723
Dr. North's Cherokee Liniment, 30, 723
Dr. North's Cholera Drops, 723
Dr. North's Cough Candy, 723
Dr. North's Expectorant Cough Candy, 722
Dr. North's Fever & Ague Killer, 34, 722
Dr. North's Mother's Balm, 723
Dr. North's Pain Killer, 723
Dr. North's Pectoral Balsam of Horehound,
 723
Dr. North's Sarsaparilla and Blood Pills,
 722, 723
Dr. P. Johnson's Golden Eye Salve, 314
Dr. Paoli's Sparkling Persian Sherbet,
 412–413, 413f
Dr. Penoyer's Extract from the Hygean
 Springs, 741–742, 741f
Dr. Peter Fahrney & Sons (Chicago), 251
Dr. Pierce's Golden Medical Discovery,
 96–97, 96f
Dr. Richardson's Peruvian Balsam, 744
Dr. Richardson's Wahoo Bitters, 743–744, 744f
Dr. Streeter's Magnetic Liniment, 495–496
Dr. Sykes New England Liver Tonic and
 Billious Annihilator, 435
Dr. Sykes' Sure Cure for Catarrh, 432–434
"Dr. Thomson" (Pittsfield), 649–650, 650f
Dr. Thomson's Galvanic Liniment, 30,
 649–650, 650f
Dr. Topping's Alterative and Cathartic
 Syrup, 717–719, 717f, 718f
Dr. U. C. Roe & Sons (Franklin Grove),
 497–498, 497f, 498f
Dr. U. C. Roe's Blood Cordial, 497–498, 497f
Dr. Van Deusen's Ague Bitters, 34
Dr. Weaver's Compound Extract of
 Fireweed, 443–446, 445f
Dr. Weaver's Humor & Liver Syrup, 443–
 446, 445f
Dr. William Beach's Wine Bitters, 222
Dr. Winchell's Teething Syrup, 243, 243f,
 247, 419–420, 420f

Dr. Wolgamott's Aperient and Blood
 Purifier, 223, 455–456, 456f
Dr. Z. Waters & Co. (Bloomington), 154,
 155, 157
Dr. Z. Waters' Family Medicines, 154–156
Dr. Z. Waters' Honey Balsam, 154–156, 157
Dr. Z. Waters' Itch Renovator, 150, 154, 156
Dr. Z. Waters' Liver and Cathartic Pills,
 154–156
Dr. Z. Waters' NE Plus Ultra Liniment,
 154–156, 155f
Dr. Z. Waters' Raspberry Balsam, 154–156
Drake, John B., 240–241, 240f, 241f, 377
 threaded glass stoppers, 29
Drake, Parker & Co. (Chicago), 241, 241f
Drake Hotel (Chicago), 377
Dresch, Henry, 712
Drown, S. DeWitt, 619
Drs. C. M. Fitch & J. W. Sykes, 434–436, 434f
druggist bottles. *See* pharmacy bottles
Dunbar's Bethesda, 208
Dunn, Frank, 363
Dunn, Thomas O., 303, 303f
Dunning, Andrew J., 703, 703f
Dunning's Celebrated Orange Bitters, 703
Durant, Jaques & Atwood (Chicago), 313
Durholt, Catherine, 663, 665f
Durholt, Henry, 663–666f, 663–667, 672, 672f
 variable neck lengths, 663, 664f
Durholt & Co. (Quincy), 658, 659
Dye, Shack, 613f
Dye's Practical Horse Shoer (Peoria), 613f

E
E & Z (Pekin), 571
E. A. Floyd's Liniment Depot (Greenville), 511
E. Anheuser & Co. (St. Louis)
 brewery-bottling industry growth, 35
 Brewing Association, 35
E. C. Cook & Co. (Chicago), 224
E. E. Eaton Co. (Chicago), 196, 196f
E. Hess & Co. (Chicago), 292
E. R. Stege Brewery (Chicago), 106
E. T. Flanagan & Co. (Belleville), 117–118
Eagle, George (New York), 677
Eagle Brewery (Peoria), 630, 631f
Eagle Hotel (Chicago), 188
Dr. Easterly, 721
Eastern Saratoga Springs Mineral Water Spa
 (New York), 741
Eaton, Amasa M., 195–196, 196f
Eaton, Charles, 196
Eaton, Daniel, 196
Eaton, Emma, 196
Eaton, George E., 241–242, 509
Eaton, Hiram, 509
Eaton, James E., 551, 602, 607, 635, 643
 Chicago (IL), 241–242, 241f
 Galesburg (IL), 507, 508–509, 508f, 509
 Peoria (IL), 611–613f, 611–614
Eaton, Julia, 613
Eaton, Sephreness, 509
Eaton & Abbey (Chicago), 196
Eaton/Hickey "clan," 509, 539, 551–552,
 607, 611, 635
Ebey, John Neff, 586
Eclectic Advertiser, 536
Eclectic Medical College (Cincinnati), 122,
 132, 150
Eclectic Medical Depot (Jerseyville), 536, 538
eclectic medical movement, 536–537
Eclipse Glass Works (Temperanceville, PA),
 63
Edds, William T., 602
Eddy, A. S., 148

Eddy, Louis, 150
Edgar A. Hall & Co. (Chicago), 271
Edinburgh (Scotland)
  Robin, McMillan & Co., 422
Edward H. Everett Co (Newark, OH), 55t, 62, 68, 158, 330, 330f
Edward R. Stege (Chicago), 106
Edward T. Flanagan & Co. (Belleville), 119
Edwards, "E.", 242, 242f
Edwards, Dr. Edward W., 242
Edwards, Francis C., 242, 242f
Edwards, George, 509–510, 509f
Edwards, [Isaac?] R., 510, 510f
Edwards House (Knox County), 510
Edwin Hayes & Bro. (DuQuoin), 477
Eger mineral waters, 88
"Egypt," Illinois area, 544
Egyptian Anodyne, 543–544, 544f
Egyptian Collyrium, 544
Egyptian Salve, 154, 662, 662f
Ehrhardt, Frederick W., 107–109, 107f, 108f
Ehrman & Co. (Chicago), 261
Eilert, Jacob K., 242–243, 242f, 243f, 245, 419–420, 420f
Elder, John, 690, 690f
Elgin Eagle Brewery (Elgin), 488
Elgin Ink Co. (Elgin), 489–490, 490f
Eliel, Gustavus, 550
Eliel, Jeremiah, 549–550
Eliel, Louis, 549–550, 549f, 550f
Eliel & Co. (Chicago), 550
Eliel & Co. (LaSalle), 550
Eliel Brewery (LaSalle), 549
embossed symbols
  4-bottle logo. See John A. Lomax (Chicago)
  AC monogram. See Christin, Arthur
  apple. See Barrett & Barrett (Chicago)
  CBCo monogram. See George Hofmann & Brothers (Chicago)
  clock face 11:00. See Van Buskirk & Henry (Chicago)
  cloverleaf. See Griggs, Edward Y.
  crescent moon and star. See Brophy, Dennis P.
  crown. See Henry H. Shufeldt & Co. (Chicago)
  diamonds (5) in star shape embossing. See Paul Rouze & Co.
  E. Y. C. monogram. See Cronk, Edward Y.
  eagle, spread-wing. See J. Grossenheider & Co. (Chicago)
  footprints and stars. See Connell, J. M.
  G (embossed backwards). See Green Brothers (Aurora)
  H (in embossed circle). See Hutmacher, Edward
  HAS monogram. See H. A. Stone & Co. (Chicago)
  HB monogram. See Hayes Brothers (Chicago)
  horseshoe. See Hayes Brothers (Chicago)
  JB monogram. See Junge & Beck (Rock Island)
  Maltese Cross. See Wolgamott, Dr. George W.
  "O" double-embossed oval. See Sass & Hafner (Chicago)
  palm tree. See Garrick & Cather (Chicago)
  propeller. See Quigley, Hopkins, and Lea (Alton)
  rings and X
    Lemont (IL). See Bolton, John G.
  SH monogram. See Haldeman, Samuel
  sheaf of wheat. See DeSteiger Glass Co. (LaSalle)
  shield. See Croskey, Abraham F.; Gillet, McCulloch & Co. (Chicago); Landsberg, Moses G.
  soda bottle. See Muller, Frederick W. "Pop"; Sass & Muller (Dunton)
  star. See Star Glass Works (New Albany, IN); Star Glass Works (Newark, OH)
    Beardstown (IL). See Ehrhardt, Frederick W.
    Chicago (IL). See A. Leibenstein & Co. (Chicago); Chapin & Gore (Chicago); Connell, J. M.; Gillet, McCulloch & Co. (Chicago); Hutchinson & Co. (Chicago); Keeley, Michael; King, Thomas; Northwestern Chemical Manufacturing Co. (Chicago); Paul Rouze & Co.
    Nokomis (IL). See Brophy, Dennis P.
    Sheldon (IL). See Miller, Jacob S.
  star and crescent moon. See Brophy, Dennis P.
  stars and footprints. See Connell, J. M.
  triangle with GJZ or P. See Zerwekh, Gottleib J.
  wreath. See Westerfield, John G.
  X and 2 rings
    Lemont (IL). See Bolton, John G.
embrocations, 43
  Dr. North's Cherokee Liniment, 30, 723
  Myers' Embrocation, 721
Emmert, Charles F., 243–244, 244–247
Emmert, Harris L., 243–244, 244–247
Emmert, John S., 243–244, 244–247, 245f, 246f, 247f, 420
Emmert & Burrell (Freeport), 244, 245, 420
Emmert Proprietary Co. (Chicago), 243–244, 244–247, 245f, 246f, 247f, 420
Empire mineral water, 208
English Horse Liniment, 662, 662f
Ennis, August, 161, 161f
Enoch Woods company, 196
Ensel & Mayer (Springfield), 726
Entwistle, Ann, 361
Entwistle, Joseph, 189, 201, 237, 238, 247–250, 248f, 249f, 334, 360–361
Entwistle & Bebbington (Chicago), 248–249, 249f, 334, 361
Entwistle & Lomax (Chicago), 249, 249f, 334, 361, 362
Epstein, Goodman & Co. (Chicago), 289–290
Epstein, Henley & Co. (Chicago), 289–290
Epstein, Henry, 289–290
equine medicines. See veterinary medicines
Eqyptian Rheumatic and Strengthening Salve, 676
Erl, Joseph, 121, 571, 602
Essence of Jamaican Ginger, 42, 96–97, 267
etched seltzer-water dispensers, study exclusion rule, 37, 40f
Evans, David H., 503
Evans, Evan, 503
Evans, Joseph, 503–505, 504f
Everybody's Liniment, 713–714, 714f
Everybody's Mill (Springfield), 713
Everybody's Store (Springfield), 713
Excelsior
  Carlinville (IL)
    Zaepffel, August, 174
Excelsior Brewing Co. (Red Bud), 687–688, 687f
Excelsior Brewing Co. (St. Louis), 688
Excelsior Liniment, 210–212, 210f, 211f
Excelsior Medicine, 210–212, 211f
Extract of Tar & Wild Cherry, 242–243, 245
Eye Infirmary (Quincy), 660
eye water, 43, 79
  Egyptian Collyrium, 544
  Kalb's Eye Water, 675–677, 675f

F

F. Binz brewery (Chicago), 319
F. C. Lang (Chicago), 330
F. C. Lang & Co. (Chicago), 330
F. J. Finkler & Co. (Dixon), 473, 473f
F. Molln & Co. (Mendota), 569, 569f, 570
F. Scammon & Co. (Chicago), 187
F. W. Muller (Arlington Heights), 103–104, 103f, 104f, 476
F. Weber & Brother (Pana), 599–600, 599f
F. X. Reisch (Springfield), 712
Fahnestock, Albree & Co. (Pittsburgh), 55, 55t, 62–63, 67
  Chicago (IL), 293f, 294
  Keokuk (IA), 675
  Lacon (IL), 97, 547, 547f
  LaSalle (IL), 53, 53f
  Peoria (IL), 626, 627f, 641, 641f
  Springfield (IL), 707f, 708
  St. Louis (MO), 505
Fahnestock, B. L., 62–63
Fahrney, Peter, 250–251, 250f, 251f
Fair Play Foundry and Machine Shop (Springfield), 713
Faloon, Dr. Mathew, 122–123
Faloon and Geltmacher (Bloomington), 122–123
Faloon's medicines, 46–47, 46f
Faloon's Rosin Weed Balsam, 46, 64
Family Medicine Depot (Springfield), 723
Farmers & Farriers Liniment, 473–475, 474f
Farrell, DeWitt C., 622, 622f
Farrell, Hiram G., 254, 614–621, 614–621f, 622
  earliest embossed bottles, 29
Farrell, William B., 251–255, 252f, 253f, 254f, 615–620
  earliest embossed bottles, 29
Farrell & Cox (Peoria), 622, 622f
Farrell's Arabian Liniment, 251–254, 252f, 253f, 254f, 511, 614–621, 615–621f, 715
  earliest embossed bottles, 29
  early liniment, 30
  Pioneer era (1840s), 75
Farrell's Arabian Nerve and Bone Liniment, 618
Farrell's Cough Syrup, 620, 621f
Farrell's Fever & Ague Drops, 614–621, 614f
  as cholera medicine, 34
  earliest embossed bottles, 29
  Pioneer era (1840s), 75
Farrell's Indian Black Ointment, 251, 252f, 254
Farrrell's Tooth-Ache Drops, 620, 621f
Farwell, James F., 255, 255f
Farwell and Brace (Chicago), 255, 255f
Federal Hill and Spring Garden Glass Works (Baltimore), 59
Fellows, John A., 651
Fellows & Caldwell (Pontiac), 651
Fellrath, Hubert, 622–625, 623–625f
  Matthews gravitating stoppers, 27
Fever Specifics, 34
Field, Marshall, 376
Field, Platus A., 172, 172f
Fife, John, 568, 568f
Filer, Charles, 255–257, 256f, 257, 257f
Filer, Henry, 255–257, 256f, 257f
Finch, Edward, 277
Findlay Bottling Co (Findlay, OH), 55t, 63, 290
Finkler, Alexander, 53, 53f, 473, 500, 550–551, 550f, 594–595, 594f, 731, 732
Finkler, Alexander C., 595
Finkler, Frank J., 52–53, 473, 550, 594, 732

Finkler, John A., 53, 473, 550, 594–595, 641, 641f, 731–732, 731f
fire grenades
  assemblage patterns, 75
  study exclusion rule, 37
Fischer, Augusta, 181
Fischer, Charles, 181–182
Fischer, Christian, 181–182, 181f, 182f
Fischer, Henry, 658, 667, 667f, 677
Fischer & Boschulte (Quincy), 658, 667, 667f
Fisher, Joseph (aka Fischer), 45, 45f, 63, 110, 113–115, 113f, 114f, 115–116, 115f, 116–117, 116f
Fisher, Williamson P., 377
Fisher & Ab Egg (Belleville), 115–116, 115f
Fisher & Rogger (Belleville), 63, 116–117, 116f
Fitch, Dr. Calvin M., 434–436, 434f
Fitzgerald, James, 221, 221f
Fitzgeralds (Jerseyville), 534
Fitzpatrick's grocery store (Springfield), 721
Fixmer, John P., 704–705, 704f
Flack, Gustaf, 678
Flagg, Edward H., 257–258, 257f, 432
Flagg & Sweet (Chicago), 258
Flanagan, Edward T., 117–118, 117f, 118f
flared lips
  Chicago (IL)
    Buck, George, 205–206, 206f
    Buck & Rayner, 206–208, 207f
    Butt, Charles, 210–212, 211f
    Gillett Chemical Works, 265–266, 266f
    Redlich, Henry O., 389, 389f
    Sanford Manufacturing Co., 393–394, 393f
    Walker & Taylor, 443–446, 444f
  Peoria (IL)
    Farrell, Hiram G., 614–621, 617f
    Farrell & Cox, 622, 622f
flared ring below lip
  Chicago (IL)
    Perrine, Charles O., 377–381, 377f
flared rounded lips
  Chicago (IL)
    Gillett Chemical Works, 265–266, 266f
  Pontiac (IL)
    Caldwell & McGregor, 651, 651f
flared tapered pharmacy-style lip
  Quincy (IL)
    Schroeder, Herman, 685, 685f
flared-ring lips
  Bloomington (IL)
    Wakefield & Co., 137–150, 140f
flask-shaped bottles. See also whiskey flasks
  Chicago (IL)
    Wolgamott, Dr. George W., 455–456, 456f
flat lozenge-blop-top soda
  Danville (IL)
    Galtermann, William, 468, 468f
flattened blob top
  Nelson (IL)
    Allison, James J., 586, 586f
flat-topped oval-blob sodas
  De Kalb (IL)
    Dee, Richard, 52, 52f
Fleischman, John, 470, 470f
Flower, George, 85
Floyd, Edward A., 511–512, 512f
Floyd's American Penetrating Liniment, 30, 511–512, 512f
Fluid Extract Buchu, 262, 312
Fluid Lightning, 227–228, 227f
fluted bottles
  Chicago (IL)
    Chapin & Gore, 214–216, 216f
    Paul Rouze & Co., 391–393, 392f
  definition, 13, 13f, 23, 23f

Galesburg (IL)
  C. Brechwald & Co. (Galesburg), 507–508, 507f
Peoria (IL)
  William A. Callender & Son, 609
Flynn, James, 668
Flynn, John J., 668–671, 668–671f
Focus of study, 1–2
Foerster, Theobald, 258, 258f
Foerster's Teutonic Bitters, 79
Folsom, Charles A., 428
Food and Drug Administration, 278, 385
footprints and stars embossed. See Connell, J. M.
Fosselman, J. B., 718, 720
Fosselman's drug store (Springfield), 704, 720, 721f
Foster, Charles M., 259–260, 259f, 260f, 705–706
Foster's Child's Relief, 259, 705, 706
Foster's Indian Health Renewer, 259–260, 259f, 260f, 705, 706
4-bottle logo, embossed. See John A. Lomax (Chicago)
Francis, John H., 626, 626f
Francis, Lewis N., 626
Francis, Willis, 626
Francis & Spier (Peoria), 626, 626f
Franks, Himan, 215
  threaded internal stopper, 29, 70
Frankstown Glass Works (Pittsburgh)
  Chicago (IL), 214f, 215–216
  threaded internal stopper, 29
Dr. Frazier, 724
Frechette, John, 260–261, 261f
Frederick Lorenz, Jr., & Thomas Wightman (Pittsburgh), 55t, 66–67
Frederick R. Lorenz, Jr. (Pittsburgh), 55, 55t, 62, 63, 626
  Belleville (IL), 116
Frederick U. Roe & Sons (Franklin Grove), 498
Frederick Z. Lang (Chicago), 330
Freeburg Hotel (Freeburg), 499
Freeman, Dr. John D., 535–537f, 535–538
  Eclectic Advertiser, 536
Freeman & Co. (Springfield), 717
Freeman Graham & Co. (Rockford), 689
Freeman's Alterative Balsam and Mother's Relief, 537
Freeman's Anti-Ague and Fever Pills, 537
Freeman's Aperient Anti-Bilious Pills, 537
Freeman's Compound Emetic Powders, 537
Freeman's Cough Balsam, 535–538, 536f
Freeman's Diarhea Syrup or Cholora Mixture, 535–538
Freeman's Eclectic Liniment, 535–538, 537f
Freeman's Liver Pills, 537
Freeman's Sudorific or Sweating Drops, 535–538, 535f
Freeman's Sweet Balsam of Life, 537
Freeman's Vermifuge, 535–538, 535f
French square bottles
  as earliest pharmacy bottle design, 73
  study exclusion rule, 37, 38f
fruit (preserve) jars
  assemblage patterns, 75
  body shape definition, 18, 21f
  Chicago (IL)
    Liebenstein, Abraham L., 330–331, 331f
  Evanston (IL)
    Westerfield, John G., 492–493, 493f
    Wheeler & Bayless (Chicago), 452–453, 453f
Fuller, Charles H., 472, 707
Fuller, Finch & Fuller (Chicago), 131, 277
Fuller, Harvey B., 472, 707

Fuller, Oliver, 277
Fuller & Fuller (Chicago), 287
The Fuller Mfg. Co. (Chicago), 471–472, 706–707
Fuller's Banner Ink (Decatur), 471–472, 471f
Fuller's Banner Ink (Springfield), 471–472, 706–707, 706f
Fuller's Liquid Fish Glue, 472, 707
Funk, Francis M., 123–124, 124f
Funk & Lackey (Bloomington), 123–124, 124f, 129, 131, 189

G
G. & L. embossing, 714
G. C. Paoli & Co. (Chicago), 413
G. H. Hausburg (Blue Island), 158–160, 159f, 280
G. J. Arnold Bottler Supplies, 311
G. J. Zerwekh & Sons Co. (Pekin), 602
G. Lomax & Co. (Chicago), 334
G. N. Spannagel (East St. Louis), 481
G. W. Barnard, 431
G. W. Young & Co. (St. Louis), 179
Gabler, Franklin, 646
Gale, Edwin O., 261, 261f
Gale, William H., 261, 261f
Gale Brothers (Chicago), 261, 261f
Galloway & Snooks (Freeport), 500
Galtermann, William, 468, 468f
  Matthews gravitating stoppers, 27
Garden City Chemical Works (Chicago), 261–262, 262f, 312
Garden City Rheumatic Institute (Chicago), 455–456
Gardiner, Charles H., 262–263, 263f
Gardiner's Rheumatic & Neuralgia Compound, 262–263, 263f
Garrick, John, 263, 263f
Garrick & Cather (Chicago), 263, 263f
Garrison, Herod, 194
Garrison & Co. (Chicago), 194
Garwood, Samuel, slug-plate patent, 73
Gaskill, Tyler, 241
Geer, Schubael, 238, 264, 264f, 392
Geer and Dorman (Chicago), 238, 264
Geltmacher, John, 122
Geo. H. Betts & Co. (Alton), 87
Geo W. Smyth, Pittsburgh Ale, 94
George H. Betts & Co. (Alton), 86–87, 87f
George Hofmann & Brothers (Chicago), 292–293, 292f
George Powell & Co. (Chicago), 387–389, 388f, 389f
Gerhardt, Mrs. William, 3
Gerhardt, William, 559–560, 559f, 560, 560f
German Cologne, 262
German Itch Ointment, 579
German Liquid Polish, 255, 255f
Gesner, William G., 558, 558f
Gesner & Lipp (Lincoln), 558, 558f
Gettysburg Katalysine Water, 208
Geyser mineral water, 208
Gillet, Egbert W., 267–268, 267f, 268f
Gillet, McCulloch & Co. (Chicago), 266, 267–268, 267f, 268f, 313
Gillet, Paul W., 267, 267f
Gillett, Egbert W., 266
Gillett, Paul W., 265–266, 265f, 266f
Gillett Chemical Works (Chicago), 265–266, 265f, 266f
Gillig, Charles E., 626–627, 627f, 643
Gillig, Karl G., 613, 627–629, 628f, 635, 643
Gillig & Singer (Peoria), 626–627, 627–629, 627f, 628f
Gillig's Wine & Beer Hall (Peoria), 628f
Gilt-Edge Tonic, 609

gin
  Chicago (IL)
    Croskey, Abraham F., 230–231, 231f
    Henry H. Shufeldt & Co., 410–411, 411f
  Columbian Gin, 230–231, 231f
  Imperial Gin, 410–411, 411f
  Quincy (IL)
    Schrader, Frank, 684
ginger ale
  Arlington Heights (IL)
    Muller, Frederick W., 103–104, 103f, 104f
  Aurora (IL)
    Green Brothers, 106
  Beardstown (IL)
    Ehrhardt, Frederick W., 107–109, 107f, 108f
  Black Hawk Ginger Ale, 691, 693
  Carlinville (IL)
    Zaepffel, August, 174
  Chicago (IL)
    Christin, Arthur, 219–221, 219f, 220f, 221f
    Cook, Egbert C., 224
    Dempsey & Hennessey, 234–236, 235f
    Hausburg Brothers, 281–282, 281f
    John A. Lomax, 364, 365
    Lang Brothers, 325–328, 326f
  De Kalb (IL)
    Dee, Richard, 52
  embossed-glass labeling, 42
  Industrial Expansion era (1870s), 79
  Jacksonville (IL)
    Buell & Schermerhorn, 524
  Peoria (IL)
    Fellrath, Hubert, 622–625, 625f
    Hickey, Ransom E., 607, 633
    Singer, Charles J., 643
  Rock Island (IL)
    Carse & Lamont, 691
    Carse & Ohlweiler, 693
ginger beer
  Chicago (IL)
    Adams, Thompson W., 188, 188f
    Dr. Cronk, 229
Gipps, Cody & Co. (Peoria), 58, 629–630, 629f
Gipps, Howe & Co. (Peoria), 630
Gipps, John M., 629–632, 629f, 631f
Gipps & Co. (Peoria), 629–630, 630–632, 631f
Gipps & Shurtleff (Peoria), 630
"Glass and Glass Makers" (Seymour), 56
glass annealing lehr, 92
glass bottle dealers
  Chicago (IL)
    Bode, 55t, 56, 59
    Croskey, Abraham F., 230
  New York (NY)
    Matthews Co., 27, 56, 67
    Putnam, 56, 68, 348
glass bottle manufacturers. See glassworks
glass gravitating stoppers
  definition, 26–27, 28f
  "Glassmaker's Marks found on Bottles" (Whitten), 59
glassworks, 54–70, 55t, 61f, 65f. See also specific companies
  maker's marks, 54–70, 55t, 61f, 65f
  website database (Whitten), 47
Glen Flora springs (Waukegan), 208
Glen Flora Water, 208
Glidden & Co. (Springfield), 720
Globe Spice Mills (Springfield), 727
Glueck, John F., 506, 506f
Godfrey, Charles C., 422
Godfrey's Cordial, 30, 715
Goggin, William L., 449–450, 449f
Golden Gate pharmacy bottles, 73
Golden Lion (Springfield), 699

Gore, James, 214–216, 214f, 215f, 216f
Goss & Abbot (Chicago), 360, 361
Gottleib Wurster & Co. (Chicago), 312, 456–461, 457–460f
Grabs, William, 468, 469
Grafe, P. Herman, 512–513, 513f
Graham, Byron, 689
Graham, David, 689, 689f
Graham, Freeman, 689, 689f
Graham, Julius, 689
Graham Bros. Distillery (Rockford), 689
Graham's Distillery (Rockford), 689, 689f
Grand Opera House (Chicago), 278
Grand Pacific Hotel (Chicago), 241
Grape Bitters, 418, 418f
Grassly, Charles W., 387
gravitating stoppers. See glass gravitating stoppers
Gray, George, 679, 680
Great Lakes-Riverine area
  immigrant density, 30, 31f, 32f
  wealth distribution, 33f
Great Western Medical Depot (Bloomington), 152, 450
Great Western Wholesale and Retail Patent Medicine Depot (Chicago), 217
Green, George W., 105–106, 105f
Green, James W., 105–106, 105f
Green, William H., 190, 190f, 193, 194f
Green Brothers (Aurora), 105–106, 105f
Greenville Building and Savings Association (Greenville), 513
Griffith, Bush & Schrader (Quincy), 684
Griffith, Miss Hannah G., 713
Griffith & Schrader (Quincy), 684
Griggs, Edward Y., 595–597, 595f, 596f
Griggs, Oakley, 595
Griggs & Co. (Ottawa), 595
Griggs & Co. (Streator), 595
Griggs & Loveland (Brighton), 538
Griggs Sarsaparilla, 595
Grimshaw, John, 649
Grimshaw, Thomas C., 649–650, 650f
Grone, Caspar H., 667, 672, 672f
Grone, Catherine, 663
Grone, Henry, 663, 672
Grone, Henry, Jr., 672, 672f
Grone & Durholt (Quincy), 663, 672, 672f
Gross, Christian, 469–470, 470f
Gross & Fleischman (Darmstadt), 470, 471
Grossenheider, Julius, 68, 236, 268–270
Grossenheider, M., 269
Grossenheider & Dickinson (Chicago), 270
Grossenheider & Kruse (Chicago), 270
ground lips
  Chicago (IL)
    A. Leibenstein & Co., 330–331, 331f
    Perrine, Charles O., 377–381, 377f
ground sheared lips
  Bloomington (IL)
    Wakefield & Co., 137–150, 145f
Grubb & Lewis (Springfield), 714
Grube, Peter, 441–442
Guide to Health, and Family Physician (Freeman), 537
gun oil
  Chicago (IL)
    Bailey & Eaton (Chicago), 195–196, 196f
Gundlach, John, 465–466, 466f
Gundlach, Philip, 465
Gunn, John S., 174–175, 175f

H
H & Bro's (Chicago), 279–281, 280f, 281f
H. A. Stone & Co. (Chicago), 426–427, 427f

H. B. Mathews Sons (Chicago), 369
H. C. Myers & Son (Springfield), 721
H. Durholt & Co. (Quincy), 667, 673
H. E. Bucklen & Co. (Chicago), 208–210, 209f
H. E. Bucklen & Co. (St. Louis), 209
H. E. Rupert (Alton), 97, 97f
H. F. & Co. (Chicago), 255–257, 256f, 257f
H. F. L. Rodemeyer & Bro. (Chicago), 391
H. J. Witt (IA), 69
H. M. Wooster & Co. (Norwalk, OH), 217
H. W. Ashley, Lehman, Upham & Co. (Chesterfield), 538
H. W. Blood (Braidwood), 162–163, 162f
H. W. Blood & Co. (Wilmington), 162, 742–743, 742f
H. W. Blood & Sons (Braidwood), 162
Haas, Christian, 649
Haas, Kasper, 648–649, 648f
Hafner, William, 396–399, 396–399f, 404
Haines, W. S., 366
Hair Cream, 662, 662f
Haldeman, Benjamin, 565
Haldeman, Samuel, 565, 565f
Hale, Martin, 412
Haley, Charles C., 393
  patent, California Pop (pop beer), 586
Haley, James, 509
half-pint shoofly flasks
  LaSalle (IL)
    DeSteiger Glass Co., 548–549
Mr. Hall, 653
Hall, Edgar A., 271
Hall, Zebulon M., 271, 271f
Hall & Winch (Chicago), 360
Haller's Patent preserve jars, 331, 331f
Hamilton, George W., 529
Hamilton, Dr. William, 525–529, 525–529f
Hamilton & Co. (Belvidere), 120
Hamilton's Anti-bilious Pills, 529
Hamilton's Indian Liniment, 525–529, 529f
Hamilton's [Never Failing] Ague Tonic, 529
Hamilton's Syrup of Blackberry & Sassafras, 525–529, 527f, 528f
Hamilton's Vegetable Ague Pills, 529
Hamilton's Vegetable Cathartic Sugar-Coated Pills, 529
Hamilton's Vegetable Cough Balsam, 525–527f, 525–529
Hamlin, John A., 271–279, 272–279f
Hamlin, Lysander B., 275–279
Hamlin, William, 274
Hamlin Theater (Chicago), 278
Hamlin's Cough Balsam, 274, 278
Hamlin's Wizard Oil, 271–279, 272–279f
Handled Chestnut Flask
  body shape definition, 13, 14f
handmade characteristic
  and measurements, 10, 10f
Handy, John, 585–586, 585f
Handy, Thomas, 585
Hansen, P. L., 450
Hansen & Welch (Chicago), 450
Happy Home Blood Purifier, 246, 247f
hard-rubber gravitating-stopper blob-top closure
  Chicago (IL)
    John A. Lomax, 349, 350f
Mr. Harkness, 742
Harkness, E. J., 427
Harlem Oil, 30
Harles, Frank, 487, 487f
Harles, Henry, 487
Harries Dayton Ale, 94
Mr. Harris (pharmacist), 367
Hartmann, Charles, 314–315, 314f, 404

Hartt Manufacturing Co. (Chicago), 399
Hatfield's (Jacksonville), 522
Hathorn mineral water, 208
Haunschild, Henry, 407–408, 408f
Hausburg, Charles, 158
   Chicago (IL), 279–281, 280f, 281–282
   Elgin (IL), 490–491, 491f
Hausburg, Gustave H.
   Blue Island (IL), 158–160, 159f, 160, 160f, 161, 161f, 489
   Chicago (IL), 279–281, 280f, 281, 281f
   Elgin (IL), 490–491, 491f
Hausburg, William A., 158
   Chicago (IL), 279–281, 280f, 281–283, 281f, 282f, 283f
   Elgin (IL), 490–491, 491f
   Kelly-patent bottles, 29, 49
Hausburg & Dettmer (Blue Island), 158, 160, 160f, 280, 489
Hausburg & Ennis (Blue Island), 158, 160, 161, 161f
Hausburg & Innes (Blue Island), 280
Hausburg Brothers (Chicago), 158, 280–281, 281–282, 281f, 282f
Hausburg Brothers (Elgin), 158
Hausburg Brothers & Co. (Elgin), 490–491, 491f
Hawes, David, 695
Hawley Glass Co. (Hawley, PA), 63
   threaded internal stopper, 29
Hayden, Albert, 718
Hayes, Edwin, 476–477, 476f, 477f, 576
Hayes, Michael, 284–287, 284f, 285f, 286f, 287f
Hayes, Patrick, 284–287, 284f, 285f, 286f, 287f
Hayes, Thomas, 477, 477f
Hayes & Sullivan (Chicago), 287
Hayes Brothers (Chicago), 69, 284–287, 284f, 285f, 286f, 287f, 310
Headley, Dr. William S., 632, 632f
Heberer, Adam, 118
Heberer, Henry, 118
Heberer, Thomas, 118–119, 119f
Heck, John, 106, 106f
Hedlund, John M., 287–289, 288f, 289f
Hedlund & Co. (Chicago), 287–289, 288f, 289f
height, variations in, 10, 10f
Heinskill, Matthew, 363
Heiss, Henry (aka Hess), 180, 180f. See also Hess, Henry (aka Heiss)
Heiss & Hutter (Centralia), 175, 180, 180f
Heller, Abraham, 324–325
Heller, Adolph, 324–325
Heller Co. (New York), 324–325
Helmbold, Henry T., 42
Helmbold's Extract of Buchu, 42
Henderson, Thomas W., 191–192, 192f
Henley, Dr. William A., 289–290
Henn & Gabler (Chicago), 583
Hennessey, Thomas, 63, 233–234, 234–236, 235f, 290–291, 290f
Henry, David, 442–443, 442f, 443f, 490, 490f
Henry & Van Buskirk (Elgin), 490, 490f
Henry Durholt & Co. (Quincy), 664, 666–667, 666f, 677
Henry H. Shufeldt & Co. (Chicago), 410–411, 411f
Henry Korf & Co. (Springfield), 707–708
Herold, Ferdinand, 566–567, 567f, 568
Herrick, George L., 473–475, 474f, 496
Herrmann, F. W., 428
Herron, W. A., 622
Hess, Charles, 292
Hess, Ernst, 291–292, 291f, 404–405, 405f
Hess, Henry (aka Heiss), 175f. See also Heiss, Henry (aka Hess)
Hess & Co. (Carlyle), 175, 175f, 176, 180

Hess & Hutter (Carlyle), 175, 176, 180
Hewit & Co. (Upper Alton), 538
Hewitt, Dr. George, 455–456
Hickey, Alex, 613f
Hickey, Hiram M., 509, 539, 539f, 552, 607, 633
Hickey, John E., 509, 539, 539f, 540, 552, 607
Hickey, Ransom E., 125, 509, 611, 613, 627
   LaSalle (IL), 551–552, 551f
   Peoria (IL), 606–607, 607f, 632–635, 632–635f
Hickey, Stephen M., 121, 129, 135, 509, 552, 607, 633
   Bloomington (IL), 124–125, 125f
   LaSalle (IL), 552–553, 552f
Hickey Bottling Works (Peoria), 242, 509, 627, 629
Hickey Peoria Springs bottling factory, 635, 643
Hickey/Eaton "clan," 509, 539, 551–552, 607, 611, 635
Highland House and Highland Beer Saloon (Highland), 520
high-shouldered bottles
   Alton (IL)
      Weisbach, Christian, 98–100, 100f
      Yoerger, Augustin, 100–102, 102f
   Chicago (IL)
      Alfred & William Strickland Co., 428–429
      Barothy & Cook, 197, 197f, 198f
      Cronk, Edward Y., 228–230, 229f
      Farwell and Brace, 255, 255f
      Gillett Chemical Works, 265–266, 265f
      Gottleib Wurster & Co., 458–461, 460f
      H & Bro's, 279–281, 280f
      Hayes Brothers, 284–287, 284f
      Hutchinson & Co., 307, 308, 308f
      John A. Lomax, 349, 350f
      Keeley, Michael, 317–320, 320f, 321f
      Lomax, George, 332–334, 332f, 333f
      Miller, Andrew J., 372–373, 372f
      Paul Rouze & Co., 391–393, 393f
      Sass & Hafner, 396–399, 398f
      Victor Barothy & Co., 196–197, 197f
      Warner, Dr. C. D., 446–449, 446f, 447f
   Galesburg (IL)
      Edwards, [Isaac?] R., 510, 510f
   Jacksonville (IL)
      Buell & Schermerhorn, 522, 523–525, 524f
      Schermerhorn, Charles, 533–534, 533f
   Joliet (IL)
      Paige, J. D., 540–542, 541f, 542f
   Naperville (IL)
      Keith, Aylmer, 579
   Nelson (IL)
      Allison, James J., 586, 586f
   Ottawa (IL)
      Finkler, Alexander, 594–595, 594f
   Peoria (IL)
      Fellrath, Hubert, 622–625, 623–625f
high-shouldered cylinders
   Chicago (IL)
      Moench & Reinhold, 373–374, 374f
      Paul Rouze & Co., 391–393, 393f
      Sykes, Charles R., 432–434, 433f
      Warner, Dr. C. D., 446–449, 446f, 447f
   Peoria (IL)
      Headley, Dr. William S., 632, 632f
high-shouldered sodas
   Aurora (IL)
      Green Brothers, 105–106, 105f
   body shape definition, 15, 17f
   Lemont (IL)
      Bolton, John G., 554–557, 555f
Hilbert, William, 134
Hill, William P., 125–126, 126f
Hill & Vanatta (Bloomington), 125–126, 126f

Mr. Hille, 733–734, 733f, 734f
Hine, W. H., 630
Hine & Co. (Peoria), 630
Historic Glass Bottle Identification and Information Website (Lindsay), xvii, 59, 71
History of Chicago (Andreas), 359, 362
Hobbs, S. J., 362
Hockenhull, John, 530–531, 530f, 531f
Hockenhull, King & Elliott (Jacksonville), 531
Hockenhull, Robert, 530–531, 530f, 531f, 717
Hockenhull & Young (Jacksonville), 531
Hockenhull's Extract of Ginger, 531
Hoffman, Frederick C., 664
Hoffman, H. H., 649–650
Hoffmann, Christian, 730, 730f
Hofman, George, Sr., 293
Hofmann, Alois, 292–293
Hofmann, George, Jr., 292–293, 292f
Hofmann, Valentine, 292–293
Hogeboom, Samuel, 293
Hogeboom, Wolf, & Co. (Chicago), 293
Hogg, William, 132
Holton, W. A., 725
Holton's Hair Restorer, 725
Honoré, Bertha, 376
Hopkins, George K., 96–97
Hopkins, Horace, 298
Horr, James, 363
Horse Ointment, 85
Horsey, Edwin H., 220
"Hotel Oval" Flask
   body shape definition, 13, 14f
Howe, Bliss S., 126–128, 155, 156–157
Howe, H., 630
Hoy, Luman (L. T.), 743
Hoyt, Pierce & Co. (Chicago), 431
Mr. Hubbard, 740
Hubert, Anthony, 175–176, 175f, 176
Huck, John, 436
Huckhardt, J. M., 528
Humboldt House hotel (Chicago), 315
Humboldt's German Bitters, 387, 387f
Humiston, Lenson S., 419, 697–698, 697f
Humphreys, E. J., 610, 635–636, 635f
Hunt, Robert L., 293–296, 293f, 294f, 295f
Hunt, Thomas M., 390
Hunt & Kenyon (Chicago), 293
Hunt & Salpaugh (Chicago), 293, 295–296
Hurlbut, Horace A., 390
Hurlbutt & Edsall (Chicago), 391
Hutchings, William A., 296–297, 296f
Hutchins, John W., 296, 296f
Hutchins Magnetic Oil, 296, 296f
Hutchinson, Andrew J., 297, 298
Hutchinson, Charles G., 297, 309–310, 310f
   bottle-stopper patent, 36f, 71, 309–310, 310f
   Hutchinson-style-bottle filling machine, 311
   star (maker's mark), 69
Hutchinson, George C., 303, 309, 309f
Hutchinson, Jane, 297
Hutchinson, Joseph, 56
Hutchinson, Joseph M., 297
Hutchinson, William A., 297, 303, 309
Hutchinson, William H., 56, 60, 297–311, 335
Hutchinson & Co. (Chicago), 297, 299f, 300
Hutchinson & Co. (Pittsburgh, PA), 303
Hutchinson & Dunn (Chicago), 303, 303f
Hutchinson & Sons (Chicago), 287, 309–310, 310f, 314, 404
Hutchinson-style internal stoppers
   Alton (IL)
      Max Kuhl & Co., 89
   Blue Island (IL)
      Hausburg, Charles, 282

bottling industry changes, 310–311
  Chicago (IL)
    Christin, Arthur, 221
    Hayes Brothers, 287
    Hennessey, Thomas, 291
    John A. Lomax, 348
    Morrison, William, 51
  as chronological marker, 71
  expanded use of, post-study, 35, 36*f*
  and lip style chronology, 24
  patent date, 36*f*, 309–310, 310*f*
  W. H. Hutchinson & Sons (Chicago), 309–311, 310*f*
Hutchinson-style soda bottles
  Beardstown (IL)
    Ehrhardt, Frederick W., 109
  Bloomington (IL)
    Kreis, Peter, 129
  Blue Island (IL)
    G. H. Hausburg, 158
  Braidwood (IL)
    Blood, Horace W., 162
  Cairo (IL)
    Kump, John H., 168
  Carlinville (IL)
    Zaepffel, August, 174
  Chester (IL)
    Wenda, John, 185
  Chicago (IL)
    A. Mette & Bro., 371
    Gottleib Wurster & Co., 460
    John A. Lomax, 366
    Seibt, Frederick (Fritz), 409
    Stenson, James, 426
  Colehour (IL)
    Kassens, Henry C., 463
  Danville (IL)
    Shean, Charles, 469
  Decatur (IL)
    Kuny, Fredrick D., 472
  East St. Louis (IL)
    Schroeder, Edward, 486
  expanded use of, post-study, 35, 36*f*
  Highland (IL)
    Mueller & Beck, 518
  Jacksonville (IL)
    Schermerhorn, Charles, 533
  Kankakee (IL)
    Kammann, [Dietrich] H., 545
  Lebanon (IL)
    Reuter, Charles J., 553
  Lemont (IL)
    Bolton, John G., 557
  Litchfield (IL)
    Weber, Frederick, 562
  Mascoutah (IL)
    Koob, August G., 568
  Ottawa (IL)
    Finkler, Alexander, 595
  paper label–free, 41*f*
  Peoria (IL)
    Eaton, James E., 614
    Fellrath, Hubert, 625
    Singer, Charles J., 643
  Springfield (IL)
    Charles J. Peterson Co., 712
    Johnson & Peterson Co., 711
  Staunton (IL)
    Stille, Rudolph, 731
  study exclusion rule, 37
  Waterloo (IL)
    Boeke, Henry, 735
Hutmacher, Edward, 667, 673–674, 673*f*
Hutmacher, Rudolph ("Rud"), 674
Hutmacher & Kreitz (Quincy), 674

Huttenlocher, John, 311–312, 311*f*, 460
Hutter, William, 175, 180, 180*f*
Huyck, J. H., jr., 262
Huyck, John H., 261–262, 262*f*, 312, 312*f*
Huyck & Knox (Chicago), 262, 312
Huyck & Randall (Chicago), 262, 312, 312*f*
Huyck Bros. & Esmay (Chicago), 262
Huyck's Full Measure Triple Flavoring Extracts, 261–262, 262*f*
Huyck's Standard Flavoring Extracts, 262
Hygean Spring Seminary (Western Saratoga), 741–742

I
I. & G. W. Lackey (Bloomington), 131
Ihmsen, Christian, 55*t*, 64
Ihmsen Glass Co. (Pittsburgh), 55*t*, 64, 450
Illinois Glass Co. (Alton, IL), 55*t*, 58, 63–64, 65*f*, 86, 89–92, 90*f*, 91*f*
  100th anniversary replica bottles, 100–102
  Beardstown (IL), 107*f*, 109
  East St. Louis (IL), 480*f*, 481
  Greenville (IL), 513, 513*f*
  Normal (IL), 151
  Waterloo (IL), 735, 735*f*
Illinois Hominy and Spice Mills (Springfield), 726–727, 726*f*, 727*f*
Illinois Infirmary (Charleston), 184
Illinois Trust & Savings Bank (Chicago), 241
Ilsley, John C., 396
immigrant densities
  and bottlers, 6*f*, 29–30, 31*f*, 32*f*
Imperial Gin, 410–411, 411*f*
Imperial Ink, 375
Indian Balm, 554, 554*f*
Indian Life Pills, 662, 662*f*
Indian Medical Depot (Chicago), 221–222
Indian Salve, 725
Indianapolis Glass Works Co. (IN), 47, 55*t*, 64
Industrial Expansion era (1870s), 34, 79–80
Inglis, Fred, 92–93, 92*f*
Inglis & Lowe (Alton), 92
Inhaling Fluid, 434–436, 434*f*
inks
  Bloomington (IL)
    J. W. Maxwell & Co., 132–133, 132*f*, 133*f*
  body shape definitions, 18, 19*f*
  Chicago (IL)
    Gillett Chemical Works, 265–266, 265*f*
    Northwestern Chemical Co., 69
    Northwestern Chemical Manufacturing Co., 375, 375*f*
    Redlich, Henry O., 389, 389*f*
    Sanford Manufacturing Co. (Chicago), 393–394, 393*f*
    Thomas, Levi H., 439–440
  Cox, Lewis, 226–227, 226*f*
  Decatur (IL)
    Fuller's Banner Ink, 471–472, 471*f*
  Elgin (IL)
    Elgin Ink Co., 489–490, 490*f*
  Pontiac (IL)
    Caldwell & McGregor, 651, 651*f*
  Quincy (IL)
    Miller, George A., 682–683, 682*f*, 683*f*
  Springfield (IL)
    Fuller's Banner Ink, 471–472, 706–707
inside screw-thread lip
  Chicago (IL)
    Wolford, Jacob A., 453–455, 454*f*
Instant Relief, 413, 414*f*, 417, 445
Instantaneous Relief, 257–258, 257*f*
internal screw threads
  Chicago (IL)
    Chapin & Gore, 214–216, 214*f*, 215*f*

    Drake, John B., 240–241, 240*f*, 241*f*
    Kirchhoff Brothers, 322–323, 322*f*
    Palmer House Hotel, 376–377, 376*f*
Ioerger. *See* Yoerger, Augustin
Ira Lackey & Brother (Bloomington), 123, 131, 132
Ira N. Malin & Son (Quincy), 684–685
Iron Spring water, Perry Springs, 648
iron swing stopper. *See* wire swing stoppers
Ironworkers' Addition (Chicago), 463
Irving's Giant Pain Curer, 446
Israel, Stephen, 312–313, 313*f*
IXL Bitters, 289–290

J
J. & A. Vatter (Monee), 572, 572*f*, 573*f*
J. & H. Hickey (Joliet), 539, 539*f*
J & J. F. Trippe & Co. (New York), 217
J. A. Hamlin & Bro. (Chicago), 278
J. A. Montgomery & Co. (Chicago), 419, 698
J. B. Bolton & Co. (Lemont), 557
J. B. Brown & Co. (Springfield), 699, 699*f*
J. B. St. Clair & Co. (Peoria), 641, 641*f*
J. Bollin & Co. (Lincoln), 558–559, 558*f*
J. Bryant & Co. (Rochester, NY), 187
J. C. Barnes & Co. (Chicago), 259, 705
J. D. Bastow's Cough Cure, 603, 603*f*
J. D. Vail & Co. (St. Louis), 541
J. E. Bradbury & Co. (Jacksonville), 521–522
J. E. Eaton & Co. (Chicago), 241–242, 241*f*
J. E. Eaton & Co. (Peoria), 613
J. Fife & Co. (Mattoon), 568, 568*f*
J. G. Nattinger (Ottawa), 595
J. Grossenheider & Co. (Chicago), 268–269
J. H. Huyck (Chicago), 262
J. H. Huyck & Co. (Chicago), 262
J. H. Reed & Co. (Chicago), 188, 206, 390–391, 390*f*, 436
J. I. Beaumont, 213
J. J. Flynn & Co. (Quincy), 668–671, 668–671*f*
J. Jeffley (Belvidere), 120
J. Johnson & Co. (Springfield), 707–709
J. Kirchhoff Brothers & Co. (Chicago), 323
J. M. Menkhaus (Carlyle), 176, 176*f*
J. W. Maxwell & Co. (Bloomington), 132–133
Jacksonville Bottling Works (Jacksonville), 524, 533
Jacques Chemical Co. (Chicago), 313
James Hammond (Vienna), 544
James Morgan (Murphysboro), 544
Jansson, Cornelius A., 704–705
Jaques, Atwood & Co, 313
Jaques, Forester F. (aka Frank), 313
Jauncey, William J., 432
Jaundice Bitters, 221–222, 222*f*
Jerseyville Ale and Porter Depot (Jerseyville), 534–535
Joerger. *See* Yoerger, Augustin
John A. Lomax (Chicago), 335–367, 336–339*f*, 341–347*f*, 349–355*f*, 357–360*f*, 363*f*, 365*f*, 366*f*
  brand names, embossed, 347–348
  buying out A. Mette & Bro., 371
  fires, 362–364, 366–367
  "Largest Bottling House in the U.S.", 1, 34, 359, 360*f*, 365
  overview, 335
John Anderton & Co. (Chicago), 191, 192*f*
John B. Drake & Co. (Chicago), 241
John Cairns & Co. (East St. Louis), 170, 486
John Cairns & Co. (St. Louis), 170, 486
John E. Brakbury & Co. (Jacksonville), 203
John H. Brown & Bro. (Springfield), 699
John Heck's Chamomile Tonic, 106, 106*f*
John Hodges (Thebes), 544

John Johnson & Co. (Keokuk, IA), 709
John N. Lipp & Brother (Lincoln), 558
John W. Harries, 191
John W. Maxwell & William Hogg (Bloomington), 132
Johnson, Dr. [Hosmer A.], 313–314, 313f
Johnson, Isaac, 288–289
Johnson, John, 135, 188, 707–712, 707–712f
Johnson, John W., 713
Mr. Johnson (pharmacist), 367
Johnson & Burk (Keokuk, IA), 709
Johnson & Cox (Peoria), 622
Johnson & Peterson Co. (Springfield), 59, 708, 709–712, 709–712f
Joseph Apple & Co. (Chicago), 192, 193f
Joss, John, 314–315, 314f, 398, 403–404, 404f
Joss & Hartmann (Chicago), 314–315, 314f
Joss & Taylor (Chicago), 314
JPF monogram, 705
Julius L. Barnstack (Edwardsville), 538
Junge, Julius, 694, 694f
Junge & Beck (Rock Island), 694, 694f
Just, Samuel, 181

K
Kadgihn, Otto, Sr., 128
Kadish, Charles J., 381, 381f
Kagy, Aaron A., 597
Kagy, John M., 597, 597f
Kagy & Trumbo (Ottawa), 597
Kagy's Superior Stomach Bitters, 597, 597f
Kaiser, Daniel, 674–675, 675f, 680
Kalb, Absalom J., 675–677, 675f, 676f
Kalb's Eye Water, 675–677, 675f
Kalb's Vegetable Quick Relief, 675–677, 676f
Kammann, [Dietrich] H., 67, 545–546, 546f
   Matthews gravitating stoppers, 27
Kane, Thomas, 315, 315f
Kaspar, Amihien, 363
Kassens, Henry C., 462–463, 462f
Kassens, Mrs. Maria, 463
Kearns, Herdman & Gorsuch (Zanesville, OH), 55t, 64, 389
   Murphysboro (IL), 576–577, 576f
   Prentice (IL), 652f, 653
Keeley (Chicago), 69
Keeley, Michael, 93–96, 93f, 315–320, 315f, 316f, 319–321f
Keeley, Thomas, 93–96, 93f, 315–317, 315f, 316f
Keeley & Brother (Alton), 93–96, 93f, 191
Keeley & Brother (Chicago), 93, 96, 191, 315–317, 315f, 316f, 320
Keeley Bottling Works (Chicago), 317
Keeley Brewing Co. (Chicago), 319
Keith, Aylmer, 579–580, 580f
Keith, Charles W., 580
Keith, William F., 580
Keith & Kimball (Naperville), 580
Keith's Persian Liniment, 30, 579–580, 580f
Kelleher, Jeremiah A., 712–713, 712f
Kelleher & Co. (Springfield), 712–713, 712f
Keller, Jacob, 584
Kelly, William H.
   gravitating stopper patent, 29, 49, 283, 283f
Kelly patent gravitating stoppers
   Chicago (IL)
      Gottlieb Wurster & Co., 458–461, 460f
      Hausburg, William A., 282–283, 283f
      Lang, Frederick C., 48–49, 49f
   definition, 26, 28f, 29
Kelly-patent Codd-style gravitating-stopper sodas
   Chicago (IL)
      Gottlieb Wurster & Co., 456–461, 460f
      Hausburg, William A., 282–283, 283f

Kentucky "Blue Lick" Water, 208
Kenyon, David, 293–294, 294–295, 294f, 295f
Keokuk Soda Water Manufactory (Keokuk, IA), 709
Kershaw, Albert, 3, 522–523, 524, 531, 532, 532f, 533
Kerwin, Michael, 319
   Richardson, Dr. Holland W., 743–744, 744f
Keystone House (Chicago), 259, 705
Keystone pharmacy bottles, 73
Keystone Tonic Bitters, 473–475, 496, 474f
kick-ups. See also central base kick-up; deep basal kick-up
   shape definitions, 22
Killian, John, 523, 523f
King (Chicago), 69
King, Thomas, 320–322, 321f
King, Dr. Z. L. (Elkhart, IN), 209
Kirchhoff, F. Gustavus, 322–323, 322f
Kirchhoff, H. August, 322–323, 322f
Kirchhoff, John Diedrich, 322–323, 322f
Kirchhoff, Julius, 322–323, 322f
Kirchhoff Brothers (Chicago), 322–323, 322f
   threaded glass stoppers, 29
Kirk & Adams Co. (Chicago), 186
Kirkpatrick, Dr. J. E., 660
Kirkpatrick family pottery, 50, 373
Kissengen mineral waters, 88
   Chicago (IL)
      John A. Lomax, 364
Kissinger mineral waters
   Quincy (IL)
      Schroeder, Herman, 685
Mr. Klohr, 472, 472f
Knackstead, William, 323, 323f
Knauf, Adam, 658, 659, 667, 668
Knemoeller, Henry, 731
Knemoeller, Rudolph, 730–731, 731f
Knemoeller & Stille (Staunton), 730–731, 731f
Knickerbocker Mineral & Soda Water, 418–419, 418f
Knickerbocker pharmacy bottles, 73
Knox, Charles M., 262, 312
Koch, Charles, 47
Koemsker, Charles, 466
Koob, August G., 46, 46f, 112, 119, 566–567, 567–568, 568f
Koob, August G., Jr., 568
Korf, Henry, 707–708
Korf & Peterson (Springfield), 708
Kreis, Peter, 121, 125, 129, 129f, 135
Krohe, Fred, 110
Krohe, Henry W., 109–110, 109f
Kruse, Henry, 269, 270, 270f
Kruse, Dr. Hero, 636–637, 636f
Kuehle, Frederick A., 576–577, 576f
Kuehle, Frederick A. C., 576
Kuehle & Son (Murphysboro), 576
Kuhl, Max, 88–89
Kump, John H., 167–168, 167f
Kuny, Fredrick D., 472–473, 472f, 473f
Kuny & Klohr (Decatur), 472, 472f

L
L. B. McCreary (Metropolis), 544
L. Eliel & Co. (Chicago), 550
L. Eliel & Co. (LaSalle), 549–550, 549f, 550f
Labbe, Charles, 172–173
Labbe & Son (Carlinville), 172–173
Lackey, George W., 129–131, 130f, 131f
Lackey, Ira, 123–124, 124f, 129–131, 130f, 131f
Lackey, John A., 129–131, 130f, 131f
Lackey and Brothers (Bloomington), 129–131, 130f, 131f
Lackey's Iron Bitters, 129–131, 130f, 131f

Lackey's Justly Celebrated Blackberry Cordial, 123–124, 124f
lady's leg neck shape
   Cairo (IL)
      Lohr, Andrew, 168–170, 168f, 169f
   Chicago (IL)
      DeMor & Co., 232, 232f
      John A. Lomax, 355f, 356
      Steel & Co., 421–422, 421f
   definition, 23, 24f
   Springfield (IL)
      Fixmer, John P., 705
lager and weiss beer
   Chicago (IL)
      J. Grossenheider & Co., 268–269, 269f
      John A. Lomax, 356–358, 357–359f, 364, 365
   LaSalle (IL)
      Eliel, Louis, 549–550, 549f, 550f
   Naperville (IL)
      Stenger, John, 582–584, 584f
   Peoria (IL)
      Central Brewery, 646
      Parent, Anthony, 639
   Rock Island (IL)
      Carse & Ohlweiler, 693
   Springfield (IL)
      Kelleher, Jeremiah A., 712
Lamb's store (Springfield), 724
Lammers, H., 646
Lamont, George, 690–691, 690f
Lampe, Lewis(?), 658, 677, 677f
Lampe & Boschulte (Quincy), 677, 677f
Lancaster Glass Works (NY), 56, 298, 348
Landsberg, Moses G., 323–325, 324f, 325f
Landsberg's Century Bitters, 323–325, 324f, 325f
Landsberg's Pure Blackberry Brandy, 324–325, 325f
Lang, August, 48, 48f, 325–328, 326f, 327f, 330
Lang, Frederick C., 48–49, 48f, 49f, 325–328, 326f, 327f, 328–330, 328–330f
   Kelly-patent bottles, 29
Lang, John A., 327
Lang, William, 48, 48f, 325–328, 326f, 327f
Lang Bros. & Miller (Chicago), 327
Lang Brothers (Chicago), 48, 48f, 325–328, 326f, 327f, 330
"Largest Bottling House in the U.S.", 1, 34, 359, 360f, 365
Larson, Anders, 678. See also Lawson, Andrew
larynx node ulcers, 30
Lauterbach & Reisch (Springfield), 712
Lawson, Andrew, 430
Lawson & Sundell (Chicago), 430
Lea, Charles G., 96–97
leather treatment
   Chicago (IL)
      Farwell and Brace, 255, 255f
      Sloan's Tannin Paste, 416
Lee, Dr. Charles M., 575–576, 575f
Lee's Pills, 30
Mr. LeGuere (pharmacist), 367
Lehman, Frederick, 409–410
Lemke, Albert J., 393
lemon beer. See small beer
Lemp, William
   St. Louis brewery-bottling industry, 35
Leonard, Richard H., 213
LeRoy, Dr. David, 574–575, 574f
Levis, Edward, 89
Levison's Red Drops, 205
Lewis, J. K., 713–714
Lewis, Joseph, 716
Lewis, Thomas, 713–714, 714f
Lewis' Everybody's Liniment, 30
Lichtenthaler, George W., 153

Liebenstein, Abraham L., 330–331, 331*f*
Liebenstein, Joseph, 331
Life Ointment, 222
Life Syrup, 414, 415*f*, 417
Lightning stopper
  study exclusion rule, 37
Lincoln, Abraham, 132, 720
Lincoln Monument poem, 719
Lindeborough Glass Works (NH), 68
Lindell Glass Co. (St. Louis), 55*t*, 64, 66
  St. Louis brewery-bottling industry, 35
Lindsay, Alexander, 75, 714–717, 715*f*, 716*f*
  as earliest embossed bottle, 29, 715
Lindsay, Bill, 59, 71
Lindsay, Morris, 75, 714–717, 715*f*, 716*f*
  as earliest embossed bottle, 29, 715
Lindsay's Nerve and Bone Pills, 716
Lipp, John N., 558, 558*f*
lipping tool
  and lip style chronology, 24, 25*f*
  use as chronological marker, 71–72
Liquid Tonic, 429
liquor
  Chicago (IL)
    Thomas & Co. (Chicago), 440–441, 440*f*
    J. H. Reed & Co. (Chicago), 390–391, 390*f*
    Uncle Marb's Old Bourbon Bitters, 651–653, 652*f*, 653*f*
"Little Egypt," Illinois area, 544
Lockport Glass Works (NY), 55*t*, 64, 66, 298, 348
Lohr, Andrew, 47*f*, 166, 167, 168–170, 421
  variable neck lengths, 10*f*
Lomax, Ann, 294, 334
Lomax, George, 189, 201, 248, 249–250, 249*f*, 332–334, 332*f*, 333*f*, 335, 359–361, 557
Lomax, George, Jr., 334
Lomax, James, 334
Lomax, James H. L., 363
Lomax, James L., 363
Lomax, John A., 188–189, 201, 239, 297, 334, 335–367, 392, 557
  full-pint sodas, 48
  gravitating stoppers, 29
  Industrial Expansion era (1870s), 79
  "Largest Bottling House in the U.S.", 1, 34, 359, 360*f*, 365
Lomax & Meagher (Chicago), 334
London, England
  Alexander & Austin, 58
  Alfred Alexander & Co., 55*t*, 58, 69, 366
  glass factories, 55*t*, 58
long necks
  Alton (IL)
    Yoerger, Augustin, 100–102, 102*f*
  Belleville (IL)
    Fisher, Joseph (aka Fischer), 113–115, 114*f*
    Thomas Heberer and Brothers, 118–119
  Bloomington (IL)
    Wakefield & Co., 137–150, 142–144*f*
    Woodard & Howe, 156–157, 157*f*
  Cairo (IL)
    Kump, John H., 167, 167*f*
    Lohr, Andrew, 168–170, 168*f*, 169*f*
    Schutter, William H., 171, 171*f*
  Carrollton (IL)
    Smith, Dr. Alexander H., 176–179
  Chicago (IL)
    Alfred & William Strickland Co., 428–429
    Butt, Charles, 210–212, 211*f*
    Cook & Co., 224–225, 224*f*
    DeMor & Co., 232, 232*f*
    Dinet, Joseph, 237–238, 237*f*, 238*f*
    Gardiner, Charles H., 262–263, 263*f*
    Gillett Chemical Works, 265–266, 266*f*
    Hamlin, John A., 271–279, 272*f*, 274*f*
    Hess, Ernst, 291–292, 291*f*
    Hutchings, William A., 296–297, 296*f*
    Hutchinson & Sons, 310, 311*f*
    J. Grossenheider & Co., 268–269, 269*f*
    John A. Lomax, 345, 346*f*, 347, 347*f*
    Landsberg, Moses G., 323–325, 324*f*, 325*f*
    P. W. Gillet & Son, 267, 267*f*
    Paul Rouze & Co., 391–393, 392*f*
    Schonwald, Francis (Franz), 402–403
    Schulz & Hess, 404–405, 405*f*
    Thomas & Co., 440–441, 440*f*
    Warner, Dr. C. D., 446–449, 446*f*, 447*f*
  DuQuoin (IL)
    Hayes, Edwin, 476–477, 476*f*
  Galesburg (IL)
    C. Brechwald & Co. (Galesburg), 507–508, 507*f*
    Edwards, [Isaac?] R., 510, 510*f*
  Jacksonville (IL)
    Hamilton, Dr. William, 525–529, 529*f*
  LaSalle (IL)
    L. Eliel & Co., 549–550, 550*f*
  Lemont (IL)
    Bolton, John G., 554–557, 556*f*
  Litchfield (IL)
    Weber, Frederick, 560–563, 561*f*
  Pekin (IL)
    Zerwekh, Gottleib J., 600–602, 601*f*
  Peoria (IL)
    Fellrath, Hubert, 622–625, 623*f*
    Hickey, Ransom E., 632–634*f*, 632–635
    J. B. St. Clair & Co., 641, 641*f*
    Kruse, Dr. Hero, 636–637, 636*f*
  Pittsburgh (PA)
    W. H. H., 303, 303*f*
  Princeton (IL)
    Althoff, Friedrich, 653–654, 654*f*
  Quincy (IL)
    M. R. & H. W. Lundblad, 677–681, 677*f*, 678*f*
  Rockford (IL)
    Graham's Distillery, 689, 689*f*
  Springfield (IL)
    Diller, Roland W., 700–702, 700*f*
    J. B. Brown & Co., 699, 699*f*
    J. Johnson & Co., 707–709, 707*f*
    Lewis, Thomas, 713–714, 714*f*
    Myers, Henry C., 721
long necks with central ring
  Springfield (IL)
    Myers, Henry C., 721
"Long Nine"
  North, Alfred A., 722
long tapered necks
  Springfield (IL)
    J. B. Brown & Co., 699, 699*f*
Lord, James J., 717–719, 717*f*, 718*f*
  "patent medicine poet," 717, 719
Lord, Thomas, 277
Lord & Co. (Springfield), 718
Lord & Smith (Chicago), 277
Mr. Lorens, 613, 635
Lorentz & Co. (Peoria), 613, 627, 635
Lorenz, Frederick, Jr., 56
Lorenz, Frederick, Sr., 56, 67
Lorenz, Moses, 62, 67, 69
Lorenz, Wightman & Co. (Pittsburgh), 547
Lorenz & Wightman (Pittsburgh), 55*t*, 62–63, 66–67, 69, 626
  Alton (IL), 97
  Chicago (IL), 247, 313, 313*f*, 387*f*, 388
  closure of, 603
  Peoria (IL), 603, 604*f*
Louis Eliel Brewery (LaSalle), 549–550
Louis Rodemeyer & Co. (Chicago), 391, 391*f*
Louisiana (MO)
  M. R. & H. W. Lundblad accounts, 681
Louisville Glass Works (KY), 64, 66, 348
Low, Edward, 361
low shoulder bottles
  Chicago (IL)
    John A. Lomax, 355*f*, 356
lozenge-shaped blob-top
  Chicago (IL)
    Gottleib Wurster & Co., 456–461, 459*f*
  Danville (IL)
    Galtermann, William, 468, 468*f*
  Litchfield (IL)
    Weber, Frederick, 560–563, 562*f*
lozenge-style oval blob-top
  Peoria (IL)
    Gipps, Cody & Co., 629–630, 629*f*
lozenge-top bottles
  Belleville (IL)
    Stoltz, Francis (Franz), 119, 119*f*
  Chicago (IL)
    Doty, Harvey C., 239, 239*f*
    definition, 24, 25*f*
  Springfield (IL)
    Slemmons & Conkling, 726–727, 726*f*
Lundblad, Caroline, 679
Lundblad, Henry W., 677–681, 677*f*, 678*f*
  earliest embossed soda bottles, 677–678
Lundblad, Mathias R., 677–681, 677*f*, 678*f*
  earliest embossed soda bottles, 677–678
Lung Syrup, 222
Lurton, Dr. Lycurgus L., 367, 367*f*
Lurton & Stewart (Chicago), 367
Lutt, Christian, 479–481, 479*f*, 480*f*, 538
Lutz, N. L.
  paper-labeled pharmacy bottle, 30*f*

M

M. Brand & Co. (Chicago), 202–203, 204*f*
M. R. & H. W. Lundblad (Quincy), 66, 677–681, 677*f*, 678*f*
  business details, 679–681
  earliest embossed soda bottles, 677–678
Magic Oil, 275
Magnesia Spring water, Perry Springs, 648
magnetic fluid
  Chicago (IL)
    Christie, Dr. Abel H., 216
    Hutchins, John W., 296, 296*f*
Magnolia Bitters, 92–93, 92*f*
Main "Poland" Water, 208
Majestic Theatre (Chicago), 215
Major, Dr. John M., 131–132
maker's marks, 54–70, 55*t*, 61*f*, 65*f*
Malin, Ira N., 684
Malin, Jas. D., 684
Malin & Schrader (Quincy), 684
malt liquor
  Peoria (IL)
    Pletscher, John C., 640
Mammoth Tea House (Chicago), 439
"man or beast," 30, 580, 618, 619*f*, 650, 675, 713, 716
Manhatten pharmacy bottles, 73
Mansion House Inn (Williamsville, NY), 297
manufacturing-technology terms, 11–12
Manzanita Bitters, 259, 427–428, 427*f*, 705
Marienbad mineral waters, 88
Married Woman's Private Medical Assistant (Brown), 600
Mars Cologne, 208
Mason, Parker R., 367–368, 367*f*
Masonic Temple (Chicago), 455

master inks
  Bloomington (IL)
    J. W. Maxwell & Co., 132–133, 133f
  body shape definition, 18, 19f
  Decatur (IL)
    Fuller's Banner Ink, 471–472, 471f
  Pontiac (IL)
    Caldwell & McGregor, 651, 651f
  Springfield (IL)
    Fuller's Banner Ink, 706–707, 706f
Mathews, David, 369
Mathews, Edwin B., 368–369
Mathews, H. B., Jr., 368–369
Mathews, Henry B., 368–369, 368f
Mattei, Angelo, 392
Matthew Rinehardt & Co. (Springfield), 725–726, 725f
Matthews, Henry B. *See* Mathews, Henry B.
Matthews, John
  gravitating stoppers, 27
Matthews gravitating stoppers
  Alton (IL)
    Jacob Buff and Max Kuhl, 88–89, 89f
  Barothy & Cook (Chicago), 47–48
  Chicago (IL)
    John A. Lomax, 345
  definition, 26–27, 28f
  Kankakee (IL)
    Kammann, [Dietrich] H., 545–546, 546f
  Olney (IL)
    Bower, William, 591–592, 591f
Matthews-style flat-topped-blob soda
  Alton (IL)
    Jacob Buff and Max Kuhl, 88–89, 88f, 89f
  Chicago (IL)
    Barothy & Cook, 47–48
  Danville (IL)
    Galtermann, William, 468, 468f
  Kankakee (IL)
    Kammann, [Dietrich] H., 545–546, 546f
  Nelson (IL)
    Allison, James J., 586, 586f
  Olney (IL)
    Bower, William, 591–592, 591f
Matthies, Adolph L., 637, 637f
Maxwell, Samuel A., 132
Mayer, Adolph, 726
Mayfield, John, 363
McAlisters All-Healing Ointment, 414
McCabe, John, 363
McCulloch, Carlton G., 266, 267–268, 267f, 268f
McCully, William, 55, 56
McDill, Dr. Andrew C., 592, 592f
McDill, John H., 592
McDill, T. W., 592
McGregor, Charles, 651, 651f
McKillop & Sprague's "Commercial Agency Report," 3
McLain, Alexander, 496
McLain, Andrew, 496
McLain, Jesse H., 496
McLain, John, 496
McLain, Samuel, 496, 496f
McLain & Brother (Forreston), 475, 496, 496f
McLean's Volcanic Oil Liniment, 511
McNeil, Francis, 719f
McNeil, William W., 719, 719f
M&Co Soda Mfg. (Highland), 515, 515f, 518
McQuaid, Edward, 383, 405–406, 406f
McQuaid & Monheimer Bros. (Chicago), 374, 385, 406
Meagher, John, 189, 201, 248, 249–250, 334, 360, 361
measurements, in description, 10, 10f
medical/pseudo-medical terms, 30, 43

medicated aerated water
  body shape definition, 15, 17f
  Chicago (IL)
    Barothy & Cook, 197, 198f
    Hayes Brothers, 284–287, 284f
    John A. Lomax, 352, 353f, 354f, 366
  embossed-glass labeling, 42
  Industrial Expansion era (1870s), 79
medicated beverages
  Hayes Brothers (Chicago), 284–287, 284f
Medicated Horehound Candy, 702
medicine bottles. *See* patent medicine bottles
The Medicine Co. (Lewistown), 557–558, 557f
medicines. *See also* bitters; patent medicine bottles; veterinary medicines
  A. Lindsay's "Anti-Billious" Pills, 716
  A. Lindsay's "Anti-Despeptic" Tonic and Pills, 716
  A. Lindsay's "Aramatic" Tooth Powder, 716
  A. Lindsay's Western Nerve Tonic, 716
  Ague Killer, 222
  Ague Tonic, 676
  All-Healing Salve, 676
  Althrop's Constitutional Tonic, 189–190, 189f
  Anti-Bilious Physic, 222
  Asthma Relief, 371–372, 371f
  Badeau's Pure Blood Maker and Liver Cure, 194–195, 195f
  Bateman's Drops, 715
  Baumann's Liniment, 587, 587f
  Betts' Ginger Brandy, 86–87, 87f
  Bligh's Tonic, 201–202, 202f
  Blood Purifier, 222, 662, 662f, 676
  Bradner's Honey Balsam, 124, 124f, 131, 131f
  Brandeth's Pills, 715
  Bristol's Nerve & Bone Liniment, 120, 120f
  Brown's Oriental Hair Renewer, 443–446
  Budlong, William C., 104–105, 105f
  Bunnel's German Liniment, 659–660, 659f
  Burleighs' Electric Balm, 660–662, 661f, 662f
  Carmen's Bitter Sweet, 183–184, 183f
  Charleston (IL)
    Van Meter, Dr. Samuel, 184, 184f
  Chicago (IL)
    A. W. Sargent & Co., 394–396, 395f
  Cholera Tincture, 222
  Chologogue, 406, 407f
  Christie's Ague Balsam, 216–219, 217f, 218f
  cider cure, 199, 199f
  Clarke's Compound Mandrake Bitters, 571–572
  Clarke's European Cough Remedy, 571
  Clarke's Excelsior Panacea, 571
  Clarke's Family Vegetable Pills, 571
  Clarke's Red Drops, 205, 205f
  Clarke's Rheumatic Elixir, 571
  Clark's Female Pills, 655
  Columbian Cathartic Bitters, 717–719, 717f
  Condition Powders, 662, 662f
  Congo Liniment, 254
  Connell's East Indian Brahminical Moonplant Remedies, 222–223, 223f
  Connell's East Indian Remedies, 222–223
  Corn & Wart Salve, 676
  Corneau & Diller's Castor Oil, 702, 702f
  Corneau & Diller's Diarrhoea Mixture, 702
  Cough Balsam, 662, 662f
  Cough Killer, 257–258
  Cough Syrup, 676
  Daniels' Tanta Miraculous, 488–489
  Davis Vegetable Pain Killer, 676
  Davis's Inimitable Hair Renewer, 610–611
  Day-Light Liver Pills, 243, 243f
  Denie's Electro-Magnetic Plaster, 154
  Diarrhea and Dysentery Cordial, 222

  Diarrhoea Syrup, 725
  Dr. A. Brown's Blood Purifier, 600, 600f
  Dr. A. Brown's Female Alterative or Women's Friend, 600
  Dr. A. C. McDill Vermifuge, 592, 592f
  Dr. A. H. Smith's Celebrated Ague Cure, 177, 178f
  Dr. A. L. Adams Liver Balsam, 186–187, 186f
  Dr. Banning's Female Pills, 722
  Dr. Bartram's Dysentary and Diarrhoea Cordial, 722
  Dr. Bartram's Nerve and Bone Liniment, 722
  Dr. Bodley's Fever & Ague Balsam, 34, 654–655, 655f
  Dr. Clark Johnson's Indian Blood Syrup, 314
  Dr. F. A. Sabine's Carbolic Ointment, 393
  Dr. F. A. Sabine's Harvest Bitters, 393
  Dr. F. A. Sabine's World's Remedy, 393
  Dr. Fahrney's Blood Vitalizer, 250–251, 251f
  Dr. Fahrney's Panacea, 250–251, 250f
  Dr. Fahrney's Uterine, 250–251, 250f
  Dr. Faloon's Blackberry Diarrhoea Specific, 122, 123f
  Dr. Faloon's Infant Cordial, 123
  Dr. Faloon's Instant Relief, 122–123, 123f
  Dr. Faloon's Rosin Weed Balsam, 46–47, 46f, 121–122, 121f, 122f, 123f
  Dr. Faloon's Tonic, 122, 123f
  Dr. Hoffmann's Red Drops, 393
  Dr. Howe's Blackberry Balsam, 127, 127f
  Dr. Howe's Egyptian Salve, 127, 127f
  Dr. Howe's Honey Balsam, 127, 127f
  Dr. Howe's Instant Pain Relief, 127, 127f
  Dr. Howe's Liniment, 127, 127f
  Dr. Howe's Vegetable Liver Pills, 127, 127f
  Dr. J. C. Wills' World Worm Candy, 563, 563f
  Dr. J. S. Gunn's Golden Vermifuge, 174–175
  Dr. J. Stickels Compound Syrup of Cephalanthus, 597–598, 598f
  Dr. Jacques German Worm Cakes, 246
  Dr. Johnson's Balsam for the Lungs, 314
  Dr. Johnson's Liniment Mixture, 314
  Dr. Jones Red Clover Tonic, 595–597
  Dr. Kruse's Malaria Antidote, 636–637, 636f
  Dr. Lee's Rheumatic Liniment, 575–576, 575f
  Dr. LeRoy's Antidote to Malaria, 574–575
  Dr. LeRoy's Cough Syrup, 574–575, 574f
  Dr. Lurton's Cholagogue, 367, 367f
  Dr. Major Quinine Substitute, 129, 130f
  Dr. Miller's Infallible Asthma Remedy, 371–372, 371f
  Dr. North's Ague Pills, 723
  Dr. North's Balsam of Horehound and Naptha, 723
  Dr. North's Cherokee Balsam of Horehound, 723
  Dr. North's Cherokee Liniment, 723
  Dr. North's Cholera Drops, 723
  Dr. North's Cough Candy, 723
  Dr. North's Expectorant Cough Candy, 722
  Dr. North's Fever & Ague Killer, 722
  Dr. North's Mother's Balm, 723
  Dr. North's Pain Killer, 723
  Dr. North's Pectoral Balsam of Horehound, 723
  Dr. North's Sarsaparilla and Blood Pills, 722
  Dr. P. Johnson's Golden Eye Salve, 314
  Dr. Penoyer's Extract from the Hygean Springs, 741–742, 741f
  Dr. Richardson's Peruvian Balsam, 744
  Dr. Streeter's Magnetic Liniment, 495–496, 495f
  Dr. Sykes New England Liver Tonic and Billious Annihilator, 435
  Dr. Sykes' Sure Cure for Catarrh, 432–434

Dr. Thomson's Galvanic Liniment, 30, 649–650, 650f
Dr. Topping's Alterative and Cathartic Syrup, 717–719, 717f, 718f
Dr. U. C. Roe's Blood Cordial, 497–498, 497f
Dr. Weaver's Compound Extract of Fireweed, 443–446, 445f
Dr. Weaver's Humor & Liver Syrup, 443–446, 445f
Dr. Winchell's Teething Syrup, 243, 243f, 419–420, 420f
Dr. Wolgamott's Aperient and Blood Purifier, 223, 455–456, 456f
Dr. Z. Waters' Family Medicines, 154–156
Dr. Z. Waters' Honey Balsam, 154–156
Dr. Z. Waters' Itch Renovator, 154, 156
Dr. Z. Waters' Liver and Cathartic Pills, 154–156
Dr. Z. Waters' NE Plus Ultra Liniment, 154–156, 155f
Dr. Z. Waters' Raspberry Balsam, 154–156
Egyptian Anodyne, 543–544, 544f
Egyptian Collyrium, 544
Egyptian Salve, 154, 662, 662f
English Horse Liniment, 662, 662f
Eqyptian Rheumatic and Strengthening Salve, 676
Essence of Jamaican Ginger, 42, 96–97, 96f, 267, 267f
Everybody's Liniment, 713–714, 714f
Excelsior Liniment, 210–212, 210f, 211f
Excelsior Medicine, 210–212, 211f
Extract of Tar & Wild Cherry, 242–243, 242f, 243f
Farmers & Farriers Liniment, 473–475, 474f
Farrell's Arabian Liniment, 251–254, 252f, 253f, 254f, 511, 614–621, 615–621f
Farrell's Arabian Nerve and Bone Liniment, 618
Farrell's Cough Syrup, 620, 621f
Farrell's Fever & Ague Drops, 614–621, 614f
Farrell's Indian Black Ointment, 251, 252f, 254
Farrrell's Tooth-Ache Drops, 620, 621f
Floyd's American Penetrating Liniment, 511–512, 512f
Fluid Lightning, 227–228, 227f
Foster's Child's Relief, 259
Foster's Indian Health Renewer, 259–260, 259f, 260f
Freeman's Alterative Balsam and Mother's Relief, 537
Freeman's Anti-Ague and Fever Pills, 537
Freeman's Aperient Anti-Bilious Pills, 537
Freeman's Compound Emetic Powders, 537
Freeman's Cough Balsam, 535–538, 536f
Freeman's Diarhea Syrup or Cholora Mixture, 535–538
Freeman's Eclectic Liniment, 535–538, 537f
Freeman's Liver Pills, 537
Freeman's Sudorific or Sweating Drops, 535–538, 535f
Freeman's Sweet Balsam of Life, 537
Freeman's Vermifuge, 535–538, 535f
Gardiner's Rheumatic & Neuralgia Compound, 262–263, 263f
German Itch Ointment, 579
Gilt-Edge Tonic, 609
Griggs Sarsaparilla, 595
Hamilton's Anti-bilious Pills, 529
Hamilton's Indian Liniment, 525–529, 529f
Hamilton's [Never Failing] Ague Tonic, 529
Hamilton's Syrup of Blackberry & Sassafras, 525–529, 527f, 528f
Hamilton's Vegetable Ague Pills, 529

Hamilton's Vegetable Cathartic Sugar-Coated Pills, 529
Hamilton's Vegetable Cough Balsam, 525–527f, 525–529
Hamlin's Cough Balsam, 274, 278
Hamlin's Wizard Oil, 271–279, 272–279f
Happy Home Blood Purifier, 246, 247f
Heck, John, 106, 106f
Hockenhull's Extract of Ginger, 531
Holton's Hair Restorer, 725
Horse Ointment, 85, 85f
Indian Balm, 554, 554f
Indian Life Pills, 662, 662f
Indian Salve, 725
Inhaling Fluid, 434–436, 434f
Instant Relief, 413, 414f, 417, 445
Instantaneous Relief, 257–258, 257f
Iodine Mineral Spring Waters, 366
Iron Spring water, Perry Springs, 648
Irving's Giant Pain Curer, 446
J. D. Bastow's Cough Cure, 603, 603f
Jaundice Bitters, 221–222, 222f
Kalb's Eye Water, 675–677, 675f
Kalb's Vegetable Quick Relief, 675–677, 676f
Keith's Persian Liniment, 579–580, 580f
Lackey's Iron Bitters, 129–131, 130f, 131f
Lackey's Justly Celebrated Blackberry Cordial, 123–124, 124f
Life Ointment, 222
Life Syrup, 414, 415f, 417
Lindsay's Nerve and Bone Pills, 716
Liquid Tonic, 429
Lung Syrup, 222
Magic Oil, 275
Magnesia Spring water, Perry Springs, 648
magnetic fluid, 216
Magnolia Bitters, 92–93, 92f
McLean's Volcanic Oil Liniment, 511
Medicated Horehound Candy, 702
Miner's Erasive Solution, 433
Morrison's Hygean Pills, 715
Mrs. Winslow's Soothing Syrup, 247
Myers' Embrocation, 721
Nerve & Bone Liniment, 715, 725
Nerve Tonic, 284, 284f
New Discovery for Consumption, 208–210, 209f
Norman's Chalybeate Cough Syrup, 118, 118f
Norman's Magic Liniment, 117–118, 117f
Norman's Pine Knot Cordial, 117–118, 117f
North American Specific, 714
North's Pectoral Balsam of Hoarhound, 721–723, 722f, 723f
Oil of Life, 125–126, 126f
Old Style Bitters, 177–179, 178f
Owen's Extract of Buchu, 724–725, 724f, 725f
Pain Exterminator, 394–396, 395f, 719, 719f
Panacea, 30
Perry Springs water, 648
Pfifferling's Trionadraphia, 170
Pile Liniment, 676
Pratt & Butcher Magic Oil, 443–446, 444f
Ready Relief, 662, 662f
Rheumatic Liniment, 719
Rheumatic Mixture, 222
Rheumatic Specific, 254
Roe's Ague Pills, 498
Roe's Balm of Gilead Ointment, 498
Roe's Cough Drops, 498
Roe's Physic, 498
Sanative Vegetable Pills, 676
Sawyer's Fluid Extract of Bark, 34, 399–400, 399f
Schroeder's Freckletine, 685
Schroeder's Imperial Balm, 685, 685f

Schroeder's Imperial Cologne, 685
Schroeder's Improved Tooth Powder, 685
Shinn's Panacea, 30
Sloan's Condition Powder, 416, 445
Sloan's Family Ointment, 416, 445
Sloan's Horse Ointment, 417, 445
Sloan's Ointment, 416, 443–446, 444f, 445f
Smith's Cholera Mixture, 714
Springfield (IL)
    McNeil, William W., 719, 719f
Stewart's Improved Medical Compound, 605–606
Stomachic Bitters, 225–226, 225f
Strickland's Wine of Life, 428–429, 428f
Sulphur Spring water, Perry Springs, 648
Swaim's Panacea, 30
Sweet Elixir for Diarrhoea, 557f, 558
Sweet's Blood Renewer, 258, 432, 432f
Sweet's Cholera Drops, 258, 432, 432f
System Builder and Blood Purifier, 151
Tonic Queen, 172, 172f
Turner's Balsam, 702
Turner's Fever & Ague Cure, 702
Turner's Gonyza & Styllingia, 702
Turner's Sarsaparilla, 702
Uncle Sam's Condition Powder, 246, 246f
Uncle Sam's Nerve & Bone Liniment, 246
Van Deusen's Ague Bitters, 727–729
Van Deusen's Cholera Specific, 728
Vermifuge, 676
Vigor of Life, 313–314, 313f
Wakefield's Ague & Fever Pills, 149
Wakefield's Blackberry Balsam, 68, 146, 148–149, 150
Wakefield's Cough Syrup, 146, 148, 149, 150
Wakefield's Egyptian Liniment, 30, 146, 148, 149
Wakefield's Fever Specific, 34, 146, 148, 149
Wakefield's Golden Ointment, 149
Wakefield's Magic Pain Cure, 148, 149
Wakefield's Nerve and Bone Liniment, 146, 148, 149
Wakefield's Strengthening Bitters, 146, 148, 149–150
Wakefield's Wine Bitters, 149–150
Wakefield's Worm Medicine, 146
Wallace & Diller's Cholera Mixture, 702
Wallace & Diller's Western Tonic or Fever and Ague Bitters, 702
Wallace's Tonic Stomach Bitters, 387–389
Warner's Cough Balsam, 428, 448
Warner's Dyspepsia Tonic, 428, 448
Warner's Emmenagogue, 428, 448
Warner's Pile Remedy, 428, 449
Warner's White Wine & Tar Syrup, 68, 446–449, 446f, 447f
Warner's Wine of Life, 428, 429, 446–449, 446f
Wells' Blackberry Cordial, 152
Wells' Genuine Liniment, 152, 153, 450
Wells' German Condition Powder, 152, 154, 452
Wells' German Liniment, 30, 152, 153, 153f, 450–452, 450f, 451f
Wells' Great Western Family Medicines (St. Louis), 153
Wells' Great Western Pills, 152
Wells' Great Western Vegetable Pills, 152, 154, 452
Wells' Pectoral Syrup of Wild Cherry, 152–153, 450–452, 451f
Western Cough Syrup, 725
Whitlow's Golden Age Cough Mixture, 606
Whooping Cough Alleviator, 676
Wild Cherry Pulmonic, 725
Woodard's Instant Relief, 155, 156, 157

(continued)

medicines (continued)
    Worm Candy, 662, 662f
    Younces Indian Cure Oil and Pain
        Destroyer, 565
    Ypsilanti Mineral Spring Water, 461–462, 461f
Meier, John C., 327
Melbourne, Australia
    William Witt & Co., 223
Melcher, Dr. Samuel H., 228
Melvin, Samuel H., 713, 719–721, 720f, 721f
Melvin & Glidden (Springfield), 720
Merchant's Hotel (Chicago), 188
Merrill patent paneled stoneware
    body shape definition, 18, 21f
    Chicago (IL)
        Adams, Thompson W., 188, 188f
        Amberg & Green, 190, 190f
        Arnold & Green, 193, 194f
        Hutchinson & Co., 298, 299f, 300, 301f
        John A. Lomax, 348
        Keeley, Michael, 320, 321f
        Sass & Hafner, 396–399, 398f
        Shure & Bro., 412, 412f
    Dr. Cronk, 229
    Joliet (IL)
        Paige, J. D., 542–543, 543f
        Paige & Vail, 541, 542f
    Springfield (IL)
        Kelleher & Co., 712–713, 712f
    St. Louis (MO)
        J. D. Vail & Co., 541
Merz, Charles, 572–573, 573f
Metric pharmacy bottles, 73
Mette, August, 49–50, 49f, 369–371, 370f
Mette, Henry, 49–50, 49f, 369–371, 370f
Mette, Louis, 49–50, 369, 371
Mr. Metz, 106
Metz, George, 637–638, 638f, 641
Metz, Johann, 637
Metz, John, 637
Metz, Rudolph, 637
Metz & Pletscher (Peoria), 637–638, 638f
Metz & Stege Brewery (Chicago), 45, 638
Meyer, John, 688
Meyer Bros. & Co. (St. Louis), 118
Meyers, Simon G., 374
Michigan Avenue Hotel (Chicago), 241
Middlewood Cincinnati Ale, 94
mid-neck ring
    Peoria (IL)
        Cleveland, Frederick, 609–610, 609f
Miles, Dr. B. F., 632
milk bottles
    study exclusion rule, 37, 39f
Miller, Dr. [Adam], 371–372, 371f
Miller, Andrew J., 50, 50f, 327, 372–373
Miller, Dr. Benjamin C., 371–372
Miller, Dr. DeLaskie, 371
Miller, George A., 682–683, 682f, 683f
Miller, Jacob S., 698, 698f
Miller, Joel
    patent internal rubber stopper, 348
Miller, Montgomery & Co. (Quincy), 683
Miller, Nicholas, 182–183, 182f
Miller, Dr. Truman W., 371–372
Miller & Clements (Chicago), 372
Miller patent internal rubber stopper
    Chicago (IL)
        John A. Lomax, 348
Millgrove Glass Co. (IN), 67
Millville round bottles
    as chronological marker, 73
    study exclusion rule, 37, 38f
Milwaukee Glass Co. (WI), 67
Miner, C. A., 433

Miner & Sykes (Chicago), 433
Miner's Erasive Solution, 433
Misinformation, sources of, 3
Mississippi Glass Co. (St. Louis), 55t, 66, 67–68
    St. Louis brewery-bottling industry, 35
Missouri Glass Co. (St. Louis), 67, 680, 681f
Mix, James, 212
Modes, William F., 67
Modes Glass Co. (IN), 67
Moeller, Henry, 315
Moench, Charles, 373–374, 374f
Moench & Reinhold (Chicago), 373–374, 374f
Molln, Frederick, 54, 54f, 569, 569f, 570
Molln & Co. (Mendota), 54
Monheimer, Isadore, 374–375, 385
Monheimer, Leonard, 374, 383, 385, 405, 406
Monheimer, Levi, 374, 385, 406
Monheimer & Co. (Chicago), 374–375, 385
Monheimer & Cook (Chicago), 374
Monheimer Bros. (Chicago), 374
monograms, embossed. See embossed symbols
Monongahela, V. A. Brown, 212
Monroe Brewery (Columbia), 465–466
Moore, Alexander, 467
Morgan Brewery (Jacksonville), 532
Morgan House (Jacksonville), 529
Morris, Nathan E., 662
Morrison, William, 50–51, 51f
Morrison's Hygean Pills, 715
Mountain Root Bitters, 419, 419f, 697–698, 697f
moving oven technology, 92
Mrs. Winslow's Soothing Syrup, 247, 420
mucilage
    Chicago (IL)
        Thomas, Levi H., 439–440
        Sanford Manufacturing Co. (Chicago),
            393–394, 393f
Mueller, Anton (aka Miller), 174, 517–519,
    518f, 520, 520f
Mueller, Gustavus, 134, 134f, 174
Mueller, John A., 173–174, 173f
Mueller & Beck (Highland), 516, 517–519, 518f
Mueller & Stein (Bloomington), 121, 125,
    129, 134, 134f, 135
Muller, Frederick W. "Pop," 103–104, 103f,
    104f, 475–476, 475f, 476f
multiple-paneled bottles. See paneled bottles
Muncie Glass Co. (IN), 67
Murphy, Anderson, 743
Murphy & Hoy (Woodstock), 743
Murphysboro Steam Bottling Works
    (Murphysboro), 476
Murray, Allen, 194
Myers, Frank, 721
Myers, Henry C., 721
Myers, Samuel, 374, 374f
Myers' Embrocation, 721

N
N. Bassler & Co. (Mascoutah), 567
narrow blob top bottles
    Chicago (IL)
        John A. Lomax, 357–359f, 358
    Peoria (IL)
        Gipps & Co., 630–632, 631f
narrow blob top quarts
    Chicago (IL)
        John A. Lomax, 350, 351f, 355f, 356
narrow blob tops. See also narrow blob top
        quarts
    Chicago (IL)
        Davies, George F., 231–232, 231f
narrow cylindrical Selters-style
    East St. Louis (IL)
        Schroeder, Edward, 482–486, 486f

narrow shouldered blob-top quarts
    Chicago (IL)
        Hausburg, William A., 282–283, 283f
    Elgin (IL)
        Althen, Casper, 487–488, 488f
narrow-shouldered blob top
    Chicago (IL)
        George Hofmann & Brothers, 292–293, 292f
National Brewing Co. (San Francisco), 506
National City Bank (Ottawa), 595
"near beer," 410
neck shapes. See also specific shapes
    definition of terms, 23, 24f
    length as time markers, 23
neckless bottles. See no neck bottles
Nectar Bitters, 212–213, 213f
Nerve & Bone Liniment, 715, 725
Nerve Tonic, 284, 284f
Neubarth, Theo, 323
Neuber, Joseph F., 560, 560f
Neuerburg, Hubert, 128
New Discovery for Consumption, 208–210
New Orleans mead
    Chicago (IL)
        Cook, Egbert C., 224
no neck bottles
    Chicago (IL)
        John A. Lomax, 343, 343f
    Peru (IL)
        Haas, Kasper, 648–649, 648f
Norman, Dr. George, 117, 119
Norman's Chalybeate Cough Syrup, 118, 118f
Norman's Magic Liniment, 117–118, 117f
Norman's Pine Knot Cordial, 117–118, 117f
North, Alfred A., 721–723, 722f, 723f
    "Long Nine," 722
North & Bristol (Springfield), 723
North American Specific, 714
North Western Hotel (Chicago), 188
North's Pectoral Balsam of Hoarhound,
    721–723, 722f, 723f
Northwestern Chemical Co. (Chicago), 69
Northwestern Chemical Manufacturing Co.
    (Chicago), 375, 375f

O
O. F. & G. A. Miller (Quincy), 682
O. K., Jr. (Bloomington), 128–129, 128f
octagonal bottles
    Quincy (IL)
        Miller, George A., 682–683, 682f
Oetter, Anton, 505–506, 505f
Ohlweiler, Jacob, 693
Ohlweiler, John, 691–693f, 691–694
Oil of Life, 125–126, 126f
ointments
    Dr. F. A. Sabine's Carbolic Ointment, 393
    Farrell's Indian Black Ointment, 251, 252f,
        254
    Life Ointment, 222
    Roe's Balm of Gilead Ointment, 498
    Sloan's Family Ointment, 416, 445
    Sloan's Ointment, 416
Old Style Bitters, 177–179, 178f
Olmsted, John, 390
"One-for-a-Man, Two-for-a-Horse," 414
onion-shaped bottles
    body shape definition, 13, 14f
    Chicago (IL)
        Thomas & Co., 79, 440–441, 440f
    Foerster's Teutonic Bitters, 79
Opedeldoc, 30
Orange Nectar Co. (Chicago), 428
organization, of text, 4
Orr, T. C., 428

Orr, W. H., 428
Orrin P. Bissel & Co. (Peoria), 605
Osborn, Thomas W., 296
Mr. Ostrom, 678
Oswego "Deep Rock," 208
Ottawa mineral waters
    Chicago (IL)
        John A. Lomax, 364, 366
Otto Kadgihn & Son (Bloomington), 128
Otto Kadgihn v. The City of Bloomington (1871), 128
Ouilmette cabin (Wilmette), 493
outside screw threads
    Chicago (IL)
        A. Leibenstein & Co., 330–331, 331*f*
oval blob-top
    Peoria (IL)
        Weber, George, 643–645, 644*f*
        Wichmann, John, 645–646, 645*f*
oval-blob-top sodas
    Alton (IL)
        H. E. Rupert, 97, 97*f*
        Weisbach, Christian, 98–100, 99*f*, 100*f*
        Yoerger, Augustin, 100–102, 102*f*
    Aurora (IL)
        Green Brothers, 105–106, 105*f*
        Stege, Albert [?], 44, 44*f*
    Beardstown (IL)
        Ehrhardt, Frederick W., 107–109, 107*f*, 108*f*
    Belleville (IL)
        Ab Egg, Louis, 110–112, 111*f*, 112*f*
        Fisher, Joseph (aka Fischer), 113–115, 114*f*
    Bloomington (IL)
        Kreis, Peter, 129, 129*f*
        Petterson & Bro., 135, 135*f*
    Blue Island (IL)
        G. H. Hausburg, 158–160, 159*f*
        Hausburg & Dettmer, 160, 160*f*
        Hausburg & Ennis, 161, 161*f*
        R. Boil & Co., 158, 158*f*
    Braidwood (IL)
        Blood, Horace W., 162–163, 162*f*, 163*f*
    Breese (IL)
        Dorries & Bro., 163, 163*f*
    Cairo (IL)
        Breihan, Henry, 10*f*, 69, 165–167, 165*f*, 166*f*
        Lohr, Andrew, 168–170, 168*f*, 169*f*
    Carlinville (IL)
        Zaepffel, August, 174, 174*f*
    Centralia (IL)
        Wehrheim, Henry G., 180–181, 180*f*
    Centreville (IL)
        Fischer, Christian, 181–182, 181*f*, 182*f*
    Chicago (IL)
        A. Mette & Bros., 369–371, 370*f*
        Barothy & Cook, 197, 198*f*
        Bower & Bryan, 202, 203*f*
        Dempsey, Daniel, 232–234, 234*f*
        Dickinson, George, 236, 236*f*
        Edwards, "E.", 242, 242*f*
        Edwards, Francis C., 242, 242*f*
        Filer, Henry, 255–257, 257*f*
        Gottlieb Wurster & Co., 456–461, 458*f*, 459*f*
        H & Bro's, 279–281, 280*f*
        Hausburg Brothers, 281–282, 281*f*
        Hayes Brothers, 284–287, 284*f*, 285*f*, 286*f*
        Hess, Ernst, 291–292, 291*f*
        Hunt & Kenyon, 294–295, 294*f*, 295*f*
        Hutchinson & Co., 304, 306–308*f*, 307, 308
        Hutchinson & Sons, 310, 311*f*
        Huttenlocher, John, 311–312, 311*f*
        John A. Lomax, 338, 339*f*, 340, 341*f*, 342*f*, 343, 343*f*, 344*f*
        Joss & Hartmann, 314–315, 314*f*
        Keeley, Michael, 317–320, 320*f*
        Lang Brothers, 325–328, 326*f*, 327*f*
        Lomax, George, 332–334, 332*f*
        Morrison, William, 50–51, 51*f*
        Paul Rouze & Co., 391–393, 393*f*
        Sass & Hafner, 396–399, 397*f*, 398*f*
        Schriber & Joss, 403–404, 404*f*
        Seibt, Frederick (Fritz), 408–409, 409*f*
        Seibt & Haunschild, 407–408, 408*f*
        Shure & Bro., 412, 412*f*
        Stenson, James, 422–426, 424*f*, 425*f*
        W. H. Hutchinson & Son, 309, 309*f*
    Colehour (IL)
        Kassens, Henry C., 462, 462*f*
    Collinsville (IL)
        Baierlein, Henry, 464, 464*f*
    Columbia (IL)
        Gundlach, John, 465–466, 466*f*
    Danville (IL)
        Shean, Charles, 468–469, 468*f*
    Darmstadt (IL)
        Gross, Christian, 469–470, 470*f*
    De Kalb (IL)
        Dee, Richard, 52, 52*f*
    definition, 24, 25*f*
    Dixon (IL)
        F. J. Finkler & Co., 473, 473*f*
    Dunton (IL)
        Sass & Muller, 475–476, 475*f*, 476*f*
    DuQuoin (IL)
        Hayes, Edwin, 476–477, 476*f*
    East St. Louis (IL)
        C. Lutt & Co., 479–481, 480*f*
        Schroeder, Edward, 482–486
        Thomas Schoebein & Co., 481–482, 482*f*
    Edwardsville (IL)
        Harles, Frank, 487, 487*f*
    Elgin (IL)
        Henry & Van Buskirk, 490, 490*f*
        Shure & Bro., 491–492, 492*f*
    Freeburg (IL)
        Darmstatter, Frederick, 499, 499*f*
    Galena (IL)
        Evans, Joseph, 503–505, 504*f*
        Oetter, Anton, 505–506, 505*f*
    Galesburg (IL)
        Edwards, [Isaac?] R., 510, 510*f*
    Galva (IL)
        Ayres, Clarence M., 510–511, 511*f*
    Greenville (IL)
        Grafe, P. Herman, 512–513, 513*f*
    Highland (IL)
        Amsler Brothers, 516–517, 516*f*
        Schott, Christian, 519–520, 519*f*
    Jacksonville (IL)
        Braun & Killian, 523, 523*f*
        Buell & Schermerhorn, 522, 523–525, 524*f*
        Kershaw, Albert, 531, 532*f*
        P. Braun & Co., 522–523, 522*f*
        Ricks, John W., 532, 532*f*
        Schermerhorn, Charles, 533–534, 533*f*
    Jerseyville (IL)
        Boyle, Charles M., 534–535, 534*f*
        Spannagel, Johanna, 538–539, 538*f*
    Joliet (IL)
        Paige, J. D., 540–542, 540*f*
    Kankakee (IL)
        Kammann, [Dietrich] H., 545–546, 546*f*
    Kewanee (IL)
        Spencer, Daniel H., 546, 546*f*
    Lemont (IL)
        Bolton, John G., 554–557, 556*f*
    Litchfield (IL)
        Gerhardt, William, 559–560, 559*f*, 560*f*
        Neuber, Joseph F., 560, 560*f*
        Weber, Frederick, 560–563, 561*f*
    O'Fallon (IL)
        Bayet & Williams, 589–590, 589*f*
        Bischoff, Benard, 590–591, 590*f*
    Pekin (IL)
        Zerwekh, Gottleib J., 600–602, 601*f*, 602*f*
    Peoria (IL)
        Gillig & Singer, 626–627, 627–629
        Hickey, Ransom E., 632–635, 633*f*, 635*f*
        Singer, Charles J., 642–643, 642*f*, 643*f*
    Peru (IL)
        Haas, Kasper, 648–649, 648*f*
    Quincy (IL)
        Durholt, Henry, 663–667, 664*f*
        Hutmacher, Edward, 673–674, 673*f*
        J. J. Flynn & Co., 668–671, 668–671*f*
    Red Bud (IL)
        Excelsior Brewing Co., 687–688, 687*f*
    Rock Island (IL)
        Carse & Lamont, 690–691, 690*f*
    Sheldon (IL)
        Miller, Jacob S., 698, 698*f*
    Springfield (IL)
        Johnson & Peterson Co., 709–712, 710*f*, 711*f*
    Staunton (IL)
        Knemoeller & Stille, 730–731, 731*f*
    Streator (IL)
        Finkler, John A., 731–732, 731*f*
    Virden (IL)
        Achilles & Hille, 733–734, 733*f*, 734*f*
    Waterloo (IL)
        Boeke, Henry, 735, 735*f*
        S. & J. Ruf, 737–738, 737*f*
    Wilmington (IL)
        H. W. Blood & Co., 742–743, 742*f*
Owen, E., 713
Owen, Thomas J. V., 724–725, 724*f*, 725*f*
Owen, William F., 725
Owen's Extract of Buchu, 724–725, 724*f*, 725*f*
Ozonized Ox Marrow, 208

P
P. Baumann & Brothers (New Athens), 587
P. Braun & Co. (Jacksonville), 522–523, 522*f*, 523, 524, 531, 532, 533
P. C. Higgins (Bunker Hill), 538
P. F. McQuillan, 213
P. McIntyre & Co. (Chicago), 422
P. W. Gillet & Son (Chicago), 267, 267*f*
Paige, J. D., 69, 162, 539, 540–543, 743
Paige, Nathanial, 400
Paige & Vail (Joliet), 541, 542–543, 542*f*, 543*f*
Pain Exterminator, 394–396, 395*f*, 719, 719*f*
Paist, Marmon & Co. (Bloomington), 131
Paist & Elder (Bloomington), 131
Pale India Ale, 422
palm tree, embossed. *See* Garrick & Cather (Chicago)
Palmer, Charles, 229
Palmer, Potter, 376–377, 376*f*
    threaded glass stoppers, 29
Palmer House Hotel (Chicago), 37–377, 376*f*
    threaded glass stoppers, 29
Panacea, 30
panaceas
    Clarke's Excelsior Panacea, 571
    Dr. Fahrney's Panacea, 250–251, 250*f*
    Panacea, 30
    Shinn's Panacea, 30
    Swaim's Panacea, 30
Paoli, Dr. Gerhard, 412–413, 413*f*
paper bottles, 440
paper labels
    Bloomington (IL)
        Dr. Faloon's Rosin Weed Balsam, 122, 122*f*
        Dr. H. S. Woodard, 155, 155*f*

(continued)

paper labels (continued)
Bloomington (IL) (continued)
Funk & Lackey, 123, 124f
J. W. Maxwell & Co., 132–133, 133f
Wakefield's Wine Bitters, 144
Woodard & Howe, 157, 157f
and bottle shape, 13, 41–42
Chicago (IL)
Bennett Pieters & Co. Red Jacket Bitters, 385f
Christie's Ague Balsam, 217, 218f
Connell, J. M., 222
Drs. C. M. Fitch & J. W. Sykes, 434f
Gillet's Chinese Liquid Bluing, 267, 268f
Hamlin's Wizard Oil, 278, 279f
Instantaneous Relief, 257–258, 257f
Old Jim Gore Burbon Whiskey, 216
Wallace's Tonic Stomach Bitters, 388f
Warner's White Wine & Tar Syrup, 447f
Elgin (IL)
Daniels' Tanta Miraculous, 489, 489f
before embossed bottles, 30, 30f
versus embossing, 74
Franklin Grove (IL)
Dr. U. C. Roe's Blood Cordial, 497f
Lewistown (IL)
The Medicine Co., 557f
Normal (IL)
System Builder and Blood Purifier, 151
Ottawa (IL)
Dr. Jones Red Clover Tonic, 595–597, 596f
Peoria (IL)
Bissell's Tonic Bitters, 605f
Callender's Liver Bitters, 609, 609f
Farrell's Cough Syrup, 620, 621f
Parent, Anthony, 638, 639f
Quincy (IL)
M. R. & H. W. Lundblad, 681
Schroeder's Imperial Balm, 685, 685f
Springfield (IL)
Bonekamp Stomach Bitters, 705
Corneau & Diller's Castor Oil, 702, 702f
St. Louis (MO)
Union Medicine Co., 179
and sunken panels, 13
paper-thin flared lip bottles
as chronological marker, 71
definition, 24, 24f
Jacksonville (IL)
Hamilton, Dr. William, 525–529, 526f
Parent, Anthony, 638–640, 639f, 640f
Park House (Jacksonville), 521
Parker, Samuel W., 241
Parks, John D., 217
Parks, Dr. Luther K., 543–544, 544f
Parshall & Co. (Chicago), 433
pasteurization
brewery versus local bottling, 74
and brewery-product expanded use, 34, 584
milk bottle study exclusion rule, 37, 39f
quart beer bottle study exclusion rule, 37
patent medicine bottles. See also bitters; medicines; pharmacy bottles; stone-bottled medicine; veterinary medicines
advertising, 42–43
Albion (IL)
Stewart, Alexander, 85, 85f
Alton (IL)
Quigley, Hopkins, and Lea, 96–97, 96f
Antebellum era (1850s), 75, 78f, 79
assemblage patterns, 74
Aurora (IL)
Budlong, William C., 104–105, 105f
Heck, John, 106, 106f

Belleville (IL)
Flanagan, Edward T., 117–118, 117f, 118f
Belvidere (IL)
Bristol, Solomon W., 120, 120f
Bloomington (IL)
C. B. Castle & Co., 46–47, 46f, 121–122, 121f, 122f
Funk & Lackey, 123–124, 124f
Hill & Vanatta, 125–126, 126f
Howe, Bliss S., 126–128, 126f, 127f, 128f
Lackey and Brothers, 129–131, 130f, 131f
Wakefield & Co., 137–147f, 137–150
Wells, Edwin M., 151–154, 151f, 152f, 153f
Woodard, Henry S., 154–156, 154f, 155, 155f, 156–157, 157f
Woodard & Howe, 126–127, 150, 154, 155, 156–157, 157f
body shape definitions, 18, 19f
Capron (IL)
Field, Platus A., 172, 172f
Carlyle (IL)
Gunn, John S., 174–175, 175f
Carrollton (IL)
Smith, Dr. Alexander H., 176–179
Charleston (IL)
Cunningham & Barnes (Charleston), 183–184, 183f
Chester (IL)
Van Meter, Dr. Samuel, 184, 184f
Chicago (IL)
Adams, Dr. A. L., 186–187, 186f
Alfred & William Strickland Co. (Chicago), 428–429, 428f, 429f
Althrop, Thomas, 189–190, 189f
Badeau, William C., 194–195, 195f
Bligh, Thomas, 201–202, 202f
Bryan, Frederick A., 205, 205f
Buck, George B., 205–206, 205f, 206–207
Butt, Charles, 210–212, 210f, 211f
Christie, Dr. Abel H., 216–219, 217f, 218f
Connell, J. M., 222–223, 223f
Drs. C. M. Fitch & J. W. Sykes, 434–436
Eilert, Jacob K., 242–243, 242f, 243f
Emmert Proprietary Co. (Chicago), 244–247, 245f, 246f, 247f
Fahrney, Peter, 250–251, 250f, 251f
Farrell, William B., 251–255
Flagg, Edward H., 257–258, 257f
Foster, Charles M., 259–260, 259f, 260f
Gale, William H., 261, 261f
Gardiner, Charles H., 262–263, 263f
Gillet, McCulloch & Co. (Chicago), 266, 267–268, 267f, 268f
Hutchins, John W., 296, 296f
Johnson Dr. [Hosmer A.], 313–314, 313f
Lurton, Dr. Lycurgus L., 367, 367f
Miller, Dr. [Adam], 371–372, 371f
Penton, Fisher & Co. (Chicago), 377
Sawyer, Sidney, 399–400, 399f
Sears & Smith, 406, 407f
Sloan, Walter B., 413–418, 414–416f
Snyder & Eilert, 419–420, 420f
Sweet, Henry J., 258, 432, 432f
Sykes, Charles R., 432–434, 433f
Walker & Taylor, 443–446, 444f, 445f
Warner, Dr. C. D., 446–449, 446f, 447f
Wells, Alvin T., 450–452, 450–452f
Wells, Edwin M., 450–452, 450–452f
Wolgamott, Dr. George W., 223, 455–456
Cram, Dr. John S., 227–228, 227f
DeMor & Co. (Chicago), 232, 232f
Dixon (IL)
Herrick, George L., 473–475, 474f
Elgin (IL)
C. M. Daniels & Co., 488–489, 488f, 489f

eye water, 43
Farmington (IL)
Bristol, Tyler & Co., 495–496, 495f
Franklin Grove (IL)
Dr. U. C. Roe & Sons (Franklin Grove), 497–498, 497f, 498f
Geneseo (IL)
"puff"-style pharmacy bottle, 30f
Greenville (IL)
Floyd, Edward A., 511–512, 512f
H. E. Bucklen & Co., 208–210, 209f
Hamlin, John A., 271–279, 272–279f
Helmbold's Extract of Buchu, 42
Jacksonville (IL)
Hamilton, Dr. William, 525–529
Hockenhull, Robert, 530–531, 530f, 531f
Jerseyville (IL)
Freeman, Dr. John D., 535–537f, 535–538
Jonesboro (IL)
Parks, Dr. Luther K., 543–544, 544f
LaSalle (IL)
Babcock, Dr. A. L., 548, 548f
Lebanon (IL)
Smith, Henry H., 554, 554f
Lewistown (IL)
The Medicine Co., 557–558, 557f
lip style definitions, 24, 24f, 25f
Lord, James J., 717–719, 717f, 718f
Maryland (IL)
Haldeman, Samuel, 565, 565f
Morris (IL)
Lee, Dr. Charles M., 575–576, 575f
LeRoy, Dr. David, 574–575, 574f
Naperville (IL)
Keith, Aylmer, 579–580, 580f
New Athens (IL)
Baumann, Peter, 587, 587f
Oquawka (IL)
McDill, Dr. Andrew C., 592, 592f
Ottawa (IL)
Griggs, Edward Y., 595–597, 595f, 596f
Stickels, Dr. J., 597–598, 598f
Pekin (IL)
Brown, Dr. Alfred, 600, 600f
Peoria (IL)
Bastow, James D., 603, 603f
Davis, William H., 610–611, 611f
Farrell, Hiram G., 614–619f, 614–621
Farrell & Cox, 622, 622f
Headley, Dr. William S., 632, 632f
Kruse, Dr. Hero, 636–637, 636f
Pioneer era (1840s), 75, 76f
Pittsfield (IL)
Thomas C. Grimshaw & Co., 649–650, 650f
Princeton (IL)
Bodley, Dr. A. R., 654–655, 655f
Quincy (IL)
Bunnel, Dr. D[avid], 659–660, 659f
Burleigh & Brothers, 660–662, 661f, 662f
Kalb, Absalom J., 675–677, 675f, 676f
Schroeder, Herman, 685, 685f
Red Drops, 43, 75
Springfield (IL)
A. & M. Lindsay (Springfield), 714–717, 715f, 716f
Lewis, Thomas, 713–714, 714f
Melvin, Samuel H., 719–721, 720f, 721f
Myers, Henry C., 721
North, Alfred A., 721–723, 722f, 723f
Owen, Thomas J. V., 724–725, 724f, 725f
study inclusion rule, 37
Thompson (Chicago), 441, 441f
Topping, Ebenezer H., 717–719, 717f, 718f
vermifuges, 43, 75

Western Saratoga (IL)
   Penoyer, Dr. Hiram, 741–742, 741f
Willard, Joseph, 453
Wills, Dr. John C., 563, 563f
Winchell, 69
Wm. McCully & Co. (Pittsburgh), 70
Wolgamott, Dr. George W., 455–456, 456f
"patent medicine poet," 717, 719
patterns of use, 74–80, 76f, 77f, 78f
Paul Rouze & Co. (Chicago), 391–393
Peabody, Frank H., 224
Pecan (Peoria), 603
Penn Glass Works, 62–63, 67
Penoyer, Dr. Hiram, 741–742, 741f
Penton, Fisher & Co. (Chicago), 377
Penton, Thomas B., 377
Penton & Co. (Chicago), 377
Pentroleum Centre (PA)
   Kump, John H., 167
People's Drug Store (Princeton), 654–655
The People's Drug Store (Bloomington), 125
People's Favorite Bitters, 581–582, 581f, 582f
Peoples National Bank (Rock Island), 695
Peoria Mineral Spring, 607, 627, 633
Pepper, Benjamin, 538
peppersauce bottles
   assemblage patterns, 75
perfumes
   assemblage patterns, 75
   Chicago (IL)
      Garden City Chemical Works, 262
      Gilet, McCulloch & Co., 262
      Gillet, McCulloch & Co., 267
      Huyck & Randall, 312
Perrine, Charles O., 377–381, 377f
Perrine honey bottles, 377–381, 377f
   bottle shape definition, 18, 21f
   Chicago (IL)
      Gillett Chemical Works, 265–266, 266f
Perry Springs (Perry), 646–648, 647f, 729, 742
   Joliet marble, 648
Perry Springs Hotel (Perry), 648
Persian Sherbet
   embossed-glass labeling, 42
petaled shoulder panels
   Belleville (IL)
      Fisher, Joseph (aka Fischer), 113–115, 114f
   Peoria (IL)
      Cairns & Hickey, 606–607, 607f
Peter, T., 175
Peterson, Alfred, 135, 135f, 711
Peterson, Charles, 135, 135f, 707, 709–712
Petterson & Bro. (Bloomington), 121, 125, 129, 135, 135f
Peuser, Otto, 381, 381f
Peuser & Kadish (Chicago), 381, 381f
Pfeifer, Philip, 566–567
Pfeifer, Philip, Jr., 567
Pfifferling, Charles, 170, 170f
Pfifferling's Trionadraphia, 170
pharmacy bottles
   Antebellum era (1850s), 79
   Chicago (IL) expanded use, post-study, 35
   as chronological marker, 73
   citrate bottles, study exclusion rule, 37, 40f
   embossing rare, 42
   flint glass, 73f
   Geneseo (IL)
      "puff"-style, 30f
   intentionally colored, 73f
   patented slug-plate
      study exclusion rule, 37
   Springfield (IL)
      Diller, Roland W., 700–702, 700–702f
      J. B. Brown & Co. (Springfield), 699, 699f

Springfield (IL) expanded use, post-study, 35
   Willard, Joseph, 453
pharmacy-style flared-lips
   Chicago (IL)
      Farwell and Brace, 255, 255f
   Quincy (IL)
      Schroeder, Herman, 685, 685f
Philadelphia (PA)
   Warner, William R., 448
Philadelphia oval bottles
   as chronological marker, 73
   study exclusion rule, 37, 38f
Phoenix (LaSalle), 62
Phoenix Distillery (Peoria), 626
Phoenix pharmacy bottles, 73
pickle bottles
   assemblage patterns, 75
   body shape definition, 18, 21f
Pieters, Bennet, 57, 79
Pieters, Bennett, 293, 374, 381–386, 382–386f
Pile Liniment, 676
pinched neck Codd marble–style sodas
   Rock Island (IL)
      Carse & Ohlweiler, 691–694, 692f
pint ale
   body shape definition, 15, 15f
   Chicago (IL)
      Carpenter, Andrew, 212, 212f
      Keeley & Brother, 315–317, 315f
      Taylor & Bro., 436–438, 436f, 437f
   Peoria (IL)
      Parent, Anthony, 638–640, 639f
pint porter
   Chicago (IL)
      Dempsey, Daniel, 232–234, 233f
      Hayes Brothers, 284–287, 287f
      John A. Lomax, 346f, 347, 347f
   Pittsburgh (PA)
      W. H. H., 303, 303f
pint soda
   Chicago (IL)
      John A. Lomax, 343, 344f
      Joseph Apple & Co., 192, 193f
      Schulz & Hess, 404–405, 405f
   Perry (IL)
      Perry Springs, 646–648, 647f
pint stoneware
   Chicago (IL)
      Keeley, Michael, 320, 321f
pint whiskey
   Chicago (IL)
      Drake, John B., 240–241, 240f, 241f
      Drake, Parker & Co., 241, 241f
      Samuel Myers & Co., 374, 374f
   LaSalle (IL)
      DeSteiger Glass Co., 548–549, 548f
pints, as term, 10
Pioneer era (1840s), 34
   product use patterns, 75, 76f
Pittsburgh ale
   Chicago (IL)
      Doty & Waugh, 239
Pittsburgh do, 94
Pittsburgh stock
   Peoria (IL)
      Parent, Anthony, 639
plain oval pharmacy bottles, 73
Plantation Bitters, 207
Plautz, Charles H., 387, 387f
Pletscher, John C., 637–638, 638f, 640–641, 640f
Plows, Harris & Upham (Chicago), 419, 698
Plows, Wm. J., 418, 419, 698
Plummer, J. B., 695
poet of patent medicine, 717, 719

poison bottles. *See also* rat poisons
   assemblage patterns, 75
   study exclusion rule, 37, 40f
pop beer. *See* California pop and pop beer
porter
   Alton (IL)
      Keeley & Brother, 94–95, 317
   Antebellum era (1850s), 79
   assemblage patterns, 74
   and black glass, 9
   body shape definition, 15, 16f
   Chicago (IL)
      Cook, Egbert C., 224
      Dempsey, Daniel, 232–234, 233f
      Doty, Harvey C., 239, 239f
      Gottlieb Wurster & Co. (Chicago), 456–461, 457f
      H. F. & Co., 257
      Hayes Brothers, 284–287, 286f, 287f
      Hutchinson & Co., 303
      Hutchinson & Dunn, 303
      Hutchinson & Sons, 310, 311f
      J. Grossenheider & Co., 269
      John A. Lomax, 336, 337f, 346f, 347, 347f, 364, 365
      John Anderton & Co., 191
      Keeley, Michael, 317–320, 318–320f
      Keeley & Brother, 315–317, 315f, 316f
      Keeley Bottling Works, 317
      Sass & Hafner, 398
      Sproat, John, 420–421, 420f, 421f
      Taylor & Bro., 436–438, 437f
      Tivoli Bottling Co., 442
   De Kalb (IL)
      Dee, Richard, 52
   Galena (IL)
      Evans, Joseph, 503–505, 504f
      Volz & Glueck, 506, 506f
   Jacksonville (IL)
      J. E. Bradbury & Co., 521
   Jerseyville (IL)
      Boyle, Charles M., 534
   Lemont (IL)
      Bolton, John G., 554–557, 555f, 556f
   Peoria (IL)
      Gipps & Co., 630
      Parent, Anthony, 638–640, 639, 639f, 640f
      Pletscher, John C., 640–641, 640f
   Pittsburgh (PA)
      W. H. H., 303, 303f
   Rock Island (IL)
      Carse & Ohlweiler, 693
   Springfield (IL)
      Kelleher & Co., 712–713, 712f
   study inclusion rule, 35
Post, George W., 194
Post & Badeau (Chicago), 194
pouring-harness, 22
Powell, George, 64, 387–389, 388f, 389f
Powell, Morris B., 581–582, 581f, 582f
Powell & Stutenroth (Naperville), 581–582
Pratt & Butcher Magic Oil, 443–446, 444f
preservation of contents
   by black glass, 9
   storage techniques, 22
Price, J., 390
Prima pharmacy bottles, 73
product categories, in study, 41–43
product jar
   Chicago (IL)
      Gillett Chemical Works, 265–266, 266f
Professor Lennord's Nectar Bitters, 212–213
Professor Leonard's Nectar Bitters, 212
propeller embossed symbol. *See* Quigley, Hopkins, and Lea (Alton)

pseudo-medical/medical terms. *See*
  medical/pseudo-medical terms
Pullna mineral waters, 88
pumpkin seed flask
  body shape definition, 13, 14*f*
  Chicago (IL)
    Alfred & William Strickland Co., 428–429
punty rod, 11
Pure Food and Drug Act (1906), 247
Putnam, 56, 68, 348. *See also* Lindeborough Glass Works (NH)
Putnam, Henry W., 68
Putnam Glass Works (Putnam, OH), 68
Putnam's Lightning fruit jars, 68
Pyrmont mineral waters, 88

Q
quart ale
  Alton (IL)
    Keeley & Brother, 93–96, 93*f*
  base bumps and dimples, 22, 22*f*
  Belleville (IL)
    Thomas Heberer and Brothers, 118–119
  Bloomington (IL)
    O. K., Jr., 128–129, 128*f*
    Schausten, William, 135–137, 136*f*
  body shape definition, 15, 15*f*
  Cairo (IL)
    Breihan, Henry, 10*f*, 69, 165–167, 165*f*, 166*f*
    Lohr, Andrew, 168–170, 168*f*, 169*f*
    Schutter, William H., 171, 171*f*
  Champaign (IL)
    Miller, Nicholas, 182–183, 182*f*
  Chester (IL)
    Wenda, John, 184–185, 185*f*
  Chicago (IL)
    Amberg & Stenson, 191, 191*f*
    Arnold, Caleb, 193, 193*f*
    Dempsey, Daniel, 232–234, 233*f*
    Filer, Henry, 255–257, 256*f*
    Gottlieb Wurster & Co., 456–461, 457*f*
    Hayes Brothers, 284–287, 286*f*, 287*f*
    Hennessey, Thomas, 290–291, 290*f*
    Hunt & Kenyon, 294–295, 294*f*, 295*f*
    Hutchinson & Co., 300, 301, 302*f*, 304, 305*f*
    Hutchinson & Sons, 310, 311*f*
    John A. Lomax, 336, 337*f*, 345, 346*f*, 355*f*, 356
    John Anderton & Co., 191, 192*f*
    Kane, Thomas, 315, 315*f*
    Keeley, Michael, 317–320, 318*f*, 319*f*
    Keeley & Brother, 315–317, 316*f*
    King, Thomas, 320–322, 321*f*
    Lang, Frederick C., 328–330, 328*f*
    Sass & Hafner, 396–399, 396*f*, 397*f*
    Seibt & Haunschild, 407–408, 408*f*
    Stenson, James, 422–426, 423*f*, 424*f*
    Van Buskirk & Henry, 442–443, 443*f*
  Danville (IL)
    Carey, Patrick, 467, 467*f*
  East St. Louis (IL)
    C. Lutt & Co., 479–481, 479*f*
    Schroeder, Edward, 482–486, 483*f*
  Galena (IL)
    Evans, Joseph, 503–505, 504*f*
    Volz & Glueck (Galena), 506, 506*f*
  Lemont (IL)
    Bolton, John G., 554–557, 556*f*
  Naperville (IL)
    Stenger, John, 582–584, 583*f*
  Peoria (IL)
    Parent, Anthony, 638–640, 639*f*
    Pletscher, John C., 640–641, 640*f*
  Springfield (IL)
    Matthew Rinehardt & Co., 725–726, 725*f*

quart beer
  Bloomington (IL)
    Schausten, William, 135–137, 136*f*
  Chicago (IL)
    Tivoli Bottling Co., 441–442, 441*f*
  Danville (IL)
    Stein, John, 469, 469*f*
  Jacksonville (IL)
    J. E. Bradbury & Co., 521–522, 521*f*
  study exclusion rule, 37
quart blob-top beer
  body shape definition, 15, 16*f*
quart blob-top sodas
  Bloomington (IL)
    Bottling House, 120–121, 121*f*
  body shape definition, 15, 17*f*
  Chicago (IL)
    Hausburg, William A., 282–283, 283*f*
    Hayes Brothers, 284–287, 285*f*, 286*f*
    Hess, Ernst, 291–292, 291*f*
    Hutchinson & Sons, 310, 311*f*
  Monee (IL)
    Vatter, Adam, Jr., 572–573, 573*f*
quart fruit jars
  Chicago (IL)
    A. Leibenstein & Co., 330–331, 331*f*
quart soda/cider/mineral water
  Chicago (IL)
    John A. Lomax, 355*f*, 356
    Stenson, James, 422–426, 425*f*
    Ypsilanti Springs Co., 461–462, 461*f*
  Colehour (IL)
    Smith, William S., 462–463, 463*f*
  East St. Louis (IL)
    C. Lutt & Co., 479–481, 480*f*
    Schroeder, Edward, 482–486, 483*f*
  Galesburg (IL)
    Eaton, James E., 509
  Industrial Expansion era (1870s), 79
  Kankakee (IL)
    Kammann, [Dietrich] H., 545–546, 546*f*
  Monee (IL)
    Vatter, Adam, Jr., 572–573, 573*f*
  Peoria (IL)
    Eaton, James E., 611–614, 611*f*
  Quincy (IL)
    Kaiser, Daniel, 674–675, 675*f*
  Rock Island (IL)
    Carse & Ohlweiler, 691–694, 693*f*
quart whiskey
  LaSalle (IL)
    DeSteiger Glass Co., 548–549
  Rockford (IL)
    Graham's Distillery, 689, 689*f*
quart whiskey cylinder
  Chicago (IL)
    DeMor & Co., 232, 232*f*
    Garrick & Cather, 263, 263*f*
Quigley, Hopkins, and Lea (Alton), 96–97, 96*f*
Quigley, Hopkins & Co. (Alton), 96
Quigley, Webb C., 96–97
Quigley Bros. & Co. (Alton), 96

R
R. Boil & Co. (Blue Island), 158, 158*f*
R. E. Hickey's Soda Water Factory (Peoria), 611, 612*f*, 613*f*
R. Hickey & Co. (LaSalle), 551–552, 551*f*
R. T. Thomas & Co. (Chicago), 440–441, 440*f*
R. Thompson & Co. (Bloomington), 132
Racoczy mineral waters, 88
Railroad Exchange Hotel (Peoria), 626
railroads
  directories, as data sources, 3, 9
  Hillsboro (IL), 600

  Pana (IL), 600
  Peoria (IL), 630
  Prentice (IL), 653
  refrigerated cars and brewery-product expanded use, 34, 35
  West Urbana (IL), 740
Randall, Amos, 262, 312, 312*f*
Ranft, John, 541
Ranken, John, 331
rat poisons
  Browen's Rat Killer (Bloomington), 144, 145*f*, 148, 149
  Peoria (IL)
    Davis, Robert, 611
Rayner, James B., 206–207, 206*f*, 207*f*
Read, Francis A., 500–501, 500*f*
Read & Bartlett (Freeport), 500–501, 500*f*
Ready Relief, 662, 662*f*
rectangular bottles
  Aurora (IL)
    Budlong, William C., 104–105, 105*f*
  Belleville (IL)
    E. T. Flanagan & Co., 117–118, 117*f*, 118*f*
  Bloomington (IL)
    C. B. Castle & Co., 46–47, 46*f*, 121–122
    Dr. H. S. Woodard, 154–156, 154*f*, 155*f*
    Funk & Lackey, 123–124, 124*f*
    Hill & Vanatta, 125–126, 126*f*
    Howe, Bliss S., 126–128, 126*f*
    Lackey and Brothers, 129–131, 130*f*
    Wakefield & Co., 137–139*f*, 137–150, 141–144*f*
    Wells, Edwin M., 151–154, 152*f*
    Woodard & Howe, 156–157, 157*f*
  Cairo (IL)
    Pfifferling, Charles, 170, 170*f*
  Capron (IL)
    Field, Platus A., 172, 172*f*
  Carrollton (IL)
    Smith, Dr. Alexander H., 176–179
  Chicago (IL)
    A. W. Sargent & Co., 394–396, 395*f*
    Adams, Dr. A. L., 186–187, 186*f*
    B. S. Cory and Son, 225–226, 225*f*
    Badeau, William C., 194–195, 195*f*
    Bailey & Eaton, 195–196, 196*f*
    Bligh, Thomas, 201–202, 202*f*
    Butt, Charles, 210–212, 210*f*, 211*f*
    Cassily & Co., 212–213, 213*f*
    Christie, Dr. Abel H., 216–219, 217*f*, 218*f*
    Collins, Dr. Silas F., 221–222, 222*f*
    Cory's and Collins' bitters, 79
    Cram, Dr. John S., 227–228, 227*f*
    Drs. C. M. Fitch & J. W. Sykes, 434–436
    Eilert, Jacob K., 242–243, 242*f*, 243*f*
    Fahrney, Peter, 250–251, 250*f*
    Flagg, Edward H., 257–258, 257*f*
    Foster, Charles M., 259–260, 259*f*, 260*f*
    Garden City Chemical Works, 261–262
    Gardiner, Charles H., 262–263, 263*f*
    Gillett Chemical Works, 265–266, 266*f*
    H. A. Stone & Co., 426–427, 427*f*
    H. E. Bucklen & Co., 208–210, 209*f*
    Hamlin, John A., 271–279, 273*f*, 274*f*, 275*f*
    Henry H. Shufeldt & Co., 410–411, 411*f*
    Hutchins, John W., 296, 296*f*
    Huyck & Knox, 312
    Huyck & Randall, 312, 312*f*
    Johnson, Dr. [Hosmer A.], 313–314, 313*f*
    Lurton, Dr. Lycurgus L., 367, 367*f*
    Miller, Dr. [Adam], 371–372, 371*f*
    P. W. Gillet & Son, 267, 267*f*
    Sawyer, Sidney, 399–400, 399*f*
    Sears & Smith, 406, 407*f*
    Sloan, Walter B., 413–418, 414*f*, 415*f*

Sweet, Henry J., 432, 432f
Taylor, Newton S., 438–439, 438f
Thompson, 441, 441f
Walker & Taylor, 443–446, 444f, 445f
Warner, Dr. C. D., 446–449, 446f, 447f
Wells, Edwin M., 450–452, 451f
Willard, Joseph, 453
Dixon (IL)
    Herrick, George L., 473–475, 474f
Franklin Grove (IL)
    Dr. U. C. Roe & Sons, 497–498, 497f, 498f
Jacksonville (IL)
    Hamilton, Dr. William, 525–529, 525f, 528f
    Hockenhull, Robert, 530–531, 530f, 531f
Jerseyville (IL)
    Freeman, Dr. John D., 535–538, 535f, 536f
Jonesboro (IL)
    Parks, Dr. Luther K., 543–544, 544f
LaSalle (IL)
    Babcock, Dr. A. L., 548, 548f
Lebanon (IL)
    Smith, Henry H., 554, 554f
Lewistown (IL)
    The Medicine Co., 557–558, 557f
Louden City (IL)
    Wills, Dr. John C., 563, 563f
Maryland (IL)
    Haldeman, Samuel, 565, 565f
medicine bottles
    body shape definition, 18, 20f
Mendota (IL)
    Viera, Manuel J., 569, 570f
Morris (IL)
    Lee, Dr. Charles M., 575–576, 575f
    LeRoy, Dr. David, 574–575, 574f
Naperville (IL)
    Keith, Aylmer, 579–580, 580f
New Athens (IL)
    Baumann, Peter, 587, 587f
Ottawa (IL)
    Griggs, Edward Y., 595–597, 595f, 596f
Pekin (IL)
    Brown, Dr. Alfred, 600, 600f
Peoria (IL)
    Bastow, James D., 603, 603f
    Cleveland, Frederick, 609–610, 609f
    Davis, William H., 610–611, 611f
    Kruse, Dr. Hero, 636–637, 636f
Quincy (IL)
    Bunnel, Dr. D[avid], 659–660, 659f
    Burleigh & Brothers, 660–662, 661f, 662f
    Kalb, Absalom J., 675–677, 675f, 676f
    Schroeder, Herman, 685, 685f
ring-neck tapered collars
    as chronological marker, 71
    and slug plates, 12, 72
Springfield (IL)
    Diller, Roland W., 700–702, 700f
    Lewis, Thomas, 713–714, 714f
    Myers, Henry C., 721
    North, Alfred A., 721–723, 722f
    Owen, Thomas J. V., 724–725, 724f
    Slemmons & Conkling, 726–727, 726f
study inclusion rule, 37
sunken-panel
    and slug plates, 72
Waukegan (IL)
    B. S. Cory & Son, 738–739
Western Saratoga (IL)
    Penoyer, Dr. Hiram, 741–742, 741f
Red Cloud Bitters, 385, 439, 439f
Red Drops, 43, 75
Red Jacket Bitters, 57, 70, 79, 374–375, 381–386, 382–386f, 405–406, 406f
Redington, Frederick, 393–394

Redlich, Henry O., 389, 389f
Redman, James B., 715
Reed, George, 609
Reed, Josiah H., 390–391, 390f
Reed & King (Jacksonville), 530
Reen, Augustus W. H., 641, 641f
Reid, Dr. John
    medicinal mixtures ad, 30
Reiland & Ranney (Chicago), 259, 706
Reinhardt, Mayer & Co. (Springfield), 726
Reinhold, William, 373–374, 374f
Religious Philosophy Publishing
    Association, 384
Reuter, Charles J., 553–554, 553f
Reuter, Fritz, 641
Rheumatic Liniment, 719
Rheumatic Mixture, 222
Rheumatic Specific, 254
ribbed barber-pole spiral embossing
    Naperville (IL)
        Powell & Stutenroth, 581–582, 581f, 582f
Richards & Hudson, 718, 718f
Richardson, Dr. Holland W., 743–744, 744f
Mr. Richter, 673
Ricks, Henry, 532
Ricks, John W., 3, 522, 523, 524, 531, 532, 533
ridged upper shoulder
    Decatur (IL)
        Fuller's Banner Ink, 471–472, 471f
    Springfield (IL)
        Fuller's Banner Ink, 706–707, 706f
ridged-barrel shaped
    Chicago (IL)
        Wolford, Jacob A., 453–455, 454f
Rinehardt, Matthew, 725–726, 725f
ringed barrel shape bottles
    Chicago (IL)
        Chapin & Gore, 214–216, 214f
ring-neck bottles
    Aurora (IL)
        Budlong, William C., 104–105, 105f
    Cairo (IL)
        Breihan, Henry, 10f, 69, 165–167, 165f, 166f
        Lohr, Andrew, 168–170, 168f, 169f
        Schutter, William H., 171, 171f
    Carrollton (IL)
        Smith, Dr. Alexander H., 176–179
    Champaign (IL)
        Miller, Nicholas, 182–183, 182f
    Chester (IL)
        Wenda, John, 184–185, 185f
    Chicago (IL)
        A. Mette & Bros., 49–50, 49f
        Alfred & William Strickland Co., 428–429
        Amberg & Stenson, 191, 191f
        Buck & Rayner, 206–207, 206f, 207f
        Carpenter, Andrew, 212, 212f
        Chapin & Gore, 214–216, 214f, 215f, 216f
        Christin, Arthur, 219–221, 221f
        DeMor & Co., 232, 232f
        Dempsey, Daniel, 232–234, 233f
        Drake, John B., 240–241, 240f, 241f
        Filer, Henry, 255–257, 256f
        Flagg, Edward H., 257–258, 257f
        Garrick & Cather, 263, 263f
        Gillett Chemical Works, 265–266, 266f
        Gottleib Wurster & Co., 456–461, 457f, 458–461, 460f
        Hamlin, John A., 271–279, 272f
        Hausburg, William A., 282–283, 282f
        Hayes Brothers, 284–287, 286f, 287f
        Hennessey, Thomas, 290–291, 290f
        Hunt & Kenyon, 294–295, 294f, 295f
        Hutchinson & Co., 300, 301, 301f, 302f, 304, 305f, 308, 308f, 309f

Israel, Stephen, 312–313, 313f
John A. Lomax, 336, 337f, 345, 346f, 347, 347f, 355f, 356
John Anderton & Co., 191, 192f
Joseph Apple & Co., 192, 193f
Keeley, Michael, 317–320, 318f, 319f
Keeley & Brother, 315–317, 315f, 316f
King, Thomas, 320–322
Kirchhoff Brothers, 322–323, 322f
Landsberg, Moses G., 323–325, 324f, 325f
Lang, Frederick C., 328–330, 328f
Lang Brothers, 325–328, 327f
Palmer House Hotel, 376–377, 376f
Perrine, Charles O., 377–381, 377f
Samuel Myers & Co., 374, 374f
Sass & Hafner, 396–399, 396f, 397f
Seibt & Haunschild, 407–408, 408f
Steel & Co., 421–422, 421f
Stenson, James, 422–426, 423f, 424f, 426f
T. Foerster & Co., 258, 258f
Taylor & Bro., 436–438, 436f, 437f
Van Buskirk & Henry, 442–443, 443f
Warner, Dr. C. D., 446–449, 446f, 447f
Wolford, Jacob A., 453–455, 454f
Ypsilanti Springs Co., 461–462, 461f
Danville (IL)
    Carey, Patrick, 467, 467f
definition, 23, 24f
East St. Louis (IL)
    C. Lutt & Co., 479–481, 479f, 480f
    Schroeder, Edward, 482–486, 483f
Elgin (IL)
    Hausburg Brothers & Co., 490–491, 491f
LaSalle (IL)
    DeSteiger Glass Co., 548–549, 548f
Lemont (IL)
    Bolton, John G., 554–557, 556f
Naperville (IL)
    Stenger, John, 582–584, 583f
Peoria (IL)
    Parent, Anthony, 638–640, 639f
    Pletscher, John C., 640–641, 640f
    Weber, George, 643–645, 644f
quart ciders
    Industrial Expansion era (1870s), 79
Rockford (IL)
    Graham's Distillery, 689, 689f
Springfield (IL)
    Matthew Rinehardt & Co., 725–726, 725f
    Slemmons & Conkling, 726–727, 726f
squared ring collar
    Alton (IL), 96–97, 96f
study inclusion rule, 35
tapered-collar
    as chronological marker, 71
    definition, 24, 25f
whiskey flasks, 24
ring-neck cider finish. See also cider-finish
    ring-neck quarts
body shape definition, 18, 18f
Chicago (IL)
    A. Mette & Bros., 49–50, 49f
Elgin (IL)
    Hausburg Brothers & Co., 490–491, 491f
Lemont (IL)
    Bolton, John G., 554–557, 556f
Peoria (IL)
    Gipps, Cody & Co., 629–630, 629f
    Metz & Pletscher, 637–638, 638f
    Wichmann, John, 645–646, 645f
Prentice (IL)
    Berry, John M., 651–653, 652f, 653f
Rock Island (IL)
    Sargent, William B., 694–695, 695f
Rising Sun Hotel (Cairo), 170

Robin, McMillan & Co. (Edinburgh), 422
Robinson, Archibald M., 377
Robinson, Leslie, 630
Roby, Edward, 463
Rock Island Bottling Works (Rock Island), 692–693
Rodemeyer, Henry, 391
Rodemeyer, Louis, 391, 391*f*, 402
Rodman House (Rock Island), 695
Roe, Frederick, 498
Roe, Dr. John, 497, 498
Roe, Dr. Malcom C., 497
Roe, Nathaniel, 498
Roe, Dr. Uriah C., 497–498, 497*f*, 498*f*
Roe Family Medicine Co. (Franklin Grove), 498
Roemheld, Julius, 374
Roemheld's Chemical Laboratorium (Chicago), 374
Roe's Ague Pills, 498
Roe's Balm of Gilead Ointment, 498
Roe's Cough Drops, 498
Roe's Hair Balsam, 498
Roe's Physic, 498
Rogers, Hiram, 680
Rogger, Ernst, 45, 114, 116–117, 116*f*
rolled lip
　Belvedere (IL)
　　Bristol, Solomon W., 120, 120*f*
　Bloomington (IL)
　　Wakefield & Co., 137–150, 137*f*, 139*f*, 140*f*
　　Wells, Edwin M., 151–154, 151*f*
　Carlyle (IL)
　　Gunn, John S., 174–175, 175*f*
　Chicago (IL)
　　A. W. Sargent & Co., 394–396, 395*f*
　　Bryan, Frederick A., 205, 205*f*
　　Farrell, William B., 251–255, 252*f*, 253*f*
　　Hamlin, John A., 271–279, 272*f*
　　Sloan, Walter B., 413–418, 414*f*
　　Wells, Edwin M., 450–452, 450*f*
　as chronological marker, 71
　definition, 24, 24*f*
　Greenville (IL)
　　Floyd, Edward A., 511–512, 512*f*
　Jacksonville (IL)
　　Hamilton, Dr. William, 525–529, 526*f*, 527*f*
　Jerseyville (IL)
　　Freeman, Dr. John D., 535–538, 535*f*, 537*f*
　Jonesboro (IL)
　　Parks, Dr. Luther K., 543–544, 544*f*
　Lebanon (IL)
　　Smith, Henry H., 554, 554*f*
　Naperville (IL)
　　Keith, Aylmer, 579–580, 580*f*
　Oquawka (IL)
　　McDill, Dr. Andrew C., 592, 592*f*
　Peoria (IL)
　　Farrell, Hiram G., 614–617*f*, 614–621
　Quincy (IL)
　　Kalb, Absalom J., 675–677, 675*f*
　Springfield (IL)
　　Diller, Roland W., 700–702, 700*f*
　　Lewis, Thomas, 713–714, 714*f*
root beer. See also small beer
　Chicago (IL)
　　Cronk, Edward Y., 228–230, 229*f*, 230*f*
　　Hutchinson & Co., 297
　　John A. Lomax, 362, 365
　　Miller, Andrew J., 50, 50*f*, 373, 373*f*
　Quincy (IL)
　　Flynn, John J., 668
　stoneware definitions, 18, 21*f*, 41–42
roped corners
　Chicago (IL)
　　Gillett Chemical Works, 265–266, 266*f*

Rose Gin, 212, 213
round bottles
　body mold lines, 12
　Chicago (IL)
　　Sanford Manufacturing Co., 393–394, 393*f*
　Millville Round
　　study exclusion rule, 37, 38*f*
　Ottawa (IL)
　　Cavarly, Dr. William W., 593, 593*f*
　and slug plates, 12
　as chronological marker, 72–73
rounded shoulder bottles
　Chicago (IL)
　　Buck, George, 205–206, 205*f*
　　Butt, Charles, 210–212, 211*f*
　　Lord, James J., 717–719
　Quincy (IL)
　　Schroeder, Herman, 685, 685*f*
　Springfield (IL)
　　North, Alfred A., 721–723, 722*f*
round-shouldered cylinders
　Springfield (IL)
　　Melvin, Samuel H., 719–721, 720*f*
Rouze, Paul, 239, 391–393, 392*f*, 393*f*
Royal Cleveland and Price Companies, 610
Royal Crown ink, 394
Royal Palm (Chicago), 263
Ruf, John, 533, 737–738, 737*f*
Ruf, Sofia, 533, 737, 738
Ruff, John, 533, 737, 738
Ruff, Sofia, 533, 737
rum
　Quincy (IL)
　　Schrader, Frank, 684
Rupert, [Henry] E., 97, 97*f*, 547, 547*f*
Rush Medical College (Chicago)
　Haines, W. S., 366
　Johnson, Dr. [Hosmer A.], 314
　Miller, Dr. DeLaskie, 371
　Siebel, J. E., 366
　Stone, Dr. Reuben R., 427
Ryan, Cornelius, 234, 236
Rylands, Ben, 68–69
Rylands, Dan, 68–69
Rylands & Codd (Barnsley, England), 55*t*, 68–69

S
S. & J. Ruf (Waterloo), 737–738, 737*f*, 738
S. A. Maxwell & Co. (Mundelein), 132
S. Hickey (LaSalle), 551
S. J. Smith & Co. (Chicago), 419, 419*f*, 697–698
S. M. Bartlett & Co., 681
S. M. Hickey & Co. (Bloomington), 124–125
Sabine, F. A., 393
Sabine Medicine Co. (Chicago), 393
salt-glaze quart stoneware
　Springfield (IL)
　　Kelleher & Co., 712–713, 712*f*
salve jars
　body shape definition, 18, 20*f*
　Chicago (IL)
　　Sloan, Walter B., 413–418, 414*f*
Samuel Myers & Co. (Chicago), 374, 374*f*
Sanative Vegetable Pills, 676, 715
Sandburg, Carl, 279
Sanders, J. H., 470, 471, 471*f*
Sanders & Co. (Darmstadt), 470, 471, 471*f*
Sands Brewery, 322
Sands' Pale Cream Ale, 171
Sanford, William, Jr., 393–394
Sanford Ink Corp. (Bellwood), 394
Sanford Manufacturing Co. (Chicago), 393–394, 393*f*, 440
　fires, 393, 394

Saratoga mineral water
　Chicago (IL)
　　Israel, Stephen, 312–313, 313*f*
Saratoga Springs water
　Quincy (IL)
　　Schroeder, Herman, 685
Saratoga "Vichy," 208
Saratoga-style mineral water bottles
　body shape definition, 15, 18*f*
　Chicago (IL)
　　Buck & Rayner, 206–207, 206*f*
　　Peuser & Kadish, 381, 381*f*
　　Ypsilanti Springs Co., 461–462, 461*f*
　last use of, 461
　Perry (IL)
　　Perry Springs, 646–648, 647*f*
Sargent, Abraham W., 394–396, 395*f*
Sargent, Ezekiel H., 396
Sargent, Nute, 695
Sargent, William B., 694–695, 695*f*
Sarkander, Fritz, 292
sarsaparilla, 229. See also small beer
　Alton (IL)
　　Weisbach, Christian, 99
　Arlington Heights (IL)
　　Muller, Frederick W., 103–104, 103*f*, 104*f*
　Dr. Cronk's, 229, 230
　East St. Louis (IL)
　　Schroeder, Edward, 486
　Jacksonville (IL)
　　Buell & Schermerhorn, 524
sasparilla. See root beer
Sass, Louis, 103, 396–399, 396–399*f*, 404, 475–476, 475*f*, 476*f*
Sass & Bro (Dunton), 103, 475
Sass & Hafner (Chicago), 103, 314–315, 396–399, 396–399*f*, 404, 475
Sass & Muller (Dunton), 475–476, 475*f*, 476*f*
Saulpaugh, David, 293, 295–296
Sawyer, Nathanial, 399, 400
Sawyer, Paige & Co. (Chicago), 400
Sawyer, Sidney, 399–400, 399*f*
Sawyer's Fluid Extract of Bark, 34, 399–400, 399*f*
Schafberg & Jarrait (Detroit, MI), 229
"Mrs. Schafer," 400, 401*f*
Schaplin's Select Powders, 400
Schauffhausten, William, 137
Schausten, William, 135–137, 136*f*
Schermerhorn, Charles, 523–525, 533–534
Schiffer, Tobias, 118
Schlitz Beer
　Chicago (IL)
　　Welch & Goggin, 450
　Peoria (IL)
　　Weber, George, 645
Schlitz Milwaukee Export and Bottled Beer, 450
Schmidt & Knecht (Alton), 89
schnapps
　Chicago (IL)
　　Van Buskirk & Henry, 442–443, 442*f*
Schneider, George, 109–110, 109*f*
Schneider & Krohe (Beardstown), 109–110, 109*f*
Schnitzler, Martin, 206, 207
Schoenbein, Thomas, 481–482, 482*f*
Schonwald, Francis (Franz), 391, 402–403, 402–403*f*
schoolhouse ink bottles
　Bloomington (IL)
　　J. W. Maxwell & Co., 132–133, 132*f*
　body shape definition, 18, 19*f*
　Chicago (IL)
　　Cox, Lewis, 226–227, 226*f*
　　Gillett Chemical Works, 265–266, 265*f*

Schott, Christian, 516, 519–520, 519f
Schott, Gerhardt, 519–520
Schott, Martin, 519–520
Schrader, August, 684, 684f
Schrader, Frank, 684–685
Schrader's Bitters, 684–685
Schrader's Highland Bitters, 684–685
Schrader's Kentucky Whisky, 684, 684f
Schriber, Charles, 314, 398, 403–404, 404f
Schriber & Co. (Chicago), 398
Schriber & Joss (Chicago), 314, 403–404, 404f
Schroeder, Edward, 59, 168, 170, 482–486
Schroeder, George, 486
Schroeder, Herman, 685, 685f
Schroeder, Noble, 432
Schroeder's Camphorated Cream, 685
Schroeder's Cold Cream, 685
Schroeder's Drug Store (Quincy), 685
Schroeder's Freckletine, 685
Schroeder's Imperial Balm, 685, 685f
Schroeder's Imperial Cologne, 685
Schroeder's Imperial Cologne Water, 685
Schroeder's Improved Tooth Powder, 685
Schroeder's Sparkling Soda Water, 685
Schultze & Eggers (St. Louis), 538
Schulz, [Otto], 404–405, 405f
Schulz & Hess (Chicago), 404–405, 405f
Schutter, William H., 171, 171f
Schwab, Charles H., 383, 385, 405–406, 406f
Schwab, McQuaid & Co. (Chicago), 374, 383, 385, 405–406, 406f
Schwab, Pieters & Co. (Chicago), 374, 383, 385, 386f, 406
Scotch Ale, 94
Scott, George E., 186–187
Scovill, Harvey (Harry), 438, 444
Scovill Co. (Chicago), 438, 444
screw caps
  embossed star, 214, 216f
  glass
    Chicago (IL), 214, 215f, 216f, 240, 376
  rubber, 214, 215f
Sears, John, Jr., 406, 407f
Sears & Bay (Chicago), 406
Sears & Smith (Chicago), 406, 407f
The Sebastopol (Bloomington), 128
Seibt, Frederick (Fritz), 407–409, 408f, 409f
Seibt & Haunschild (Chicago), 407–408, 408f
Seidlitz Powder, 207
Seipp, Conrad, 409–410, 410f
Seipp & Lehman (Chicago), 409–410
Seipp Brewing Co. (Chicago), 410
Mr. Sell, 56, 66
Sell, Hale & Co. (St. Louis), 680
Selters mineral waters, 88, 115
  embossed-glass labeling, 42
  Peoria (IL)
    Singer, Charles J., 643
Selters-style bottles
  Alton (IL)
    Weisbach, Christian, 98–100, 100f
  Belleville (IL)
    Fisher, Joseph (aka Fischer), 113–115, 114f
  body style definition, 15, 18f
  East St. Louis (IL)
    Schroeder, Edward, 482–486, 486f
  Galesburg (IL)
    Edwards, [Isaac?] R., 510, 510f
  and petal-paneled bottles, 23
Selters-style neck
  Chicago (IL)
    Moench & Reinhold, 373–374, 374f
seltzer
  Aurora (IL)
    Green Brothers, 106

Cairo
  Breihan, Henry, 167, 421
  Lohr, Andrew, 167, 421
Chicago (IL)
  Hayes Brothers, 287
  John A. Lomax, 364, 365
  Seibt & Haunschild, 407, 408
  Sproat, John, 421
  W. P. Brazelton & Co., 203, 204f
East St. Louis (IL)
  Schroeder, Edward, 486
Litchfield (IL)
  Weber, Frederick, 562
Pana (IL)
  Weber, Frederick, 562
Peoria (IL)
  Fellrath, Hubert, 622–625, 625f
Rock Island (IL)
  Carse & Ohlweiler, 693
seltzer-water dispensers
  study exclusion rule, 37, 40f
semi-cathedral shape
  Chicago (IL)
    Gillett Chemical Works, 265–266, 266f
serifed embossed lettering, 9
  Bloomington (IL)
    Wakefield & Co., 137–150, 142f
  Chicago (IL)
    Miller, Andrew J., 50, 373
  Peoria (IL)
    Farrell, Hiram G., 614, 614f
    Fellrath, Hubert, 622, 623f
Seymour, H. J., 56–57
Shackelford & McCoy (Marion County, IN), 563
shampoo
  Mendota (IL)
    Viera, Manuel J., 569, 570f
  Viera's Toilet Shampoo, 569, 570f
sharp slightly rounded shoulders
  Chicago (IL)
    Gardiner, Charles H., 262–263, 263f
sharp-shouldered blob tops
  Rock Island (IL)
    Carse & Ohlweiler, 68, 691–694, 692f
sharp-shouldered cylinders
  Chicago (IL)
    Buck, George, 205–206, 206f
    Farrell, William B., 251–255, 252f, 253f
    John A. Lomax, 358, 358f, 359f
  Oquawka (IL)
    McDill, Dr. Andrew C., 592, 592f
sharp-sided "coffin" flask
  body shape definition, 13, 14f
Shawneetown (IL)
  Docker & Ferring, 538
  Lee, Dr. Charles M., 575
Shean, Charles, 468–469, 468f
Sheboygan Mineral Water, 208
  Codd Marble gravitating stopper sodas, 693
Sherman House (Chicago), 384, 435
shield-shaped panels
  Chicago (IL)
    Drs. C. M. Fitch & J. W. Sykes, 434–436
    Lomax, George, 332–334, 333f
    Wolford, Jacob A., 453–455, 454f
  definition, 13
Shinn, John
  Panacea ad, 30
Shinn's Panacea, 30
Shonwaldt, Francis (aka Franz Schonwald), 57
shoofly flasks
  Alton (IL)
    Alton Glass Co., 86, 86f
  body shape, 13

Chicago (IL)
  Drake, Parker & Co., 241, 241f
  Palmer House Hotel, 376–377, 376f
  Samuel Myers & Co., 374, 374f
LaSalle (IL)
  DeSteiger Glass Co., 548–549, 548f
Prentice (IL), 64
  Berry, John M., 651–653, 652f, 653f
shoofly flasks, pint
  Prentice (IL)
    Berry, John M., 651–653, 653f
shoofly flasks, quart
  Prentice (IL)
    Berry, John M., 651–653, 652f, 653f
Short Blake pharmacy bottles, 73
short bottles
  Chicago (IL)
    Farrell, William B., 251–255, 252f
short flared neck w/ lip
  Springfield (IL)
    A. & M. Lindsay, 714–717, 715f, 716f
short lozenge-style blob-top sodas
  Litchfield (IL)
    Weber, Frederick, 560–563, 562f
short necks. See also short necks w/ mid-neck ring
  Alton (IL)
    George H. Betts & Co., 86–87, 87f
    Jacob Buff and Max Kuhl, 88–89, 88f, 89f
  Belleville (IL)
    Fisher, Joseph (aka Fischer), 113–115, 114f
    Koob, August G., 46, 46f
  Bloomington (IL)
    Hill & Vanatta, 125–126, 126f
    Wakefield & Co., 137–150, 143f
  Cairo (IL)
    Lohr, Andrew, 168–170, 168f, 169f
  Carlyle (IL)
    Gunn, John S., 174–175, 175f
  Chicago (IL)
    A. Mette & Bros., 369–371, 370f
    Barothy & Cook, 47–48
    Bennett Pieters & Co., 381–386, 382f
    Bryan, Frederick A., 205, 205f
    Dempsey, Daniel, 232–234, 234f
    Farrell, William B., 251–255, 254f
    Gottleib Wurster & Co., 456–461, 459f
    Hamlin, John A., 271–279, 272f–274f
    Hayes Brothers, 284–287, 284f
    John A. Lomax, 340, 342f, 345, 345f, 348, 349f
    Morrison, William, 50–51, 51f
    Sass & Hafner, 396–399, 397f
    Sears & Smith, 406, 407f
    Sloan, Walter B., 413–418, 414f
    Snyder & Eilert, 419–420, 420f
    Van Buskirk & Henry, 442–443, 442f
  Danville (IL)
    Galtermann, William, 468, 468f
    Shean, Charles, 468–469, 468f
  Dunton (IL)
    Sass & Muller, 475–476, 475f, 476f
  East St. Louis (IL)
    C. Lutt & Co., 479–481, 480f
    Schroeder, Edward, 482–486, 485f, 486f
  Greenville (IL)
    Floyd, Edward A., 511–512, 512f
  Jacksonville (IL)
    Hamilton, Dr. William, 525–529
  Kankakee (IL)
    Kammann, [Dietrich] H., 545–546, 546f
  Lebanon (IL)
    Smith, Henry H., 554, 554f
  Maryland (IL)
    Haldeman, Samuel, 565, 565f

(continued)

short necks (continued)
   Mattoon (IL)
      J. Fife & Co., 568, 568f
   Nokomis (IL)
      Brophy, Dennis P., 588, 588f
   Olney (IL)
      Bower, William, 591–592, 591f
   Ottawa (IL)
      Finkler, Alexander, 594–595, 594f
      Kagy, John M., 597, 597f
   Pana (IL)
      F. Weber & Brother, 599–600, 599f
   Peoria (IL)
      Fellrath, Hubert, 622–625, 624f
      Francis & Spier, 626, 626f
      Gillig & Singer, 627–629, 628f
      Headley, Dr. William S., 632, 632f
      Singer, Charles J., 642–643, 642f, 643f
      Strehlow, Robert, 643, 644f
   Pontiac (IL)
      Caldwell & McGregor, 651, 651f
   Princeton (IL)
      Bodley, Dr. A. R., 654–655, 655f
   Quincy (IL)
      Bunnel, Dr. D[avid], 659–660, 659f
      Burleigh & Brothers, 660–662, 661f, 662f
      Schroeder, Herman, 685, 685f
   Springfield (IL)
      A. & M. Lindsay, 714–717, 715f, 716f
      Diller, Roland W., 700–702, 700f
      Dunning, Andrew J., 703, 703f
      Fixmer, John P., 704–705, 704f
      Johnson & Peterson Co., 709–712, 710f, 711f
      Lewis, Thomas, 713–714, 714f
      Lord, James J., 717–719, 717f
      McNeil, William W., 719, 719f
      Melvin, Samuel H., 719–721, 720f
      North, Alfred A., 721–723, 722f
      Owen, Thomas J. V., 724–725, 724f, 725f
      Van Deusen, Martin M., 727–729, 727f
   Staunton (IL)
      Hoffmann, Christian, 730, 730f
      Swain, Calvin H., 430–431, 431f
short necks w/ mid-neck ring
   Springfield (IL)
      Owen, Thomas J. V., 724–725, 724f
short shouldered cylinders
   Chicago (IL)
      Redlich, Henry O., 389, 389f
shoulder decorations
   applied seals
      definition, 23, 23f
      Kirchhoff Brothers (Chicago), 322–323
   bumps
      definition, 23, 23f
   fluted
      definition, 23, 23f
   petal-paneled
      definition, 23, 23f
   Quincy (IL)
      M. R. & H. W. Lundblad, 677–681
   raised rings
      definition, 23, 23f
   ridges
      definition, 23, 23f
shoulder double-ring
   Quincy (IL)
      M. R. & H. W. Lundblad, 677–681
shoulder shapes. *See also specific shapes*
   definition of terms, 22–23, 23f
shouldered blob-top bottles. *See also shouldered blob-top quarts; shouldered blob-top sodas*
   Bloomington (IL)
      Bottling House, 120–121, 121f
      Schausten, William, 135–137, 136f
   Carlyle (IL)
      J. M. Menkhaus, 176, 176f
   Chester (IL)
      Wenda, John, 184–185, 185f
   Chicago (IL)
      Amberg & Stenson, 191, 191f
      Barrett & Barrett, 199, 199f
      Dorman & Co., 238–239, 239f
      Geer, Schubael, 264, 264f
      J. Grossenheider & Co., 268–269, 269f
      Victor Barothy & Co., 196–197, 196f
   definition, 24, 25f
   Joliet (IL)
      Paige, J. D., 540–542, 541f, 542f
   LaSalle (IL)
      L. Eliel & Co., 549–550, 550f
   Peoria (IL)
      Metz & Pletscher, 637–638, 638f
shouldered blob-top quarts
   Chicago (IL)
      Dorman & Co., 238–239, 239f
      Hutchinson & Co., 307, 308f
      John A. Lomax, 345, 345f, 346f
   East St. Louis (IL)
      Schroeder, Edward, 482–486, 483f
shouldered blob-top sodas
   Alton (IL)
      Weisbach, Christian, 98–99, 99f
      Weisbach, Jacob, 98, 98f
      Yoerger, Augustin, 100–102, 101f, 102f
   Beardstown (IL)
      Schneider & Krohe, 109–110, 109f
   Belleville (IL)
      Ab Egg, Louis, 110–112, 111f
      Fisher, Joseph (aka Fischer), 113–115, 113f, 114f
      Fisher & Ab Egg, 115–116, 115f
      Fisher & Rogger, 116–117, 116f
      Koob, August G., 46, 46f
   Bloomington (IL)
      Mueller & Stein, 134, 134f
      S. M. Hickey & Co., 124–125, 125f
   Cairo (IL)
      Kump, John H., 167–168, 167f
      Lohr, Andrew, 168–170, 168f, 169f
      Schroeder, Edward, 170, 170f
   Carlinville (IL)
      Weber & Mueller, 173–174, 173f
   Centralia (IL)
      Besant, Daniel J., 179–180, 180f
   Centreville (IL)
      Fischer, Christian, 181–182, 181f, 182f
   Chicago (IL)
      A. Mette & Bros., 369–371, 370f
      Ainsworth & Lomax, 188–189, 188f
      Amberg & Green, 190, 190f
      Bebbington, Edward K., 189, 200–201
      Bower & Bryan, 202, 202f
      Clowry & Fitzgerald, 221, 221f
      Dempsey, Daniel, 232–234, 234f
      Dempsey & Hennessey, 234–236, 235f
      Dickinson, George, 236, 236f
      Eaton, James E., 241–242, 241f
      Entwistle, Joseph, 247–248, 248f
      Gottleib Wurster & Co., 456–461, 458–461, 459f, 460f
      Grossenheider & Dickinson, 270, 270f
      Grossenheider & Kruse, 270, 270f
      Hausburg Brothers, 281–282, 281f
      Hayes Brothers, 284–287, 284f, 285f
      Hedlund & Co., 287–289, 288f, 289f
      Hess, Ernst, 291–292, 291f
      Hunt, Robert L., 293–294, 293f
      Hutchinson & Co., 298, 299f, 300, 300f, 301, 302f, 304, 306–308f, 307
      Hutchinson & Dunn, 303, 303f
      J. Grossenheider & Co., 268–269, 268f
      John A. Lomax, 335–336, 336–338f, 338, 340, 342f, 345, 345f, 348
      King, Thomas, 320–322, 322f
      Lang, Frederick C., 328–330, 329–330f
      Lang Brothers, 48, 48f, 325–328, 326f
      Lomax, George, 332–334, 332f
      Louis Rodemeyer & Co., 391, 391f
      Miller, Andrew J., 372–373
      Moench & Reinhold, 373–374, 374f
      Paul Rouze & Co., 391–393, 393f
      Sass & Hafner, 396–399, 398f
      Schonwald, Francis (Franz), 402–403
      Simonds, F. W., 412–413, 413f
      Stenson, James, 422–426, 425f
      Sundell & Co., 429–430, 429f, 430f
      W. P. Brazelton & Co., 203–204, 204f
      William Knackstead & Co., 323, 323f
   Colehour (IL)
      Smith, William S., 462–463, 463f
   Collinsville (IL)
      Bischoff, Benard, 464–465, 464f
   Darmstadt (IL)
      Gross & Fleischman, 470, 470f
      Sanders & Co., 471, 471f
   De Kalb (IL)
      Dee, Richard, 52, 52f
   Decatur (IL)
      Kuny, Fredrick D., 472, 472f
      Kuny & Klohr, 472, 472f
   Dixon (IL)
      Frank J. Finkler & Co., 52–53, 52f
   DuQuoin (IL)
      T. H. & Bro., 477, 477f
   East St. Louis (IL)
      Schroeder, Edward, 482–486, 483f, 484f
   Elgin (IL)
      Dettmer, William, 489, 489f
      Hausburg Brothers & Co., 490–491, 491f
   Freeport (IL)
      Crotty Brothers, 499, 500f
      Read & Bartlett, 500–501, 500f
   Frogtown (IL)
      Burger, Henry, 501, 501f
   Galena (IL)
      Evans, Joseph, 503–505, 504f
      Oetter, Anton, 505–506, 505f
   Galesburg (IL)
      Aley, Jacob H., 507, 507f
      Eaton, James E., 508–509, 509f
      Edwards, George, 509–510, 509f
   Highland (IL)
      Beck & Brother, 517, 517f
      Mueller & Beck, 517–519, 518f
      Weber & Mueller, 520, 520f
   Joliet (IL)
      J. & H. Hickey, 539, 539f
      Paige, J. D., 540–542, 540f
      Paige & Vail, 542–543, 543f
   Lacon (IL)
      Rupert, [Henry] E., 547, 547f
   LaSalle (IL)
      A. Finkler & Co., 550–551, 550f
      Hickey, Stephen M., 552–553, 552f
      R. Hickey & Co., 551–552, 551f
   Lebanon (IL)
      Reuter, Charles J., 553–554, 553f
   Lemont (IL)
      Bolton, John G., 554–557, 555f
   Lincoln (IL)
      Gesner & Lipp, 558, 558f

Litchfield (IL)
   Weber, Frederick, 560–563, 561*f*, 562*f*
Mascoutah (IL)
   Bassler, Nikolaus J., 566–567, 566*f*
   Herold, Ferdinand, 567, 567*f*
   Koob, August G., 567–568, 568*f*
Mattoon (IL)
   J. Fife & Co., 568, 568*f*
Mendota (IL)
   F. Molln & Co., 569, 569*f*
   Molln, Frederick, 54, 54*f*
   Wohlers, Edward, 569–570, 570*f*
Metamora (IL)
   Bottling House, 570–571, 571*f*
Monee (IL)
   J. & A. Vatter, 572, 572*f*, 573*f*
   Vatter & Merz, 572–573, 573*f*
Murphysboro (IL)
   Kuehle, Frederick A., 576–577, 576*f*
Ottawa (IL)
   B. W. Coutant & Co., 593–594, 593*f*
Pana (IL)
   F. Weber & Brother, 599–600, 599*f*
Pekin (IL)
   Zerwekh, Gottlieb J., 600–602, 601*f*, 602*f*
Peoria (IL)
   Cairns & Hickey, 606–607, 607*f*
   Fellrath, Hubert, 622–625, 623*f*, 625*f*
   Gillig & Singer, 626–627, 627*f*
   Hickey, Ransom E., 632–635, 632*f*, 634*f*
   Humphreys, E. J., 635–636, 635*f*
   J. B. St. Clair & Co., 641, 641*f*
   Matthies, Adolph L., 637, 637*f*
   Reen, Augustus W. H., 641, 641*f*
   Singer, Charles J., 642–643, 642*f*
Princeton (IL)
   Althoff, Friedrich, 653–654, 654*f*
Quincy (IL)
   Boschulte & Brother, 657–658, 657*f*
   Boschulte & Co., 658–659, 658*f*
   Durholt, Henry, 663–667, 663*f*, 664*f*
   Fischer & Boschulte, 667, 667*f*
   Grone & Durholt, 672, 672*f*
   Kaiser, Daniel, 674–675, 675*f*
   Lampe & Boschulte, 677, 677*f*
Rock Island (IL)
   Carse & Ohlweiler, 691–694, 691*f*, 693*f*
Rockford (IL)
   Chisholm, John, 688, 688*f*
Springfield (IL)
   J. Johnson & Co., 707–709, 707*f*, 708*f*
   Johnson & Peterson Co., 709–712
Waterloo (IL)
   Ruf, John, 737–738, 737*f*
West Urbana (IL)
   Doane, William, 740
shouldered blob-top stoneware
  Springfield (IL)
   Kelleher & Co., 712–713, 712*f*
shouldered bottles
  Quincy (IL)
   Bunnel, Dr. D[avid], 659–660, 659*f*
shouldered broad cylinder bottles
  Ottawa (IL)
   Cavarly, Dr. William W., 593, 593*f*
shouldered cylinders
  Belvedere (IL)
   Bristol, Solomon W., 120, 120*f*
  Bloomington (IL)
   Wakefield & Co., 137–150, 143*f*
   Wells, Edwin M., 151–154, 151*f*
  Carlyle (IL)
   Gunn, John S., 174–175, 175*f*
  Chicago (IL)
   A. Mette & Bros., 369–371, 370*f*

Buck & Rayner, 206–208, 207*f*
Lang Brothers, 325–328, 326*f*
Penton, Fisher & Co., 377
Wells, Edwin M., 450–452, 450*f*, 451*f*
Greenville (IL)
   Floyd, Edward A., 511–512, 512*f*
Jerseyville (IL)
   Freeman, Dr. John D., 535–538, 537*f*
Ottawa (IL)
   Cavarly, Dr. William W., 593, 593*f*
Peoria (IL)
   Farrell, Hiram G., 614–621, 614*f*, 615*f*, 616*f*, 617*f*, 618*f*
Springfield (IL)
   McNeil, William W., 719, 719*f*
shouldered oval cylinders
  Peoria (IL)
   Farrell, Hiram G., 614–621, 617*f*
   Farrell & Cox, 622, 622*f*
shouldered ovals
  Alton (IL)
   Quigley, Hopkins, and Lea, 96–97, 96*f*
  Chicago (IL)
   Hamlin, John A., 271–279, 272*f*, 273*f*
   J. H. Reed & Co., 390–391, 390*f*
shouldered quarts
  Cairo (IL)
   Lohr, Andrew, 168–170, 168*f*, 169*f*
  Chicago (IL)
   Schulz & Hess, 404–405, 405*f*
shouldered squat cylinders
  Chicago (IL)
   Hall, Zebulon M., 271, 271*f*
shouldered tall pints
  Chicago (IL)
   Schulz & Hess, 404–405, 405*f*
shouldered-blob-top quarts
  Chicago (IL)
   Barrett & Barrett, 199, 199*f*
shoulder-embossed bottles. *See also* shoulder-embossed cylinders
  Alexander Arbogast maker's mark bottles, 57
  Chicago (IL)
   Alfred & William Strickland Co., 428–429, 428*f*, 429*f*
   Chapin & Gore, 214–216, 216*f*
   Cook & Co., 224–225, 224*f*
   DeMor & Co., 232, 232*f*
   Egbert C. Cook & Co., 224–225, 224*f*
   Garrick & Cather, 263, 263*f*
   Hausburg, William A., 282–283, 282*f*
   Hayes Brothers, 284–287, 286*f*, 287*f*
   Hunt & Kenyon, 294–295, 294*f*, 295*f*
   Hutchinson & Co., 300, 301, 301*f*, 302*f*, 304, 305*f*, 308, 308*f*, 309*f*
   J. Grossenheider & Co., 268–269, 268*f*, 269*f*
   John A. Lomax, 336, 337*f*
   Keeley & Brother, 315–317, 315*f*, 316*f*
   Landsberg, Moses G., 323–325, 324*f*, 325*f*
   Lang, Frederick C., 328–330, 329*f*
   Stenson, James, 422–426, 423*f*
   Taylor & Bro., 436–438, 436*f*, 437*f*
   Warner, Dr. C. D., 446–449, 446*f*, 447*f*
  Galena (IL)
   Evans, Joseph, 503–505, 504*f*
   Volz & Glueck (Galena), 506, 506*f*
  Naperville (IL)
   Stenger, John, 582–584, 583*f*
  Rockford (IL)
   Graham's Distillery, 689, 689*f*
shoulder-embossed cylinders
  Bloomington (IL)
   Wakefield & Co., 137–150, 142*f*

shoulder-embossed only
  Chicago (IL)
   Cook & Co., 224–225, 224*f*
   DeMor & Co., 232, 232*f*
   Keeley & Brother, 315*f*, 316, 316*f*
   Strickland Co., 428, 428*f*
   Warner, Dr. C. D., 446, 446*f*
  Rockford (IL)
   Graham's Distillery, 689, 689*f*
Shufeldt, George A., Jr., 410
Shufeldt, Henry H., 410–411, 411*f*
Shufeldt, William T., 410
Shure, Henry, 412, 412*f*, 491–492, 492*f*
Shure, Joseph, 412, 412*f*, 491–492, 492*f*
Shure & Bro. (Chicago), 412, 412*f*, 491
Shure & Bro. (Elgin), 491–492, 492*f*
Shurtleff, N., 630
Siebel, J. E., 366
Mr. Sieber, 643
Simms, Wm. H. (Lyons, NY), 187
Simonds, F. W., 412–413, 413*f*
Mr. Singer, 635
Singer, Charles J., 626–629, 635, 642–643
siphon bottles, 461
Slemmons, Conkling & Co. (Springfield), 727
Slemmons, Samuel A., 726–727, 726*f*, 727*f*
Slemmons & Conkling (Springfield), 726–727
slightly domed shoulder
  Pontiac (IL)
   Caldwell & McGregor, 651, 651*f*
Sloan, Oscar B., 417
Sloan, Walter B., 413–418, 414–416*f*, 445
   "One-for-a-Man & Two-for-a-Horse," 414
Sloan's Complete Farrier and Cattle Doctor (Sloan, W. B.), 445
Sloan's Condition Powder, 416, 445
Sloan's Family Ointment, 416, 417, 445
Sloan's Garden City (journal), 417
Sloan's Horse Liniment, 417
Sloan's Horse Ointment, 417, 445
Sloan's Ointment, 415*f*, 416, 443–446
Sloan's Tannin Paste, 416
Sloan's Veterinary and Family Medicines (Chicago), 414
slope-shouldered bottles, 340, 342*f*, 549*f*. *See also* slope-shouldered quarts; slope-shouldered sodas; steeply sloped shoulders
  Chicago (IL)
   A. Leibenstein & Co., 330–331, 331*f*
   Davies, George F., 231–232, 231*f*
   Dinet, Joseph, 237–238, 237*f*, 238*f*
   Emmert Proprietary Co., 244–247
   Farrell, William B., 251–255, 254*f*
   Gale Brothers, 261, 261*f*
   George Hofmann & Brothers, 292–293
   Gillett Chemical Works, 265–266
   Gottleib Wurster & Co., 458–461, 460*f*
   H & Bro's, 279–281, 280*f*
   Hutchinson & Co., 308, 308*f*, 309*f*
   J. Grossenheider & Co., 268–269, 269*f*
   John A. Lomax, 343, 343*f*, 345, 345*f*, 348, 349, 349*f*, 350*f*, 355*f*, 356, 357*f*, 358
   Paul Rouze & Co., 391–393, 392*f*
   Perrine, Charles O., 377–381, 377*f*
   Sass & Hafner, 396–399, 398*f*
   Schriber & Joss, 403–404, 404*f*
   Seipp, Conrad, 409–410, 410*f*
   Stenson, James, 422–426, 425*f*
   Taylor, Newton S., 438–439, 438*f*
   Tivoli Bottling Co., 441–442, 441*f*
   Victor Barothy & Co., 196–197, 196*f*
   Welch & Goggin, 449–450, 449*f*
  Danville (IL)
   Stein, John, 469, 469*f*

(continued)

slope-shouldered bottles (continued)
　Decatur (IL)
　　Fuller's Banner Ink, 471–472, 471f
　Dunton (IL)
　　Sass & Muller, 475–476, 475f, 476f
　DuQuoin (IL)
　　Hayes, Edwin, 476–477, 476f
　　T. H. & Bro., 477, 477f
　East St. Louis (IL)
　　C. Lutt & Co., 479–481, 480f
　　Schroeder, Edward, 482–486, 483f, 484f, 485f
　　Thomas Schoebein & Co., 481–482, 482f
　Edwardsville (IL)
　　Harles, Frank, 487, 487f
　Elgin (IL)
　　Hausburg Brothers & Co., 490–491, 491f
　　Henry & Van Buskirk, 490, 490f
　　Shure & Bro., 491–492, 492f
　Freeburg (IL)
　　Darmstatter, Frederick, 499, 499f
　Freeport (IL)
　　Read & Bartlett, 500–501, 500f
　Frogtown (IL)
　　Burger, Henry, 501, 501f
　Galena (IL)
　　Oetter, Anton, 505–506, 505f
　Galesburg (IL)
　　Aley, Jacob H., 507, 507f
　　Edwards, George, 509–510, 509f
　Galva (IL)
　　Ayres, Clarence M., 510–511, 511f
　Greenville (IL)
　　Grafe, P. Herman, 512–513, 513f
　Highland (IL)
　　Amsler Brothers, 516–517, 516f
　　Weber & Mueller, 520, 520f
　Jacksonville (IL)
　　Braun & Killian, 523, 523f
　　Kershaw, Albert, 531, 532f
　　P. Braun & Co., 522–523, 522f
　　Ricks, John W., 532, 532f
　Jerseyville (IL)
　　Boyle, Charles M., 534–535, 534f
　　Spannagel, Johanna, 538–539, 538f
　Kankakee (IL)
　　Kammann, [Dietrich] H., 545–546, 546f
　Kewanee (IL)
　　Spencer, Daniel H., 546, 546f
　Lemont (IL)
　　Bolton, John G., 554–557, 555f
　Litchfield (IL)
　　Gerhardt, William, 559–560, 559f, 560f
　　Neuber, Joseph F., 560, 560f
　　Weber, Frederick, 560–563, 562f
　Naperville (IL)
　　Stenger, John, 582–584, 584f
　Peoria (IL)
　　Cairns & Hickey, 606–607, 607f
　Rock Island (IL)
　　Junge & Beck, 694, 694f
　Springfield (IL)
　　Fuller's Banner Ink, 706–707, 706f
slope-shouldered quarts
　Aurora (IL)
　　Stege, A., 106, 106f
　Blue Island (IL)
　　G. H. Hausburg, 158–160, 159f
　Chicago (IL)
　　Christin, Arthur, 219–221, 220f, 221f
　　Geer, Schubael, 264, 264f
　　Hutchinson & Co., 307, 307f, 308f
　　John A. Lomax, 338, 339f
　　Lang, Frederick C., 328–330, 329f, 330f
　　M. Brand & Co., 202–203, 204f

Elgin (IL)
　Althen, Casper, 487–488, 488f
　Dettmer, William, 489, 489f
　Hausburg Brothers & Co., 490–491, 491f
LaSalle (IL)
　L. Eliel & Co., 549–550, 550f
Peoria (IL)
　Gipps, Cody & Co., 629–630, 629f
　Gipps & Co., 630–632, 631f
　Metz & Pletscher, 637–638, 638f
　Weber, George, 643–645, 644f
　Wichmann, John, 645–646, 645f
Rock Island (IL)
　Junge & Beck, 694, 694f
　Sargent, William B., 694–695, 695f
slope-shouldered sodas. See also slope-shouldered bottles
　Alton (IL)
　　Weisbach, Jacob, 98, 98f
　Aurora (IL)
　　Stege, Albert [?], 44, 44f
　Belleville (IL)
　　Stoltz, Francis (Franz), 119, 119f
　body shape definition, 15, 17f
　Chicago (IL)
　　Hutchinson & Co., 301, 302f, 304, 306–308f, 307, 308
　　Lang, Frederick C., 328–330, 329f, 330f
　　Lomax, George, 332–334, 332f
　　Sundell & Co., 429–430, 429f, 430f
　Peoria (IL)
　　Cairns & Hickey, 606–607, 607f
　　Fellrath, Hubert, 622–625, 623–625f
　Quincy (IL)
　　Durholt, Henry, 663–667, 664f
　　Kaiser, Daniel, 674–675, 674f
slope-shouldered stoneware
　Chicago (IL)
　　A. J. Miller, 50, 50f
slope-shouldered tall pints
　Peoria (IL)
　　Gipps & Co., 630–632, 631f
slug-plate embossed bottles
　Alton (IL)
　　Weisbach, Christian, 98–100, 100f
　Blue Island (IL)
　　G. H. Hausburg, 158–160, 159f
　Cairo (IL)
　　Lohr, Andrew, 168–170, 168f, 169f
　Chicago (IL)
　　Hayes Brothers, 284–287, 286f
　　Hutchinson & Co., 307, 308f
　　Lang, Frederick C., 328–330, 330f
　　Palmer House Hotel, 376, 376f
　　Stenson, James, 422–426, 425f
　　Tivoli Bottling Co., 441–442, 441f
　　Welch & Goggin, 449–450, 449f
　　Ypsilanti Springs Co., 461–462, 461f
　as chronological marker, 72–73
　Colehour (IL)
　　Smith, William S., 462–463, 463f
　Cunningham & Ihmsen (Pittsburgh), 60
　Danville (IL)
　　Stein, John, 469, 469f
　De Kalb (IL)
　　Dee, Richard, 52
　definition, 12
　Elgin (IL)
　　Althen, Casper, 487–488, 488f
　　Dettmer, William, 489, 489f
　flasks, study exclusion rule, 37, 40f
　Jacksonville (IL)
　　J. E. Bradbury & Co., 521–522, 521f
　Monee (IL)
　　Vatter, Adam, Jr., 572–573, 573f

Naperville (IL)
　Stenger, John, 582–584, 584f
Peoria (IL)
　Eaton, James E., 611–614, 611f
　Gipps & Co., 630–632, 631f
pharmacy citrate, study exclusion rule, 37, 40f
and post-mold technology, 72
Rock Island (IL)
　Junge & Beck, 694, 694f
Springfield (IL)
　Van Deusen, Martin M., 727–729, 727f
study inclusion rule, 37
small beer
　Chicago (IL)
　　A. J. Miller, 50, 50f
　　Dempsey, Daniel, 234
　　Hutchinson & Co., 297–298, 298f, 299f, 300, 301f
　　John A. Lomax, 348, 363f, 366f
　　Miller, Andrew J., 372–373, 373f
　description of, 50
　Joliet (IL)
　　Paige, J. D., 542, 543, 543f
　Lemont (IL)
　　J. B. Bolton & Co., 556f, 557
　Milwaukee (WI)
　　Taylor & Brother, 438, 438f
　St. Louis (MO)
　　Vail, John D., 542
stoneware bottle definitions, 18, 21f, 39f
Smedley, John D., 383
Smith, Abel, 410, 411
Smith, Dr. Alexander H., 176–179, 177f, 178f
Smith, Charles Gilman, 406, 407f
Smith, Charles R., 418, 418f
Smith, David, 377
Smith, Edward, 179
Smith, Henry H., 554, 554f
Smith, John B., 383, 405–406, 406f
Smith, John F., 455
Smith, Lafayette, 277
Smith, Orrin, 419, 697–698
Smith, Samuel, 418–419, 418f
Smith, Sylvester J., 419, 419f, 697–698
Smith, William, 89
Smith, William S., 462–463, 463f
Smith & Duncan (Chicago), 255
Smith & Plows (Chicago), 418
Smith & Videto (Boston, MA), 223
Smith patent paneled stoneware
　body shape definition, 18, 21f
　Dr. Cronk, 229
Smith's Cholera Mixture, 714
snap-on swing stoppers
　and lip style chronology, 24
snuff bottles
　assemblage patterns, 75
Snyder & Eilert (Chicago), 243, 244, 245, 419–420, 420f
soda and mineral water
　Alton (IL)
　　A. & F. X. Yoerger, 87
　　Alton Soda Water Co., 87
　　Geo. H. Betts & Co., 87
　　H. E. Rupert, 97, 97f
　　Jacob Buff and Max Kuhl, 88–89, 88f, 89f
　　Keeley & Brother, 94–95, 317
　　Weisbach, Christian, 98–100, 99f, 100f
　　Weisbach, Jacob, 98, 98f
　　Yoerger, Augustin, 100–102, 101f, 102f
Antebellum era (1850s), 75, 77f
Arlington Heights (IL)
　Muller, Frederick W., 103–104, 103f, 104f
assemblage patterns, 74

Aurora (IL)
    Green Brothers, 105–106, 105f
    Stege, Albert [?], 44–45, 44f, 106f
base-embossed only, 500, 594
Beardstown (IL)
    Ehrhardt, Frederick W., 107–109, 107f, 108f
    Schneider & Krohe, 109–110, 109f
Belleville (IL)
    Ab Egg, Louis, 110–112, 111f, 112f
    Clark, George N., 112, 112f
    Fisher, Joseph (aka Fischer), 45, 45f, 113–115, 113f, 114f
    Fisher & Ab Egg, 115–116, 115f
    Fisher & Rogger, 116–117, 116f
    Stoltz, Francis (Franz), 119, 119f
blob-top
    inclusion in study, 35
    paper label–free, 41f
    and study focus, 2
Bloomington (IL)
    Bottling House, 120–121, 121f
    Kreis, Peter, 129, 129f
    Mueller & Stein, 134, 134f
    Petterson & Bro., 135, 135f
    S. M. Hickey & Co., 124–125, 125f
Blue Island (IL)
    G. H. Hausburg, 158–160f, 159f
    Hausburg & Dettmer, 160, 160f
    Hausburg & Ennis, 161, 161f
    R. Boil & Co., 158, 158f
body shape definitions, 15, 16–17f
"bottle cap"
    study exclusion rule, 37
Braidwood (IL)
    Blood, Horace W., 162–163, 162f, 163f
Breese (IL)
    Dorries & Bro., 163, 163f
Breihan, Henry, 10f, 69, 165–167, 165f, 166f
Cairo (IL)
    Kump, John H., 167–168, 167f
    Lohr, Andrew, 47, 47f, 168–170, 168f, 169f
    Schroeder, Edward, 170, 170f
Carlinville (IL)
    Labbe & Son, 172–173
    Weber, John, 173
    Weber & Mueller, 173–174, 173f
    Zaepffel, August, 174, 174f
Carlyle (IL)
    Hess, Henry (aka Heiss), 175, 175f
    Hubert, Anthony, 175–176, 175f
    J. M. Menkhaus, 176, 176f
Centralia (IL)
    Besant, Daniel J., 179–180, 180f
    Heiss & Hutter, 180, 180f
    Wehrheim, Henry G., 180–181, 180f
Centreville (IL)
    Fischer, Christian, 181–182, 181f, 182f
Champaign (IL)
    Miller, Nicholas, 182–183
Chester (IL)
    Wenda, John, 184–185, 185f
Chicago (IL)
    A. J. Miller, 50, 50f
    A. Mette & Bros., 49, 369–371, 370f
    Ainsworth & Lomax, 188–189, 188f
    Amberg & Green, 190, 190f
    Amberg & Stenson, 191, 191f
    Barothy & Cook, 47–48, 197, 197f
    Bebbington, Edward K., 189, 200–201, 200f, 201f
    Bower & Bryan, 202, 202f
    Buck & Rayner, 206–207, 206f, 207f
    Christin, Arthur, 219–221, 219f, 220f, 221f
    Clowry & Fitzgerald, 221, 221f
    Cook, Egbert C., 224–225, 224f

Dempsey, Daniel, 232–234, 234f
Dempsey & Hennessey, 234–236, 235f
Dickinson, George, 236, 236f
Dinet, Joseph, 237–238, 237f, 238f
Eaton, George E., 241–242
Eaton, James E., 241–242, 241f
Edwards, "E.", 242, 242f
Edwards, Francis C., 242, 242f
Entwistle, Joseph, 247–248, 248f
Entwistle & Bebbington, 248–249, 249f
Entwistle & Lomax, 249, 249f
Filer, Henry, 255–257, 256f, 257f
Geer, Schubael, 264
Gottleib Wurster & Co., 456–461, 457f–460f
Grossenheider & Dickinson, 270, 270f
Grossenheider & Kruse, 270, 270f
H & Bro's, 279–281, 280f, 281f
Hausburg, William A., 282–283, 283f
Hausburg Brothers, 281–282, 281f, 282f
Hayes Brothers, 284–287, 285f, 286f
Hedlund & Co., 287–289, 288f, 289f
Hess, Ernst, 291–292, 291f
Hunt, Robert L., 293–294, 293f
Hunt & Kenyon, 294–295, 294f, 295f
Hunt & Salpaugh, 295–296
Hutchinson & Co., 298, 299f, 300, 300f, 301, 302f, 303–304, 306–309f, 307–308
Hutchinson & Dunn, 303, 303f
Hutchinson & Sons, 310, 311f
Huttenlocher, John, 311–312, 311f
J. Grossenheider & Co., 268–269, 268f, 269f
John A. Lomax, 58, 335–336, 336–339f, 338, 340, 341f, 342f, 343, 344–346f, 348, 349–350, 349–352f, 364, 365
Joss & Hartmann, 314–315, 314f
Keeley, 69
Keeley, Michael, 317–320, 320f, 321f
Keeley Bottling Works, 317
King, Thomas, 320–322, 322f
Knackstead, William, 323, 323f
Lang, Frederick C., 48–49, 49f, 328–330, 329–330f
Lang Brothers, 48, 48f, 325–328, 326f, 327f
Lomax, George, 332–334, 332f, 333f
Lomax, John A., 58
Louis Rodemeyer & Co., 391, 391f
Miller, Andrew J., 372–373, 372f, 373f
Moench & Reinhold, 373–374, 374f
Morrison, William, 50–51, 51f
Paul Rouze & Co., 391–393, 392f, 393f
Peuser & Kadish, 381, 381f
Sass & Hafner, 396–399, 397f, 398f
Schonwald, Francis (Franz), 391, 402–403, 402–403f
Schriber & Joss, 403–404, 404f
Schulz & Hess, 404–405, 405f
Seibt, Frederick (Fritz), 408–409, 409f
Seibt & Haunschild, 407–408, 408f
Shure & Bro., 412, 412f
Simonds, F. W., 412–413, 413f
Smith, Samuel, 418, 418f
Stenson, James, 422–426, 424f, 425f, 426f
Van Buskirk & Henry, 442–443, 443f
Ypsilanti Springs Co., 58, 461–462, 461f
Chicago-style quart
    Industrial Expansion era (1870s), 79
chronological markers, 71
Colehour, 69
Colehour (IL)
    Kassens, Henry C., 462, 462f
    Smith, William S., 462–463, 463f
collared-blob style
    and Hutchinson-style stoppers, 35
    study exclusion rule, 37
    study inclusion rule, 37

Collinsville (IL)
    Baierlein, Henry, 464, 464f
    Bischoff, Benard, 464–465, 464f
Columbia (IL)
    Gundlach, John, 465–466, 466f
Corneau & Diller's Bottled Soda, 702
Danville (IL)
    Galtermann, William, 468, 468f
    Shean, Charles, 468–469, 468f
Darmstadt (IL)
    Gross, Christian, 469–470, 470f
    Gross & Fleischman, 470, 470f
    Sanders & Co., 471, 471f
De Kalb (IL)
    Dee, Richard, 52, 52f
Decatur (IL)
    Kuny, Fredrick D., 472, 472f
    Kuny & Klohr (Decatur), 472, 472f
disclaimer embossing, 73
Dixon (IL)
    F. J. Finkler & Co., 473, 473f
    Frank J. Finkler & Co., 52–53, 52f
Dr. Cronk, 229
Dunton (IL)
    Sass & Muller, 475–476, 475f, 476f
DuQuoin (IL)
    Hayes, Edwin, 476–477, 476f
    Hayes, Thomas, 477, 477f
    T. H. & Bro., 477, 477f
East St. Louis (IL)
    C. Lutt & Co., 479–481, 480f
    Schroeder, Edward, 482–486, 483f–486f
Edwardsville (IL)
    Harles, Frank, 487, 487f
Elgin (IL)
    Dettmer, William, 489, 489f
    Hausburg Brothers & Co., 490–491, 491f
    Henry & Van Buskirk, 490, 490f
    Shure & Bro., 491–492, 492f
flavored
    Alton (IL), 98
    Arlington Heights (IL), 103
    Aurora (IL), 106
    Jacksonville (IL), 524
    Olney (IL), 592
    Peoria (IL), 607, 633, 636
Freeburg (IL)
    Darmstatter, Frederick, 499, 499f
Freeport (IL)
    Crotty Brothers, 499, 500f
    Read, Francis A., 500–501, 500f
    Read & Bartlett, 500–501, 500f
Frogtown (IL)
    Burger, Henry, 1, 501, 501f
Galena (IL)
    Evans, Joseph, 503–505, 504f
    Oetter, Anton, 505–506, 505f
Galesburg (IL)
    Aley, Jacob H., 507, 507f
    Eaton, James E., 508–509, 508f
    Edwards, George, 509–510, 509f
    Edwards, [Isaac?] R., 510, 510f
Galva (IL)
    Ayres, Clarence M., 510–511, 511f
Greenville (IL)
    Grafe, P. Herman, 512–513, 513f
hand-blown crown-top
    study exclusion rule, 37
height measurements, in description, 10
Highland (IL)
    Amsler Brothers, 516–517, 516f
    Beck & Brother, 517, 517f
    Mueller & Beck, 517–519, 518f
    Schott, Christian, 519–520, 519f
    Weber & Mueller, 520, 520f

(continued)

soda and mineral water (continued)
　Hutchinson-style
　　expanded use, post-study, 35, 36f
　　paper label–free, 41f
　　study exclusion rule, 37
　Jacksonville (IL)
　　Braun & Killian, 523, 523f
　　Buell & Schermerhorn, 523–525, 524f
　　Kershaw, Albert, 3, 531, 532f
　　P. Braun & Co., 522–523, 522f
　　Ricks, John W., 3, 532, 532f
　　Ruf, John, 533
　　Schermerhorn, Charles, 533–534, 533f
　Jerseyville (IL)
　　Boyle, Charles M., 534–535, 534f
　　Spannagel, Johanna, 538–539, 538f
　Joliet (IL)
　　J. & H. Hickey, 539, 539f
　　Paige, J. D., 540–542, 540f
　　Paige & Vail, 542–543, 543f
　Kankakee (IL)
　　Kammann, [Dietrich] H., 545–546, 546f
　Kewanee (IL)
　　Spencer, Daniel H., 546, 546f
　Lacon (IL)
　　Rupert, [Henry] E., 547, 547f
　LaSalle (IL)
　　A. Finkler & Co., 550–551, 550f
　　Alexander Finkler & Co., 53, 53f
　　Doll, Ben, 549
　　Hickey, Stephen M., 552–553, 552f
　　R. Hickey & Co., 551–552, 551f
　Lebanon (IL)
　　Reuter, Charles J., 553–554, 553f
　Lemont (IL)
　　Bolton, John G., 554–557, 555f, 556f
　Lincoln (IL)
　　Gesner & Lipp, 558, 558f
　lip style definitions, 24, 25f
　Litchfield (IL)
　　Gerhardt, Mrs. William, 3
　　Gerhardt, William, 559–560, 559f, 560f
　　Neuber, Joseph F., 560, 560f
　　Weber, Frederick, 560–563, 561f, 562f
　M. R. & H. W. Lundblad, 66
　Mascoutah (IL)
　　Bassler, Nikolaus J., 566–567, 566f
　　Herold, Ferdinand, 567, 567f
　　Koob, August G., 567–568, 568f
　Mattoon (IL)
　　J. Fife & Co., 568, 568f
　Mendota (IL)
　　F. Molln & Co., 569, 569f
　　Molln, Frederick, 54, 54f
　　Wohlers, Edward, 569–570, 570f
　Metamora (IL)
　　Bottling House, 570–571, 571f
　Milwaukee (WI)
　　Taylor & Brothers, 438, 438f
　Monee (IL)
　　Vatter, Adam, Jr., 572–573, 573f
　　Vatter & Merz, 572–573, 573f
　murder schemes, 185
　Murphysboro (IL)
　　Kuehle, Frederick A., 576–577, 576f
　neck length, 10, 10f, 23, 663, 664f
　Nelson (IL)
　　Allison, James J., 586, 586f
　O'Fallon (IL)
　　Bayet & Williams, 589–590, 589f
　　Bischoff, Benard, 590–591, 590f
　Olney (IL)
　　Bower, William, 591–592, 591f
　Ottawa (IL)
　　B. W. Coutant & Co., 593–594, 593f

　　Crotty Brothers, 594, 594f
　　Finkler, Alexander, 594–595, 594f
　Pana (IL)
　　F. Weber & Brother, 599–600, 599f
　Pekin (IL)
　　Zerwekh, Gottleib J., 600–602, 601f, 602f
　Peoria (IL)
　　Cairns & Hickey, 606–607, 607f
　　Eaton, James E., 611–613f, 611–614
　　Fellrath, Hubert, 622–625, 623f, 624f
　　Gillig & Singer, 626–627, 627–629, 627f, 628f
　　Hickey, Ransom E., 632–635, 632–635f
　　Humphreys, E. J., 635–636, 635f
　　J. B. St. Clair & Co. (Peoria), 641, 641f
　　Matthies, Adolph L., 637, 637f
　　Reen, Augustus W. H., 641, 641f
　　Singer, Charles J., 642–643, 642f, 643f
　Perry (IL)
　　Perry Springs, 646–648, 647f
　Peru (IL)
　　Haas, Kasper, 648–649, 648f
　Pioneer era (1840s), 75
　Princeton (IL)
　　Althoff, Friedrich, 653–654, 654f
　Quincy (IL)
　　Boschulte & Brother, 657–658, 657f
　　Boschulte & Co., 658–659, 658f
　　Durholt, Henry, 663–667, 663f, 664f
　　Fischer & Boschulte, 667, 667f
　　Grone & Durholt, 672, 672f
　　Henry Durholt & Co. (Quincy), 666–667, 666f
　　Hutmacher, Edward, 673–674, 673f
　　J. J. Flynn & Co., 668–671, 668–671f
　　Kaiser, Daniel, 674–675, 674f, 675f
　　Lampe & Boschulte (Quincy), 677, 677f
　　M. R. & H. W. Lundblad, 677–681, 677f, 678f
　Red Bud (IL)
　　Excelsior Brewing Co. (Red Bud), 687–688, 687f
　Rock Island (IL)
　　Carse & Elder, 690, 690f
　　Carse & Lamont, 690–691, 690f
　　Carse & Ohlweiler, 691–693f, 691–694
　　Sargent, William B., 694–695, 695f
　Rockford (IL)
　　Chisholm, John, 688, 688f
　Schroeder's Sparkling Soda Water, 685
　Sheldon (IL)
　　Miller, Jacob S., 698, 698f
　Springfield (IL)
　　J. Johnson & Co., 707–709, 707f, 708f
　　Johnson & Peterson Co., 709–712, 709–712f
　　W. W. Watson & Son, 729–730
　Staunton (IL)
　　Hoffmann, Christian, 730, 730f
　　Knemoeller & Stille, 730–731, 731f
　Streator (IL)
　　Finkler, John A., 731–732, 731f
　　Streator Bottle & Glass Co., 69
　and study focus, 2
　study inclusion rule, 35
　Sundell & Co. (Chicago), 429–430, 429f, 430f
　swing stoppers
　　definitions, 26, 26f
　thick walls of, 26
　Thomas Schoebein & Co. (East St. Louis), 481–482, 482f
　Victor Barothy & Co., 196–197, 196f, 197f
　Virden (IL)
　　Achilles & Hille, 733–734, 733f, 734f
　W. H. Hutchinson & Son, 309, 309f
　W. P. Brazelton & Co., 203–204, 204f

　Waterloo (IL)
　　Boeke, Henry, 735, 735f
　　DePuyt, George, 736, 736f
　　Ruf, John, 737–738, 737f
　　S. & J. Ruf, 737–738, 737f
　West Urbana (IL)
　　Doane, William, 740, 740f
　Wilmington (IL)
　　H. W. Blood & Co., 742–743, 742f
　Wm. McCully & Co. (Pittsburgh), 70
soda bottle embossed symbol. See Muller, Frederick W. "Pop"; Sass & Muller (Dunton)
Soda Water Manufactory (Litchfield), 562–563
Soda Water Manufactory (Mascoutah), 567
Sour Mash 1867 whiskey, 215–216, 455
South Chicago bottles
　Blue Island (IL)
　　Hausburg, Charles, 282, 491
Spaa mineral waters, 88
Spanish Tonic Bitters, 260–261, 261f
Spannagel, George N., 481
Spannagel, Johanna, 479–481, 479f, 480f, 538–539, 538f
Spence & Bradford (Caledonia), 544
Spencer, Daniel H., 546, 546f
spices
　Chicago (IL)
　　Hall, Zebulon M., 271, 271f
　Springfield (IL)
　　Slemmons & Conkling, 726–727
Spier, James G, 626, 626f
Spier & Co. (Peoria), 626
Sproat, John, 172, 420–421, 420f, 421f
spruce beer, 42. See also small beer
　Braidwood (IL)
　　Blood, Horace W., 69, 162
　Cairo (IL)
　　Breihan, Henry, 69
　Champaign (IL)
　　Miller, Nicholas, 182
　Chicago (IL)
　　Amberg & Green, 190
　　Arnold & Green, 190, 191, 193, 193f, 194f
　　Eaton, George E., 241–242
　　John A. Lomax, 348, 364, 365
　　Sass & Hafner, 396–399, 398f
　Joliet (IL)
　　Paige, 69
　　Paige, J. D., 540–542, 541f
　　Paige & Vail, 541, 542–543, 542f, 543f
　Quincy (IL)
　　Flynn, John J., 668
　Springfield (IL)
　　Kelleher & Co., 712–713, 712f
　St. Louis (MO)
　　J. D. Vail & Co., 541
square bitters
　body shape definition, 15, 16f
　Carrollton (IL)
　　Smith, Dr. Alexander H., 176–179
　Chicago (IL)
　　Bennett Pieters & Co., 381–386, 382f, 383f, 384f
　　George Powell & Co., 387–389, 388f, 389f
　　Mason, Parker R., 367–368, 367f
　　Mathews, Henry B., 368–369, 368f
　　Monheimer & Co., 374–375
　　Plautz, Charles H., 387, 387f
　　Schwab, McQuaid & Co., 405–406, 406f
　　"T. Pieters & Co.", 386–387
　　Taylor & Wright, 439, 439f
　Peoria (IL)
　　Bissell, Orrin P., 603–606, 604f, 605f

Francis & Spier, 626, 626f
Strehlow, Robert, 643, 644f
William A. Callender & Son, 607–609
Springfield (IL)
  Van Deusen, Martin M., 727–729, 727f
Swain, Calvin H., 430–431, 431f
Woodstock (IL)
  Richardson, Dr. Holland W., 743–744, 744f
square bottles
  Alton (IL)
    George H. Betts & Co., 86–87, 87f
  Charleston (IL)
    Cunningham & Barnes, 183–184, 183f
  Chicago (IL)
    "Mrs. Schafer," 400, 401f
    Althrop, Thomas, 189–190, 189f
    C. R. Smith & Co., 418, 418f
    Croskey, Abraham F., 230–231, 231f
    Hamlin, John A., 271–279, 275f
    Hogeboom, Wolf, & Co., 293
    Landsberg, Moses G., 323–325, 324f, 325f
    S. J. Smith & Co., 419, 419f
    Sloan, Walter B., 413–418, 415f
    Sykes, Dr. John W, 435
    Van Buskirk & Henry, 442–443, 442f
    Walker & Taylor, 415f, 443–446
    Warner, Dr. C. D., 446–449, 448f
  Elgin (IL)
    C. M. Daniels & Co., 488–489, 488f, 489f
  Evanston (IL)
    Westerfield, John G., 492–493, 493f
  French Square
    study exclusion rule, 37, 38f
  Jacksonville (IL)
    Hamilton, Dr. William, 525–529
  Pittsfield (IL)
    Thomas C. Grimshaw & Co., 649–650, 650f
  and slug plates, 12, 72
  Springfield (IL)
    J. B. Brown & Co., 699, 699f
    Stickels, Dr. J., 597–598, 598f
  study inclusion rule, 37
  sunken-panel
    definition, 13
    and slug plates, 72
square case bottles
  Charleston (IL)
    Cunningham & Barnes (Charleston), 183–184, 183f
  Chicago (IL)
    Althrop, Thomas, 189–190, 189f
    Cassily & Co., 212–213, 213f
    Frechette, John, 260–261, 261f
    Stone, Dr. Reuben R., 427–428, 427f
    Warner, Dr. C. D., 446–449, 448f
  Dixon (IL)
    Herrick, George L., 473–475, 474f
  Forreston (IL)
    McLain & Brother, 496, 496f
  Jerseyville (IL)
    Freeman, Dr. John D., 535–538, 536f
  Nokomis (IL)
    Brophy, Dennis P., 588, 588f
  Ottawa (IL)
    Kagy, John M., 597, 597f
  Peoria (IL)
    Bissell, Orrin P., 603–606, 604f, 605f
    Francis & Spier, 626, 626f
    William A. Callender & Son, 607–609
  Quincy (IL)
    Schrader, Frank, 684–685
  Sandwich (IL)
    Abel, Humiston & Co., 697–698, 697f

Springfield (IL)
  Dunning, Andrew J., 703, 703f
  Fixmer, John P., 704–705, 704f
  Lord, James J., 717–719, 717f
  Van Deusen, Martin M., 727–729, 727f
Woodstock (IL)
  Richardson, Dr. Holland W., 743–744, 744f
square flared-ring lip
  Chicago (IL)
    Walker & Taylor, 443–446, 444f
square medicine bottles, body shape definition, 18, 20f
square pickle/preserve bottles
  Evanston (IL)
    Westerfield, John G., 492–493, 493f
square shoulders
  Chicago (IL)
    Eilert, Jacob K., 242–243, 242f, 243f
squared flared-ring lips
  Chicago (IL)
    Badeau, William C., 194–195
    Flagg, Edward H., 257–258
    Gillett Chemical Works, 265–266, 265f
squat ale
  Chicago (IL)
    H & Bro's, 279–281, 281f
    Hutchinson & Sons, 310, 311f
    Kane, Thomas, 315, 315f
    Sass & Hafner, 396–399, 396f, 397f
squat Chicago-style quarts
  Blue Island (IL)
    Hausburg & Ennis, 161, 161f
  body shape definition, 15, 17f
  Chicago (IL)
    Hess, Ernst, 291–292, 291f
    Hutchinson & Sons, 310, 311f
squat cylindrical liquor bottles
  Rockford (IL)
    Graham's Distillery, 689, 689f
squat lozenge bottles
  Chicago (IL)
    Doty, Harvey C., 239, 239f
squat pint porter
  Chicago (IL)
    Sproat, John, 420–421, 420f, 421f
squat porter
  body shape definition, 15, 16f
  Chicago (IL)
    Doty, Harvey C., 239, 239f
    Sproat, John, 420, 421f
squat quarts
  Joliet (IL)
    Paige, J. D., 540–542, 541f, 542f
squat salve jars
  Chicago (IL)
    Sloan, Walter B., 413–418, 414f, 415f
    Walker & Taylor, 415f, 443–446
squat sodas
  Alton (IL)
    Yoerger, Augustin, 100–102, 102f
  Chicago (IL)
    A. Mette & Bros., 369–371, 370f
    Gottleib Wurster & Co., 458–461, 460f
    H & Bro's, 279–281, 280f
    Schulz & Hess, 404–405, 405f
  Danville (IL)
    Shean, Charles, 468–469, 468f
  Quincy (IL)
    Durholt, Henry, 663–667, 664f
    M. R. & H. W. Lundblad, 677–681, 677f, 678f
St. Clair, J. B., 731
St. Louis Exchange Hotel (Peoria), 626
St. Louis Glass Works (St. Louis), 56, 66, 681f

St. Louis Medical College, 544
St. Louis Soda Co. (St. Louis), 95
St. Nicholas Hotel (Springfield), 538
Stafford, John F., 383
stamped stoneware bottles. See stoneware bottles
stamping depiction, in text, 5
Star Glass Works (New Albany, IN), 69, 304, 305f, 319, 320, 321f
Star Glass Works (Newark, OH), 55t, 68
Stebbins & Reed (Chicago), 390
Stedlin, Mary, 179
Steel, [William B.], 421–422, 421f
Steel & Co. (Chicago), 421–422, 421f
steep decorative shoulders
  Chicago (IL)
    Gillett Chemical Works, 265–266, 266f
steeply sloped shoulders
  Chicago (IL)
    Eilert, Jacob K., 242–243, 242f, 243f
Stege, Albert [?], 44–45, 44f, 106, 106f
Stege, Edward, 44
Stege, Edward A., 106
Stege, George, 106
Stege, Richard, 106
Stein, John, 469, 469f
Stein, Louis, 134, 134f, 174
Stein & Helm Brewery (Amboy), 469
Stenger, John, 582–584, 583f, 584f
Stenger, Nicholas, 583
Stenger, Peter, 583
Stenger Naperville Brewery (Naperville), 583, 584f
Stenson, James, 191, 191f, 422–426, 422f–426f
Steward, Dugal, 443
Stewart, Alexander, 85, 85f
Stewart, Dr. James T., 605–606
Stewart, O. A. T., 151
Stewart's Improved Medical Compound, 605–606
Stickels, Dr. J., 597–598, 598f
Stille, Ernst Rudolph, 731
Stille, Herman W., 731
Stille [Ernst R.], 730–731, 731f
Stoelze, Fidel, 465
Stoltz, Francis (Franz), 112, 119, 119f
Stomachic Bitters, 225–226, 225f
Stone, G. H., 427
Stone, Hedding A., 426–427, 427f
Stone, Nathaniel F., 426
Stone, R. R., 259
Stone, Dr. Reuben R., 427–428, 427f, 705
stone-bottled medicine
  Naples (IL)
    Handy, John, 585–586, 585f
stoneware, blob top
  Naples (IL)
    Handy, John, 585–586, 585f
stoneware, hand-turned quart
  Naples (IL)
    Handy, John, 585–586, 585f
stoneware, sharp-shouldered
  body shape definition, 18, 21f
  Chicago (IL)
    A. J. Miller, 50, 50f
    Arnold & Green, 193, 194f
  Naples (IL)
    Handy, John, 585–586, 585f
stoneware, sharp-shouldered small beer
  body shape definition, 18, 21f
stoneware bottles. See also stone-bottled medicine
  beer, study exclusion rule, 37, 39f
  body shape definitions, 18, 21f
  bottle attrition, 43–44

(continued)

stoneware bottles (continued)
  Cairo (IL)
    Lohr, Andrew, 47, 47f, 169f, 170
  Chicago (IL)
    A. J. Miller, 50, 50f
    Adams, Thompson W., 188, 188f
    Arnold & Green, 193, 194f
    Cronk, Edward Y., 230, 230f
    Dempsey, Daniel, 234
    Dinet, Joseph, 238, 238f
    Hutchinson & Co., 297, 298f, 300, 301f
    John A. Lomax, 348, 362, 363f, 365, 366f
    Keeley, Michael, 320, 321f
    Miller, Andrew J., 372–373, 373f
    Sass & Hafner, 396–399, 398f
    Shure & Bro., 412, 412f
  stoneware bottles study, xx
  county distribution of bottlers, 4–5t, 6f
  Dr. Cronk, 229
  Lemont (IL)
    J. B. Bolton & Co., 557
  Merrill-patent paneled
    as chronological marker, 71
  Milwaukee (WI)
    Taylor & Brother, 438, 438f
  Naples (IL)
    Handy, John, 585–586, 585f
  recently documented, 47, 50
  small beer. See small beer
  Smith-patent paneled
    as chronological marker, 71
  Springfield (IL)
    Kelleher & Co., 712–713, 712f
  stamping depiction, in text, 5, 7, 9–10
  as study focus, 2, 4
  Upper Alton (IL)
    stoneware pottery vessel style study, xx
  Vermillionville (IL)
    stoneware production study, xx
stoppers. See closures
storage, and body shape, 22
Storm, Gustavus
  slug plate patent, 72
straight brandy finish
  Chicago (IL)
    Palmer House Hotel, 376–377, 376f
straight-sided blop-top
  Chicago (IL)
    Seipp, Conrad, 409–410, 410f
straight-sided flask shape
  Chicago (IL)
    Wolgamott, Dr. George W., 455–456, 456f
straight-sided shouldered-blop top
  Chicago (IL)
    Geer, Schubael, 264, 264f
strap-sided bottles
  Aurora (IL)
    Heck, John, 106, 106f
  Charleston (IL)
    Van Meter, Dr. Samuel, 184, 184f
  Chicago (IL)
    H. A. Stone & Co., 426–427, 427f
    Palmer House Hotel, 376–377, 376f
  Springfield (IL)
    Owen, Thomas J. V., 724–725, 725f
strap-sided oval
  Springfield (IL)
    Owen, Thomas J. V., 724–725, 725f
strap-sided Union-style flasks
  Quincy (IL)
    Schrader, August, 684, 684f
strap-sided whiskey flasks
  body shape definition, 13, 14f
  Chicago (IL)
    Chapin & Gore, 214–216, 215f

Quincy (IL)
  Schrader, August, 684, 684f
Streator Bottle & Glass Co. (Streator), 55t, 67, 69
Streeter, W. E., 495
Strehlow, Robert, 643, 644f
Strehlow's German Wine Bitters, 643, 644f
Strickland, Alfred, 428–429, 428f, 429f, 448
Strickland, Herrmann, and Strickland (Chicago), 428
Strickland, William, 428–429, 428f, 429f, 448
Strickland's International Cased Good Co. (Chicago), 429
Strickland's Wine of Life, 428–429, 428f
stubby-necked sodas
  Chicago (IL)
    Keeley, Michael, 317–320, 320f, 321f
  Ottawa (IL)
    Finkler, Alexander, 594–595, 594f
Stutenroth, Charles W., 581–582, 581f, 582f
sudorific patent medicines, 43
Sullivan, John J., 287
Sulphur Spring water, Perry Springs, 648
Sundell, C. J., 678
Sundell, Charles, 429–430, 429f, 430f
Sundell & Co. (Chicago), 429–430, 429f, 430f
Sundell & Lawson (Chicago), 678–679
sunken-panel bottles
  Aurora (IL)
    Budlong, William C., 104–105, 105f
  Belleville (IL)
    E. T. Flanagan & Co., 117–118, 117f
  Bloomington (IL)
    C. B. Castle & Co., 46–47, 46f, 121–122, 121f, 122f
    Dr. H. S. Woodard, 154–156, 154f, 155f
    Funk & Lackey, 123–124, 124f
    Hill & Vanatta, 125–126, 126f
    Howe, Bliss S., 126–128, 126f
    Lackey and Brothers, 129–131, 130f
    Wakefield & Co., 137–139f, 137–150, 141–144f
    Wells, Edwin M., 151–154, 152f
    Woodard & Howe, 156–157, 157f
  Cairo (IL)
    Pfifferling, Charles, 170, 170f
  Capron (IL)
    Field, Platus A., 172, 172f
  Carrollton (IL)
    Smith, Dr. Alexander H., 176–179, 177f, 178f
  Charleston (IL)
    Cunningham & Barnes, 183–184, 183f
    Van Meter, Dr. Samuel, 184, 184f
  Chicago (IL)
    "Mrs. Schafer," 400, 401f
    A. W. Sargent & Co., 394–396, 395f
    Adams, Dr. A. L., 186–187, 186f
    Althrop, Thomas, 189–190, 189f
    B. S. Cory and Son, 225–226, 225f
    Badeau, William C., 194–195, 195f
    Bailey & Eaton, 195–196, 196f
    Bligh, Thomas, 201–202, 202f
    Butt, Charles, 210–212, 210f, 211f
    Christie, Dr. Abel H., 216–219, 217f, 218f
    Collins, Dr. Silas F., 221–222, 222f
    Cory's and Collins' bitters, 79
    Cram, Dr. John S., 227–228, 227f
    Croskey, Abraham F., 230–231, 231f
    Drs. C. M. Fitch & J. W. Sykes, 434–436
    Eilert, Jacob K., 242–243, 242f, 243f
    Fahrney, Peter, 250–251, 250f
    Flagg, Edward H., 257–258, 257f
    Foster, Charles M., 259–260, 259f, 260f
    Garden City Chemical Works, 261–262
    Gardiner, Charles H., 262–263, 263f

    Gillett Chemical Works, 265–266, 266f
    H. E. Bucklen & Co., 208–210, 209f
    Hamlin, John A., 271–279, 273f, 274f, 275f
    Hutchins, John W., 296, 296f
    Huyck & Knox, 312
    Huyck & Randall, 312, 312f
    Johnson, Dr. [Hosmer A.], 313–314, 313f
    Miller, Dr. [Adam], 371–372, 371f
    P. W. Gillet & Son, 267, 267f
    Sears & Smith, 406, 407f
    Sloan, Walter B., 413–418, 415f
    Sweet, Henry J., 432, 432f
    Taylor, Newton S., 438–439, 438f
    Thompson, 441, 441f
    Walker & Taylor, 443–446, 444f, 445f
    Warner, Dr. C. D., 446–449, 446f, 447f
    Wells, Edwin M., 450–452, 451f
    Willard, Joseph, 453
  Dixon (IL)
    Herrick, George L., 473–475, 474f
  Elgin (IL)
    C. M. Daniels & Co., 488–489, 488f, 489f
  Franklin Grove (IL)
    Dr. U. C. Roe & Sons, 497–498, 497f, 498f
  Jacksonville (IL)
    Hamilton, Dr. William, 525–529, 528f
    Hockenhull, Robert, 530–531, 530f, 531f
  Jerseyville (IL)
    Freeman, Dr. John D., 535–538, 536f
  Jonesboro (IL)
    Parks, Dr. Luther K., 543–544, 544f
  Lebanon (IL)
    Smith, Henry H., 554, 554f
  Lewistown (IL)
    The Medicine Co., 557–558, 557f
  Maryland (IL)
    Haldeman, Samuel, 565, 565f
  Mendota (IL)
    Viera, Manuel J., 569, 570f
  Morris (IL)
    Lee, Dr. Charles M., 575–576, 575f
  Naperville (IL)
    Keith, Aylmer, 579–580, 580f
  New Athens (IL)
    Baumann, Peter, 587, 587f
  Ottawa (IL)
    Griggs, Edward Y., 595–597, 595f, 596f
  and paper labels, 13
  Pekin (IL)
    Brown, Dr. Alfred, 600, 600f
  Peoria (IL)
    Bastow, James D., 603, 603f
    Cleveland, Frederick, 609–610, 609f
    Davis, William H., 610–611, 611f
    Kruse, Dr. Hero, 636–637, 636f
  Pittsfield (IL)
    Thomas C. Grimshaw & Co., 649–650, 650f
  Princeton (IL)
    Bodley, Dr. A. R., 654–655, 655f
  Quincy (IL)
    Burleigh & Brothers, 660–662, 661f, 662f
    Kalb, Absalom J., 675–677, 675f, 676f
  rectangular bottles
    and slug plates, 72
  shield-shaped panels. See shield-shaped panels
  Springfield (IL)
    Diller, Roland W., 700–702, 700f
    J. B. Brown & Co., 699, 699f
    Lewis, Thomas, 713–714, 714f
    Myers, Henry C., 721
    Owen, Thomas J. V., 724–725, 724f
    Slemmons & Conkling, 726–727, 726f
  square bottles
    and slug plates, 72

Stickels, Dr. J., 597–598, 598f
   Waukegan (IL)
      B. S. Cory & Son, 738–739
   Western Saratoga (IL)
      Penoyer, Dr. Hiram, 741–742, 741f
Swaim's Panacea, 30
Swain, Calvin H., 430–431, 431f
Swain Manufacturing and Distilling Co. (Chicago), 431
Swain's Bourbon Bitters, 430–431, 431f
Swartz's Livery Stable (Jerseyville), 534
Sweet, Henry J., 258, 432, 432f
Sweet and Schroeder (Chicago), 258
Sweet Elixir for Diarrhoea, 557f, 558
Sweet's Blood Renewer, 258, 432, 432f
Sweet's Cholera Drops, 258, 432, 432f
swing stoppers. *See also specific stoppers*
   definition of terms, 26–28f, 26–29
Sykes, Dr. Charles J., 436
Sykes, Charles R., 432–434, 433f
Sykes, Dr. John W, 434–436, 434f
symbols, embossed. *See* embossed symbols
Synnott, T. W., 440
syphilis, 30
syrups and extracts
   Chicago (IL)
      Hutchings, William A., 296–297, 296f
System Builder and Blood Purifier, 151

T
T. Foerster & Co. (Chicago), 258, 258f
T. H. & Bro. (DuQuoin), 477, 477f
T. J. V. Owen & Brother (Springfield), 725
T. Keeley & Brother, 94
"T. Pieters & Co." (Chicago), 386–387
tall beer
   Chicago (IL)
      Seipp, Conrad, 409–410, 410f
tall cylindrical sodas
   Chicago (IL)
      Gottleib Wurster & Co., 456–461, 459f
   Jacksonville (IL)
      Buell & Schermerhorn, 522, 523–525, 524f
      Schermerhorn, Charles, 533–534, 533f
   Nelson (IL)
      Allison, James J., 586, 586f
tall pint
   LaSalle (IL)
      L. Eliel & Co., 549–550, 549f
   Rock Island (IL)
      Junge & Beck, 694
tall shouldered cylinder
   Chicago (IL)
      W. H. Hutchinson & Son, 309, 309f
   Nelson (IL)
      Allison, James J., 586, 586f
tapered bodies
   Beardstown (IL)
      Ehrhardt, Frederick W., 107–109, 107f
   Chicago (IL)
      Henry H. Shufeldt & Co., 410–411, 411f
tapered heels
   Chicago (IL)
      Paul Rouze & Co., 391–393, 393f
Tapered Stone Sodas
   body shape definition, 18, 21f
target balls
   assemblage patterns, 75
   study exclusion rule, 37
Taylor, Frank C., 439, 439f
Taylor, George, 410, 411
Taylor, James, 436
Taylor, John, 436–438, 436–438f
Taylor, Joseph, 436–438, 436–438f
Taylor, Newton S., 438–439, 438f, 443–446

Taylor, Robert J., 323
Taylor & Bro. (Chicago), 436–438, 436–438f
Taylor & Wright (Chicago), 439, 439f
Taylorville (IL)
   McNeil, William W., 719
Tazewell, Charles, 488
technology terms. *See* bottle technology terms
tenpin-shaped bottles
   base shape definition, 22, 22f
   Chicago (IL)
      Barothy & Cook, 197, 198f
      Dinet, Joseph, 237–238, 237f, 238f
      Simonds, F. W., 412–413, 413f
   Naperville (IL)
      Powell & Stutenroth, 581–582, 581f, 582f
10-sided bottles
   Albion (IL)
      Stewart, Alexander, 85, 85f
   Alton (IL)
      Yoerger, Augustin, 100–102, 101f
   Beardstown (IL)
      Ehrhardt, Frederick W., 107–109, 107f
ten-sided sodas
   Chicago (IL)
      Smith, Samuel, 418–419, 418f
Thebes (IL)
   John Hodges, 544
thickened flared lip
   Bloomington (IL)
      Wakefield & Co., 137–150, 139f
   Chicago (IL)
      Thompson, 441, 441f
   Jacksonville (IL)
      Hamilton, Dr. William, 525–529, 529f
   Springfield (IL)
      McNeil, William W., 719, 719f
      Melvin, Samuel H., 719–721, 720f
      Owen, Thomas J. V., 724–725, 724f
thickened squared flare-ring lips
   Chicago (IL)
      Gillett Chemical Works, 265–266, 265f
thickened-ring lip
   Bloomington (IL)
      Wakefield & Co., 137–150, 140f
   Chicago (IL)
      Hutchins, John W., 296, 296f
   Louden City (IL)
      Wills, Dr. John C., 563, 563f
thin flared lip
   Pittsfield (IL)
      Thomas C. Grimshaw & Co., 649–650
   Quincy (IL)
      Bunnel, Dr. D[avid], 659–660, 659f
Thomas, Dr. E., 153
Thomas, Frederick A., 441
Thomas, Lancaster
   slug-plate patent, 72
Thomas, Levi H., 439–440
Thomas, R. T., 440–441, 440f
Thomas & Co. (Chicago), 440–441, 440f
   Antebellum era (1850s), 79
Thomas C. Carter and John Johnson (Chicago), 188
Thomas C. Grimshaw & Co. (Pittsfield), 649–650, 650f
Thomas Co. (Chicago), 440
Thomas Heberer and Brothers (Belleville), 79, 118–119, 119f
Thomas Lynch Distillery (Chicago), 449
Thomas Schoebein & Co. (East St. Louis), 481–482, 482f
Thomas Wightman & Co. (Pittsburgh), 55t, 69
   Cairo (IL), 166f, 167
   Chicago (IL), 245f, 247
Thompson (Chicago), 441, 441f

Thompson, Alexander M., 441
Thompson, Daniel D., 441
Thompson, Eliza A., 441
Thompson, Hugh B., 441
Thompson, James, 441
Thompson, Mary H., 441
Thompson, Merritt W., 441
Thompson, R., 132
Thompson, Robert, 145
Thompson, Robert C., 441
Thompson & Taylor (Chicago), 441
Thompson W. Adams' Saloon and Restaurant (Chicago), 188
Thomson, John D., 649
Thomson, Thomas, 649
threaded ground lip
   Chicago (IL)
      Wheeler & Bayless, 452–453, 453f
thumb knob glass caps
   Drake, John B., 240–241, 240f
time markers. *See* chronological markers
time span of study, 29–35
Timmerman, Gerard, 95
Tivoli Bottling Co. (Chicago), 441–442, 441f
Tolle, Charles A., 562
Tonic Queen, 172, 172f
tonics, 43
   A. Lindsay's "Anti-Despeptic" Tonic and Pills, 716
   A. Lindsay's Western Nerve Tonic, 716
   Ague Tonic, 676
   Althrop's Constitutional Tonic, 189–190, 189f
   Aurora (IL)
      John Heck's Chamomile, 106, 106f
   Badeau's Pure Blood Maker and Liver Cure, 194–195, 195f
   beer
      Chicago (IL), 49–50, 49f
   Bininger's Wine Bitters, 431
   Bissell's Great Western Tonic Bitters, 603–606, 604–606f
   Bissell's Tonic Bitters, 603–606, 604–606f
   Bligh's Tonic, 201–202, 202f
   Carmen's Bitter Sweet, 183–184, 183f
   Daniels' Tanta Miraculous, 488–489, 488f
   Dr. Faloon's Tonic, 122, 123f
   Dr. Jones Red Clover Tonic, 595–597
   Dr. Sykes New England Liver Tonic and Billious Annihilator, 435
   Gilt-Edge Tonic, 609
   Great Western Wholesale and Retail Patent Medicine Depot (Chicago), 217
   Hamilton's [Never Failing] Ague Tonic, 529
   Keystone Tonic Bitters, 473–475, 474f
   Key-stone Tonic Bitters, 496, 496f
   Liquid Tonic, 429
   Magnolia Bitters, 92–93, 92f
   Nerve Tonic, 284, 284f
   Old Style Bitters, 177–179, 178f
   People's Favorite Bitters, 581–582, 581f, 582f
   Sawyer's Fluid Extract of Bark, 34, 399–400
   Spanish Tonic Bitters, 260–261, 261f
   Stewart's Improved Medical Compound, 605–606
   Swain's Bourbon Bitters, 430–431, 431f
   Tonic Queen, 172, 172f
   Wallace & Diller's Western Tonic or Fever and Ague Bitters, 702
   Warner's Wine of Life, 428, 429, 446–449
tooled blob-top lips
   Chicago (IL)
      Monheimer & Co., 374–375
tooled cider-style finish
   Springfield (IL)
      Fixmer, John P., 705

tooled lip
  Peoria (IL)
    William A. Callender & Son, 609
tools
  bottler's mask, 26, 28f
  lipping tool
    and lip style chronology, 24, 25f, 71–72
Topping, Ebenezer H., 717–719, 717f, 718f
Torbet, Dr. N. H., 544
Townsend, Dr. C. M., 274–275
Tremont House hotel (Chicago)
  threaded glass stoppers, 29
triangle, embossed, GJZ or P. *See* Zerwekh, Gottleib J.
Triton mineral water, 208
Trommer Extract of Malt, 59
Trower, Dr. T. B., 184
Turner, Samuel, 241
Turner's Balsam, 702
Turner's Bitters, 702
Turner's Fever & Ague Cure, 702
Turner's Gonyza & Styllingia, 702
Turner's Lotion, 700–702, 700f
Turner's Sarsaparilla, 702
Tyler, Charles, 495–496, 495f

U
ulcerated or putrid soar throat, 30
ulcerous diseases, 30
Uncle Sam's Condition Powder, 246, 246f
Uncle Sam's Nerve & Bone Liniment, 246, 246f
unguent jars. *See also* wide unguent style
  body shape definition, 18, 20f
unguent/spice
  Chicago (IL)
    Hall, Zebulon M., 271, 271f
Union Brewery (Peoria), 645
Union Coal and Oil Co. (Maysville, KY), 196
Union Medicine Co. (Carrollton), 179
Union Medicine Co. (St. Louis), 179
Union pharmacy bottles, 73
upside-down storage, 22, 51, 500, 594
U.S. Census data
  as data source, 8–9
  on immigrant densities, 30, 31f, 32f
  on wealth distribution, 30, 33f
use patterns, 74–80, 76f, 77f, 78f
uterine patent medicines, 43

V
V. A. Brown's Monongahela, 212, 213
Vail, Delos W., 541
Vail, John D., 541–542, 542–543, 543f
Van Buskirk, John A., 442–443, 442f, 443f, 490, 490f
Van Buskirk & Henry (Chicago), 442–443, 442f, 443f
Van Deusen, Abraham, 729
Van Deusen, Martin M., 727–729, 727f, 728f
Van Deusen's Ague Bitters, 727–729, 727f, 728f
Van Deusen's Cholera Specific, 728, 729
Van Meter, Dr. Samuel, 184, 184f
Vanatta, Sanford K., 125–126, 126f
Vatter, Adam, 572, 572f, 573, 573f
Vatter, Adam, Jr., 572–573, 573f
Vatter, Jacob, 572, 572f, 573, 573f
Vatter & Merz (Monee), 572–573, 573f
Vermifuge, 676
vermifuges, 43
  Dr. A. C. McDill Vermifuge, 592, 592f
  Dr. J. C. Wills' World Worm Candy, 563, 563f
  Dr. J. S. Gunn's Golden Vermifuge, 174–175, 175f
  Freeman's Vermifuge, 535–538, 535f
  Great Western Wholesale and Retail Patent Medicine Depot (Chicago), 217
  Pioneer era (1840s), 75
  Vermifuge, 676
veterinary medicines
  A. & M. Lindsay's Liniment, 30, 75, 714–717, 715f, 716f
  Baumann's Liniment, 587, 587f
  Bristol's Nerve & Bone Liniment, 120, 120f
  Chicago (IL)
    Taylor, Newton S., 438–439, 438f
  Dr. A. H. Smith's Golden Liniment, 176, 177f
  Dr. North's Cherokee Liniment, 723
  Dr. Streeter's Magnetic Liniment, 495–496, 495f
  Dr. Thomson's Galvanic Liniment, 30, 649–650, 650f
  English Horse Liniment, 662, 662f
  Farrell's Arabian Liniment, 251–254, 252f, 253f, 254f, 511, 614–621, 615–621f
  Farrell's Arabian Nerve and Bone Liniment, 618
  Floyd's American Penetrating Liniment, 30, 511–512, 512f
  Kalb's Vegetable Quick Relief, 675–677, 675f, 676f
  Keith's Persian Liniment, 30, 579–580, 580f
  Sloan's Condition Powder, 416, 445
  Sloan's Horse Liniment, 417
  Sloan's Horse Ointment, 417, 445
  Sloan's Ointment, 416, 443–446, 444f, 445f
  Stewart, Alexander, 85, 85f
  Wakefield's Egyptian Liniment, 149
  Wells' German Liniment, 152
  Woodard's Instant Relief, 156
Vichy mineral waters, 88
  Quincy (IL)
    Schroeder, Herman, 685
Victor Barothy & Co. (Chicago), 196–197
Viera, Manuel J., 569, 570f
Viera's Toilet Shampoo, 569, 570f
Vigor of Life, 313–314, 313f
Volz & Glueck (Galena), 505, 506, 506f
Vredenburgh's Old Stand (Springfield), 723

W
W. A. Holton & Co., 87
W. B. & H. G. Farrell's Linseed (Peoria), 619f
W. H. Hutchinson & Son (Chicago), 309, 309f
W. P. Brazelton & Co. (Chicago), 203–204, 204f
W. Quigley & Sons (Alton), 96
W. Steele (Brisbane, Australia), 223
W. T. S. & Co. (Chicago), 410
W. T. Shufeldt & Co. (Chicago), 410
W. W. Watson & Son (Springfield), 729
Wade, Zachariah, 646
Wahrlich, Louis, 374
Wakefield, Cyrenius, 132, 137–147f, 137–150
Wakefield, Homer, 150
Wakefield, Oscar, 148, 150
Wakefield, Dr. Zera, 132, 144–146, 148
Wakefield & Co. (Bloomington), 137–147f, 137–150
Wakefield & Thompson (Bloomington), 132, 145
Wakefield Family Medicine Co. (Bloomington), 150
Wakefield's Ague & Fever Pills, 144, 145f, 149
Wakefield's Blackberry Balsam, 68, 137, 137f, 138f, 146, 148, 150
  ingredients, 150
Wakefield's Cough Syrup, 137, 139, 139f, 146, 148, 149, 150
Wakefield's Egyptian Liniment, 30, 139, 140f, 146, 148, 149
Wakefield's Fever & Ague Pills, 34
Wakefield's Fever Specific, 34, 140–141, 141f, 142f, 146, 148, 149
Wakefield's Golden Ointment, 141, 142f, 148, 149, 150
Wakefield's Magic Pain Cure, 141, 142f, 143f, 148, 149
Wakefield's Nerve and Bone Liniment, 141, 143f, 146, 148, 149
Wakefield's Strengthening Bitters, 143f, 144, 144f, 146, 148, 149–150
Wakefield's Wine Bitters, 144, 148, 149–150
Wakefield's Worm Medicine, 146
Walker, T. Irving, 443–446, 444f, 445f
Walker & Taylor (Chicago), 414, 415f, 417–418, 439, 443–446, 444f, 445f
Wallace, J. H., 388
Wallace & Diller (Springfield), 700, 702, 721
Wallace & Diller's Cholera Mixture, 702
Wallace & Diller's Fever & Ague Bitters, 34
Wallace & Diller's Western Tonic or Fever and Ague Bitters, 702
Wallace's Tonic Stomach Bitters, 64, 387–389, 388f, 389f
Walter B. Sloan's Medical Depot (Chicago), 416
Walters, Jacob M., 597
Walters & Church (Ottawa), 597
Walters & Co. (Ottawa), 597
Warner, Dr. C. D., 428, 446–448f, 446–449
Warner, Hubert H. (Rochester, NY), 448
Warner, Major L., 449
Warner, William R. (Philadelphia), 448
Warner's Cough Balsam, 428, 448
Warner's Dyspepsia Tonic, 428, 448
Warner's Emmenagogue, 428, 448
Warner's Imported English Gin, 446–449, 448f
Warner's Pile Remedy, 428, 449
Warner's Proprietary Medicine Co. (Chicago), 428, 448
Warner's White Wine & Tar Syrup, 68, 446–449, 446f, 447f
Warner's Wine of Life, 428, 429, 446–449, 446f
waste glass, 43–44
Waters, Henry, 150
Waters, Orin, 150
Waters, Dr. Zera, 126–128, 150–151, 155, 157
Waters & Howe (Bloomington), 126, 154, 157
Waters' Abdominal and Uterine Supporter, 150
"Waters' Family Medicines" (Waters, Z.), 150
Water's Golden Ointment, 149, 150
Watson, Benjamin, 646–648, 647f, 729–730
Watson, William W., 729–730
Waukesha Mineral Rock Water
  Chicago (IL)
    John A. Lomax, 365
wax-seal jars
  Chicago (IL)
    Gillett Chemical Works, 265–266, 266f
weak-shouldered bottles
  Bloomington (IL)
    Wakefield & Co., 137–150, 145f
wealth distribution, 30, 32f
Weatherbee fastener
  Chicago (IL)
    John A. Lomax, 345
  definition, 26, 26f
  patent date, 26
Weber, Frederick, 174, 559, 560–563, 561f, 562f, 599–600, 599f, 645
Weber, George, 643–645, 644f
Weber, Jacob, 174, 519, 520, 520f, 599–600, 599f